13TH TEXAS SYMPOSIUM ON RELATIVISTIC ASTROPHYSICS

13TH TEXAS SYMPOSIUM ON
RELATIVISTIC ASTROPHYSICS

Chicago, Illinois, USA
December 14–19, 1986

Editor: Melville P Ulmer

World Scientific

Published by

World Scientific Publishing Co Pte Ltd.
P. O. Box 128, Farrer Road, Singapore 9128

Library of Congress Cataloging-in-Publication data is available.

13TH TEXAS SYMPOSIUM ON RELATIVISTIC ASTROPHYSICS

ISBN 9971-50-307-7
9971-50-310-7 pbk.

Printed in Singapore by Kyodo-Shing Loong Printing Industries Pte Ltd.

INTERNATIONAL ORGANIZING COMMITTEE

N. Bahcall, P. G. Bergmann, A. Cameron, J. Ehlers, E. J. Fenyves, R. Giacconi, F. C. Jones, S. A. Korff, M. Livio, L. Mestel, L. Motz, Y. Ne'eman, I. Ozvath, D. Pines, W. Priester, R. Ramaty, I. Robinson, R. Ruffini, A. Salam, D. N. Schramm, E. Schuding, M. M. Shapiro, G. Shaviv, L. C. Shepley, H. J. Smith, J. Stachel, M. P. Ulmer, S. Weinberg, J. A. Wheeler, J. C. Wheeler.

LOCAL ORGANIZING COMMITTEE

W. D. Arnett, E. Carlson, D. Hegyi, R. Hildebrand, E. W. Kolb, R. G. Kron, D. Q. Lamb, Jr., P. Meyer, D. Pines, D. N. Schramm, J. Truran, M. S. Turner, M. P. Ulmer, P. Vandevoort, R. M. Wald.

SCIENTIFIC ADVISORY COMMITTEE

R. D. Blanford, M. Davis, S. Glashow, J. E. Grindlay, J. Gunn, A. H. Guth, J. B. Hartle, D. J. Helfand, J. Ostriker, W. Press, M. J. Rees, M. A. Ruderman, J. Silk, K. S. Thorne, R. J. Weymann, R. Weiss, D. T. Wilkinson.

WORKSHOP ORGANIZERS

J. Grindlay, J. Truran, J. Gallagher, D. Wilkinson, J. Hartle, J. Bahcall, M. Davis, J. C. Wheeler, D. York, P. Michelson, E. Becklin, D. Lamb, M. Ruderman, J. Simpson, G. Cassiday, M. Geller, M. Ruderman, D. Pines, S. Shapiro, L. Smarr, B. Burke, M. Rees, G. Steigman, T. Weekes, R. Wald, M. Turner.

Financial assistance was received from:

- Department of Energy
- Fermi National Accelerator Laboratory
- National Aeronautics and Space Administration
- National Science Foundation
- Northwestern University
- University of Chicago

PREFACE

The Thirteenth Texas Symposium on Relativistic Astrophysics was held in Chicago from December 15 to 19, 1986. The organizers of the meeting tried several experiments. One was to organize the meeting with morning plenary sessions and afternoon sessions devoted to workshops that were organized by workshop chairmen. Another experiment was to organize dialogue/debate sessions on current topics of controversy. One such session was held between G. Tammann and J. Huchra. This session was most appreciated and we are fortunate that both authors have contributed texts that are included here. The other experiment was to hold a conference in Chicago in December. Here we were also successful. Over six hundred people attended and the weather cooperated as well. As the Mayor of Chicago noted, Chicago is a "world class city", and the cultural excitement of the city greatly enhanced the conference.

MELVILLE P. ULMER
Department of Physics and Astronomy
Northwestern University
Illinois
U.S.A.

CONTENTS

Part III NEW TELESCOPE AND DETECTOR TECHNOLOGY

Part IV SUPERSTRINGS/SUPERSYMMETRY

Part V QUANTUM GRAVITY

Part VI RELATIVITY AND NUMERICAL ASTROPHYSICS

Part IX 3K BACKGROUND

Part X LARGE SCALE STRUCTURE OF UNIVERSE AND DARK MATTER

Part XI QSO ABSORPTION LINES

Part XII GRAVITATIONAL LENSES

Part XIII ACTIVE GALACTIC NUCLEI AND JETS

Part XIV OUR GALACTIC NUCLEUS

Part XV SUPERNOVAE

Part XVI GALACTIC EVOLUTION AND CHRONOLOGY

Part XVIII COMPACT OBJECTS

Part XIX GAMMA RAY BURSTS

Part XX COSMIC RAYS

Part XXI ULTRA HIGH-ENERGY COSMIC RAYS

Part XXII SUPERCONDUCTING COSMIC STRINGS

13TH TEXAS SYMPOSIUM ON RELATIVISTIC ASTROPHYSICS

THE COSMOLOGICAL DISTANCE SCALE

JOHN P. HUCHRA

Center for Astrophysics
60 Garden Street
Cambridge, Massachusetts 02138

1. THE PROBLEM

Almost by definition, the cosmological distance scale impacts on all fields of extragalactic astronomy and cosmology. The intrinsic sizes and masses of galaxies, the expansion age of the universe, the luminosities of upper main sequence stars, predictions of cosmic nucleosynthesis, and the amount of "excess" or dark matter in galaxies and galaxy clusters, to name just a few problems, all depend on H_0. Even more important for cosmologists is the use of the expansion age as a valuable test of cosmological models — if the ages of the oldest objects in the universe (eg. gobular clusters) are really different from (especially *greater than*) the expansion age derived from the Hubble constant, significant modifications must be made to the simple Big Bang.

Unfortunately, we experimental astrophysicists have a problem. There exist today two camps of very stubborn observers who each believe they know the Hubble Constant to better than 10%. One group gets approximately 50 km/s/Mpc and the other gets 100 km/s/Mpc! Although other experimenters have derived intermediate values, by and large these two disparate values are the ones most often adopted for astrophysical calculations. (For references see Rowan-Robinson[1]), Madore and Tully [2]), and the long series of articles referenced therein by Sandage and Tammann and by de Vaucouleurs and his collaborators).

These two values have come to be known as the long (50) and short (100) distance scale. Each, taken at face value, is not unreasonable. Nothing terribly special in observational extragalactic astronomy must happen for either of these values to be correct — although members of each camp will try to tell you that the other is all hogwash!

In this talk, I would like to 1. compare the derivations of the two scales,

2. compute realistic *external* error budgets for the steps involved in the distance ladder by way of discussing some of the problems of deriving H_0, and 3. describe the best current independent derivation of H_0 and its uncertainty.

2. THE LONG AND SHORT OF IT

The differences between Sandage and Tammann (ST), de Vaucouleurs (DV), and other workers in the field are differences in detail and philosophy. Following the work of Hubble himself, Sandage has adopted what is called the 'Principle of Precision Indicators,' essentially picking what one believes is the best standard candle at each rung of the distance ladder and using only that technique. De Vaucouleurs and van den Bergh adopt the alternative approach of 'Spreading the Risks,' using as many possible techniques at each rung and averaging the results.

The differences in the scales of ST and DV are given in Table 1 as a function of distance. The differential and cumulative differences are in percentages. As can be seen, the differences between ST and DV start out quite small.

TABLE 1

DIFFERENCES IN THE SHORT AND LONG SCALE

Where	Differential	Cumulative
Galactic Extinction	0.12	12 %
Local Group	0.15	29 %
to M101	0.20	55 %
to VIRGO	0.1 – 0.3	~ 80 %
to COMA	0.1 – 0.2	~ 95 %

Differences in the two scales start out inside our own galaxy. ST adopt a polar cap free model for absorption by dust in our galaxy, while DV uses a plane parallel model with approximately a 20% loss at the poles. Both are wrong. The best current estimates of galactic absorption derived from the IRAS satellite studies and from galactic HI maps give intermediate values. Stepping to the Local Group galaxies increases the difference; ST and DV use slightly different calibrations of the Cepheid period-luminosity relation – again different calibrating stars in our own galaxy plus differences in dust absorption corrections. By Andromeda, the difference in scale is nearly 30%.

Past the nearest galaxies even larger differences begin to appear. These are primarily due to the different standard candles applied by each group. ST now rely primarily on the brightest stars in galaxies to cover out to the Virgo Cluster and on Supernovae to measure further out. DV uses brightest stars, the blue Tully-Fisher relation, luminosity classes, novae and several other techniques to cover the same range.

Each step is NOT major; the difference in distance *at* any distance is well within the errors, but the total effect of all of the slightly different choices is a factor of 2!

3. THE PROBLEMS IN GETTING H_0: AN ERROR BUDGET

Derivation of the global value of H_0 is fraught with many difficulties. A list of some of the problems is given below.

1. Virgo Flow (and larger scale velocity perturbations)
2. Galactic Obscuration in a) Our Galaxy, b) other Galaxies
3. Confusion - Identification of Single Objects in Crowded Fields
4. Metallicity Effects
5. Environmental Effects
6. Population Effects (variation of galaxy properties)
7. Statistical Selection Effects – Malmquist, Scott, etc.

All steps in the distance ladder suffer from selection effects. Most of these have been well known for years, and some, like #3, have produced monumental blunders. Some problems, like #1 were not even recognized a few decades ago, and only have become tractable in the last few years. Some problems are quite insidious, like #'s 4, 5 and 6, because calibrating them out almost requires and answer ahead of the time. Fortunately, some, like the Malmquist Bias, can be dealt with relatively easily[3)].

All of these problems introduce errors into the calibration of distance scales. At each rung of the ladder, distances become more and more inaccurate. Unfortunately, most workers in the field sweep errors in the local calibration under the rug. For this reason, it is useful to try to list the errors in each step and to try to derive a "cumulative" quadrature external error estimate for the determination of H_0. I will essentially follow the steps taken by Rowan-Robinson[1)] and others.

Table 2 gives estimates of the errors in the determination of the zero points for a variety of distance indicators and/or steps in the distance ladder. The errors are quoted as the percentage uncertainty in the distance. These zero point errors are not cumulative, ie. they ***do not*** include the errors in earlier steps necessary to derive them. For secondary indicators, they are only the errors in the mean resulting from the scatter (either observational or cosmic) in the properties of the objects themselves. **The error in H_0 is the quadrature sum of the errors in the steps used in its derivation.**

TABLE 2
RMS ERROR BUDGETS IN ZERO POINTS

TECHNIQUE OR ITEM		FRACTIONAL ERROR
Hyades Modulus		0.05
Extinction in our Galaxy		0.05
Cepheid P-L,P-L-C		0.12[a]
Globular Clusters		dead ?
HII Regions		dead!
Brightest Stars –	Red	0.15[a]
	Blue	0.20[a]
Type I SN		0.16
IR-Tully Fisher		~ 0.15
Virgo Distance (error due to)		
Velocity uncertainty		0.07
Virgo Flow V		0.07
Coma Distance (" ")		
Large Scale Flows		0.1 (!)

[a] Neglecting problems with metallicities!

Most distance determination start with the distance to the Hyades open star cluster; a conservative estimate of the error in its distance is 5%. As stated above, the error in distance due to inadequate galactic extinction models is of order 5%. Only 12 (ST) or 18 (DV) galactic Cepheids are used to derive the zero point of the P-L-C relation - the *scatter* in the zero point is 12%!

As we step to secondary indicators the errors increase. Globular cluster luminosity functions and properties of giant HII regions are currently not used in distance scale determinations – although the GCLF may be making a comeback. Brightest stars come dearly - there is, after all, only one per galaxy and each galaxy used must have a distance determined with Cepheids. There are not many nearby relatively luminous galaxies, so the scatter in brightest star luminosities is 30-40%. There are only 3 type I SN in galaxies with measured distances; the scatter in their luminosities would imply distance errors of 16%. To make matters worse, the SN I galaxies only have distances measured via secondary indicators. The scatter in the *calibrators* of the IR-Tully Fisher relation is 15% in distance.

Now come the real nasty problems. Most people forget that the Virgo Cluster velocity is only known to about 5%. The correction for the Virgo infall is also only known to about 7% in distance. Then there are the newly discovered large scale motions (Dressler et al.[4]), which may bolix up H_0 even beyond the Coma Cluster.

As they say in the trade, PICK YOUR POISON. No matter what sequence of rungs or standard candles you use to get to large enough distances, the *external* error in your derivation of H_0 *cannot be less than about 25%!*

4. THE MIDDLE GROUND

A group of us have been trying some new approaches to the distance scale problem. In particular, several of the problems mentioned above can be minimized by a careful choice of measurements. In 1976, R. B. Tully and R. Fisher proposed a new, objective technique for measuring the relative distances of galaxies[5]. They capitalized on M. Robert's earlier discovery of a correlation between the rotation rates and absolute luminosities of spiral galaxies. Their technique was significantly improved upon by the measurement of luminosities in the near infrared[6]. Observations near 2μ have the advantage of nearly eliminating the effects of both galactic extinction and absorption internal to the galaxy measured. Infrared observations also help to minimize the effects of metallicity and variance in stellar populations. The galaxian light in the near-IR is dominated by the red giant stars that are the best tracer of the galaxy's mass.

The first major effort to apply the IR-Tully Fisher (IRTF) technique was aimed at measureing the perturbation of the local velocity field caused by the massive Virgo Cluster of galaxies[7]. That study produced the best determination of the infall velocity of the Local Group into Virgo. Unfortunately, the sample of galaxies used for that study was selected by a variety of relatively poorly defined and distance dependent criteria – HI flux, velocity limits, etc. – that make it totally unsuitable for the determination of H_0.

The second major effort applying IRTF was aimed specifically at the determination of H_0.[8] Ten moderately distant clusters of galaxies were selected and HI and IR observations were made of ~ 150 edge-on spiral galaxies. Each cluster was sampled to approximately the same depth, about 5 magnitudes down from the brightest spiral, to essentially eliminate problems with Malmquist bias.

Distances to each cluster were derived using direct calibrations of IRTF from nearby bright spiral galaxies with Cepheid distances. This minimizes the number of steps required to reach directly to the Coma Cluster and beyond (where we hope the velocity field is quiet enough for a determination of the *global* value of H_0). The motion of the local supercluster w.r.t. the microwave background was also detected and its effect was removed from H_0.

Table 3 shows the values of H_0 derived with three different calibrations of Cepheid distances to nearby galaxies. The ST and DV calibrations (by definition!) produce different values of H_0. We believe that the best Cepheid distances, however, come from the recent infrared work of Madore and collaborators.[9] IR observations of Cepheids have all of the inherent advantages mentioned above for galaxies as well as being at wavelengths where the star's amplitude is small.

TABLE 3
IRTF ESTIMATES OF H_0

Adopted Zero Point Calibration	H_0
IR-Cepheids (MFMM)	92 $km\ s^{-1}\ Mpc^{-1}$
Sandage and Tammann	74
de Vaucouleurs	100

Reasonable bounds on the probable error might be given by the ST and dV calibrations. This value of H_0 corresponds to an expansion age of ~ 11 billion years for an open universe ($\Omega \sim 0.1$), with errors that still overlap the lower bounds of age determinations for the globular clusters.

5. THE BOTTOM LINE

If we are lucky (ie. if Coma isn't shuckin' and jivin' in a large scale flow), we may now know the Hubble Constant to ~ 25 % — the one sigma *external* error.

The best value of H_0 from the IR-Tully Fisher work on moderately distant clusters is ~ 90 km $s^{-1}Mpc^{-1}$. Values between 50 and 100 cannot be honestly excluded. Keep your fingers crossed for the Hubble Space Telescope.

I would like to thank my collaborators Rob Kennicutt, Marc Aaronson, Greg Bothun, John Hoessel, Jeremy Mould, Paul Schechter, Bob Schommer, and Brent Tully as well as my friends and sometimes adversaries Allan Sandage, Gustave Tammann and Gerard de Vaucouleurs for our wonderful continuing dialog on H_0. This work is supported by the Smithsonian Institution.

REFERENCES

1) Rowan-Robinson, M. *The Cosmological Distance Ladder*, (San Francisco: Freeman and Co.), (1985).

2) Madore, B. and Tully, R. B. eds. *Galaxy Distances and Deviations from Universal Expansion*, (Dordrecht: Reidel), (1986).

3) Eddington, A. S., *M.N.R.A.S.* **73**, 359 (1913).

4) Dressler, A., Burstein, D., Davies, R., Faber, S., Lynden-Bell, D., Terlevitch, R. and Wegner, G. Ap. J. in press, (1987).

5) Tully, R. B. and Fisher, R. *Astron. and Ap.* **54**, 661 (1977).

6) Aaronson, M., Huchra, J. and Mould, J. *Ap. J.* **229**, 1 (1979).

7) Aaronson, M., Huchra, J., Mould, J., Schechter, P. and Tully, R. B. *Ap. J.* **258**, 64 (1982).

8) Aaronson, M., Bothun, G., Mould, J., Huchra, J., Schommer, R. and Cornell, M. *Ap. J.* **302**, 536, (1986).

9) Madore, B., McAlary, C., McLaren, R., Welch, D., Neugebauer, G. and Matthews, K. *Ap. J.* **294**, 560 (1985) and references therein.

THE VALUE OF THE HUBBLE CONSTANT

G.A.Tammann

Astronomisches Institut der Universität Basel, Binningen
European Southern Observatory, Garching

ABSTRACT

Distances to nearby galaxies from primary distance indicators are quite secure; they define an uncontroversial zero point for the calibration of the Hubble constant H_o. They can be used to calibrate distance indicators reaching to recession velocities of ~5000 km s^{-1}, necessary to find the global value of H_o. Unfortunately the mean properties of nearby (calibrating) galaxies deviate systematically from those of distant galaxies, because the latter are discriminated by our catalogues against their underluminous/undersized analogues. Neglect of this selection bias yields too high values of H_o (90 - 100 km s^{-1} Mpc^{-1}). Several bias-free methods to derive sufficiently large distances are discussed; they are recognized by their conformity with a linear expansion field and their insensitivity against the sample selection, i.e. they agree for field and cluster galaxies. A solid net of unbiased distance determinations requires H_o = 50 $\pm$ 10 km s^{-1} Mpc^{-1}.

I. Introduction

The current literature on the Hubble constant flaunts a bimodal distribution characterized by values of either H_o = 90 - 100 [km s^{-1} Mpc^{-1}] (de Vaucouleurs and Corwin[1]; Aaronson and Mould[2]) or H_o = 50 (Sandage and Tammann[3]). Since the Universe expands undoubtedly in only one mode and since the published error ranges of the two values of about $\pm$10 are mutually exclusive, one or both determinations must carry a large systematic error. To trace its origin it is necessary to procede step by step going from small distances out to recession velocities of ~5000 km s^{-1}, where the influence of peculiar velocities becomes negligible, and to observe the gradual divergence of the "short distance scale" (H_o = 90 - 100) and the "long distance scale" (H_o = 50). The divergent distance scales can be tested objectively against the average expansion field, which from independent evidence can be shown to be linear . In principle it is therefore possible to decide from present data which of the two distance scales is more nearly correct.

In Section II the distances to nearby galaxies with primary distance indicators are discussed; they are uncontroversial and agree on average within $0^m_.3$ between different authors. Unfortunately they reach to only ~7 Mpc, where the recession velocities (~350 km s^{-1}) may still be dominated by local perturbations. After a demonstration in Section III of the linearity of the local expansion field, two distance scales are illustrated in Section IV which are in clear conflict with this linearity. The cause of the non-linearity is identified in Section V with the effect of selection bias. Several bias-free routes to H_o are tabulated and briefly discussed in Section VI. The conclusions are given in Section VII.

II. Nearby Calibrators

Among the primary distance indicators Cepheid variables are the most powerful ones. The zero point of their period-(color)-luminosity relation comes from Cepheids in galactic clusters; the distances of the latter are tied to the Hyades cluster and in two cases directly to trigonometric parallax stars corrected for metallicity differences. The reliability of the Cepheid distances to within ~$0^m.2$ is confirmed in several galaxies by RR Lyrae variables, Mira variables, color-magnitude diagrams of main-sequence stars, the brightes blue and red stars and by novae. Much progress has been made with these additional primary distance indicators in recent years. In addition the advent of the infrared P-L relation of Cepheids has reduced considerably remaining problems with absorption and metallicity differences.

A compilation of the relevant data since 1980 leads to impartial mean distances of 13 "calibrators" (plus 11 additional fellow group members)[4]. The 13 calibrators define a mean zero point of the extragalactic distance scale which is in perfect agreement with that adopted from independent data by Sandage and Tammann in 1974[5]; it is only $0^m.3$ brighter than de Vaucouleurs' 1979 zero point[6] and $0^m.2$ brighter than the zero point (yet defined by only three galaxies) of Aaronson and Mould[7]. This agreement to within 15% provides clear evidence that the discrepancies in H_o are <u>not</u> due to very local problems. The near constancy of the zero point over the last 12 years provides in addition confidence that also future revisions will remain at the 15% level.

III. The Linearity of the Local Expansion Field

Pushing the distance scale further out requires some guide line against which any longer-range distance indicator can be tested. The guide line is provided by the true overall form of the expansion field.

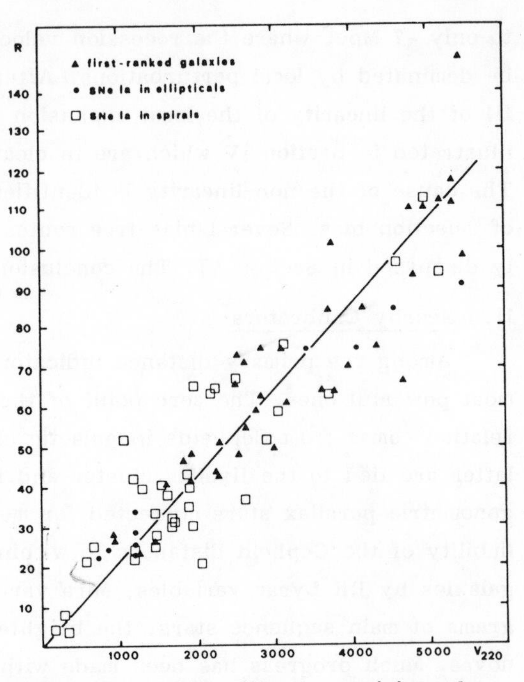

Fig.1. The distance-velocity diagram of first-ranked galaxies in groups and clusters and of supernovae of type Ia at maximum light. For the derivation of the distances R (in Mpc) it was arbitrarily assumed that $M_B(1)=-23\overset{m}{.}5$ and $M_B(\max)=-20\overset{m}{.}0$; the resulting relative distances are suitable for mapping the shape of the velocity field. The velocities are corrected for a selfconsisting Virgocentric flow model (with a Local Group infall of v_{VC} = 220 km s^{-1})[8)9)]. The scatter in the diagram is due to peculiar velocities, observational errors and - to a lesser extent - internal luminosity scatter.

The latter can be mapped by standard candles without requiring absolute distances. First-ranked galaxies and supernovae Ia are for this purpose ideal because their discovery is independent of any luminosity scatter for the relevant distance range (~ 5500 km s^{-1}). These objects reveal the local expansion field to be linear in good approximation (Fig.1).

IV. Evidence for Non-linear Expansion?

All distance indicators which seemingly require a high value of H_o ~90-100 imply a non-linear expansion field in contradiction with Section III. Atypical example is the infrared Tully-Fisher method, which combines infrared magnitudes of spiral galaxies with 21cm line widths as luminosity indicator. If this method is consistently applied to a sample of field galaxies by Aaronson et al.[10)] one finds an expansion field as shown in Fig.2; it requires an increase of the Hubble constant with distance of up to H_o= 120 at v = 3000 km s^{-1}. The same method yields for a fainter sample of cluster galaxies[7)] H_o = 90 at v = 3000 km s^{-1}. The non-linearity of the expansion and the sample-dependent value of H_o suggest a selection bias (cf. Section IV).- A parallel test for de Vaucouleurs´main distance indicator, the so-called Λ_c-index[11)], reveals an even stronger deviation from a linear expansion field[4)].

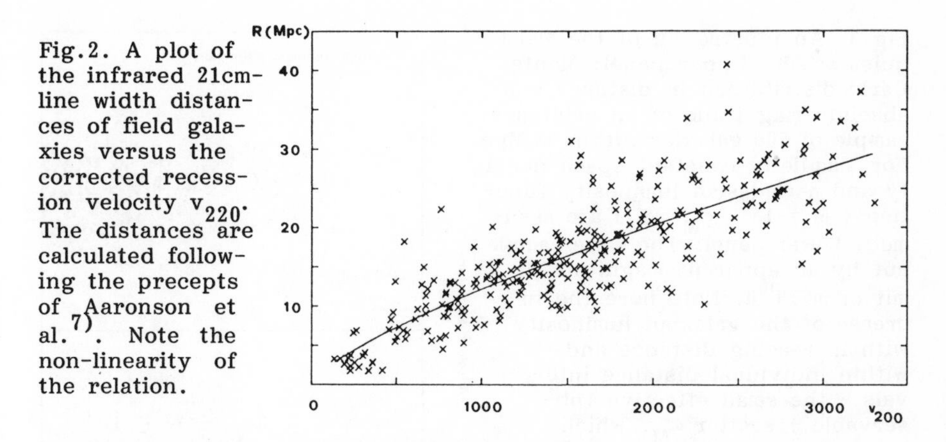

Fig.2. A plot of the infrared 21cm-line width distances of field galaxies versus the corrected recession velocity v_{220}. The distances are calculated following the precepts of Aaronson et al.[7]. Note the non-linearity of the relation.

V. Selection Bias - the Key to All Discrepancies

The extragalactic distance scale is beset by biasses. Most important is the so-called Malmquist effect resulting from the fact that all available galaxy samples are more or less strictly limited by apparent magnitude and that therefore galaxies at large distances enter our catalogues only if they are overluminous. Therefore the mean luminosity of a galaxy sample increases with distance, and distant galaxies are brighter than their nearby calibrator "twins". For the ensemble of all galaxies the effect is overwhelming; for galaxies of a specific property the size of the effect depends on the intrinsic luminosity scatter. With an intrinsic magnitude scatter of $0^{m}.6-0^{m}.8$ at constant 21cm-line width or Λ_c-index, these luminosity indicators are rather weak and still strongly vulnerable to selection bias. The Malmquist bias always leads to too short a distance scale and to a non-linear expansion field as found in Section IV. An analytical correction is not possible because of the patchy space distribution of galaxies. A remedy would be to use the inverse relation (luminosity indicator versus absolute magnitude), if galaxy samples were fair samples of the luminosity indicator, which is generally not the case[12]. Another possibility, yet difficult to realize systematically, is to sample to fainter apparent magnitudes as one goes to larger distances. Since field galaxies are generally taken from the bright Shapley-Ames catalogue, while cluster galaxies come from fainter catalogues, the Malmquist bias enters differently into the two types of samples and tends to give smaller values for H_o from cluster galaxies than from field galaxies. Exactly this discrepancy was found in Section IV for the infrared Tully-Fisher distances.

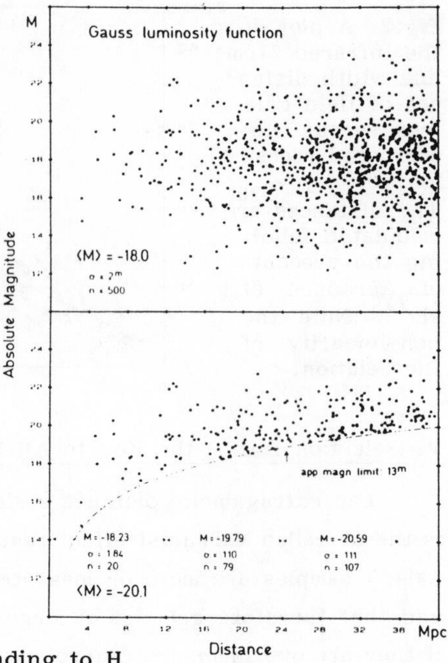

Fig.3. An illustration of the Malmquist effect. Upper panel: Monte Carlo distribution in distance and absolute magnitude of an arbitrary sample of 500 galaxies within 38 Mpc. For simplicity constant space density and a Gaussian luminosity function ($M = -18^m$, $\sigma_M = 2^m$) are assumed. Lower panel: The same sample cut by an apparent-magnitude limit of $m = 13\overset{m}{.}0$. Note here the increase of the galaxian luminosity with increasing distance and - within individual distance intervals - the small effective (observable) scatter σ_M, which tends to underestimate, if taken at face value, the size of the Malmquist bias. (By permission of A. Spaenhauer).

VI. The Net Work of Distances Leading to H_o

Table 1 exhibits the best distance determinations leading to the global value of H_o. They are either insensitive to selection bias or are corrected as well as possible for bias. They employ population I and II distance indicators whose distance scales are independent of each other, they comprise field and cluster galaxies, and yet they yield at all levels $H_o = 55$ within the errors, which implies a nearly linear expansion field, the latter being a prerequisite of any valid distance scale. Sandage[13] has in fact shown from additional dwarf galaxies that the linearity after subtraction of the Virgocentric flow extends down to distances of ~4 Mpc where the Local Group deceleration becomes important. The distance scale reaches into the optimum range of v ~ 10 000 km s^{-1} for the determination of H_o(global). The large-scale streaming motions of galaxies of ~600 km s^{-1} become here negligible. At still larger distances one finds strictly not the present value of H_o, but a combination of H_o and the deceleration q_o.

At this conference John Huchra has given an impressive list of errors which plague the determination of H_o. However, they are not cumulative for any given distance indicator. The different, mutually supporting distance indicators, each of which carries some specific errors, define therefore H_o to within an estimated mean error of $0\overset{m}{.}3$ (15%).

Table 1: The routes to the global value of H_o via field and cluster galaxies

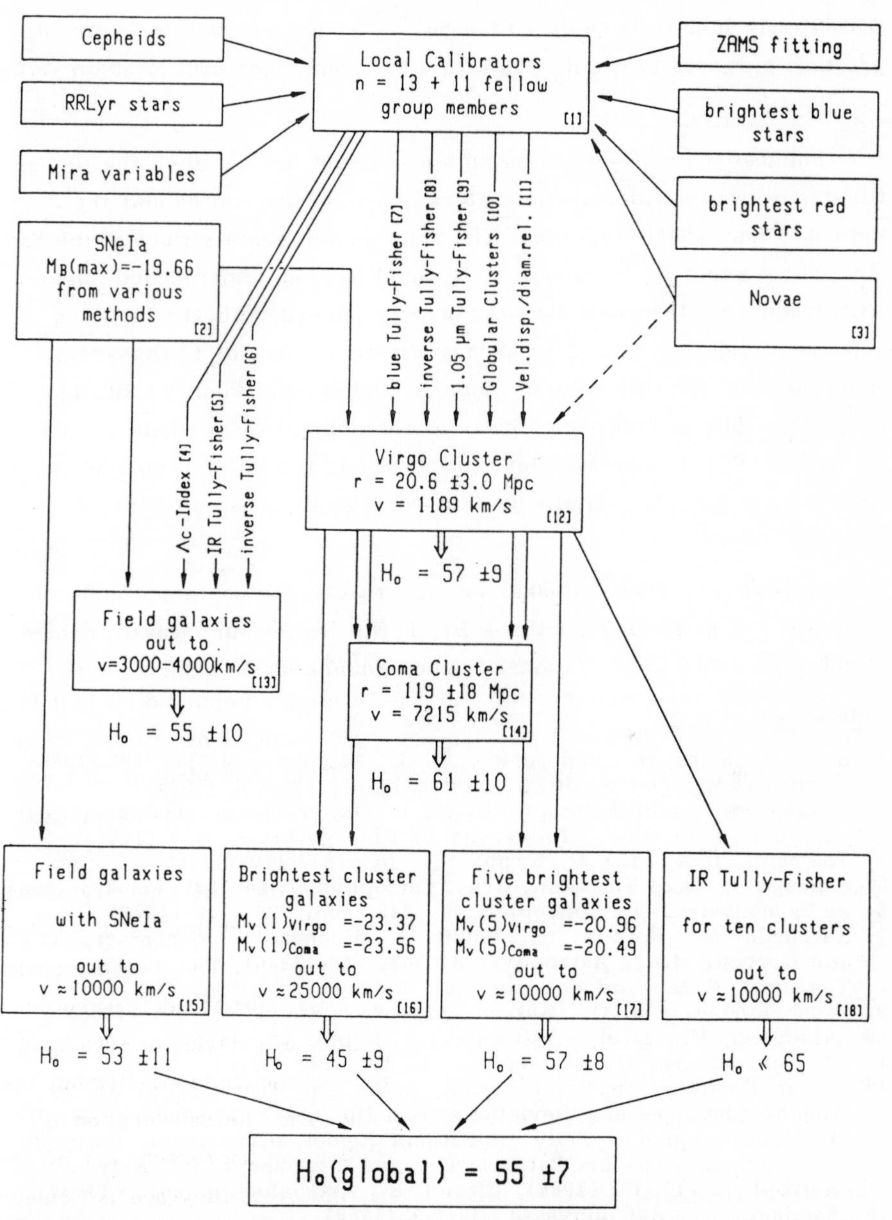

Sources. 1, 2, 4-8, 12-18: ref. 4; 3: S. van den Bergh, private communication (1986); 9: N. Visvanathan, Astrophys. J. 288, 182; 10: S. van den Bergh, C. Pritchet, and C. Grillmair, preprint (1986); 11: A. Dressler, preprint (1986).

The straight mean from Table 1 is $H_o = 55 \pm 7$. History shows that it is easier to overestimate H_o than to underestimate it. Indications exist that H_o may indeed be smaller than 50[14]. At present it is best for all practical purposes to use $H_o = 50$ with a 99% confidence limit of $30 < H_o < 70$.

VII. Conclusions

Suggestions of H_o being as high as 90 - 100 are based on the unjustified optimism that distance indicators like 21cm-line widths and the Λ_c-index are free of selection bias. The value would create serious problems: our Galaxy would be oversized[15], it would have an unparallelled supernova frequency, the baryonic mass density would fall short of binding clusters of galaxies, and the maximum Friedmann time of 11 Gigayears could not meet the time scale of globular clusters and would hardly be compatible with the nucleosynthesis of the actinides.- A rounded value od $H_o = 50$ ($\pm$ 7) avoids the above problems and is - after proper allowance is made for selection effects - in very good agreement with all available evidence.

Acknowledgment. These remarks are the result of a continued collaboration with Dr. A. Sandage. I thank Dr. J. Huchra for our pleasant debate. Support from the Swiss National Science Foundation is acknowledged.

References:

1. de Vaucouleurs, G., and Corwin, H. G., Astrophys. J. 308, 487 (1986).
2. Aaronson, M., and Mould, J., Astrophys. J. 303, 1 (1986).
3. Sandage, A., and Tammann, G.A., in: Inner Space - Outer Space, eds. E.W. Kolb et al., University of Chicago Press, p.3 (1986).
4. Tammann, G.A., I.A.U. Symp. 124, in press (1987).
5. Sandage, A., and Tammann, G.A., Astrophys. J. 190, 525; 194, 223 (1986).
6. de Vaucouleurs, G., Astrophys. J. 223, 730; 224, 710 (1978).
7. Aaronson, M., Bothun, G., Mould, J., Huchra, J., Schommer, R.A., and Cornell, M.E., Astrophys. J. 302, 536 (1986).
8. Tammann, G.A., and Sandage, A., Astrophys. J. 294, 81 (1985).
9. Kraan-Korteweg, R.C., Astron. Astrophys. Suppl. 66, 255 (1986).
10. Aaronson, M., et al., Astrophys. J. Suppl. 50, 241.
11. de Vaucouleurs, G., Astrophys. J. 227, 380 (1979).
12. Kraan-Korteweg, R.C., Cameron, L.M., and Tammann, G.A., in: Galaxy Distances and Deviations from Universal Expansion, eds. B. F. Madore and R.B. Tully, Dordrecht: Reidel, p.65 (1986); Bottinelli, L., Gouguenheim, L., Paturel, G., and Teerikorpi, P., Astron. Astrophys. 156, 157 (1986); Giraud, E., Astrophys. J. 301, 7 (1986).
13. Sandage, A., Astrophys. J. 307, 1 (1986).
14. Sandage, A., private communication (1987).
15. van der Kruit, P.C., Astron. Astrophys. 157, 230 (1986).

INTERFEROMETRIC GRAVITATIONAL WAVE DETECTION AT MIT

P. Saulson, R. Benford, M. Burka, N. Christensen, M. Eisgruber,
P. Fritschel, A. Jeffries, J. Kovalik, P. Linsay, J. Livas, and R. Weiss
Massachusetts Institute of Technology
Cambridge, Massachusetts, USA

1. INTRODUCTION

A gravitational wave passing through a set of freely falling test masses will perturb them by a (usually) minute amount. The experimental task in detecting gravitational waves is to build a set of test masses which are sufficiently isolated from non-gravitational effects that they are freely falling to a good approximation, and simultaneously to measure the relative displacements of the masses with high enough precision to detect the effect of weak gravitational waves. Thus experimenters must fight both stochastic noise forces on the masses, and transducer noise in the measuring system.

A second orthogonal classification of noise terms is into the categories of fundamental noise sources and ubiquitous ones. Fundamental noise terms are those whose magnitude can be written down in an expression that contains a fundamental constant, such as Planck's constant or Boltzmann's constant. Ubiquitous noise terms are all of the important noise terms which aren't fundamental. For example, the fundamental transducer noise in an interferometric system is shot noise in the light, which is proportional to $(hc\lambda/P)^{1/2}$, where λ is the wavelength and P is the available light power. A fundamental stochastic noise source is thermal excitation of modes of vibration of the test masses or their suspensions, which goes like $(kTf_0/mQ)^{1/2}$, where f_0, m, and Q characterize the resonant mode. A fundamental noise term which blurs the distinction between transducer noise and stochastic force is the fluctuating radiation pressure on the test masses from the light in the interferometer. The fact that it grows with P, while shot noise shrinks with P, means that there is an optimum light power, for a given choice of mass and signal frequency. This is the mechanism which enforces the uncertainty principle in interferometers.

Ubiquitous noise terms are often the ones which give experimenters the most difficulty, especially as their magnitudes depend more on engineering details of an instrument rather than on simple physical arguments. The most ubiquitous stochastic force is the coupling of the laboratory ambient vibration spectrum to the test masses. Isolation by pendulums and springs is in principle straightforward, but subtle in practice. A ubiquitous transducer noise, at least in interferometers with delay-line optical folding schemes, is spurious interference from scattered light in the instrument. Perhaps naively, it was not even anticipated until it was seen in the Munich 3-m interferometer. Another ubiquitous transducer noise is the second order sensitivity of interferometers to laser injection angle (beam jitter noise). Tracking down and taming these noise terms is often what takes most of the experimental effort to improve sensitivity.

2. THE MIT 1.5-METER INTERFEROMETER

At MIT, we have a Michelson interferometer with arms 1.5 meters long. Light makes 28 round trips through the arms by means of a Herriott delay line. A Spectra-Physics Model 2020 laser illuminates the interferometer with 300 mW of single mode power at λ=514.5 nm. We do not frequency-stabilize the light, as the Michelson interferometer is (in principle) insensitive to the optical frequency, so long as the arms are sufficiently close in length. Instead, we add a wide-band phase modulation to decrease the coherence length of the light to reduce the effects of scattering in the interferometer. Light in the two arms receives a differential 5.38

MHz phase modulation, so that our fringe interrogation is not affected by excess amplitude noise in the laser light.

The four spherical mirrors which form the optical cavities are clamped to the three test masses of mass 8 kg (end masses) and 15 kg (central mass). Tungsten wires 1 meter long hung from the top of the vacuum cans support the masses. We control the six rigid-body modes of each mass by servos which use RF capacitive displacement sensors and electrostatic actuators. The interferometer is locked on the dark point of a fringe with two nested servo loops, a fast loop which drives a Pockels cell, and a wide dynamic range loop which drives one of the masses. The interferometer sits in a vacuum system pumped to below 10^{-6} torr by ion pumps.

Figure 1 is a typical noise spectrum of our instrument. Tests show that it is dominated almost entirely by ubiquitous noise terms rather than fundamental ones. The white noise at high frequencies is close to shot noise, but fails to fall as we increase the power further. At intermediate frequencies (1 to 4 kHz), ambient vibrations (coupled through the injection optics as beam jitter) dominate, appearing as a forest of resonant peaks. Below 1 kHz, ambient vibration coupled through the string modes of the suspensions dominate the appearance of the spectrum. In between these modes is a broad band noise which we have not diagnosed yet.

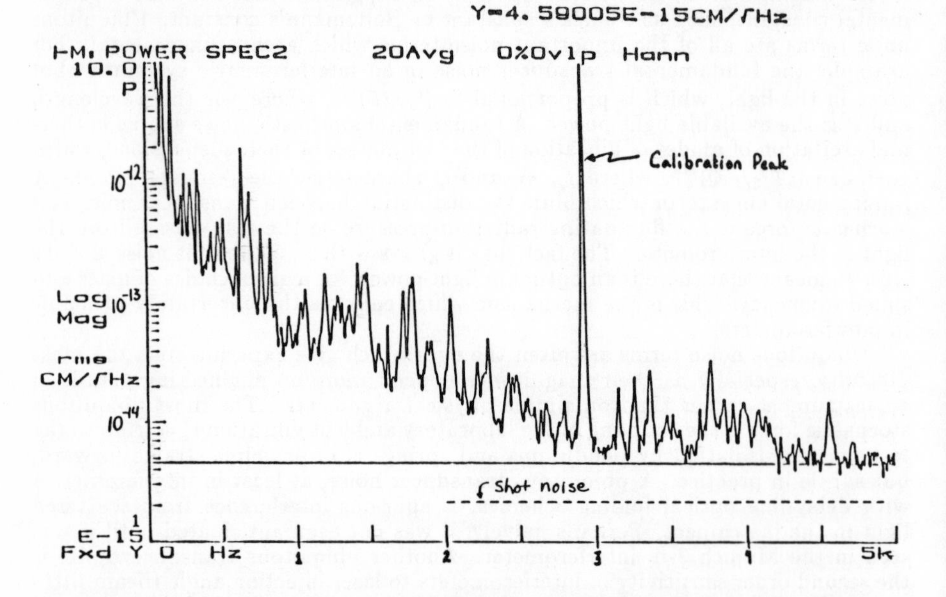

Figure 1. Displacement noise spectrum of the MIT 1.5-m interferometer

Our efforts have concentrated on reducing extra noise due to scattered light. The approach of adding wide-band phase noise to the light should allow us make scattering unimportant compared with shot noise, were it not for the fact that our phase modulator adds some amplitude noise to the light as well, which limits the improvement. Current work is concentrating on characterizing the scattering mechanism in the interferometer, and on implementing a fiber optic output coupler to improve the fringe contrast. This last device is useful because the beam-splitter on the central mass can not be aligned from outside the apparatus, nor is it easy to open the vacuum system and readjust it. With the fiber coupler, we can adjust the contrast to better than 99.7% with a coupling efficiency of 60%. At this higher contrast we can use a smaller dither about the dark fringe, thus reducing sensitivity to excess amplitude noise and also reducing the shot noise level. Results from these tests should be available soon.

We expect to make progress on our 1.5-meter prototype in the high frequency part of the spectrum where we are dealing (presumably) with transducer noise only. (To the extent that our scattering problem is due to upconversion of low frequency motion of the masses, this separation of transducer noise from mechanical noise may not be correct.) Progress on this instrument at lower frequencies will be more difficult, because mechanical changes are quite awkward to make in it, where there is room for them at all.

3. A NEW 5-METER INTERFEROMETER

For this reason, we have undertaken to build a new prototype interferometer, taking advantage of what we have learned so far. It will be constructed in a high bay in our laboratory, which is equipped with an overhead hoist of 3 ton capacity. The space is large enough to accomodate an interferometer with a baseline of 5 meters. More important, the new antenna will have vacuum tanks which are 8 feet tall and 4.5 feet in diameter at the two ends, and 8 feet by 7 feet at the vertex. Further, the masses will be suspended by a compound suspension from sturdy pillars anchored to the ground. The pillars will pass through holes in the base of the tanks, isolated by compliant bellows from the acoustically driven vibrations of the tanks themselves. The portion of the tank above the base will be removable without disturbing the internal parts. This design will simultaneously give us better isolation for the masses and allow us ready access for changes which we have not had on the 1.5-meter device.

The large tanks will be connected by beam pipes 30 inches in diameter. The scale is such that we should be able to test components appropriate in size to the 4 km LIGO system which the Caltech and MIT groups are proposing. The vacuum system will consist of a rotary pump and turbo pump for roughing, and ion pumps for quiet maintenance of the high vacuum. An additional feature is a fourth tank, identical to the two end tanks, pumped independently by the same set of pumps. This will be used for parallel development of improved suspension designs and components, such as magnetic suspensions and interferometric fiber optic displacement sensors. The vacuum system is under construction at NuVac Systems, Inc. of Kingston, Mass., with delivery scheduled for March 1987.

The design of the interferometer which will fill the vacuum system is still under discussion, but some features have been specified. The light will come from a Nd:YAG laser operating at $\lambda = 1.06\mu$. The mirrors will be 8-inch diameter by 4-inch thick cylinders of silicon, with plasma sputter coatings. Each mirror will be the final mass in a compound pendulum, which is itself further isolated by mechanical (or eventually magnetic) springs. Damping and control will be implemented with optical shadow sensors and a mix of electromagnetic and electrostatic actuators. At the vertex of the interferometer, separate suspensions will carry each mirror, the beamsplitter, and other optical components. Optical fibers will bring the light through the vacuum chamber wall, both on input and output. Construction and commissioning of this instrument will be a major task over the next year or two. Until it is completed, the 1.5-meter interferometer will be kept operating for a continuing program of improvements of the high frequency noise.

OPERATION OF THE 2270 kg GRAVITATIONAL WAVE RESONANT ANTENNA OF THE ROME GROUP

E.Amaldi, P.Bonifazi, P.Carelli, M.G.Castellano, G.Cavallari, E.Coccia, C.Cosmelli, V.Fogliettí, S.Frasca, R.Habel, I.Modena, R.Onofrio, G.V.Pallottino, G.Pizzella, P.Rapagnani, F.Ricci

Dipartimento di Fisica, Universita' "La Sapienza", Roma
Dipartimento di Fisica, Universita' "Tor Vergata", Roma
Istituto Nazionale di Fisica Nucleare
Istituto di Fisica dello Spazio Interplanetario, CNR
Istituto di Elettronica dello Stato Solido, CNR
European Organization for Nuclear Research
ENEA, Centro Ricerche Energia Alternativa

The gravitational wave antenna of the Rome group, installed at CERN in Geneva, was operated from November 1985 until July 1986. The antenna consists of a 5056 alluminium cylinder of L=3 m lenght and M=2270 kg mass cooled with liquid helium to T=4.2 K. The mechanical vibrations are detected by means of a resonant capacitive transducer with equivalent mass m_t=0.35 kg, electrical capacity C_t=3891 pF and a gap, between the two plates, d=51 μm. The resonant frequency of the transducer is very close to that of the bar: the system bar + transducer has resonant modes at ν_- =907.116 Hz and ν_+ =923.083 Hz. The corresponding merit factors are $Q_-=3.2\times10^6$ and $Q_+=5.6\times10^6$.

The electrical signal from the transducer is amplified by means of a dc SQUID. For adapting the high impedance of the transducer to the low impedance of the SQUID we use a superconducting transformer with N=1250 turn ratio.

During the nine months of operation the antenna has produced data for about 70% of the time. The rest of the time has been spent for refilling with cryogenic liquids, setting and calibration of the apparatus and searching for causes of unexpected disturbances that occasionally appeared. We stopped the operation of the antenna on 21

July 1986 in order to improve both the mechanical filtering and the electrical matching.

During the period of operation the noise temperature T_W was on the average between 15 and 20 mK. In certain periods (lasting a few days) was as low as 12 mK. The data analysis has been fully completed for the period 15 April 1986 through 21 July 1986 which includes the periods when the Louisiana and Stanford antennas were also in operation. Short bursts of gravitational radiation were searched. Assuming the duration of these possible bursts to be of the order of τ_g=1 ms we calculate the minimum (SNR=1) value of h(t), for the most favourable conditions of direction and polarization, from the formula $h(t) = \frac{1}{\tau_g} \frac{L}{v^2} \sqrt{kT_W/M}$, v being the sound velocity in alluminium. The behaviour of the hourly average of h(t) versus time is shown in the figure. A preliminary coincidence analysis between the antennas at LSU and Rome and LSU and Stanford was done for a period of about one month with a threshold h(t) $\sim 10^{-17}$. Between LSU and Rome, we found one coincidence at zero time delay; for random data the probability for such a coincidence is about 1/7. Between LSU and Stanford no statistically significant coincidences at zero time delay were found above background. A detailed analysis is under way.

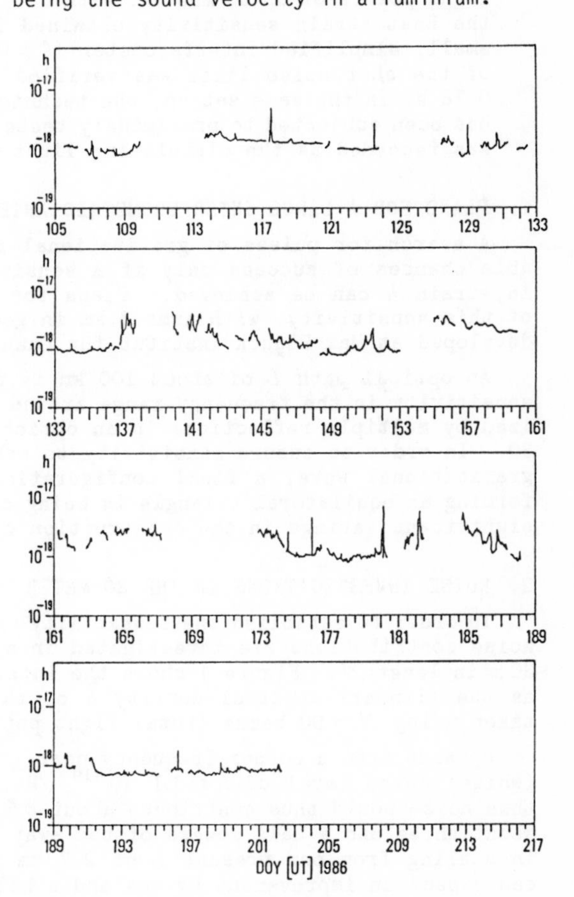

THE GARCHING 30-METER PROTOTYPE AND PLANS FOR A LARGE GRAVITATIONAL WAVE DETECTOR

A. Rüdiger, R. Schilling, L. Schnupp, D. Shoemaker
W. Winkler, G. Leuchs, K. Maischberger

Max-Planck-Institut für Quantenoptik
D - 8046 Garching, FRG

ABSTRACT

The 30-meter prototype of a laser interferometer for detecting gravitational radiation has practically reached the fundamental limit of shot noise in a spectral range from 500 Hz to 6 kHz, at light powers of up to 250 mW. With N = 90 beams in the delay line, the best strain sensitivity obtained is $\tilde{h} = 1.1 \cdot 10^{-19}/\sqrt{\text{Hz}}$. In a small, simplified interferometer ($\ell = 0.30$ m, $N = 2$) the attainment of the shot noise limit was verified even up to light powers of 0.75 W. In the same set-up, the technique of 'recycling' the light has been subjected to preliminary tests. A reduction of shot noise was recorded as the circulating light power was enhanced.

1. PLANS FOR A LARGE INTERFEROMETRIC ANTENNA

A search for pulses of gravitational radiation will have reasonable chances of success only if a sensitivity of better than 10^{-21} in strain h can be achieved. Plans for an interferometric antenna of this sensitivity, with arms 3 km in geometric length ℓ, are being developed at Max-Planck-Institut für Quantenoptik[1].

An optical path L of about 100 km is planned to achieve the best sensitivity in the frequency range around 1 kHz. This is to be realized by multiple reflections in an optical delay line: $L = N \cdot \ell$, $N \approx 30$. In order to ensure sensitivity to arbitrary polarization of the gravitational wave, a final configuration of three interferometers forming an equilateral triangle is being considered, which will allow significant savings in the construction of tunnels and end houses.

2. NOISE INVESTIGATIONS IN THE 30-METER PROTOTYPE

In order to establish the feasibility of a large antenna, the major noise contributions are investigated in a prototype with arms of $\ell =$ 30 m in length[2]. Figure 1 shows the interferometer noise, expressed as the (linear) spectral density $\tilde{h}$ of the strain $h = \delta\ell/\ell$. It was taken using $N = 90$ beams (total light path L about 2.7 km).

Upwards from a corner frequency near 1 kHz, there is a rather flat (white) noise level of $\tilde{h} \approx 1.1 \cdot 10^{-19}/\sqrt{\text{Hz}}$. In a bandwidth of 1 kHz this noise would thus contribute about $3.5 \cdot 10^{-18}$ to the signal, still about three and a half powers of ten away from our goal of $h \approx 10^{-21}$. In scaling from our present L of 2.7 km to the planned 100 km, one can expect an improvement by one and a half powers of ten. The more difficult part will be the further reduction of the noise sources that limit the present displacement sensitivity.

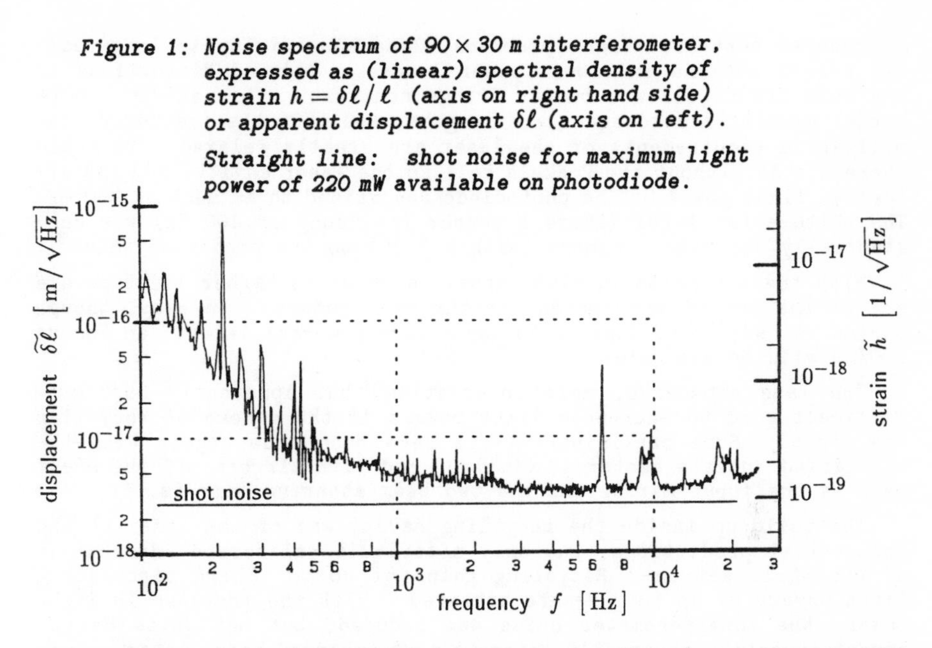

Figure 1: Noise spectrum of 90×30 m interferometer, expressed as (linear) spectral density of strain $h = \delta\ell / \ell$ (axis on right hand side) or apparent displacement $\delta\ell$ (axis on left).

Straight line: shot noise for maximum light power of 220 mW available on photodiode.

The main noise contribution seems to be the shot noise in the photodiode. The theoretical value, taking into account the finite fringe contrast, is shown as the horizontal line; it is calculated for the photocurrent maximum (70 mA), corresponding to a maximum light power of 220 mW impinging on the photodiode.

The structures seen in Figure 1 are significant, since the spectra, representing large numbers of averages, are quite reproducible. The peaks between 6 kHz and about 12 kHz, all well above our frequency range of interest, clearly stem from mechanical resonances. All resonances are at such high frequencies owing to the simple construction and suspension of the main optical components [3].

The corner frequency at which the very steep rise towards low frequencies begins has been moved further down. The remaining low frequency noise has still not been fully explained. The difficulty in identifying its origins is due to a similarly steep rise in all laser noise mechanisms as well as in the seismic background. Seismic noise is quite effectively filtered by a double-pendulum suspension, as borne out by calculated *and* measured isolation transfer functions [3].

It may well be significant that the shot noise limit is not exactly reached. An excess noise of about 3 dB remained unaccounted for. A further reduction of the (theoretical) shot noise by an increase of light power will make other possible noise contributions stick out more clearly and thus allow better identification.

3. NOISE INVESTIGATIONS IN A SIMPLIFIED 0.30-METER INTERFEROMETER

A step in this direction was taken with an extremely simplified interferometer with only 2×0.30 m of light path: the beam is immediately reflected from the *near* mirrors ($N = 2$). This set-up results in reduced scattered-light problems, less sensitivity to longitudinal mirror motions and lower power losses. Reduced distortions of the wave fronts allow a better fringe contrast. This set-up can be better symmetrized to equal arm lengths, and thus the frequency stabilization requirements of the laser are greatly relaxed. We could therefore do without the Pockels cell in the laser cavity, and the effective light power on the photodiode was raised to as much as 750 mW. The white-noise level (above a corner frequency of 400 Hz) was substantially decreased, almost (within 3 dB) to the predicted value[3].

With these results in mind, attempts to go to higher light powers are at the top of the agenda for the near future. An *easy*, though expensive, way to do that is to use a more powerful laser: an Innova 100/20 will be available in early 1987.

The less expensive, more interesting, but apparently *much more difficult* road to increased light powers is the scheme of recycling the light. Some preliminary tests have been made, again with the simplified interferometer (2×0.30 m), and with mirrors off the shelf used as the input mirror and the two beam-steering mirrors.

The build-up inside the recycling cavity was of the order of the theoretical predictions, but only after quite elaborate adjustments of the whole set-up. Recycling gains of up to 15 and circulating light powers of up to 2 W were achieved. With the increase in light power, the interferometer noise was reduced, but not quite by the amount expected, apparently owing to some enhanced beam jitter. Further investigations are under way to clarify this point.

4. SQEEZED STATES

Following the proposal by Caves[4], the prospect of increasing the shot noise limited sensitivity using sqeezed states has been theoretically investigated for a non-ideal interferometer[5] taking into account the specific aspects of the stabilization technique commonly used. It apears that the possible improvement is not only limited by the quantum efficiency of the photo detector and other linear losses, but it also depends critically on other imperfections like wavefront distortion, depolarization and unequal losses in the interferometer arms all leading to a non-zero intensity in a dark fringe. For linear losses of 5% and a fringe visibility of 99% a 2.5-fold improvement in sensitivity seems possible, corresponding to a 6-fold increase in laser power.

REFERENCES

1) W. Winkler, et al.: *Plans for a large gravitational wave antenna in Germany*, Fourth Marcel Grossman Meeting (MG4), Rome 1985

2) D. Shoemaker, et al.: *Progress with the 30 meter prototype for an interferometric gravitational wave detector*, MG4, Rome 1985

3) D. Shoemaker, et al.: *Noise behavior of the Garching 30 meter prototype gravitational wave detector*, to be published (1987)

4) C.M. Caves: *Quantum-mechanical noise in an interferometer*, Phys. Rev. **D 23** (1981) 1693

5) J. Gea-Banacloche and G. Leuchs: *Applying squeezed states to non-ideal interferometers*, to be published (1987)

SOME IDEAS ABOUT (I) DETECTING LOW-FREQUENCY GRAVITATIONAL RADIATION AND (II) X-RAY FAST TIMING OBSERVATIONS TO SEARCH FOR GRAVITY WAVE PULSARS

Peter F. Michelson

Department of Physics, Stanford University
Stanford, California 94305

ABSTRACT

In recent years it has become apparent that it is desirable to extend the sensitivity of gravitational radiation detectors to lower frequencies. The sources of interest that may be detectable at lower frequency with terrestrial detectors include stochastic gravitational radiation. In part 1 of this paper the sensitivity to stochastic background radiation of a cross-correlation experiment with widely separated detectors is discusssed. In part 2 of this paper the possibility of detecting millisecond x-ray pulsars in low-mass accreting x-ray binary systems is discussed. This is especially relevant to gravitational wave astronomy because these sources could be continuous emitters of gravitational radiation.

1. SENSITIVITY OF CORRELATION EXPERIMENTS TO STOCHASTIC BACKGROUND RADIATION

Although there are no firm predictions about the strength of stochastic-background gravitational radiation in the frequency range likely to be accessible by terrestrial detectors ($f > 10$ Hz), it is fair to say that $\Omega_g < \Omega_{rad}$ ($\sim 2.4 \times 10^{-5}$) for the region of plausible signals[1]. Ω_{rad} is the density parameter for the photon background. A number of sources that include cosmological phase transitions, cosmic strings, primordial scale-invariant metric fluctuations, and overlapping bursts from collapse and coalescence of massive Population III objects , have been discussed in the literature[1]. Rather than discussing the possibility of detecting any particular source, I will make the simplifying assumption that the background can be characterized by a broadband scale-invariant spectrum with an rms dimensionless wave strain, in a bandwidth $\Delta f = f$, of

$$< h > = h_g(f)\sqrt{f} = 10^{-22} (\Omega_g / 10^{-8})^{1/2} (1 \text{ Hz} / f) . \quad (1)$$

For a cross-correlation experiment with two identical detectors that are not necessarily colocated or aligned[2], the signal-to-noise ratio (SNR) is given by

$$S/N = \pi^2 (\Delta f\, t)^{1/2} \frac{\int g(fx, \Omega_1, \Omega_2)\; h_g^2(f)\, df}{\int h_n^2(f)\, df} , \quad (2)$$

where t is the integration time and h_n^2 is the strain noise spectral density of the detectors. The function g depends on the frequency, the detector configurations, the separation of the detectors and their angular orientation. The SNR of eq.(2) is obtained assuming that the output of one of the detectors is filtered by a filter with transfer function H(f) = sgn(g). In the limit of a local correlation experiment eq.(2) reduces to the usual expression [3]. Fig. 1a shows the angular dependence of g for different values of the relative separation. This is the behavior expected of the cross-correlation function R for a narrowband measurement. Fig. 1b shows the dependence of R for a broadband measurement with $\Delta f = f$. These curves are computed for the case of two linear quadrupole detectors with axes oriented along ε_1 and ε_2.

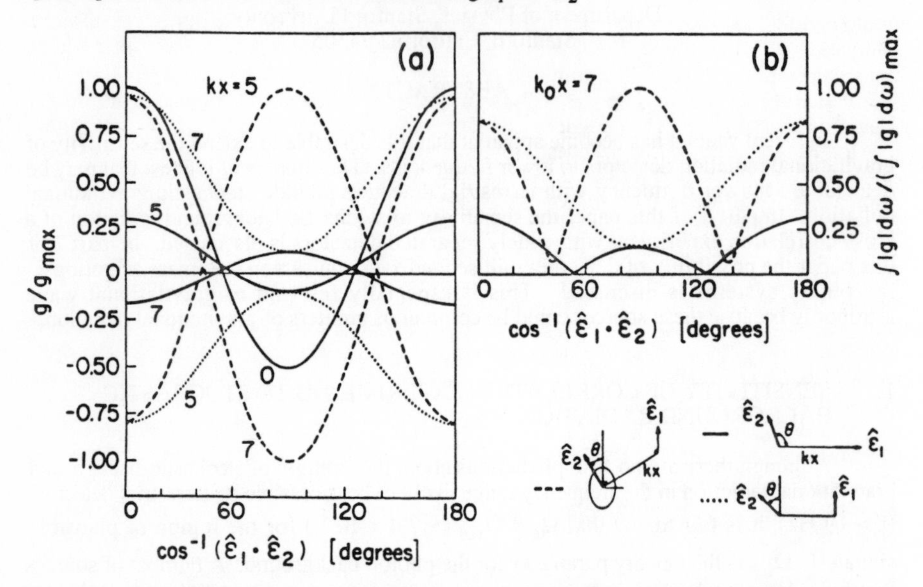

Figure 1: (a) Angular dependence of g expected for two linear quadrupole detectors; (b) the angular dependence of R for a broadband measurement ($\Delta f = f$).

The solid curve of Fig. 2 illustrates the reduction in SNR as detector separation is increased. This is the SNR for a broadband measurement derived assuming optimum relative orientation of the detectors. The dotted curve is derived with the same assumptions except that a filter with transfer function sgn(g) has not been applied. An important conclusion to be drawn from a detailed analysis is that the (energy) SNR falls off roughly as (detector separation)$^{-1}$.[2]

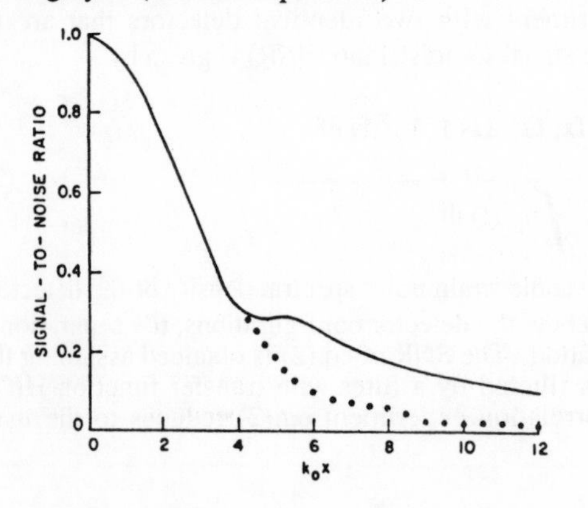

Figure 3: Signal-to-noise ratio (SNR) as a function of detector separation. The solid curve is derived assuming optimum relative orientation of the two detectors and a measurement bandwidth equal to frequency. The SNR is normalized to unity at zero separation. The dotted curve is derived with the same assumptions except that a filter with transfer function sgn(g) has not been applied.

2. X-RAY FAST TIMING OBSERVATIONS TO SEARCH FOR GRAVITY WAVE PULSARS

Among the candidates for millisecond x-ray pulsars are weakly-magnetized neutron stars in low-mass x-ray binaries. A neutron star in such a system may be spun-up by accretion to a period near a millisecond. In this region the star is unstable to nonaxisymmetric modes which grow by gravitational radiation reaction[4]. A stability analysis shows that the star's viscosity, which damps the modes, will determine which nonaxisymmetric mode will dominate. Wagoner[5] showed that in the steady-state it is likely that an $m = 3$ or 4 mode will dominate and the system will emit continuous gravitational radiation in the frequency range between about 200 Hz and 1 kHz. Without a priori knowledge of the frequency of this radiation its detection will, at best, be difficult. The detection problem is not aided by the binary motion of the source (usually unmeasured). However, we can expect the x-ray emission from the system, generated by the accretion process, to be weakly modulated at the same frequency as the gravity wave signal. The sensitivity of a search for such an x-ray pulsar is limited by the detector collecting area and is subject to the same degradation in signal-to-noise caused by the binary motion effects. The sensitivity to a weakly-modulated flux can be improved dramatically using a large detector. Then, the integration time can be kept short enough to avoid significant degradation in signal-to-noise from orbital Doppler effects. Because most x-ray instruments flown to date have lacked the area and telemetry rate to do a sensitive search, the best chance to detect such an x-ray/gravity-wave pulsar may be with the X-Ray Large Array (XLA), a new concept for a space station based 100 m^2 x-ray detector array proposed by the Naval Research Laboratory, Stanford University, and the University of Washington[6]. The dual channel detection of the source in both x-rays and gravity waves would provide two complementary measures of the neutron star's properties.

3. ACKNOWLEDGEMENTS

I gratefully acknowledge useful discussions about all aspects of this work with W.M. Fairbank, M.S. McAshan, J.C. Price, T. Stevenson and R.C. Taber. Many people have contributed to the XLA concept in ways too numerous to discuss here; in particular K.S. Wood, W.M. Fairbank, H. Gursky, H. Friedman, P. Boynton, R. Wagoner, M.R. Yearian, J. Norris, A. Bryson, E. Reeves and P. Hertz. The work discussed here has been supported in part by the National Science Foundation.

4. REFERENCES

1. C. J. Hogan, "Stochastic gravitational radiation from cosmological phase transitions", *Mon. Not. R. Astr. Soc.* **218**, 629 (1986).
2. P.F. Michelson, "On detecting stochastic background gravitational radiation with terrestrial detectors", submitted to *Mon. Not. R. Astr. Soc.* (January, 1987).
3. R. Weiss, in *Sources of Gravitational Radiation*, edited by L. Smarr (Cambridge University Press, Cambridge, 1979).
4. J.L. Friedman and B.F. Schutz, "Secular instability of rotating Newtonian stars", *Ap. J.* **222**, 281 (1978).
5. R.V. Wagoner, "Gravitational radiation from accreting neutron stars", *Ap. J.* **278**, 345 (1984).
6. K.S. Wood (NRL), P.F. Michelson (Stanford), P. Boynton (Washington) *et al.*, *An X-ray Large Array (XLA) for the NASA Space Station* , proposal to the National Aeronautics and Space Administration (December, 1985).

STATUS REPORT ON THE LASER INTERFEROMETRIC GRAVITATIONAL WAVE DETECTOR AT GLASGOW UNIVERSITY

N.A. Robertson, J. Hough, G.A. Kerr, N.L. Mackenzie, J. Mangan, B.J. Meers, G.P. Newton, D.I. Robertson and H. Ward

Department of Physics and Astronomy
University of Glasgow, Glasgow G12 8QQ, Scotland

Over the past 6 years we have been developing the 10 metre baseline prototype laser interferometric gravitational wave detector at Glasgow[1]) and have been improving its displacement sensitivity by a factor of approximately 4 per year. Currently we are investigating possible noise sources which have been limiting the sensitivity of the detector above the level dictated by photon shot noise in the amount of light at present being used. Earlier this year a sensitivity of $(2 - 3)\times10^{-17}$ m/$\sqrt{\text{Hz}}$ around 1 kHz had been achieved, and it was thought that the piezoelectric transducers used to control the length of one of the Fabry-Perot cavities might be contributing excess noise to the system. Continuing investigations led us to realise that the type of bonding being used between the cavity mirrors, the piezoelectric transducers and the test masses was a dominant factor in the level of noise perceived. In particular our best results to date have been achieved by using a cyano-acrylate glue between the mirrors and test masses, and removing the piezoelectric transducers altogether to reduce the number of bonds. This has necessitated the development of a different method to control the cavity length which is now achieved using a coil and magnet system, the coil being bonded onto the rear of the test mass. A sensitivity of $\sim 5 \times 10^{-18}$ m/$\sqrt{\text{Hz}}$ at 2 kHz has been achieved using cyano-acrylate glue. By comparison, the use of silicone oil or a softer glue raises the noise level by greater than 10 db. However photon shot noise has not yet been achieved, and we believe that we are still limited by excess noise associated with the bonding and by mirror distortion introduced by the bonding. We plan soon to investigate the use of thermally matched mirrors and test masses, with the mirrors optically contacted to the masses.

Our plans for a detector of km arm length are well advanced, a design study having been completed earlier this year[2]) . Funding for this project is being sought at present from the SERC in Britain.

REFERENCES

1. Ward, H., Hough, J., Newton, G.P., Meers, B.J., Robertson, N.A., Hoggan, S., Kerr, G.A., Mangan, J.B., and Drever R.W.P., IEEE Transactions on Instrumentation and Measurement, IM-34, 261 (1985).
2. Hough, J. et al., 'A British Long Baseline Gravitational Wave Observatory', Design Study Report, May 1986 (GWD/RAL/86-001)

PROSPECTS FOR USING THE NAVSTAR/GPS CONSTELLATION AS A DETECTOR OF GRAVITATIONAL RADIATION

Peter M. Hansen
The Aerospace Corporation

SUMMARY

The Navstar/GPS constellation will include 18-21 satellites in inclined, semi-synchronous orbits. Each satellite will continuously transmit navigational data on two separate carrier frequencies whose stability is controlled by atomic standards. The atomic clocks will be cesium-beam ($\Delta\nu/\nu \sim 10^{-13}$) and eventually hydrogen-masers ($\Delta\nu/\nu \sim 10^{-15}$). It was hoped that Doppler tracking of these carriers could provide continuous (24 hr.), and global (4π sr.), gravity wave detection data for unlimited periods in the future.

A major difficulty in using these satellites is tracking them over extended intervals through variable, and scintillating, regions of the ionosphere ($\Delta\nu/\nu \sim 10^{-11}$). A computer simulation and detection experiment performed at The Aerospace Corporation showed that data processing and analysis techniques are capable of detecting gravity wave effects for $h_g \sim 10^{-14}$, or at the frequency stability level of improved cesium standards. This detection limit was determined by the ability to reconstruct, and remove, earth model ephemeris effects from the Doppler tracking data.

Correlated State Gravitational Radiation Antennas*

J. Weber

University of Maryland, College Park, Maryland 20742

and

University of California, Irvine, California 92717

Some years ago B. S. DeWitt[1], Thorne[2], and Braginsky[3] proved that gravitational radiation antennas can measure the Riemann tensor exactly, without any quantum limit.

For antennas in thermal equilibrium, the energy absorption is proporational to the square of the Riemann[4] tensor, and the fluctuations are those of systems with known temperatures.

Antennas may be prepared, in well defined[4] quantum states, very different from thermal equilibrium states. If quality factors are high, such states will persist for long periods required for the antenna to approach thermal equilibrium.

For antennas in correlated states the energy absorption is a linear function[4] of the Riemann tensor, and the fluctuations may be much smaller[4] than for systems in thermal equilibrium. This provides, then, one way to achieve sensitivity greater than the limits for systems in thermal equilibrium, with amplifiers limited by spontaneous emission.

*Research supported in part by the National Science

Foundation.

1. B. S. DeWitt, in Gravitation: An Introduction to Current Research, L. Witten (Ed.), J. Wiley and Sons, New York (1962).

2. Thorne, Drever, Caves, Zimmermann, and Sandberg, Phys. Rev. Lett. 40, 667, 1978.

3. V. B. Braginsky, in Topics on Theoretical and Experimental Gravitation Physics, edited by V. de Sabbata and J. Weber, Plenum Press, New York, Page 105, 1977.

4. J. Weber, Phys. Rev. A, 23, 2, 761, 1981, Physics Letters 81A, 542, 1981.

Sir Arthur Eddington Centenary Symposium, Vol. 3, pages 1-77, World Scientific 1986.

PERTURBED BINARY SYSTEMS - AN UNUSUAL SOURCE OF GRAVITATIONAL RADIATION

Michael Fitchett
Canadian Institute for Theoretical Astrophysics,
University of Toronto, Toronto, M5S 1A1 CANADA

Binary systems are an important source of gravitational radiation. Most theoretical estimates of the likely amplitude of gravitational radiation from binary systems have assumed the orbits of the components to be *circular*, in which case the close binary systems have the highest gravitational wave luminosity. However, if a binary system is perturbed, the resultant orbit (if bound) may be very eccentric. Since $L_{gw} \propto (1 - e^2)^{-7/2}$, perturbed systems may have a very high gravitational wave luminosity. This paper briefly summarizes a detailed calculation[1)] of the case of a velocity perturbation to one component of a binary system. Expected velocity perturbations are not large and so the perturbations will be most effective for *wide* binary systems. Assuming the direction of the velocity perturbation to be isotropically distributed, it is possible to show that:

(i) such perturbed systems are on average as luminous a source of gravitational waves as close unperturbed binaries.

(ii) the duration of emission prior to coalescence is very similar to that of the close binaries.

(iii) the characteristics of the gravitational radiation emitted are very interesting. The most luminous perturbed sources generate power at many multiples of the orbital frequency. The peak in the distribution corresponds to frequencies similar to those at which the close circular systems radiate. The spectral distribution is similar in shape to that of a *burst* source, but these sources are also *continuous*.

(iv) orbital evolution due to gravitational radiation energy and angular momentum losses causes these sources to become *more* luminous, despite the fact that their orbits are circularizing. In fact all of their interesting properties are maintained until coalescence.

This effect will be of most significance for the binary VMO (very massive object) systems postulated by Bond and Carr[2)]. However such perturbations could also be important for low mass systems. On a purely speculative note, maybe the unusual character of the gravitational radiation emitted by these systems would enable the ingenious workers in the field of gravitational radiation detection to come up with some way of detecting them.

1 Fitchett,M.J., 1987, to appear in *Monthly Notices of the Royal Astronomical Society*

2 Bond,J.R., & Carr,B.J., 1984, *Monthly Notices of the Royal Astronomical Society*, **207**, 585

LIMITS ON A STOCHASTIC GRAVITATIONAL WAVE BACKGROUND FROM OBSERVATIONS OF EARTH NORMAL MODES

S. P. Boughn, Haverford College

J. R. Kuhn, Michigan State University

The lowest quadrupole mode, ${}_0S_2$, of the earth has a relatively large cross-section for gravitational waves with a frequency of 0.31 Hz. Previous anslyses of the excitation of this mode have led to upper limits to the energy spectral density of gravitational radiation at this frequency which are several orders of magnitude larger than the closure density of the universe per octave.[1,2] The seismic background at this frequency is apparently due to local atmospheric pressure load variations[3] and not to actual excitation of the normal mode. In fact, the ${}_0S_2$ mode has only been observed after huge earthquakes.[4] Therefore, by averaging seismic power spectra at quiet locations much lower limits can be expected.

We have begun to analyze approximately 200 station-years of data collected by the Project IDA global network of vertical seismometers.[5] Programs[6] were written to automatically remove tidal accelerations, earthquakes and, data dropouts. Thirty day power spectra of the cleaned data are computed and the individual power spectra averaged. Three station-years of data have been analyzed and no excitation of the ${}_0S_2$ is observed. This implies an upper limit on the gravitational wave background of 2×10^6 erg/s cm^2 Hz or about closure density/octave. Fundamental modes with spherical harmonic order higher than 8, i.e., ${}_0S_\ell$ $\ell > 8$, are observed in the data and are probably the residual excitation of small earthquakes near the earth's surface.

By averaging the remainder of the IDA data the seismic noise should be reduced by about an order of magnitude. If no resonant excitation is observed, then the above upper limit will be reduced by a factor of ten. Rough calculations indicate that the background excitation of the ${}_0S_2$ mode by earthquakes is another three orders of magnitude below this; however, this sensitivity could only be achieved if much quieter sites are located or if a method to correct for local seismic noise is found. Local barometric pressure load fluctuations, although globally incoherent, excite the earth modes at some level. The magnitude of this effect has not yet been estimated.

1) Weber, J., Phys. Rev. Letters, 18, 498 (1967).
2) Boughn, S. P. and Kuhn, J. R., Ap. J., 286, 387 (1984).
3) Agnew, D. C. and Berger, J., Jour. Geo. Res., 83, 5420 (1978).
4) Aki, K. and Richards, P. G., Quantitative Seismology, vol. 1, (W. H. Freeman, San Francisco, 1980), pp 363-364.
5) Agnew, D. C., Berger, J., Farrell, W., and Gilbert, F., Eos Trans. AGU 57, 180 (1976).
6) Programs written by G. Zweibel and C. O'Neal, Princeton University.

Topology of the Universe: Motivation for the Study of Large–Scale Structure

Adrian L. Melott
Department of Physics and Astronomy
University of Kansas
Lawrence, Kansas 66045 U.S.A.

ABSTRACT :

The large–scale structure of the Universe has been characterized as spongy, or like isolated clusters, sheets, filaments, cells, bubbles, and other analogs. The use of a computer program to study isodensity surfaces separating over– and under–dense regions can quantify the topology. Gaussian distributions can be distinguished from, for example, bubble distributions using this test. Improved data from redshift surveys and QSO absorption line studies will help test the random phase hypothesis in cosmology.

1. INTRODUCTION

Recently, interest has grown in the topology of large–scale structure in the Universe. These topological studies concentrate on the connectedness of structures, rather than their size. Thinking has considered, for example, whether the supercluster–void distribution resembles a "cell structure." In a cell structure presumably there are isolated voids surrounded by cell walls, with all the walls connected. One could equally imagine isolated clumps of galaxies surrounded by empty space.

The question can be quantified by smoothing the discrete data with a Gaussian, and examining the connectedness of regions bounded by an isodensity surface. Hereafter I shall save space by letting G = Gaussian. It has been argued[1)] that for a G distribution the regions bounded by the mean density should be a sponge — with both "supercluster" and "void" regions mutually interlocking and equivalent. (I will refer to regions denser than the chosen threshold as "superclusters" and regions less dense as "void".) It has been argued that observational data is compatible with this hypothesis[1,2,3)].

Still, showing that the data is compatible with a sponge structure does not prove the random phase (G) hypothesis. More detailed examination of various possibilities is needed. I will describe recent work with J. R. Gott and D. Weinberg on testing this idea, as well as work in progress.

2. MATHEMATICAL BASIS

When we look at galaxies in space, we are not interested in them as isolated points but as tracers of structure. We therefore construct a smoothed density field by convolving the data with a window $W = e^{-r^2/\lambda^2}$. The density field (sampled in cells) can then be divided into regions above or below a given threshold.

The structure of the "superclusters" can be characterized by the genus, or "number of holes" in the structure. The genus (g) is the number of cuts which can be made without dividing the surface into two pieces. A torus (donut) has g = 1; a sphere has g = 0. Two spheres have g = −1, since one cut "undone" could join them. It has been argued that a random phase distribution should be spongy, (have large positive g), at the median density[1]. In a G distribution, this can be shown analytically, and the genus calculated for arbitrary density thresholds[5,6]

3. NUMERICAL RESULTS

The solid curve in Fig. 1 shows the generic shape of the g-ν curve for any G distribution. The amplitude of this curve is determined by an integral over the power spectrum; we used the CDM spectrum here[7]. We use ν to indicate how many standard deviations the threshold is from the mean. (We use this notation even in non–G distributions.) Also in Fig. 1 we show visual realizations of a cold dark matter model. The notation "7% high" means the contours enclose the 7% most dense portion of the volume; "7% low" means the 7% most empty portion.

On the genus curve, we compare the initial, final (evolved), and biased cold dark matter CDM distributions to the theoretical value based on the power spectrum we tried to fit, with good agreement. Of course, the initial, final, and biased topologies need not agree — but they do here. A CDM model should look very G. Theories such as cosmic strings or explosions in which structure arises from non–G perturbations need not fit this curve and most do not[8]. It is important to examine the genus curve because many of these do have g(0) > 0, which is "spongelike", but they do not follow the G genus curve. Examining models in redshift space[9] has shown only a slight effect on the topology in most cases; even if large–scale velocity fields exist, this should be true since no one seems to suggest flows where $\Delta v/v_H > 0.1$.

Non–G behavior also can arise gravitationally from G perturbations. We know this is true because many numerical studies have shown nonzero three–point correlations. I want to show results from a "hot dark matter"[10] model, sometimes called the "pancake theory". The initial conditions are G, (see the dashed line below) but the result of dynamical evolution is not[9]. This particular asymmetry is characteristic of bubble distributions, or of systems of sheets. By the way, it does not resemble a filamentary net, because the latter does not keep isolated holes dominant for $\nu > -1$ as do bubble, sheet, and HDM systems[11]. The genus curve and pictures in Fig. 2 indicate the behavior of this model.

Note that 16% low looks very different from 16% high — it has isolated bubbles, even when periodic boundary conditions are taken into account. It is time to stop talking in general terms about whether models or galaxy surveys look like specific visual analogies, and quantify these things.

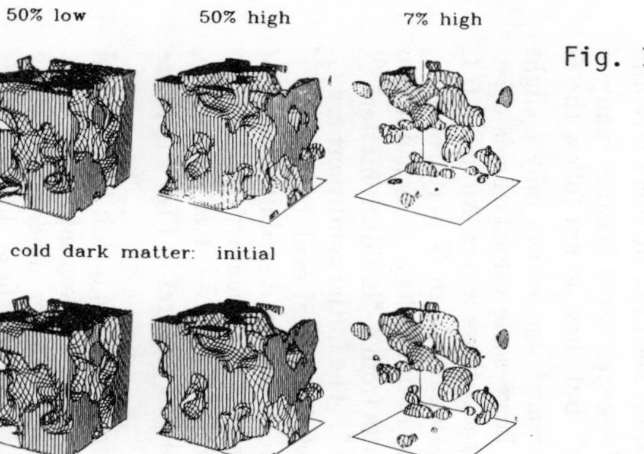
7% low
50% low
50% high
7% high
cold dark matter: initial
cold dark matter: final

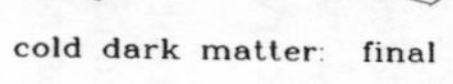

Fig. 1

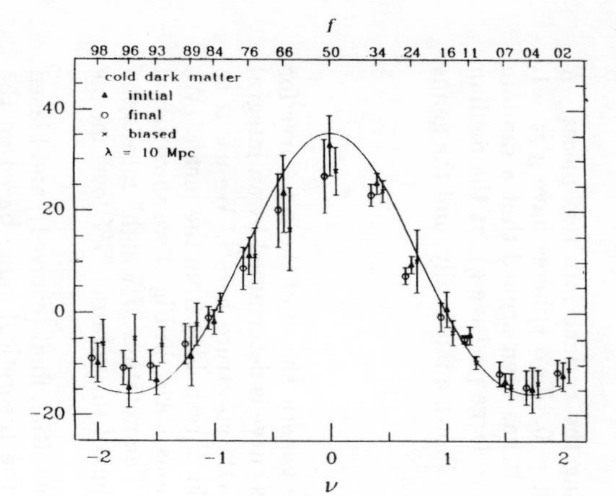
f
98 96 93 89 84 76 66 50 34 24 16 11 07 04 02
cold dark matter
initial
final
biased
λ = 10 Mpc
genus
40
20
0
−20
−2
−1
0
1
2
ν

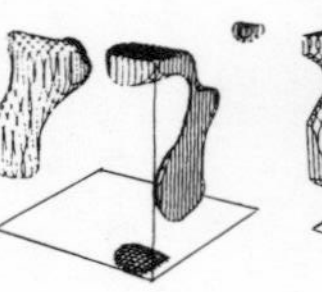
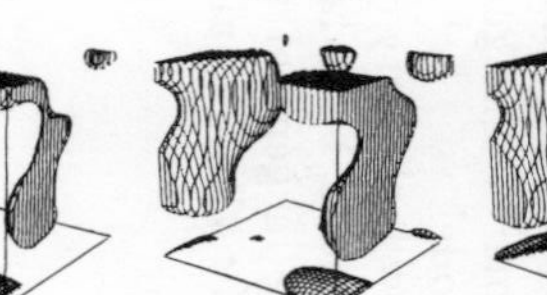
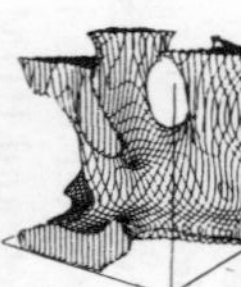
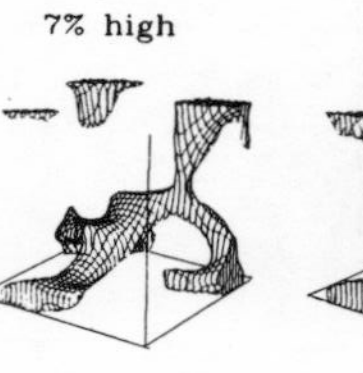
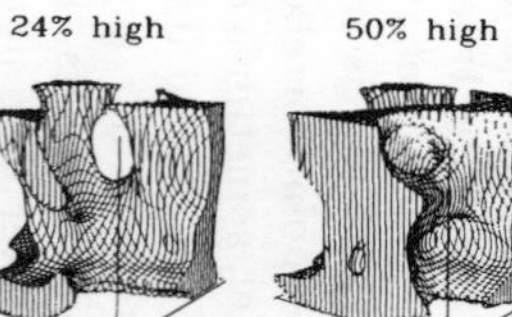
7% low
16% low
24% low
50% low
7% high
16% high
24% high
50% high
neutrino model: final

Fig. 2

final
λ = 3
5
0
−5
−2
−1
0
1
2

4. FUTURE WORK

We made an error early on[1] by failing to do that. We said that one could not distinguish between sponges and bubbles by a cross–sectional view because one could be seeing cross–sections of tunnels or holes. However, a cross–section of a G distribution should be (when properly smoothed to remove discreteness) light–dark symmetric, like the famous yin–yang symbol (Fig. 3) of Taoism. This is different from light holes on a dark background, or vice–versa. Thus it may be possible to discriminate quantitatively based on a slice of the Universe[12], and such work is underway[13]. Even much of the mathematical formalism goes over to 2D[14], with some modification. But 3D surveys over as much volume as possible are still needed as well, beginning with our neighborhood[15]. Resolution is not a problem in implementing this work. Tests have shown that smoothing over λ equal to the mean intergalaxy separation is adequate to remove discreteness. This scale is much smaller than the size of superclusters and our studies show one can easily distinguish, e.g. a bubble from a G distribution with such smoothing. The real problem is sample volume. Our early analysis of the CFA survey[1] was hampered by our need to cut out a cube and use periodic boundary conditions. We now have eliminated these constraints, and also plan to use other data with greater linear scale[13,15].

It is also important to look back in time to study the evolution of the topology. We can more reliably test the random phase hypothesis by looking at larger scales and earlier times. QSOs can be used as probes of large as well as small–scale structure, by carefully designed surveys.[16] Large ground–based telescopes coupled with the Hubble Space Telescope will be needed to optimally explore all redshift regimes. Even if this were only "mapping the Universe", it would be worthwhile. We also have a chance to learn something about its initial conditions.

5. ACKNOWLEDGEMENTS

I am grateful to my collaborators for their talents and permission to discuss common work, as well as to the NSF for supercomputer time under grants AST–83–13128, AST–86–13897, a Dudley Observatory travel grant, and University of Kansas General Research Allocation 3868–20–0038.

REFERENCES

1) Gott, J. R. Melott, A., and Dickinson, M., *Ap.J.* **306**, 341 (1986).

2) Einasto, J., Einasto, M., Melott, A., and Saar, E., *M.N.R.A.S.*, in press (1987).

3) S. Gregory, personal communication (1986); R. Giovanelli and M. Haynes, personal communication (1986).

4) Gott, J. R., Weinberg, D. H., and Melott, A. L., *Ap.J.*, submitted (1987).

5) Bardeen, J. M. Bond, J. R., Kaiser, N. and Szalay, A. S., *Ap.J.* **304**, 15 (1986).

6) Hamilton, A. J. S., Gott, J. R., and Weinberg, D. H., *Ap.J.* **309**, 1 (1986).

7) Peebles, P. J. E., *Ap.J.Lett.* **263**, L1 (1982); Blumenthal, G., Pagels, H., and Primack, J., *Nature* **299**, 37 (1982).

8) Weinberg, D. H., Charlton, J. Dekel, A. Gott, J. R., Melott, A. L., Ostriker, J. P., and Scherrer, R., in preparation (1987).

9) Melott, A. L., Weinberg, D. H., and Gott, J. R., *Ap.J.*, submitted (1987).

10) Bisnovatyi–Kogan, G. S., and Novikov, I. D., *Sov.Astron.* **24**, 516 (1980); Doroshkevich, A. G., Zel'dovich, Ya. B., Sunyaev, R. A., and Khlopov, M. Yu, *Sov.Astron.Lett.* **6**, 252 (1980); Bond, J. R., Efstathiou, G., and Silk, J., *Phys.Rev.Lett.* **45**, 1980 (1980).

11) Weinberg, D. H., Gott, J. R., and Melott, A. L., *Ap.J.* submitted (1987).

12) de Lapparent, V., Geller, M., and Huchra, J. P., *Ap.J.Lett.* **302**, L1 (1986).

13) Hudson, J. P., Melott, A. L., Weinberg, D. H., and Gott, J. R., in preparation (1987).

14) Hamilton, A. H., Gott, J. R., and Melott, A. L., in preparation (1987).

15) Bhavsar, S., Giovanelli, R., Gott, J. R., Haynes, M., Melott, A. L., and Weinberg, D. H., in preparation (1987).

16) Crotts, A. P. S., Melott, A. L., York, D. G., and Fry, J. N. *Phys.Lett.* **155B**, 251 (1985).

LIQUID MIRROR TELESCOPES

Ermanno F. Borra

Département de Physique, Université Laval, Québec, Canada G1K 7P4

ABSTRACT

I describe the liquid mirror telescope concept and the results of tests performed on 1-m diameter prototype liquid mirrors. The concept appears sounds. I speculate briefly on the future of the liquid mirror telescope.

1. Liquid mirror telescopes.

The basic concept has been discussed by Borra[1]. It is rather straightforward to show that, in a rotating fluid, adding the vectors of the centrifugal and gravitational accelerations gives a surface that has the shape of a parabola . Using mercury one gets therefore a reflecting parabola that could be used as the primary mirror of a telescope. The focal length of the mirror L is related to the acceleration of gravity g and the angular velocity of the turntable ω by

$$L = g/2\omega^2 \qquad (1)$$

For mirrors of astronomical interest the periods of rotation are about 6 seconds and the linear velocities of the rims range between 5 and 20 km/h.

The characteristics of mercury (such as reflectivity) are discussed in [1] and [3]. Metallic mercury is a toxic metal but not overly so. Mercury evaporates very slowly; even a large liquid mirror telescope will not be a threat to the environment and minimal ventilation will keep the telescope room safe.

Liquid mirror telescopes can only observe the zenith and cannot therefore track like conventional telescopes. Liquid mirror telescopes were never taken seriously in the past because of this major limitation. On the other hand, modern technology gives us now alternate tracking techniques. For imagery or slitless spectroscopy one can track with a CCD operating in the driftscan mode and clock the pixels at the same rate as the image drifts in the focal plane. This technique is currently used at Steward Observatory to obtain wide-band direct images with a fixed transit telescope [2]. Although a night yields integration times of the order of a minute per object, coadding the nightly images gives longer effective integration times. High or medium resolution slit spectroscopy is also possible with fiber optics. Tracking could be done mechanically , in the focal plane, with the tip of a fiber. The light could then be piped to a spectrograph on the ground. This technique has not been demonstrated but the mechanical precisions needed are not very high and optical fibers are used for astronomical spectroscopy ; there is thus every reason to believe that the technique is feasible.

2. Tests and scaling up considerations.

The results of optical shop tests with a 1-m diameter f-1.65 liquid mirror have been published[3]. Knife-edge tests showed that the central part of the mirror was of good quality and that the focal length was stable. The tests also showed the presence of low-amplitude ($< \lambda/10$) concentric ripples caused by vibrations originating in the laboratory. These ripples can be dampened and should not be present on an isolated mountain top. Imaging of a resolution test chart confirmed the results of the knife-edge test. Hartmann tests showed that the overall shape of the surface is a parabola, within the uncertainties of the tests ($\sim \lambda/10$).

Borra, Beauchemin and Lalande[4] have reported on a small number of observations of star trails obtained with a 1-m liquid mirror telescope. They obtained images of the order of 2 arcseconds, consistent with the seeing expected at the location of the telescope. They also discussed some scaling-up considerations (curvature of the earth, coriolis forces), concluding that mirrors 30-m mirrors should be feasible.

3. Recent developments

We have built a 1-m diameter f-4.7 LMT on the campus of Laval University and operated it during every clear night of the Summer and Fall of 1986. We used a programable 35-mm camera as detector to look for optical flashes and flares in stars and, especially, to gain practical experience with the operation of a LMT. The camera was programmed to take consecutive frames of 2 minute duration. We did not have any tracking. Within the 2 minute duration of every 35-mm frame, stars show up as long trails . Flashes in the sky would show up as trails of short duration. The observatory was very simple. It was very similar to the set-up shown in [4]. We did not build any foundations; the telescope frame and the mirror rested directly on the ground.

Star trails give a very good and simple way to evaluate the behavior of the mirror, for the time dimension runs along the length of the trails. Any abnormal behavior (focus changes, oscillations, etc..) can readily be detected by a simple visual examination. The results were very satisfactory, especially if one considers the simplicity of the telescope. Focus was generally stable, although it sometimes tended to drift slowly. On several nights we obtained 2 arcsecond wide trails that stayed in sharp focus during the 1 hour 12 minutes duration of a 35 mm roll of film. We experienced at first some difficulty with the wind but the problems were solved by adding a thin mylar cover above the container. The mylar cover is very thin (8 microns thick), does not appear to affect the quality of the images but shelters effectively the surface of the mirror from the wind. We observed on 42 clear nights and obtained 200 hours of data on film. On a few nights we detected oscillations of the images, with low amplitude (a few arcseconds) and periods of the order of a few seconds. They are probably due to imperfect leveling of the mirror. A full evaluation of the data is in progress and will appear elsewhere. We found evidence of concentric ripples on the mirror as reported in [3] and [4]. The concentric rings, seen around bright stars, caused by these ripples seemed to be much fainter than the ones seen in [4]. This may be due to the fact that the telescope was located in a slightly different place. The ripples probably originate from a nearby building.

4. Astronomy with liquid mirror telescopes.

Liquid mirror telescopes make sense only if they can be built to large dimensions and at moderate costs. All our results seem to indicate that this can be accomplished. In particular, I would like to emphasize that making liquid mirrors has been very easy. Setting up the testing facilities and understanding the results of the tests done on very fast parabolas took most of our energies. The entire project has also been very cheap as it has cost about 30,000 dollars. These facts are obviously very important if we hope to build very large liquid mirror telescopes at low cost.

Given that Liquid mirror telescopes should have the expected advantages of cost and size one can foresee building a few 30-m class LMT's dedicated to the study of very faint objects. If a 30-m LMT can be produced at a cost of the order of 10 million dollars than one can then envision the possibility of building 10 LMT's next to each other for an equivalent collecting area of 100-m. However a major improvement to the way astronomical observations are carried out could be brought up by building many 8-meter class LMT's at different latitudes. If they can be produced, individually, for about 2 million dollars, many institutions could presumably afford such instruments. This would give a coverage of most of the sky as there are plenty of dark sites between -40 and + 50 degrees of latitude. For an investment of a couple of hundred million dollars (spread among several countries) one would have access to most of the sky, removing therefore the sky coverage limitation of the zenith telescope. An even greater benefit of this idea is that a very large number of people would have access to a substantial amount of telescope time. Because telescope time would be less valuable than it is now, researchers would be more free to experiment with instrumental set ups and to repeat crucial observations. This array of telescopes could first be used for an all sky very deep multicolor survey. After the survey, they could be scheduled much as telescopes are scheduled nowadays, with telescope time granted to investigators of other institutions.

Finally, looking into the next century, one can envision a 30-m LMT on the moon or even perhaps a 100-m multiple mirror LMT. That could be the ultimate telescope.

REFERENCES

1) Borra, E.F. , J. Roy. Astron. Soc. 76, 245 (1982)

2) McGraw ,J.T., Cawson, M.G.M., Keane, M.J., and Cornell, M.E., Bull. Am. Astron. Soc. 18, 942 (1987)

3) Borra, E.F., Beauchemin, M. , Arsenault, R., and Lalande, R., Publ. Astron. Soc. Pac. 97, 454 (1985).

4) Borra, E.F., Beauchemin, M., and Lalande, R., Astrophys. J. 297, 846 (1985)

GROUND-BASED INTERFEROMETRY

F. Roddier

National Optical Astronomy Observatories
ADP Division, Tucson, Arizona 85726

ABSTRACT

The results of high angular resolution interferometric observations are reviewed and prospects of future developments with ground-based telescopes are discussed. Stellar sources up to magnitude m = 16 have been resolved at the diffraction limit of large telescopes. Complete image reconstruction of turbulence degraded images is now currently achieved and new promising techniques are being developed for this purpose. Submilli-arcsecond resolution has been achieved with long baseline interferometers using intensity interferometry and milli-arcsecond resolution is now achieved with much higher sensitivity using direct Michelson interferometry.

1. Introduction

As shown in table I below, the resolution of ground-based telescopes is severely limited by atmospheric turbulence. It improves only slightly in the infrared where it becomes limited by the telescope size. For an 8-m telescope under excellent seeing conditions, the diffraction limit of 0.3" could be achieved at 10 microns. The Hubble Space Telescope will provide a factor 10 improvement in the visible, but little or none in the IR due to its limited size. In the UV range, its resolution limit will be comparable to that in the visible because the mirror figure is not accurate enough for the diffraction limit to be reached.

wavelength (microns)	0.25	0.5	2.2	5	10
8-m ground-based telescope (excellent seeing)		0"5	0"4	0"34	0"3
2.4-m Space Telescope	0"05	0"05	0"22	no resolution gain compared to ground	

Table I: Resolution limits at optical wavelengths

Because telescopes can be built on the ground much larger than in space, it is worth developing techniques that will allow imaging at the diffraction limit of such telescopes. This limit is given in table II for an 8-m telescope.

wavelength (microns)	0.5	2.2	5	10
angular resolution	0"015	0"07	0"15	0"3

Table II: Diffraction limit of an 8-m telescope

As we shall show, the techniques for such high angular resolution imaging are already available. Even higher resolution has been obtained by coupling telescopes together interferometrically. Because imaging in the infrared is less sensitive to seeing and requires cryogenics, long baseline interferometry at IR wavelengths is expected to be developed on the ground whereas long baseline interferometry in the visible will probably be most productive in space. One more reason to develop interferometry on the ground is to learn how to do it before going to space.

2. A short historical review

In 1868, Fizeau first proposed to use interferometric techniques to measure stellar diameters [1]. In 1873, following Fizeau's ideas Stephan [2,3] put a mask with two apertures on the 1-m telescope of the Observatory in Marseille (France) and was able to detect fringes. However no star could be resolved with such a small baseline. In 1891, using independently the same technique, Michelson [4] in the U.S. was able to measure accurately the size of Jupiter's

Galilean satellites. Following Michelson, Hamy [5] in France first determined interferometrically the size of a small planet (Vesta). In 1920, Michelson [6] and Anderson [7] resolved the binary star Capella finding an angular separation of one twentieth of an arcsecond. Finally, in 1921, the first stellar disk was resolved by Michelson and Pease [8]. They found that the angular diameter of Betelgeuse was 0.04". In 1931, Pease [9] had already measured nine stellar diameters.

However, the technique was considered to be very difficult and for a long time nobody attempted to repeat Pease's observations. After the last world war radioastronomy was developing very fast. Because of the low resolution of their telescopes, radioastronomers were compelled to use interferometric techniques and did it with great success. It was then realized that some radiotechniques such as homodyne detection might also apply to optical wavelengths as well. This led to the development of intensity interferometry by Hanbury Brown and his coworkers (1956-1974). They were able to use baselines up to 200m and to measure 32 stellar diameters as small as 0.5 milli-arcsecond with an accuracy better than 5% yielding empirical effective temperatures and bolometric corrections for early type stars [10]. They were also able to observe very close spectroscopic binaries such as Gamma2 Velorum [11] and Alpha Virgo (Spica) [12]. They have shown that one of the components of Gamma2 Velorum fills the Roche lobe, whereas the distance of Spica (84 parsecs) was determined with a 3% accuracy. The main drawback of intensity interferometry is its limiting magnitude: about 2. Because of that, the technique is no longer in use.

In 1970, Labeyrie [13] proposed a very efficient technique to obtain information at the diffraction limit of a single large telescope. The technique called speckle interferometry is based on statistical analysis of the granular structure of short exposure images. Since its discovery, it has been extensively used by several groups and produced a large harvest of double star measurements yielding new and more accurate stellar masses [14]. It was used to determine the orbit of Pluto's satellite Charon [15], the orientation of the rotation axis of asteroids [16], and to resolve central objects in strong HII regions such as R136a in 30 Dauradus [17] or HD 97950 in NGC 3603 [18] which were earlier considered as possible supermassive single objects. The faintest object resolved up to now is the A component of quasar PG 1115+08 (visual magnitude 16) produced by a gravitational lens effect [19].

Speckle interferometry was also applied with great success in the infrared where it led to the discovery of cool companions [20] and possibly brown dwarfs [21]. It was applied to study as a function of wavelength of young pre-main-sequence stars and protostellar candidates in regions of stellar formation and led to the discovery of complex structures [22-28], some of which may be planets in formation [29]. It was also applied to the study of dust envelopes around red giants, supergiants and Mira type stars, in view of understanding the mass loss mechanism by which most of the matter is recycled in the universe [30,31].

3. Prospects in single apertue imaging

Speckle interferometry has three important limitations. First, it does not yield an image of the object but only its autocorrelation. Second, each observation has to be calibrated using a reference source under the same seeing conditions. Time variations of seeing seriously limit the accuracy of the measurements. Third, it is not optimal for bright sources, i.e. when atmospheric noise dominates over photon or detector noise. Techniques have been developed to overcome these limitations. We will mainly describe three of them which we believe are the most promising: triple correlation, pupil-plane (or amplitude) interferometry and phase closure.

Speckle interferometry consists of computing the image average energy spectrum

$$<|I(f)|^2> = |O(f)|^2 \cdot <|S(f)|^2> \qquad (1)$$

where I(f), O(f), and S(f) are the Fourier transforms of the brightness distribution in the instant image, in the object and in the instant point spread function respectively. The brackets < > denote averages over a large number of images. The so-called speckle transfer function $<|S(f)|^2>$ is obtained by observing a point source (unresolved star) under the same seeing conditions. Dividing the image spectrum by the speckle transfer function yields the squared modulus of the object Fourier transform (the Fourier transform of which is the object autocorrelation). The phase of O(f) is lost in the process and therefore cannot be used to recover the object.

Triple correlation, also called bi-spectral analysis or speckle masking consists of computing a third order moment, the image average bi-spectrum given by [32]

$$<I(f_1)I(f_2)I^*(f_1+f_2)>=O(f_1)O(f_2)O^*(f_1+f_2) \, . \, <S(f_1)S(f_2)S^*(f_1+f_2)> \quad (2)$$

Eq. (2) shows that the image bi-spectrum is the product of the object bi-spectrum times a transfer function which is theoretically real. Hence the phase of the image average bi-spectrum gives the phase of the object bi-spectrum. Moreover, the phase of the object Fourier transform O(f) can be determined from the phase of its bi-spectrum but for linear terms which relate only to the location of the object. Hence the shape of the object can be entirely recovered using bi-spectral analysis.

The method has been successfully demonstrated and used to reconstruct images of multiple stars and objects such as Eta Carinae [33] or HD 97950 [34]. The main drawback of the technique is that it is computer time consuming and requires large computer memories. Moreover, as speckle interferometry, it is not optimal for bright sources when atmospheric noise dominates over photon or detector noise.

Pupil-plane interferometry consists of recording fringes in a plane conjugate to the telescope pupil through a shearing interferometer. Using a 180 degree rotational shear, one can map entirely the fringe visibility in two dimensions up to the telescope resolution limit. This technique is more efficient than speckle interferometry on bright sources since it minimizes the effect of atmospheric fluctuations. Moreover calibration is little sensitive to seeing conditions. It has been successfully demonstrated and led to the discovery and mapping of the inner part of the dust shell of Alpha Ori (Betelgeuse) [35,36]. However, as speckle interferometry, it does not provide the phase of the object Fourier transform.

Phase closure is a means for recovering phase information which automatically cancels out the effect of atmospheric disturbances as well as optical aberrations. It is therefore much more efficient than other techniques which all require averaging. It was discovered by a radioastronomer [37] and it is now widely used in aperture synthesis at radiowavelengths [38]. It was first applied in optics by Goodman and his coworkers [39,40] and was only recently tested on a telescope [41]. Up to now, it was believed to require a mask on the telescope pupil which seriously limits its optical efficiency. Very recently methods have been proposed to combine phase closure with pupil-plane interferometers avoiding the use of any mask [42,43]. Progress is expected in this field in the near future.

4. Long baseline interferometry

Intensity interferometry is no longer in use because of its low sensitivity. Another radiotechnique, heterodyne detection has been successfully applied to infrared interferometry yielding both astrophysical as well as astrometric results [44,45] and is expected to give useful results at wavelengths of 10 microns or longer [46]. However Michelson interferometry is clearly the most sensitive technique at shorter wavelengths. It has been successfully demonstrated in the visible with small independent telescopes by Labeyrie [47] in 1975, by Shao and Staelin [48] in 1980 and by Davis and Tango [49] in 1985.

Up to now, the accuracy with which stellar diameters have been estimated has been limited by seeing calibration to 10-15% [50]. The related uncertainty on empirical effective temperatures, about 300 K, is still too large to put serious constraints on stellar models. The situation is much better in the infrared where stellar diameters have been determined with a 5% average accuarcy [51,52], yielding effective temperatures often within 100 K. Moreover, secondary maxima have been detected in the visibility function putting constraints on limb darkening models. Using larger telescopes, future long baseline interferometry in the infrared could also yield fundamental astrophysical results on circumstellar dust shells and stellar formation [53].

Because the coherent area and the isoplanatic patch are larger in the infrared, it will soon become possible to cophase large telescopes at wavelengths of 5 microns and above, using nearby reference stars in the visible and adaptive mirrors [54]. Used in a long baseline interferometric mode, these telescopes will become very powerful tools. This is the case of the future European Very Large Telescope (VLT) and, with a smaller baseline, of the future double telescope of the University of Arizona.

References

[1] Fizeau, C. R. Acad. Sc. (Paris) 66, p. 932 (1868).
[2] Stephan, C. R. Acad. Sc. (Paris) 76, p. 1008 (1873).
[3] Stephan, C. R. Acad. Sc. (Paris) 78, p. 1008 (1874).
[4] Michelson, A. A., Nature 45, p. 160 (1891).
[5] Hamy, M., Bull. Astr. (Paris) 16, p. 257 (1899).
[6] Michelson, A. A., Ap. J. 51, p. 257 (1920).
[7] Anderson, J. A., Ap. J. 51, p. 263 (1920).
[8] Michelson, A. A., Pease, F. G., Ap. J. 53 p. 249 (1921).
[9] Pease, F. G., Ergebn. Exacten. Naturwiss. 10, p. 84 (1931).
[10] Hanbury Brown R. et al., M.N.R.A.S. 167, p. 121 (1974).
[11] Hanbury Brown R. et al., M.N.R.A.S. 148, p. 103, (1970).
[12] Herbison-Evans D. et al., M.N.R.A.S. 151, p. 161 (1971).
[13] Labeyrie, A., A. & A. 6, p. 85 (1970).
[14] McAllister, H. A., Hartkopf, W. I., Catalog of interferometric measurements of binary stars. Georgia State University, Atlanta.
[15] Hetterich, N., Weigelt, G., A. & A. 125, p. 246 (1983).
[16] Drummond, J. D. et al, Icarus 61, p. 132 & p. 232 (1985).
[17] Weigelt, G., Baier, G., A & A. 150, p. L18 (1985).
[18] Baier, G. et al., A & A. 151, p. 61 (1985).
[19] Hege, E. K., Ap. J. 248, p. L1 (1981).
[20] Dyck, H. M. et al., Ap. J. 255, p. L103 (1982).
[21] McCarthy, D. W. et al., Ap. J. 290, p. L9 (1985).
[22] Chelli, A. et al., Nature 278, p. 143 (1979).
[23] Howell, R. R. et al, Ap. J. 251, p. L21 (1981).

[24] Chelli, A. et al., A. & A. 117, p. 199. (1983).
[25] Dyck, H. M., Howell, R. R., Astron. J. 87, p. 400 (1982).
[26] McCarthy, D. W., Ap. J. 257, p. L93 (1982).
[27] Chelli, A. et al., Ap. J. 280, p. 163, (1984).
[28] Beckwith, S. et al., Ap. J. 287, p. 793 (1984).
[29] Mariotti, J. M. et al., A. & A. 120, p. 237 (1983).
[30] Dyck, H. M. et al., Ap. J. 287, p. 801 (1984).
[31] Lohmann, A. W. et al., Applied Optics 22, p. 4028 (1983).
[32] Weigelt, G. Ebersberger, J., A. & A. (submitted).
[33] Hofmann, K.-H., Weigelt, G., A. & A. 167, p. L15 (1986).
[34] Roddier, C., Roddier, F., Ap. J., 270, p. L23 (1983).
[35] Roddier, F., Roddier, C., Ap. J., 295, p. L21 (1985).
[36] Jennisson, R. C., M.N.R.A.S. 118, p. 276 (1958).
[37] Rogers, A. E. E., Proc. SPIE 231, p. 10 (1980).
[38] Russel, F. D., Goodman, J. W., J. Opt. Soc. Am. 61, p. 182 (1971).
[39] Rhodes, W. T., Goodman, J. W., J. Opt. Soc. Am. 63, p. 647 (1973).
[40] Baldwin, J. E. et al., Nature 320, p. 595 (1986).
[41] Hofmann, K.-H., Weigelt, G., Applied Optics 25, p. 4280 (1986).
[42] Roddier, F., Roddier, C., Optics Comun. 60, p. 350 (1986).
[43] Sutton, E. C. et al., Ap. J. 217, p. L97 (1977).
[44] Sutton, E. C. et al., A. & A. 110, p. 324 (1982).
[45] Townes, C. H., J. Astrophys. Astr. 5, p. 111 (1984).
[46] Labeyrie, A., Ap. J. 196, p. L71 (1975).
[47] Shao, M., Staelin, D. H., Applied Optics 19, p. 1519 (1980).
[48] Davis, J., Tango, W. J., Proc. ASA 6, p. 34 (1985).
[49] Faucherre, M. et al., A & A. 120, p. 263 (1983).
[50] Di Benedetto, G. P., Conti, G., Ap. J. 268, p. 309 (1983).
[51] De Benedetto, G. P., A. & A. 148, p. 169 (1985).
[52] Dyck, H. M., Kibblewhite, E. J., Pub. Astr. Soc. Pac. 98, p. 260 (1986).
[54] Beckers, J. M. et al, Proc. SPIE 628, p. 290 (1986).

DESIGN CONSIDERATIONS FOR A FUTURE LARGE OPTICAL/NEAR-IR TELESCOPE IN SPACE

Pierre Y. Bely*
Space Telescope Science Institute
3700 San Martin Drive, Baltimore, Maryland, 21218

Abstract

Site and design options for a successor to the Hubble Space Telescope are discussed. It is argued that a telescope with a 6 to 8 meter monolithic mirror, launched fully assembled, and placed into geosynchronous orbit or at one of the stable earth-moon Lagrange points, offers the greatest feasibility and effectiveness.

1. Introduction

It takes from ten to fifteen years from inception to completion of a large astronomical telescope, whether on the ground or in space. Although the Hubble Space Telescope (HST) is still to be launched, its limited operational lifetime of fifteen years means that it soon will be time to start making serious plans for a successor.

It is difficult to predict what the exact observational requirements will be fifteen years from now. Our premise will therefore be that, following a four-century-old trend, the next generation optical space telescope should offer a significant increase over HST's capabilities in some or all of the following apects: flux-collecting power, resolution, wavelength coverage, and observational efficiency. In this paper we briefly discuss some of the main design and site options and make an attempt at defining a feasible and effective approach.

2. Optical Design

Reflecting telescopes have traditionally been built around monolithic primary mirrors. When larger than 5 to 6 meters in diameter, monolithic mirrors become increasingly heavy and difficult to cast, handle and support. Recently, the need for very large aperture telescopes on the ground has promoted two new approaches:

- the telescope array, or multi-mirror telescope (MMT), composed of a number of parallel pointing telescopes with their beams combined to a common focus,

- and the segmented primary mirror telescope (SMT), where the primary mirror is a mosaic of common-focus mirrors supported by actuators which maintain optical alignment.

Typical applications of these new approaches to a large space-borne telescope are shown schematically in Figure 1. The traditional concept, in the form of a scaled-up HST, and a short-focal-length monolithic mirror with reduced baffling are also shown for comparison. Advantages and drawbacks of these various configurations are briefly discussed below.

---

* Affiliated to the Astrophysics Division, Space Science Department, European Space Agency.

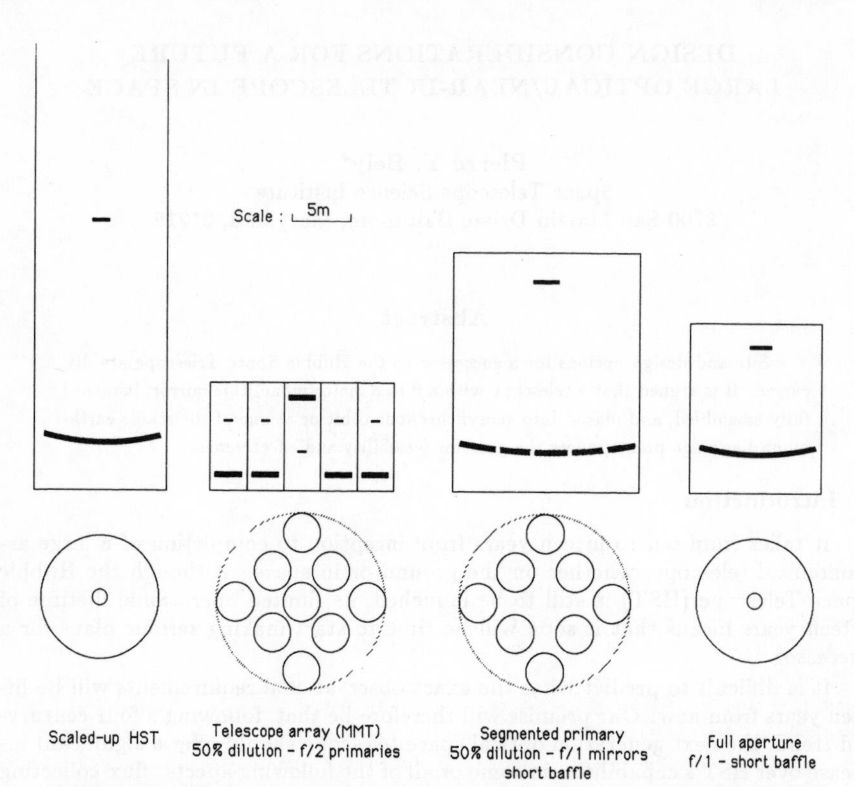

Fig. 1 Comparative sizes and configurations of possible options for a large telescope in space. All are drawn to the same scale, on the assumption that the impinging flux is equivalent to that of a 10-meter full-aperture mirror.

Telescope arrays are very compact and lends themselves to modular construction. Component telescopes could be launched independently and connected once in space. Spreading out the component telescopes also allows for higher resolution with the same optical area. Unfortunately, the necessary beam transfer and the recombining telescope increase the number of optical reflections from the original two, to six or more, with a resulting considerable loss in throughput. In the far ultraviolet as much as 90% of the light would be lost in the successive reflections. Other drawbacks include a very small field and the challenging task of combining all the individual beams in phase.

The segmented primary mirror retains the throughput and field advantages of the traditional two-mirror configuration while benefiting from a modular construction. Indeed, a segmented mirror telescope could be launched as separate elements and assembled in space. As in the case of the telescope array, the individual mirror segments could be spread out to increase resolution. However, the gain is marginal and may not justify the loss in image point-spread function uniformity. The main drawbacks of this solution are the off-axis figure of the optical segments and the mechanical complexity. Hundreds of extremly precise mirror actuators and sensors would be needed, resulting in poor overall system reliability.

In the final analysis, if these new approaches make it possible to reach very large apertures, they come at the expense of throughput, design simplicity, and reliability. As long as it remains feasible and launchable, the traditional Cassegrain telescope with a passive monolithic primary mirror remains unsurpassed.

3. Near-infrared coverage

Active cooling of large optics to reduce the background emission seen by near infrared detectors is impractical, but it is possible to reach a respectable temperature, in the 130°K range, by purely passive means[1,2]. Operation at low temperatures imposes numerous constraints. In particular, the mirror would have to be tested in vacuum during the figuring process. However, unrestricted access to the 1 to 20 micron range would surely be scientifically decisive, and would bridge the gap with dedicated IR observatories such as the Large Deployable Reflector (LDR).

4. Tracking

HST relies on "body pointing" and has no articulating elements. Since the telescope forms an integral part of the spacecraft, tracking accuracy is obtained by constraining disturbances to a very low level, and pointing the whole assembly by means of very quiet attitude control devices. However, attitude control performance has reached a technological plateau and near-future improvements will only be marginal[3]. To cope with an inherently more massive spacecraft and avoid degrading the resolution of an instrument with several times the aperture of HST, a multi-level attitude control will have to be used. The most obvious solution consists of augmenting the traditional body tracking system with adaptive optics to correct for long time scale deformations due to orbital effects, and beam steering for image drift and jitter compensation.

In addition extreme care will have to be exercised to reduce disturbances to a very low level. As an example, steerable solar panels will have to be abandoned. Solar panel wings are highly flexible and are usually the source of the lowest natural frequencies in spacecraft. In addition to being a source of jitter, they restrict slewing speed and impose long stabilization times.

Preference should also be given to very high orbits where disturbances due to atmospheric drag and gravity gradient are minimal to nonexistent.

5. Site Options

The main characteristics and evaluation criteria for possible sites of a spaceborne astronomical observatory are summarized in Table 1.

Among these possible sites, the geosynchronous orbit (GEO), and the L4 or L5 earth-moon Lagrange points, affords overwhelming advantages:
- they do not have the short period thermal cycling of low earth orbits (LEO), a prime condition for accurate pointing and tracking and a stable optical figure,
- the demand on attitude control is minimal since gravity gradient and aerotorque are practically nonexistent, an important consideration for a low level of internal mechanical excitations,
- the observing efficiency can be very high since there is virtually no target eclipsing by the earth, and bright-earth induced stray light is less constraining,

Table 1.

Location	Launch energy (MJ/kg)	Angle subtended by earth	Aerotorque and gravity disturbance	Thermal variation	Observation efficiency	Ease of maintenance	Scheduling flexibility
LEO	33(1)	135°	high	high(1)	30%(1)	high	poor
GEO	58	17°	very low	low	90 %	fair (2)	high
Moon	90	2°	very high	high	40 % (3)	good (4)	good (4)
L4/L5	62	2°	ultra low	low	90 %	poor	high (5)

(1) a sun-synchronous orbit would require higher launch energy, but would have less thermal variation and higher observational efficiency.
(2) assumes servicing to geosynchronous orbit, a likely possibility in the early 21st century.
(3) assumes no observing during moon day because of stray light.
(4) assumes moon base, unlikely before year 2020.
(5) assumes no dedicated relay satellites.

- long exposures can be obtained without interruption,
- real-time control of the observatory can be continuous (GEO only), simplifying scheduling and permitting astronomers to intervene during the course of their observations, thus optimizing data gathering. Also, a direct link to the ground imposes fewer restrictions on the high level of data transfer required by future large-sized electronic detectors.

Clearly, the operational advantages of GEO or the L4/L5 points far outweigh the weight penalty, and they appear indispensable to ensure the technical feasibility and economic justification of any major future astronomical facility in space.

6. Launching and Assembly on orbit

Table 2 lists the payload capability and failure rate of the present and near future launchers. Two conclusions can be drawn from this data.
- The heavy lift vehicles (HLV) which will become available in the 1990's will be capable of launching to geosynchronous orbit a fully assembled telescope in the 6 to 8 meter range. This should alleviate the need for any on-orbit assembly up to that diameter. Although on-orbit assembly should become fairly common in the next decade, it will remain costly and applicable only to the simplest tasks. In view of the mechanical complexity, alignment accuracy and contamination risks associated with astronomical instruments, a telescope launched fully assembled and tested appears a far preferable solution.
- The failure rate of launchers is still fairly high. This, combined with the difficulty in servicing the geosynchronous orbit, should favor simple and relatively low-cost space observatories as opposed to unique and complex systems like HST. It may even be economically justifiable to build two almost identical telescopes, as was done for the OAO series, to guard against a major failure which would otherwise cripple space astronomy for several years.

Table 2.

Launcher	Capability to GEO (tons)	Payload max diam. (m)	Failure rate %
Present			
Delta + PAM	1.2	2.5	7
Ariane 4	2.2	3	22
Titan 34D	1.9	3	18
H-II (Japan)	2	3	-
Proton (SL-12)	2.1	3.5	7
Shuttle + OTV(future)	7	4.6	4
Planned			
Ariane 5 (enhanced)	8	4	-
US HLV - Jarvis (enhanced)	20 ?	10.5	-
US HLV - UPC (enhanced)	22?	8	-
USSR HLV	30?	8?	-

7. Conclusion

In summary, the following rationale may proposed for the design of the next large space telescope:

- site the observatory in geosynchronous orbit for minimal disturbance, high observing efficiency and real time control,
- select the telescope aperture to be compatible with the payload mass and size capability of the upcoming heavy lift launchers (say 6 to 8 meters in diameter and an overall mass of 20 tons),
- passively cool the optics to about 130° K in order to cover the near-IR,
- use a multi-level attitude control (i.e. beam steering) to guarantee tracking accuracy,
- launch the spacecraft fully assembled and tested,
- keep design and fabrication simple and relatively low-cost and build a standby clone to palliate possible launch or on-orbit failure.

A very preliminary concept based on the above criteria is shown in Figure 2.

Although the proposed approach is not extendable to the truly large telescopes of the 15-meter size and above, it appears to be the most justifiable and practical for a first-generation successor to HST.

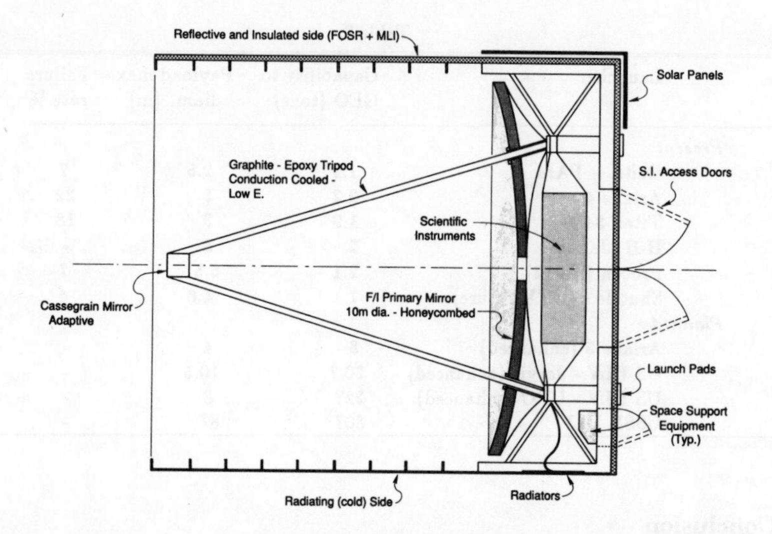

Fig. 2 Schematic section of an observatory designed along the proposed rationale. The primary mirror is monolithic and lightweighted, with an f/1 ratio. To further reduce overall length, baffling is short, as permitted by GEO or Lagrange point siting. The primary mirror and the secondary mirrors are passively cooled . Solar panels are directly mounted on the spacecraft body to reduce jitter. The secondary mirror is adaptive and steerable for fine guiding and jitter compensation.

References

[1] Swanson, P.N., "A lightweight low cost large deployable reflector (LDR)", JPL D-2283 report, 1985.

[2] Bely P.Y., Bolton J. F., Neeck S. P., Tulkoff P. J., "Space Ten-Meter Telescope (STMT), Structural and thermal feasibility study of the primary mirror", SPIE Proc. 751, 1987.

[3] Laskin R.A. and Sirlin S.W." Future payload isolation and pointing system technology", Journal of Guidance, Control, and Dynamics, Vol. 9, 4, p 469, 1986.

CRYSTAL DIFFRACTION LENSES FOR IMAGING GAMMA-RAY TELESCOPE

Robert K. Smither

Argonne National Laboratory, Argonne, IL 60439

Recent developments[1-5] in the design of crystal diffraction lenses has made it possible to build a gamma-ray telescope that produces a real image of a distant object. This can be done both with surface diffraction optics and with transmission diffraction optics. The mathematical solution to both cases is the same but they differ in the way the crystals are cut and bent. Until recently, there was no solution for the transmission case. It is this last case that will be highlighted in this paper, because it is new and because it makes much more efficient use of the crystal material when high energy photons (>50 keV) are being focused. A transmission crystal lens is shown in fig. 1. It consists of many concentric rings of crystal subassemblies or individual

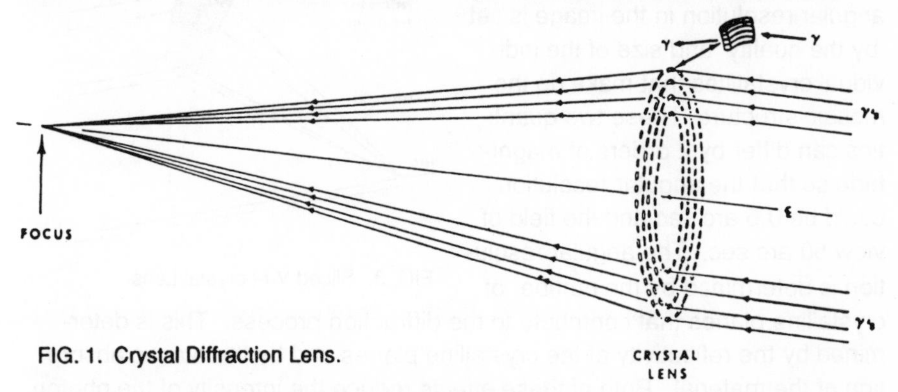

FIG. 1. Crystal Diffraction Lens.

crystals, orientated so that they all diffract the incident flux of monochromatic photons so as to form an image at the focal point. Each ring uses a different set of crystal planes and/or a different material for it's crystals. The radius of each ring is adjusted so that the Bragg condition, $n\lambda = 2d\sin\Theta$, is satisfied. The first type of lens uses normal crystals with uniform crystal lattice spacings. This results in a condenser lens that can collect photons from a large aperture and focus them on a small area. Direct imaging is limited in this case because the focal spot can never be smaller than the radial thickness of the diffraction crystal. The angular resolution of this type of lens can be a few arc sec, however, so an image of a distant object can be generated by scanning the object, forming the image line by line. In order to obtain direct, point to point focusing and thus high resolution imaging, it is necessary to use crystals in which the lattice spacing changes as a function of position in the crystal. This type of crystal shall be referred to as a "Variable-Metric" or V-M crystal. The variation of the lattice spacing can be accomplished by either appling a thermal gradient or growing a two-element crystal and varing the concentration of the elements.[1] By bending the V-M crystal and adjusting the lattice spacing to match this bending, it is possible to obtain a line focus of a point source (fig. 2). If the crystal is now bent around the axis of the lens then a point focus is ob-tained. The bending of the crystal in this unnatural way is achieved by slicing up the crystal into thin sections and then bending it as one would a deck of cards (fig. 3).

FIG. 2. Bent V-M crystal lens.

The field of view of the lens is determined by the mosaic structure of the crystals, while the limit on the angular resolution in the image is set by the quality and size of the individual crystallites that make up the mosaic structure. These two quantities can differ by 2 orders of magnitude so that the angular resolution could be 0.5 arc sec and the field of view 50 arc sec. The angular resolution is determined by the number of crystalline planes that contribute to the diffraction process. This is determined by the reflectivity of the crystalline planes and by the atomic absorption of the material. Both of these effects reduce the intensity of the photon beam as it passes through the crystal and limit the number of planes that contribute. Fig. 4 shows separately the limit set on the angular resolution by the reflectivity of the planes(Darwin) and by the atomic absorption, as a function of photon energy for the [400] planes in Si. The two effects are plotted separately so one can see how they compare at different photonenergies. The two curves are combined to obtain the expected limit.For highly reflecting planes like this one, the Darwin width is the limiting effect at low

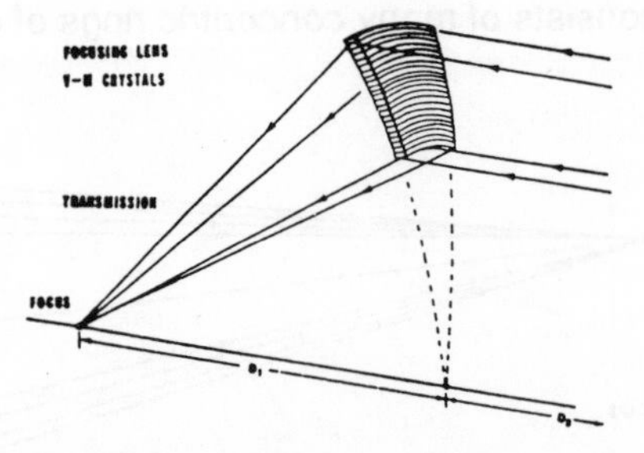

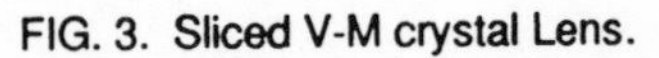

FIG. 3. Sliced V-M crystal Lens.

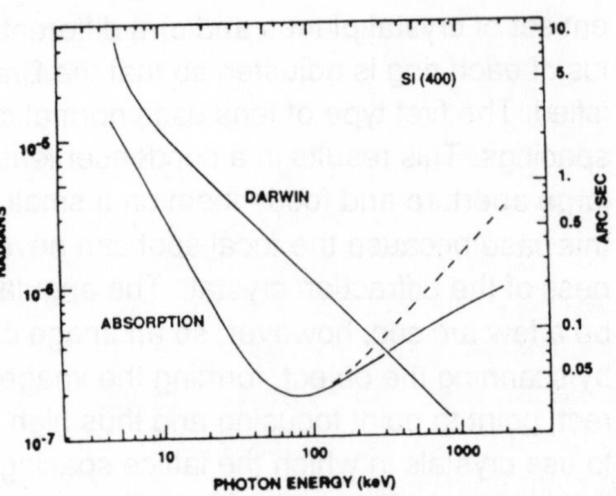

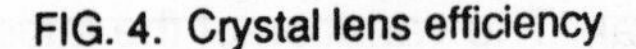

FIG. 4. Crystal lens efficiency

energies. At some higher energy the atomic absorption always takes over as the limiting effect. The lower limit for the angular resolution decreases with increasing energy from a few arc sec to a few tenths of an arc sec for photon energies near the the 511 keV annihilation radiation. The dashed line is an aditional limit produced by the finite size of the crystillites.

Both types of crystal lenses can be used to construct an gamma ray telescope. Since the strength of astrophysics gamma-ray sources tend to be very weak (10^{-3}/sec to 10^{-7}/sec) the lens will have to be quite large to collect enough photons in a reasonable time to detect the source. The efficiency of the lens is defined as the produce of the reflectivity of the crystalline planes and the transmission of the photons through the crystal. At higher energies the reflectivity is down but the atomic absorption is also down so one can maximize the efficiency by choosing the right thickness. A plot of efficiency verses crystal thickness for different energy photons, using the [400] planes in Ge appears in fig. 5. Efficiencies of 30 to 40% are possible for the energies up 300 keV and 27% for the 511 keV annihilation line. This could be of special interest for investigating the elusive 511 keV line observed near the center of our galaxy. This source as been as strong as 2×10^{-3}photons/cm^2/sec and as weak as 0.5×10^{-3}/cm^2/sec. If one constructed a large lens (4 m dia.), tuned to 511 keV, and 80 % of its area filled with crystals 25% efficient, the 2×10^{-3} photons/cm^2/sec flux would result in 50 photons/sec at the focal spot. The 0.5×10^{-3}/cm^2/sec flux would give13 photons/sec. Both of these signal strengths are very much larger than the background signal incident on the 5 mm dia. focal spot. The major source of 511 keV background is the shield around the detector which could be 10^{-3} photons/sec for a noninteractive shield. For an interactive shield this background could be as low as 10^{-5}photons/sec. A distant source with a flux of 5×10^{-7}photons/cm^2/sec at the lens would result in a flux at the detector of 1.2×10^{-2} photons/sec. This weak signal is

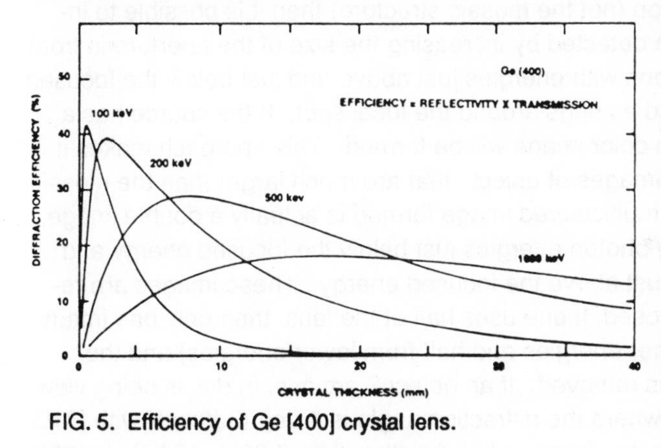

FIG. 5. Efficiency of Ge [400] crystal lens.

still a factor of 100 to 1000 times the background. The main drawback with the observation of a weak source is the long observation time needed to form a good image. A day's observation of this 5×10^{-7} source with the 4 m dia. lens would collect 1068 photons. If the detector is relatively efficient, this would be enough to identify not only the centroid of the image but also it's size and general shape. The efficiency of the lens will be higher for lower energy photons with a corresponding higher sensitivity for weak sources.

The major drawback of the crystal lens is that it is difficult to change the energy being focused. Varies schemes have been proposed to change the energy by 10 to 20 % but larger changes appear quite difficult. A number of lenses will be needed to cover the full spectrum. The monochromatic nature of the lens presents problems in the detection of weak astrophysical sources where most of the radiation appears as a continuum spectrum with little or no strength in sharp lines. In this case only a small fraction of the radiation is focused in the focal spot. If the source has a finite angular size larger than imaging resolution (not the mosaic structure) then it is possible to increase the bandwidth detected by increasing the size of the aperture in front of the detector. Photons with energies just above and just below the focused energy will be focused as rings around the focal spot. If the source has a shape then a multiple color image will be formed. This approach makes it possible to generate images of objects that are much larger than the mosaic structure width. The multicolored image formed is actually a double image, one generated by the photon energies just below the focused energy and one by the energies just above the focused energy. These images are reversed and superimposed If one uses half of the lens, then only one image occurs (half from higher energies and half from lower energies) and the confusion of images is removed. If an object 2 arc min. in dia. is being viewed at a wave- length where the diffraction angle is 1 deg.(300 keV with [310] planes in quartz) then the detected bandwidth will be 3.33%, 10 keV for 300 keV photons and 40 keV for the 600 keV photons. When the large collection area of the lens is taken into account, this modest bandwidth results in a quite usable signal for many distant objects. As the photon energy increases, the bandwidth increases and almost compensates for the reduced intensity per energy interval. For the Crab source, the 4 m dia. lens focus gives 18 photons/sec for the 300 keV continuum and 9 photons/sec for 600 keV.

Some experiments have been preformed on crystal elements very similar to what would be used in a large lens. One of the bent-crystal elements tested had a radius of curvature (around the axis of the lens) is 95 cm. This crystal was 1 mm thick in the radial direction, 4 mm deep in the beam direction, and 5 cm wide perpendicular to the beam. This results in a focal spot 1 mm in diameter for a point source. Fig. 8 shows the focused energy spectrum in a Ge detector for a source of 59.5 keV gammas (^{241}Am) located 20 m away. Fig. 8 shows the response of the detector as the source was moved in the vertical direction. This is equivalent to the lens making an angular scan of the source. The detector aperture was wide open so the acceptance width of

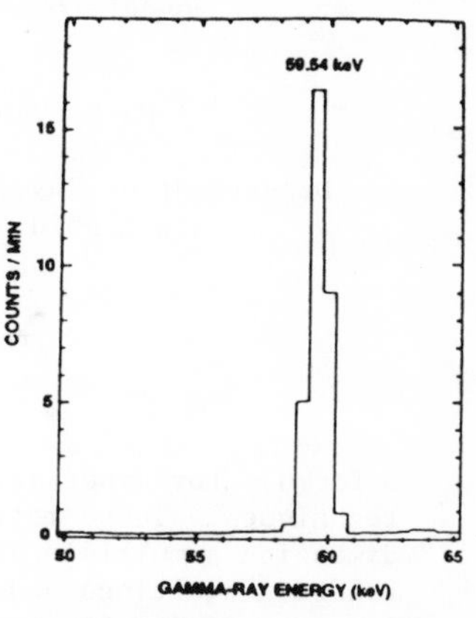

FIG. 7. Energy spectrum in detector.

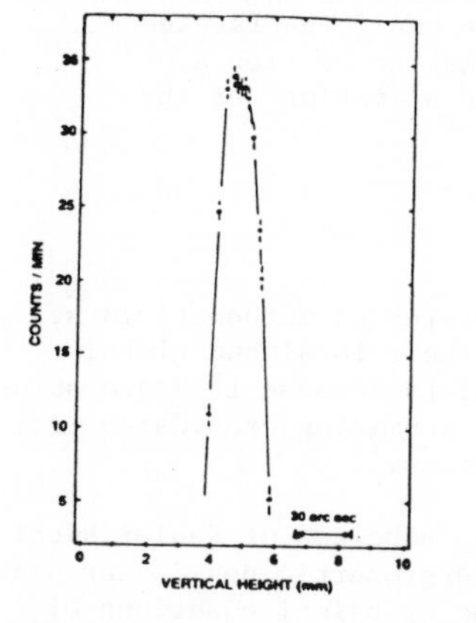

FIG. 8. Vertical scan of source.

the crystal was determined by the source size(2 mm dia.), the radial thickness, and the mosaic structure of the crystal. The flat top of the response reflects the source width. The sharp rise and fall reflects the narrow width of the mosaic structure. The count rate was 10% higher than predicted. This is due to a small increase in the width of the mosaic structure by the binding. This work was supported by the U.S. Department of Energy, BES-Material Science under Contract No. W-31-109-ENG-38.

References:

[1] Smither, R. K. , New Method for Focusing X-rays and Gamma-rays, Rev, Sci. Instrum., 44,131 (Feb. 1982).

[2] Smither, R.K., A New Method for Focusing and Imaging X-Rays and Gamma-Rays with Diffraction Crystals, Sym. on Future X-Ray Experiments,X-Rays in the 80's, GSFC, Oct. 1981. NASA Tech. Mem. No. 83848 (Nov. 1981).

[3] Smither, R.K., Gamma-Ray and X-Ray Telescopes Using Variable-Metric Diffraction Crystals, 11th Texas Sym. on Relativistic Astrophysics, Austin, Texas, Dec. 1982. Annal of New York Acad. Sci., 44, 673 (1983).

[4] Lund, N. and Smither, R.K., A Bragg Crystal Flux Concentrator for Annihilation Radiation, 16th Inter. Cosmic Ray Conf., Banggalore, India. (1983).

[5] Smither, R.K. (U.S. Patent No. 4,429,411, entitled "Instrument and Method for Focusing X-Rays, Gamma-Rays and Neutrons", Inventor: Robert K. Smither Issued: Jan., 1984.)

CALABI-YAU METRICS FROM SUPERSPACE*

Mark Evans† and Burt A. Ovrut

Department of Physics, University of Pennsylvania
Philadelphia, Pennsylvania 19104-6396

ABSTRACT

A formula for hyperkahler metrics is derived using superspace techniques. These metrics are Kahler, Ricci flat and have dimension a multiple of 4. Thus they satisfy the Vacuum Einstein equations and are analogs of the manifolds favored for superstring compactification. Hyperkahler metrics are also solutions to the tree level equation of motion for the superstring.

1. Introduction

Kahler Ricci flat manifolds have long interested mathematicians, in part because of the beautiful interplay of their local and global properties.[1] These manifolds interest physicists because they are solutions of the vacuum Einstein equations and are promising candidates for compactifying superstrings from 10 to 4 dimensions.[2]

Hyperkahler manifolds form an interesting subclass of Kahler Ricci flat manifolds. They correspond to finite supersymmetric non-linear sigma models[3,4] and so are exact solutions to the classical equations of motion for the superstring.[5]

Recently, we have derived a formula for hyperkahler metrics using superspace techniques.[6] In this talk I shall describe the problem in more detail and briefly sketch our solution.

2. Differential Geometry

From general relativity we are all familiar with the notion of an ordinary real manifold. I shall describe a chain of refinements of this notion that will end with hyperkahler manifolds. The first refinement is purely topological in nature. A complex manifold[7] has coordinate charts diffeomorphic to C^n and transition functions between charts that are holomorphic in the local coordinates. Clearly every complex manifold is a real manifold of even dimension on which the real coordinates can

†Invited talk at 13th Texas Symposium on Relativistic Astrophysics, Chicago, Illinois, December 14-19, 1986.

*Work supported in part by the Department of Energy Contract number DOE-AC02-76-ERO-3071.

be paired to form complex coordinates. The fact that the transition functions are holomorphic means that this structure lifts to the tangent space, where there exists a mapping ("multiplication by i") that can, by a suitable real coordinatisation, everywhere be put in the form

$$J(X)^{\mu}{}_{\nu} = \begin{pmatrix} & -\mathbf{1} \\ \mathbf{1} & \end{pmatrix} \tag{1}$$

I shall refer to the tensor in equation (1) as the complex structure of the manifold. A tensor that, like a complex structure, squares to give minus the identity is called an almost complex structure. However, coordinate transformations do not in general suffice to put an almost complex structure in the form of equation (1). The obstruction is the Nijenhuis tensor

$$N^{\mu}_{\nu\lambda} = (\partial_{\sigma}J^{\mu}{}_{\lambda} - \partial_{\lambda}J^{\mu}{}_{\sigma})J^{\sigma}{}_{\nu} - (\partial_{\sigma}J^{\mu}{}_{\nu} - \partial_{\nu}J^{\mu}{}_{\sigma})J^{\sigma}{}_{\lambda} \tag{2}$$

An almost complex structure for which the Nijenhuis tensor vanishes is termed integrable. A famous theorem states that a real manifold is a complex manifold if it admits an integrable almost complex structure.

All of these conditions are highly non-trivial topological properties of a manifold. For example, the spheres S^n do not admit almost complex structures for $n = 4$ or $n \geq 8$ (this is something like the statement that you cannot put a non-vanishing vector field on S^2). On the other hand, S^6 admits an almost complex structure but is widely believed not to admit one that is integrable.

So far we have not introduced a metric. Let us now do so, and demand that the complex structure generate an isometry

$$g_{\mu\nu}J^{\mu}{}_{\lambda}J^{\nu}{}_{\sigma} = g_{\lambda\sigma} \tag{3}$$

Such a metric is said to be Hermitean, and it is Kahler if its Riemannian (i.e. torsion free) connection leaves the complex structure covariantly constant

$$\nabla_{\mu}J^{\nu}{}_{\lambda} = 0 \tag{4}$$

A hyperkahler metric is Kahler with respect to three distinct complex structures which satisfy a quaternionic algebra

$$\{J_{(a)}, J_{(b)}\} = -2\delta_{ab}\,\mathbf{1} \qquad a,b = 1...3 \tag{5}$$

This definition is very restrictive. To represent the algebra of equation (5) the manifold must be of dimension 4k (k an integer). Equation (4) says that the holonomy commutes with each complex structure, and so is at most Sp(k). Since $Sp(k) \subseteq SU(2k)$, hyperkahler manifolds are Ricci flat.

3. Non-Linear Sigma Models

A non-linear sigma model is a theory of fields which take their values in some manifold called the target space. By constructing a sigma model with a target space we know to be hyperkahler we obtain a formula for such a metric. The appropriate sigma models are those with (4,0) supersymmetry.

In a 1 + 1 dimensional space-time the (p,0) supersymmetry algebra[8] is

$$\{Q_{+i}, Q_{+j}\} = 2\,\delta_{ij}\,P_{+} \qquad 1 \leq i,j \leq p \tag{6}$$

The supercharges Q_{+i} are real one-component Majorana-Weyl fermions, and P_+ is the generator of right translations (we use light-cone coordinates).

The most general sigma model with (1,0) supersymmetry has action

$$S = \int d^2x[(G_{\mu\nu}(X_o) + B_{\mu\nu}(X_o))\partial_+ X_o^{\mu}\, \partial_- X_o^{\nu}$$

$$-i\, G_{\mu\nu}\psi^{\mu}(\partial_-\psi^{\nu} + \Gamma^{\nu}_{\lambda\sigma}\partial_- X_o^{\lambda}\psi^{\sigma})] \tag{7}$$

The scalar fields X_o^{μ} are to be thought of as coordinates for the target space, and ψ^{μ} are their fermionic superpartners. The couplings $G_{\mu\nu}$, $B_{\mu\nu}$ are functions of the scalar fields X_o^{μ}; $G_{\mu\nu}$ is symmetric in its indices and is interpreted as the metric for the target space while $B_{\mu\nu}$ is antisymmetric. $\Gamma^{\nu}_{\lambda\sigma}$ is a connection that is the usual Christoffel symbol constructed out of $G_{\mu\nu}$, plus a torsion piece constructed out of $B_{\mu\nu}$.

The supersymmetry transformations that leave (7) invariant are

$$Q_{+1} X_o^{\mu} = i\, \psi^{\mu} \tag{8a}$$

$$Q_{+1}\psi^{\mu} = -\, \partial_+ X_o^{\mu} \tag{8b}$$

(1,0) supersymmetry places no restriction on the geometry of the target space. However, imposing additional supersymmetries alters this situation in a very interesting way.[3,9]

What would transformation Q_{+i} look like when acting on X_o^{μ}? Presumably just like (8a), except that the right hand side would be a different combination of fermions

$$Q_{+i} X_o^{\mu} = i\, J_{(i)}(X_o)^{\mu}{}_{\nu}\psi^{\nu} \tag{9a}$$

$$Q_{+i}(J_{(i)}{}^{\mu}{}_{\nu}\psi^{\nu}) = -\, \partial_+ X_o^{\mu} \tag{9b}$$

Given that (8) is a symmetry, (9) will be as well provided the chiral transformations

$$\psi^{\mu} \to J_{(i)}(X_o)^{\mu}{}_{\nu}\psi^{\nu} \qquad (10)$$

are also symmetries. If we demand that (10) is indeed a symmetry of (7) and that transformations (8) and (9) satisfy the algebra (6) we learn the following remarkable facts (we have assumed $B_{\mu\nu} = 0$): each $J_{(i)}$ is an integrable almost complex structure with respect to which the metric is Kahler. They also satisfy a Clifford algebra like equation (5).

Thus the target space for a (2,0) supersymmetric sigma model (with $B_{\mu\nu} = 0$) is Kahler, and for a (4,0) model is hyperkahler. By constructing actions for these models in the appropriate superspace we can derive expressions for Kahler or hyperkahler metrics. The Kahler case is straightforward[10] but yields nothing new, and so I shall sketch the hyperkahler construction. Be warned that the proofs of many of the statements I shall make are algebraically tedious. More details appear elsewhere.[6]

3. Superspace Construction

Superspace[11] is the group manifold corresponding to the supersymmetry algebra (6). It has coordinates (x^+, x^-, $\theta^{+1},\ldots,\theta^{+p}$), where the θ's are odd elements of a Grassmann algebra. Functions defined on superspace are superfields, and transform under the algebra (6) through translations on superspace induced by left or right group multiplication. The generators of these motions are denoted Q_{+i} and D_{+i} respectively.

$$Q_{+i} = \partial_{+i} - i\theta^{+i}\partial_{+} \qquad (11a)$$

$$D_{+i} = \partial_{+i} + i\theta^{+i}\partial_{+} \qquad (11b)$$

For our purposes, the important facts about Q_{+i} and D_{+i} are that they separately satisfy the algebra (6) and mutually anti-commute

$$\{D_{+i}, Q_{+j}\} = 0 \qquad (12)$$

We may therefore take the Q's to represent the supersymmetry generators, and use the D's to constrain superfields in a supersymmetric way.

Real scalar (4,0) superfields have the component expansion

$$X^{\mu} = X^{\mu}_{o} + i\theta^{+i}X_{+i} + \frac{i}{2}\theta^{+i}\theta^{+j}X^{\mu}_{+i+j} - \frac{1}{3!}\theta^{+i}\theta^{+j}\theta^{+k}X^{\mu}_{+i+j+k} - \frac{1}{4!}\theta^{+i}\theta^{+j}\theta^{+k}\theta^{+\ell}X^{\mu}_{+i+j+k+\ell} \qquad (13)$$

Each superfield contains sixteen independent real component fields, only two of which (the lowest component and its fermionic superpartner) appear in the action (7). It is therefore necessary to impose constraints on the superfields X^{μ} such that the lowest component and one superpartner are unconstrained and all other components are determined in terms of these physical fields.

An appropriate set of constraints is

$$D_{+i}X^{\mu} = L_{(a)ij}\, J_{(a)}(X)^{\mu}{}_{\nu}\, D_{+j}X^{\nu} \qquad a = 1\ldots 3 \tag{14}$$

where the $L_{(a)}$ are constant 4 x 4 antisymmetric matrices that satisfy the Clifford algebra (5). Note that the index (a) is not summed over in equation (14). The superfields will in fact be overconstrained by (14) unless the $J_{(a)}$ also satisfy the Clifford algebra (5) and are integrable complex structures.

We now have the desired supersymmetry multiplet, but cannot yet write an action in superspace for a reason peculiar to (p,0) supersymmetry: the superspace integration measure carries a Lorentz charge -p/2. Thus for a (4,0) action constructed out of scalar superfields we would have to introduce ∂_-^2 into the Lagrangian to get a Lorentz invariant action. This, however, would lead to an unacceptable higher derivative theory. We therefore introduce a new superfield V^+, with Lorentz charge +1, and impose constraints on it so that no new degrees of freedom are introduced into the theory. Suitable constraints are

$$h_{(a)}(X) = -\, iL_{(a)ij}\, D_{+i} D_{+j} V^{+} \tag{15a}$$

$$g(X) = \partial_{+} V^{+} \tag{15b}$$

Constraints (15) are consistent if

$$\partial_{\mu} g = \frac{1}{4} J_{(a)}{}^{\lambda}{}_{\mu}\, \partial_{\lambda} h_{(a)} \qquad a = 1\ldots 3 \tag{16}$$

which says that $g + \frac{1}{4}\, ih_{(a)}$ is holomorphic with respect to $J_{(a)}$.

The action may now be written

$$S = \int d^2x d^4\theta\, f_{\mu}(X)\, \partial_- X^{\mu} V^{+} \tag{17}$$

Expanded in components, action (17) does not always yield a non-linear sigma model. However, if we take

$$f_{\mu} = \partial_{\mu} g \tag{18}$$

we do indeed get a non-linear sigma model with $B_{\mu\nu} = 0$. We may read off from this action our hyperkahler metric

$$G_{\mu\nu} = 2\Big[\sum_{b=1}^{3} J_{(b)}{}^{\lambda}{}_{\mu} J_{(b)}{}^{\sigma}{}_{\nu} + \delta^{\lambda}{}_{\mu}\delta^{\sigma}{}_{\nu}\Big]\partial_{\lambda} g\, \partial_{\sigma} g \tag{19}$$

We have here a formula for a hyperkahler metric in terms of the three complex structures $J_{(a)}$, and a function g, introduced in equation (15) and satisfying equation (16).

It is perhaps natural to ask whether this metric is non-trivial. If we take the J's to be constant antisymmetric matrices then the metric

(19) is conformally flat, and hence flat (a manifold that is conformally flat and Ricci flat is flat). Thus if we can make the J's simultaneously constant by coordinate transformation the metric (19) is flat. Furthermore, if we cannot make the J's simultaneously constant then, from the covariant constancy of the J's, we cannot make the Christoffel symbols vanish, and so the metric cannot be flat. It is therefore crucial to know whether the complex structures can be made simultaneously constant by coordinate transformation.

The J's can be made constant one at a time because they are integrable. Thus we may choose a frame in which $J_{(2)}$ (say) takes the form of equation (1). Then $J_{(3)}$ can be written as the coordinate transform of some analogous constant, canonical matrix. Does there exist a coordinate transformation that leaves $J_{(2)}$ invariant and takes $J_{(3)}$ back to its canonical form? The answer is, in general, no. We have constructed (in the four dimensional case) the most general linear algebra transformation that does this, and shown that it does not, in general, have enough freedom to satisfy the integrability condition that would make it a coordinate transformation. Hence we have been able to construct explicit J's which we can prove cannot be made constant by coordinate transformation. Thus the metric in equation (19) is, in general, not flat.

REFERENCES

1) Calabi, E. in "Algebraic Geometry and Topology: A Symposium in Honor of S. Lefschetz", Princeton, 1957.
   Yau, S.-T., Proc. Nat. Acad. Sci. 74 (1977) 1798.

2) Candelas, P., Horowitz, G., Strominger, A. and Witten, E. Nucl. Phys. B258 (1985) 46.

3) Alvarez-Gaume, L., and Freedman, D., Commun. Math. Phys. 80 (1981) 443.

4) Galperin, A., Ivanov, E., Ogievetsky, V., and Sokatchev, E., Class. Quant. Grav. 2 (1985), 601, 617;
   Alvarez-Gaume, L. and Ginsparg, P., Commun. Math. Phys. 102 (1985) 311;
   Morozov, A., and Perelmov, A., Nucl. Phys. B271 (1986), 620;
   Grundberg, J., Karlhede, A., Lindstrom, U. and Theodoridis, B., Stockholm preprint (1986);
   Hull, C., Nucl. Phys. B260 (1985), 182;
   Sokatchev, E. and Stelle, K., Imperial College preprint TP/85-86/37.

5) Lovelace, C., Phys. Lett. 135B (1984), 75.

6) Evans, M. and Ovrut, B., Pennsylvania preprint UPR-0321T (1986).

7) Chern, S., "Complex Manifolds without Potential Theory", Van Nostrand, 1967.

8) Sakamoto, M., Phys. Lett. 151B (1985) 115;
   Siegel, W., Nucl. Phys. B238 (1984) 307;
   Alvarez-Gaume, L., Commun. Math. Phys. 90 (1983), 262;
   Friedan, D., and Windey, P., Nucl. Phys. B235 (1984), 395;
   Evans, M., and Ovrut, B., Phys. Lett. 171B (1986), 177.

9) Gates, J., Hull, C., and Rocek, M., Nucl. Phys. B248 (1984), 157;
Hull, C., and Witten, E., Phys. Lett. 160B (1985), 398;
Sen, A., SLAC preprint 3919 (1986);
Braden, H., and Frampton, N., N. Carolina preprint (1986).

10) Evans, M. and Ovrut, B., Phys. Lett. 175B (1986), 145.

11) Wess, J., and Bagger, J., "Supersymmetry and Supergravity", Princeton (1983).
Gates, J., Grisaru, M., Rocek, M., and Siegel, W., "Superspace", Benjamin (1983).

COSMOLOGICAL STABILITY OF QUANTUM COMPACTIFICATION

Marcelo Gleiser†

Theoretical Astrophysics Group
Fermilab, P.O.B. 500
Batavia, Illinois 60510 U. S. A.

ABSTRACT

We discuss the cosmological stability of higher dimensional models that feature internal manifolds given by the product of two spheres. In particular, we consider the case when the total number of dimensions is even. After we obtain the vacuum energy coming from one-loop fluctuations of scalars and spin-$\frac{1}{2}$ fermions, we show how a realistic cosmological scenario can arise by balancing the quantum energy with monopole-like contributions.

1. Introduction

The quest for a unified theory of the fundamental interactions based on higher-dimensional spacetimes has been the focus of much attention in recent years[1]. From the upsurge of the traditional Kaluza-Klein theories in the mid- seventies to superstring theories nowadays, the inclusion of "some" extra dimensions in the description of nature seems to be imperative if we are to unify gravity with the gauge interactions.

Once we accept the possible physical relevance of extra-dimensions, two questions come immediately to our minds; why do the scales characterizing the physical four-dimensional spacetime and the internal compact space differ by roughly 60 orders of magnitude nowadays and what determines the stability of the internal space. The physical motivations for both questions should be obvious. We find no trace of extra dimensions on scales between $10^{-16}cm$ to $10^{28}cm$ and also have very strict limits on the time variation of the fundamental couplings that would be induced by a time variation of the internal space. To put it concisely, the internal space is extremely small and static (or very nearly so)[2].

On view of the above comments, it is important to study the stability of the various compactification mechanisms proposed to date in order to select the proper ground-state to the physical theory coming from extra-dimensions. Different compactification schemes will provide different ground-state geometries with

† Talk presented at the XIII Texas Symposium in Relativistic Astrophysics, December 1986. The work reported was done in collaboration with P. Jetzer and M. A. Rubin.

different stability properties. Although we expect the correct ground-state to be univocally determined once a better understanding of string theory is achieved, for the moment a stability analysis of compact manifolds is certainly useful.

To our knowledge, there are presently three possible contributions to the energy-momentum tensor that can play a role in stabilizing the internal space. One can introduce a $4+D$-dimensional cosmological constant and use it to try to balance the curvature of the internal space, one can calculate 1-loop quantum corrections that give rise to an attractive force coming from imposing boundary conditions to the fields in the effective action, very much like the Casimir effect in QED[3) , or one can use the fact that most actions include anti-symmetric tensor fields that can assume non-trivial topological configurations on the compactified background[4) .

All the above possibilities have been extensively analysed in the literature, including some combinations of these effects with each other and with finite temperature[5). Without going into details, it has been shown that models with a cosmological constant are semi-classically unstable against quantum tunneling; for large values of the internal radius the cosmological constant dominates and the system behaves like a $4+D$-dimensional de Sitter spacetime. (There is some apparent controversy as to whether the tunneling actually occurs. The root of the discussion is in the Weyl rescaling of the metric that, according to some authors, is fundamental to the proper interpretation of the potential. We sustain that physical results will depend on the rescaling since the theory is not conformally invariant and that the absence of tunneling that occurs with the rescaled metric has to be reinterpreted in the light of the uncertainty principle, since the rescaling changes the definition of time. In any case, the necessity of rescaling has not been clarified yet. Work in these lines is on progress.)[6)

Thus, the more interesting possibilities seem to include the quantum and monopole effects. Accordingly, we have studied the possibility of obtaining a stable compactification with these effects taken into account[7). We used, in ref. 7, an approximate expression for the Casimir effect in even dimensions and showed that, for certain theories in ten dimensions, it is possible to obtain a stable background. The promising model comes from $N=1$ supergravity with fermionic condensation and with an internal space given by the product of two 3-spheres. It is yet far from being related to the phenomenologically more promising Calabi-Yau manifolds but it represents an example where the calculations can be presently performed and that will be of relevance once a realistic compactification is achieved.

In the next section, we will present the results of a more complete calculation of the vacuum energy for even dimensional spaces with two internal spheres.[8) In particular, we will present the results for the $M^4 \times S^3 \times S^3$ geometry that is of relevance to our previous stability analysis of ref. 7. In section 3 we will use these results within a cosmological context and write Einstein's equations for the ten-dimensional case including monopole-like contributions. These latter terms come from the presence of the Kalb-Ramond 3-form in the action of $N=1$ supergravity and can be used to induce the $S^3 \times S^3$ compactification of the internal space. We will then integrate the equations numerically to study the possibility of finding solutions that exhibit both stability of the internal space and a physical rate of expansion for the four-dimensional spacetime.

2. Vacuum Energy of $M^4 \times S^M \times S^N$ in Even Dimensions

We are interested in calculating the 1-loop quantum fluctuations of scalars and spin-$\frac{1}{2}$ fields in the background $M^4 \times S^M \times S^N$ geometry in even dimensions. We refer the reader to ref. 8 for details. In principle higher-spin fields should be included but we will concentrate on the simplest case at this level. The 1-loop quantum potential is given in general by

$$V_Q = bV_Q^{(0)} - 4fV_Q^{(1/2)} , \tag{1}$$

where b and f are the number of bosons and fermions, respectively.

If we use the zeta-function regularization, we can write V_Q as

$$V_Q^{(i)} = i\left(ln\bar{\mu}\varsigma^{(i)}(0) + \frac{1}{2}\varsigma'^{(i)}(0)\right), \quad i = 0, 1/2 , \tag{2}$$

where $\bar{\mu}$ is a constant coming from the measure of the path integral. By taking into account the eigenvalues and degeneracies of the Laplacian operator of scalars and fermions in the above product manifold, we end up with an extremely complicated expression for the quantum potential. We will thus make two simplifications; we will only consider the case with two internal 3-spheres (which is the case of interest anyway) and will restrict ourselves to small deviations from equality in the ratio of the two internal radii. Accordingly, we define ε, the deviation from equality of the radii, as

$$\frac{C}{B} \equiv 1 + \varepsilon , \tag{3}$$

where B (C) is the radius of the M (N) -sphere (here, $M = N = 3$) and retain, in what follows, terms of no higher than linear degree in ε. With this approximation we obtain, after some algebra,

$$V_Q = \frac{V_4}{B^4}[b(3.639 \times 10^{-4} - \varepsilon 6.053 \times 10^{-4} + 3.315 \times 10^{-4}(1 - 2\varepsilon)ln(B\bar{\mu}))+ \\ + f(3.657 \times 10^{-6} - \varepsilon 5.414 \times 10^{-5})]. \tag{4}$$

This is the quantum potential we were looking for. Note that the original symmetry under the interchange of the two radii was lost due to our asymmetric choice of expansion. The next step is to try to solve Einstein's equations in the presence of this potential. Following Candelas and Weinberg (see ref.3), we should be able to find a constant solution for the internal radii by balancing the quantum correction with the classical curvature term. We can write Einstein's equation for the $M^4 \times S^3 \times S^3$ geometry whithin our approximation, as,

$$\frac{6(1-\varepsilon)}{B^2} = \frac{8\pi G_4}{V_4} V_Q \tag{5.1}$$

$$\frac{6(2-3\varepsilon)}{B^2} = \frac{8\pi G_4}{V_4} B \frac{\partial V_Q}{\partial B} \tag{5.2}$$

$$\frac{6(2-3\varepsilon)}{B^2} = \frac{8\pi G_4}{V_4}\frac{\partial V_Q}{\partial \varepsilon} \ , \tag{5.3}$$

where we used that $G_D = G_4 \Omega_M \Omega_N$, with Ω_M being the volume of the M-dimensional sphere.

Now we impose that $V_Q(\bar{\mu}, B_0, \varepsilon_0) = 0$ and that B_0 and ε_0 are critical points of the quantum potential. The first condition is equivalent to fine-tuning the 4-dimensional effective cosmological constant to zero and allows one to solve for $\bar{\mu}$ in terms of B_0 and ε_0 . We note that in odd-dimensional theories the 1-loop potential is independent of the renormalization parameter $\bar{\mu}$ and one is forced to introduce a D-dimensional cosmological constant in order to obtain a flat 4-dimensional spacetime.

By using the expression of the quantum potential in equation 5, we find

$$\varepsilon_0 = 1.447 \times 10^{-1} + 6.871 \times 10^{-2}\frac{f}{b} \tag{6}$$

$$\frac{B_0^2}{8\pi G_4} = b10^{-6}(8.14 - 5.26 \times 10^{-1}\frac{f}{b}) \ . \tag{7}$$

In particular, if we take $b \sim f \sim 10^4$, we find $B_0 \simeq 1.428 L_P$, where L_P is the Planck length. Thus, for a sufficiently large number of matter fields, the 1-loop calculation is valid. Note also that ε_0 is small enough to justify our approximate solution. We have checked if this approximate solution is stable by explicitly calculating the second derivative of the effective potential obtained by taking the quantum and classical terms together. As expected, the solution given by equations (6) and (7) is not stable, representing a saddle point of the effective potential. This result, together with many others in the literature, seems to suggest that quantum effects alone are not enough to stabilise the internal space. We note however that the effective potential that will come from a realistic solution of a higher dimensional theory will be certainly more complex than the prototypes analysed so far. Hopefully, it will also exhibit the necessary properties for stability.

3. Cosmological Evolution of Quantum Compactification

Now that we have the expression for the quantum potential we are in a position to study the cosmological evolution of our model. Here, we will neglect the corrections that come from performing the quantum calculation in a time-varying background. These corrections may be of relevance in determining the sign of the kinetic term for the internal radius and thus may affect conclusions about the stability of perturbations.[9] As usual, we consider a generalized Robertson-Walker metric for the product space $M^4 \times S^3 \times S^3$. We will also include a general monopole-like term in the energy-momentum tensor, as in ref.7. It is the balance between this term and the quantum contribution that can make stability possible.

By writing the scale factors for the physical spacetime and for the internal space as $A(t)$, $B(t)$ and $\varepsilon(t)$ and by writing the constants in the monopole contribution as f_0^B and f_0^ε , Einstein's equations become,

$$\frac{\ddot{A}}{A}+\frac{\dot{A}}{A}\left(2\frac{\dot{A}}{A}+3\dot{\varepsilon}+6\frac{\dot{B}}{B}\right)=-\frac{1}{32\pi^4B^{10}}\Big\{b\big[8\alpha(1-3\varepsilon)+\beta(1-10\varepsilon)-\gamma(1-5\varepsilon)+ 2\gamma(5-22\varepsilon)ln(B\bar{\mu})\big]+f\big[8\delta(1-3\varepsilon)+\xi(1-10\varepsilon)\big]\Big\}+ -\frac{1}{2B^6}\left[(f_0^B)^2+(f_0^\varepsilon)^2(1-6\varepsilon)\right] \quad (8.1)$$

$$\frac{\ddot{B}}{B}+\frac{2}{B^2}+\frac{\dot{B}}{B}\left(5\frac{\dot{B}}{B}+3\dot{\varepsilon}+3\frac{\dot{A}}{A}\right)=\frac{1}{96\pi^4B^{10}}\Big\{b\big[32\alpha(1-3\varepsilon)-\beta(3+26\varepsilon)-5\gamma(1-5\varepsilon)+ 2\gamma(13-74\varepsilon)ln(B\bar{\mu})\big]+f\big[32\delta(1-3\varepsilon)-\xi(3+26\varepsilon)\big]\Big\}+ +\frac{1}{2B^6}\left[3(f_0^B)^2-(f_0^\varepsilon)^2(1-6\varepsilon)\right] \quad (8.2)$$

$$\ddot{\varepsilon}+\frac{2(1-2\varepsilon)}{B^2}+\frac{\ddot{B}}{B}+\frac{\dot{B}}{B}\left(9\dot{\varepsilon}+5\frac{\dot{B}}{B}+3\frac{\dot{A}}{A}\right)+3\dot{\varepsilon}\frac{\dot{A}}{A}=\frac{1}{96\pi^4B^{10}}\Big\{b\big[3\gamma(1-5\varepsilon)+ 5(1-2\varepsilon)(\beta+2\gamma ln(B\bar{\mu}))\big]+5f(1-2\varepsilon)\xi\Big\}+ +\frac{1}{2B^6}\left[-(f_0^B)^2+3(f_0^\varepsilon)^2(1-6\varepsilon)\right] \quad (8.3)$$

where α, β, γ, δ and ξ are, in this order, the numbers appearing in the result for the quantum potential, eq.(4). The 00 equation was not written and can be used as a constraint equation.

The usual procedure now is to find the constant solutions for the internal radii by fine tuning the effective cosmological constant (the r.h.s. of eq. (8.1)) to zero and by setting the r.h.s. of eqs.(8.1) and (8.2) to zero. By doing this we can also express the regularization parameter $\bar{\mu}$ in terms of the constant values for the internal radii, as we mentioned earlier. This was done for the static case without the monopole term in ref.8. Note that the usual interpretation of eq.(8.2) as an equation for a scalar field $\phi = ln(\frac{B}{B_0})$ is not possible here since the potential is a function of the two internal radii. A detailed analysis of the constant solution and its stability is in progress[10]. For now, we show in fig.1 the result of a numerical integration of eq.(8). Note that the perturbation ε oscillates around zero while the internal radius B oscillates around a constant value. The physical radius expands with a rate that is sensitive to the initial conditions used. This example suffices to show that, within certain approximations, it is possible to obtain a cosmological scenario coming from higher-dimensional theories that can be compatible with the usual Friedmann cosmology.

REFERENCES

1) For a review see Duff, M.J., Nilsson, B.E.W. and Pope, C.N., Phys. Rep. **130**, 1(1986).

2) Gleiser, M., Rajpoot, S. and Taylor, J.G., Ann. Phys. **160**, 299(1985).
3) For a review see Chodos, A., "Quantum Aspects of Kaluza-Klein Theories" in "An Introduction to Kaluza-Klein Theories" ed. by Lee, H.C. (World Scientific, 1984); Candelas, P. and Weinberg, S., Nucl. Phys. **B237** 397(1984).
4) See ref.1).
5) See ref.7) and Accetta, F.S. in his contribution to this conference.
6) Gleiser, M. , Holman, R. and Kolb, E. W., work in progress. Maeda, K. ICTP preprint IC/86/316, September 1986.
7) Accetta, F.S., Gleiser, M., Holman, R. and Kolb, E.W., Nucl. Phys. **B276**501(1986).
8) Gleiser, M., Jetzer, P. and Rubin, M.A., Fermilab preprint 87/25, January 1987.
9) Gilbert, G., Mclain, B. and Rubin, M.A., Phys. Lett. **142** 28(1984).
10) Gleiser, M., Jetzer, P. and Kolb, E.W., work in progress.

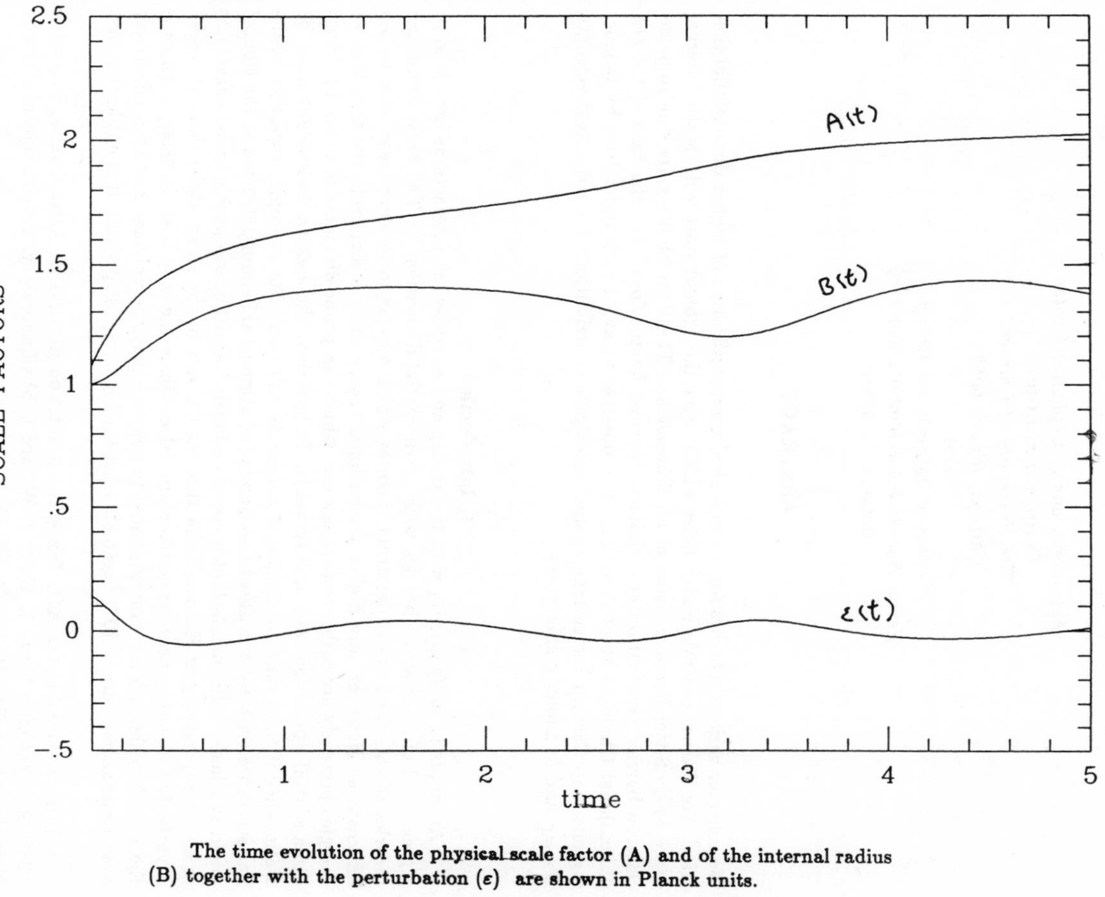

The time evolution of the physical scale factor (A) and of the internal radius (B) together with the perturbation (ε) are shown in Planck units.

INSTABILITIES OF HIGHER DIMENSIONAL COMPACTIFICATIONS

Frank S. Accetta†

Astronomy and Astrophysics Center
Enrico Fermi Institute
The University of Chicago
Chicago, Illinois 60637
and
Theoretical Astrophysics Group
Fermi National Accelerator Laboratory
Batavia, IL 60510

ABSTRACT

We consider various schemes for cosmological compactification of higher dimensional theories. We discuss possible instabilities which drive the ground state with static internal space to de Sitter-like expansion of all dimensions. These instabilities are due to semi-classical barrier penetration and classical thermal fluctuations. For the case of the ten dimensional Chapline-Manton action, it is possible to avoid such difficulties by balancing one-loop Casimir corrections against monopole contributions from the field strength H_{MNP} and fermionic condensates.

1. Introduction

Attempts to unify gravity with the strong and electro-weak interactions have lead to a great deal of interest in theories with extra spatial dimensions[1]. The most promising theories of this type are superstring theories which appear to be consistent only in ten dimensions. However, any higher dimensional theory must incorporate the fact that at energies presently accessible to accelerators, which can probe distances of order 10^{-16}cm, extra spatial dimensions are unobservable. In addition, these extra dimensions must be static since if they vary, fundamental constants will vary. For example, variation in the fine structure constant can affect the amount of primordial helium produced at the time of nucleosynthesis[2]. Requiring that these abundances lie within acceptable limits constrains in either superstring or Kaluza-Klein theories the size of the extra dimensions at nucleosynthesis to be very near its equilibrium value. Since the only scale in these theories is the Planck scale, it is not unreasonable to suppose that the universe has been effectively four dimensional since $\sim 10^{-42}$ seconds after the big bang. At present, it is not known how

† Talk presented at the XIII Texas Symposium on Relativistic Astrophysics, December 1986. Address after Sept. 1, 1987: J. Willard Gibbs Laboratory, Department of Physics, Yale University, New Haven Ct. 06520

the universe evolved from say ten dimensions to four dimensions plus some presumably compact internal space. It appears that such an evolution would require a change in the topology of space-time, a highly non-perturbative effect. However, once the universe has this product space structure it is possible to study in a cosmological setting how it evolves so that the extra dimensions presently form a small, static internal space.

Our approach to studying the cosmological evolution of the extra dimensions is to begin with matter fields defined on a $4+D$ dimensional manifold with the *ansatz* that this manifold has the product space structure $M^{4+D} = R^1 \times Q^3 \times \prod_{i=1}^{\alpha} S_i^d$, where Q^3 is the physical 3-space with radius a, $D = \alpha d$ and the internal radii are $b_1, \ldots, b_\alpha$. If we introduce a $4+D$ dimensional cosmological constant Λ^{4+D}, and for the present take $\alpha = 1$, then by balancing Λ^{4+D} against the vacuum stress energy of the matter fields (gravity will be treated classically here) the internal D-sphere is stable against small perturbations around some equilibrium value, b_0, of the internal radius. This configuration comes about by requiring that the minimum energy state be static and have a vanishing Λ^4.

Compactification stabilized due to the vacuum stress energy of quantum fluctuations due to non-trivial boundary conditions, is analagous to the Casimir effect in quantum electrodynamics[3)], while compactification due to classical stress energy can arise from the existence of monopole configurations for gauge and matter fields[4)].

Though these stabilization schemes are perturbatively stable, it has been demonstrated that the ground state manifold is semiclassically unstable–there is a nonzero probability for decay via quantum tunneling through a potential barrier[5)]. In addition, at non-zero temperature there exists the possibility of classically rolling over the barrier due to thermal fluctuations[6)]. In both cases, the instability is characterized by a de Sitter-like expansion of all dimensions.

The semiclassical instability is the result of adding a cosmological constant to the action. However for higher dimensional supergravity theories, such the field theoretic limit of the heterotic string, one cannot have a cosmological constant since this explicitly breaks supersymmetry. One can achieve a stable compactification in such theories by balancing Casimir-like one-loop quantum effects against monopole configurations which include contributions from fermionic condensates[7)].

In section 2 we will discuss stabilization of the internal space using the higher dimensional cosmological constant and Casimir contribution (monopole contributions give similar results). In section 3 we will discuss the situation for ten dimensional supergravity where stabilization of the internal space is brought about by balancing Casimir and monopole contributions.

2. Semi-classical and Thermal Instabilities

The free energy for non-interacting spinless matter fields in thermal equilibrium at temperature T is

$$\beta T = \frac{1}{2} \ln \det(-\Box_{4+D} + \mu^2). \tag{1}$$

Here the product space manifold, $S^1 \times S^3 \times S^D$, is Euclidean with the time direction compactified to a circle of radius $\beta/2\pi$. After regularization, and generalizing to a set of spinless, noninteracting fields, the free energy can be approximated[8] in the "flat-space" limit, $a >> b$, as

$$F = \frac{\Omega_3}{b^4}\left[c_1 - c_2(2\pi bT)^4 - c_3(2\pi bT)^{4+D}\right]. \tag{2}$$

Here Ω_3 is the volume of the physical 3-sphere, c_1 is the Casimir coefficient c_N of Candelas and Weinberg, while c_2 and c_3 are thermal terms. Equation (2.2) has the correct high $(T > 1/2\pi b)$ and low $(T < 1/2\pi b)$ temperature limits for the free energy. For our product space metric, the stress-energy tensor has the form $T_{MN} = \mathrm{diag}(\rho,\ p_3\tilde{g}_{ij},\ p_D\tilde{g}_{mn})$ and the components of T_{MN} can be obtained from Eq. (1) using standard thermodynamic relations generalized to higher dimensions. Plugging these results into Einstein's equations, one finds that the equation of motion for the b scale factor can be written

$$\begin{aligned}\frac{\ddot{b}}{b} + (D-1)\frac{\dot{b}^2}{b^2} + 3\frac{\dot{b}\dot{a}}{ba} &= -\frac{(D-1)}{b^2} + \frac{D(D-1)}{(D+4)} \\ &\times\left[\left[b_0^{-2} + \frac{4}{D}\frac{b_0^{D+2}}{b^{D+4}}\right] + \frac{b_0^{D+2}}{b^{D+4}}\frac{c_3}{c_1}(2\pi bT)^{(D+4)}\right].\end{aligned} \tag{3}$$

This can be recast as an equation of motion for a scalar field minimally coupled to gravity in four dimensions, with potential

$$\begin{aligned}V(\Phi,T) =& \frac{(D-1)\Lambda m_{Pl}^2}{8\pi(D+2)}\left[\frac{(D+4)}{(D-2)}(\Phi^{(2/D)(D-2)} - 1) + \Phi^{-8/D}\right. \\ &\left. - \left[1 + \frac{c_3}{c_1}(2\pi b_0 T)^{D+4}\right]\Phi^2 + \frac{c_3}{c_1}(2\pi b_0 T)^{D+4}\right].\end{aligned} \tag{4}$$

At temperatures less than some critical temperature, V is unbounded from below for large values of Φ and has a barrier which separates this region from the vacuum $(\Phi = \Phi_0)$ with static internal space of radius $b = b_0$ and zero cosmological constant. The lifetime of the compactified state can be estimated in a straightforward fashion. The semi-classical decay rate per unit 4-volume is $\Gamma = m^4\exp(S_4)$ where S_4 is the four dimensional euclidean action for the field Φ and m is a determinant with mass scale m_{Pl}. For $T = 0$ and $D = 7$, we can approximate V with $V(\bar{\Phi}) \approx 0.093\Lambda\bar{\Phi}^2 - 0.159\Lambda\bar{\Phi}^3/m_{Pl}$ (here Λ is Λ^{4+D}) and the tunnel action is $S_4 \approx 165 m_{Pl}^2/\Lambda$. The decay amplitude becomes of order one when $\tau \approx m_{Pl}^{-1}\exp(41 m_{Pl}^2/\Lambda)$ so that the compactification lifetime will be longer than the present age of the universe, $\tau > H_0^{-1}$, for values of $\Lambda \leq 0.3 m_{Pl}^2$ which corresponds to values of $b_0 \geq 11 l_{Pl}$.

At finite temperature, $V(\Phi,T)$ has a local minimum Φ_0 when $T < T_{crit}$ while for $T > T_{crit}$, $V(\Phi,T)$ monotonically decreases. The barrier height drops as T increases, vanishing at $T = T_{crit}$. As might be expected, classical thermal fluctuations of the compactified space are important for temperatures $T > 1/2\pi b$. If we require that $\Lambda^{4+D} \leq 0.3 m_{Pl}^2$, then $T_H = 1/2\pi b \leq 1.49 \times 10^{-2} m_{Pl}$, while $T_{crit} \leq 2 \times 10^{-2} m_{Pl}$. This narrow region of interesting temperatures should not be surprising since $V(\Phi,T)$ is such a strong function of T. The finite temperature vacuum decay rate is $\Gamma \approx \beta^{-4}\exp[-\beta S_3(\Phi,T)]$. Now the relevent scale for the determinant is $1/\beta$ and the fact that at finite temperature euclidean time is periodic in β allows us to write $S_4 = \beta S_3$ and S_3 is the three dimensional euclidean action (free energy).

We see that if $\Phi = \Phi_0$ when $T > T_{crit}$, stabilization of the extra dimensions is impossible. In the region $T_{crit} \geq T_{comp} \geq T_H$, where T_{comp} is the temperature at which $\Phi = \Phi_0$, then the decay rate is large only for $T_{comp} \sim T_{crit}$ and to avoid a destabilizing thermal fluctuation, the initial entropy must be made small. This corresponds to a small value for Λ^{4+D} which in turn implies a larger radius for the internal space. Though the decay rate is large only for $T_{comp} \sim T_{crit}$, we should note that $T_{crit} << m_{Pl}$ so that if compactification takes place near the Planck scale, it seems difficult to have hot initial conditions in these models.

3. Stability for Ten-dimensional Supergravity

Though the instabilities discussed in the last section can be avoided, or at least postponed by adjusting parameters, it would be preferable if such fine tuning were not necessary. That $V(\Phi, T)$ is unbounded from below for large values of Φ is a consequence of including a cosmological constant in the higher dimensional action. Since we wish to consider compactifications in more realistic supersymmetric models, alternate compactification schemes which do not include Λ^{4+D} and so may not contain instabilities should be investigated.

Type I or heterotic string theories contain $N = 1$ supersymmetry coupled to $N = 1$ super-Yang-Mills in ten dimensions. The action contains an antisymmetric rank-2 tensor with an accompanying three-index field strength H. Returning to our product space metric with 2 internal 3-spheres, we can use the Freund-Rubin *ansatz*[9] for the field strength H_{MNP}, giving it a monopole configuration on each of the $i = 1$, 2 internal 3-spheres:

$$H_{MNO} = \sqrt{g^{(3)}}\epsilon_{m_i n_i p_i} f^{(i)}(t), \tag{5}$$

and setting it to zero on the external space. The Bianchi identities then tell us that $f^{(i)}(t) = f_0^{(i)}/b_i^d(t)$. The vacuum stress energy due to monopole configurations will scale as $1/b_i^{2d}$.

For a manifold $R \times S^3 \times S^D$, with $a \to \infty$, the Casimir contribution to the vacuum stress energy has the form

$$F = \Omega_3 \left[\frac{A + A' \ln(2\pi\rho^2)}{b^4} \right] \tag{6}$$

Here, A and A' are calculable coefficients, $\rho^2 = \mu^2 b^2$, and μ is a regularization scale. In odd dimensions, A' vanishes so that F does not explicitly depend on an undetermined parameter. For our purposes, we can neglect the logarithmic dependence on the radius and set the numerator equal to a constant.* Using our product space *ansatz*, we write the Casimir free energy as

$$F = \Omega_3 \sum_{i=1}^{2} \frac{A^{(i)}}{b_i^4}. \tag{7}$$

---

* A calculation of one-loop effects in even dimensions is discussed by M. Gleiser in his contributions to the proceedings.

Since the monopole and Casimir energies scale differently, there will be non-trivial values of the b_i for which the internal spaces are static. However, for models which do not contain fermionic condensates, Minkowski space appears as a perturbatively unstable point for the equations of motion.

In the case of the ten dimensional Chapline-Manton action[10)], it is possible to obtain a stable compactification. Consider the bosonic part of this action including gluino and subgravitino couplings. We set the Yang-Mills field strength to zero and the dilaton to a constant $\sigma = \sigma_0$. The internal space is a product of two 3-spheres. In addition we impose the Freund-Rubin condition for H_{MNP} and fermionic condensates. These are related to each other in a non-trivial fashion through the dilation field equation

$$e^{-\sigma_0}(H_{MNP})^2 = \frac{3}{2}e^{-\sigma_0/2}H_{MNP}(Tr\bar{\chi}\Gamma^{MNP}\chi) \tag{8}$$

After adding Casimir terms, setting $b_1 = b_2 = b$, and rescaling coefficients we find that the b equation of motion can be written

$$\frac{\ddot{b}}{b} + 5\frac{\dot{b}^2}{b^2} + 3\frac{\dot{a}\dot{b}}{ab} = -\frac{2}{b^2} + \frac{4A}{3b^{10}} + \frac{c'}{b^6} \tag{9}$$

The coefficient c' is a function of the monopole strengths of H and the fermionic condensates. In terms of an effective four dimensional scalar field $\phi = ln(b/b_0)$, we can define an equation of motion with potential

$$V(\phi) = b_0^{-2}\left[-e^{-2\phi} + \frac{c'}{6b_0^4}e^{-6\phi} + \frac{(2b_0^4 - c')}{10b_0^4}e^{-10\phi} + \frac{12b_0^4 - c'}{15b_0^4}\right]. \tag{10}$$

The critical points for this potential are $\phi_1 = 0$ and $\phi_2 = \frac{1}{4}\ln[-2b_0^4/(2b_0^4 - c')]$ with $2b_0^4 < c'$, $c' > 0$. For ϕ_1 there exists a minimum at b_0 when $4b_0^4 > c'$, $c' > 0$ or for $c' < 0$. For ϕ_2 we find that Minkowski space is once again a maximum. To realize $\phi_1 = 0$, set the gluino monopole strenth equal to the negative of the H monopole strength. Then $c' = 6b_0^4/5$ and the effective four dimensional cosmological constant vanishes. No fine tuning is needed to realize a stable compactification but in this approach, stability away from the $b_1 = b_2$ line in phase space is unknown.

REFERENCES

1) For a review of Kaluza-Klein Supergravity see Duff, M. J., Nilsson, B. E. W. and Pope, C. N., Phys. Rep. **130**, 1 (1986); for a pedagogical discussion of superstrings see Green, M. B., Schwarz, J. H. and Witten, E. *Superstring Theory, Vol. I, II* (Cambridge University Press, 1987); for a review of the role of extra dimensions in cosmology see Kolb, E. W., "Particle Physics and Cosmology" in Proceedings of the 1986 TASI, Santa Cruz, California, ed. Haber, H. (World Scientific, 1987).
2) Kolb, E. W., Perry, M. J., Walker, T. P., Phys. Rev. D **33**, 869 (1986).
3) For a review, see Chodos, A., in *High Energy Physics 1985* ed. Gürsey, F. and Bowick, M. (World Scientific, 1986); Appelquist, T. and Chodos, A., Phys. Rev. Lett. **50**, 141 (1983); Candelas, P. and Weinberg, S., Nucl. Phys. **B237**, 397 (1984).

4) Horvath, Z., Palla, L., Cremmer, E., and Scherk, J., Nucl. Phys. **B127**, 57 (1977); Randjbar-Daemi, S., Salam, A., and Strathdee, J., *ibid.* **B214**, 491 (1983).
5) Frieman, J. A. and Kolb, E. W., Phys. Rev. Lett. **55**, 1435 (1985).
6) Accetta, F. S. and Kolb, E. W., Phys. Rev. D **34**, 1798 (1986).
7) Accetta, F. S., Gleiser, M., Holman, R., and Kolb, E. W., Nucl. Phys. **B276**, 501 (1986).
8) Randjbar-Daemi, S., Salam, A., and Strathdee, J., Phys. Lett. **135B**, 388 (1984); Okada, Y., Nucl. Phys. **B264**, 197 (1986).
9) Freund, P. G. O. and Rubin, M. A., Phys. Lett. **97B**, 233 (1980).
10) Chapline, G. F. and Manton, N. S., Phys. Lett. **120B**, 105 (1983).

Initial Conditions and Quantum Cosmology

James B. Hartle

Department of Physics
University of California
Santa Barbara, California 93106

ABSTRACT

A theory of initial conditions is necessary for a complete explanation of the presently observed large scale structural features of the universe, and a quantum theory of cosmology is probably needed for its formulation. The kinematics of quantum cosmology are reviewed and some candidates for a law of initial conditions discussed. The proposal that the quantum state of a closed universe is the natural analog of the ground state for closed cosmologies and is specified by a Euclidean sum over histories is sketched. When implemented in simple models this proposal is consistent with the most important large scale observations.

1. INTRODUCTION

The evidence of the observations is that the universe was a simpler place earlier than it is now — more homogeneous, more isotropic, more nearly in thermal equilibrium.[1] In this early simplicity we can find explanations of much of the apparent complexity of the present universe by understanding the dynamical processes that happened over its history. For example we hope to understand stars and galaxies by understanding the evolution of an initial spectrum of perturbations on the prevailing homogeneity and isotropy through the action of gravitational attraction. We hope to understand the abundances of the elementary particles and nuclear species by understanding the evolution of an initial democracy of thermal equilibrium through expansion and fundamental particle interactions. The successes of this program, which I do not need to review here,[1] have naturally raised the question of what is the explanation of the simplicity of the early universe? What is the reason for the homogeneity and isotropy? What is the origin of the primordial spectrum of density fluctuations? In short, can we find a compelling law which specifies the initial conditions of the universe. I want to discuss some proposals for this law here , but first I would like to make a few remarks about this enterprise in general.

(1) A law specifying initial conditions is a different kind of law from those we are used to in physics. In physics we are used to laws which specify evolution. Such laws require boundary conditions and the laws — Einstein's equation and the matter field equations — which govern cosmological evolution are no exception. For most of physics the boundary conditions are determined by observation of the rest of the universe outside the system whose evolution is being considered. If we don't see any incoming radiation, we calculate with no incoming radiation boundary conditions. In cosmology, by definition, "the system" is everything and there is no "rest of the universe" to pass the specification of the boundary conditions off to. As different as it is from our usual occupation, the nature of the cosmological problem forces us to consider the specification of the boundary conditions for cosmological evolution as part of the laws of physics.

(2) Inflation is not enough. We are used to the assertion that inflation explains the homogeneity and isotropy of the universe. Starting from sufficiently regular initial conditions it certainly does that. However, it cannot be a complete explanation. One can certainly imagine irregular universes today very unlike the one we see. Evolved backwards in time, whatever the physics, whether it is classical or quantum, whether there is a phase transition or not, these irregular universes came from *some* initial conditions and inflation, therefore, cannot rule these out. The initial conditions which lead to irregular universes are generic, at least as counted today, because there are many more irregular possibilities then the rather regular universe we observe. Further, if the initial conditions are sufficiently irregular, sufficiently anisotropic, sufficiently inhomogeneous, one suspects that inflation will not even to act to dominate the expansion of the universe. Such highly irregular initial conditions, however, are likely to be generic. For, following the argument of Penrose,[2] consider a generic irregular universe today. This generic universe is likely to become more irregular in the future because of the attractive nature of gravity and, therefore, because of the approximate time reversal invariance of the laws of physics, likely to have had a similarly more irregular past. Thus as critical as inflation is in explaining the absence of monopoles, the size of the horizon and the evolution of density perturbations it is unlikely to be an explanation of the homogeneity and isotropy in the context of all initial conditions. A law for initial conditions is still needed.

(3) Quantum Gravity is Important. Classical cosmological spacetimes can be singular, for example the Friedman models, or non-singular, for example, de Sitter space. Depending on the physics of the matter there could either have been a

big bang or a small bounce. Singularities are not easy to avoid in general relativity. The singularity theorems of classical general relativity suggest that, with reasonable assumptions on the matter physics, the classical extrapolation of any universe similar to our own into the past will encounter a big bang singularity.[3] If the classical evolution is singular, then curvatures which vary significantly on the scale of a Planck length will occur and quantum gravitational effects will be significant for the early universe. This is one reason why quantum gravity is the appropriate context in which to search for a theory of initial conditions.

The second reason is that it is difficult to imagine a classical theory of initial conditions. What classical principle for example would single out the *particular* density fluctuations we have today? By contrast, fluctuations occur naturally and fundamentally in a quantum mechanical theory. Of course, one could always imagine a *statistical* classical theory but such a law of initial conditions would contrast with the deterministic character of classical evolutionary laws.

For these two reasons quantum cosmology seems to be the appropriate context to search for a theory of initial conditions. Conversely if one has a theory of quantum gravity the construction of a theory of initial conditions for cosmology would seem to be one of its most fitting applications.

2. KINEMATICS OF QUANTUM COSMOLOGY

2.1 The wave function of the universe.

In quantum mechanics we describe the state of a system by giving its wave function. The wave function enables us to made predictions about observations made on a spacelike surface; it thus captures quantum mechanically the classical notion of the "state of the system at a moment of time." The arguments of the wave function are the variables describing how the system's history intersects the spacelike surface. For example, for the quantum mechanics of a particle, the histories are particle paths $x(t)$. We write for the wave function

$$\psi = \psi(x,t). \tag{2.1}$$

The t labels the hypersurface and the x specifies the intersection of the history with it.

In the quantum mechanics of a closed cosmologies with fixed (for simplicity) spatial topology, say, that of a 3-sphere S^3, the histories are the 4-geometries on

$S^3 \times \mathbf{R}$. The appropriate notion of a 4-geometry fixed on a spacelike surface is the 3-geometry, ${}^3\mathcal{G}$, induced on that surface. One can think of this as specified by a 3-metric h_{ij} on the fixed spatial topology. Thus for the quantum mechanics of a closed cosmology we write[4]

$$\Psi = \Psi[{}^3\mathcal{G}] = \Psi[h_{ij}(\mathbf{x})]. \tag{2.2}$$

Note that there is no additional "time" label. This is because a generic 3-geometry will fit in a generic 4-geometry at locally only one place if it fits at all. The 3-geometry itself carries the information about its location in spacetime. This labeling of the wave function correctly counts the degrees of freedom. Of the six components of h_{ij}, three are gauge. If one of the remaining three is time, there are left two degrees of freedom — the correct number for a massless, spin-2 field.

The space of all three geometries is called superspace. Each "point" represents a different geometry on the fixed spatial topology. In the case of pure gravity that we have been describing, the wave function is a complex function on superspace. With the inclusion of matter fields the wave function depends on their configurations on the spacelike surface as well, and we write typically

$$\Psi = \Psi[h_{ij}(\mathbf{x}), \phi(\mathbf{x})]. \tag{2.3}$$

A law for initial conditions in quantum cosmology is a law which prescribes this wave function.

2.2 Interpretation

To make contact with observations we must specifiy the observational consequences of the state of the universe being described by this or another wave function. This is usually called an "interpretation" of Ψ. There is little doubt that what I can say here will not address every issue which can be raised on this fascinating topic and even less doubt that it will not satisfy many who have thought about the subject. I would like, however, to offer some minimal elements of an interpretation which I believe will enable an attribution of Ψ to the universe to be confronted by cosmological observations. These elements are an example of "an Everett interpretation" although the words and emphasis may be different from other interpretations in this broad catagory.[5]

The idea is to take quantum mechanics seriously. One assumes that there is one wave function Ψ defined on a preferred configuration space which contains all the predictable information about observations in the universe. If Ψ is sufficiently peaked about some region in the configuration space we predict that we will observe the correlations between the observables which characterize this region. If Ψ is small in some region we predict the observations of the correlations which characterize this region are precluded. Where Ψ is neither small nor sufficiently peaked we don't predict anything. That's it.

The natural reaction to such an interpretations is to ask "Where is probability?" In response, I can say two things:

First, probabilities for single systems have no direct observational interpretation and the universe by definition is a single system. Second, this interpretation implies the usual probability interpretation of quantum mechanics when applied to ensembles of identically prepared systems: Suppose, for example, the configuration space of a single system is $\mathcal{C}$. The configuration space of an ensemble of N systems is $\mathcal{C}^N$. The wave function of an ensemble of N identically prepared systems is

$$\Psi(q_1, ..., q_N) = \psi(q_1)\psi(q_2)\cdots\psi(q_N), \tag{2.4}$$

where $\psi(q)$ is the wave function of a single system. It was pointed out independently by a number of people[6] in the late 60's that, for large N, such a wave function is sharply peaked in the variable which is the frequency, f_a, that a measurement of q on each member of the ensemble yields the result a. The value of f_a about which Ψ is peaked is $|\psi(a)|^2$.

A more precise statement of this result for ensembles can be given in the familiar Hilbert space formulation of quantum mechanics. Consider a single system described by a wave function ψ. Possible observations correspond to operators in the Hilbert space of states. For the physical interpretation of ψ for a single system assume only the following: *If ψ is an eigenfunction of an observable A then an observation of A will yield (with certainty) the eigenvalue. For those observables of which ψ is not an eigenfunction there is no prediction for the outcome of an observation.* We can then derive the probability interpretation of ψ as follows:

An ensemble of identically prepared systems can be viewed as single system with wave function (2.4). On the Hilbert space on $\mathcal{C}^N$ there is an operator $\hat{f}_a$ corresponding to observing q on the first system, q on the second, etc., and then computing the frequency that a given value a occurs. For an infinitely large ensemble of identical systems, each in a state ψ, it is a mathematical fact that the product wave function (2.4) is an eigenfunction of this operator

$$\hat{f}_a \Psi = |\psi(a)|^2 \Psi. \tag{2.5}$$

In this way we deduce the probability interpretation of quantum mechanics from its predictions about individual systems.

An interpretation of this kind means that ones ability to predict in quantum cosmology is limited as it is for any single quantum system. We do not expect, for example, that the wave function will be sharply peaked about the *particular* arrangement of galaxies in the universe. We do hope that it might be peaked about the form of the galaxy-galaxy correlation function at the present epoch.

2.3 The Wheeler-DeWitt Equation

In a theory of quantum spacetime we are not free to propose any wave function as a theory of initial conditions. It must satisfy the constraints which implement the dynamics of the particular theory of quantum gravity we have in mind. For example, in pure Einstein gravity three of the constraints are[4]

$$iD_i\left(\frac{\delta\Psi}{\delta h_{ij}(\mathbf{x})}\right)=0, \tag{2.6}$$

where D_i is the covariant derivative of the metric h_{ij}. These constraints enforce gauge invariance in the spacelike surface. The additional dynamical constraint is

$$\left[l^2\nabla^2+l^{-2}h^{\frac{1}{2}}\left({}^3R-2\Lambda\right)\right]\Psi=0, \tag{2.7}$$

where $l=(16\pi G)^{\frac{1}{2}}$ is the Planck length (in units where $\hbar=c=1$) and ∇^2 is a wave operator in the 3-metric. Explicitly

$$\nabla^2=G_{ijkl}\frac{\delta^2}{\delta h_{ij}(\mathbf{x})\delta h_{kl}(\mathbf{x})}\quad+\quad\left(\begin{matrix}\text{linear terms depending}\\ \text{on factor ordering}\end{matrix}\right), \tag{2.8a}$$

with

$$G_{ijkl}=\frac{1}{2}h^{-\frac{1}{2}}\left(h_{ik}h_{jl}+h_{il}h_{jk}-h_{ij}h_{kl}\right). \tag{2.8b}$$

This constraint, called the Wheeler-DeWitt equation, reflects the dynamics of general relativity. In a different theory of quantum gravity there would be a different constraint.

One can think of the Wheeler-DeWitt equation as a kind of wave equation in superspace. Finding a law for initial conditions may be viewed as the problem of specifying the boundary conditions which select from its many solutions the one which is the wave function of our universe.

2.4 Semiclassical Approximation

In the present universe we would expect a semiclassical approximation to Ψ to be good, and even where it is not precisely valid this approximation often gives useful qualitative information about the behavior of the wave function. In the semiclassical approximation the wave function is given by

$$\Psi \sim A\cos(S). \tag{2.9}$$

in regions of superspace which are classically allowed, and by

$$\Psi \sim Ae^{I} + Be^{-I}. \tag{2.10}$$

in regions which are classically forbidden. Here S is a real Lorentzian action and I a real Euclidean action.

The utility of this approximation is that it allows direct contact with classical physics because the correlations it predicts are just those of classical physics. Put differently, semiclassically in the classically allowed region the wave function corresponds to an ensemble of classical trajectories determined by the action function S. What we can learn from the gradient of S is the tangent vector to the classical trajectory. What we do not know is the initial position. The various possibilities define the ensemble. Further, in the semiclassical approximation we recover a notion of time — the time of these classical trajectories.[7,11] These properties make the interpretation and comparison with observation a good deal easier in this approximation.

3. PROPOSALS FOR A LAW OF INITIAL CONDITIONS

In recent years a number of different proposals for a law of initial conditions have been put forward. There is Roger Penrose's time asymmetric proposal that the Weyl tensor vanish on initial singularities but not on final singularities.[2] There are the proposals that the universe nucleates spontaneously "from nothing" worked out most explicitly and clearly by Alex Vilenkin.[8] This proposal can be implemented concretely by solving the Wheeler-DeWitt equation with the boundary condition that "at the boundaries of superspace the wave function includes only outgoing modes." There is the proposal of Narlikar and Padmanabhan[9] that one should quantize only the conformal factor and enforce the constraints only semiclassically. There is the proposal of Fischler, Ratra and Susskind[10] that the wave function of the universe is singled out from other solutions of the Wheeler-DeWitt equation by the requirement that the energy in the matter field not diverge at the singularity. Frank Tipler[11] has a proposal in which the universe "explodes from nothing". There is the proposal of Stephen Hawking and his collaborators[12] that the universe is in the cosmological analog of its ground state, implemented concretely by specifying the wave function as a Euclidean sum over histories. There are no doubt many more.

It would be difficult to review all these proposals, not least, because they are not all sufficiently developed to compare their predictions. To illustrate the ideas I will therefore focus on the ground state wave function proposal because it is the most concrete and the one with which I am most familiar.

The proposal is that the quantum state of the universe is the analog for closed cosmologies of the ground state, or state of minimum excitation. To understand what this means note first that what is meant is not a state of minimum energy. For closed cosmologies there is no natural notion of time, therefore no natural notion of energy, therefore no natural Hamiltonian, and therefore no natural notion of a state with the lowest eigenvalue of the Hamiltonian. For systems which possess a Hamiltonian, however, finding the lowest energy eigenstate is not the only way of finding the ground state wave function. One can also calculate it directly as a Euclidean sum over histories. For example, for a particle in a potential $V(x)$ the ground state wave function is

$$\psi_0(x) = \sum_{\text{paths}} \exp\left(-I[x(\tau)]\right), \tag{3.1}$$

where $I = \int dt(m\dot{x}^2/2 + V(x))$ is the Euclidean action and the sum is over all paths which start at the argument of the wave function at $\tau = 0$ and proceeed to a configuration of minimum action in the infinite past. This construction generalizes to the quantum mechanics of closed cosmologies.

For a closed cosmology we shall mean by the ground state wave function:

$$\Psi[{}^3\mathcal{G}, \phi(\mathbf{x})] = \sum_{{}^4\mathcal{G}, \phi(\mathbf{x},t)} \exp\Big(-I[{}^4\mathcal{G}, \phi(x)]\Big) \tag{3.2}$$

The sum over ${}^4\mathcal{G}$ is over all compact, Euclidean 4-geometries which have a boundary on which the induced three geometry is the argument of the wave function and *no other boundary.* The sum over $\phi(x)$ is over all field configurations which match the argument $\phi(\mathbf{x})$ on the boundary and are otherwise regular. These conditions are the analogs in the particle case of the conditions that the paths start at the argument of the wave function and proceed to a configuration of minimum action in the past. Such a construction, properly implemented, should automatically yield a solution of the Wheeler-DeWitt equation.

Since the remarkable simplicity of the early universe suggests that it is in a state of low excitation, it is a natural conjecture that this is the wave function of the universe.

Although it has been stated here for spacetime theories of gravity, and for the dynamics of general relativity in particular, this proposal can be framed in a large class of gravitational theories. This is because it is essentially a topological proposal – " the boundary condition for the universe is that there is no boundary". To the extent that a gravitational theory, for example string theory, incorporates the idea of manifold there will be a similar proposal for the quantum state of the universe with that gravitational dynamics also.

How do we test this proposal? As with any theory we need to compare its predictions with the observations. The cosmological observations which we might hope to explain in a theory of initial conditions might be summarized in a few big cosmological facts.[1]

1) Spacetime is homogeneously four dimensional with Euclidean topology on familiar scales. We usually take this for granted but it is important to remember that all aspects of geometry have an observational basis.

2) The universe is large, old and expanding. This is the observation that the scales of cosmology are not the scales of quantum gravity or of elementary particle physics. The age of the universe, for example, is approximately 10^{60} times larger than the Planck scale. An important test for any theory of initial conditions is whether it explains how the universe can be as big as it is.

3) The universe contains matter and radiation distributed homogeneously and isotropically on the largest scales. A theory of initial conditions should explain the approximate large scale simplicity.

4) The spatial geometry is approximately flat.

5) There is a spectrum of density fluctuations which produced the galaxies. A theory of initial conditions should certainly be expected to produce a spectrum of initial fluctuations which, in amplitude and spectrum, serve as a suitable input for galaxy formation.

6) The entropy of the matter in the universe is low compared to what it might have been and increasing in the direction of expansion. This is another characterization of the fact that the universe is apparently more ordered earlier than it is now.

I would now like to describe how the ground state wave function proposal compares with these observations. To do so I will describe in rough terms the calculations which have been made. Essentially all of them are minisuperspace models or linear perturbations of minisuperspace models. In a minisuperspace model one truncates the infinite number of degrees of freedom to a finite number to obtain a tractable quantum mechanical model. This is usually done by imposing symmetries such as homogeneity and isotropy on the geometries and matter fields which contribute to the sum over histories. One is thus at best obtaining approximate information about the wave function on an infinitesimal region of superspace. The region can be enlarged by considering the linear deviations from exact symmetry. This is a large enlargement from the point of view of number of degrees of freedom, but still small compared to the whole of superspace. The hope is that although this region is small it will contain the region which is compatible with observations. If there is consistency with observations here, the rest of the task will be to show the wave function is small everywhere else. A typical model is the homogeneous, isotropic massive scalar field model of Hawking[12] later extended to include linear perturbations by Halliwell and Hawking.[13] The minisuperspace is obtained by restricting to geometries which are homogeneous and isotropic whose metrics have the form

$$d\hat{s}^2 = -dt^2 + a^2(t)d\Omega_3^2, \tag{3.3a}$$

where $d\Omega_3^2$ is the metric on the unit 3-sphere, and restricting to homogeneous field configurations for which

$$\phi = \phi(t). \tag{3.3b}$$

Euclidean geometries with the same symmetry are obtained by setting $t = -i\tau$ in (3.3a). Geometry and field on a spacelike hypersurface are characterized by just two numbers a and ϕ. The minisuperspace is thus just two dimensional and we write

$$\Psi = \hat{\Psi}(a, \phi). \tag{3.4}$$

The semiclassical approximation to $\hat{\Psi}$ is obtained by making a steepest

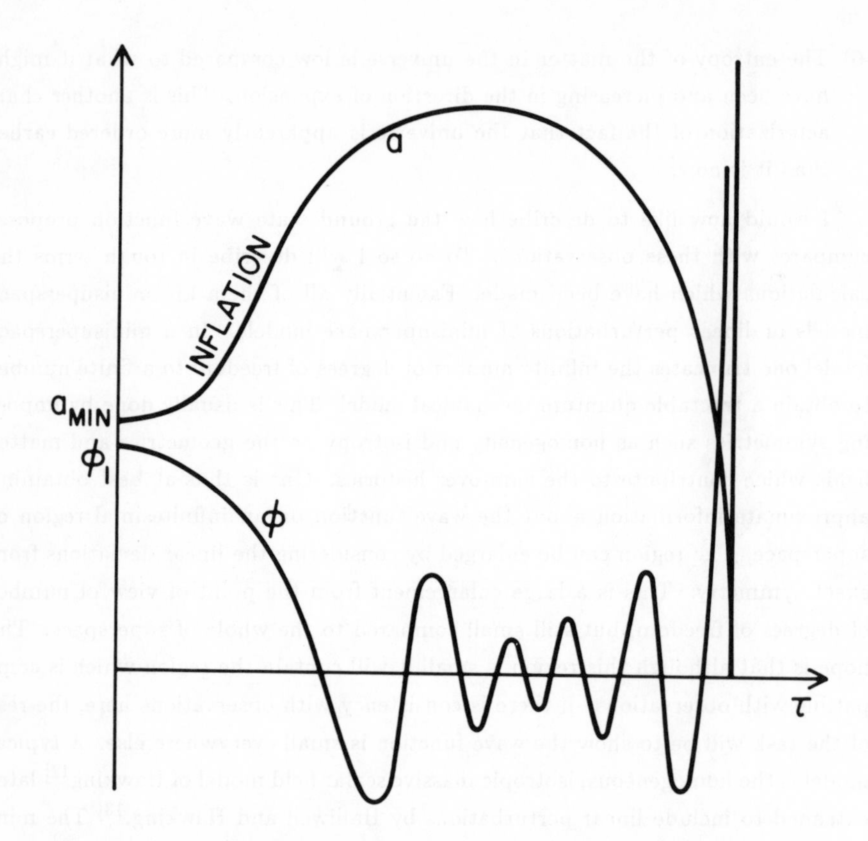

Figure 1: A schematic graph of a typical Lorentzian extremizing configuration. The ground state wave function prescription implies that the solutions start at a minimum radius with $\dot{\phi} \approx 0$. In the domain where ϕ varies slowly the universe follows a de Sitter like inflationary expansion with $H = m\phi_1$. Later the scalar field begins to oscillate and the universe evolves approximately as though matter dominated. Eventually a maximum expansion is reached, the universe recollapses and matter and geometry become singular. A sufficiently large $m\phi_1$ would provide a long enough inflationary period to explain the present large size of the universe and its approximate spatial flatness. The oscillation of the scalar field models the creation of matter.

descents approximation to the defining Euclidean functional integral (3.2). This singles out the appropriate extrema to compute the classical action in either (2.9) or (2.10). In the classically allowed region these extrema solve the classical equations of motion

$$\ddot{\phi} + \frac{3\dot{a}}{a}\dot{\phi} + m^2\phi = 0, \tag{3.5a}$$

$$\left(\frac{\dot{a}}{a}\right)^2 = -\frac{1}{a^2} + \dot{\phi}^2 + m^2\phi^2. \tag{3.5b}$$

They are the possible classical trajectories in the ensemble determined semiclassically by $\hat{\Psi}$. In this way an ensemble of possible classical behaviors for the universe is fixed.

The analysis of which extrema contribute to the semiclassical approximation depends on a detailed analysis of (3.2) which I cannot develop here. Very roughly, however, it proceeds as follows: For small a the steepest descents approximation to (3.2) is given by a compact Euclidean geometry which is a portion of a distorted 4-sphere bounded by a 3-sphere boundary of radius a. The extremizing field configuration is the regular solution of the field equations which assumes the homogeneous value ϕ on the boundary. As a is increased, for sufficiently large ϕ, the Euclidean geometry tries to close and a maximum radius is encountered beyond which it does not exist. Beyond this boundary is the classically allowed region. At this boundary $\dot{a} \approx 0$, $\dot{\phi} \approx 0$ so that the extremizing trajectory enters the the classically allowed region with these boundary conditions. The near zero rate of change of ϕ implied by this proposal has important consequences.

A typical classical trajectory in the ensemble fixed by the ground state wave function is shown in Figure 1. For small t, while $\dot{\phi} \approx 0$ the $m^2\phi^2$ term in (3.5a) acts as an effective cosmological constant. The universe inflates and according to Hawking and Page[14] the ensemble is peaked about inflations of arbitrarily long duration. This is how the universe can get big and still be in its ground state and how it can become arbitrarily close to spatially flat. Eventually the field begins to oscillate and acquires kinetic energy. In the crude terms of the model, this is how the matter of the universe is created.

Halliwell and Hawking[13] have analyzed the linear fluctuations about this minisuperspace model. These are specified by metrics and field configurations of the form

$$ds^2 = d\hat{s}^2 + \epsilon_{\alpha\beta}(\mathbf{x},t)dx^\alpha dx^\beta, \tag{3.6a}$$

$$\phi = \phi(t) + f(\mathbf{x},t), \tag{3.6b}$$

where ϕ and $d\hat{s}^2$ are as in (3.3). The linear fluctuations can be expanded in harmonics on the 3-sphere

$$\epsilon_{\alpha\beta}(\mathbf{x},t) = \sum_n \epsilon^{(n)}(t)Q^{(n)}_{\alpha\beta}(\mathbf{x}), \tag{3.7a}$$

$$f(\mathbf{x},t) = \sum_n f^{(n)}(t)Q^{(n)}(\mathbf{x}). \tag{3.7b}$$

The $h^{(n)}$ and $f^{(n)}$ can be thought of as local coordinates in superspace. We write

$$\Psi = \Psi\left(a, \phi, \epsilon^{(1)}, f^{(1)}, \epsilon^{(2)}, f^{(2)}, \ldots\right). \tag{3.8}$$

The linear modes decouple in linearized perturbation theory. Correspondingly the wave function becomes a product

$$\Psi = \hat{\Psi}(a, \phi) \prod_n \psi^{(n)}\left(\epsilon^{(n)}, f^{(n)}\right). \tag{3.9}$$

In the semiclassical approximation for $\hat{\Psi}$, the Wheeler-DeWitt equation for the $\psi^{(n)}$ becomes an ordinary Schrodinger equation

$$i\frac{\partial \psi^{(n)}}{\partial t} = H^{(n)}\psi^{(n)}. \tag{3.10}$$

The time t being the time of the classical trajectory of the unperturbed problem. The theory of fluctuations in this approximation is thus effectively a field theory in the curved background spacetime which gives the semiclassical approximation to $\hat{\Psi}$. Not surprisingly, the ground state prescription for the wave function requires each mode to start out in its ground state. The wave function is sharply peaked about the observed homogeneity and isotropy and this remains true at later times. There are still quantum fluctuations and these fluctuations, evolved through the inflationary expansion, provide a plausible spectrum of density fluctuations from which to grow galaxies. The physics of this evolution is essentially classical and essentially the same as in familiar inflationary stories. What the ground state wave function proposal provides is the initial condition for the evolution.

The results of this and some other representative* minisuperspace calculations are summarized in Table 1. There one sees that, while the wave function has been calculated on only a limited region of superspace, plausible mechanisms have been provided by which the ground state wave function can explain most of the observed large scale features of the universe. While the enterprise is very different from what we are used to in physics, it is just possible that the remarkable simplicity of the early universe is described by this simplest of all possible quantum states and that all the complexity we see around us arises from its quantum fluctuations and the attractive force of gravity.

---

* While I have attempted to be helpful in citing some typical calculations, I have not been able to be exhaustive in what already is a large literature.

TABLE 1

OBSERVED PROPERTY	MINISUPERSPACE MODELS	RESULTS	SELECTED REFERENCES
Spacetime is 4-dimensional with Euclidean topology	Local properties largely unstudied but some Kaluza-Klein models	There are preferred compactifications in Kaluza-Klein models	15–18
The universe is large and old	Homogenous and isotropic geometry with either massive scalar field or in $(\text{curvature})^2$ theories	There are trajectories along which the universe "inflates" to a large size	12, 19–22
Matter and geometry are nearly homogeneous and isotropic	Homogeneous but anisotropic models	The wave function is sharply peaked about zero anisotropy	13, 23–25
Space is nearly flat	Homogeneous and isotropic geometry with either massive scalar field or in $(\text{curvature})^2$ theories	The distribution of Ω_0 is sharply peaked about $\Omega_0 = 1$	12, 14
The spectrum of density fluctuations	Homogeneous and isotropic geometry with massive scalar field plus linear inhomogeneous perturbations in matter and geometry	The wave function predicts initial quantum fluctuations which evolve classically during an inflationary epoch to give a scale free spectrum with a plausible amplitude	13
The entropy of the universe is low and increasing in the direction of expansion	Homogeneous and isotropic geometry with massive scalar field plus linear inhomogeneous perturbations in matter and geometry	Order when the universe is small evolves to disorder when the universe is big	13, 26, 27

ACKNOWLEDGEMENTS

Substantially similar versions of this talk were presented at the SWOGU-/ICOBAN '86 conference, the 26th Liège International Astrophysical Colloquium, and the VIIIth International Congress on Mathematical Physics and will appear in the proceedings of those conferences. Part of the work for the preparation of this report was done while the author was in residence at the Institute for Theoretical Physics at the University of California, Santa Barbara. He is grateful for the Institute's hospitality. The preparation of this report was supported in part by NSF Grant PHY85-06686 and at the Institute for Theoretical Physics by NSF Grant PHY82-17853, supplemented by funds from NASA.

REFERENCES

1. For reviews of the current status of cosmological observations and models see *Physical Cosmology: Les Houches 1979*, ed. by Balian, R., Audouze, J., and Schramm, D. (North-Holland Amsterdam, 1980); and *Inner Space/Outerspace*, ed. by Kolb, E.W., Turner, M.S., Lindley, D., Olive, K and Seckel, D. (Chicago University Press, Chicago, 1986).

2. Penrose, R. in *General Relativity: An Einstein Centenary Survey,* ed. by Hawking, S.W., and Israel, W., (Cambridge University Press, Cambridge, 1979).

3. See, *e.g.,* Hawking, S.W., and Ellis, G.F.R., *Ap. J.* **152**, 25 (1968) and Geroch, R., and Horowitz, G.T. in *General Relativity: An Einstein Centenary Survey*, ed. by Hawking, S.W., and Israel, W., (Cambridge University Press, Cambridge, 1979).

4. For further details on the canonical quantum mechanics of gravity see, *e.g.,* Kuchař, K. in Relativity, Astrophysics, and Cosmology, ed. by Israel, W. (D. Reidel, Dorchecht, 1973).

5. See, *e.g.,* Everett, H. *Rev. Mod. Phys.* **29**, 454 (1957), the many articles reprinted and cited in *The Many Words Interpretation of Quantum Me-*

chanics, ed. by DeWitt, B., and Graham, N. (Princeton University Press, Princeton, 1973), and the lucid discussion in Geroch R. *Noûs,* **18** 617 (1984).

6. Finkelstein, D. *Trans. N.Y. Acad. Sci.* **25**, 621 (1963); Graham, N. unpublished Ph.D. dissertation, University of North Carolina 1968 and in *The Many Worlds Interpretation of Quantum Mechanics*, ed. by DeWitt, B. and Graham, N. (Princeton University Press, Princeton, 1973); and Hartle, J.B. *Am. J. Phys.* **36**, 704 (1968).
7. Banks, T. *Nucl. Phys.* **B249**, 332 (1985).
8. See, *e.g.,* Vilenkin, A. *Phys. Lett.* **B117**, 25 (1983), *Phys. Rev.* **D27**, 2848 (1983), *Phys. Rev.* **D30**, 509 (1984), *Phys. Rev.* **D32**, 2511 (1985), TUTP preprint 85-7.
9. See, *e.g.,* Narlikar, J.V. and Padmanabhan, T. *Physics Reports* **100**, 151 (1983); Padmanabhan, T. "Quantum Cosmology — The Story So Far," (unpublished lecture notes).
10. Fischler, W., Ratra, B., andd Susskind, L. *Nucl. Phys.* **B259**, 730 (1985).
11. Tipler, F., *Physics Reports*, **137**, 231, (1986).
12. See, *e.g.,* Hawking, S.W. in *Astrophysical Cosmology: Proceedings of the Study Week on Cosmology and Fundamental Physics,* ed. by Brück, H.A., Coyne, G.V. and Longair, M.S. (Pontificial Academiae Scientiarum Scripta Varia, Vatican City, 1982) and Hawking, S.W. *Nucl. Phys.* **B239**, 257 (1984).
13. Hawking, S.W. and Halliwell, J., *Phys. Rev.* **D31**, 1777 (1985).
14. Hawking, S.W. and Page, D.N. *Nucl. Phys.* **B264**, 185 (1986).
15. Wu, Z.C., *Phys. Lett.* **146B**, 307 (1984), *Phys. Rev.* **D31**, 3079 (1985), Hu, X.N. and Wu, Z.C., *Phys. Lett.* **149B**, 87 (1984).
16. Halliwell, J.J., "Quantum Cosmology of the Einstein Maxwell Theory in 6-Dimensions." (preprint)
17. Okada, Y. and Yoshimura, M., *Phys. Rev.* **D323**, 2164 (1986).
18. Hartle, J.B., *Class Quan. Grav.* **2**, 707 (1985).
19. Hawking, S.W. and Luttrell, J.C. *Nucl. Phys.* **B247**, 250 (1984).
20. Horowitz, G., *Phys. Rev.* **D31**, 1169 (1985).

21. González-Diaz, P.F., *Phys. Lett.* **159B**, 19 (1985).

22. Carow, U. and Watamura, S., *Phys. Rev.* **D32**, 1290 (1985).

23. Hawking, S.W. and Luttrell, J.C., *Phys. Lett.* **B146**, 307 (1984).

24. Wright, W. and Moss, I., *Phys. Lett.* **154B**, (1985).

25. Amsterdamski, P., *Phys. Rev.* **D31**, 1169 (1985).

26. Hawking, S.W., *Phys. Rev.* **D32**, 2489 (1985).

27. Page, D.N., *Phys. Rev.* **D32**, 2496 (1985).

STABILITY OF DE SITTER SPACE AND THE COSMOLOGICAL CONSTANT PROBLEM

L.H. Ford

Department of Physics and Astronomy, Tufts University
Medford, Massachusetts 02155

ABSTRACT

A review is given of work on the stability of de Sitter space and its possible relevance to the cosmological constant problem. It is argued that there is no compelling evidence for an intrinsic instability of de Sitter space which could explain the vanishing of the cosmological constant. However, specific field theories in de Sitter space can lead to instability. The most promising models for cosmological constant damping are those in which the instability does not depend upon the unique properties of de Sitter space, but also occurs in a wider class of metrics.

1. INTRODUCTION

Consideration of field theories with spontaneous symmetry breaking in the early universe leads to a serious cosmological constant problem. As the universe cools, one or more phase transitions occur and change the effective value of the cosmological constant Λ. The value after the last transition must be extremely small ($< 10^{-30}$ g/cm^3) compared to the typical energy scales at the time of the transition, yet there is no natural explanation why this should be so. If de Sitter space is unstable for some reason, then this might provide a natural mechanism for the effective value of Λ to relax toward zero. This possibility has stimulated a considerable amount of discussion on the issue of the stability of de Sitter space. Here a brief review of this work will be given.

2. INTRINSIC INSTABILITY OF DE SITTER SPACE

The first possibility is that de Sitter space is intrinsically unstable, that is, has an instability that does not depend upon special characteristics of the matter fields present. Instability under classical metric perturbations would fall into this category. However, Abbott andd Deser[1] have shown that de Sitter space is classically stable, at least under metric perturbations which are localized within one horizon volume. Thus, it seems that any instability must have a quantum origin.

An example of a quantum instability was given by Ginsparg and Perry[2]. This is the nucleation of black holes in de Sitter space, which is analogous to the formation of black holes at finite temperature in flat spacetime[3]. However, the nucleation rate is negligably small except when the spacetime curvature is characterized by Planck dimensions. In any case, this instability is not of the type which would lead to a solution of the cosmological constant puzzle.

It was recently suggested by Antoniadis, et al[4] that the graviton propagator in de Sitter space has an infrared singularity. This would mean that the space is unstable when linearized perturbations are quantized. However, Allen[5] has shown that this apparent singularity is an artifact of the choice of gauge, and that gravitons in de Sitter space are well behaved.

An instability could conceivably also arise at the semiclassical level of quantized fields on a classical background. To be classed as intrinsic, it should be relatively independent of the nature of the matter fields. Parker[6] has noted that nonvanishing of the scalar curvature leads to creation of nonconformally coupled particles. Although this need not imply that de Sitter space is unstable, it does suggest that particle creation could be linked to a cosmological constant damping mechanism.

Mottola[7,8] has argued that noninteracting quantum fields in de Sitter space will exhibit particle creation and consequent cosmological constant damping. However, in order for this to happen, particles would have to be created at a rate which more than compensates for their redshifting by the exponential expansion. Calculations[9] of particle creation in models with a de Sitter phase of finite duration indicate that the net number of particles created is independent of the length of the de Sitter phase, rather than increasing monotonically as would be required if de Sitter space were to be unstable.

Myhrvold[10] has suggested that particle interactions could lead to runaway particle creation. For example, a $\lambda\phi^4$ self-coupled scalar field has a finite amplitude in an expanding universe for one particle to split into three; these three then become nine particles, and so on. However, because this is a tree level process, it is essentially classical and there should be growing solutions of the classical wave equation. The absence of such solutions indicates that this process is a redistribution of energy among modes rather than an instability. Consequently, it is unlikely that there is any intrinsic instability of de Sitter space associated with particle creation.

3. INSTABILITY OF PARTICULAR FIELD THEORIES IN DE SITTER SPACE

It is much easier to devise specific models in which instabilities arise in de Sitter space. The various versions of the inflationary universe model[11-14] all incorporate a means for halting the inflationary expansion which involve some form of instability. Starobinsky[15] has shown that de Sitter space is unstable in the semiclassical approximation in which conformal anomaly corrections are introduced into the Einstein equations. A massless, minimally coupled scalar field in de Sitter space has the peculiar property that it does not possess a de Sitter invariant vacuum state[16,17]. This leads to linear growth of $\langle\phi^2\rangle$. If it is coupled to other fields, then the energy-momentum tensor, $\langle T_{\mu\nu}\rangle$, can also grow linearly in time,[18] causing de Sitter space to be unstable. Traschen and Hill[19] have recently discussed a short time scale instability of massive scalar fields in de Sitter space.

An instability of de Sitter space cannot be counted as an acceptable solution to the cosmological constant problem unless it leads to a realistic cosmology. In particular, it must permit sufficient reheating to allow cosmological nucleosynthesis to proceed. Hiscock[20] has shown that this will not happen in an adiabatically decaying de Sitter space. This issue has also been discussed by Freese, et al[21]. If the instability is unique to de Sitter space, then it can only act when the metric remains close to the de Sitter metric. Thus successful cosmological constant damping requires an instability that continues to act in the post-de Sitter phase.

A model which fulfills this condition was discussed by Dolgov[22]. This involves a scalar field coupled to curvature with a $\xi R\phi^2$ coupling with $\xi<0$; the equation of motion is

$$\Box\phi + \xi R\phi = 0 . \tag{1}$$

Initially, ϕ grows exponentially in time and reacts against the de Sitter metric. At late times ϕ continues to grow linearly in time in such a way that its energy-momentum tensor asymptotically cancels the original cosmological constant term. The residual correction is of order t^{-2}, where t is the age of the universe. This is within the acceptable limits at the present time. It is also desirable that the time scale for the instability of de Sitter space be sufficiently long to allow adequate inflation[11]. This will occur in this model if $|\xi|$ is sufficiently small. This bound is discussed in detail in Ref. 23; however, for inflation at the GUT scale one needs $|\xi| < 0.1$. The scale factor at late times expands as a power law:

$$a(t) \propto t^{\alpha}, \qquad \alpha = (2|\xi|+1)/(4|\xi|), \tag{2}$$

so small $|\xi|$ leads to $\alpha\gg1$, which is undesirable.

This problem is avoided in a model in which ϕ satisfies

$$\Box\phi + \kappa R\ln(R\ell^2_{Pl})\phi = 0, \tag{3}$$

where ℓ_{Pl} is the Planck length. This coupling to curvature arises naturally as a result of one loop quantum corrections in interacting theories. This model is essentially equivalent to the model Eq.(1) with a slowly varying effective value of ξ; $\xi_{eff} = \kappa\ \ln(R\ell^2_{Pl})$. The advantage of this model is that the value of ξ_{eff} which governs the onset of instability can be smaller than the present value. If $|\xi_{eff}| \approx 10^{-1}$ at the time of inflation, then $|\xi_{eff}| \approx O(1)$ today. Another model[23,24], arising from one loop quantum corrections in scalar electrodynamics, is a scalar field which satisfies the equation

$$\Box\phi + \lambda R_{\mu\nu}t^{\mu}t^{\nu}\phi = 0 \tag{4}$$

where $R_{\mu\nu}$ is the Ricci tensor and t^{μ} is the Robertson-Walker conformal Killing vector. This model leads to behavior very similar to that of the $R\phi^2$ coupling model. One might be concerned that unstable fields coupled to curvature could make localized distributions of matter, such as a neutron stars, unstable in the present universe. This seems, however, not to happen[23]. So long as the matter distribution is outside of its own Schwarzschild radius and the strength of the coupling is not too great ($|\xi|<1$ in the a case of Eq.(1)), no unstable modes arise. This can be understood physically in the following way: only modes whose wavelengths are of the order of the radius of the star feel the curvature within the star, longer wavelength modes are too delocalized; however, short wavelength modes are not destabilized as easily as are long wavelength modes.

The main flaw in all of these models based on unstable scalar fields is that they lead to difficulties with gravitation on scales small compared to the cosmological horizon. In the case of the $R\phi^2$ model, the effective value of Newton's constant vanishes at late times[22]. In the models of Eqs. (3) and (4) there are terms in the scalar field energy-momentum tensor proportional to $R_{\mu\nu}\phi^2$ which produce similar problems. Thus it seems that these models are not a complete solution to the cosmological constant problem. Nonetheless, they do show that cosmological constant damping behavior through 100 orders of magnitude is possible. This holds out hope that a better model can be found which does fully solve the cosmological constant problem through instability of de Sitter space.

Acknowledgement: This work was supported by the National Science Foundation under Grant PHY-8506593.

References

1. Abbott, L.F. and Deser, S., Nucl. Phys. B195, 76 (1982).

2. Ginsparg, P. and Perry, M.J., Nucl. Phys. B222, 245 (1983).

3. Gross, D.J., Perry, M.J., and Yaffe, L.G., Phys. Rev. D25, 330 (1982).

4. Antoniadis, I., Iliopoulos, J., and Tomaras, T.N., Phys. Rev. Lett. 56, 1319 (1986).

5. Allen, B., Phys. Rev. D34, 3670 (1986).

6. Parker, L., Phys. Rev. Lett. 50, 1009 (1983).

7. Mottola, E., Phys. Rev. D31, 754 (1985).

8. Mottola, E., Phys. Rev. D33, 1616 (1986).

9. Ford, L.H., "Gravitational Particle Creation and Inflation", submitted to Phys. Rev. D.

10. Myhrvold, N., Phys. Rev. D28, 2439 (1983).

11. Guth, A.H., Phys. Rev. D23, 347 (1981).

12. Albrecht, A. and Steinhardt, P.J., Phys. Rev. Lett. 48, 1220 (1982).

13. Linde, A., Phys. Lett. 108B, 389 (1982).

14. Linde, A., Phys. Lett. 129B, 177 (1983).

15. Starobinsky, A.A., Phys. Lett. 91B, 99 (1980).

16. Vilenkin, A. and Ford, L.H., Phys. Rev. D26, 1231 (1982).

17. Allen, B., Phys. Rev. D32, 3136 (1985).

18. Ford, L.H., Phys. Rev. D31, 710 (1985).

19. Traschen, J., and Hill, C.T., Phys. Rev. D33, 3519 (1986).

20. Hiscock, W., Phys. Lett. 166B, 285 (1986).

21. Freese, K., Adams, F., Frieman, J., and Mottola, E., "Cosmology with Decaying Vacuum Energy", to be published.

22. Dolgov, A.D., "An Attempt to Get Rid of the Cosmological Constant", in The Very Early Universe, G.W. Gibbons, S.W. Hawking, and S.T.C. Siklos, eds. (Cambridge University Press, Cambridge, 1983) p. 449.

23. Ford, L.H., "Cosmological Constant Damping by Unstable Scalar Fields" submitted to Phys. Rev. D.

24. Ford, L.H., Phys. Rev. D31, 704 (1985).

Sigma Model Corrections to the Effective Field Theory of Superstrings

K. S. Stelle*
The Institute for Advanced Study
Princeton, New Jersey 08540, U.S.A.

Abstract

Corrections to the effective field equations of superstring theories starting at cubic order in the string tension α' require modifications to the lowest order vacuum solutions. Nonetheless, the symmetries of the lowest order vacuum solutions are preserved at higher orders in α'. This preservation occurs through corresponding modifications to the spacetime supersymmetry transformations and to the vacuum relation between the spin connection and the gauge connection.

The effective field theory for the massless modes of string theories may be derived either from string perturbation calculations directly or from the requirement that two-dimensional conformal invariance be preserved. The latter approach [1] examines the consistency conditions necessary for conformal invariance to be preserved for a string propagating in a set of classical background fields that correspond to the massless modes of the string. Essentially, this amounts to setting the β-function to zero for the two-dimensional generalized non-linear σ-model represented by the string propagating on the given background. In particular, from the metric β-function condition $\beta_{ij} = 0$, at order $(\alpha')^0$ one obtains Einstein's equation for the background spacetime metric coupled to the other background fields. This happens because the one-loop σ-model counterterm for the metric is just the Ricci tensor [2] plus appropriate stress tensor expressions for the other backgrounds.

For the ten-dimensional heterotic string theory [3] coupled to a transverse d=8 background metric $g_{ij}(x)$ (with the other two dimensions taken to be flat), and to an antisymmetric tensor $B_{ij}(x)$ and a Yang-Mills connection $A_i(x)$, the lowest-order σ-model action is [4]

---

* On leave of absence from The Blackett Laboratory, Imperial College, London SW7 2BZ, England. Work supported in part by the U.S. National Science Foundation under Grant No. PHY-8217352.

$$I = \frac{1}{4\pi\alpha'}\int d^2\sigma\,\{g_{ij}\partial_\mu x^i\partial_\mu x^j + \varepsilon^{\mu\nu}B_{ij}\partial_\mu x^i\partial_\nu x^j$$
$$+ i\lambda^a(\partial_+\lambda^a + (\omega_k{}^{ab} + \tfrac{1}{2}H_k{}^{ab})\lambda^b\partial_+x^k)$$
$$+ i\psi^A(\partial_-\psi^A + A_k{}^{AB}\psi^B\partial_-x^k) + \tfrac{1}{2}F_{ij}^{AB}\lambda^i\lambda^j\psi^A\psi^B\} \qquad (1)$$

where λ^a is a one component (Majorana-Weyl) worldsheet spinor, transverse spacetime vector and ψ^A is a one-component spinor transforming in the $(1,16)\oplus(16,1)$ representation of $SO(16)\times SO(16)$. This is the linearly realized part of the gauge symmetry of the heterotic string. The partial derivatives ∂_+ and ∂_- are taken with respect to $\sigma^+=\tau+\sigma$ and $\sigma^-=\tau-\sigma$, the light-cone variables on the worldsheet. Inclusion of a dilaton background $\phi(x)$ is made at order $(\alpha')^0$ via coupling to the worldsheet Euler density $\int\phi(x)R(\gamma)\sqrt{-\gamma}\,d^2\sigma$, where $\gamma_{\mu\nu}$ is the worldsheet metric (which has been gauged to $\eta_{\mu\nu}$ in (1)). H_{ijk} is the field strength for B_{ij} : $H_{ijk} = \partial_{[i}B_{jk]}$ to lowest order. The heterotic string action (1) possesses a (1,0) worldsheet supersymmetry [4] with one Majorana-Weyl generator.

At higher orders in α', the backgrounds $g_{ij}(x)$, $B_{ij}(x)$ and A_i^{AB} must be renormalized - they represent the coupling constants of the two dimensional theory. In addition, in order to cancel anomalies in the background general coordinate and gauge invariances, it is necessary to add $\alpha'(\omega_{3Y} - \omega_{3L})$, where ω_{3Y} and ω_{3L} are the Yang-Mills and Lorentz Chern-Simons 3-forms [5]. Setting the β-functions for the various background fields to zero, one obtains the effective field equations. To lowest order, these equations admit solutions with $R_{ij} = H_{ijk} = 0$ and with the spin connection embedded in the Yang-Mills gauge group ($\omega_i{}^{ab}=A_i{}^{ab}$; the other $A_i{}^{AB} = 0$) [6]. Keeping only terms that contribute at orders up to $(\alpha')^3$ in corrections to the lowest order solution, the corrected effective field equations for g_{ij}, ϕ and B_{ij} are [7,8]

$$R_{ij} - \nabla_i\nabla_j\phi - \tfrac{1}{4}H_{ikl}H_j{}^{kl} + \tfrac{\alpha'}{2}[R_{iklm}R_j{}^{klm} - F_{ik}\cdot F_j{}^k]$$
$$- \tfrac{\alpha'}{2}e^{-\phi}\nabla_m[e^\phi H_{kl(i}R_{j)}{}^{mkl}] - \alpha' e^{-\phi}\nabla^k\nabla^l[e^\phi R_{k(ij)l}]$$
$$- \xi(\alpha')^3 W_{ij} = 0 \qquad (2)$$

$$R - 2\nabla^2\phi - \partial_i\phi\partial^i\phi - \tfrac{1}{12}H_{ijk}H^{ijk} - \tfrac{\alpha'}{4}[F_{ij}\cdot F^{ij} - R_{ijkl}R^{ijkl}] + \xi(\alpha')^3 Y = 0 \qquad (3)$$

$$\nabla^i(e^\phi H_{ijk}) = 0\,, \qquad (4)$$

where $\xi = \frac{1}{24}\zeta(3)$,

$$\mathcal{Y} = \frac{1}{64} t^{i_1 \cdots i_8} t_{j_1 \cdots j_8} \mathcal{R}_{i_1 i_2}{}^{j_1 j_2} \cdots \mathcal{R}_{i_7 i_8}{}^{j_7 j_8} \qquad (5)$$

$$t^{ijk\ell mnpq} M_{ij} M_{k\ell} M_{mn} M_{pq} = 24 \operatorname{tr} M^4 - 6(\operatorname{tr} M^2)^2 + \frac{1}{2} \varepsilon^{ijk\ell mnpq} M_{ij} M_{k\ell} M_{mn} M_{pq},$$

for arbitrary $M_{ij} = -M_{ji}$ and

$$W_{ij} = \frac{\delta}{\delta g^{ij}} \int dx\, \mathcal{Y}\,. \qquad (6)$$

Equations (2-6) induce a number of changes in the vacuum compared to the $(\alpha')^0$ solutions of ref [6]. Most prominently, the Ricci tensor no longer vanishes. However, we may still set H = 0 to this order. Keeping only contributions of order $(\alpha')^3$ and less in deviations from the $(\alpha')^0$ solution (which has ϕ = const., H = 0 and $A_i{}^{ab} = \omega_i{}^{ab}$), one must solve the corrected gravitational and dilaton equations

$$\mathcal{R}_{ij} = \nabla_i \nabla_j \phi + \xi(\alpha')^3 W_{ij} \qquad (7)$$

$$\mathcal{R} = 2\nabla^2 \phi + \xi(\alpha')^3 \mathcal{Y}\,. \qquad (8)$$

Equation (8) simplifies greatly if one also requires the transverse spacetime to be Kähler, i.e. to have a covariantly constant complex structure J_{ij}, $J_i{}^j J_j{}^k = -\delta_i{}^k$. For Ricci-flat Kähler spaces, $\mathcal{Y} = 0$, so (8) just becomes $\mathcal{R} = 2\nabla^2\phi$. The Kähler condition is also necessary in order for the vacuum solution to admit a rigid spacetime supersymmetry [6]. In that case, one finds that

$$W_{ij} = \nabla_{\hat{i}} \nabla_{\hat{j}} S \qquad (9)$$

$$\nabla_{\hat{i}} \equiv J_i{}^j \nabla_j\,, \qquad S = 2\mathcal{R}^{\hat{i}}{}_j{}^{\hat{k}}{}_\ell \mathcal{R}^j{}_m{}^\ell{}_n \mathcal{R}^m{}_i{}^n{}_k\,. \qquad (10)$$

The corrected solution to the equations (7,8) has

$$\phi = \text{const.} + \xi(\alpha')^3 S \qquad (11)$$

$$\mathcal{R}_{ij} = \xi(\alpha')^3 (\nabla_i \nabla_j S + \nabla_{\hat{i}} \nabla_{\hat{j}} S)\,. \qquad (12)$$

Equation (12) is precisely the condition for conformal invariance of a (2,2) worldsheet supersymmetric σ-model [9] (i.e. one with two Majorana-Weyl and two Majorana-anti-Weyl supersymmetry generators). This occurs because the Kähler, H=0 and A=ω conditions reduce the interacting part of (1) to a (2,2) worldsheet supersymmetric σ-model.

Although the modified vacuum condition (12) is inconsistent with the $\mathcal{R}_{ij}$ = 0 requirement of the the original rigid spacetime supersymmetry of the vacuum [6], it is nonetheless of precisely the correct form to allow a modified rigid supersymmetry to be preserved [8]. If the bosonic terms in the gravitino transformation that are non-vanishing for H=0, A=ω configurations are modified to read

$$\delta\psi_M = \nabla_M\varepsilon + 6\xi(\alpha')^3\, \tau_M{}^{PQ}\Gamma_{PQ}\varepsilon \qquad (13)$$

$$\tau_M{}^{PQ} = \nabla^S \mathcal{R}_{MK}{}^{[P|R}\mathcal{R}_{STRN}\mathcal{R}^{TKN|Q]}\,, \qquad (14)$$

then upon specialization to a space $R^2 \otimes \mathcal{K}$ where $\mathcal{K}$ is Kähler, one finds that setting $\psi_M = 0$ in the vacuum requires precisely (12) as the integrability condition. Thus, the corrected effective equations (2-4) are consistent with maintaining N=1 supersymmetry of the four dimensional vacuum. In particular, they are consistent with maintaining N=1 supersymmetry in a vacuum $R^4 \otimes \mathcal{K}_6$ corresponding to the direct product of a flat four dimensional Minkowski space with a compact 6-dimensional Kähler "internal" space.

The effective Yang-Mills field equations also will receive σ-model corrections. Since the kinectic $F_{\mu\nu}\cdot F^{\mu\nu}$ term in the effective Lagrangian occurs at order α', as can be seen on dimensional grounds or from the dilaton equation (6), it would be necessary to know the order $(\alpha')^4$ corrections to the effective field equations. Nonetheless, since the effective field equations should be covariant under the full gauge symmetry of the heterotic theory, and not just under the SO(16)×SO(16) that can be linearly realized on the σ-model variables in (1), one can deduce by consideration of all covariant possibilities that the $A=\omega$ vacuum condition must receive corrections [8]. In the case of the $E_8 \times E_8$ heterotic string, gauge covariance requires that this condition be modified to read

$$A_i{}^{ab} = \omega_i{}^{ab} - (\alpha')^3\tau_i{}^{ab} \qquad (15)$$

where $\tau_i{}^{ab}$ is just the specialization to the internal space of the tensor $\tau_M{}^{PQ}$ (14) that occured in the gravitino transformation.

The modified embedding condition (15) has the effect of giving a vacuum value to the Yang-Mills field that lies in an SU(3) subgroup of E_8, even though the holonomy group of the spin connection $\omega_i{}^{ab}$ is U(3). The residual Yang-Mills symmetry of the vacuum is therefore undisturbed with respect to the lowest order solutions in α', and remains $E_6 \times E_8$. Thus, despite the complications induced by the higher order σ-model corrections to the effective field equations, the symmetries to the vacuum remain undisturbed. These detailled considerations using the explicit forms of the effective equation corrections are consistent with other discussions based upon the non-renormalization of the compactified theory's superpotential [10] and with consideration of the worldsheet superconformal algebra [11].

References

1. E.S. Fradkin and A.A. Tseytlin, Phys. Lett. **158B** (1985), 316;
A. Sen, Phys. Rev. **D32** (1985) 2102;
C.G. Callan, D. Friedan, E.J. Martinec and M.J.Perry, Nucl. Phys. **B262** (1985), 593.

2. D. Friedan, Phys. Rev. Lett. **45** (1980), 1057.

3. D.J. Gross, J. Harvey, E. Martinec and R. Rohm, Phys. Rev. Lett. **54** (1985), 502; Nucl. Phys. **B256** (1985) 253.

4. C. M. Hull and E. Witten, Phys. Lett. **160B** (1985), 398.

5. M.B. Green and J. H. Schwarz, Phys. Lett. **149B** (1984), 117.

6. P. Candelas, G. Horowitz, A. Strominger and E. Witten, Nucl. Phys. **B256** (1985), 46.

7. M.T. Grisaru, A.E.M. van de Ven and D. Zanon, Phys. Lett. **173B** (1986), 199; Nucl. Phys. **B277** (1986), 388, 409;
D. J. Gross and E. Witten, Nucl. Phys. **B277** (1986),1;
M.D. Freeman and C.N. Pope, Phys. Lett. **174B** (1986), 48;
D. Nemeschansky and A. Sen, Phys. Lett. **178B** (1986), 365;
A. Sen, Phys. Lett. **178B** (1986), 370;
C. G. Callan, I.R. Klebanov and M.J. Perry, Nucl. Phys. **B278** (1986), 78.

8. C.N. Pope, M.F. Sohnius and K.S.Stelle, Nucl. Phys. **B** (in press);
P. Candelas, M.D. Freeman, C.N.Pope, M.F. Sohnius and K.S. Stelle, Phys. Lett. **177B** (1986), 341;
M.D. Freeman, C.N. Pope, M.F. Sohnius and K.S. Stelle, Phys. Lett. **178B** (1986), 199;
M.D.Freeman, C.N. Pope, C.M. Hull and K.S. Stelle, Phys. Lett. **B** (in press).

9. P.S. Howe, G. Papadopoulos and K.S. Stelle, Phys. Lett. **174B** (1986),405.

10. M. Dine, N.Seiberg, X.G. Wen and E. Witten, Nucl. Phys. **B278** (1986),769.

11. A. Sen, preprint SLAC-PUB-4071

RECENT DEVELOPMENTS IN GENERAL RELATIVITY

Robert M. Wald
Enrico Fermi Institute and Department of Physics,
University of Chicago
Chicago, Illinois 60637, U.S.A.

ABSTRACT

We briefly review the status of some of the issues involving general relativity which played a role in the early Texas Symposia. Some recent theoretical developments in classical general relativity are then discussed, with particular attention to work of Friedrich and Christodoulou and Klainerman on global existence of exact solutions describing weak gravitational waves.

It is fair to say that theoretical developments in classical general relativity have not played a central role at this Texas Symposium nor at the ones in the recent past. This has not always been the case. Indeed, the Texas Symposia came into existence 23 years ago largely because the discovery of quasars indicated the presence of gravitational collapse phenomena in astrophysics; further development of the theory of general relativity was urgently needed to describe such phenomena. I shall begin by briefly reviewing the status of the issues which were posed at that time.

In the invitation letter to the first Texas Symposium in 1963, the following questions were posed:[1] a) The astronomers observed some unusual objects connected with radio sources. Are these the debris of a gravitational implosion? b) By what machinery is gravitational energy converted into radio waves? 3) Does gravitational collapse lead, on our present assumptions, to indefinite contraction and a singularity in space time? d) If so, how must we change our theoretical assumptions in order to avoid this catastrophe?

It is interesting to note how rapidly the views of the relativity and astrophysics communities evolved with regard to question (d). The singularity theorems of the late 1960s and early 1970s showed that singularities are an inevitable feature of general relativity. Furthermore, it became clear to theorists that one could continue to make physically sensible predictions even when singularities are produced, provided that they are causally confined within black holes. Thus, singularities have become an accepted feature of collapse phenomena (at least at the classical level), and question (d) appears strikingly anachronistic today.

A modern reformulation of questions (a), (b), and (c) -- which uses some hindsight but fairly captures all the issues raised by the original questions -- might read as follows: (1) Is gravitational collapse inevitable for sufficiently massive and dense objects? (2) If so, then (i) are singularities inevitable and (ii) what is the final state resulting from gravitational collapse? (3) Given that the final state is a black hole, can energy be extracted from it in such a way as to plausibly power quasars and other highly energetic astrophysical sources?

Theoretical developments in general relativity -- mainly during the late 1960s and early 1970s -- were remarkably successful in providing definitive answers to most of the above questions. The answer to question (1) is a clear "yes." Given a physically reasonable equation of state of matter below a density ρ_0, and given only minimal assumptions about the equation of state at densities above ρ_0, rigorous upper mass limits for spherical bodies can be derived.[2)] More recently, other results have been obtained[3)] which establish the inevitability of collapse for any sufficiently compacted body (with no assumption of symmetries).

The answer to question (2i) also is a clear "yes," as the singularity theorems already alluded to above have shown. However, part of the answer to (2ii) remains open. It is generally believed that the "cosmic censor hypothesis" holds: that all singularities of gravitational collapse are hidden within black holes, so that the final state of collapse will always be a black hole. However, the main theoretical evidence for this belief remains the examples of spherical collapse and perturbations of spherical collapse, and a number of interesting failures of clever attempts to construct counterexamples; no general proof has been given. I will return to this point towards the end of this paper. However, assuming that black holes are produced, that they "settle down" to a stationary final state and that no matter or electromagnetic fields are externally present (or, if they are, that their gravitational influence is negligible), it is known from the black hole uniqueness theorems that the final state must be a Kerr black hole. (In the case where electromagnetic fields may be present, the charged Kerr black hole is unique.[4]) Thus, modulo the cosmic censor hypothesis, the final state of collapse is described by the Kerr metric.

With regard to question (3), energy to power some astrophysical sources containing black holes can be provided simply by the accretion of matter into the black hole; the large effective gravitational potential well provided by the hole can be used to heat up the matter and produce X-rays, etc. However, in the case of some quasars, it would appear that if a black hole is present, it must play a more active role. Thus, it is interesting that the answer to question (3) is another clear "yes": Not only can rotational energy be extracted in principle from a black hole as originally shown by Penrose, but the Blanford-Znajek process[5] provides an astrophysically plausible mechanism for doing so.

The above discussion should not be interpreted as suggesting that the issues involving quasars and other energetic sources which prompted the convoking of the first Texas Symposium have been resolved. As can be seen from the review of Rees at the present Texas Symposium, many questions remain and many details (and, undoubtedly, some entirely new ideas) must be added to the currently favored models. However, essentially all of these remaining questions involve plasma astrophysics rather than general relativity. The main issues involving general relativity do appear to have been successfully resolved. Until some new astrophysical issues arise which involve general relativity in a fundamental way, I believe that theoretical developments in general relativity will continue to play a peripheral role at the Texas symposia.

Nevertheless, some research in general relativity has continued on these "old" issues involving collapse and black holes, and I wish to mention one important result which has been obtained very recently: an analytic proof of stability of the Kerr metric. If cosmic censorship holds and the Kerr black hole is the final state of gravitational collapse, then it had better be stable. The issue of stability was intensively investigated in the early 1970s, and a proof of stability was given by Press and Teukolsky.[6] However, a key step in their analysis involved a numerical search for modes with infinite superradiance. Obviously, only a finite number of modes could be checked in this manner, so the possibility of instabilities setting in at modes of high azimuthal number m -- as occurs in the Friedman instability[7] of all rotating fluid stars -- remained open. This gap has now been closed by as yet unpublished results of Whiting presented at the Relativity Workshop of this Symposium. Starting with the Teukolsky equation, Whiting uses a nonlocal transformation of each mode (involving an integral transform of the radial function) to obtain an equivalent equation for which stability can be proven in the following manner. The new equation corresponds to the equation satisfied by a mode of a wave equation in a new spacetime which has a Killing field which is null everywhere. Nonexistence of exponentially growing modes then follows from existence of a conserved, locally positive definite energy for the new field variable in the new spacetime. (It is the ergoregion of Kerr which makes the energy integrand there non-positive definite and prevents application of this approach to the original Teukolsky equation.) Thus, the belief in the stability of Kerr black holes has been put on a sounder footing.

With regard to "newer" issues, during the past ten years one of the main areas of research in classical general relativity relevant to the interests of the Texas Symposium has concerned gravitational radiation, specifically attempts to derive properties of exact solutions describing radiation phenomena in a rigorous manner. I shall discuss some significant recent developments in this area concerning global evolution of solutions describing weak gravitational waves. However, I shall begin by making a few remarks about the quadrupole formula for gravitational radiation. The main reason for making these remarks is to motivate the desire to obtain rigorous results concerning radiation from isolated systems. In addition, although the validity of the quadrupole formula has been a subject much discussed at recent Texas symposia, I believe that a brief review of the simple reasons why difficulties arise with approximation techniques in the case of self-gravitating systems may be useful for those who have not followed these discussions. However, I shall make no attempt to review the considerable literature on this subject -- all of which goes well beyond the elementary remarks made here.

In the case where gravity is weak, i.e., spacetime is nearly flat, it is reasonable to write the spacetime metric, g_{ab}, as the sum of a flat metric, η_{ab}, and a tensor, γ_{ab},

$$g_{ab} = \eta_{ab} + \gamma_{ab} \quad (1)$$

and retain only terms linear in γ_{ab} in Einstein's equation. In an appropriate gauge, the quantity $\bar{\gamma}_{ab} = \gamma_{ab} - \frac{1}{2}\gamma_{ab}$ then satisfies the ordinary wave equation in Minkowski spacetime with source given by the matter stress-energy tensor T_{ab}, so the solution for $\bar{\gamma}_{ab}$ is given by the same retarded Green's function as is used in electromagnetism. The total energy carried by gravitational waves can then be computed in this linearized gravity approximation from the second order Einstein tensor (or other appropriate pseudo-tensor); to the lowest order this agrees with the total Bondi energy flux of the exact theory.[8] The same assumptions (i.e., "slow motion," or, more precisely, typical wavelengths >> source size) as yields the dipole formula in electromagnetism then leads to the quadrupole formula here. There is no difficulty whatsoever (apart, of course, from justifying the linearized approximation and the slow motion approximation) with this derivation for problems such as radiation by a spinning rod or by masses connected with a spring.

However, a difficulty does occur when self-gravitating masses are involved -- as, for example, in the case of the binary pulsar -- even if gravity is "weak." The problem is simply that in the linearized approximation, masses do not attract each other! This fact is already manifest in Newtonian gravity via the formula $F = Gm_1m_2/r^2$, which involves the product of m_1 and m_2 and hence is of second order in an expansion about the case of no gravitational field. Thus, in linearized gravity two masses will move along straight lines of the flat metric, η_{ab}, rather than orbit around each other. One's first reaction to this problem might be to proceed simply by inserting the correct, orbiting motion "by hand." However, if one does this, there will be no solutions of the linearized Einstein equation because that equation implies the linearized conservation equation, $\partial^a T_{ab} = 0$, which, in turn, implies straight line motion of the masses. The linearized approximation thus is unsuitable for describing radiation from self-gravitating masses; one needs an approximation scheme which describes the motion of the bodies more accurately in lowest order.

Newtonian gravity properly describes the orbiting motion of self-gravitating masses and the post-Newtonian approximation scheme thus avoids the above difficulties of linearized gravity. The post-Newtonian approximation can, in essence, be viewed as an expansion of Einstein's equation in powers of $1/c$. (A better way of formulating this notion is to view constants of nature like c as fixed but consider one-parameter families of general relativistic spacetimes which, in the limit, approach a Newtonian spacetime in an appropriate manner; however, the notion of an "expansion in powers of $1/c$" will suffice for our purposes here.) However, new difficulties now arise since an expansion in powers of $1/c$ is a very poor way of solving radiation problems, as the following simple example illustrates.

Consider the scalar wave equation for a field ϕ in Minkowski spacetime with bounded source $j(\vec{x}, t)$,

$$\nabla^2\phi - \frac{1}{c^2}\frac{\partial^2\phi}{\partial t^2} = j \tag{2}$$

Let us attempt to solve this equation by an expansion in powers of $1/c$. The zeroth order approximation, $\phi^{(0)}$, satisfies

$$\nabla^2\phi^{(0)} = j \tag{3}$$

and thus is given by the instantaneous Coulomb solution with source j. The first order correction, $\phi^{(1)}$, satisfies

$$\nabla^2\phi^{(1)} = 0 \tag{4}$$

which has the obvious and natural solution $\phi^{(1)} = 0$. The second order correction, $\phi^{(2)}$, satisfies

$$\nabla^2\phi^{(2)} = \frac{\partial^2\phi^{(0)}}{\partial t^2} \tag{5}$$

Here there is a clear sign of trouble, because the source term, $\partial^2\phi^{(0)}/\partial t^2$, will not, in general, fall to zero rapidly enough at infinity to permit a solution for $\phi^{(2)}$ which vanishes at infinity. Thus, we do not appear to be generating a sensible description of scalar radiation from a bounded source. In fact, what we are doing by this procedure is generating a "near zone" expansion of a time symmetric Green's function solution with source j. (Which time symmetric Green's function we have depends on our choice of boundary conditions at infinity.) In order to get the retarded solution, we must choose a non-zero solution of eq. (4) for $\phi^{(1)}$, a step which would appear very unnatural in this procedure. Even if we chose the solutions $\phi^{(n)}$ corresponding to the retarded Green's function, it would not be straightforward to extract radiation formulas from them, since the $\phi^{(n)}$ all diverge at infinity. These difficulties would become more severe if the equation satisfied by ϕ were nonlinear.

The purpose of the above remarks has not been to suggest that there is serious reason to doubt the validity of the quadrupole formula for self-gravitating systems where gravity is weak. Indeed, it seems to me to be overwhelmingly plausible that for such systems the nonlinear terms in γ_{ab} in Einstein's equation will provide the "effective gravitational stresses" needed to make orbiting motion consistent, but that the matter energy density will still dominate the "effective gravitational energy density" and that the slow motion approximation will remain valid. If so, then the quadrupole formula will hold. (As mentioned above, I shall not attempt to review here the literature concerning derivations of the quadrupole and radiation reaction formulas.) Rather my purpose has been to graphically illustrate the serious difficulties one can encounter in attempting to extract information about radiating systems via approximation techniques even in simple contexts. It would be extremely useful to have a body of results concerning the general properties of exact solutions describing radiation systems. Such a body of results might help suggest which approximation techniques would be most useful, and unquestionably it would be of great use in justifying the validity of any approximation technique which is employed.

How do things stand with regard to results on exact solutions describing gravitational radiation from isolated systems in general relativity? Until the recent results described below, the situation stood as follows. A mathematically precise notion of an isolated system -- i.e., an asymptotically flat spacetime -- has existed for over twenty years.[9)] For such spacetimes, notions such as radiated energy are well defined. However, no exact solution was (or still is) explicitly known which satisfies all the requirements of asymptotic flatness (with no sources or singularities extending out to infinity) and has non-zero radiated energy. More importantly, no theorems were even known guaranteeing existence of such solutions!

The initial value formulation of general relativity tells us "how many" solutions of Einstein's equation exist. However, the analysis of global dynamical behavior of solutions is generally beyond the power of currently available mathematical techniques. For example, prior to the results I shall describe below, it was conceivable that even for initially very weak gravitational waves, the nonlinear terms in Einstein's equation acting over a very long time could result in singularities. If progress is to be made toward proving the cosmic censor hypothesis discussed above, it will be necessary to gain a great deal of mathematical control over the global dynamical behavior of solutions.

The first result which I wish to mention concerns a proof of existence of an appropriately large class of asymptotically flat spacetimes with radiation. Recall that the definition of asymptotic flatness (at null infinity) involves the existence of an "unphysical spacetime" $(\hat{M}, \hat{g}_{ab})$ with boundary whose interior is conformally related to the physical spacetime (M, g_{ab}),

$$\hat{g}_{ab} = \Omega^2 g_{ab} \tag{6}$$

Here the conformal factor Ω is required to vanish on the parts of the boundary representing future and past null infinity, denoted $\mathscr{I}^+$ and $\mathscr{I}^-$ respectively; see, e.g., [10] for further discussion and a complete listing of all conditions. Thus, a smooth spacetime $(\hat{M}, \hat{g}_{ab})$ with boundary which is conformal to a Ricci flat spacetime and whose conformal factor satisfies appropriate conditions is equivalent to an asymptotically flat spacetime. It would be very convenient for a proof of existence of asymptotically flat solutions to analyze evolution via the "conformal Einstein equation" in the unphysical spacetime, using $\mathcal{I}^+$ or $\mathcal{I}^-$ as an initial data surface. However, a straightforward transcription of Einstein's equation to the unphysical spacetime yields equations which become singular where $\Omega = 0$.

This problem has recently been overcome by Friedrich[11)] who reformulated the conformal Einstein equation so that it has a well posed initial value formulation and is nonsingular even when $\Omega = 0$. By using $\mathcal{I}^+$ (together with a "vertex point" i^+ representing future timelike infinity) as an initial data surface, he thereby establishes existence of an appropriately large class of spacetimes which contain radiation and are asymptotically flat at future null infinity. However, the results are "local" near $\mathcal{I}^+$, i.e., the solutions thus obtained could, when evolved backward into the past, develop singularities and/or fail to be asymptotically flat at $\mathcal{I}^-$.

An even more significant step toward obtaining global dynamical information about solutions describing weak gravitational waves appears to have been taken very recently in as yet unpublished work of Christodoulou and Klainerman. Their arguments involve a large number of highly technical steps, and the details of the proofs of some of these steps remain to be filled in, but it appears that the following result will be established: For any initial data which is sufficiently close to that of flat spacetime as measured by a certain norm, Q, the evolved solutions are geodesically complete and are asymptotically flat at both past and future null infinity (although not necessarily with the degree of smoothness usually assumed there).

In very brief and rough outline, their method of proof involves using maximal hypersurfaces labeled by a time function, t, and constructing vector fields in the spacetime which are "nearly" conformal isometries. Bel-Robinson-like tensors are then constructed out of Lie derivatives of the curvature with respect to these vector fields. The norm, Q(t), then is defined in terms of integrals over the maximal hypersurfaces involving these tensors. Remarkably, Q has the properties that (i) by Sobolev inequalities, boundedness of Q implies boundedness of curvature and (ii) by Einstein's equation, dQ/dt can be bounded in terms of Q and t. The theorem results from using properties (i) and

(ii) in combination: The inequality (ii) implies that if Q is sufficiently small initially, then it remains bounded, whereas (i) then implies that the curvature remains bounded, which, in turn, implies that no singularities develop so that Einstein's equation makes sense and the inequality (ii) continues to hold. Detailed fall-off behavior of the curvature at null and timelike infinity also is then obtained.

Assuming that no surprises occur as the authors fill in the details of their proof, this result will provide the first global "nonsingularity theorem" to be proven for Einstein's equation. It will show that weak gravitational waves behave in accord with prevailing beliefs (although there is an indication from their results that certain components of the curvature may fall off more slowly at infinity than is generally assumed). It will provide a small -- but very important -- step in the direction of proving the cosmic censor hypothesis.

REFERENCES

1) Robinson, I., Schild, A., and Schücking, E.L., eds., "Quasi-Stellar Sources and Gravitational Collapse", University of Chicago Press (Chicago, 1965).
2) Hartle, J.B., Phys. Rept. 46C, 201 (1978).
3) Schoen, R. and Yau, S.T., Commun. Math. Phys. 90, 575 (1983).
4) Mazur, P.O., J. Phys. A15, 3173 (1982).
5) Blanford, R.D. and Znajek, R.L., Mon. Not. R. astr. Soc. 179, 433 (1977).
6) Press, W.H. and Teukolsky, S.A., Astrophys. J. 185, 649 (1973).
7) Friedman, J.L., Commun. Math. Phys. 62, 247 (1978).
8) Habisohn, C., J. Math. Phys. 27, 2759 (1986).
9) Penrose, R., Phys. Rev. Lett. 10, 66 (1963).
10) Wald, R.M., "General Relativity", University of Chicago Press (Chicago, 1984).
11) Friedrich, H., Commun. Math. Phys. 103, 35 (1986).

THE PROBLEM OF DISSIPATION IN RELATIVISTIC FLUIDS

William A. Hiscock and Lee Lindblom

Department of Physics, Montana State University
Bozeman, Montana 59717

ABSTRACT

This paper describes the properties of the Israel-Stewart theory of dissipative relativistic fluids. The conditions needed to guarantee the stability, hyperbolicity and causality of the evolution of fluctuations about an equilibrium state are described. An unexpected relationship between these conditions is revealed: the stability conditions are satisfied if and only if the hyperbolicity and causality conditions hold. The manner in which the complicated (14 degrees of freedom) dynamics of an Israel-Stewart fluid reduces in appropriate limits to the familiar (5 degrees of freedom) dynamics of a relativistic ideal fluid or a Navier-Stokes-Fourier fluid is described.

I. INTRODUCTION. The original attempts to include the effects of viscosity and thermal conductivity in a relativistic fluid theory were made by Eckart[1)] and Landau and Lifshitz[2)]. These theories are the simplest Lorentz covariant generalizations of the Navier-Stokes-Fourier theory. Unfortunately, these theories are extremely pathological[3)]. They admit no stable equilibrium states; they do not have hyperbolic evolution equations; and they violate causality[4)].

In this paper we describe the properties of a more complicated theory of dissipative relativistic fluids proposed by Israel and Stewart[5-8)] that overcomes many of the problems found in the original theories.

II. THE ISRAEL-STEWART THEORY. The dynamical variables of an Israel-Stewart fluid include the familiar variables of an ideal relativistic fluid: the particle number density n, the energy density ρ, the pressure p, the temperature T, the entropy per particle s, and the four-velocity u^a (with $u^a u_a = -1$). In addition, however, these fluids also have as dynamical variables the scalar stress τ, the heat flux vector q^a (with $q^a u_a = 0$) and the spatial stress tensor τ^{ab} (with τ^{ab}

symmetric, traceless and $\tau^{ab}u_b = 0$). These variables are related to the particle number current N^a and the stress energy tensor T^{ab} by:

$$N^a = n\,u^a, \tag{1}$$

$$T^{ab} = \rho\, u^a u^b + (p+\tau)q^{ab} + \tau^{ab} + q^a u^b + q^b u^a, \tag{2}$$

where $q^{ab} = g^{ab} + u^a u^b$ and g^{ab} is the metric tensor which is used to raise and lower indices. The evolution equations for these fluids include the conservation laws for these quantities:

$$\nabla_a N^a = \nabla_a T^{ab} = 0. \tag{3}$$

In an ideal fluid (where $\tau = q^a = \tau^{ab} = 0$) these conservation laws together with an equation of state, $s = s(\rho,n)$, and the first law of thermodynamics,

$$d\rho = nT\,ds + [(\rho+p)/n]\,dn, \tag{4}$$

would be sufficient to determine the evolution of the fluid. In an Israel-Stewart fluid, however, these equations must be supplemented with evolution equations for the new dynamical variables τ, q^a, and τ^{ab}. Their equations for these quantities are the following:

$$\tau = -\varsigma\Big[\nabla_a u^a + \beta_0 u^a\nabla_a\tau - \alpha_0\nabla_a q^a - \gamma_0 T q^a\nabla_a(\alpha_0/T) + \tfrac{1}{2}\tau T\nabla_a(\beta_0 u^a/T)\Big], \tag{5}$$

$$\begin{aligned} q^a = -\kappa T q^{ab}\Big[&(\nabla_b T)/T + u^c\nabla_c u_b + \beta_1 u^c\nabla_c q_b - \alpha_0\nabla_b\tau - \alpha_1\nabla_c\tau^c{}_b + \gamma_2\nabla_{[b}u_{c]}q^c \\ &- (1-\gamma_0)T\tau\nabla_b(\alpha_0/T) - (1-\gamma_1)T\tau^c{}_b\nabla_c(\alpha_1/T) + \tfrac{1}{2}Tq_b\nabla_c(\beta_1 u^c/T)\Big], \end{aligned} \tag{6}$$

$$\begin{aligned} \tau^{ab} = -\eta(q^{ac}q^{bd} + q^{ad}q^{bc} - \tfrac{2}{3}q^{ab}q^{cd})\Big[&\nabla_c u_d + \beta_2 u^e\nabla_e\tau_{cd} - \alpha_1\nabla_c q_d \\ &- \gamma_1 T q_c\nabla_d(\alpha_1/T) + \tfrac{1}{2}T\tau_{cd}\nabla_e(\beta_2 u^e/T) + \gamma_3\nabla_{[c}u_{e]}\tau^e{}_d\Big]. \end{aligned} \tag{7}$$

In these equations ς, η, and κ are the viscosity coefficients and the thermal conductivity; the α_i, β_i, and γ_i are new "second order" coefficients to be computed from a microscopic theory of the fluid (e.g., kinetic theory[6,7,9]) or determined empirically. This choice of equations is motivated by the need to enforce the second law of thermodynamics. If the entropy current, s^a, is defined by,

$$Ts^a = snTu^a + q^a - \tfrac{1}{2}(\beta_0\tau^2 + \beta_1 q^b q_b + \beta_2\tau^{bc}\tau_{bc})u^a + \alpha_0\tau q^a + \alpha_1\tau^a{}_b q^b, \tag{8}$$

then it follows from the evolution equations (3) - (7) that the total entropy of the fluid will be a non-decreasing function of time, since

$$T\nabla_a s^a = \tau^2/\varsigma + q^a q_a/(\kappa T) + \tau^{ab}\tau_{ab}/(2\eta). \tag{9}$$

The equilibrium states (those satisfying $\nabla_a s^a = 0$) of a fluid satisfying these equations are identical to the equilibrium states of the simpler Eckart and Landau-Lifshitz theories. In particular, the vector field u^a/T must be a Killing vector field, each of the thermodynamic variables must be constant along the integral curves of u^a, and the thermodynamic variable, $\theta = -s + (\rho + p)/nT$, must have vanishing gradient.

III. STABILITY, CAUSALITY AND HYPERBOLICITY. To determine whether or not the Israel-Stewart theory suffers from the same problems found in the Eckart and the Landau-Lifshitz theories, we undertook a systematic study of the evolution of small perturbations about an arbitrary equilibrium state[10]. To investigate the stability of these perturbations, an energy functional was constructed. By diagonalizing this functional, we determined that certain conditions, $\Phi_A > 0$, are necessary and sufficient for the stability of these perturbations; the Φ_A are defined by:

$$\Phi_1 = (\partial p/\partial \rho)_s, \tag{10}$$

$$\Phi_2 = (\partial \rho/\partial s)_p (\partial p/\partial s)_\theta, \tag{11}$$

$$\Phi_3 = \beta_0, \tag{12}$$

$$\Phi_4 = \beta_2, \tag{13}$$

$$\Phi_5 = (\rho + p)(1 - \Phi_1) - [1/\beta_0 + 2/(3\beta_2) + K^2/\Phi_7], \tag{14}$$

$$\Phi_6 = \rho + p - [2\beta_2 + 2\alpha_1 + \beta_1]/[2\beta_1\beta_2 - (\alpha_1)^2], \tag{15}$$

$$\Phi_7 = \beta_1 - [(\alpha_0)^2/\beta_0 + 2(\alpha_1)^2/(3\beta_2) + (\partial T/\partial s)_n/(nT^2)], \tag{16}$$

where $$K = 1 + \alpha_0/\beta_0 + 2\alpha_1/(3\beta_2) - (n/T)(\partial T/\partial n)_s. \tag{17}$$

Therefore, by restricting the values of the second order coefficients α_i and β_i and by imposing the usual thermodynamic constraints on the specific heats, etc., implied by eqs. (10) and (11) it is possible to have stable equilibrium configurations in the Israel-Stewart theory.

We also investigated the conditions under which the equations governing the evolution of the perturbations are hyperbolic. The following conditions are sufficient (perhaps not necessary) to assure that the perturbation equations are a symmetric hyperbolic system:

$$(\partial n/\partial\theta)_T > 0, \quad (18) \qquad (\partial\rho/\partial T)_\theta > 0, \quad (19)$$

$$(\partial\rho/\partial T)_n > 0, \quad (20) \qquad \beta_0 > 0, \quad (21)$$

$$\beta_2 > 0, \quad (22) \qquad \beta_1(\rho + p) > 1, \quad (23)$$

For a hyperbolic system of equations, the propagation of information is controlled by the characteristic velocities of the differential operator. We computed the characteristic velocities for the equations governing the perturbations away from an equilibrium state. Six characteristic velocities are zero, four characteristic velocities are given by (each occurs twice),

$$(v_T)^2 = [(\rho + p)(\alpha_1)^2 + 2\alpha_1 + \beta_1]/\{2\beta_2[\beta_1(\rho + p) - 1]\}, \tag{24}$$

and four more are given by the roots of the quartic equation,

$$A(v_L)^4 + B(v_L)^2 + C = 0, \tag{25}$$

where

$$A = \beta_0\beta_2[\beta_1(\rho + p) - 1], \tag{26}$$

$$B = A(\Phi_7/\beta_1 - 2) + \beta_0\beta_2\left[\beta_1\Phi_5 + (\beta_1 K - \Phi_7)^2/(\beta_1\Phi_7)\right], \tag{27}$$

$$C = - A - B + \beta_0\beta_2\Phi_5\Phi_7. \tag{28}$$

The perturbations are guaranteed to propagate causally as long as these characteristic velocities are less than the speed of light (one in our units) and the system of equations is symmetric hyperbolic.

A remarkable relationship exists between the conditions for the stability, causality and hyperbolicity of these perturbation equations. The stability conditions for these perturbations, eqs. (10) - (16), are satisfied if and only if both the hyperbolicity conditions, eqs. (18)-(23), are satisfied and the characteristic velocities are real and less than the speed of light[10]. A slightly stronger proposition is false. The stability conditions are not equivalent to the characteristic velocities being real and less than the speed of light[11].

IV. <u>THE CLASSICAL FLUID LIMIT</u>. A potential difficulty for the Israel-Stewart theory is the embarrassing complexity of its dynamical structure. These fluids have fourteen dynamical fields, while an ideal relativistic fluid or a non-relativistic Navier-Stokes-Fourier fluid have only five. If this theory is capable of describing ordinary laboratory fluids, how do these additional degrees of freedom disappear? To understand this question we have analyzed the dispersion relations for the plane wave perturbation solutions to the Israel-Stewart theory[4]. When these dispersion relations are examined in the "classical" (i.e., long wavelength compared to the mean free path) limit we find that nine of the modes are strongly damped at lowest order in the wavenumber k:

$$\omega_1 = -\, i/(\varsigma\beta_0), \qquad (29) \qquad\qquad \omega_{2-6} = -\, i/(2\eta\beta_2), \qquad (30)$$

$$\omega_{7-9} = -\, i(\rho + p)/\{\kappa T[\beta_1(\rho + p) - 1]\}. \qquad (31)$$

The remaining five modes have dispersion relations which are simply the relativistic generalizations of the five modes of a Navier-Stokes-Fourier fluid in this long wavelength limit:

$$\omega_{10,11} = \pm k(\partial p/\partial \rho)_s^{1/2} - \tfrac{1}{2} i k^2 \left[\tfrac{4}{3}\eta + \varsigma + \kappa(\partial p/\partial \rho)_s(\partial \rho/\partial s)_p^2/n^2 T\right](\rho + p)^{-1}, \qquad (32)$$

$$\omega_{12} = -\, i\kappa k^2(\partial T/\partial s)_p/(nT), \qquad (33)$$

$$\omega_{13,14} = -\, i\eta k^2/(\rho + p). \qquad (34)$$

Thus the complicated dynamical structure of an Israel-Stewart fluid does reduce to the familiar dynamics of ordinary fluid mechanics in the regime where experimental data are most prevalent.

This research was supported by NSF grants PHY85-05484 and PHY85-18490.

REFERENCES

1. Eckart, C., Phys. Rev. 58, 919 (1940)
2. Landau, L. and Lifshitz, E.M., Fluid Mechanics (Addison-Wesley, Reading, Mass.) Section 127 (1958)
3. Hiscock, W.A. and Lindblom, L., Phys. Rev. D31, 725 (1985)
4. Hiscock, W.A. and Lindblom, L., "Linear Plane Waves in Dissipative Relativistic Fluids", Phys. Rev. D (submitted), (1987)
5. Israel, W., Ann. Phys. (N.Y.) 100, 310 (1976)
6. Stewart, J.M., Proc. Roy. Soc. (London) A357, 59 (1977)
7. Israel, W., Stewart, J.M., Proc. Roy. Soc. (London) A365, 43 (1979)
8. Israel, W. and Stewart, J.M., Ann. Phys. (N.Y.) 118, 341 (1979)
9. Grad, H., Comm. Pure Appl. Math. 2, 331 (1949)
10. Hiscock, W.A. and Lindblom, L., Ann. Phys. (N.Y.) 151, 466 (1983)
11. Hiscock, W.A., Lindblom, L., "Stability in Dissipative Relativistic Fluid Theories", in Contemporary Mathematics: Mathematics in General Relativity edited by J. Isenberg (1987)

DO CLOSED UNIVERSES RECOLLAPSE?*

Frank J. Tipler
Department of Mathematics and Department of Physics
Tulane University
New Orleans, Louisiana 70118 USA

ABSTRACT

It is widely believed that closed universes - those with a compact Cauchy hypersurface - behave globally as the dust-filled Friedmann universe with S^3 spatial topology: start at an all-encompassing initial singularity, expand to a maximal hypersurface, and recollapse to an all-encompassing final singularity. In reality, it is not known if the generic S^3 closed universe recollapses. In fact, I shall show that there are even S^3 *Friedmann* universes satisfying all the standard energy conditions (and with zero cosmological constant) that expand forever. However, if a generic closed universe at some point in its history attains a maximal hypersurface, then it does originate at an initial all-encompassing initial singularity, and does recollapse to an all-encompassing final singularity. But only certain spatial topologies admit maximal hypersurfaces, and hence permit recollapse: roughly speaking, the only closed universes which can ever evolve maximal hypersurfaces are those whose Cauchy hypersurfaces have topology S^3 or $S^2 \times S^1$, or a more complicated topology formed from these two basic types by connected summation and certain identifications. All known solutions to Einstein's vacuum equations with S^3 or $S^2 \times S^1$ Cauchy hypersurface topology recollapse, so I conjecture that *all* vacuum solutions with these Cauchy hypersurface topologies recollapse. I shall also state a recollapse conjecture for matter-filled spatially homogeneous closed universes, and give a general recollapse theorem for Friedmann universes: if the positive pressure criterion, the dominant energy condition and the matter regularity condition hold, then an S^3 Friedmann universe originates at an initial singularity, expands to a maximal hypersurface, and recollapses to a final singularity. Counter-examples indicate that this Friedmann recollapse theorem is more or less the most general recollapse theorem for the Friedmann universe.

1. WHO CARES IF CLOSED UNIVERSES RECOLLAPSE?

Since this Symposium on Relativistic Astrophysics consists more of astrophysicists than relativists, I should like to provide a justification for investigating the Recollapse Problem to the former, who generally think the business of science is finding an explanation of observed past or present phenonmena, and who often think that the behavior of the universe in the far future is therefore irrelevant to science in general and to their work in particular. There are at least three reasons for regarding the Recollapse problem as important to physical cosmology.

The first is rather trivial: at some point in her career, every astrophysicist teaches an elementary astronomy course, including some cosmology. The elementary texts almost uniformly assert that closed universes, by which the texts usually mean universes with S^3 spatial topology, recollapse. In fact, it is not known if "realistic" S^3 closed

*Work supported in part by the NSF under grants PHY-8409672 and PHY-8603130.

universes recollapse, where "realistic" means a universe which is inhomogeneous and which contains the matter fields of contemporary particle physics. Furthermore, only closed universes with certain very special spatial topologies can have a maximal hypersurface and hence recollapse: in particular, a closed universe with a T^3 spatial topology cannot admit a maximal hypersurface and hence (probably) must expand forever.

The second reason is that the early universe behavior of many quantum cosmological models depend on their global temporal structure. For example, the Hartle-Hawking model[1] in its simplest form postulates that the wave function of the universe is defined globally on S^4, and only when the densities are significantly below the Planck density -- that is, far away from what classically would be the initial and final singularities -- can the universe be described as topologically $S^3 \times R^1$, and metrically spacetime. The reason for limiting the domain of the wave function to a compact 4-manifold is to eliminate the necessity for global boundary conditions; as Hawking puts it, the most plausible boundary condition is that there is no boundary. But this compactness in 4 dimensions requires recollapse, for an ever-expanding universe is open in the future temporal direction, and thus would require boundary conditions at future temporal infinity. In a sense, these future boundary conditions are avoided in the Hartle-Hawking model by identifying the high density future with the high density past (though "past" and "future" lose their meaning in these high density regions[1], in part because it is no longer meaningful to talk about the trajectory of a single classical universe); the early universe depends on the future because quantum mechanically (but not classically - see Hawking's paper in these Proceedings), the past *is* the future. This identification can be carried out only if we have recollapse. To the extent we are interested in whether the Hartle-Hawking model correctly describes the early universe, we are interested in the Recollapse Problem.

The third reason is that the plausibility of the inflationary model of the universe depends in part on whether closed universes recollapse. The inflationary model derives its appeal by purporting to show that certain major features of the visible universe (the fact that it is nearly flat, for example) are nearly independent of the initial conditions. However, inflationary models will in fact be nearly independent of the initial conditions only if they asymptotically approach the de Sitter state during the accelerating phase. Whether or not this approach occurs is termed the "cosmic no hair conjecture". As pointed out by Barrow[2,3], attempts to prove no-hair theorems have assumed that the spatial 3-curvature scalar is non-positive on the grounds that universes with positive 3-curvature scalar recollapse. But in fact, I shall give below S^3 Friedmann models (all S^3 Friedmann models necessarily have positive 3-curvature scalar) which expand forever. On the other hand, if more realistic S^3 closed models *do* recollapse, it is possible that *generically* they recollapse too soon for inflation to occcur, suggesting that inflation is unlikely to occur closed universes.

My conventions will be those of Hawking and Ellis[4]. The cosmological constant will be assumed to be zero. I shall in large part be summarizing work done jointly by myself and J.D. Barrow[5] and by myself, J.D. Barrow and G.J. Galloway[6].

2. IMPLICATIONS OF MAXIMAL HYPERSURFACES

The first theorem establishes the necessity of all-encompassing initial and final singularities in a universe with a compact maximal hypersurface. Recall that a spacelike hypersurface **S** is said to be a *maximal hypersurface* if $z^a{}_{;a} = 0$ everywhere on **S**, where z^a is the unit normal to **S**. A singularity to the past of a spacetime point set **P** will be said to be *all-encompassing* if every inextendible timelike curve λ in $I^-(\mathbf{P})$ - i.e., in the past of

P - has a proper time length less than a universal constant L (i.e., the length of $\lambda \cap I^-(\mathbf{P})$ is less than L). An all-encompassing final singularity is defined analogously.

Theorem 1: Let **S** be a compact maximal Cauchy hypersurface. Then there is an all-encompassing singularity to the past of **S** and an all-encompassing singularity to the future of **S**, and further the length of *every* timelike curve in the entire spacetime is less than a universal constant L, provided

(1) $R_{ab}V^aV^b \geq 0$, for all timelike vectors V^a;

(2) At least one of the tensors $z^cz^dz_{[a}R_{b]cd[e}z_{f]}$, $z_{a;b}$, or $R_{ab}z^az^b$, is non-zero on **S**, where z^a is the normal vector to **S.**

Theorem 1 was first stated and proved by Marsden and Tipler[7]. Condition (1), the timelike convergence condition, merely says that gravity is always attractive. Condition (2) says that somewhere on the maximal hypersurface, the gravitational tidal forces are non-zero, or at least the hypersurface is not a hypersurface of time symmetry. For vacuum spacetimes, a hypersurface **S** of time symmetry ($z_{a;b} = 0$ everywhere on **S**) would imply that the future and past of **S** are identical. It is very unlikely that the gravitational tidal forces are identically zero everywhere on **S**, so condition (2) is a generic condition.

The next theorem shows that a maximal hypersurface will never evolve in some universes with compact Cauchy hypersurfaces; only certain topologies admit maximal hypersurfaces.

Theorem 2: If **S** is a spacelike compact orientable maximal hpersurface, then it must have the topology

$$[S^3]_1 \# [S^3]_2 \#...\# [S^3]_n \# k(S^2 \times S^1)$$

(where $[S^3]_i$ is a manifold which is covered by a homotopy 3-sphere, "#" denotes the connected sum, and $k(S^2 \times S^1)$ means the connected sum of k copies of $S^2 \times S^1$), provided the following hold:

(1) The Einstein equations without cosmological constant hold,
(2) the weak energy condition holds, and
(3) the induced metric on S is not flat.

In particular, since T^3 cannot be so written, a closed universe whose Cauchy hypersurface has topology T^3 cannot evolve a maximal hypersurface. Theorem 2 was first proved in [5] (see also [8]). The Theorem is an application of a theorem of Schoen and Yau[9], later generalized by Gromov and Lawson[10]. Witt[11] has recently applied the Schoen-Yau theorem to the existence of maximal hypersurfaces in asymptotically flat space. The hypotheses and conclusions in Theorem 2 are weaker that those of the equivalent theorem in [5] and [8] : in the latter, the manifold $[S^3]_i$ is S^3/P_i, the quotient of S^3 with P_i , a subgroup of O(4) which acts standardly on S^3. To obtain this stronger conclusion, an additional hypothesis ruling out exotic differentiable structures was made; in effect this added hypothesis ruled out homotopy spheres which are not spheres (i.e., it explicitly ruled out manifolds which violate the Poincaré Conjecture), and it ruled out more exotic identifications of S^3 than S^3/P_i. (I am grateful to J. Friedman for discussions on this point.) Schoen and Yau[9] give other hypotheses which reduce $[S^3]_i$ to

S^3/P_i. The important point is that the only recollapsing universes with*simple* spatial topologies are those with either S^3 or $S^2 \times S^1$ spatial topology.

3. RECOLLAPSE IN S^3 FRIEDMANN UNIVERSES

The archetypical S^3 closed universe is the closed Friedmann universe, so it is interesting that there are S^3 Friedmann models that expand forever but in which the matter obeys all the standard energy conditions. To see this, recall that if the matter is a perfect fluid with equation of state $p = (\gamma - 1)\mu$, conservation of energy implies that $\mu = (3M/8\pi G)R^{-3\gamma}$, where M is a constant and R is the usual Friedmann scale factor. The Friedmann constraint equation is thus

$$(R'/R)^2 = M/R^{3\gamma} - 1/R^2 \tag{1}$$

The homogeneity, isotropy, and S^3 spatial topology imply $\mu > 0$ and $M > 0$. Clearly the universe expands forever if $\gamma \leq 2/3$ (if $\gamma = 2/3$, $M > 1$, since the LHS of (1) is positive if the universe is expanding initially). It is easily checked that if $\gamma = 2/3$, the weak, strong, and dominant energy conditions are satisfied.

If $\gamma = 2/3$, the generic condition is not satisfied, but we can add dust satisfying $\mu = (3M_D/8\pi G)R^{-3}$ to the fluid having $\gamma = 2/3$, with M and M_D chosen so that the Friedmann constraint equation is

$$(R'/R)^2 = -M_D/R + 1 \tag{2}$$

which is the Friedmann equation for dust, but with $k = -1$ (the open Friedmann universe); clearly such an S^3 Friedmann universe expands forever, and it is easily checked that the generic condition is satisfied, together with all the other above mentioned energy conditions. What happens is this: when $\gamma < 1$, the pressure is negative, and in general relativity, negative pressure generates a *repulsive* gravitational force (this is why the inflationary universe inflates; some of the fields now being considered by particle physicists have strong negative pressures[6]). When $\gamma \leq 2/3$, this repulsion overwhelms the attractive force due to $\mu > 0$; i.e., the attractive force due to the positive spatial curvature.

If negative pressures are eliminated, we can prove S^3 Friedmann universes recollapse:

Theorem 3: If the positive pressure criterion, the matter regularity condition, and the dominant energy condition hold, then a Friedmann universe with S^3 spatial topology expands from an initial singularity to a maximal hypersurface, and then recollapses to a final singularity.

Theorem 3 is proved in [6]. Counter-examples[6] indicate that the hypotheses of Theorem 3 cannot be significantly weakened. The *positive pressure critierion* of Collins and Hawking[12] says that $\Sigma p_i \geq 0$, where p_i are the principal pressures of the stress-energy tensor. In the Friedmann universe, the 3 principal pressures are all equal, so the criterion reduces to $p \geq 0$. In more general spacetimes, the positive pressure criterion is much weaker than $p_i \geq 0$, $i = 1,2,3$; in fact, the latter condition is violated in certain Bianchi type IX S^3 universes containing electromagnetic fields[6], but the postive pressure criterion holds, and recollapse occurs. The *matter regularity condition*[6] -- which, roughly speaking, asserts that the stress-energy tensor is well-behaved except at a p.p. curvature singularity[4] -- and the dominant energy condition are required to ensure that the pressure doesn't blow up and stop the evolution before the maximal hypersurface is reached. (Don't laugh -- this can actually happen in S^3 Friedmann universes[6].)

4. CLOSED UNIVERSE RECOLLAPSE CONJECTURES

All known vacuum solutions to Einstein's equations with Cauchy hypersurface topology S^3 or $S^2 \times S^1$ are known[6] to recollapse, so I propose

Conjecture 1: All globally hyperbolic vacuum C^2 maximally extended closed universes with S^3 or $S^2 \times S^1$ spatial topology expand from an all-encompassing initial singularity to a maximal hypersurface, and recollapse to an all-encompassing final singularity.

Examples indicate[6] that the conditions on the matter tensor in Theorem 3 are sufficient to obtain recollapse, at least in homogeneous universes, so I therefore propose

Conjecture 2: All globally hyperbolic C^2 maximally extended spatially homogeneous closed universes with S^3 or $S^2 \times S^1$ spatial topology, and with stress-energy tensors which obey

(1) the strong energy condition,
(2) the positive pressure criterion,
(3) the dominant energy condition, and
(4) the matter regularity condition,

expand from an all-encompassing initial singularity to a maximal hypersurface, and recollapse to an all-encompassing final singularity.

I challenge the reader to prove these conjectures, or to give counter-examples.

5. REFERENCES

[1]Hartle, J. and Hawking, S.W., Phys. Rev. **D28**, 2960 (1983); see also the papers by Hartle and Hawking in these Proceedings.

[2]Barrow, J.D., Phys. Lett. **B180**, 335 (1986).

[3]Barrow, J. D., "Cosmic No-Hair Theorems and Inflation", preprint.

[4]Hawking, S.W. and Ellis, E.F.R., *The Large-Scale Structure of Space-Time*, (Cambridge University Press, 1973).

[5]Barrow, J.D. and Tipler, F.J., Mon. Not. Roy. astr. Soc. **216**, 395 (1985).

[6]Barrow, J.D., Galloway, G.J., and Tipler, F.J., Mon. Not. Roy. astr. Soc. **223**, 835 (1986).

[7]Marsden, J.E. and Tipler, F.J., Phys. Rep. **66**, 109 (1980).

[8]Barrow, J.D. and Tipler, F.J., *The Anthropic Cosmological Principle* (Oxford University Press, 1986), chapter 10.

[9]Schoen, R. and Yau, S.-T., Manuscripta Math. **28**, 159 (1979).

[10]Gromov, M. and Lawson, H.B., Inst. Hautes Etudes Sci. Publ. Math. **58**, 83 (1983) I am grateful to J. Friedman for this reference.

[11]Witt, D.M., Phys. Rev. Lett. **57**, 1386 (1986).

[12]Collins, C.B. and Hawking, S.W., Astrophys. J. **239**, 317 (1973)

GRAVITATIONAL RADIATION DAMPING IN SYSTEMS WITH COMPACT COMPONENTS

James L. Anderson

Stevens Institute of Technology
Hoboken, New Jersey 07030

ABSTRACT

Balance equations and the radiation reaction force are derived for slow-motion, gravitationally bound systems with compact components such as neutron stars and black holes. To obtain these results, use is made of the surface integral method developed by Einstein, Infeld and Hoffmann. As a consequence, all quantities involved in the derivation are finite and hence no renormalization procedure is required. Approximate expressions for the fields required to evaluate the surface integrals are obtained using the methods of matched asymptotic expansions and multiple time scales. The results obtained are the same as those derived previously for systems with non-compact components.

1. Introduction

It is a remarkable fact that the equations of motion of sources of electric and gravitational fields are contained within the field equations for these quantities in general relativity and hence do not need to be postulated separately as they do in special relativity. This fact was first exploited by Einstein-Infeld-Hoffmann (EIH)[1] to derive the so-called post-Newtonian equations of motion for gravitating sources. It is equally remarkable that, to the best of our knowledge, this method has never been used to derive expressions for the effects of radiation damping in such systems in spite of its superiority over other methods of derivation, especially if these systems contain compact sources such as black holes or neutron stars with strong gravitational fields.

Previous attempts to derive the effects of radiation reaction in systems with compact sources have all suffered from one or more of the following deficiencies:

i) source model not applicable. Neither δ-function or continuous matter models work. The former bear no obvious relation to either neutron stars or black holes and the latter cannot be used in the case of black holes. For neutron stars, the continuous matter models lead to difficulties when several such sources are in interaction because the fields in the neighborhoods of the sources are in general not weak and cannot be adequately treated by weak-field perturbation methods.

(ii) mathematical inconsistencies. A number of approximation schemes lead to divergent integrals when carried to sufficiently high order. Others require some form of renormalization to obtain finite results. Furthermore, in most derivations of the radiation reaction force, the resultant equations of motion admit unphysical solutions, e.g., runaway solutions.

(iii) incomplete or inappropriate derivation. Any attempt to use the results of linear perturbation theory in the case of compact sources is obviously unjustified. Several derivations assume that one can equate the flux calculated by the so-called quadrupole formula to the time rate of change in the Newtonian energy of the source. Other derivations assume the validity of the geodesic equations of motion. All such assumptions are in fact unnecessary in general relativity since the equations of motion, including reaction effects, are a consequence of the field equations.

The derivation we will outline here avoids, we believe, these difficulties and as such is the first derivation that can, with some confidence, be applied to a system such as the binary pulsar PSR1913+16. Since the results we derive here are the same as those that have been applied to this system in the past, they agree with the observed orbital period change to within three percent.

2. Approximation Methods

In order to evaluate the surface integrals that arise in the EIH method, it is necessary to find approximate expressions for the fields present. Because of the singular nature of the problem it is necessary to use several different techniques in order to avoid non-uniform expansions. In our work we have employed the method of matched asymptotic expansions (MAE), first introduced into the subject by Burke,[2] as well as the method of multiple time scales (MTS).[3] In both cases, the expansion parameter ε is the ratio of the light travel time across the system divided by the longest relevant time scale, usually the Newtonian orbit period, associated with the system.

In using the MAE, the matching region is the overlap between the inner zone of dimensions of the order of a few wave lengths and the outer or radiation zone that extends to infinity. In the inner zone one must solve a Poisson equation for the fields at each stage of the approximation where the sources are functionals of lower order approximate fields. Since the EIH surfaces can be taken to lie entirely in the weak field region when dealing with slow-motion systems, it is not necessary to solve these equations everywhere but rather only in the weak field region. As a consequence, the solutions at each stage of the approximation are only determined in the inner zone up to a harmonic function. These functions in turn are fixed in part by matching to an outer zone solution ala Burke, in part by the imposition of coordinate conditions such as the deDonder conditions and in part by the EIH surface integrals. In principle the method can be carried to any desired order of approximation but in practice the required computations become exceedingly lengthy beyond the lowest orders.

Because of the existence of secular effects in gravitationally bound systems such as periastron advance and orbital period change, it is necessary to employ MTS in order to avoid non-uniformities in time in the approximation expansions employed. In this method one assumes that the unknowns in the problem depend on time through their dependence on several times t_1, t_2, t_3, ... where usually $t_i = \varepsilon^{n_i} t$. The dependence on these multiple times is determined in such a way that the coefficients of terms in the equations that would lead to secularly growing terms in the solution are made to vanish.

Unfortunately, neither of these methods of approximation so far has been made completely rigorous and so one might doubt their validity. At present the best one can say is that in other fields their application leads to results that agree very well with experiment and there is no reason to suppose that they will not work equally well in general relativity.

3. The EIH Method

The field equations of general relativity can be put into the Landau-Lifshitz form [4)]

$$U^{\mu\nu\rho}{}_{,\rho} = \Theta^{\mu\nu} \tag{1}$$

where

$$U^{\mu\nu\rho} = (1/16\pi)\{(-g)(g^{\mu\nu}g^{\rho\sigma} - g^{\mu\rho}g^{\nu\sigma})\}_{,\sigma} \tag{2}$$

and $\Theta^{\mu\nu} = (-g)(T^{\mu\nu} + t_{LL}{}^{\mu\nu})$. Here $T^{\mu\nu}$ is the stress-energy tensor of any other fields in the problem and $t_{LL}{}^{\mu\nu}$ is the Landau-Lifshitz pseudo-tensor.

The essence of the EIH method is to integrate the field equations (1) over the surface of a small sphere surrounding the A'th compact source in the system to obtain the equation of motion for that source. Because $U^{\mu\nu\rho} = -U^{\mu\rho\nu}$, the spatial derivatives of $U^{\mu\nu\rho}$ do not contribute to the integrals since the integral of a curl over a closed surface vanishes identically. One can also integrate Eqs. (1) over a large sphere whose surface lies in the far radiation zone to obtain a balance equation for "energy" and in a similar manner one can obtain a balance equation for "angular momentum." The actual results follow from the fact that the values of the surface integrals must be independent of the size of the sphere and do not depend upon any interpretation of $t_{LL}{}^{\mu\nu}$ as a stress-energy complex for the gravitational field.

4. Results

The balance equations for "energy" and "angular momentum" in lowest order of approximation obtained by this method agree completely with those obtained previously[5]. The lowest order contribution to the radiation reaction force in deDonder coordinates is quite different in form from the so-called standard expression

$$F^{i}_{A\ react} = -(2/5)x_{Aj}d^{5}\mathrm{I\!\!\!-}^{ij}/dt_{1}^{5} \qquad (4)$$

where $\mathrm{I\!\!\!-}^{ij}$ is the reduced mass quadrupole moment associated with the source and $t_1 = \varepsilon t$. Nevertheless it yields the same balance equations as does (4) and solutions of the equations of motion can be transformed into solutions of the equations of motion with the reaction force given by Eq. (4).

This work was supported by the National Science Foundation, Grant No. PHY-85 03879.

----

1. Einstein, A., Infeld, L. and Hoffmann, B., Ann. of Math., 39, 66 (1938).

2. Burke, W. L., J. Math. Phys. 12, 401 (1971).

3. For a discussion of these methods see, for example, Nayfeh, A. "Perturbation Methods" (Wiley, New York, 1973).

4. Landau, L. D. and Lifshitz, E. M., "The Classical Theory of Fields" (Addison-Wesley, Reading, Mass., 1962).

5. Peters, P. C., Phys. Rev. 136, B1224 (1964); Anderson, J. L., Phys. Rev. Lett. 45, 1745 (1980).

QUANTUM STRESS TENSORS AND THE WEAK ENERGY CONDITION

Thomas A. Roman

Physics/Earth Sciences Department
Central Connecticut State University
New Britain, Connecticut 06050, USA

ABSTRACT

The stress-energy tensors associated with certain processes predicted by quantum field theory violate the weak energy condition and are shown to possess, as a result, one or both of the following properties: T_{ab} is non-diagonalizable by a local Lorentz transformation, and $T_{ab}U^aU^b$ is not bounded below for all unit timelike vectors U^a. It is also shown that Penrose's singularity theorem will still hold if the weak energy condition is replaced by an "averaged" (i.e., nonlocal) weak energy condition.

1. Introduction

All singularity theorems proved in general relativity require some type of "energy condition"; i.e., some restriction on the behavior of matter and energy present in spacetime. The weakest such condition:

$$T_{ab}U^aU^b \geq 0 \tag{1}$$

for all timelike vectors U^a, is known as the "weak energy condition" (WEC). This condition implies that the energy density is positive for all observers. (By continuity, eq. (1) will also hold when U^a is null vector.) It is used in Penrose's singularity theorem[1-2] as a sufficient condition for the focusing of null geodesics in order to prove null geodesic incompleteness in open universes containing closed trapped surfaces.

The energy conditions are _local_ conditions in the sense that they hold in a neighborhood of each point of spacetime. It is possible that for extreme situations involving high density and large spacetime curvature, such as the late stages of stellar collapse or recollapse of the universe, these conditions no longer hold.

The lead in relaxing the energy conditions was taken by Tipler[3] in a seminal 1978 paper. He showed that many of the results of the original singularity theorems could be preserved if the "strong energy condition":

$$(T_{ab} - 1/2\, g_{ab}T)U^aU^b \geq 0 \tag{2}$$

for all timelike vectors U^a, was replaced with the WEC and the assumption that the strong energy condition holds only "on the average." The average is taken over the entire history of a causal geodesic.

In considering whether the WEC is violated, Tipler proved the following:

Proposition (Tipler[3]). If $T_{ab}U^aU^b$ is bounded below for all unit timelike vectors U^a in T_p (the tangent space of a point p) and if T_{ab} is type I, then $T_{ab}K^aK^b \geq 0$ at p for all null vectors K^a in T_p.

The assumption that T_{ab} is type I (in the language of Hawking and Ellis[2]) means that T_{ab} is diagonal; i.e., there always exists a frame of reference in which the "flux" term T_{10}, can be made to vanish. The assumption that $T_{ab}U^aU^b$ be bounded below essentially means that this quantity cannot be made arbitrarily negative by Lorentz-boosting the unit timelike vector U^a. The physical implications of Tipler's proposition is that if T_{ab} has the form of most known classical matter fields and if $T_{ab}U^aU^b$ is bounded below for all unit timelike vectors, then the WEC still holds for all null vectors, even though it may be violated for some timelike vectors. Therefore, any singularity theorem which uses the WEC to prove null geodesic incompleteness (such as Penrose's theorem) will still hold, if the WEC is violated for some timelike vectors provided Tipler's proposition holds.

2. New Results (details may be found in Roman[4])

We have shown that certain processes predicted by quantum field theory violate one or both of the assumptions in Tipler's proposition. More specifically, we have shown that:

(i) T_{ab} for a spherically symmetric evaporating black hole violates the WEC and, as a result, is non-diagonalizable by a local Lorentz transformation in the vicinity of the apparent horizon - i.e., T_{ab} is type IV[2] in this region. (Here and in what follows T_{ab} means $\langle T_{ab}\rangle$, the renormalized vacuum expectation value in a suitable vacuum state.)

(ii) T_{ab} for massless scalar radiation emitted by a moving mirror in a 2-D spacetime is type II[2] and can violate the WEC for certain accelerations of the mirror.

(iii) The stress tensors for (i) and (ii), as well as the type I stress tensor for the (experimentally verified) Casimir effect, have the property that $T_{ab}U^aU^b$ is not "bounded below" for all unit timelike vectors U^a.

It is possible that similar, but as yet unknown, quantum processes in which the WEC is locally violated could prevent the formation of a singularity in the gravitational collapse of a star.

Therefore, it is important to prove a singularity theorem using a weaker energy condition. We prove that:

(iv) The singularities predicted by Penrose's theorem will still occur if there exists at least one closed trapped surface in spacetime for which the WEC holds only on the average along every null geodesic making up the boundary of the future of that surface.

The physical implication of this theorem is that once a trapped surface has formed in a stellar collapse, a small localized violation of the WEC is insufficient to prevent the subsequent formation of a singularity.

References

(1) Penrose, R., Phys. Rev. Lett. 14, 57 (1965).
(2) Hawking, S. W., and Ellis, G. F. R., "The Large Scale Structure of Spacetime" (Cambridge University Press, London, 1973).
(3) Tipler, F. J., Phys. Rev. D 17, 2521 (1978).
(4) Roman, T. A., Phys. Rev. D 33, 3526 (1986).

DIRAC'S LARGE NUMBERS HYPOTHESIS-- A BRIDGE FOR UNIFYING GRAVITATION AND QUANTUM THEO.

Ronald Gautreau
Physics Department
New Jersey Institute of Technology
Newark, New Jersey 07102

The Large Numbers hypothesis put forth by Dirac is a profound and elegant attempt to find a link between the atomic world of quantum physics and the cosmological world of General Relativity. The thought that there might be some physical significance in the similarities of large dimensionless numbers of the order 10^{39} that arise from certain ratios of cosmological and atomic quantities is a reflection of the manner in which Dirac viewed physics.

In the nearly half century from 1938 until his death in 1984, Dirac continuously advocated the LNh, but he was never able to reconcile his LNh with General Relativity. After various earlier formulations of the LNh[1,2,3], his last version[4] involves two separate metrics. In one metric ds_E times and distances are measured in cosmological ephemeris units, while in the other metric ds_A times and distances are measured in separate atomic units. The atomic metric ds_A is not determined from the Einstein field equations, so that Dirac's last formulation of the LNh lies outside the scope of General Relativity. Dirac was not completely satisfied with this formulation. As he remarked[3]:

> The foregoing work is all founded on the Large Numbers hypothesis, in which I have great confidence. It also requires the assumption of two metrics, which is not so certain. The only reason for believing in two metrics is that up to the present no alternative way of bringing in the Einstein theory has been thought of. But this situation could change.

I have recently developed a method for incorporating Dirac's LNh into the standard Einstein theory of General Relativity[5]. In my approach, I follow Dirac and assume that along the world line of a galaxy of the cosmological fluid comprising the Universe there are separate times measured by coincident cosmological and atomic clocks:

<u>ephemeris time T</u> is measured by mechanical processes such as the motion of an oscillating pendulum, a planet orbiting the sun, or a galaxy moving through the Universe;

<u>atomic time t_A</u> is measured by atomic processes such as the decay rate of radioactive atoms.

It is usually tacitly assumed that in a galaxy the evolution of stars proceeds synchronously with the evolution of atoms. Dirac's point is that this assumption need not necessarily be correct. In principle it is possible that cosmological processes may evolve at a different rate from atomic processes, so that the ratio dT/dt_A of the "tickings" of the clocks may vary with the age of the Universe.

Cosmological evolution is governed by General Relativity, while the evolution of atomic processes is determined separately by quantum theory. At present, these two theories stand independent and unrelated, and there is no a priori reason why the two theories should yield a constant value for the ratio dT/dt_A as the Universe evolves.

If indeed it is the case that T and t_A are not equal to each other, there will be some functional relationship $T = f(t_A)$ along the world line of each galaxy. From the cosmological principle this will be the same for all galaxies.

The relationship $T = f(t_A)$ acts simply as a coordinate transformation between the two times. A metric can be expressed equally well in terms of either time, and $ds(x_i,T) = ds(x_i,t_A)$. Further, since T = constant and t_A = constant surfaces coincide, the proper distance between T-simultaneous events will be the same as the proper distance between t_A-simultaneous events:

$$R = \int_{T=const} ds(x_i,T) = \int_{tA=const} ds(x_i,t_A) \qquad (1)$$

In this manner all calculations can be done within the context of the standard Einstein field equations. It is not necessary to introduce separate metrics nor separate ephemeris and atomic units of length as Dirac has done. Proper length measured by R in (1) is determined from Einstein's standard theory.

The functional relationship $T = f(t_A)$ is found from the LNh. The details are given in Ref. 5. When a quantity in a Large Number is formulated in terms of Newtonian time, the transition to the LNh is made by taking this time to be atomic time t_A. The time variation of the various LN's is then worked out for the Einstein-de Sitter (ES) Universe, which Dirac claims is in agreement with his LNh[4].

In his LNh, Dirac requires that large dimensionless numbers in Nature of the order 10^{39} must be interrelated by equations where the coefficients are close to unity. This requires that the various LN's should be proportional to each other, so they all have the same time variation. When the LN's, as worked out for the ES Universe, are set proportional, one finds for all combinations of LN's the unique relationship $t_A \propto T^2$.[5]

This is different from what Dirac obtained in his last paper on the LNh.[4] From dimensional arguments built around his two metrics ds_E and ds_A, Dirac determined that $T \propto t_A{}^2$.

Ephemeris clocks such as the evolution of galaxies and atomic clocks like decaying atoms are physical systems. As such, the behavior of the "ticking" of these clocks should be determinable first hand from physical principles, and physicists should be able to work out the functional relationship $T = f(t_A)$, and correspondingly the ratio dT/dt_A. In this case there would be no need for a Large Numbers hypothesis, for the Large Numbers would be whatever they came out to be when the calculations were done.

However, we do not as yet possess an understanding of the relationship between General Relativity and quantum theory, so that $f(t_A)$ can not be worked out from present knowledge of physics. In the grand scheme of physics it is hoped that the two theories can be merged into a single unified theory, but we are at present far removed from this.

The usual approach towards unification has been to try to quantize General Relativity. Inherent in this is the assumption that quantum mechanics is a "superior" theory that General Relativity must be subservient to. A little reflection, though, shows that this may not be the most prudent course. Quantum theory rests on an intrinsically incomplete substructure. As Dirac notes in discussing "The Future of Atomic Physics,"[6] and in one of his last publications appropriately titled "The Inadequacies of Quantum Field Theory,"[7] a set of working rules has evolved that artificially removes infinities from quantum field theory by a renormalization procedure.

Although the results of quantum field theory happen to be in surprisingly good agreement with observation, this does not prove that the theory is correct. Dirac points out that the Bohr theory gave very good answers in simple cases, but it was basically wrong conceptually. Similarly, according to Dirac, the renormalized kind of quantum theory presently followed is not justifiable. Even though there is good agreement with experiment, the quantum mechanics presently in use is not based on strict mathematics. It is a set of working rules, and not a complete dynamical theory.

On the other hand, Einstein's theory of gravitation is a completely self consistent theory. Furthermore, the theory is intrinsically beautiful.

So it seems quite logical that in trying to unite gravitation and quantum mechanics, one should use General Relativity as the starting point, and try to modify quantum theory appropriately to fit into the scheme of General Relativity. This is exactly 180° removed from the usual approaches taken towards unification.

The next question is how then to proceed? The way is unclear. But now that we have a scheme by which Dirac's LNh can be incorporated into Einstein's standard theory of General Relativity, perhaps the LNh might serve as a bridge across the gulf between General Relativity and quantum theory. In following this or any other lead, though, it seems that the approach should not be to try to "quantize" General Relativity. Rather, it seems that instead one should attempt to "General Relativize" quantum theory--whatever that may mean.

References

1,2,3,4. Dirac, P.A.M., Proc. Roy. Soc. Lon. Ser. A: **165**, 199 (1938); **333**, 403 (1973); **338**, 439 (1974); **365**, 19 (1979).
5. Gautreau, R., Int. J. Theor. Phys., **24**, 877 (1985).
6. Dirac, P.A.M., Int. J. Theor. Phys., **23**, 677 (1984).
7. Dirac, P.A.M., "The Inadequacies of Quantum Field Theory," preprint.

STRINGS AND OTHER DISTRIBUTIONAL SOURCES IN GENERAL RELATIVITY

Robert Geroch
Enrico Fermi Instiute and Deptartment of Physics
University of Chicago, Chicago, Il 60637

Jennie Traschen
Enrico Fermi Institute and Department of Astronomy
University of Chicago, Chicago, Il 60637

ABSTRACT

This paper deals with two broad issues: the formulation of a mathematical framework for concentrated sources in general relativity, and its application to strings. We isolate a class of those metrics whose curvature tensors are well-defined as distributions. It is shown that shells of matter -- but neither point particles nor strings -- can be described by metrics in this class. This conclusion is examined in more detail for the case of strings. We estimate the errors inherent in certain determinations of the mass per unit length of a cosmic string, and in certain calculations of the gravitational radiation from such a string.

SUMMARY[1)]

A useful idealization in physics is that in which some smooth distribution of sources is replaced by a "concentrated source". In the case of electromagnetism there is a natural mathematical framework for this idealization; one allows both the Maxwell field and the charge-current density to be, rather than smooth tensor fields, distributions. As an example, let the charge density in electrostatics be the distribution of strength σ confined to some 2-dimensional shell in space. Then the Maxwell equations require that the jump in the normal component of the electric field on crossing this shell be equal to σ. A similar idealization to concentrated sources would be useful in general relativity[2)]. But here the mathematical framework cannot be as simple as for electromagnetism, for Einstein's equation, being nonlinear, does not make sense as an equation on a distributional metric.

There have been attempts to introduce into general relativity sources to represent gravitating point particle, *i.e.*, sources concentrated on 1-dimensional surfaces in space-time. It now appears, however, that the metrics in these constructions may not be physically realistic; one expects that such a concentration of matter would result in collapse through a horizon, and that inside this horizon there will be further structure. By contrast, sources concentrated on a 3-dimensional submanifold S have been successfully introduced in general relativity[3]. The magnitude of the curvature (and hence the stress-energy) concentrated on S is given in terms of the jump in the first derivatives of the metric across S. Therefore, while there is apperently no viable treatment of point particles as concentrated sources in general relativity, there is a satisfactory treatment of thin shells of matter or radiation.

The intermediate case is that of strings, *i.e.*, of sources concentrated on 2-dimensional surfaces in space-time. Is there, or is there not, a satisfactory treatment in relativity of such concentrated sources?

We introduce[1] a mathematical framework for concentrated suorces in general relativity. The key step is the definition of a class of metrics, called *regular* metrics, for which the curvature tensor makes sense as a distribution. It is the regular metrics that can arise from a distributional source. The definition requires that the metric be locally bounded with locally bounded inverse, and have locally square-integrable weak first derivative. It turns out that the metrics for thin shells of matter or radiation, described above, are regular. But the class is much more general than this, admitting, *e.g.* certain metrics that are not even continous. The main theorem asserts that, for a regular metric with source concentrated on some submanifold of space-time, that submanifold must be of dimension three[1]. Thus, point particles and strings -- which correspond to dimensions one and two -- are not permitted as sources. This result is closely related to the fact that the energy density of the Newtonian gravitational field of a massive shell is locally integrable, but not for a point particle or string. We also obtain an approximation theorem, which gives the sense in which a smooth metric must approximate a regular metric in order that the curvature tensor of the former be close to the curvature tensor of the latter[1].

A particular case of a concentrated source is a string, that is, a source concentrated on 2-dimensional surfaces in space-time. Consider the cylindrically symmetric static space-time[4]

$$-dt^2 + dz^2 + dr^2 + \beta^2(r)d\phi^2 \tag{1}$$

where z,r,ϕ are cylindrical coordinates. Let

$$\beta = \frac{l}{\gamma}\sin(\gamma r/l) \quad , r \leq l \tag{2a}$$

$$\beta = (r - l + \frac{l}{\gamma}\tan\gamma)\cos\gamma \quad , r > l, \tag{2b}$$

where $l>0$ and $\gamma \epsilon (0,\pi/2)$ are constants. For $r \leq l$ the nonzero components of stress energy are $-T^t_t = T^z_z = \gamma^2/l^2$. For $r>l$ the space-time is Minkowski space-time with angular deficit $2\pi(1-\cos\gamma)$. Note that the mass per unit length, defined by the integral of $-T^t_t$ over surfaces of constant t and z, is equal to $2\pi(1-\cos\gamma)$, and hence is equal to the deficit angle.

Now fix γ, and consider the limit $l \rightarrow 0$. Then the exterior metric becomes Minkowski space-time with angular deficit $2\pi(1-\cos\gamma)$; and one can think of the source as approaching a "line mass density " of mass per unit length also $2\pi(1-\cos\gamma)$.

One might thus regard this limit as representing a source in general relativity concentrated on a 2-dimensional surface in space-time. The equality here between the angular deficit and the mass per unit length would be analogous to that, in electrostatics, between the jump in the normal component of the electric field and the surface charge density. Both relate the external field to the strength of the source.

However, Minkowski space-time with angular deficit – the limit of the family (1)-(2) – turns out not to be a regular metric. Thus, we cannot in any natural way regard this as the external metric for a distributional source. What happens, then, if one introduces some general source for the external spacetime (2), and then takes the limit as that source becomes concentrated on the axis? Examples[1] suggest that, if one wishes to recover in the limit a fixed relationship between source and external field, then some restrictions on the character of the source or the nature of the limit must be imposed. We argue however, that there are *no* such restricitons that are fully satisfactory. If, for example, one merely imposes an energy condition, then for a given fixed angular deficit one can obtain in the limit a variety of values for the mass per unit length of the source. Even fixing an equation of state for the matter will not do. It turns out, for example, that for almost every choice of equation of state there exists *no* source whatever made of that matter, and having for its external metric Minkowski space-time with angular deficit. And even when there does exist such a source, like the matter of (1)-(2), further difficulties arise. For example, if the static and cylindrical symmetries are relaxed, then it is questionable whether one can recover any simple relation between source and field.

There are at least two situations in work on cosmic strings, in which the string is idealized as a concentrated, gravitating source. The first is when the mass per unit length of a string is inferred from external geometrical effects, *e.g.*, from the formation of a double image of a quasar or from a step discontinuity in the microwave background radiation. radiation from such a string is computed. These situations involve precisely the idealization that appears to be problematic in relativity. Errors will be introduced into the cosmic strings computations as a result of using this idealization. For example an external mass which introduces a perturbing gravitational field of order λ will imply corrections to the mass per unit length equals missing angle formula of at least order λ, though they may be much larger.

References

1) For full discussion, see Geroch,R. and Traschen,J., Enrico Fermi Inst. preprint *"Strings and other Distributional Sources in General Relativity"*,(1987) University of Chicago, Chicago, Il. 60637.

2) See *e.g.*, Taub,A., *J. Math. Phys.*,21,1423 (1980).

3) Israel,W., *Nuovo Centima*, *44B*, 1 (1966) ; Chandrasekhar,S. and Xanthopoulos,B., *Proc. Royal Soc. London A*, 398, 223 (1985).

4) Gott,R., *Astrophysical Journal*, 282, 4221 (1985) ; Hiscock,W., *Phys. Rev D* 31, 3288 (1985).

RECENT PROGRESS IN JET MODELING

Michael L. Norman
National Center for Supercomputing Applications
and Department of Astronomy
University of Illinois, Urbana-Champaign
Champaign, IL 61820

ABSTRACT

Results of supercomputer simulations are presented which address two long-standing problems surrounding the interpretation of high luminosity extragalactic jets: 1) propagational stability and 2) confinement. The effects of both passive and dynamic magnetic fields on the structure and appearence of jets is illustrated. Theoretical and computational challenges posed by constructing self consistent radio source models are discussed.

1. INTRODUCTION

Coincident with the hypothesis that the extended lobes of radio galaxies are powered by collimated plasma jets[1,2] were also raised the major theoretical problems that this hypothesis posed. Namely, the generation, collimation, propagation and termination of the jets, which, at that time, had not yet been observed. A decade of intense observational effort using radio interferometers has produced well over one hundred clear detections of jets in radio galaxies and quasars[3,4] and has served to sharpen and constrain these questions. A useful generalization to emerge from these observations is that jets appear to come in two distinct "flavors" depending upon whether or not the radio core power at 5 GHz exceeds $\approx 10^{24}$ W/Hz[5]. Jets in low power sources are generally two-sided, smooth in appearance, rapidly spreading and feed edge-darkened plume-like radio tails. As a rule, the projected magnetic field is oriented perpendicular to the jet axis. Jets in high power sources are generally one-sided, clumpy in appearance, highly collimated and feed edge-brightened radio lobes with hot spots. The projected magnetic field in these jets is almost always oriented along the jet axis.

The current theoretical view is that the first class of jets are turbulent, transonic and essentially gas dynamical, where their observed spreading rates and brightness distributions are governed by mass entrainment and bouyancy[6]. The second class of jets are thought to be quasi-laminar, hypersonic and possibly relativistic flows of extremely diffuse plasma (compared to ambient) in which the observed compact knots and hotspots locate large-scale internal shocks[7]. Our understanding of the dynamical role of the magnetic fields in both classes of jets is incomplete at this time.

Numerical simulations have focussed exclusively on the second class of jets because they are both theoretically interesting and technically tractible with the current generation of codes and computers. The purpose of this paper is to present some new results of supercomputer simulations which address the problems of jet stability and confinement in high power radio sources incorporating magnetic fields. In addition to permitting a study of their influence on jet structure and stability, the inclusion of magnetic fields into the dynamical models allows for the first time a more realistic computation of their radio surface brightness distribution, with tantalizing results, and forshadows the move to construct ever more detailed models of extragalactic radio sources.

2. HYDRODYNAMIC STABILITY

How is it that extragalactic jets propagate as far as they do? Their stability is impressive either in terms of distance (hundreds of kiloparsecs) or in terms of jet diameters (hundreds). A related theoretical question is what determines how far a supersonic jet may propagate? Whereas it is known that the deceleration and spreading that accompany mass entrainment in subsonic and transonic jets limit their propagation, it is thought that large-scale disruptive instabilities limit the propagation of supersonic jets in which entrainment rates are greatly reduced[7)].

Linear perturbation theory of hydrodynamic Kelvin-Helmholtz instabilities have shown that the lowest-order nonaxisymmetric mode (helical, kink) have large spatial and temporal growth rates at all Mach numbers[8,9)]. Since these modes displace the jet, their nonlinear growth could disrupt the jet by destroying its directionality. In a nonlinear simulation of temporal growth of kink modes in a 2-D planar "slab" jet, Woodward[10)] identified a disruptive mode driven by a resonance mechanism involving the constructive interference of weak compression waves reflecting between the channel walls. A difficulty with this work is that one does not know how to relate the temporal growth of an isolated mode in periodic geometry to the *spatial* growth of the same mode as it convects out along the jet, appropriate to the astrophysical case.

As shown in Fig. 1, the nonlinear spatial growth of kink instabilities contains important new features which escaped the temporal simulations of Woodward. In this calculation, a 2-D planar "slab" jet is initialized across the computational domain. The jet is initially in pressure balance with the surrounding uniform gas. The jet gas is admitted continuously on the left and exits on the right through an outflow boundary condition. A one percent sinusoidal transverse velocity perturbation is continuously applied at the jet inlet at the resonant frequency of the fundamental kink mode as determined by perturbation theory[11)].

The three snapshots illustrate three phases of evolution of the slab: 1) spatial growth of Woodward's resonant kink mode to perceptible amplitude, 2) subsequent disruption of jet channel due to continued growth of this mode and the formation of a propagating jet head (here

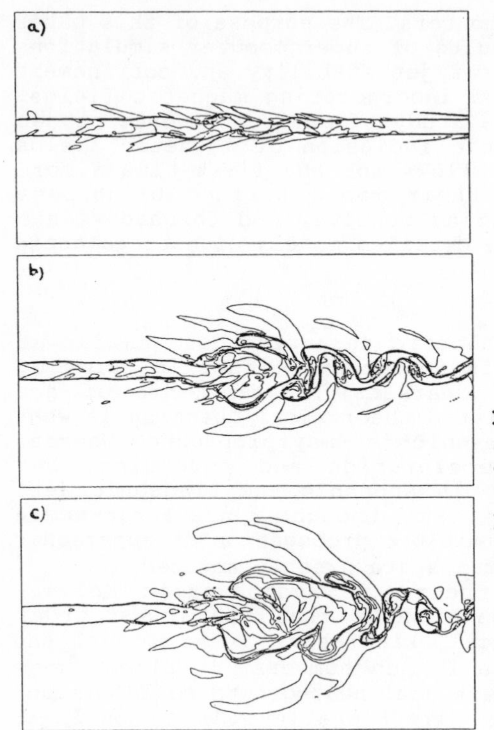

Fig. 1 2-D Hydrodynamical simulation of the spatial growth of a disruptive kink instability on a supersonic planar jet (η=0.1,M=3). The gas stream enters continuously at left and is wiggled at the resonant frequency of the the fundamental kink mode. The three snapshots (a-c) show the subsequent evolution as described in the text.

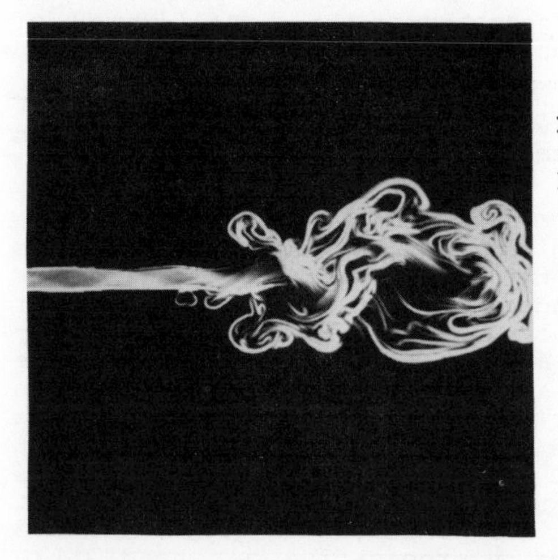

Fig. 2 Simulated radio surface brightness distribution of the jet in Fig. 1 at late times. The emissivity is taken to be B^2P, where B and P are the field strength and gas pressure. The magnetic field is initially uniform and longitudinal and is of negligible strength. The filamentary structure is the result of mass entrainment and flux stretching.

two-thirds the way across the box), and 3) growth of gaseous lobe and emergence of well-defined bow shock as the jet head advances through the ambient gas. The lobe is supplied by the jet, which develops a large amplitude sinusoidal structure and flaps through several jet widths while maintaining its collimation. The wavelength and phase velocity of the sinusoid agree well with theoretical predictions of the fundamental mode[11].

This evolution represents one point in a parametric study of the nonlinear response of slabs of different densities and Mach numbers to different driving frequencies. The results of this study can be summarized as follows. The slab is disrupted at early times by a resonant kink instability in the manner shown above independent of driving frequency. The disruption length scales linearly with the Mach number, but is insensitive to the density ratio. A gaseous lobe forms at the disruption point, which is fed by a flapping jet. The size of this lobe is sensitive to density ratio and the amplitude of flapping at the jet head. For perturbations at or below the resonant frequency for the *fundamental* kink mode, this amplitude is large, although wavelengths and phase speeds agree well with linear theory. For perturbations above the resonant frequency, the slab does not flap appreciably, but rather pinches as it enters the lobe.

Although these simulations are unrealistic models of extended radio source formation, due to the assumed initial conditions, I would expect a similar outcome from a pure propagating jet simulation. If this is the case, then large-scale instabilities may indeed control the formation site of radio lobes and detailed jet structures within them, such as sudden jet deflections and multiple hot spots.

3. MAGNETOHYDRODYNAMIC STABILITY

One cannot help wondering whether the longitudinal magnetic fields characteristic of high power radio jets are somehow responsible for their remarkable stability. Some of these jets are straight as an arrow over tens jet diameters (e.g., 3C175). As we have seen above, a hydrodynamic explanation would necessitate very large Mach numbers in order to push out the disruption length appropriately. This may well be the answer. Nevertheless, simulations entirely analogous to the above incorporating a strong longitudinal magnetic field have also been performed in order to see whether magnetic tension can stabilize kinking under certain circumstances.

The expectation on the basis of linear perturbation theory[12] is that stability can be acheived for $M_{ms}<1$, where M_{ms} is the magnetosonic Mach number in the jet. We have confirmed this expectation numerically by varying M_{ms} while holding all other jet and perturbation parameters fixed. In a simulation with $M_{ms}=1.2$, just above the stability threshold, kinking is observed, although at much reduced amplitude compared to the hydrodynamic case at similar times. A simulation with $M_{ms}=0.5$, well below the stability threshold, shows no displacement whatsoever. A simulation of an "equipartition jet", where the gas and magnetic field pressures were initially equal and $M_{ms}=2.2$, exhibits kinking as expected, and produces a lobe of folded flux ropes.

Although this work is preliminary, it is clear that magnetic fields change dramatically the structural and stability properties of supersonic jets, forcing us to revamp our notions based on hydrodynamic models.

4. CONFINEMENT

As reviewed elsewhere in this volume[4)], jets in the most luminous radio galaxies and quasars are inferred to be overpressurized with respect to the surrounding intergalactic gas, yet are highly collimated. This is the source of the so-called confinement problem, which dates back to the first detection of a radio jet in a quasar[13)]. Magnetic confinement has traditionally been invoked to explain this, however there is virtually no direct evidence in support of the hypothesis. This does not necessarily rule out the hypothesis, for no one knows what a magnetically confined jet looks like. Numerical magnetohydrodynamic simulations of pinched jets are beginning to be performed[14,15)], however, and should provide observers with magnetic signatures to look for.

An alternate possibility, appreciated only recently, is that, for sufficiently high Mach numbers of advance, the jet may be ram pressure confined along its entire length[16,17)]. In such cases, the bow shock is hypersonic, and closely envelops the jet in a high pressure bubble. A numerical simulation of this is shown in Burns et al., this volume. An attractive feature of this model is a natural explanation for why only high luminosity jets should be overpressurized. It can be shown that the degree of overpressure within the hypersonic bubble should scale as $\eta^{1/3}\mathcal{L}^{2/3}$, were η and $\mathcal{L}$ are the density ratio and kinetic luminosity of the jet, respectively.

5. TOWARD SELF-CONSISTENT RADIO SOURCE MODELS

It is almost self-evident that to understand extragalactic radio sources, we have to model them. The numerical simulations that have been done to date represent, at best, incomplete models of radio sources as none have adressed all of the observed parameters simultaneously. These are polarized and unpolarized intensity, E-vector position angle, and spectral index as function of frequency and position projected onto the sky. However, unless this level of contact with observations is made, we cannot hope to decide between the range of physically plausible interpretations of radio source morphology. To do this will require two things: 1) a model for the energetic evolution of the relativistic electron population coupled self-consistently to the underlying dynamical flow, and 2) numerical model mapping codes which translate from physical to observed variables, allowing for various telescope response characteristics (resolution, dynamic range) and viewing angle to be factored in. Both tasks represent considerable code development efforts. By virtue of the MHD codes that are now being developed, we are well on our way to achieving 1), although our current lack of understanding of relevant particle acceleration mechanisms constitute a roadblock. Perhaps an inferential approach will be required for progress here, much as the current models are being used to infer radio sources' underlying dynamical laws.

The inclusion of magnetic fields into the dynamical model is an essential first step toward self-consistent radio source models since now the entire class of observational features related to polarization and field orientation can be addressed. Even if the fields are passive, one is able, for the first time, to evaluate their contribution to the synchrotron emissivity formula directly, rather than through ad hoc assumptions. As an example of such features, Fig. 2 presents a simulated radio surface brightness distribution of the late stages of evolution of the unstable jet in Fig. 1. revealing considerable filamentary structure. The emissivity is taken to be the product of the square of the local magnetic field strength and the gas pressure. The magnetic field is initially longitudinal and uniform in strength, and is chosen to be dynamically unimportant. The bright filaments track the magnetic field, which is highly tangled and nonuniforn due to flux winding and stretching. Polarized filaments are observed in highly resolved radio lobes, such as in Cygnus A[18).

<u>REFERENCES</u>

1. Blandford, R.D. and Rees, M.J., M.N.R.A.S., <u>169</u>, 395 (1974).
2. Scheuer, P.A.G., M.N.R.A.S., <u>166</u>, 513 (1974).
3. Bridle, A.H. and Perley, R.A., Ann.Rev.A., <u>22</u>, 319 (1984).
4. Burns, J.O., Clarke, D.A. and Norman, M.L., these proceedings.
5. Bridle, A.H., Astron. J., <u>89</u>, 979 (1984).
6. Bicknell, G.V., Ap. J., <u>286</u>, 68 (1984).
7. Norman, M.L., in *Radiation Hydrodynamics in Stars and Compact Objects*, ed. D. Mihalas, (Springer:New York) (1986).
8. Hardee, P.E., Astrophys. J., <u>287</u>, 523 (1984).
9. Hardee, P.E., Astrophys. J., <u>303</u>, 111 (1986).
10. Woodward, P.R., in *Astrophysical Radiation Hydrodynamics*, eds. K.-H. Winkler and M.L. Norman, (Reidel:Dordrecht), (1986).
11. Hardee, P.E., Norman, M.L. and Clarke, D.A., in preparation.
12. Ferrari, A., Trussoni, E. and Zaninetti, L., M.N.R.A.S.,<u>196</u>, 1051.
13. Potash, R.I. and Wardle, J.F.C., 1980,Ap. J., <u>239</u>, 42 (1980).
14. Clarke, D., Norman, M. and Burns, J., Ap. J Lett. <u>311</u>, L63 (1986).
15. Lind, K., Ph.D. dissertation, Caltech (1986).
16. Norman, M.L. and Burns, J.O. in preparation.
17. Begelman, M.C. and Cioffi, D. in preparation.
18. Perley, R., Dreher, J. and Cowan, J., Ap. J. Lett <u>285</u>, L35 (1984).

Computer Simulations of Relativistic Star Clusters: The Movie

STUART L. SHAPIRO AND SAUL A. TEUKOLSKY
Center for Radiophysics and Space Research, and
Departments of Astronomy and Physics,
Cornell University, Ithaca NY 14853.

Abstract

We have constructed a new numerical code which solves Einstein's equations for the dynamical evolution of a *collisionless* gas of particles in general relativity. Our initial investigation is restricted to spherically symmetric systems, but the gravitational field can be arbitrarily strong and particle velocities can be arbitrarily close to the speed of light. Our computational scheme combines the tools of numerical relativity with those of N-body particle simulation. We solve the Vlasov equation in general relativity by particle simulation and determine the gravitational field using the ADM $3+1$ formalism. Physical applications include the stability of relativistic star clusters, the binding energy criterion for stability, the collapse of star clusters to black holes, and relativistic violent relaxation. Astrophysical applications include the possible origin of quasars and active galactic nuclei via the collapse of dense star clusters to supermassive black holes. We find that our method is extremely accurate, even in the case of black hole formation. It provides a unique proving ground for testing different computational algorithms and gauge choices for the construction of numerical spacetimes.

I. Introduction

We have recently shown that it is possible to integrate the full Einstein equations for the dynamical motion of an arbitrary spherical, *collisionless* configuration in general relativity (Shapiro and Teukolsky 1985a, b, c, 1986, hereafter Papers I, II, III and IV, respectively). We can track the dynamical evolution of a relativistic star cluster on the computer by combining the techniques of numerical relativity with those of N-body particle simulations. Our resulting method is even able to handle epochs characterized by total gravitational collapse leading to the formation of a black hole. Our general relativistic Vlasov integrations follow the formation and growth of the black hole accurately without the appearance of numerical or physical singularities.

In Paper I we presented some historical background to this problem, together with a detailed discussion of the computational method and a battery of test-bed calculations that help assess the reliability of our code. In Paper II we applied the code to address several longstanding theoretical issues in relativistic stellar dynamics, including the stability of relativistic star clusters in dynamical equilibrium, the collapse of unstable star clusters to black holes, and relativistic "violent relaxation". Based on our findings, we proposed in Paper III a plausible scenario for the origin of quasars and active galactic nuclei (AGNs) via the collapse of dense star clusters embedded in galactic nuclei to supermassive black holes. In Paper IV we explored alternative coordinate choices, ideally suited for evolving the very centrally condensed clusters most likely to form in Nature.

II. The Movie

These computations are an example of a problem that requires extensive use of graphical display to visualize the dynamical behavior predicted by the simulation. We have recently produced a computer-generated color movie that dramatically illustrates the collapse of unstable clusters to black holes.

The Einstein equations for this process were solved on the Cornell University Supercomputer (an assembly of Floating Point Systems' FPS 264 Array Processors hosted by an IBM 3090-400). The computer tracked the motion of a representative sample of stars in the cluster, and the positions of these stars at successive times are portrayed on color film. The film itself lasts 8 minutes and was made on a Cray XMP at Digital Productions in Los Angeles.

The movie consists of 6 scenes. At the beginning of Scene 1, the stars orbit about their mutual center at speeds close to the speed of light with a Maxwell-Boltzmann distribution. Their motion exactly counterbalances the inward pull of their combined gravity. If Newton's theory of gravity were correct, the cluster would remain in this equilibrium state forever. However, gravity is stronger in General Relativity and the cluster is unstable to catastrophic collapse. As time advances, the stellar orbits are seen to spiral inward toward the center. At the very center the concentration of mass becomes so great that a black hole forms. During the collapse, the computer sends out light rays from the center to determine when this occurs. Before the black hole forms, the light rays have no problem traveling outward forever, escaping from the cluster entirely. However, once the black hole forms the light rays are permanently trapped within. All of these features are portrayed in the movie.

Scene 2 is a cartoon to illustrate that time is slowed down by the strong gravitational field that develops near a black hole. In this scene clocks that are far from the black hole are seen to advance more rapidly than clocks close up. The lapse function α gives the ratio of time measured by a local clock to time measured very far away.

Scene 3 shows how α plunges to zero in the center of the cluster when the black hole forms.

Scene 4 is a comparison of the same Maxwell-Boltzmann collapse as seen by two different sets of observers. For one set ("maximal"), clocks advance further than for the other set ("polar"). Hence the collapse proceeds further for the maximal observers, and the stars get closer to the singularity at the center of the black hole. The singularity is a region of infinite density and infinite spacetime curvature caused by the pile-up of matter at the center. All the presently-known laws of physics break down at a singularity. One of the goals of numerical relativity is to study black hole formation on the computer without encountering the spacetime singularity, otherwise the computer code will crash before the evolution is complete. For both maximal and polar observers near the center, time slows down sufficiently as the gravitational field gets stronger that neither set gets crushed by the singularity. However, polar observers avoid the singularity to a greater extent, as expected from theoretical considerations.

Most astrophysicists believe that quasars (and Active Galactic Nuclei) are powered by supermassive black holes, with a mass more than $10^8 M_\odot$. Scene 5 demonstrates how such a black hole can form in a galaxy whose core is made largely of neutron stars. It may take up to 10^{10} years for a galaxy to reach the point at which its core is sufficiently relativistic for catastrophic collapse. Once it reaches this point, the implosion takes only a few days. It is this implosion that is depicted in the film for an $n = 4$ relativistic polytrope. When the collapse is complete, the black hole contains 5% of the cluster mass. The rest of the stars continue to orbit about the black hole in dynamical equilibrium.

Scene 6 shows 4 typical orbits of stars near the black hole of scene 5. The first star swings around the center on a nearly elliptical orbit. By its second orbit the hole has grown sufficiently that it swallows the star. The second star orbits as close as it can (about 2 Schwarzschild radii) without actually being captured, undergoing a very large perihelion precession.

The third star moves on an almost circular orbit, which is unperturbed by the formation of the black hole (Birkhoff's theorem). The last star starts far from the center, but happens to be on a very radial orbit and plunges into the black hole.

ACKNOWLEDGEMENTS

This work has been supported in part by National Science Foundation grants AST 84-15162 and PHY 83-05288 at Cornell University. Computations supporting the research were performed on the Cornell National Supercomputer Facility which is supported in part by the National Science Foundation and IBM corporation.

References

Shapiro, S. L., and Teukolsky, S. A. (1985*a*). *Astrophys. J.*, **298**, 34 (Paper I).
———. (1985*b*). *Astrophys. J.*, **298**, 58 (Paper II).
———. (1985*c*). *Astrophys. J. (Letters)*, **292**, L41 (Paper III).
———. (1986). *Astrophys. J.*, **307**, 575 (Paper IV).

GRAVITATIONAL RADIATION FROM COLLISIONS OF COMPACT STARS

Charles R. Evans
Theoretical Astrophysics 130 - 33
California Institute of Technology
Pasadena, CA 91125

I. INTRODUCTION

The effects of strong-field gravity on the generation of gravitational radiation are being investigated by modeling a sequence of collisions of relativistic compact stars. The sequence ranges from the Newtonian limit, $M/R \to 0$, to stars with the maximum stable mass, giving $M/R \sim 0.21$ (in units with $G=c=1$). A two-dimensional finite difference code is employed[1)] which self-consistently solves the equations of general relativity and relativistic hydrodynamics for axisymmetric configurations. Through a modification of the code, analogous Newtonian collisions are calculated. These form the sequence baseline ($M/R=0$) and we give some results here. Collisions in the present study are restricted to zero impact parameter. Such collisions are likely to be exceedingly rare, though some may occur in the cores of massive, dense, highly evolved star clusters. From a gravitational physics standpoint however, collisions represent an important model problem in which to study gravitational radiation emission from a violent, strong-field source. In addition, it might be hoped that the detailed results and intuition developed from this study could provide some insight to the problem of the final coalescence of compact binaries. Strong fields may be expected to significantly affect the emission of gravitational radiation during the final stages in the coalescence of binaries such as PSR 1913+16.

In the next few years it will be increasingly important to produce precise numerical models for potential sources of gravitational radiation and provide a survey of possible waveforms to be seen in planned detectors. Not only are planned detector sensitivities[2)] reaching levels where sources may confidently be seen, but rapid advances in supercomputer technology are making possible detailed modeling of such sources with numerical relativity techniques. It is widely believed that the strongest emitters of gravitational radiation, and those perhaps most likely to be first detected, are coalescing compact binaries such as binary neutron stars or black hole binaries.[2)] Hence realistic astrophysical sources (an exception would be axisymmetric rotating stellar collapse associated with type-II supernovae) involve full three dimensional (3D), asymmetric dynamics (i.e., three nontrivial spatial dimensions plus time). It is this, plus bulk velocities close to that of light and strong gravitational fields, that is required for copious gravitational radiation emission.

It is currently possible to produce high resolution numerical simulations in 2D. 3D calculations using presently available resources are possible, but much cruder and will require considerable computing costs. In pursuing 3D predictions, a modeler faces a choice: utilize a simpler calculation employing Newtonian gravity, with use of the quadrupole formula for gravity waves, or make a more self-consistent, but vastly more complicated calculation of the full general relativistic field (i.e., the techniques of numerical relativity). The most efficient emitters of gravitational waves, and ones of most astrophysical interest, are those that have strong internal fields and high velocities, precisely where the results should differ most. If the field is sufficiently strong a black hole forms and there is no question; numerical relativity must be used. In other circumstances it is clearly important to know to what level of accuracy a simpler solution will suffice. In the fully nonlinear regime only a direct comparison between two calculations can quantitatively settle this question.

Direct collisions between stars have been investigated by Seidl and Cameron[3], who considered collisions of main sequence stars, by Smarr and Wilson[4], who modeled a relativistic collision of $\gamma=2$ neutron stars, and by Gilden and Shapiro[5], who treated the Newtonian collisions of $\gamma=2$ and $\gamma=5/3$ polytropes. In multidimensional simulations it is important, though rare, to have direct comparisons between two different numerical methods applied to the same problem. In the Newtonian case, a comparison with the present effort is provided by the previous work of Gilden and Shapiro.[5]

II. NUMERICAL METHOD

The numerical method is a two dimensional Eulerian finite difference scheme which utilizes an adaptive mesh and explicit time integration.[1] The gravitational field equations are split into first-order, Hamiltonian form using the ADM formalism.[6] The maximal time slicing condition and quasi-isotropic spatial gauge are used to fix the spacetime coordinates. A numerical regularization technique has been developed to handle finite differencing near the origin of coordinates. The relativistic hydrodynamic equations are written in a form closely resembling Eulerian Newtonian hydrodynamics. A second-order monotonicity transport scheme is used.

Accurate simulation of hydrodynamic phenomena is crucial for the correct calculation of gravitational radiation produced in the collision. The code must be able to model correctly density gradients, shocks and rarefactions, adiabatic compression and expansion, and circulation. All of these effects play a role in producing a complex integrated signal from the collision and coalescence. To simulate these hydrodynamic effects accurately requires sufficient zoning to resolve details in the core. At the same time, in the full relativistic calculation, the mesh must extend out to the wave zone in order to yield accurately the emitted radiation. In these initial Newtonian hydrodynamic simulations meshes with 150 radial zones and 90 angular zones over a quadrant were employed.

III. COLLISION DYNAMICS

In all these calculations a polytropic equation of state is utilized (with $\gamma=2$). The idealized nature of the collisions (zero impact parameter) precludes considering more detailed microphysics. Indeed there is a compelling reason to use a polytropic relation since with it the Newtonian equations are scale free. In principle, for a free-fall collision one calculation (for a given γ) yields the waveform for stars of different masses and radii

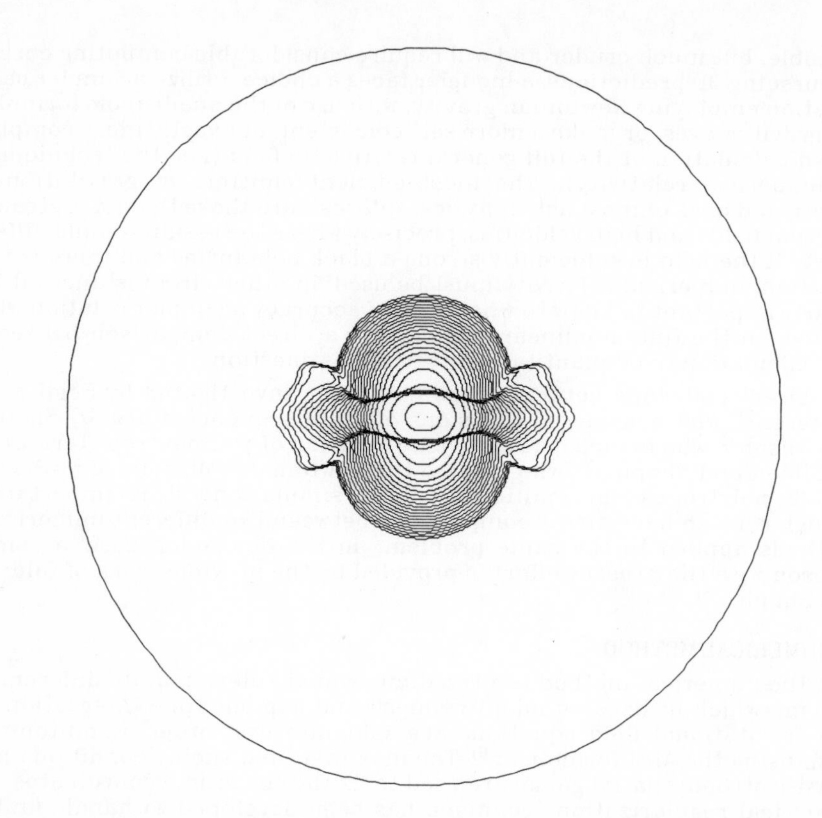

Figure 1. Snapshot of collision depicted as contours of pressure as stellar centers are passing through the shock fronts. Collision axis is vertical. Shock is evident erupting from the surface at the equatorial plane. Circle indicates radius of outer boundary.

up to scalings of the amplitude and time coordinate. When relativity is included the scaling is broken by the addition of the relativity parameter M/R.

The initial Newtonian calculations can be compared to previous work by Gilden and Shapiro.[5] The most important comparison is between the gravitational radiation waveforms produced by the collision and subsequent coalescence. The waveform is a complex product of the process of stellar merger and post-collision hydrodynamics. Part of the collision dynamics can be seen in the first two figures. The two stars merge by passing through a strong shock wave (see Figure 1). This strong recoil shock ejects between 5 and 10 percent of the mass of the system. The remaining 90 to 95 percent of the material is bound. This strong shock and ejecta can be seen in Figure 2. Also evident is the merged, hot, oblate central region undergoing a period of reexpansion. This is followed by recollapse of the core, which then undergoes large amplitude oscillations and fluid circulation. The motion of the core is damped due to acoustic energy

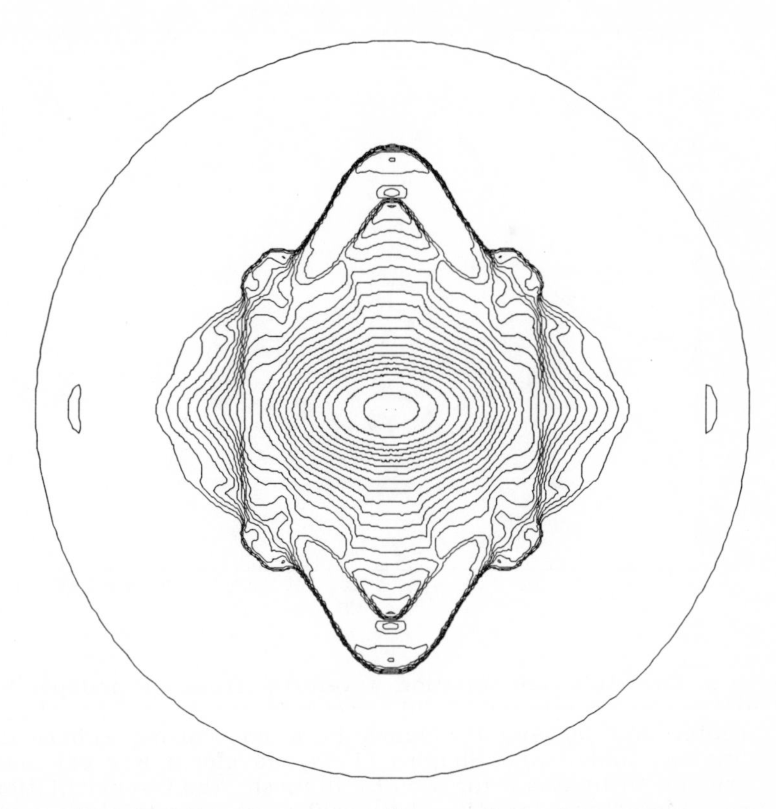

Figure 2. Shortly after the merger, the recoil shocks can be seen moving outward with shock ejected material trailing behind. Some 5 to 10 percent of the outer layers of the stars are ejected from the system. The denser central region, now oblate, is seen reexpanding. This material subsequently recollapses, undergoes large amplitude oscillations and fluid circulation.

losses to the hot, newly-formed atmosphere (no other dissipation mechanism is included).

The waveforms from the present study and from Gilden and Shapiro[5] are shown in Figure 3. There are broad elements of agreement between the two calculations despite having employed very different numerical schemes and coordinate systems. Work is in progress on extending the calculation away from the Newtonian limit.

ACKNOWLEDGEMENTS

Numerical computations were performed on the Cray X-MP at the National Center for Supercomputing Applications, University of Illinois, Urbana, Illinois. This research is supported in part by NSF grants PHY83-08826 and AST85-14911.

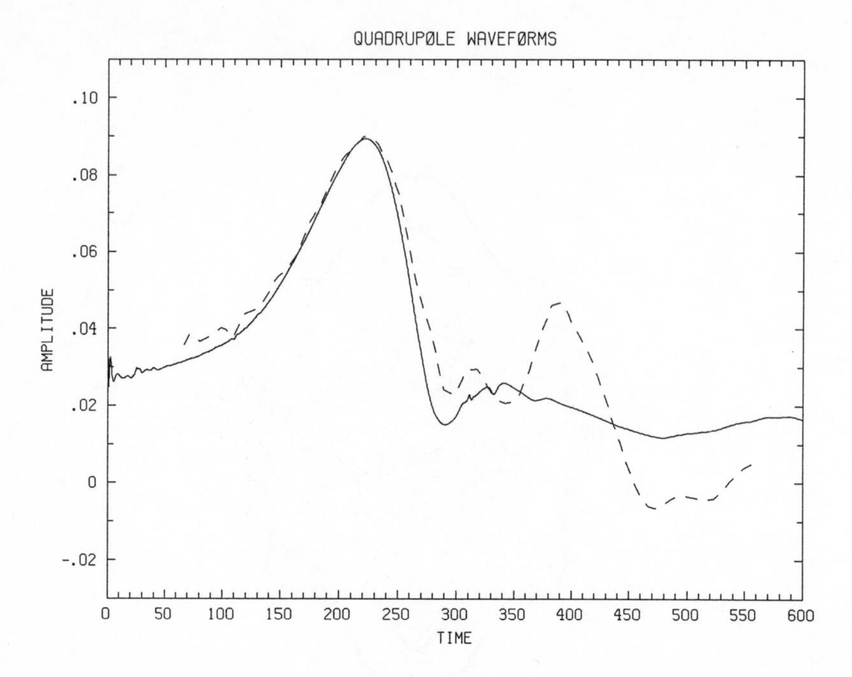

Figure 3. Gravitational radiation waveforms from two independent 2D calculations are compared. Solid curve is from recent calculation using 150 radial and 90 angular zones in a code using spherical-polar coordinates. Gilden and Shapiro (1984) waveform was calculated in cylindrical coordinates using a 50 by 40 mesh. The two calculations are in firm agreement for the main pulse of gravitational radiation (produced during the initial merging; see Figure 1) but differ at later times associated with the reexpansion and recollapse phases.

REFERENCES

[1] Evans, C. R. "An Approach for Calculating Axisymmetric Gravitational Collapse." *In*: Dynamical Spacetimes and Numerical Relativity, ed. J. Centrella. Cambridge University Press: Cambridge (1986).

[2] Thorne, K. S. "Gravitational Radiation." *To appear in*: Newtonian Principia Tercentenary, ed. S. W. Hawking and W. Israel. Cambridge University Press: Cambridge (1987).

[3] Seidl, F. G. P. and Cameron, A. G. W. Ap. Space Sci., 15, 44 (1972).

[4] Wilson, J. R. "A Numerical Method for Relativistic Hydrodynamics." *In*: Sources of Gravitational Radiation, ed. L. Smarr. Cambridge University Press: Cambridge (1979).

[5] Gilden, D. L. and Shapiro, S. L. Ap. J. 287, 728 (1984).

[6] Arnowitt, R., Deser, S. and Misner, C. W. "The Dynamics of General Relativity." *In*: Gravitation: An Introduction to Current Research, ed. L. Witten. John Wiley: New York (1962).

3D NUMERICAL RELATIVITY

Takashi NAKAMURA
Department of Physics, Kyoto University
Kyoto JAPAN

ABSTRACT

A method for solving initial data and dynamical equations of 3-geometry in 3D numerical relativity is given. Numerical results on 3D time evolution of pure gravitational waves are also presented.

1. 3D TIME EVOLUTION OF PURE GRAVITATIONAL WAVES

From the results of 2D numerical relativity, phase cancellation effects and gravitational radiation emitted by a particle falling into a Kerr or Schwarzschild black hole[1)], 3D processes should be necessary for a strong emitter of gravitational radiation. In fact Miyama, Nagasawa and Nakamura[2)] showed in 3D processes such as fragmentation the efficiency of emission of gravitational radiation increases at least factor 10 by using the Landau-Lifshitz formula to estimate the energy. It is now desirable to construct a fully general relativistic code to answer the questions such as 1) What is the final structure of space time after the collapse in general? 2) What kind of information can we extract from the gravitational radiation emitted during the collapse? The answers to the second questions are urgent because it is expected that the gravitational waves from the collapse will be detected by bars and/or beam detectors in early 1990s[3)].

Fortunately the recent speed of super computers is fast enough to construct such a code. As a first step of 3D code in numerical relativity, I will show some numerical results of 3D time evolution of pure gravitational radiation. Namely we consider only metric and neglect matters degree of freedom.

We adopt the (3+1)-formalism of the Einstein equations. As for the coordinate condition we adopt ;

$$\alpha=1 \text{ and } \beta^i=0$$

where α is a lapse function and β^i is a shift vector. Then the basic equations in the (3+1)-formalism become ;

$$^{(3)}R+K^2=K_{ij}\, K^{ij}, \quad (1) \qquad K_i{}^j{}_{;j} = K_{;i}, \qquad (2)$$

$$(\partial/\partial t)\, K_{ij} = {}^{(3)}R_{ij} + K\, K_{ij} - 2K_i{}^m K_{mj}, \qquad (3)$$

$$(\partial/\partial t)\, \gamma_{ij} = -2K_{ij}, \qquad (4)$$

where $K=TR(K_{ij})$ and ; is a covariant derivative with respect to γ_{ij}.

We shall write down the Einstein equations in (x,y,z) coordinate. Then γ_{xx}, γ_{yy},etc are the basic variables. Let us assume every quantity Q in (x,y,z) has an Taylor expansion as,

$$Q=\sum_{p,q,r}^{\infty} a_{pqr}(t)x^p y^q z^r/p!q!r!. \qquad (5)$$

Then it is very easy to show that Eq.(5) can be reexpressed as,

$$Q=\sum_{l,m} r^l Q_{lm}(r^2,t) Y_{lm}(\theta,\phi). \qquad (6)$$

We need $(\partial/\partial z)Q$, $(\partial^2/\partial^2 z)Q$ etc to calculate R_{ij} in r.h.s. of Eq.(3). As $(\partial/\partial z)Q$, $(\partial^2/\partial^2 z)Q$ etc should be expressed like Eq.(5), they have also the form as Eq.(6), that is, for example ,

$$(\partial/\partial z)Q=\sum_{l,m} r^l (\partial/\partial z)Q_{lm}(r^2) Y_{lm}(\theta,\phi). \qquad (7)$$

It is easy to determine the relation between $(\partial/\partial z)Q_{lm}(r^2)$ and $Q_{lm}(r^2)$ as,

$$(\partial/\partial z)Q_{lm}(r^2) = C1z(l,m)\partial/\partial w Q_{l-1\ m}$$

$$+C2z(l,m)(2w\partial/\partial w Q_{l+1\ m}+(2l+3)Q_{l+1\ m}), \qquad (8)$$

where

$C1z(l,m)=sqrt((l+m)(l-m)/(2l+1)/(2l-1))$,

$C2z(l,m)=sqrt((l+m+1)(l-m+1)/(2l+3)/(2l+1))$,

and

$$w=r^2.$$

We have similar formula for $(\partial/\partial xQ)_{lm}(r^2)$ and $(\partial/\partial yQ)_{lm}(r^2)$. However for the second derivatives of Q, formulae become complicated. Therefore we use REDUCE to derive the FORTRAN source program.

We expand K_{ij} and γ_{ij} as Eq.(6). Then Eq.(4) has a simple form for (l m) component. However as Eq.(3) is nonlinear, we solve this equation as follows

$$(\partial/\partial t)K_{ij}{}^{lm}(r^2,t)=\int({}^{(3)}R_{ij}+K\,K_{ij}-2K_i{}^nK_{nj})\;Y_{lm}d\Omega/r^l, \qquad (9)$$

In Eq.(9) we use 41 points Gauss quadrature in θ-direction and 16 points Discrete Fourier Transform for ϕ-direction. The number of grids in r-direction is 100 with $r_{max}=7r_0$. For each r, m ranges from -8 to 8 and l does from $|m|$ to $|m|+10$.

For weak amplitude case the numerical solution should essentially agree with the linearized one. In Fig.1, we show time evolution of $\gamma_{xx}-1$ in the equatorial plane for l=2 and m=2 localized initial wave packet. At early steps, we can see two peaks for constant r, which is the characteristics of l=2 and m=2 waves. The wave propagates outwards and finally leaves from the numerical boundaries without showing any artificial reflection of the waves. Computing time for this evolution takes three hours for 700 time steps by using FACOM VP200 which is a Japanese super computer with 400MFLPS.

Now we compare the numerical result with that derived from the linearized solution which is given in ref 1). For comparison we should express the solution in (x,y,z) coordinate. Moreover we need Y_{lm} component of the linearized solution (Q_{lm}) because it is our basic variable in numerical simulations. This process is very tedious and complicated. However if we use REDUCE, for example, it is very easy to obtain FORTRAN list. It is found that the relative error(or difference) between calculated Q_{lm} and that of the linearized solution is at most a few percent up to the time when the wave leaves from the numerical boundary. As we take into account the nonlinear terms in the numerical calculations, we can expect a few percent difference(error) for $A=10^{-2}$ where A is the amplitude of the wave. The difference of ADM energy flux between the calculated and linearized solutions is found to be at most a few percent. This suggests us a possibility of estimating the energy of the gravitational waves from the numerical results within an error of a few percent even in 3D numerical relativity. We have also done the time evolution of rather high amplitude waves with A=0.2. In this case, by the effects of nonlinearity an central peak with propagating waves near the numerical boundaries can be seen.

As is shown in Appendix of ref 1), the basic equations in 3D numerical relativity is very complicated. So use of algebraic computing software such as REDUCE and MACSYMA is indispensable in constructing codes as well as checking and analyzing numerical results as shown in this paper.

This work was in part supported by the Grant-in-Aid for Scientific Research of Ministry of Education, Science and Culture (61740206).

References

1) Nakamura, T. (1986) "Gravitational radiation and 3D numerical relativity", in H.Sato and T. Nakamura(eds) "Proceedings of the 14-th Yamada Conference on gravitational collapse and relativity" World Scientific Publishing Co, Singapore.

2) Miyama, S.M.(1986) Nagasawa, M. and Nakamura, T. "Gravitational radiation in the 3D processes", in H.Sato and T. Nakamura(eds) "Proceedings of the 14-th Yamada Conference on gravitational collapse and relativity" World Scientific Publishing Co, Singapore.

3) Thorne, K.S. "Gravitational Radiation" This Volume.

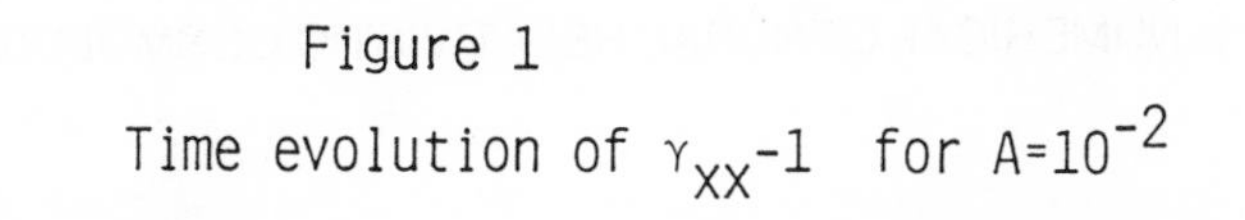

Figure 1

Time evolution of $\gamma_{xx}-1$ for $A=10^{-2}$

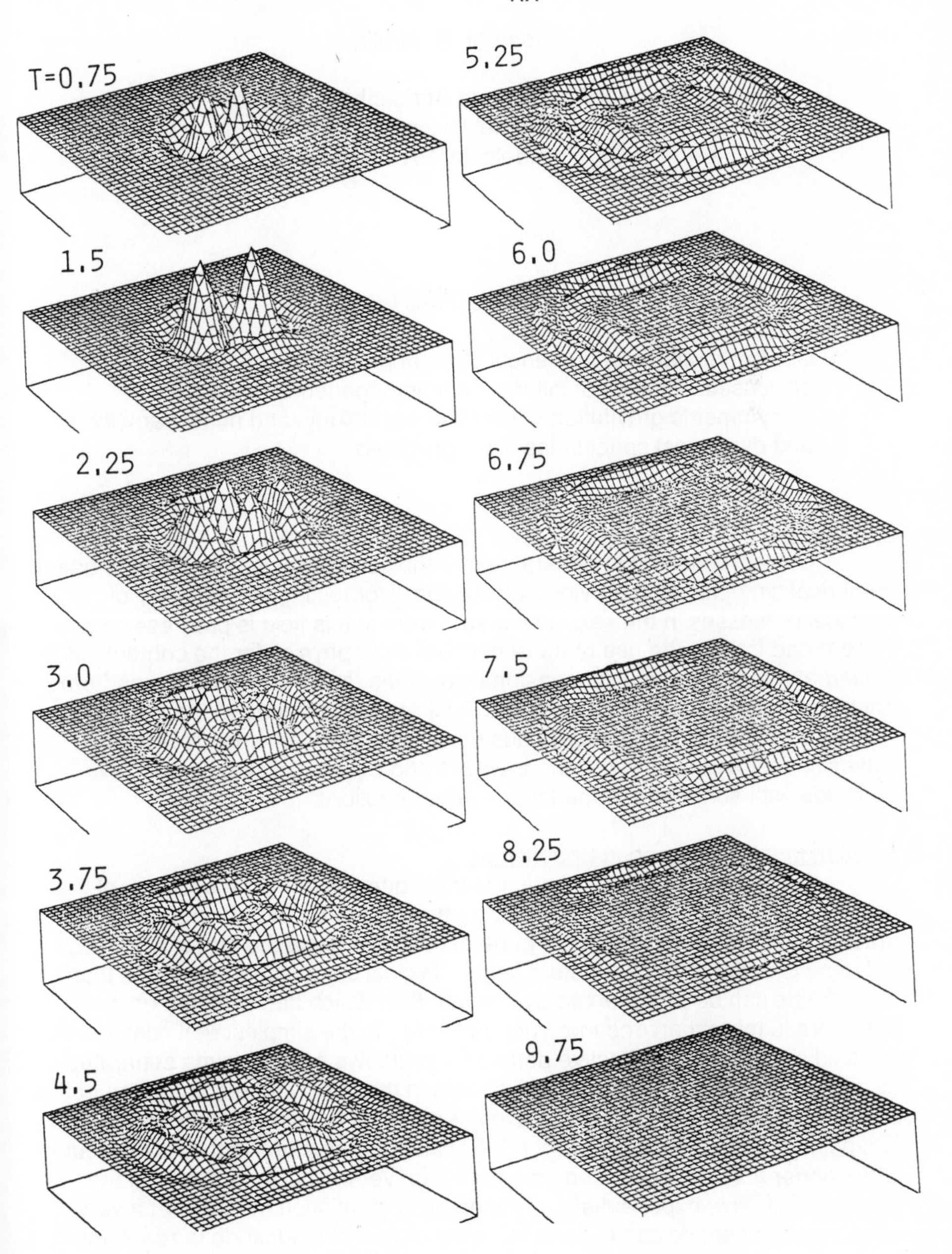

NUMERICAL GENERAL RELATIVISTIC COSMOLOGY

Joan M. Centrella

Dept. of Physics and Atmospheric Science
Drexel University
Philadelphia, PA 19104

ABSTRACT

Current research in numerical general relativistic cosmology is discussed. Studies of inflation in inhomogeneous universes, axisymmetric gravitational waves in cosmology, and nonlinear waves and dynamical spacetimes are highlighted.

Introduction

Current research in general relativistic cosmology is fueled by the use of numerical simulations to provide "laboratories" for testing the behavior of physical processes in the early universe. Work in this field is progressing along three broad fronts: the use of more detailed input physics for the contents of the universe, the study of multidimensional solutions, and the influx of ideas from nonlinear dynamics. In this paper, we illustrate these trends with a look at models of inflation in inhomogeneous universes, axisymmetric gravitational waves in cosmology, and nonlinear waves and dynamical spacetimes. We conclude with some brief remarks on future directions.

Inflation in Inhomogeneous Spacetimes

According to current theory, the inflationary era[1,2] in the history of the universe begins at about $t \sim 10^{-35}$ s after the big bang. At that time, the gravitational field obeys classical general relativity, while the rest of the universe is in the form of quantum fields. Typical computations assume that these fields can be represented by a scalar field which has symmetries such that there is more than one true vacuum state. In the simplest real scalar case, the field possesses an effective potential having two equal minima separated by a potential barrier. The field can remain in the false vacuum at the top of this barrier since there are no gradients in the potential to push it off; however, this is a point of unstable equilibrium. If the scalar field remains in the false vacuum state longer than an expansion time for the universe, the scale factor of the universe will grow exponentially. This process of inflation means that a very small region of space can expand by many orders of magnitude to reach the size of the observable universe. Regions with strong inhomogeneities can inflate so that their gradients are small, thus producing the approximately uniform cosmos we observe today.

The initial work on the inflationary model was done within the framework of a spatially homogeneous cosmology. This leaves open the crucial question of whether inhomogeneities at t ~ 10^{-35} s can prevent inflation.

Kurki-Suonio, Matzner, Centrella and Wilson[3,4] are investigating this issue using numerical relativity in planar symmetry. Beginning with the code developed by Centrella and Wilson to solve the equations of general relativity coupled to hydrodynamics, they added a scalar field coupled gravitationally to the rest of the universe. Their work so far has concentrated on the effects of strong inhomogeneities in the fluid source. They find that such inhomogeneities tend to make inflating more difficult. Nevertheless, inflation is a robust idea in the sense that if the scalar field can be kept at the false vacuum long enough by tuning the field theory parameters to produce a flatter potential, the inhomogeneous models will inflate. The next step in their work will consider the effects of large shear inhomogeneities.

An alternative approach to the study of inflationary models is being carried out by Piran and collaborators using Regge calculus in vacuum models[5].

Axisymmetric Gravitational Waves

Another aspect of general relativistic cosmology is the study of waves in more than one spatial dimension. Holcomb and Matzner[6,7] have undertaken a project to study the behavior of axisymmetric cosmological models numerically, A major physical motivation for this effort is to examine the behavior of collapsing objects in an expanding, shearing universe, and to determine the circumstances under which black holes will form.

A number of interesting technical issues have arisen. Their 2-D numerical relativity code is a descendant of the one developed by Evans and collaborators[8] for the study of asymptotically flat spacetimes. In the cosmological case, the background universe itself has curvature, and so the boundary conditions must be evolved. In addition, the meaning of a minimal shear gauge needs to be re-evaluated in the case of a model with background shear.

Holcomb and Matzner are currently looking at the implosion of quadrupolar gravitational waves in a Kasner background. Work continues to make this code as accurate as those used in the study of asymptotically flat spacetimes.

Ove[9] is also carrying out numerical studies of 2-D cosmological models, with the aim of shedding light on the question of cosmic censorship.

Dynamical Spacetimes and Nonlinear Gravitational Waves

A deeper understanding of cosmology near the big bang requires an understanding of the basic physics of general relativity in the nonlinear regime. The initial conditions of the universe, even at the end of the quantum gravity era, are unknown. To explore the early universe we must be able to study strong dynamical gravitational fields.

Recent years have seen a trend toward exploring the nonlinear regime of dynamical general relativistic systems, stimulated by ideas and techniques from nonlinear dynamics[10,11]. This work is part of the growing recognition of the importance of nonlinear phenomena in many branches of physics. We note that numerical relativity has produced a number of spacetimes that are dynamical in the sense that gravitational waves are generated: the collision of two black holes[12], the collision of two neutron stars[13,14], and rotating collapse[15]. However, these spacetimes are only weakly dynamical in the sense that the background is flat and therefore not dynamical itself. The dynamics comes only from the motions of the sources and the curvature effects they produce. In contrast, cosmological spacetimes can have backgrounds that are highly dynamical themselves, a situation that can lead to a variety of interesting physical effects.

One aspect of the influence of nonlinear dynamics on general relativity is the study of the "generic" singularity in the Einstein equations by Belinsky, Lifshitz and Khalatnikov (BKL)[16]. By studying the behavior of the general solution to the Einstein equations near the initial singularity, they found that all observers falling into the singularity see the same pattern of evolution. This pattern is the same as in the spatially homogeneous Mixmaster model[17] and can be understood in terms of the tidal acceleration along principal axes. As an observer falls into the singularity there is an infinite series of cycles, in which one axis compresses and the other two axes oscillate between stretching and compression. These cycles are grouped into eras, with a change of era marked by a change in the axis which undergoes compression. This is a highly dynamical spacetime. It exhibits chaotic behavior in the sense that slightly different initial conditions lead to arbitrarily different states later[18].

Another means of studying nonlinear behavior in general relativity is to look at "wave-like" phenomena, which we characterize as traveling signals varying in both space and time. One category of such phenomena are the "soliton" solutions[19-21]. We note that although these solutions are obtained using techniques that yield solitons in other physical systems, they differ from those solitons in that their amplitudes change as they propagate. A different means of studying nonlinear waves is being pursued by Centrella and Matzner[22]. They are using numerical relativity to evolve standing plane waves on flat spacetime with constant mean curvature slices. Beginning with the linearized solution, they gradually increase the amplitude of the waves on the initial slice. The solution evolves away from the linear solution with a change in both amplitude and phase. This behavior is being modeled analytically with a second order perturbative solution.

Future Directions

Future studies of numerical general relativistic cosmologies promise to be both exciting and demanding. Our growing ability to handle multidimensional spacetimes, coupled with the continued cross-fertilization of ideas from field theory, will lead to studies of the effects of inflation in models with more degrees of freedom. Concepts such as mode-mode coupling, dispersion, and the tendency to steepen and shock in gravitational waves will be studied through the synergistic use of numerical simulations and analytic modeling. Eventually these phenomena will be considered in the arena of dynamical spacetimes like the BKL solution. Such investigations of the early universe will hopefully lead to a deeper understanding of both the basic physics of general relativity in the nonlinear regime, as well as which classes of initial conditions lead to the universe we observe today.

Acknowledgments

I would like to thank my colleagues and collaborators R. Matzner, H. Kurki-Suonio, C. Evans, J. Wilson, and K. Holcomb for many stimulating discussions. I am also indebted to R. Matzner for many useful and interesting comments on the manuscript. I am pleased to acknowledge support from NSF grants PHY84-17918 and PHY84-51732, Drexel University, Lawrence Livermore National Laboratory, Cray Research, RCA, and Sun Microsystems.

References

1. Guth, A., Phys. Rev. D23, 347 (1981).

2. Linde, A., Phys. Lett. 108B, 389 (1982).

3. Kurki-Suonio, H., Centrella, J., Matzner, R., and Wilson, J., Phys. Rev. D, in press (1987).

4. ----------, preprint (1987).

5. Piran, T., private communication (1986).

6. Holcomb, K., Ph.D. Dissertation, University of Texas at Austin, unpublished (1986).

7. ----------, in Dynamical Spacetimes and Numerical Relativity, ed. J. Centrella (Cambridge: Cambridge University Press, 1986), p. 187.

8. Evans, C., in Dynamical Spacetimes and Numerical Relativity, ed. J. Centrella (Cambridge: Cambridge University Press, 1986), p.3.

9. Ove, R., in Dynamical Spacetimes and Numerical Relativity, ed. J. Centrella (Cambridge: Cambridge University Press, 1986), p.201.

10. Thorne, K., in Nonlinear Phenomena in Physics, ed. F. Claro (New York: Springer Verlag, 1985), p. 280.

11. Barrow, J., Physics Reports 85, 1 (1982).

12. Smarr, L., in Sources of Gravitational Radiation, ed. L. Smarr (Cambridge: Cambridge University Press, 1979), p. 245.

13. Gilden, D. and Shapiro, S., Ap. J. 287, 728 (1984).

14. Evans, C., presentation at this meeting.

15. Piran, T. and Stark, R., in Dynamical Spacetimes and Numerical Relativity, ed. J. Centrella (Cambridge: Cambridge University Press, 1986), p.40.

16. Belinsky, V., Khalatnikov, I., and Lifshitz, E., Adv. Phys. 31, 639 (1982).

17. Misner, C., Phys. Rev. Lett. 22, 1071 (1969).

18. Khalatnikov, I., Lifshitz, E., Khanin, K., Shchur, L., and Sinai, Ya., in General Relativity and Gravitation, eds. B. Bertotti, F. deFelice, and A. Pascolini (Dordrecht: Reidel, 1984), p. 343.

19. Belinsky, V. and Zakharov, V., Sov. Phys. JETP 50, 1 (1980).

20. Carr, B. and Verdaguer, E., Phys. Rev. D28, 2995 (1983).

21. Ibañez, J. and Verdaguer, E., Phys. Rev. D31, 251 (1985).

22. Centrella, J., in Dynamical Spacetimes and Numerical Relativity, ed. J. Centrella (Cambridge: Cambridge University Press, 1986), p.123.

INFLATION AND THE COSMIC NO-HAIR THEOREM

Jaime A. Stein-Schabes*

Theoretical Astrophysics Group
Fermi National Accelerator Laboratory
Batavia, Illinois 60510

ABSTRACT

I review and extend the prescription for a successful new inflation paying particular attention to cases where the metric is that of an inhomogeneous and anisotropic space-time. I restate and sketch a proof for the No-Hair theorem and discuss some of the consequences for the dynamics of the scalar field and Inflation.

I. Introduction

Observations on the very large scales (100 Mpc) reveal a Universe almost featureless and flat. No dramatic differences are observed between two or more different regions in the sky that were in no causal contact in the distant past. This featureless nature of the Universe is not only present in the distribution of luminous matter but also in the microwave background radiation coming to us from the very early and hot universe. Naturally, we ask ourselves, why ? Why is the Universe around us so remarkably homogeneous and isotropic on large scales. This fascinating question has been in the minds of physicist for some time, and many different explanations have been given. I would only like to mention a few, which in my view represent the two antipodes of the issue. I would only elaborate on a few of these and will start with those that are more philosophical or even mystical in nature.

• *Initial conditions*: The initial conditions where such that the Universe started homogeneous, isotropic and flat and has remained like this ever since. This argument is perfectly consistent with observations of the Universe on the large scale, however it is then difficult to explain the existence of structure on smaller scales, like galaxies and cluster of galaxies. I should say that there are ways of introducing perturbations that will leave the Universes homogeneous and isotropic on large scales but produce suitable density perturbations on the small scale. How natural it is for this perturbations to be there is still a matter of debate.

• *Anthropic Principle*: This philosophical position has been most eloquently argued by Barrow and Tipler[2], and I would like to make two quotations, the first by Prof. Steven Hawking who captured the essence of the principle, *The fact that we have observed the Universe to be isotropic is only a consequence of our existence.* S. Hawking[3] while the second quotation comes from Collins and Hawking[4].

---

* The work reported is based on a collaboration with Lars Jensen [1]

-Since it would seem that the existence of galaxies is a necessary condition for the development of intelligent life, the answer to the question "Why is the Universe isotropic" is "because we are here." These argument are extremely difficult to overpower, however, one has the feeling that this is not enough. Something is missing. One would like to have a dynamical process by which the universe could have started in any arbitrary initial state and ended, as a consequence of this process, looking like our own. There have been at least two such proposals, the first was the *Chaotic Cosmological Program* proposed by Misner[5] where the initial singularity was thought to be of the Mixmaster Bianchi IX type. In this model the Universe started in an initial chaotic singularity that evolved through dissipative processes (mixmaster oscillations, neutrino viscosity, particle creation, etc) into our Universe. This model was soon dismissed for technical reasons[5]. The second proposal was the Inflationary Universe[6]. In this scenario the Universe undergoes a phase transition during which it supercools and expands exponentially fast, the expansion is driven by some vacuum energy which effectively behaves as a cosmological constant, the Universe finally returns to a hot radiation dominated phase due to some reheating process at the end of Inflation. The effect of this expansion is to smooth out any inhomogeneities and anisotropies that might have been present at the time (see Turner[7] for a review).

Inflation has been discussed at length in the literature so I will not go into any detail about the process, instead I will give a recipe concocted by P. Steinhardt and M. Turner[8] that will assure inflation takes place, at least when the background geometry is that of a homogeneous and isotropic Robertson-Walker (RW) model.

II. "A Prescription for Successful New Inflation"

The first ingredient for inflation is a very weakly coupled scalar (singlet) field ϕ which comes equipped with a potential $V(\phi)$ such that it has a flat region where the field can *roll slowly*. For concreteness I will write the potential has $V(\phi) = V_0 + V_1(\phi)$ with V_0 a constant or a very slowly varying function of ϕ (henceforth we will identify V_0 with the cosmological constant Λ) and such that for $\phi \in [0, \phi_1]$ it dominates the potential and $V(\phi) = V_1(\phi)$ for $\phi \geq \phi_1$. It is also necessary that $V(\phi)$ has an absolute stable minimum for some $\phi \geq \phi_1$. As we mentioned before this prescription was formulated for a RW background. What I would like to show now is that this prescription can be extended to encompass a much wider class of anisotropic and inhomogeneous cosmological models. This extension comes in the form of a theorem, the so called No-Hair theorem. For a detailed proof of the theorem in the homogeneous case we refer the reader to Wald[9] and in the inhomogeneous case to Jensen and Stein-Schabes[1] (other important references can be found in ref. [10].)

The theorem was first formulated by Gibbons and Hawking, and Hawking and Moss[11]. However, the original formulation was so general that the theorem was incorrect. The form proven[1] states the following,

The Cosmic No-Hair Theorem: Any *expanding* universe that is *not positively curved* ($^3R \equiv P \leq 0$), with a metric that can be written in a *synchronous form*, with a *positive cosmological constant* and an energy-momentum tensor satisfying the *Strong energy condition*, $T_{\mu\nu}t^\mu t^\nu \geq 0$ and $T_{\mu\nu}t^\nu$ is a non-spacelike vector, and the *Dominant energy condition*, $(T_{\mu\nu} - \frac{1}{2}g_{\mu\nu}T)t^\mu t^\nu \geq 0$, for all timelike t^ν, will approach asymptotically the de Sitter solution. For a perfect fluid these conditions become simply, $\rho \geq 0$, $\rho \geq |p|$ (all known forms of matter satisfies this condition) and $\rho + 3p \geq 0$ respectively.

Before sketching the proof I would like to stress out that if the constraint on the three-curvature is only satisfied locally, i.e. on a small patch of the manifold, then that small region will become de Sitter. In the case of homogeneous models the sign of the three-curvature does not change, so the theorem holds globally, however, in the more general inhomogeneous case, it only holds locally.

The connection between the No-Hair theorem and Inflation becomes obvious when the flat part of the potential is identified with the cosmological constant. The action is that of Gravity minimally coupled to a scalar field ϕ and some matter Lagrangian L_m

$$S = \int \sqrt{-g}\left(R - \frac{1}{2}g^{\mu\nu}\partial_\mu\phi\partial_\nu\phi - V(\phi) + L_m\right)d^4x \tag{2.1}$$

Where $g = det(g_{\mu\nu})$ and R is the Ricci scalar. Units are taken so that $8\pi G = c = 1$. By varying the action with respect to the dynamical fields the following equations are obtained,

$$R_{\mu\nu} - \frac{1}{2}g_{\mu\nu}R = T_{\mu\nu} - \frac{1}{2}\Lambda g_{\mu\nu} \tag{2.2}$$

$$\frac{1}{\sqrt{-g}}\partial_\mu\left(\sqrt{-g}\partial^\mu\phi\right) = -\frac{dV(\phi)}{d\phi} \tag{2.3}$$

with $T_{\mu\nu} = T_{\mu\nu(m)} + T_{\mu\nu(\phi)}$. The energy momentum tensor for the ϕ field is,

$$T_{\mu\nu(\phi)} = \frac{1}{2}\partial_\mu\phi\partial_\nu\phi - \left(\frac{1}{4}(\partial_\alpha\phi\partial^\alpha\phi) + \frac{1}{2}V_1(\phi)\right)g_{\mu\nu} \tag{2.4}$$

It is now clear that the problem has been reduced to finding what is the most general asymptotic solution of Einstein's equation (2.2). The No-Hair theorem tell us it is de Sitter for a large class of models.

Proof: Using the synchronous gauge we can write eq.(2.2) in components[12)], where a positive definite three-metric $h_{ab} \equiv -g_{ab}$ has been defined. Its time derivative will be denoted by $s_{ab} = \dot{h}_{ab}$,

$$R^0_0 = -\frac{1}{2}\dot{s}^a_a - \frac{1}{4}s^b_a s^a_b = T^0_0 - \frac{1}{2}T - \Lambda$$

$$R^0_a = \frac{1}{2}(s^b_{a;b} - s^b_{b;a}) = T^0_a \tag{2.5}$$

$$R^b_a = -P^b_a - \frac{1}{2\sqrt{h}}\frac{\partial}{\partial t}(\sqrt{h}s^b_a) = T^b_a - \frac{1}{2}\delta^b_a T - \delta^b_a\Lambda$$

P_{ab} is the three-Ricci tensor calculated using h_{ab} and $h = det\ h_{ab}$. ($\sqrt{h}$ can be interpreted as the volume element in three-space). Introducing the volumetric expansion rate (Hubble parameter) $K \equiv \frac{1}{2}\frac{\dot{h}}{h} = \frac{1}{2}s^a_a$ the equations read,

$$-R^0_0 = \dot{K} + \frac{1}{4}s^b_a s^a_b = -T^0_0 + \frac{1}{2}T + \Lambda \tag{2.6a}$$

$$-R^a_a = \dot{K} + K^2 + P = -T^a_a + \frac{3}{2}T + 3\Lambda \tag{2.6b}$$

If we introduce the trace free part of s_{ab}, $2\sigma_{ab} \equiv (s_{ab} - \frac{1}{3}s^c_c h_{ab})$, we find that

$$s^a_b s^b_a = s_{ab}s^{ab} = \frac{1}{3}(s^a_a)^2 + 4\sigma_{ab}\sigma^{ab} = \frac{4}{3}K^2 + 4\sigma_{ab}\sigma^{ab} \tag{2.7}$$

Substituting this into the R^a_a equation

$$\dot{K} = \Lambda - \frac{1}{3}K^2 - \sigma_{ab}\sigma^{ab} - (T^0_0 - \frac{1}{2}T) \tag{2.8}$$

Eliminating $\dot{K}$ we get,

$$\Lambda - \frac{1}{3}K^2 = -\frac{1}{2}\sigma_{ab}\sigma^{ab} - T^0_0 + \frac{P}{2} \tag{2.9}$$

Using the positivity of $\sigma_{ab}\sigma^{ab} \geq 0$, and the assumptions of the theorem we get the following inequality,

$$\dot{K} \leq \Lambda - \frac{1}{3}K^2 = -\sigma_{ab}\sigma^{ab} - T^0_0 + \frac{P}{2} \leq 0 \tag{2.10}$$

from here we get that $K^2 \geq 3\Lambda$, and by integrating the first part we get

$$K \leq \sqrt{3\Lambda}/\tanh(\sqrt{\frac{\Lambda}{3}}(t + t_0(x^i))) \tag{2.11}$$

where $t_0(x^i)$ is an arbitrary integration function*. In the large time limit we get,

$$0 \leq K - \sqrt{3\Lambda} \leq 4\sqrt{3\Lambda}e^{-2\sqrt{\frac{\Lambda}{3}}(t+t_0)} \to 0 \tag{2.12}$$

we can also deduce that, $\sigma_{ab}\sigma^{ab}, T_{00}$ and P vanish like $e^{-2\sqrt{\frac{\Lambda}{3}}(t+t_0)}$. Integrating the R_{ab} we get,

$$h_{ab}(t, x^i) = e^{2\sqrt{\frac{\Lambda}{3}}t}\tilde{h}_{ab}(x^i) \tag{2.13}$$

which has the same time behaviour has the de Sitter solution. From the dominant energy condition we can show that all components of T^ν_μ vanish exponentially fast. But what about the three-metric ? Is it really flat ? To answer this we should first notice that the time and space part of the metric have become independent,i.e. the metric is separable into its time and space parts. Now we invoke locality. Inside any given physical volume V_0, space-time rapidly becomes equal to the vacuum de Sitter space-time. Due to the exponential expansion a fluctuation of scale l is rapidly redshifted and "smoothed" over V_0, inhomogeneities are literally pushed out of the horizon and anisotropy, as well as the energy-momentum tensor, decay exponentially. It is important to keep in mind that although space-time becomes de Sitter locally there is no reason for this to happen globally. ∎

---

* P could be positive as long as it is not very large and (2.10) holds .

I would wish to point out that the theorem is valid in any number of dimensions. An $(n+1)$-dimensional cosmology under the influence of a positive cosmological constant will eventually expand at a rate of $\sqrt{\Lambda/n}$ in each of the n spatial directions. This could be catastrophic for some Kaluza-Klein type theories.

III. Scalar Field Dynamics

But what about the scalar field ? The equation of motion for ϕ is

$$\ddot{\phi} + K\dot{\phi} + (\Gamma\dot{\phi}) + \frac{1}{\sqrt{-g}}\partial_i\left(\sqrt{-g}\partial^i\phi\right) = -\frac{dV(\phi)}{d\phi} \tag{3.1}$$

where Γ is a term introduced to take into account the reheating. If we assume ϕ is smooth over a certain patch then we can neglect spatial gradient terms and the equation for ϕ is just that of a particle moving inside a potential $V(\phi)$ with friction $K\dot{\phi}$. I have just shown, that in general $K \geq \sqrt{3\Lambda} \equiv K_{isotropic}$ this implies that the friction on the field is always larger in an inhomogeneous and anisotropic model than in the standard RW case. The consequence of this is that the scalar field will, in general, roll slower and allow the Universe to inflate a few *more* e-folds. This extra inflation is insignificant, what is important is that the number of e-fold the Universe inflates is not decreased but rather increased by introducing the anisotropies and/or inhomogeneities (this had been point out earlier in ref. [13]).

IV. Conclusions

I have argued that the fact that the No-Hair theorem is valid locally for a large class of inhomogeneous and anisotropic models, makes Inflation the most attractive answer to the question of why the Universe is isotropic and homogeneous. A sketchy proof of the theorem has been given[1] and some of the consequences for the scalar field evolution have been analysed. In particular, we have shown[1] that Inflation can take place even if the Universe is highly inhomogeneous and anisotropic. It is important to mention the fact that Inflation is not just capable of smoothing out the Universe, but that it makes it with an unbelievable efficiency[14].

The fact that there are models that do not undergo inflation indicate that there is probably a whole class of them that cannot be smoothed out, either because Inflation never takes place, or there isn't enough Inflation[15]. The results I have presented here are encouraging, however, it would be nice to quantify just how large (measure) is the set of initial conditions that would or would not Inflate. If we could ever prove that the set of models that do undergo inflation, or alternatively, that satisfy the conditions set by the No-Hair theorem, is of non-zero measure in the space of all possible universes, then we could confidently say that Inflation is our best bet.

I would like to thank Mike Turner for the invitation to the Texas meeting. This work was supported by the Department of Energy and NASA.

References

1) Jensen, L.G. and Stein-Schabes, J.A., Phys. Rev. **D35**,4 (Feb 1987).
- Jensen, L.G. and Stein-Schabes, J.A., Proceedings of the first course of the International School of Particle Astrophysics, Erice,Italy 1986.

2) Barrow, J.D and Tipler, F. *The Anthropic Cosmological Principle.*, Oxford University Press, (1984).
3) Hawking, S.W. in ref. (2).
4) Collins, C.B. and Hawking, S.W., Ap.J., **180**,137 (1973).
5) Misner,C.W.Ap.J., **151**, 431 (1968).
- Matzner, R. and Misner, C.W., Ap.J., **171**, 415 (1972).
6) Guth, A.H.,Phys. Rev. **D23**, 347 (1981).
- Linde, A.D., Phys. Lett. **108B**, 389 (1982).
- Albrecht, A. and Steinhardt, P.J., Phys. Rev. Lett. **48**, 1220 (1982).
7) Turner, M.S., *The Inflationary Paradigm*, Proceedings of the Cargèse School on Fundamental Physics and Cosmology, eds. J. Audouze and J. Tran Thanh Van, Editions Frontieres, Gif-Sur-Yvette (1985).
8) Steinhardt, P. and Turner, M.S. Phys.Rev. **D29**, 2162 (1984).
9) Wald, R.W., Phys. Rev. **D28**, 2118 (1983).
10) Boucher, W., Gibbons, G.W. and Horowitz, G.T., Phys.Rev. **D30**, 2447 (1984).
- Boucher, W., in *Classical General Relativity*, ed. W.B. Bonnor *et al* Cambridge University Press, Cambridge, England, 1984).
- Starobinskii, A.A., J.E.T.P. Letters **37**, 66 (1983).
11) - Gibbons, G.W. and Hawking, S.W., Phys. Rev. **D15**, 2738 (1977).
- Hawking,S.W. and Moss,I.G., Phys. Lett. **110B**, 35 (1982).
12) Landau,L.D. and Lifshitz,E.M., *The Classical Theory of Fields*, Pergamon Press (1971).
13) Steigman,G. and Turner, M.S., Phys. Lett. **128B**, 295 (1983).
14) Jensen,L.G. and Stein-Schabes, J.A., Phys. Rev. **D34**,931 (1986).
- Turner,M.S. and Widrow,L., Phys. Rev. Lett. **57**, 2237 (1986).
15) Barrow,J.D., Phys. Lett. **180B**, 335 (1986).

BIG BANG NUCLEOSYNTHESIS: CHALLENGES AND OPPORTUNITIES

GARY STEIGMAN

Departments of Physics and Astronomy
The Ohio State University
174 W. 18th Avenue
Columbus, OH 43210, USA

ABSTRACT

Big Bang Nucleosynthesis provides the only probe of the early evolution of the Universe which is constrained by observational data. According to the "standard" hot, big bang model, the light elements (D, ^3He, ^4He, ^7Li) were synthesized in astrophysically interesting abundances during the first few minutes of the evolution of the Universe. A quantitative comparison of the predicted abundances with those derived from contemporary astronomical observations reveals the consistency of the standard model and leads to valuable constraints on cosmology and on elementary particle physics. Recently obtained observational data of very high quality provides the opportunity to test the standard model even more closely. To introduce the subsequent talks at this workshop I will place the new data on deuterium, helium and lithium in the context of the challenges posed to the standard model and - the opportunities offered.

PRIMORDIAL NUCLEOSYNTHESIS IN THE STANDARD MODEL

The so-called "standard" hot big bang model is defined by adopting the simplifying assumptions that (during the epochs of interest) the Universe is homogeneous and is expanding isotropically; that there are three families of light (<<MeV), left-handed neutrinos ($N_\nu = 3$) which are non-degenerate; and, there are no additional exotic particles. Within this framework, the yields of the light elements (D, ^3He, ^4He, ^7Li), produced during the early evolution of the Universe, depend mainly on η, the (present) ratio of nucleons to photons ($\eta = N/\gamma$; $\eta_{10} = 10^{10}(n_N/n_\gamma)_0$). For a comprehensive review of theory and observations, see Reference 1.

The present mass density in nucleons ($\rho_N = \Omega_N \rho_{crit}$) depends on η and on the present temperature ($T_{\gamma 0} = 2.7\Theta K$) of the cosmic background radiation.

$$\Omega_N h_{50}^2 = 0.00353\Theta^3 \eta_{10}. \quad (1)$$

In (1), $h_{50} = H_0/50 kms^{-1} Mpc^{-1}$ (H_0 is the present value of the Hubble parameter). For $T_{\gamma 0} = 2.76 \pm 0.05(3\sigma)^{2)}$, we are led to

$$\Omega_N h_{50}^2 = (0.015 \pm 0.001)\eta_{10}. \quad (2)$$

Bounds on η from nucleosynthesis will constrain the present mass density in nucleons (due to the dependence of Ω_N on H_0^2, the density parameter Ω_N is less well constrained than is ρ_N).

Zeroth Order

It has long been known that for $1 \lesssim \eta_{10} < 10$, the predicted abundances of the light elements are in good agreement with those derived from observational data. This "first cut" establishes the consistency of the standard model and encourages us to pursue more detailed comparisons of theory and observation. It is noteworthy that even for the nucleon-to-photon ratio in this broad range, we may test the cosmological consistency of the standard model. The lower bound $\Omega_N h_{50}^2 \gtrsim 0.01$ is consistent with the density estimated for the luminous parts of galaxies[3]; nucleons, as constrained by big bang nucleosynthesis, are more than able to account for the dynamically determined mass in the shining parts of galaxies. Of even more interest, perhaps, is the upper bound $\Omega_N h_{50}^2 \lesssim 0.16$; a critical density, nucleon-dominated ($\Omega_N = 1$) Universe is excluded within the context of the standard model.

First Order

Yang et al.[4] (see also Ref. 1) attempted a more detailed comparison of theory and observation considering, in particular, the galactic evolution of deuterium and helium-3 and, noting the uncertainties in the primordial abundance of helium-4 derived from current observational data. Such analyses narrow the range allowed for η. For example, since deuterium is destroyed in the course of galactic evolution, the present interstellar abundance[5] provides a lower bound to the primordial abundance. For $(D/H)_p \gtrsim 2 \times 10^{-5}$, we are led to an upper bound to the nucleon-to-photon ratio: $\eta_{10} \lesssim 7$. Since deuterium is burned to the less fragile helium-3, observations of deuterium and helium-3 can be used[4] to set an upper bound to the sum of the primordial abundances of these elements. For $[(D + {}^3He)/H]_p \lesssim 1 \times 10^{-4}$, we are led to a lower bound to the nucleon-to-photon ratio: $\eta_{10} \gtrsim 3$. These new bounds to η narrow the range allowed for the present density of nucleons: $0.04 \lesssim \Omega_N h_{50}^2 \lesssim 0.11$.

A more detailed comparison of theory and observation for helium-4 now becomes crucial. For $3 \lesssim \eta_{10} \lesssim 7$ and $N_\nu = 3$, the predicted primordial mass fraction of ^4He lies in the range: $0.24 \lesssim Y \lesssim 0.25$. This prediction is consistent with the range of abundances inferred from present data.[6,7] If we deviate from the standard model to permit the inclusion of a fourth family (light, left-handed) neutrino ($N_\nu = 4$), we would predict (for $\eta \geq 3$ and a neutron half-life $\tau_{1/2} \geq 10.4$ min.) a lower bound to the primordial abundance of $Y_p \geq 0.254$. If we believe the present data is accurate to 3 significant figures, this prediction is marginally excluded. A somewhat safer conclusion is that $N_\nu \leq 4$.

For our more restricted range in η, the primordial abundance of lithium is restricted to: $(^7Li/H)_p = (0.8 - 4) \times 10^{-10}$. The newly appreciated importance of lithium will be discussed shortly. Suffice it to say here that the lithium observed in Pop II stars (see Ref. 8 for details and references) is consistent with the above range.

The foregoing discussion has been in the nature of a brief review to set the stage for an exposition of the recent developments.

RECENT DEVELOPMENTS

New observational data and revised nuclear reaction rates have provided a recent stimulus to our subject. New data on lithium is reviewed by Hobbs[8] and that on helium-4 by Shields[6]. Here, I will provide an overview and context for their discussions.

Lithium

At relatively low nucleon-to-photon ratio ($\eta \lesssim 3$), primordial lithium is produced primarily directly via: $^3H(\alpha,\gamma)^7Li$; lithium is destroyed via $^7Li\ (p,\alpha)\ ^4He$. New laboratory experiments have led to revisions in each of these rates. The production rate has increased[9] and the destruction rate decreased[10] leading to the prediction of more ^7Li emerging from big bang nucleosynthesis[11] for $\eta_{10} \lesssim 4$-5.

The recent observational data, reviewed by Hobbs[8] shows that _if_ the Pop II abundances are to be associated with the primordial abundance, then $(^7Li/H)_p \leq 1.5 \times 10^{-10}$. This upper bound to the primordial abundance is consistent with a nucleon-to-photon ratio in the range[11]: $2 \lesssim \eta_{10} \lesssim 5$.

Deuterium

As emphasized earlier, the interstellar data on deuterium only provides a lower bound to its primordial abundance. To utilize ^3He to provide an upper bound[4] requires difficult observations (see Rood[12]) and involves some theoretical model dependence. Of great and fundamental value to cosmology would be observations of, or limits to, deuterium in extragalactic environments, especially those in which the gas may be very nearly primordial. York et al.[13] have searched for deuterium in Mrk 509 at a redshift $z_{abs} = 0.0333$. Deuterium was not observed and their[13] limit, $D/H \lesssim 1 \times 10^{-4}$, is consistent with that inferred earlier[4] from D plus ^3He.

Of very great interest is the recent work of Carswell et al.[14] who (may) have observed deuterium in absorption ($z_{abs} = 3.08571$) against the quasar Q0420-388. The high redshift and the low (~ 1/5 solar) abundance of the metals suggest that the gas in this system may be relatively unprocessed. An abundance $D/H \approx 4 \times 10^{-5}$ is derived[14] but, it is noted that the safest conclusion is that there is an upper bound: $D/H \lesssim 1 \times 10^{-4}$.

Helium-4

Recent observational data, along with the associated perils and pitfalls, are reviewed by Shields[6]; the extrapolation from the observational data to the primordial abundance is discussed by

Gallagher[7]. Suffice it to say that all extant high quality data is consistent with $Y_p = 0.24 \pm 0.01$. To effect a truly critical comparison between theory and observation however, we need to know Y_p to <u>three</u> significant figures.

Kunth[15] has found no evidence for a helium trend with oxygen (for low O/H) and derives: $Y_p = 0.243 \pm 0.009(3\sigma)$; $Y_p \leq 0.252$. In contrast, Pagel et al.[16] compare helium with nitrogen and find evidence for a Y versus N/H trend; they derive $Y_p = 0.237 \pm 0.015(3\sigma)$; $Y_p \leq 0.252$. Finally, Steigman et al.[17] (see Gallagher[7]) compare helium and carbon, find evidence for a trend and derive: $Y_p = 0.235 \pm 0.012(3\sigma)$; $Y_p \leq 0.247$. Current data is not inconsistent with a primordial abundance as low as $Y_p = 0.22-0.23$ which is lower than that allowed by the standard model. What is relevant however, is that this same data is consistent with a primordial abundance as large as $Y_p = 0.24 - 0.25$, consistent with the predictions of the standard model.

CHALLENGES AND OPPORTUNITIES

The new developments discussed above provide challenges to and, opportunities for the standard hot, big bang model. If recent data is confirmed and extended there will be further tests of the consistency of the standard model and the bounds to η and N_ν will be narrowed.

For deuterium the challenge is to set limits to or, observe the D/H ratio in extragalactic regions, especially gas rich regions which may have suffered minimal stellar processing. The opportunity is that if <u>upper</u> limits of the order $10^5(D/H)_p \lesssim 4-10$ can be confirmed, then <u>lower</u> bounds to N/γ will follow: $\eta_{10} \gtrsim 3-5$. Indeed, for $(D/H)_p \lesssim 1\times10^{-4}$ (and $\tau_{1/2} \gtrsim 10.4$ min.), the helium abundance is predicted to exceed,

$$Y_p \gtrsim 0.238 + 0.014(N_\nu - 3). \tag{3}$$

For $N_\nu = 4$ then, $Y_p \geq 0.252$ - barely consistent with current estimates.

Lithium provides an even greater challenge - to astrophysics as well as cosmology. There are many issues to be resolved. Is the Pop II lithium abundance the primordial abundance? If so, what is the origin of the Pop I abundance? And, why, after more than 10Gyr, is lithium undepleted in the Pop II stars (in contrast to the destruction observed for the younger Pop I stars of similar mass)? The opportunity is that if Pop II = Primordial, then the upper limit $(^7Li/H)_p \leq 1.5 \times 10^{-10}$ limits the nucleon-to-photon ratio to a narrow range[11]: $2 \lesssim \eta_{10} \lesssim 5$. The corresponding upper limit to the present mass density in nucleons, $\Omega_N h_{50}^2 \lesssim 0.08$, might suggest the existence of WIMPs - non-baryonic dark matter[18]. For η in the above range, the standard model ($N_\nu = 3$; $\tau_{½} = 10.5 \pm 0.1$ min.) predicts a helium abundance: $0.235 \lesssim Y_p \lesssim 0.247$, in agreement with current data. For $N_\nu = 4$ however, $Y_p \gtrsim 0.249$ - only marginally allowed. Finally, it is worth noting that within the content of the standard model, a primordial lithium abundance in the range $10^{10}(^7Li/H)_p = 1.2 \pm 0.3$ corresponds to a predicted deuterium abundance: $4 \lesssim 10^5(D/H)_p \lesssim 15$. Such a range will set limits to exotic models for the chemical evolution of the Galaxy.

The challenge for helium is to infer the primordial abundance to an accuracy of 5% or better. Given the difficulties of obtaining the data with that sort of accuracy, combined with the uncertain corrections for ionization, collisions and evolution, this is quite a challenge. But, the payoff is very large. For example, suppose the upper bound $Y_p \lesssim 0.247$ derived from the data by Steigman et al.[17] is confirmed. Within the standard model this would lead to the independent upper bound to N/γ: $\eta_{10} \lesssim 6$, $\Omega_N h_{50}^2 \lesssim 0.09$; again a hint of WIMPs. Furthermore, in concert with the lower bound to Y_p in equation (3), this would lead to the exclusion of a fourth generation (light) neutrino ($N_\nu \leq 3.6$).

The above examples have been chosen to illustrate the opportunities available if only the challenges can be met. Careful observational work on the abundances (extra-galactic, if possible) of deuterium, helium and lithium will yield the data necessary for a critical test of the standard model. Current data suggests $\eta_{10} \approx 4 \pm 1$ and $\Delta N_\nu \lesssim 1$;

this is consistent with $\Omega_N \gtrsim \Omega_{GAL}$ and $\Omega_N << 1$. The challenge is to confirm or refute these dramatic results.

ACKNOWLEDGEMENTS

I am grateful to Bob Rood, Lew Hobbs, Greg Shields and Jay Gallagher for their contributions to this workshop. This work is supported by the DOE at OSU.

REFERENCES

1. Boesgaard, A.M. and Steigman, G., Ann. Rev. Astron. Astrophys. 23, 319 (1985).
2. D. Wilkinson, This Volume (1987).
3. Faber, S.M. and Gallagher, J.S., Ann. Rev. Astron. Astrophys. 17, 135 (1979).
4. Yang, J., Turner, M.S., Steigman, G., Schramm, D.N. and Olive, K.A., Ap. J. 281, 493 (1984).
5. Steigman, G., In Preparation (1987).
6. Shields, G.A., This Volume (1987).
7. Gallagher, J.S., This Volume (1987).
8. Hobbs, L.M., This Volume (1987).
9. Schroder, U. Redder, A., Rolfs, C., Azuma, R.E., Buchmann, L., King, J.D. and Donoghue, T.R., Munster, Universitat Preprint, (1986).
10. Rolfs, C. and Kavanagh, R.W., Nucl. Phys. A455, 179 (1986).
11. Kawano, L., Schramm, D.N. and Steigman, G., Preprint (1986).
12. Rood, R.T., This Volume (1987).
13. York, D.G., Ratcliff, S., Blades, J.C., Cowie, L.L., Morton, D.C. and Wu, C.C., Ap. J. 276, 92 (1984).
14. Carswell, R., Irwin, M.J., Webb, J.K., Baldwin, J.A., Atwood, B., Robertson, J.G and Shaver, P.A., Preprint (1986).
15. Kunth, D., PASP 98, 984 (1986).
16. Pagel, B.E.J., Terlevich, R.J. and Melnick, J., PASP 98, 1005 (1986).
17. Steigman, G., Gallagher, J.S. and Schramm, D.N., In Preparation (1987).
18. Schramm, D.N. and Steigman, G., Ap. J. 243, 1 (1981).

GROPING FOR THE COSMOLOGICAL ^{3}He ABUNDANCE

Robert T. Rood
Astronomy Department, University of Virginia
and National Radio Astronomy Observatory,* Charlottesville, VA 22903

T. M. Bania
Department of Astronomy, Boston University, Boston, MA 02215

T. L. Wilson
Max-Planck-Institut für Radioastronomie, D5300 Bonn 1, FDR

ABSTRACT

An extensive search for emission from the 8.7 GHz hyperfine line of $^3He^+$ in 17 galactic H II regions. We detect ^{3}He in about half and have obtained significant upper limits for the others. Our detections have line-to-continuum ratios in the range 10^{-3}–10^{-4}. For such lines, systematic errors can be much larger than random noise, so we have simultaneously observed many weak recombination lines in order to estimate the magnitude of the systematic effects on our measurements. We find these effects are larger than were thought and the abundances we now derive are lower than previously reported. The largest ^{3}He/H ratio is more than six times larger than our smallest limit. This observation gives strong support for source-to-source variations. There is a marginal tendency for sources outside the solar circle to have larger abundances. For sources where the ^{3}He was found the average ^{3}He/H is 4.8×10^{-5}. Our smallest upper limits are about 10^{-5}.

1. Introduction

The abundance of ^{3}He can be used to test or constrain theories of cosmology, stellar evolution, and chemical evolution of the Galaxy. Several years ago we reported the detection of the 8.7 GHz hyperfine line of $^3He^+$ in three galactic H II regions using the NRAO* 140 ft telescope in Green Bank.[1] The abundances found at that time indicated substantial source-to-variation and gave some indication that the production of ^{3}He in low mass stars might not be occuring in the the way envisaged in the simplest models.[2] A survey of ^{3}He in a much larger number

* The National Radio Astronomy Observatory is operated by Associated Universities Inc. under contract with the National Science Foundation.

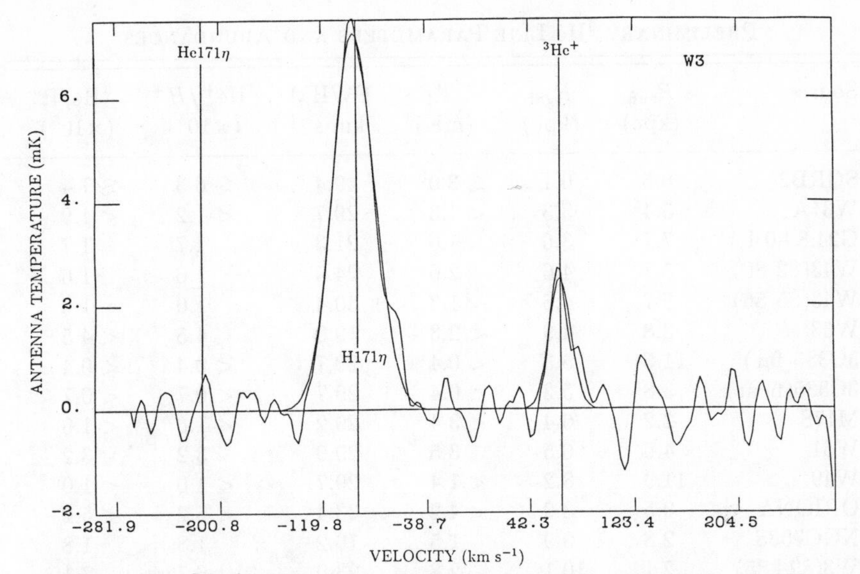

FIGURE 1. The $^3He^+$–171η region of the spectrum for W3. The vertical marks show the expected line centers. This spectrum shows data taken only during 1985 and represents 77.5 hrs integration. (We did lots of tests on W3.)

of sources was indicated so that the cosmological and stellar contributions could be sorted out. We have now completed a first step of that survey with extensive observations of 17 H II regions made with the 140 ft between March, 1984 and October, 1986. Since a detailed paper[3] is almost ready for submission, we present here only a brief summary of preliminary results.

In the earlier paper we discussed at length the problems introduced by systematic error in experiments of this sort. In particular, frequency dependent baseline structure can lead to substantial errors in spectral line parameters. Our further experiments have driven that point home. Much of our work in the past four years has been devoted to isolating and reducing these errors. Improvements have been made such that the baselines are one to two orders of magnitude better than in our earlier observations. It is now possible, for instance, to identify a quasi-sinusoidal ripple with peak-to-peak amplitude of 1.5 mK in the 10 K continuum source W3. We now feel that our earlier abundances were in some cases a factor of several to high because this "ripple" was not completely removed. Because of the difficulty in accessing such errors, we view the current results as but one step in a larger project which will include observations made with the completlely independent system at Effelsberg.

PRELIMINARY ^{3}He LINE PARAMETERS AND ABUNDANCES

Source	$R_{\rm sun}$ (kpc)	$R_{\rm gal}$ (kpc)	T_A (mK)	FWHM km s^{-1}	^{3}He$^+/H^+$ ($\times 10^5$)	^{3}He/H^a ($\times 10^5$)
SGRB2	8.5	0.1	$\lesssim 3.0^b$	49.4	$\lesssim 6.3$	$\lesssim 7.4$
W31A	5.4	3.3	< 1.3	29.7	< 1.2	< 1.9
G24.8+0.1	7.7	3.6	1.6	21.3	1.7	1.7
W43(82-86)	5.7	4.6	2.6	24.5	1.6	1.6
W43(85-86)	5.7	4.6	1.3	30.4	1.0	1.0
W33	3.8	4.9	< 2.3	19.9	< 4.5	< 4.5
3C385(far)	11.5	5.3	< 0.4	29.7	< 0.4	< 0.4
3C385(near)	3.8	5.3	< 0.4	29.7	< 0.7	< 0.7
M17S	2.2	6.4	< 3.4	29.2	< 1.6	< 1.6
W51	4.6	6.5	3.5	29.9	3.2	3.2
W49	11.9	8.2	< 1.4	29.7	< 1.0	< 1.0
ORIONA	0.5	8.9	< 4.2	33.4	< 5.2	< 5.2
NGC7538	2.8	9.9	1.5	19.2	1.8	1.8
W3(82+85)	2.1	10.1	2.8	23.0	6.7^c	7.1^c
W3(85)	2.1	10.1	2.5	20.1	5.2^c	5.5^c
G133.8+1.4	2.1	10.1	< 2.1	29.7	< 6.3	< 6.3
S162	3.5	10.3	$\lesssim 0.9^b$	29.7	$\lesssim 2.7$	$\lesssim 2.7$
S206	2.5	10.7	2.5	18.0	4.5	4.5
S311	4.1	11.0	< 0.8	29.7	< 2.2	< 2.2
S209	10.1	18.9	1.8	19.9	1.5	2.1

[a] If for the observed recombination lines ^{4}He$^+$/H$^+ < 0.085$, helium is assumed to be underionized and the ^{3}He abundance accordingly increased.

[b] Even though the fit to the line appears to be good, ambiguity in baseline fitting forces us to claim a limit rather than a detection.

[c] This value should be increased by a factor of 1.6 to correct for clumping.

2. Summary of Results

- Preliminary abundances are given in the table. When sources were observed in 1982,[1)] grand averages are presented. In a few cases the results of eliminating the earlier data are shown. For 3C385, which has a distance ambiguity, results with both distances are given to show the sort of abundance errors introduced by distance errors.
- Spectra for two of our detections are shown in Figures 1–2. These are typical (*i.e.*, the best) of our detections. Because the strength of the ^{3}He$^+$ hyperfine

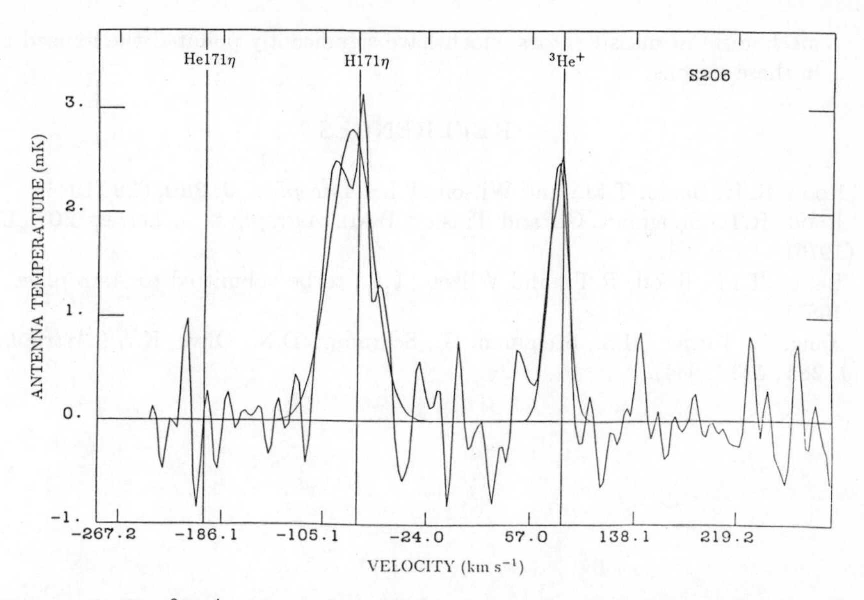

FIGURE 2. The $^{3}He^{+}$–171η region of the spectrum for S206. The total integration time is 28.4 hrs.

line depends on column depth and the continuum temperature and recombination lines depend on emission measure, many of the best candidate ^{3}He sources are rather obscure H II regions like S206.

- The fact that some sources have substantially more ^{3}He than others seems inescapable. This can be taken as evidence for stellar production of ^{3}He. Unfortunately many of the sources with high ^{3}He are in the outer galaxy where one expects less stellar processing than at, say, the location of W43. More specific indicators, such as ^{13}C, which should be produced in the same stars as ^{3}He, are if anything anticorrelated. The outer galaxy sources W3 and S206 are physically very different yet both have fairly high ^{3}He, so the apparently high abundances are probably not due to some quirk of source structure. There is still much to be understood about the distribution of ^{3}He in the galaxy and the details of its stellar production/destruction.
- If stars make a positive net contribution to ^{3}He, then our lowest upper limits (^{3}He/H $\sim 10^{-5}$) serve as an upper limit for the cosmological ^{3}He. Our limits are embarassing low for standard calculations. They require that the nucleon to photon ratio, $\eta \gtrsim 10^{-9}$. At the same time the deuterium and lithium results just discussed by Steigman (or ref. [4], for example) place *upper* limits on η several times smaller. It is premature to make much of this result for at least two reasons. (1) Upper limits are even more prone to systematic error than the detected lines. (2) Some of the strongest limits are for extremely large H II regions—to paraphrase Mayor Washington, W49 is a Galaxy Class H II region. The low abundance may be evidence for some yet unknown ^{3}He destruction

mechanism in massive stars which have significantly polluted the ionized gas in these regions.

REFERENCES

1) Rood, R.T., Bania, T.M., and Wilson, T.L., *Astrophys. J.* **280**, 629 (1984).
2) Rood, R.T., Steigman, G., and Tinsley, B.M., *Astrophys. J. Letters* **207**, L57 (1976).
3) Bania, T.M., Rood, R.T., and Wilson, T.L., to be submitted to *Astrophys. J.* (1987).
4) Yang, J., Turner, M.S., Steigman, G., Schramm, D.N., Olive, K.A., *Astrophys. J.* **281**, 493 (1984).

THE EVOLUTION OF THE GALACTIC LITHIUM ABUNDANCE

L. M. Hobbs

University of Chicago
Yerkes Observatory, Williams Bay, WI 53191-0258, USA

1) Introduction

In the interstellar medium and/or in the outer regions of sufficiently young, solar-like stars, very low abundances of approximately 10^{-9}, 10^{-11}, and 10^{-10} by number are observed for the light elements Li, Be, and B, respectively. The cosmic origin of these trace constituents of the universe is not satisfactorily understood[1]. In contrast to the abundant lighter elements H and He, the isotope ^{6}Li and all isotopes of Be and B are not products of primordial nucleosynthesis according to standard theory, although at least 10% of present-day ^{7}Li probably is. In contrast also to the abundant heavier species from C to Fe, the elements Li, Be, and B are destroyed, not produced, by progressive stellar nucleosynthesis, owing to fast (p,α) reactions which occur at the low temperatures $T \geqslant 3 \times 10^6$ K. Hence, some uncertain Galactic mechanism apparently has supplemented stellar and big-bang nucleosynthesis in building up the first few rows of the periodic table. However, there existed prior to 1982 no direct observational evidence for this presumed steady Galactic buildup of Li, Be, and B, and the available evidence remains indirect even today.

In addition, early observations[2] of lithium in stellar spectra revealed an important deficiency in the theory of stellar structure, even in simple, essentially unevolved stars like the sun. Observations reported in 1986 have strikingly extended[3] this classical puzzle, which has not yet been fundamentally explained. In possible similarity to the measured dearth of neutrinos released deep in the solar interior, the progressive destruction of fragile lithium nuclei in the convective envelopes of unevolved, cool stars provides a precise test of the actual conditions in such envelopes. It apparently emphasizes the inadequacy of the phenomenological, mixing-length theory of convection that is used in calculating some portions of essentially all stellar models.

With contemporary instruments, the generally quite weak 6707 Å resonance line of neutral Li can be measured relatively easily from the ground in the spectra even of fairly faint stars with surface temperatures $T_e \lesssim 7000$ K. There are, therefore, many more observational data available on the Galactic abundance of Li than on those of Be or B. This report focuses entirely on the Li data and on its implications for the interconnected questions of (1) big-bang nucleosynthesis, (2), Galactic nucleosynthesis, and (3) the structure of stellar convective envelopes.

2) Population I Objects

The Li abundance in the surface layers of the sun[4)] is $(Li/H) \leq 1 \times 10^{-11}$, at the present solar age of about 5 Gyr. This abundance is smaller by at least two orders of magnitude than both the current interstellar value[5)] and the presumed initial value at the birth of the solar system, which is inferred from chondritic meteorites[6)]. Working from these and other early clues, Herbig discovered[2)] that solar-like stars in the nearby Hyades cluster, which has an intermediate age of about 0.7 Gyr, also show an intermediate abundance, $Li/H \approx 3 \times 10^{-10}$. He therefore postulated that ordinary main-sequence stars with $M/M_{\odot} \approx 1$ progressively destroy their surface lithium, in contradiction to theoretical stellar models [7,8)].

Recent observations have verified this result in considerable detail and significantly augmented it. Figure 1 shows, as a function of surface temperature or mass, the Li abundances measured in 33 main sequence menbers of the Hyades[3,9)]. In the mass range shown, standard stellar models incorrectly predict that the observable surface abundance of Li should be essentially invariant at the initial (i. e. interstellar) value[8)], which is observed to be about $Li/H \approx 1 \times 10^{-9}$ with relatively large uncertainties. Instead, the interstellar abundance seems to constitute an upper limit to those of the Hyades stars. The large deficiency of Li at $T_e \approx$ 6600 K discovered in these naked-eye stars only in 1986 may be an effect distinct from the classical one observed at $T_e \lesssim 6000$ K and could arise, in sufficiently circulation-free atmospheres, from a diffusion process[10)]. Observations of stars in the much younger Pleiades cluster[11)], with an age $\lesssim$ 0.1 Gyr, are consistent with the pattern set by these interstellar, solar system, and Hyades data.

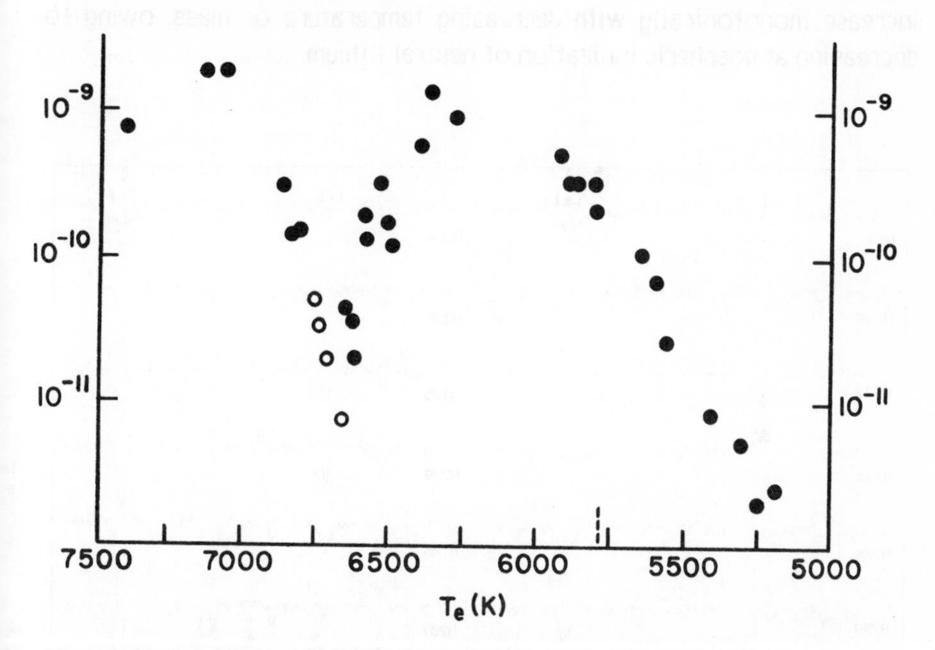

Fig. 1. The variation of Li/H with T_e along the Hyades main sequence.

Also realized in the last year was the goal[2)] of using a series of sequentially older -- and, accidentally, more distant, much fainter -- open clusters, with their accurately known relative ages, to map empirically the evolution of the Galactic lithium abundance back to cosmologically pertinent epochs, in a step-by-step fashion. Three clusters, NGC 752[12)], M67[13,14)], and NGC 188[15)], which have approximate ages[16)] of 1.7, 5, and 10 Gyr, have been observed. NGC 188, where only preliminary data for two stars are available so far, is believed to be among the oldest possible Population I clusters. The spectra of the stars in these three older clusters confirm a monotonic advance, with increasing age, of the destruction of lithium in the surface layers of cool main sequence stars. The pattern seen in the five clusters together and in the sun is a consistent one. It also indicates that the Li abundance in the Galactic gas showed no appreciable change, going back over at least the past 10 Gyr, from the present interstellar value.

For illustration, most of the spectra from NGC 752 are shown in Figure 2. At a given abundance, the strength of the Li I 6707 Å line would increase monotonically with decreasing temperature or mass, owing to decreasing atmospheric ionization of neutral lithium.

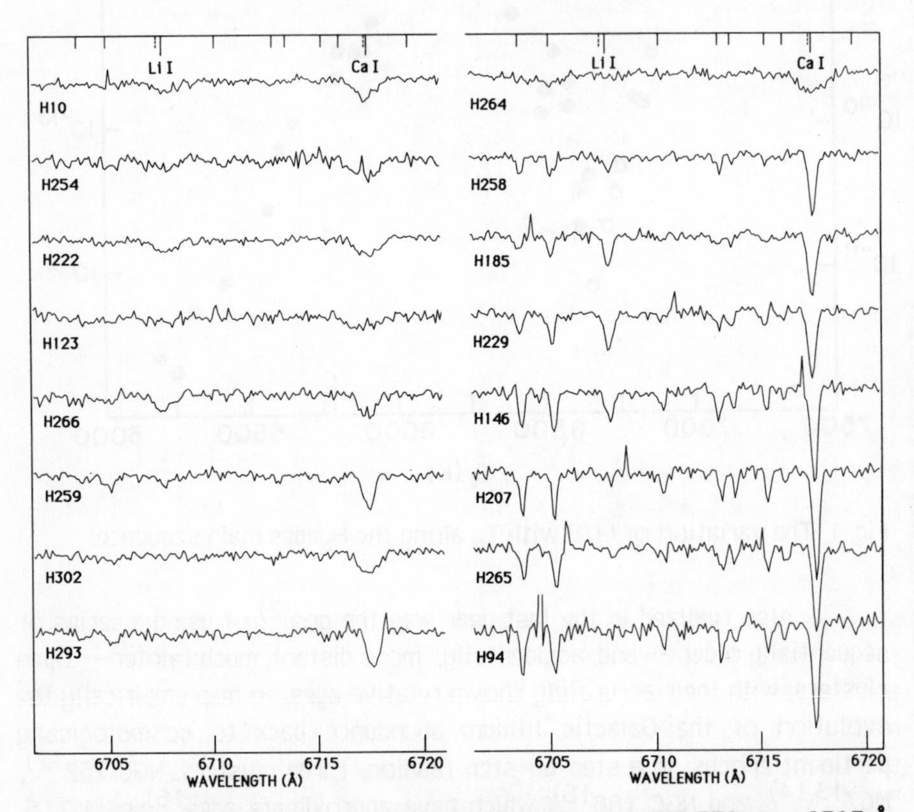

Fig. 2. The spectra of 16 main-sequence members of NGC 752 near 6707 Å. The stars are ordered according to decreasing surface temperature, which ranges approximately from 6900 to 5300 K. Star H293 shows $T_e \approx 6460$ K, for example. The positions of various stellar lines, including the Li I line, are marked at the top.

In summary, a maximum lithium abundance near Li/H $\approx 1 \times 10^{-9}$ exists in a broad range of Population I objects, including the ISM, the solar system, five open clusters with widely ranging ages, and field F and G stars. No evident change in Li/H is seen over the past 10 Gyr. We conclude that (1) this uniformity suggests an <u>upper limit</u> $(Li/H)_{BB} \lesssim 1 \times 10^{-9}$ on the Li produced in the big bang; (2) there has been negligible net Li production by Galactic mechanisms in at least the last 10 Gyr; and (3) there exists at present no satisfactory theory of the nuclear burning of Li in the convective envelopes of main-sequence stars with $T_e \lesssim 7000$ K.

3) Population II Stars

On the basis of the results for the sun and the oldest open clusters, the 6707 Å line might be expected to be undetectably weak in the still older, unevolved Population II stars with $M/M_\odot \lesssim 0.9$ or $T_e \lesssim 6200$ K. Nevertheless, in 1982 Spite and Spite made the important and surprising discovery[17)] that the Li I line can be readily detected in the spectra of nearly all of the rare, nearby, bright, very old halo stars, or subdwarfs, with $T_e \gtrsim 5600$ K. The resulting surface lithium abundance deduced for these stars is $Li/H \approx 1 \times 10^{-10}$, at least an order of magnitude greater than the solar value and in striking agreement with conventional models of big-bang nucleosynthesis[18,19,20)] for a cosmic ratio of baryons to photons, $\eta \approx 3 \times 10^{-10}$. The Spites argued qualitatively that the much lower abundance of heavy elements in the old Population II stars led to shallower convective zones, within which the surface lithium has been fully preserved over nearly the Hubble time. These atmospheric Li abundances therefore can faithfully reproduce the primordial one. For the most metal-poor subdwarfs, the near invariance of the Li/H ratio with stellar mass within the admittedly narrow mass interval corresponding to $5600 \lesssim T_e \lesssim 6200$ K lends empirical support to this hypothesis. Subsequent observational studies[21,22,23,24)] have fully confirmed and extended the original data, to a group of halo stars which now numbers about 30.

This result, $Li/H \approx 1 \times 10^{-10}$, has been found for all sufficiently hot halo dwarfs which have heavy-element abundances less than 4% of the solar ones. Much lower Li abundances are found in a small fraction of halo stars, as their metallicity increases through approximately 10%. Such a result is qualitatively expected at some metallicity threshhold, because convection may then extend to sufficiently deeper layers to destroy Li. However, in some cases, halo stars of nearly identical metallicity and mass show very different Li/H ratios[17,24]. Other stellar properties, perhaps including rotation, must be important as well.

The current lack of a reliable theoretical understanding of Li destruction in solar-like stars, which is so evident from the open clusters, precludes at present a conclusive proof of the assumption that the halo-star atmospheres are accurate tracers of the primordial Li/H ratio. The possibility that $(Li/H)_{BB} \approx 1 \times 10^{-9}$ and that the halo stars with $T_e \gtrsim$ 5600 K have subsequently suffered a uniform, ten-fold reduction of atmospheric Li/H over about 15 Gyr -- although the solar reduction in only a third of that time is ten times larger still -- cannot yet be excluded. Thus, the halo stars provide a <u>lower limit</u>, $(Li/H)_{BB} \gtrsim 1 \times 10^{-10}$.

4) Conclusions

The most widely accepted current hypothesis[17] is that of a "low" primordial Li abundance, $(Li/H)_{BB} = (1.2 \pm 0.3) \times 10^{-10}$, which corresponds to a cosmic baryon fraction $2 \times 10^{-10} \leq \eta \leq 5 \times 10^{-10}$. In this picture, the hotter halo stars must have suffered no Li destruction over a Hubble time, and about 90% of the present-day interstellar Li must have been produced earlier than at least 10 Gyr ago by some unclarified Galactic process, of which a number have been proposed[25] but not convincingly confirmed. The strongest arguments favoring these ideas are (1) the observations of ^{2}H, ^{3}He, and ^{4}He, in combination with theoretical models of big-bang element synthesis, and perhaps (2) the observed invariance with stellar mass of the Li/H ratio in the atmospheres of extreme halo stars with heavy-element abundances less than 4% of the solar values and $T_e \gtrsim 5600$ K.

The alternative possibility of a "high" original abundance, $(Li/H)_{BB} \approx 1 \times 10^{-9}$, cannot yet be excluded conclusively by the Li data alone. In this picture the hotter halo stars must have suffered a nearly uniform, ten-fold reduction of atmospheric Li, and no net Galactic production of Li at all is required. In conventional big-bang models this Li/H ratio requires either $\eta \lesssim 1 \times 10^{-10}$ or $\eta \gtrsim 10 \times 10^{-10}$, values which conflict with the combination of theory and the observed values of ^2H, ^3He, and/or ^4He. Either future observations of extragalactic Li or fundamentally improved theoretical calculations of stellar convective zones probably would fully resolve this issue.

1) Boesgaard, A. M., and Steigman, G., Ann. Rev. Astr. Ap. 23, 319 (1985).
2) Herbig, G. H., Ap. J. 141, 588 (1965).
3) Boesgaard, A. M., and Tripicco, M., Ap. J. Letters 302, L49 (1986).
4) Muller, A. E., Petrymann, E., and de la Reza, R., Solar Phys. 41, 53 (1975).
5) Hobbs, L. M., Ap. J. 286, 252 (1984).
6) Nichiporuk, W., and Moore, C. B., Geochim. Cosmochim Acta 38, 1691 (1974).
7) Bodenheimer, P., Ap. J. 142, 451 (1965).
8) D'Antona, F., and Mazzitelli, I., Astr. Ap. 138, 431 (1984).
9) Cayrel, R., Cayrel de Stroebel, G., Campbell, B., and Dappen, W., Ap. J. 283, 205 (1984).
10) Michaud, G., Ap. J. 302, 650 (1986).
11) Duncan, D. K., and Jones, B. F., Ap. J. 271, 663 (1983).
12) Hobbs, L. M., and Pilachowski, C., Ap. J. Letters 309, L17 (1986).
13) Hobbs, L. M., and Pilachowski, C., Ap. J. Letters 311, L37 (1986).
14) Spite, F., Spite, M., Peterson, R. C., and Chaffee, F. H., Astr. Ap. (in press).
15) Pilachowski, C., and Hobbs, L. M., Ap. J. (to be submitted).
16) VandenBerg, D. A., Ap. J. Suppl. 58, 711 (1985).
17) Spite, F., and Spite, M., Astr. Ap. 115, 357 (1982).
18) Yang, J. et al, Ap. J. 281, 493 (1984).
19) Kawano, L., Schramm, D. N., and Steigman, G., preprint.
20) Sale, K. E., and Mathews, G. J., Ap. J. Letters 309, L1 (1986).
21) Spite, M., Maillard, J. P., and Spite, F., Astr. Ap. 141, 56 (1984).
22) Spite, F., and Spite, M., Astr. Ap. 163, 140 (1986).
23) Beckman, J., Rebolo, R., and Molaro, P., preprint.
24) Hobbs, L. M., and Duncan, D. K., Ap. J. (in press).
25) Audouze, J., Boulade, O., Malinie, G., and Poilane, Y., Astr. Ap. 127, 164 (1983).

MEASUREMENTS OF THE PRIMORDIAL ABUNDANCE OF HELIUM

G. A. Shields

Department of Astronomy

University of Texas at Austin

ABSTRACT

Spectrophotometry of the emission lines from H II regions in dwarf irregular galaxies provides the best opportunity for measuring the primordial abundance of helium. Recent studies cast doubt on the large variations in helium abundance found in early investigations. A primordial helium mass fraction $Y_P = 0.232 \pm 0.014$ is recommended on the basis of a recent study by Pagel and co-workers[1,2] with some added allowance for systematic errors. This value is consistent with the standard hot Big Bang model of the Universe.

1. Introduction

Dwarf irregular galaxies have abundant ionized gas and strong emission lines suitable for derivation of the He/H abundance ratio. Their low abundances of heavy elements suggest that stellar evolution has added relatively little helium to the amount produced in the Big Bang. The relevant emission lines are optically thin and dominated by radiative recombination, for which the atomic data are accurately known. Therefore there is hope of achieving the precision needed for cosmologically interesting conclusions, namely to discriminate reliably between $Y_P = 0.23$ and 0.24.

Peimbert and Torres-Peimbert[3] pioneered the use of dwarf irregulars in this way. Using measurements of He/H and O/H in different objects to find $\Delta Y/\Delta Z$, the rate of enrichment in helium with increasing heavy element mass fraction, they extrapolated to $Z = 0$ to find Y_P, the "primordial" value. Subsequent studies generally have involved samples of blue compact galaxies (BCG). Examples include Lequeux *et al.*[4], French[5], and Kunth and Sargent[6]. These and other studies pointed to values of Y between 0.22 and 0.24, at the low end of the range allowed by the standard Big Bang model. A few reported values were low enough to conflict with the model[7].

Many topics related to the primordial abundance of helium are discussed in the ESO Symposium[8] and the review by Boesgaard and Steigman[9]. Matzner[10] reviews Big Bang model predictions, and Kunth[11] and Pagel *et al.*[28] review recent observations. A detailed discussion of observational uncertainties is given by Davidson and Kinman[12]. A wealth of information on dwarf irregular galaxies is contained in the volume edited by Kunth, Thuan, and Tran Tanh Van[13]. In this review, I emphasize the abundance derivation methods and systematic errors involving the physical structure of the H II regions. I conclude by recommending a value for the primordial abundance, based on the recent results of Vilchez *et al.*[2] and an allowance for systematic errors.

2. Observations

Measurements of helium in dwarf irregular galaxies are based on spectrophotometry of the optical emission-line intensities. The most commonly used He I lines are λ5876, λ6678, and λ4471. These lines are produced mainly by radiative recombination, so that their intensities relative to Hβ give the He^+/H^+ ratio. The abundances of He^0 and He^{+2} in giant extragalactic H II regions are relatively small, and therefore He/H is close to He^+/H^+ in most cases.

Uncertainties affecting the line intensities include the calibration of the spectrograph, interstellar reddening, and underlying absorption lines in the stellar continuum. (This continuum is dominated by the star cluster associated with the H II region.) The strong He I λ5876 emission line can be affected by narrow telluric H_2O absorption lines[12], and for some redshifts, by interstellar Na D absorption in our Galaxy[5]. Measurements of several members of the Balmer series, typically Hα through Hϵ, allow a simultaneous solution for the reddening coefficient $C \equiv \log_{10} I(H\beta)$ and the equivalent width in absorption of the Balmer lines in the stellar continuum. Dinerstein and Shields[14], (hereinafter called "DS") found C = 0.35 and EWA = 1 A for the bright BCG NGC 4861; and these values are typical of giant extragalactic H II regions. (An additional problem in NGC 4861 was the presence of Wolf-Rayet emission blends at λ4600 and λ5800.) At present, one must guess the He I absorption equivalent widths in terms of EWA (Hβ)[15]. Because of the uncertainties caused by underlying absorption and W-R emission, helium abundance measurements are best done for objects with Hβ emission equivalent widths well in excess of 100Å.

3. Ionic Abundances

For the conditions of giant extragalactic H II regions, most He I emission lines are optically thin and strongly dominated by radiative recombination. For a line at wavelength λ formed purely by radiative recombination, the He^+/H^+ abundance ratio can be derived from the equation

$$\frac{I(\lambda)}{I(H\beta)} = \frac{\epsilon(\lambda)N(He^+)}{\epsilon(H\beta)N(H^+)}, \qquad (1)$$

where $\epsilon = \alpha^{eff}(\lambda)h\nu$ is the emission coefficient (ergs cm^3 s^{-1}) and α^{eff} is the effective recombination coefficient, accurately known for H and He[16,17].

Some complications result from the fact that the 2 ^3S level of He I is metastable and acquires a substantial population from recombinations onto the triplet levels. One effect is to produce a large optical depth in λ3889 (2 ^{3}S - 3 ^{3}P), which leads to fluorescent conversion of λ3889 photons into λ7065. DS found that λ7065 was enhanced $\sim$ 30% relative to the other helium lines of NGC 4861; and this corresponds to $\tau(\lambda 3889) \approx 10$, consistent with the expected value for the surface brightness of the object and a typical line width of $\sim$ 30 km s^{-1} (see § 4).

Recently, Ferland[18] has called attention to the importance of allowing for collisional excitation from 2 ^{3}S as a contribution to the He I line intensities. Earlier workers were aware of this effect[15]; but the rate coefficients calculated by Berrington *et al.*[19] provide an improved basis for calculating the effect and show that collisional excitation of λ6678 (a singlet) is significant. For the relative contribution by collisional excitation to λ5876, Ferland gives

$$\gamma_{CE}(\lambda 5876) \equiv \frac{I(\lambda 5876)_{coll}}{I(\lambda 5876)_{rec}} = \frac{12.0t^{0.56}e^{-3.776/t}}{(1+3390t^{-0.21}/n_e)}, \qquad (2)$$

where $t \equiv T/10^4$K. The corresponding ratio for λ6678 is about one-third as large. For the electron densities of interest, $n_e \leq 10^3$ cm^{-3}, γ_{CE} is proportional to n_e; and $\gamma_{CE}(\lambda 5876) \simeq 0.03$ for $n_e = 100\ \text{cm}^{-3}$ and $T = 1.4 \times 10^4$ K. Electron densities usually are derived from the [S II] doublet ratio $r \equiv I(\lambda 6717)/I(\lambda 6731)$. In most giant extragalactic H II regions, this ratio is near its low density limit of $r_{low} \simeq 1.4$. For the limited spectral resolution and signal-to-noise ratio of most studies n_e is not well measured. The value of γ_{CE} depends strongly on T, and this adds further uncertainty.

Ferland[18] analyzes published observations for 5 BCG and tabulates the implied helium abundances with and without allowance for collisional excitation. Allowing for collisional excitation, he finds a helium abundance $Y = 0.207 \pm 0.016$, low enough to conflict with the standard Big Bang model. Because of the serious implications of this result, I will take time to point out some uncertainties. For CG1116+51, Ferland adopts $n_e = 900\ \text{cm}^{-3}$ on the basis of the [S II] ratio measured by DS; and allowance for collisional excitation reduces Y from 0.26 to 0.20. However, Garnett[20] has obtained high resolution spectrophotometry of CG1116+51 in the red and finds a [S II] ratio near the low density limit, implying a much smaller correction for collisional excitation. From the observations of I Zw 18 by Davidson and Kinman[12], Ferland finds $n_e = 300$ cm and $Y = 0.186$. However, Figure 4 of Davidson and Kinman suggests that the [S II] ratio may be consistent with the low density limit. Davidson and Kinman's quoted "standard error" in the $I(\lambda 5876)$ for this faint object corresponds to $\pm$ 0.015 in Y, and they suggest an upward correction of order 10% for systematic errors. Ferland[18] also derives helium abundances for II Zw 40 and Tol 35 from the observations by Kunth and Sargent[6], adopting $n_e = 300\ \text{cm}^{-3}$ and 250 cm^{-3}, respectively. However, Kunth and Sargent's quoted errors for $\lambda 6717$ and $\lambda 6731$ are such that the ratio could be near the low density limit; and for both objects, $\lambda 4471$ gives a substantially larger He/H value than $\lambda 5876$. For Tol 35, the $\lambda 5876$ intensity measured by Kunth and Sargent, implies $He^+/H^+ = 0.074$, rather than the value 0.066 quoted by Kunth and Sargent and used by Ferland. Furthermore, Tol 35 has a rather low degree of ionization, $N(O^+)/N(O^{+2}) = 0.43$; and this means that the correction for He^0 could be large, as discussed in Section 4. Thus, although Ferland[6] is correct to emphasize the need to allow for collisional excitation, his derived value Y = 0.207 probably is too low.

4. The Ionization Correction

The He^+/H^+ ratio is converted to the elemental abundance by the use of an ionization correction factor,

$$He/H \;=\; i_{CF}(He^+/H^+). \tag{3}$$

Trace amounts of He^{++} are sometimes implied by the detection of narrow emission lines of He II, and the He^{+2}/H^+ ratio can be calculated from the nebular value of I(λ4686)/I(Hβ). More serious is the problem of allowing for He^0 in the H II region. This currently is done by guesswork, guided by photoionization models and the observed O^+/O^{+2} and S^+/S^{+2} ratios.

Photoionization models generally assume a central cluster of ionizing stars with ionizing photon luminosity

$$Q(H^0) \;\equiv\; \int_{\nu_H}^{\infty} L_\nu (h\nu)^{-1} d\nu, \tag{4}$$

surrounded by an infinite volume of gas, usually consisting of clouds or filaments of atomic number density n and volume filling factor f. A filling factor $f \approx 10^{-2}$ to 10^{-1} typically is needed, in order to reconcile the observed size, luminosity, and electron density. The photoionization and thermal equilibrium equations are solved as a function of radius, from the ionizing source outward until the $H^+ \rightarrow H^0$ transition zone, at the Strömgren radius R_S, has been passed. The local ionization fraction of an ion, $X(A^{+i}) \equiv n(A^{+i})/n(A)$ is averaged with respect to volume emission measure to give a mean ionic abundance

$$\langle X(A^{+i})\rangle \;\equiv\; \frac{\int X(A^{+i}) n_e dV}{\int n_e dV} \tag{5}$$

for comparison with observed fractional abundances. Models are characterized by (1) the chemical abundances, (2) the ionizing star temperature T_*, and (3) the ionization parameter,

$$U \equiv \phi_i / nc \tag{6}$$

where $\phi_i \equiv Q(H^0)/4\pi R_s^2$.

Photoionization models with $T_* \geq 40{,}000$ K have O^{+2} throughout most of their H^+ volumes, with O^+ confined to a small volume near R_S. In the transition zone, $X(He^+)$ is larger than $X(H^+)$; photons reaching this zone have frequencies just under $4\nu_H$, for which He^0 presents a larger photoionization cross section than H^0. As a result, one finds $i_{CF} = \langle X(H^+)\rangle / \langle X(He^+)\rangle = 0.98$ to 0.99. Thus the so-called "correction for neutral helium" is actually negative and of magnitude ~ 1 or 2%[21]. The value of $\langle X(O^+)\rangle$ varies roughly as U^{-1}. In contrast, for $T \approx 38000$ K, He^+ and O^{++} are confined to small volumes near the ionizing source; and He^0 and O^+ coexist throughout most of the H^+ volume.

The dwarf irregular galaxies used for helium abundance measurements typically have $O^+/O \simeq 0.15$ to 0.3, so that relatively hot ionizing stars are involved. Peimbert, Rodriguez, and Torres-Peimbert[22] suggested a helium ionization correction formula, $i_{CF} = (1 - 0.25 O^+/O)^{-1}$. This formula was based in part on spatial variations in resolved nebulae, and its application to integrated spectra of giant extragalactic H II regions may be wrong in principle. Moreover, the following results of DS illustrate the danger of using any such formula[21].

DS constructed a successful photoionization model of NGC 4861 with a single ionizing star cluster in the center. This model had $O/H = 1.0 \times 10^{-4}$, T_* = 45,000 K, and $U = 10^{-2.6}$. Alternatively, DS explored a composite model involving two segregated ionized volumes characterized by (1) T = 55,000 K and $U = 10^{-2.4}$ and (2) T = 35,000 K and $U = 10^{-3.0}$. Region (2) contributes only 12% of the $H\beta$ emission but has $\langle X(He^+)\rangle = 0.27$, so that the composite model has $\langle X(He^+)\rangle = 0.90$. Thus, for the same O^+/O ratio, the simple model with high T_* has $i_{CF} = 0.99$ whereas the composite model has i_{CF} = 1.09. More generally, such a composite model can have a neutral helium abundance $\langle X(He^0)\rangle$ almost as large as $\langle X(O^+)\rangle$.

These results suggest that one should avoid formulae for i_{CF} and instead restrict helium measurements to H II regions with O^+/O less than a cutoff value, say 0.15, in the hope that these may have $i_{CF} \simeq 1.00$. However, one must bear in mind that i_{CF} could be nearly as large as $(1 - O^+/O)^{-1}$. This may be true in individual cases, even if $< X(He^0) >$ is typically small for a large sample of high excitation H II regions. Therefore, one should be cautious when interpreting results for objects with exceptionally low values of He^+/H^+. DS suggest that the ionization of Ar may discriminate between simple and composite models; and Peña[23)] finds that, for star formation described by an IMF, the fraction of ionizing photons contributed by stars cooler than 38,000 K is likely to be small. However, further study is needed.

Photoionization models may also have a role to play with regard to the collisional excitation correction discussed in Section 3. The relative collisional contribution to the line emission, γ_{CE}, is proportional to n_e, which may not be determined reliably from the [S II] doublet ratio. However, the ionization parameter U can be determined by fitting photoionization models to the oxygen line intensities[14,21)] Then the electron density is given by $n_e \simeq \phi_i/cU$. The mean surface brightness for a filled sphere is

$$S(H\beta) = \pi^{-1}\phi_i\epsilon(H\beta)\alpha_B^{-1},$$

where α_B is the recombination coefficient of H^0 to all excited states. Thus, from the observed surface brightness and U as determined by a photoionization model, one can estimate n_e. When this method is applied to the giant H II complex S 10 in M101, I find $n_e \simeq 6\ \mathrm{cm}^{-3}$, based on (1) taking $Q(H^0)$, R, and O/H from Shields and Searle[24)]; and (2) determining $U = 10^{-2.5}$ and $T_* = 40{,}000$ K from Figure 1 of Shields[25)]. This density is too low to be reflected in the [S II] ratio, and it gives negligible excitation of He I. Future studies of extragalactic H II regions should give the surface brightness of the area contributing most of the $H\beta$ flux, whenever possible.

There remains the possibility that an H II region consists of unresolved, high surface brightness "knots." In this case, the measured $S(H\beta)$ and the derived n_e will be underestimates. The Hubble Space Telescope will be valuable in this regard. Another constraint on the surface brightness of the emitting regions is provided by the self-absorption of He Iλ3889 (see Section 3). From the principles described by Osterbrock[17)], one readily finds that $\tau(\lambda 3889)$ is proportional to S_{rec}(He I) $/\Delta$v, where S_{rec} is the surface brightness of any optically thin He I recombination line and Δ v is the line width. For a normal helium abundance, S_{rec}(He I) is proportional to S(Hβ). If $\tau(\lambda 3889)$ can be measured, for example from the fluorescent enhancement of $\lambda 7065$[17,26], then the surface brightness characterizing the emission region can be estimated; and the electron density follows from the ionization parameter. Consistency between the observed surface brightness and $\tau(\lambda 3889)$ would be a test of the presence of unresolved structure in the surface brightness or in the velocity field.

5. Results and Discussion

Recent observations are summarized by Kunth[11] and Pagel *et al.*[28]. Pagel *et al.* notes that when Y is plotted against O/H, several BCG with unusually large N/O ratios also have relatively large values of Y; and when Y is plotted against N/H, the diagram may have somewhat less scatter than for Y *vs.* O/H (see also Vigroux *et al.*[27]). On this basis, Pagel *et al.*[28] find Y = (0.238 $\pm$ 0.005) + (2.9 $\pm$ 1.5) $\times$ 10^3 N/H from an analysis of a sample of objects having high signal-to-noise observations, low O/H, low O^+/O, and large EWE (Hβ). Revising this result to take account of collisional excitation, Vilchez *et al.*[2] find Y_P = 0.232$\pm$ 0.004 (1σ error). These authors assumed n_e = 50 cm^{-3} for most objects. To this quoted error, arising from the scatter in the Y - N/H diagram, I suggest adding $\pm$ 0.010 to allow for uncertainties in the neutral helium correction and collisional excitation. This is based on assuming, rather arbitrarily, (1) that $<X(He^0)>$ typically is $\leq$ 0.05, and (2) that electron densities typically are $\leq$ 150 cm^{-3} in BCG. This leads to a recommended value

$$Y_P \;=\; 0.232 \pm\; 0.014 \qquad (6)$$

This is consistent with the standard Big Bang model[9].

Further improvements in the accuracy of primordial helium measurements will be needed for cosmologically interesting conclusions to be drawn. Observationally, linear detectors and verifiable calibration techniques at the level of 1 or 2% accuracy are necessary. Spectrographs with a resolution matched to the intrinsic line widths ($\sim$ 30 km s^{-1}) will minimize the error associated with setting the underlying continuum level; but several hydrogen and helium lines must be observed simultaneously, requiring a large number of detector elements. A better understanding of the ionizing star clusters will help with the problem of underlying He I absorption lines and the question of He^0 zones around cooler ionizing stars. Further photoionization model studies may also reveal spectral tests relevant to the He^0 problem. Spectroscopic observations at high spatial resolution will be important as a probe of the ionization structure (*e.g.*, the correlation of O^+/O with He^+/H^+). Studies of the Small Magellanic Cloud with Space Telescope may provide a valuable opportunity. The electron density needed for the collisional excitation correction can be studied by means of sensitive observations of the [S II] and [O II] doublets at high spectral resolution as well as by a combination of surface brightness measurements, λ7065 fluorescence measurements, and photoionization modeling. These techniques should be applied to a carefully selected sample of objects having $O/H \leq 10^{-4}$, $O^+/O \lesssim 10^{-1}$, and $W(H\beta) \gtrsim 200$Å. A reasonable goal for the next five years may be to restrict Y with high confidence to an uncertainty margin of $\pm5\%$.

I am indebted to Bernard Pagel for sharing results in advance of publication, and to Richard Matzner and Gary Steigman for valuable discussions. This research was supported in part by grant F-910 from the Robert A. Welch Foundation and NSF grant AST 83-14962. During the Fall, 1986 semester, the author was Visiting Professor in the Astronomy Department and Space Sciences Laboratory of the University of California, Berkeley.

REFERENCES

1. Pagel, B. E. J., personal communication (1986).

2. Vilchez, J., Pagel, B. E. J., Terlevich, R. J., and Melnick, J., in preparation (1986).

3. Peimbert, M., and Torres-Peimbert, S., *Ap. J.*, **193**, 327 (1974).

4. Lequeux, J., Peimbert, M., Rayo, J. F., Serrano, A., and Torres-Peimbert, S., *Astr. Ap.*, **80**, 155 (1979).

5. French, H. B., *Ap. J.*, **240**, 41 (1980).

6. Kunth, D., and Sargent, W. L. W., *Ap. J.*, **273**, 81 (1983).

7. French, H. B., and Miller, J. S., *Ap. J.*, **248**, 468 (1981).

8. Shaver, P. A., Kunth, D., and Kjär, K., eds., *ESO Workshop on Primordial Helium* (Garching: ESO) (1983).

9. Boesgaard, A. M., and Steigman, G., *Ann. Rev. Astr. Ap.*, **23**, 319 (1985).

10. Matzner, R., *Publ. Astr. Soc. Pac.*, **98**, 1049 (1986).

11. Kunth, D., *Publ. Astr. Soc. Pac.*, **98**, 984 (1986).

12. Davidson, K., and Kinman, T. D., *Ap. J. Suppl.*, **58**, 321 (1985).

13. Kunth, D., Thuan, T. X., and Tran Tanh Van, J., (eds.) *Star Forming Dwarf Galaxies,* (Paris: Editions Frontières) (1985).

14. Dinerstein, H. L., and Shields, G. A., *Ap. J.*, **311**, 45 (1986).

15. Rayo, J. F., Peimbert, M., and Torres-Peimbert, S., *Ap. J.*, **255**, 1 (1982).

16. Brocklehurst, M., *M.N.R.A.S.*, **157**, 211 (1972).

17. Osterbrock, D. E., *Astrophysics of Gaseous Nebulae*, (San Francisco: Freeman) (1974).

18. Ferland, G. .J., *Ap. J. (Letters)*, **310**, L67 (1986).

19. Berrington, K. A., Burke, P. G., Freitas, L., and Kingston, A. E., *J. Phys. B.*, **18**, 4135 (1985).

20. Garnett, D. R., personal communication (1986).

21. Stasinska, G., *Astron. Ap.*, **34**, 320 (1980).

22. Peimbert, M., Rodriquez, L. F., and Torres-Peimbert, S., *Rev. Mexicana Astron. Astrof.*, **1**, 161 (1974).

23. Peña, M., *Publ. Astr. Soc. Pac.*, **198**, 1061 (1986).

24. Shields, G. A., and Searle, L., *Ap. J.*, **222**, 821 (1978).

25. Shields, G. A., *Publ. Astr. Soc. Pac.*, **98**, 1072 (1986).

26. Robbins, R. R., *Ap. J.*, **151**, 511 (1968).

27. Vigroux, L., Stasinska, G., and Comte, G., in *Star Forming Dwarf Galaxies*, D. Kunth, T. X. Thuan, and J. Tran Tanh Van, eds. (Paris: Editions Frontières), p. 425 (1985).

28. Pagel, B. E. J., Terlevich, R. J., and Melnick, J., *Publ. Astr. Soc. Pac.*, **98**, 1005 (1986).

29. Peimbert, M., *Publ. Astr. Soc. Pac.*, **198**, 1057 (1986).

PRIMORDIAL HELIUM AND GALACTIC CHEMICAL EVOLUTION

John S. Gallagher, III
Lowell Observatory
Flagstaff, Arizona

Gary Steigman
Department of Physics, The Ohio State University
Columbus, Ohio

David Schramm
Department of Astronomy and Astrophysics, University of Chicago
Chicago, Illinois

ABSTRACT

Corrections of observed ^{4}He abundances for contamination by stellar debris are reconsidered by using ^{12}C instead of ^{16}O as a tracer of galactic chemical evolution. This method suggests that corrections for stellar contributions to HII region ^{4}He abundances in metal-poor dwarf galaxies have been overestimated and that the primordial ^{4}He abundance is close to the values observed in these systems.

1. INTRODUCTION

Big bang nucleosynthesis models provide strong predictions for the primordial abundance of ^{4}He, $Y_p = 0.25 \pm 0.015$[1]. Tests of this prediction can be made from spectrophotometric measurements of emission lines of HI and HeI from ionized gas HII regions in metal-poor galaxies[2-4]. Analysis of the relative intensities of these lines in combination with theoretical models for the ionization structure of HII regions gives Y_{gal}, the present abundance of ^{4}He in the interstellar gas. The precision of these estimates approaches 5%, although uncertainties still exist in both the HII region models and the interpretation of the HeI spectrum in terms of recombination lines[5,6]. An additional uncertainty is introduced in converting from Y_{gal} to Y_p, which requires that a correction be made for ^4He that has been injected into galaxies as a result of stellar nucleosynthesis, $Y_p = Y_{gal} - Y_*$.

2. STELLAR HELIUM PRODUCTION IN GALAXIES

Historically, the size of Y_* has been traced by using ^{16}O as an indicator of the degree of galactic chemical processing[7-8]. Oxygen abundances have the practical advantage that they can be measured in HII regions from the same optical spectrophotmetric data that are used in determining Y_{gal}. Typically the ^{16}O abundances are converted to total "metal" (i.e. everything but H and He) mass fractions Z by assuming solar system-like relative abundance distributions of elements. One can then use either observations or theoretical stellar nucleosynthesis models to predict the incremental change in Y_{gal} as a function of Z. The form that is usually adopted for this model is[9]

$$\Delta Z/\Delta Y = constant \approx 1 - 3.$$

On the basis of this expression, the observed Y_{gal} are corrected for stellar contamination and Y_p is derived.

There are hints of problems with this approach. For example, it is not clear whether the data for low metallicity galaxies support a simple linear $\Delta Y/\Delta Z$ model[3]. On theoretical grounds, one can object that ^{16}O is entirely produced by massive, short lived stars, while ^4He is injected by all stars more massive than about 1.5 solar masses[10-11]. The production of ^4He by a single stellar generation thus requires more than 10^9 yr versus about 10^7 yr for ^{16}O production. Because of this difference in time scales, ^{16}O will not track the ^4He abundance in a rapidly evolving galaxy or star forming region within a galaxy.

3. CARBON AS A TRACER OF GALACTIC CHEMICAL EVOLUTION

Recently Steigman, Gallagher, and Schramm[12] have reexamined the problem of estimating stellar helium contamination levels in low metallicity dwarf galaxies. We present only a brief summary of their paper here. The best way to trace stellar ^{4}He production would be to find some other element or set of elements that mimic the ^{4}He yields as a function of stellar mass. Unfortunately, no other element is as widely synthesized as ^{4}He, but ^{12}C comes close and is therefore a very useful indicator of degree of chemical evolution in simple galaxies.

Abundances of ^{12}C in interstellar gas can be obtained from *IUE* ultraviolet spectrophotometry for only a few galaxies[13–15]. Furthermore, there are uncertainties about the amount of ^{12}C that could be locked up in interstellar grains[16]. Despite these difficulties, the ^{12}C abundances present an interesting pattern. The C/O ratio decreases to lower metallicity levels, suggesting that ^{16}O abundances overestimate the level of chemical pollution by stars in metal-poor galaxies. It then follows that Y_* will tend to be overestimated, and as a result the deduced value of Y_p is too small.

This effect qualitatively amounts to a non-linear relationship between Y_* and Z, which is nearly flat at low Z and rises once Z begins to approach the solar metallicity level that is characteristic of chemically highly evolved galaxies. The standard linear fit model for Y_* connects the low and high Z regimes, and thus leads to excessive estimates for Y_* at the low Z values that are typical of metal-poor, gas-rich dwarf galaxies. However, if ^{12}C is used as an indicator of chemical enrichment, Steigman et al. show that the size of Y_* is negligible in these types of metal poor galaxies. This conclusion agrees with the observed absence of a correlation between ^{16}O and ^{4}He abundances in these types of galaxies[3].

Using the standard nebular models with no correction for effects of collisionally excited HeI emission, Steigman et al. find

$$Y_p = 0.235 \pm 0.012$$

from the published ^{4}He abundances for metal-poor dwarf galaxies. In this model, the Small Magellanic Cloud has experienced very little chemical evolution, and so provides an excellent site in which nebular ^{4}He abundance determinations can be improved and issues such as the role of collisional excitation of HeI emission can be readily tested. The measured Y_{gal} in the Small Magellanic Cloud will then essentially be a measurement of Y_p.

The outer reaches of spiral galaxy disks are also known to contain HII regions with low metal abundances, and these have been used to help establish Y versus Z relationships[17]. However, if time dependent effects are important in galactic chemical evolution, then these regions could give misleading results. The problem is that spiral galaxy disks normally have a radial gradient in abundance that increases inwards[18]. Thus chemical abundances in outer disks of spirals result from some unknown combination of local production and radial mixing of metals[19,20]. For example, one might find in the outer disk a region in which most of the metals have been mixed from an interior zone, with low metallicity resulting primarily from dilution of this material by unprocessed gas. The abundance distribution pattern in this gas will then be typical of high metallicity environments; e.g. the C/O abundance ratio would be higher than in dwarf galaxies having a similar ^{16}O abundance. Under these hypothetical conditions the Y_* - Z relationship would be that for a high rather than low metallicity system. Intrinsically metal-poor galaxies are therefore the best places in which to estimate Y_p from HII region data.

4. CONCLUSIONS

1. Corrections for stellar ^{4}He production in chemically unevolved galaxies must take into account the wide range of stellar masses and thus stellar lifetimes that produce ^{4}He. As a result, ^{16}O is not always a reliable indicator of the level of stellar ^{4}He contamination in galaxies.

2. ^{12}C is potentially a reliable tracer of the level of stellar ^{4}He production in metal-poor galaxies. An analysis of ^{12}C abundances in a few metal-poor, gas-rich dwarfs suggests the correction factors for stellar ^{4}He are negligibly small, and that classical HII region models then imply $Y_p =$ 0.235, in good agreement with predictions of standard big bang nucleosynthesis models.

3. Not all metal-poor environments necessarily have experienced the same history of chemical enrichment. Thus outer parts of spiral galaxy disks may have different relative abundance distributions than are found in metal-poor dwarf galaxies. Primordial ^{4}He abundance determinations from HII region observations are most safely carried out in chemically isolated systems, such as dwarf galaxies.

REFERENCES

1. Boesgaard, A. M. and Steigman, G. 1985, *Ann.Rev.Astron.Ap.*, **23**, 319.

2. Peimbert, M. and Torres-Peimbert, S. 1976, *Ap.J.*, **203**, 581.

3. Kunth, D. and Sargent, W.L.W. 1983, *Ap.J.*, **273**, 81.

4. Davidson, K. and Kinman, T. 1985, *Ap.J.Suppl.*, **58**, 321.

5. Dinerstein, H.L. and Shields, G.A. 1986, *Ap.J.*, **311**, 45.

6. Ferland, G. 1986, *Ap.J.(Letters)*, **310**, L67.

7. Peimbert, M. and Torres-Peimbert, S. 1974, *Ap.J.*, **193**, 327.

8. Pagel, B.E.J. 1982, *Phil.Tran.Roy.Soc.London Ser.A*, **307**, 19.

9. Maeder, A. 1983, In *ESO Workshop on Primordial Helium*, p.89.

10. Iben, I., Jr. and Truran, J. W. 1978, *Ap.J.*, **220**, 980.

11. Mallik, D.C.V. and Mallik, S.V. 1985, *J.Astrophys.Astron.*, **6**, 113.

12. Steigman, G., Gallagher, J.S., and Schramm, D.N. 1987, in preparation.

13. Dufour, R.J., Shields, G.A., and Talbot, R.J. 1982, *Ap.J.*, **252**, 461.

14. Dufour, R. J. 1985, In *Future of UV Astronomy Based on Six Years of IUE Research*, (NASA CP-2349), p.107.

15. Dufour, R.J. Schiffer, F.H., and Shields, G.A. 1985, *ibid.*, p.111.

16. Spitzer, L., Jr. and Jenkins, E.B. 1975, *Ann.Rev.Astron.Ap.*, **13**, 133.

17. Rayo, J., Peimbert, M., and Torres-Peimbert, S. 1982, *Ap.J.*, **255**, 1.

18. Pagel, B.E.J. and Edmunds, M. 1981, *Ann.Rev.Astron.Ap.*, **19**, 77.

19. Gallagher, J.S. 1984, In *Stellar Nucleosynthesis*, ed. C. Chiosi and A. Renzini (D. Reidel, Dodrecht), p.325.

20. Lacey, C.G. and Fall, M.S. 1985, *Ap.J.*, **290**, 154.

MEASUREMENTS OF COSMIC MICROWAVE RADIATION

D. T. Wilkinson

Department of Physics
Princeton University
Princeton, N. J. 08544

ABSTRACT

The spectrum and isotropy of the 2.7 K cosmic radiation carry information about important physical processes in the early universe. Spectral measurements from $\lambda = 50$ cm to $\lambda = 1$ mm have not uncovered spectral distortions at a level of a few percent absolute accuracy. Anisotropy measurements accurately find the Galaxy's velocity with respect to the radiation, but have failed to definitely see the perturbations which should accompany the formation of large scale structure. However, two tentative detections were reported at this conference.

1. Introduction

A workshop on the 2.75 K radiation was held at the 13th Texas Symposium. This paper is intended to provide a backdrop for the new results described in the brief contributions by workshop speakers. Mainly spectral and anisotropy measurements were discussed. In the past few years a new series of spectral measurements (following a 15 year lapse) have substantially improved the accuracy of measurements across the available wavelength band (50 cm to 1 mm). Isotropy measurements are gradually approaching levels of $\Delta T/T \sim$ (a few) x 10^{-5} at angular scales from 10 arcseconds to 90 degrees. Two tentative detections of anisotropy were described in workshop talks; both are at relatively long wavelength so must be checked for contamination by foreground radio sources (Galaxy or weak radio sources). No new

This work funded in part by the National Science Foundation and the National Aeronautics and Space Administration.

results on polarization measurements were presented. However, David Meyer[1] discussed intriguing results on attempts to use CI fine structure (λ = 0.61 mm) in a quasar absorption cloud at z = 1.776 to measure the temperature of cosmic microwave radiation at that epoch. They are able to put an upper limit of 16 K on the temperature, expected to be 7.6 K at z = 1.776; the rest wavelength of 0.61 mm is near the CMR peak wavelength of about 0.7 mm then.

2. Spectral Measurements

The cosmic microwave radiation (CMR) spectrum might be distorted by several mechanisms. A large energy release (by decaying particles?) in the redshift range $10^9 > z > 10^3$ could leave a signature in the radiation today. Hot reionized plasma at more recent epochs 300 > z > 3 could distort the spectrum by adding Bremsstrahlung radiation at long wavelength and by scattering low energy photons to shorter wavelengths (Compton diffusion). Neither of these effects has yet been seen.

The current spectral measurements are summarized in Figure 1.

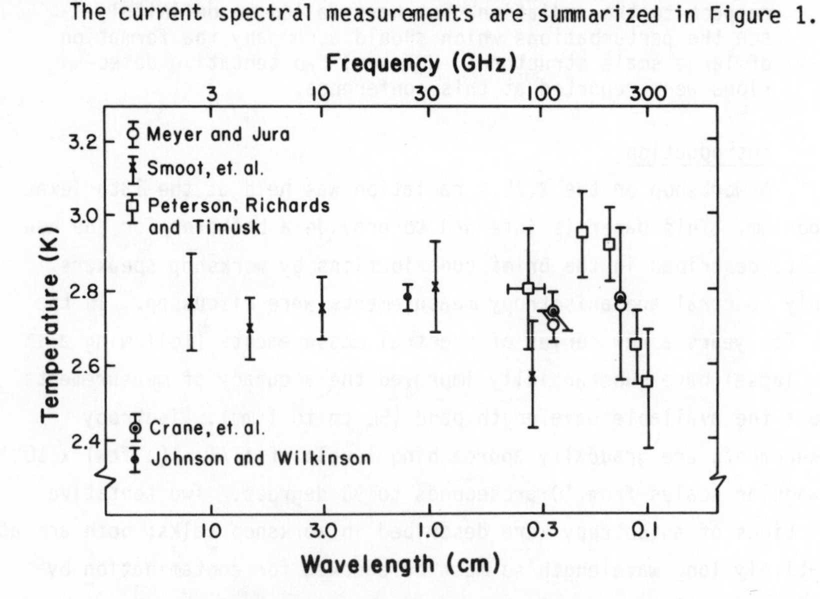

Figure 1. Recent measurements of the temperature of the Cosmic Microwave Radiation. The experiments are described in the text. A weighted mean of these points gives a temperature of 2.756 ± 0.016 K.

Using interstellar CN molecular excitation, Meyer and Jura[2] have made sensitive measurements of the CMR temperature at wavelengths of 2.64 mm and 1.32 mm -- the excitation energies of the two lowest rotational levels. They have extended[3] the work to include 3 molecular clouds, all giving temperatures consistent with 2.70 ± 0.04 K at 2.64 mm and 2.76 ± 0.20 K at 1.32 mm. At this conference Crane, et al.[4] reported similar results, and also an important new measurement of the intrinsic line width and, hence, the saturation correction. The possibility of some collisional excitation of the CN still contributes significantly to the uncertainty in these results.

The results labeled Smoot, et al.[5] were obtained from White Mountain California with radiometers which shared a common L He reference load and simultaneously measured atmospheric emission. At longer wavelengths errors are due mostly to emission from the Galaxy and warm parts of the L He load; at short wavelength subtraction of atmospheric emission is the main problem. The results agree with older ground-based measurements (see Weiss[6]) but have 2 to 3 times better accuracy. Peterson, et al.[7] (and reported at this conference) have measured the spectrum over the peak using filters and a bolometer at balloon altitudes. These results (T_{AVG} = 2.78 ± 0.11 K) do not confirm the higher temperature found by Woody and Richards[8], but the trend of the points (high from 2 mm to 1.5 mm, low at 1 mm) do mimic the distortion reported by Woody and Richards.

A new measurement[9] (reported by poster at this conference) has achieved 1% absolute accuracy at a wavelength of 1.2 cm. A sketch of the apparatus is shown in Figure 2. It is flown in a balloon at 26 km altitude to eliminate atmospheric emission. The critical parts of the radiometer and the cold load are contained in a L He bath to approximate thermal equilibrium with the CMR. Motors move the cold load out and lower the horn antenna to observe the sky. Thus, a cold reference and the CMR are observed with relatively small corrections needed to the basic measurement. The main systematic errors are due to uncertainty in the emission temperature of the Kapton window (T_{WND} = 35 ± 12 mK) and the wall radiation due to warm parts of the horn antenna (T_{HRN} = 50 ± 12 mK). When all corrections and errors are accounted for the result of this measurement is T_{CMR} = 2.783 ± 0.025 K, where this error is found by combining (many) systematic errors by quadrature; it should be regarded as a one standard deviation error.

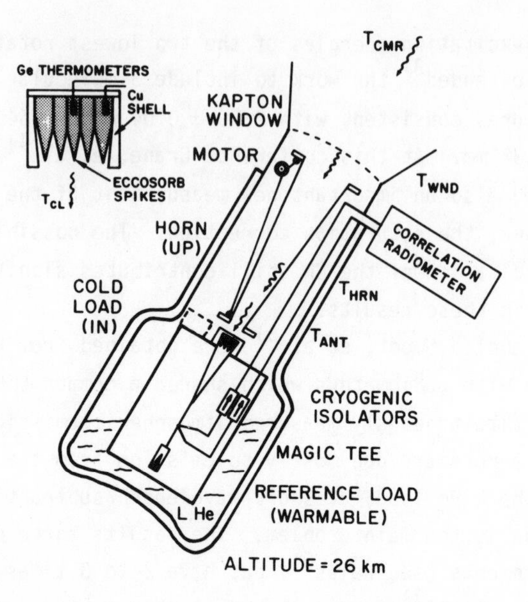

Figure 2. The balloon-borne radiometer used to obtain the point at λ = 1.2 cm in Figure 1. The critical cold (reference) load is shown expanded at the upper left. The radiometer alternately measures the radiation temperatures of the cold load and the horn antenna.

The weighted mean of all measurements shown in Figure 1 is T_{CMR} = 2.756 ± 0.016 K. This fit to a constant temperature hypothesis gives a Chi-square of 14 for 15 degrees of freedom indicating that 1) deviations are not statistically significant and 2) experimenters have done a reasonable job of estimating errors even though they are mainly systematic.

Progress on an important new measurement at λ = 50 cm was reported by Sironi[10]. Contamination by Galactic radiation (~ 7 K) and ground radiation (~ 6 K) currently limits the accuracy of these results to ± 0.7 K, but prospects are good for lowering these errors by further measurements and modeling of Galactic emission and antenna sidelobes. Such long wavelength measurements are important in looking for effects of reheated intergalactic plasma at the redshift of active star formation.

3. Anisotropy Measurements

Anisotropy is being used to study two important cosmic processes, the initial density perturbations which led to large scale structure of matter and a measurement of our velocity with respect to the comoving frame (presumed to be the same as that of the CMR). The only anisotropy measured to date is the dipole effect; the elusive primordial perturbations have not yet been identified. When this is finally accomplished, a wealth of detailed information (amplitude and angular spectrum of perturbations) will be available about the universe at $z \sim 1000$. Until then, the observed upper limits serve to constrain the many possible models[11] Figure 3 summarizes the current best

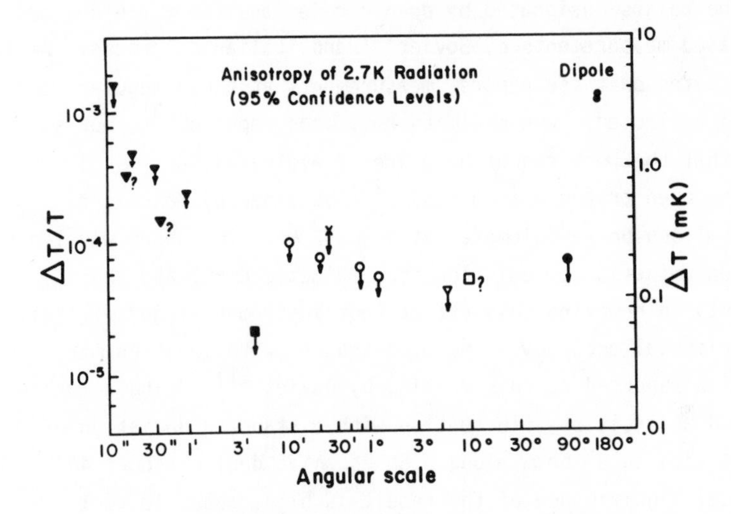

Figure 3. Experimental results on anisotropy of the Cosmic Microwave Radiation. The points marked with (?) are possible detections reported at this conference (see text). Except for these and the dipole effect, all results are upper limits.

measurments of CMR anisotropy. Starting from the left we see relatively recent measurements at angular scales less than 1 arcminute made with the VLA.[12] Using an interferometer to search for weak background fluctuations is exacting work, requiring special care in reducing the data. The techniques have now been developed to a high degree and, intriguingly, faint extended sources are being seen.[13] The filled triangles with question marks are representative of the signals being detected at $\lambda = 6$ cm. Primordial fluctuations are not expected[14] to exist at this small angular scale. Probably what is being seen is 1) the tail of the weak radio source spectrum, or 2) emission from knots of reheated plasma at quite high redshift[15] ($z \geqslant 10$). Observations at other wavelengths (preferably shorter) are needed to identify the sources from their spectral signatures.

The point at 4.5 arcminutes was obtained using the 1.2 cm maser amplifier and 140' telescope at the NRAO's Green Bank observatory.[16] The upper limit has been used to rule out models of structure formation from adiabatic, baryon-dominated, density fluctuations. Cold dark matter models[14] predict excessive anisotropy for an open universe ($\Omega <$ 0.3). The points designated by open circles and the x are due to ground-based measurements by Soviet[17] and Italian[18] groups. As seen from the dates on these papers, measurements at these angular scales have been neglected. Lower limits have been reported[19], but so briefly that the work cannot be properly evaluated.

The open triangle is a result[20] obtained by Melchiorri's group using a balloon-borne bolometer at $\lambda = 0.7$ mm. The short wavelength led to contamination by emission from Galactic dust, and the uncertainty in removing this effect (not instrumental noise) limited the experimental accuracy. The open square (with question mark) shows a detection reported at this meeting by Davies.[21] Using a maximum likelihood analysis sky fluctuation with a standard deviation of 0.1 mK are found with an 8° beam along a 50° scan at declination + 40°. The statistical significance of the result is high, about 10 to 1. However, at a frequency of 10.4 GHz Galactic emission is a serious problem, unless accurate measurements are made of the same region at another frequency. Extrapolating from long wavelength maps is exceedingly risky because of uncertainties in the spectral index β, where $T_{GAL} \propto \lambda^{\beta}$ and $\beta \approx -2.8$. For example, extrapolation from the 408 MHz map of Haslam, et al.[22] creates bumps of $\pm$ 0.1 mK in T_{GAL} at 10.4 GHz if the spectral index is patchy (on an 8° scale) to $\pm$ 0.5%, about an order of magnitude below the known constancy of β. Furthermore, the measured strip is not far from the North Polar Spur, a region known for spectral index anomalies. Until the same piece of sky is carefully measured at a nearby frequency (work underway at 5.0

GHz by Davies, et al.) the signal at 8° in Figure 3 should be regarded as probably arising from Galactic emission.

The problems encountered in the preceeding two experiments illustrate that there is an optimum wvelength range for CMR anisotropy experiments somewhere around 3 mm. At longer wavelengths radio sources and Galactic radiation create troublesome foreground contamination, at shorter wavelengths, Galactic dust emission is a problem.

The points in Figure 3 at angular scales of 90° (quadrupole)and 180° (dipole) are measured using special radiometers in balloons or satellites to avoid large scale fluctuations in atmospheric emission. Typically, small horn antennas (beamwidth ~ 7°) are used to compare the CMR temperature between two points 90° apart in the sky. Balloon-borne measurements have made accurate dipole measurements at 3 mm[22] and 1.2 cm.[23] More recently, a Soviet satellite, reaching beyond the Moon's orbit has made a large-scale anisotropy measurement[24] at a wavelength of 8 mm. Again, Galactic emission must be estimated and removed from the two longer wavelength measurements, but the 3 mm measurement has somewhat higher statistical noise. Sky coverage is good for these measurements, so good fits of the dipole model to sky maps[25] are obtained. Results for the measured dipole distribution are given in Table I. Clearly, all three measurements are in excellent agreement.

Table I.

Research Group	Dipole Amplitude (mK)	Right Ascension (hr)	Declination (degrees)	Reference
Berkeley	3.46 ± 0.17	11.3 ± 0.1	- 6.0 ± 1.4	(22)
Princeton	3.18 ± 0.17	11.2 ± 0.1	- 8 ± 2	(23)
Moscow	3.16 ± 0.12	11.3 ± 0.15	- 7.5 ± 2.5	(26)

Most, if not all, of this dipole effect is due to the Sun's motion through the CMR [Dipole Amplitude = (2.75 K) v_{sun}/c]. The motion of the Galaxy (or local group) is of cosmological interest, and to find it one simply subtracts the Sun's velocity with respect to the Galactic center from the mean velocity obtained from Table I. The Galaxy's velocity through the CMR is found to be 600 ± 50 km sec^{-1} toward galactic longitude 265° and galactic latitude + 25°, in the constellation Hydra. This speed is surprisingly large compared to typical random velocities of galaxies in clusters; also it is not directed toward the largest known mass clump nearby -- the Virgo cluster. Possible explanations of this are that 1) there are other unknown mass clumps or 2) we are part of a very large scale mass flow which is carrying the Galaxy, the Virgo cluster, and a large region of the local universe through the CMR. This last explanation would resolve the long-standing disagreement between the Galactic velocity measured with respect to the CMR and the velocity measured with respect to galaxies out to a distance of 50 Mpc.

Anisotropy measurements are about to reach a new stage of sophistication. Instrument sensitivities have improved through the use of cryogenics and wideband bolometers. Also, better strategies are developing to minimize atmospheric noise, for example symmetric scan patterns are being used to cancel gradients in the very red sky noise spectrum. The use of balloons, rockets and satellites is increasing. Experiments now being built should reach a level of $\Delta T/T = 10^{-5}$ at angular scales from 20 arcminutes to 90°. If the anisotropies are indeed just below current limits, as most of us feel they must be, the next few years should see this field turn from one of searching into one of studying.

REFERENCES

1. Meyer, D. M., Black, J. H., Chaffee, F. H., Jr., Foltz, C. B. and York, D. G., Astrophys. J. Letters 308, L37 (1986).
2. Meyer, D. M. and Jura, M., Astrophys. J. Letters 276, L1 (1984).
3. Meyer, D. M. and Jura, M., Astrophys. J. 297, 119 (1985).
4. Crane, P., Hegyi, D. J., Mandolesi, N. and Danks, A. C., Astrophys. J. 309, 822 (1986).
5. Smoot, G. F., De Amici, G., Friedman, S. D., Witebsky, C., Sironi, G., Bonelli, G., Mandolesi, N., Cortiglioni, S., Morigi, G., Partridge, R. B., Danese, L. and De Zotti, G., Astrophys. J. Letters 291, L23 (1985).

6. Weiss, R., Ann. Rev. Astron. Astrophys. 18, 489 (1980).
7. Peterson, J. B., Richards, P. L. and Timusk, T., Phys. Rev. Letters 55, 332 (1985).
8. Woody, D. P. and Richards, P. L., Astrophys. J. 248, 18 (1981).
9. Johnson, D. and Wilkinson, D. T., Astrophys. J. Letters (Feb. 1, 1987).
10. Sironi, G., Bersanelli, M., Bonelli, G., Conti, G., Marcellino, G., Limon, M. and Reif, K., This Volume.
11. Bond, J. R., This Volume.
12. Knoke, J. E., Partridge, R. B., Ratner, M. I., Shapiro, I. I., Ap. J. 284, 479 (1984). Fomalont, E. B., Kellerman, K. I. and Wall, J. V., Astrophys. J. Letters 277, L23 (1984). Partridge, R. B. and Knoke, J. E., Inner Space/Outer Space Conference, eds. Kolb, et al. (Chicago University Press, 1984).
13. Fomalont, E. B., Kellerman, K. I. and Wall, J. V., This Volume. Martin, H. M. and Partridge, R. B., submitted to Astrophys. J. Letters (1986). Partridge, R. B., IAU Symposium 124, Beijing, China (1986).
14. Bond, J. R. and Efstathiou, G., Astrophys. J. Letters 285, L45 (1984). Vittorio, N. and Silk, J., Astrophys. J. Letters 285, L41 (1984).
15. Hogan, C. J., Astrophys. J. Letters 284, L1 (1984). Ostriker, J. and Vishniac, E. T., Astrophys. J. Letters 306, L51 (1986).
16. Uson, J. M. and Wilkinson, D. T., Astrophys. J. 283, 471 (1984); Nature 312, 427 (1985).
17. Parijskij, Yu. N., Petrov, Z. E. and Cherkov, L. N., Sov. Astron. Letters 3, 263 (1977). Also see, Lasenby, A. N. and Davies, R. D., M.N.R.A.S. 203, 1137 (1983).
18. Caderni, N., De Cosmo, V., Fabbri, R., Melchiorri, B., Melchiorri, F. and Natale, V. Phys. Rev. D16 2424 (1977).
19. Berlin, A. B., Bulaenko, E. V., Vitkovsky, V. V., Kononov, V. K., Paijskij, Yu. N. and Petrov, Z. E., IAU Symposium 104, eds. Abell and Chincarini (Reidel Publishing Co., 1983).
20. Melchiorri, F., Melchiorri, B., Ceccarelli, C. and Pietranera, L. Astrophys. J. Letters 250, L1 (1981).

21. Davies, R. D., Watson, R., Daintree, E. J., Hopkins, J., Lasenby, A. N., Beckman, J., Sanchez-Almeida, J. and Rebolo, R., This Volume.
22. Lubin, P. M., Epstein, G. L. and Smoot, G. F., Phys. Rev. Letters 50, 616 (1983).
23. Fixsen, D. J., Cheng, E. S. and Wilkinson, D. T., Phys. Rev. Letters 50, 620 (1983).
24. Strukov, I. A. and Skulachev, D. P., Sov. Astron. Letters 10, 1 (1984).
25. For color pictures of the two long wavelength maps see: Wilkinson, D. T., Science 232, 1517 (1986).
26. Revised dipole fit reported by Lukash at IAU Symposium 124, Beijing, China (1986).
27. Rubin, V. C., Thonnard, N., Furd, W. K. and Roberts, M. S., Astron. J. 81, 719 (1976).

THE MISSING LINK: LYMAN ALPHA CLOUDS AS A RELIC OF PRIMORDIAL CDM DENSITY PERTURBATIONS

J.R. Bond[1], A.S. Szalay[2] and J. Silk[3]
1. CITA, Toronto; 2. Eotvos University, Budapest; 3. U.C. Berkeley

ABSTRACT: We summarize our picture[1)] of Lyman alpha clouds as expanding relics of primoridial density peaks of subgalactic mass in cold dark matter (CDM) models whose pressure-driven expansion is triggered by photoionization at $z \sim 4$. We show that the HI column density distribution as a function of N_{HI} and redshift for these clouds is compatible with observations for an ionization history of the IGM which just satisfies the Gunn-Peterson (GP) constraint.

The Biased CDM Model: In the standard normalization[2)] for biased cold dark matter assuming $\Omega = 1$, $\Omega_B = 0.1$, $h = 0.5$, the peaks of the density field of fractional amplitude $\delta\rho/\rho = \nu\sigma$ above 2.8σ which produce bright galaxies are just collapsing at $z = 3.7$. Here, $\sigma(R_f, z)$ is the *rms* amplitude of density fluctuations at redshift z (extrapolated to low z using the linear growth law for fluctuations), smoothed on a Gaussian filtering scale R_f (with mass $M \sim 10^9\ M_\odot(R_f/100\text{kpc})^3$). This normalization implies $\sigma(R_f, z = 0) \approx 5.4(R_f/100\text{kpc})^{-0.4}$, so 100 kpc scale peaks of height above about 1σ would be nonlinear at $z = 3.5$. The fate of clouds of various masses between $\sim 10^8 - 10^{11}\ M_\odot$ ($R_f \sim 50-400$ kpc) depends upon whether they have already collapsed when (and if) a strong photoionizing flux is liberated during the early phases of galaxy formation, and whether the masses of those still collapsing are above the Jeans mass at photionization: such objects presumably form dwarf galaxies. Less massive clouds still dynamically evolving can expand and be observed as the rapidly evolving population of Lyα clouds.

Photoionization History: Strong photoionization, assumed here to onset suddenly at redshift $z_{pi} \approx 3.5$, raises the temperature of the metal-deficient gas to $T \approx 3 \times 10^4$K. Prior ionization is not precluded in our model provided $T \leq 10^4$K. For the small abundance of HI allowed by the GP constraint (optical depth to the quasar ~ 0.1) to arise totally from photoionization places a severe constraint on the average number of ionizing photons per baryon Y_γ; e.g., at $z = 3.5$, $Y_\gamma \approx 1$ is the minimum required. This exceeds the expected flux from quasars by a factor of $\sim 10 - 100$, but massive stars ($\sim 30\ M_\odot$) of cosmological density Ω_{MS} generate $Y_\gamma \sim 5(\Omega_{MS}/10^{-6})$, so primeval galaxies can certainly generate such a flux. UV and optical background light constraints only restrict Y_γ to be $< 5 \times 10^3$.

Cloud Hydrodynamics: Initial conditions for the clouds at redshift z_i were taken to be spherical Gaussian profiles around a central peak in the perturbed density, $\delta(r) = \delta_c(z_i)\exp(-r^2/2r_c^2)$. Two parameters completely specify a model cloud, r_c, the initial comoving radius of the cloud, and $\nu\sigma$, the central overdensity extrapolated by linear theory to the epoch of photoionization ($\nu\sigma(R_f, z_{pi}) = (1+z_i)\delta_c(z_i)/(1+z_{pi})$, where ν is the relative height of the collapsing peak). We loosely identify clouds of size r_c with peaks of the primordial density field smoothed on the scale $R_f \approx r_c$, though this relation depends somewhat on cloud height. Neutral hydrogen density profiles were determined as a function of redshift using a spherically symmetric hydro code. We can compute the HI column density for each cloud N_{HI} $(r_c, \nu\sigma, b, z)$ as a function of r_c, $\nu\sigma$, the impact parameter of the line of sight from the center b and the redshift z (Fig.1).

N_{HI} **Distribution Function**: The comoving number density of clouds with perturbation amplitude between ν and $\nu + d\nu$ can be estimated from the differential peak density $\mathcal{N}_{pk}(\nu, R_f)d\nu$ determined from the theory of Gaussian random fields[2]. Typically $\mathcal{N}_{pk}$ is small for $\nu < 0$, approximately flat for $\nu \sim 1-2$, with a precipitous falloff for $\nu > 3$. For a given cloud scale r_c and redshift z, the relation N_{HI} (ν, b) can then be used to obtain the column density (comoving) distribution function for clouds constrained to be observed at impact parameter b: $n_{cl*}(N_{HI}|b)\ d\log(N_{HI}) \approx \mathcal{N}_{pk}(\nu(N_{HI}, b), r_c)$ $\frac{\partial \nu(N_{HI}, b)}{\partial \log(N_{HI})}\ d\log(N_{HI})$. We plot $n_{cl*}(N_{HI}|0)$ in Fig.2. The number of clouds per unit redshift follows from $\frac{dN_{cl*}}{dz d\log(N_{HI})} = H_0^{-1}(1+z)^{1/2} \int n_{cl*}(N_{HI}|b)\ 2\pi b db$.

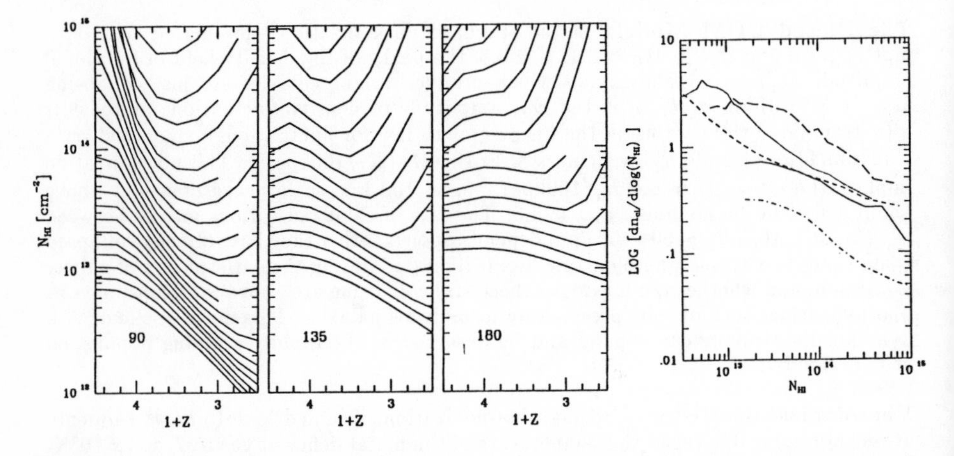

Figure 1: N_{HI} evolution, assuming $Y_\gamma(z) = 1$ after $z_{pi} = 3.5$, for $b = 0$ and $r_c = 90$, 135, and 180 kpc. The amplitudes $\nu\sigma(z_{pi})$ range (from the bottom up) between 1 and 3 in steps of 0.1. The 90 kpc model includes $\nu\sigma = 3.2$ as well. Photoionization occurs at the left boundary. N_{HI} curves for other photoionization histories are obtained by multiplying by $Y_\gamma(z)$. It is remarkable that with the minimal photoionization flux required by the GP constraint, our inferred HI column densities are in the range observed for the Lyα clouds.

Figure 2: $n_{cl*}(N_{HI}|b = 0)$ for various cloud radii and redshifts. The $r_c = 90$ kpc distribution function at redshifts $z = 3$ (long dashed), 2.5 (dotted) and 2 (short dashed) demonstrate redshift evolution for the $Y_\gamma = 1$ photoionization history. The drop in amplitude from $z = 3$ to $z = 2$ is comparable to that observed ($\sim (1+z)^{1.2\pm0.4}$). The shapes should be compared with the observed distribution ($\propto N_{HI}{}^{-0.8}$) inferred from Lyα equivalent widths. The larger scale clouds $r_c = 135$ (solid) and 180 (dot-dashed) kpc, shown at $z = 2.5$, give similar shapes. The overall cloud abundance is comparable to that inferred for Lyα systems and is similar to the dwarf galaxy abundance.

1. Bond, J. R., Szalay, A. S. and Silk, J. 1986, *Ap. J.* , submitted.
2. Bardeen, J. R., Bond, J. R., Kaiser, N. and Szalay, A.S. 1986, *Ap. J.* **304**, 15.

LARGE SCALE STRUCTURE IN THE BACKGROUND RADIATION

PHILIP M LUBIN

SPACE SCIENCES LABORATORY AND LAWRENCE BERKELEY LABORATORY
UNIVERSITY OF CALIFORNIA, BERKELEY, CA.

Introduction

One of the fundamental tests of the assumptions of isotropy and homogeneity is the measurement of the large angular scale structure of the cosmic background radiation. These measurements give us a snapshot of the universe at a time when the physical conditions were much different than now. In the twenty years since the discovery of the radiation, a substantial amount of attention has been given to developing this picture. We now have near full sky maps ($>$ 80%) from balloon measurements at 3 and 12 mm wavelength and from the Soviet Prognos 9 satellite at 8 mm.

Current Results

The dominant term visible in all of these maps is the dipole distribution with an amplitude of about 3 mK. Substantial galactic contamination is also evident in both the 8 mm and 12 mm data. Each experiment uses approximately gaussian beams and has a similar angular resolution with σ=2.5 ° (6 ° FWHM) for $P(\theta)=e^{-\theta^2/2\sigma^2}$. These experiments all use coherent detectors with about a 1 GHz bandwidth. The accuracy of the dipole amplitude is currently limited to about 5% primarily from calibration errors while the statistical error is 1 - 2%. The dipole directions of the 3mm and 12mm data have an error of about 1.5 ° primarily due to pointing reconstruction uncertainty (magnetometer errors) while the dipole directions of the 3mm and 12mm data are 1.6 ° apart. These data give a Solar velocity direction of $\alpha = 11.25 \pm 0.15$ hours, $\delta = -5.6 \pm 2.0$ ° and an average dipole amplitude of 3.26 ± 0.23 mK where the increased errors include the slight differences between the dipoles. Using a Galactic Solar velocity of 230 km s^{-1} toward l^{II} = 90 °, b^{II} = 0 °, yields a Galactic velocity of 540 ± 50 km s^{-1} towards $\alpha = 10.6 \pm 0.3$ hours, $\delta = -23 \pm 5$ °. Assuming that the velocity of the Sun relative to the Local Group is 295 km s^{-1} toward l^{II} = 97.2 °, b^{II} = -5.6 ° (Sandage, private communication 1986) gives an inferred Local Group velocity relative to the background radiation of 610 ± 50 km s^{-1} towards $\alpha = 10.8 \pm 0.3$ hours, $\delta = -25 \pm 5$ ° or $l^{II} = 272 \pm 5$ °, $b^{II} = 30 \pm 5$ °. This gives an angle of 45 ° between the center of the Virgo cluster and the Local Group velocity. Figure 1. shows the map made by combining the 3mm and 12mm data and covers 90% of the sky with a limiting sensitivity of about 0.3 mK per 6 ° field of view. The map is given in celestial coordinates as a $\cos(\delta)$ projection. Figure 2. is the same map after removal of the dipole. In both maps the galactic emission was removed from the 12mm data. Figure 3. gives the autocorrelation of the residual map (Fig. 2.) as a function of angle. Monte Carlo simulation of the maps gives an error for the autocorrelation of 0.01 mK^2. No structure is apparent in the residual map, although as discussed below this is difficult to quantify precisely.

The Upper Limits Game

Looking for an uncertain signal, of unknown signature and of unknown magnitude is difficult at best. Subtle systematic errors, atmospheric statistics, long term detector stability, varying antenna sidelobe pickup and numerous other effects may cause an otherwise observable signal to disappear into the noise. The converse is also true. Witness the difficulty in confidently detecting the cluster (Zeldovich) cooling effect which for some clusters is apparently an order of magnitude above the current sensitivity level quoted for fluctuations in the background radiation. The proper statistical interpretation of data to produce null upper limits is argued to factors of 2-3, while cosmological models are being rejected at these same levels perhaps unnecessarily. These difficulties apply to a varying degree at all angular scales. One of the fundamental problems we face is not knowing what we are looking for. It is possible that even in the present maps a signal is present but we do not know its signature. Substantial effort is needed to more fully exploit the existing and future maps so that a real signal will not escape detection.

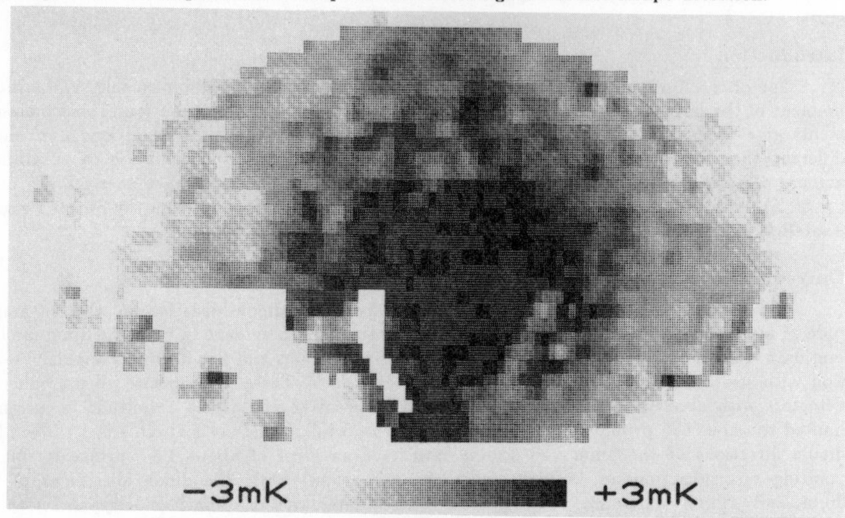

Figure 1 - Map of the sky in celestial coordinates at 3 and 12mm.

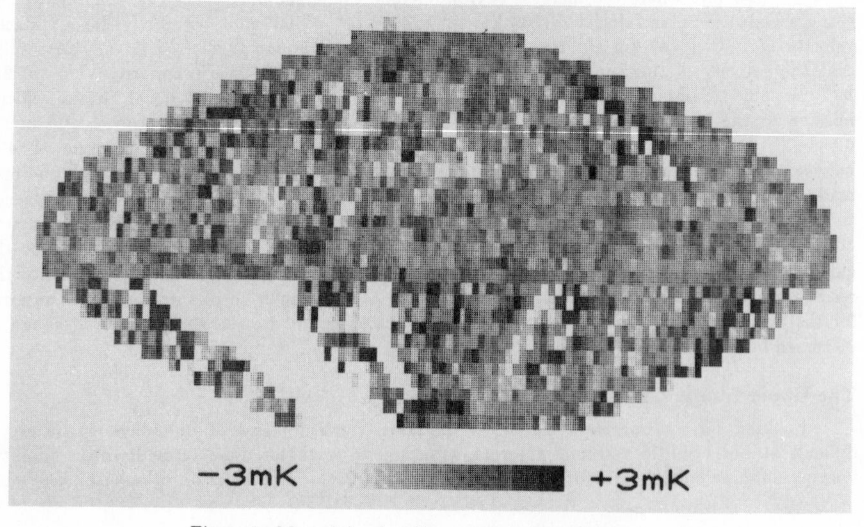

Figure 2 - Map of the sky after removal of the dipole.

Calibrators

Known calibrators or bench marks are useful as cross checks of the overall experiment performance. For large scale measurements there are several possibilities covering many order of magnitude. Known temperature targets are useful but do not test far field response. The moon is a good target for inflight calibration and tests both pointing and detector gain. The moon typically appears as about a 1 K signal in a 6° beam. This level insures high statistics in a short time while being small enough to avoid saturation effects. Unfortunately the moon is only good as a 3-5% absolute calibrator due to uncertainty in its radiometric emission. The dipole itself is a useful bench mark at the 10^{-3} level and observation of the earths orbital motion around the sun is another at the 10^{-4} level. At the 10^{-5} level planetary emission can be observed at a predictable level (factor of 2). Emission from our galaxy is another bench mark but our ability to a priori predict its magnitude is difficult particularly at millimeter wavelengths. For smaller angular scales down to tens of arc minutes, planetary emission is still a useful weak bench mark. At smaller scales, radio sources are available to some extent but cannot always be reliably predicted. Bench marks should be used whenever possible particularly if 1) they do not appreciably reduce the total observing time 2) they are predictable in value and 3) they produce a signal within an order of magnitude of the ultimate sensitivity. Such measurements would help the community judge the merit of each experiment.

Future Experiments

Currently upper limits on low order multipole moments (quadrupole, octupole) are $\sim 5\ x\ 10^{-5}$. At this level galactic emission is already a serious consideration at 8 and 12 mm. By going to systematic multi-frequency measurements perhaps another order of magnitude in sensitivity can be gained. The COBE satellite to be launched by 1990 will make a one year measurement at 3,6 and 9 mm and should reach a sensitivity of 0.1 mK per 6° field of view and a few parts in 10^{-6} to low order multipole moments. COBE will also give us a much better understanding of the diffuse galactic emission which will then allow future ground based and airborne experiments to take advantage of this increased knowledge to make more sensitive measurements.

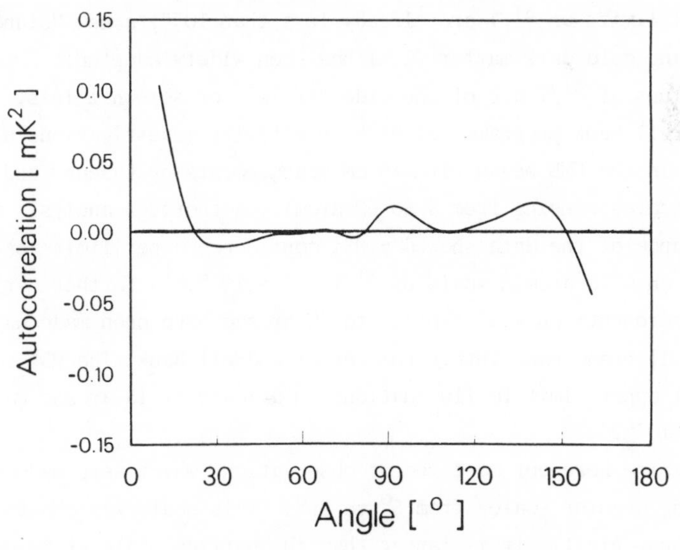

Figure 3 - Autocorrelation of residual map (Fig. 2).

FLUCTUATIONS IN THE CMB ON ANGULAR SCALES OF 5° TO 15°

R. D. Davies
University of Manchester, Nuffield Radio Astronomy Laboratories
Jodrell Bank, Macclesfield, Cheshire U.K.
A.N. Lasenby
University of Cambridge, Mullard Radio Astronomy Observatory
Madingley Road, Cambridge

ABSTRACT

High sensitivity observations at λ3 cm have detected fluctuations in the sky at a level of $\delta T/T = 3.7 \times 10^{-5}$ on an angular scale of 8°. These fluctuations are most likely intrinsic to the CMB; a component of galactic emission could also be present.

1. Introduction

The Cosmic Microwave Background (CMB) is believed to originate at the epoch ($z \sim 1000$) when electrons and protons recombine as the Universe cools subsequent to the Big Bang. At this epoch the Universe is evidently highly isotropic ($\delta T/T << 1$). Some degree of anisotropy on a wide range of angular scales is expected as a precursor of the structure we see in the present day ($z=0$) Universe. Scenarios invoking ordinary baryonic matter predict $^{\delta T}/T$ in the range 10^{-4} to 10^{-3} for accepted values of Ω, the density of the Universe relative to that required for closure. Since observational limits on $^{\delta T}/T$ are already less than 10^{-4}, a new "standard" model involving cold dark matter (CDM) has been widely adopted; predicted values of $^{\Delta T}/T$ are of the order of 10^{-5} or somewhat less.

The Jodrell Bank programme of high sensitivity investigations of fluctuations in the CMB began with λ6 cm measurements by Lasenby and Davies[1)] on scales ranging from 8 to 60 arcmin; a thorough analysis of the significance of the data showed a 95% confidence upper limit to fluctuations on a 10 arcmin scale of $^{\delta T}/T < 3 \times 10^{-4}$. A further series of λ6 cm measurements on scales of 10 to 20 arcsec have been made using the Mk IA-Mk II broad band interferometer at Jodrell Bank. The 95% confidence level upper limit to fluctuations on a scale of 15 arcsec was $^{\delta T}/T < 5.8 \times 10^{-4}$.

We summarize here our most recent observations which were made at λ3 cm covering angular scales from 5° to 15°. This scale was chosen for several reasons. Firstly it is larger than the horizon scale at recombination (2°) so that causal processes in the early Universe cannot remove anisotropy. Secondly this scale is not smeared out by scattering from the sub-degree reionization structure expected at the epoch of galaxy formation (z=3-10). Also, the CDM scenario predicts greater fluctuations on larger angular scales.

2. Observations and results

The current λ3 cm experiment[2] achieved high sensitivity by using low-noise broad-band receivers for long-integration observations of right ascension drift-scales at fixed declination. The observing mode consisted of rapid switching between two beams each 8° wide and separated by 8° with a slower wagging of the double beam by $\pm 4^o$, thereby generating a three-beam response which had a long-term stability some 10 times better than simple two-beam switching. The high altitude (2300m) of the Teide Observatory, Izana, Tenerife ensured that the observed noise in the experiment was set by internal receiver noise rather than atmospheric noise for 80% of the time.

The first aim of the observations was to survey a substantial fraction of the sky in a search for isolated high amplitude fluctuations as predicted in certain scenarios[3]. The declination range -15° to $+55^o$ was covered in 5° steps. This survey was then complemented by a deep survey at Dec = 0° and Dec = 40°.

The raw stacked data at both Dec = 0° and 40° showed significant fluctuations. A separation between the noise contribution from the receiver and that from the sky was made using the day-to-day variations and applying a maximum likelihood calculation which takes into account the geometry of the beams. This analysis showed that a non-zero variance was preferred over zero variance by a factor of approximately 10 to 1 in likelihood. Within the 8° beam we detect an excess standard deviation of 0.10 mK which is attributed to sky fluctuations. This corresponds to an observed $\delta T/T = 3.7 \times 10^{-5}$ for the CMB on this angular scale. When the observed fluctuations are corrected for the smoothing by our 8° beam we estimate that the <u>intrinsic</u> fluctuations on an 8° scale are $\delta T/T = 5.7 \times 10^{-5}$. This result applies to the Dec = 40° data.

It should be noted that our detection of sky fluctuations may contain a component of high latitude radio continuum emission from the Galaxy. If the brightness temperature spectral index were 2.8 the rms contribution would be less than 0.08 mK. Further, as there is no match between the observed signals and that expected from an extrapolation of published low-frequency maps, we believe that the spectral index is steeper than this and we are most likely detecting fluctuations in the CMB.

3. Discussion

In common with other recent measurements of the microwave background anisotropy the current results rule out baryonic models of galaxy formation[4]. This conclusion is particularly strong for our experiment because reionization is very unlikely to have erased the primordial structure, as it may do for small-scale structure.

The standard CDM model with the addition of "inflation" and the requirement that Ω=1 makes unambiguous predictions for $^{\delta T}/T$ on various angular scales. The CDM model of Bond and Efstathiou[5] when folded with our beam makes a prediction a factor of 8 lower than our observed value of $^{\delta T}/T$. If Ω<1, then the observations are closer to the predictions.

Experiments are continuing to confirm and explore the fluctuations detected in these experiments. λ3 cm observations are continuing at Teide Observatory on scales of 5° and 2°, while at Jodrell Bank λ6 cm observations are being made with an interferometer having a lobe separation of 5°. We acknowledge the contributions to the work described above of our collaborators, S.Padin, R.J.Davis,D.Waymont, R.Watson, E.J.Daintree, J.Hopkins, J.E.Beckman, J.Sanchez-Almeida and R.Rebolo.

References

1. Lasenby,A.N. and Davies,R.D., Mon.Not.R.astr.Soc., 203, 1137 (1983).
2. Davies,R.D., Watson,R., Daintree,E.J., Hopkins,J., Lasenby, A.N., Beckman,J., Sanchez-Almeida,J. and Rebolo,R., Nature submitted (1987).
3. Bajtlik,S., Juszkiewicz,R., Proszynski,M. & Amsterdamski,P., Astrophys.J., 300, 465 (1986).
4. Wilson, M.L. and Silk, J., Astrophys.J., 243, 14 (1981).
5. Efstathiou, G. and Bond, J.R., Microwave Background fluctuations and dark matter. Preprint (1986).

Cosmic Background Spectrometers

J. B. Peterson
Joseph Henry Laboratory, Princeton University,
Princeton, N. J. 08544

Abstract

Five balloon or rocket borne cosmic background spectrometers are compared. The diversity of the set of instruments reduces the chance that they have a common undetected systematic error. Two instruments have been flown successfully. The data are summarized and compared to other measured values of the cosmic background temperature.

Introduction

The development of ever more accurate absolute cosmic background photometers continues apace. Table 1 shows that the five balloon or rocket borne instruments use a wide variety of methods to measure the cosmic background radiation (CBR) temperature. I will describe the instruments below and summarize the data from the two successful flights.

Table 1. Charcteristics of the five balloon or rocket borne absolute CBR photometers.

Group	Platform	Spectro-meter	Antenna	Window	3 K Calibr. in Flight	Successful Flight
Berkeley McMaster & Princeton	Balloon	Filters	Winston Conc.	-	yes	yes
Princeton	Balloon	Receiver	Smooth Horn	Kapton	yes	yes
Berkeley & Nagoya	Rocket	Filters	Winston Conc.	-	no	no
British Columbia	Rocket	F. T. S.	Winston Conc.	-	yes	no
Queen Mary College	Balloon	F. T. S.	Corrugated Horn	Poly-ethylene	yes	no

Berkeley Balloon Borne Filter Photometer

In this liquid helium cooled photometer[1)] five filters of ~20% bandwidth are used to select bands from 90 GHz to 300 GHz. In the lowest band atmospheric emission is negligible at balloon altitudes but in the 300 GHz band the atmospheric emission exceeds the CBR flux. By observing with zenith angles θ from 20° to 55° and fitting the data to $\sec\theta$ the atmospheric emission can be measured and subtracted. A 3.2K blackbody is used for calibration in flight. The calibration Dicke switch is between the antenna and the rest of the photometer. Because window emission and reflection are difficult to measure no window is used prevent condensation of frozen air in the optics. Helium boil off gas flowing out the antenna is relied on to flush the antenna.

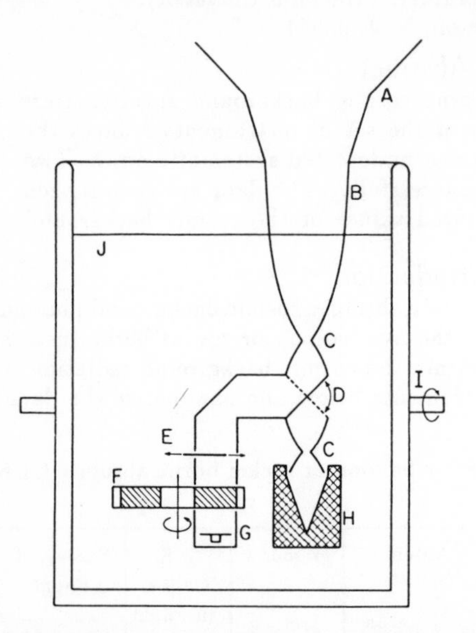

Figure 1. A schematic diagram of the Berkeley balloon borne filter photometer: A, apodizing horn; B, liquid helium cooled antenna; C, throughput limiting apertures; D, Dicke switch (dashed line shows the position during calibration); E, chopper; F, filter wheel; G, ^{3}He cooled bolometer; H, 3.2 K blackbody; I, cryostat pivot; J, superfluid liquid helium level.

The experiment was flown successfully in November 1983. The measured flux from the calibrator was stable over the five hour flight but the measured flux from the sky decreased by ~2% per hour. There are two possible explanations for this drift. First, laboratory tests show that ~50 ml of frozen air in the antenna can reduce the sensitivity of the photometer by 10%. However, photographs of the antenna taken during previous flights show no evidence of frozen air. Second, it is possible that decaying atmospheric emission is responsible for the decrease in flux. The authors have concluded that it is not possible to decide among these possibilities and have included uncertainties to accommodate both in their error analysis.

The instrument is capable of better accuracy than was achieved in the first flight so it has been prepared to fly again. Gary Bernstein and Marc Fischer have added a hot spot source to measure the transmittance of the antenna during the flight, a camera to photograph the antenna and a faster zenith drive system. A larger balloon has been requested to achieve a float pressure half that of the 1983 flight. The experiment is ready and waiting for a balloon.

Princeton Balloon Borne Receiver Radiometer

This instrument[2)] consists of a room temperature correlation receiver sensitive at 25 GHz with liquid helium cooled calibration optics. The first magic tee and the first isolators in each arm are also cooled. It is the only instrument of the five described here that carries all calibration equipment aloft. Operating at a frequency well below the other high altitude experiments offers several advantages. Even with a small balloon atmospheric emission (2 mK) is negligible. The emissivity of plastics used for window material increases dramatically with frequency. Window emission is a small correction(35 mK) at 25 GHz. The same window would be almost as bright as the CBR at 300 GHz. Because waveguide optics are used reflection coefficients and losses of the components of the radiometer can be measured easily. The small correction(50mK) for antenna emission can be calculated with confidence.

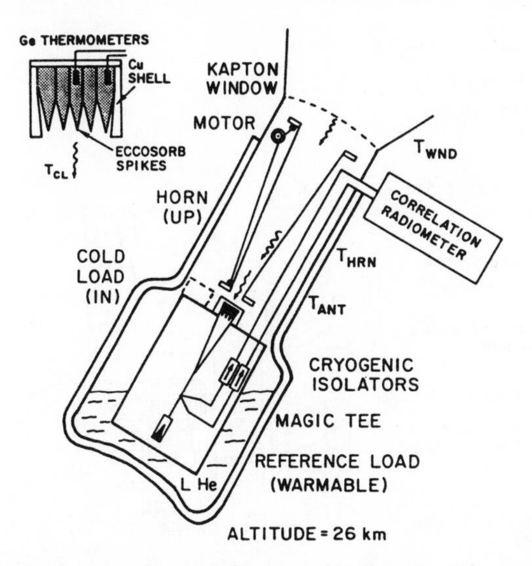

Figure 2. The Princeton Balloon Borne Receiver Radiometer.

The instrument was flown in April 1985. The authors claim to have opened the era of 1% measurements. Their result is included in Figure 4.

It is important to note that the largest uncertainties in the result come from emission in the instrument; there is room for improvement here. Much lower emissivity window material is available, the antenna can be cooled or used with a lower loss mode, cryogenic mixers can be used to reduce offsets due to detection of mixer emission. An evacuated enclosure around the black body sources would increase confidence in the precision of the thermometry.

Berkeley-Nagoya Rocket Borne Filter Photometer.

Four 20% bandwidth bandpass filters act as dicroic beam splitters dividing the incoming radiation into five channels[3)] from 300 GHz to 3000 GHz. The four lowest bands use bolometer detectors, two Ge:Ga photoconductors share the highest channel. One photoconductor is stressed to provide lower frequency response. The photometer is sensitive to CBR flux in the lowest three bands, the highest band will determine the baseline of the the data from the Infrared Astrnomical Satellite (IRAS) 100 μm band. The bands below it will be used to measure the spectrum of the cold interstellar dust detected by IRAS. The antenna is cooled by the helium bath to a temperature below that of the CBR. On the Wein side of the Planck function this guarantees that antenna emission is small. No window is needed to protect the optics from condensation. The experiment is calibrated on the ground using a blackbody inserted in the antenna. A small source in the photometer optics is heated periodically during calibration and during the flight. This source acts as a transfer standard. The photometer also carries a warm source inside the optics cover. The warm source provides a check that the sensitivity of the instrument is not altered during launch.

This photometer is the second of this design. The first was launched in September 1985. The cover failed to release and no astronomical data were taken. The photometer worked as expected. The flight of the second photometer is scheduled for February 1987.

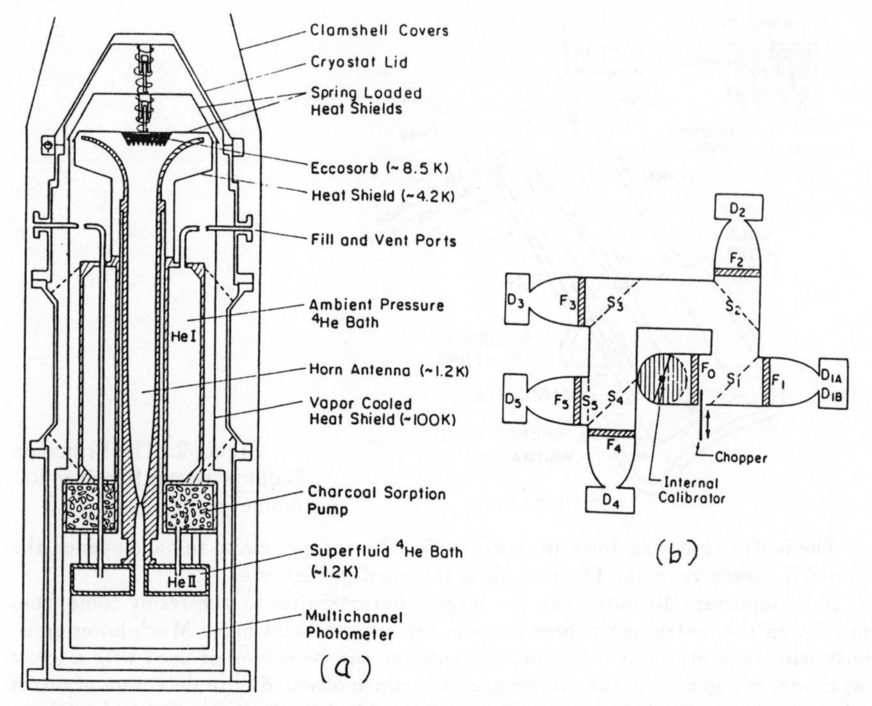

Figure 3. The Berkeley-Nagoya rocket borne filter photometer shown in vertical cross section (a) and in horizontal cross section (b) through the multi-channel photometer.

British Columbia Rocket Borne Fourier Transform Spectrometer

This instrument[4)] consists of a rapid scan polarizing interferometer, cooled Winston concentrator antenna and two ^{3}He cooled bolometers. Low pass filters between the interferometer and the detectors provide sensitivity ranges of 150-600 GHz and 150-900 GHz. The spectrometer is used in the four port geometry so each detector compares the sky to a cold blackbody reference source. The source is of unique design, it is a pyrex cone with the tip curved. The interior is coated with a thin layer of bismuth to increase absorption at low frequencies. A flight is planned within the year.

Queen Mary College Balloon Borne Fourier Transform Spectrometer

A four port polarizing interferometer is used in this instrument[5)] as well but back to back corrugated feed horns are used to define the beam on the sky and in the interferometer. The interferometer is used only over the single mode band (135-165 GHz) of the waveguide that connects the horns. An oversized Winston concentrator with small aperture at a waist of the beam directed to the sky acts as a shield against diffraction of warm radiation into the spectrometer. The optics are sealed with a polyethylene window. The entire spectrometer with its reference source is stepped in temperature in a small range near 2.78K. If the CBR has a thermal spectrum the measured difference spectrum will null when the spectrometer is at the cosmic background temperature. The instrument is designed to have 10 mK accuracy. It will be about a year before the it flies.

Data

Data from the high altitude experiments are in agreement with ground based measurements[6] and in agreement with optical measurements of the temperature of CN in interstellar molecular clouds[7]. The result of from the Princeton[2] balloon experiment T = 2.783±0.025 K agrees with the averages from other data sets and it dominates any weighted average of the temperature. In the next few years the high altitude experiments should provide a wealth of data on the largely unexplored Wein side of the CBR spectrum.

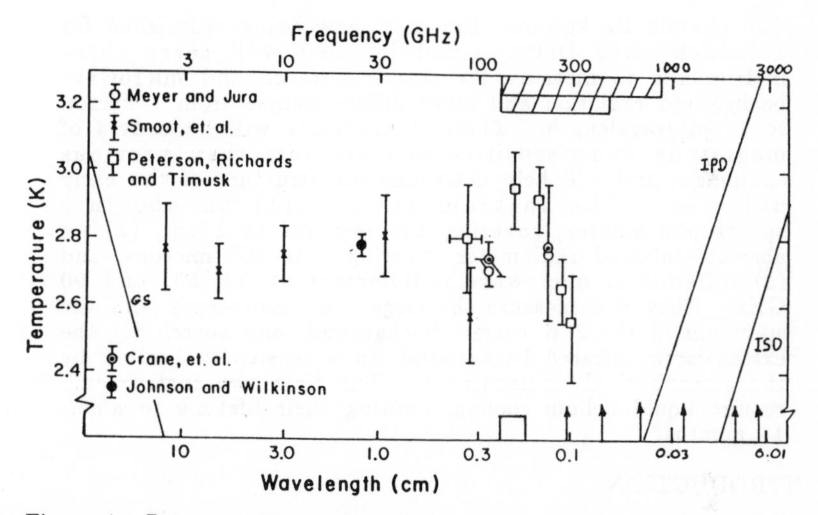

Figure 4. Recent measurements of the cosmic background temperature are shown along with the frequency ranges of the Queen Mary College F. T. S. (open box), the British Columbia F. T. S. (cross hatched boxes) and the Berkeley filter photometer (arrows). The antenna temperature due to interstellar dust(ISD) and interplanetary dust(IPD) and Galactic synchrotron radiation(GS) are also shown.

References

1) Peterson, J. B., et. al., Phys. Rev. Lett. **55**, 332 (1985)
2) Johnson, D. G. and Wilkinson, D. T. preprint, submitted to Astrophys. J. Lett. (1987)
3) Lange, A. E., et. al. *A Rocket-borne submillimeter Radiometer* Preprint (1986)
4) Gush, H. P., *Proceedings of the 1983 Conference on Space Helium Dewars* 99 (1984)
5) Martin, D. H., *Progress in Cosmology* 119, D. Reidel (1982)
6) Smoot, G. F., et. al., Astrophys. J Lett. **291** 123 (1985)
7) Meyer, D. M. and Jura, M., Astrophys. J. **297** 119 (1985)

8) This research was supported in part by a grant from the NSF

CAPABILITIES OF THE COSMIC BACKGROUND EXPLORER

J. C. Mather

Laboratory for Extraterrestrial Physics
Goddard Space Flight Center
Greenbelt, MD 20771

ABSTRACT

The Cosmic Background Explorer, now being redesigned for a launch on a Delta rocket in 1989, will carry three instruments to measure the cosmic infrared and microwave background radiation and other diffuse sources from 1 micron to 1 cm wavelength. These instruments will be orders of magnitude more sensitive and accurate than previous equipment and will help determine the structure of the early universe. The instruments are (1) an absolute spectrophotometer, covering 100 microns to 1 cm, (2) an absolute infrared radiometer covering 1 to 300 microns, and (3) differential microwave radiometers at 32, 53, and 90 GHz. They will measure the large scale anisotropy and the spectrum of the 3 K cosmic background, and search for the extragalactic infrared background, to a sensitivity limited by the astrophysical environment. The first two instruments require liquid helium cooling, limiting their lifetime to about 14 months.

1. INTRODUCTION

The Cosmic Background Explorer satellite[1)] will enable measurements of the diffuse IR and microwave cosmic background radiation at greatly improved sensitivity and accuracy. Its location above the atmosphere, its year-long observing time, and its full sky coverage are essential to pushing back the observational limits to those set by our astrophysical environment.

Measurements of this radiation are crucial for testing theories of the early universe. Imaging telescopes such as the Einstein and the Space Telescope are not expected to resolve galactic size objects past a redshift of about 10, because of confusion, small angular size, an apparent lack of sources, and potentially intervening material. However, the theory of galaxy formation is incomplete, and the number of options is large, allowing the possibility of primeval black holes, cosmic strings that explode or that seed galaxy formation, gravity waves, antimatter annihilation, reheating of the intergalactic plasma, and so forth. The accumulated diffuse radiation from such processes can be studied at much greater redshifts, up to about 10^6. Earlier processes are expected to be thermalized and could not be distinguished from each other by any observable characteristic.

The spacecraft is being redesigned for launch on a Delta rocket, as a result of the Challenger explosion and the temporary closing of the California Shuttle launch site. To obtain a complete sky survey and protection from the Sun and Earth, it will use the IRAS orbit, at an altitude of 900 km, with a 99° inclination sunsynchronous terminator orbit. This orbit permits a favorable instrument orientation, with the line of sight perpendicular to the Sun and approximately toward the zenith. The COBE will spin at 0.8 rpm, with its axis maintained approximately vertical and 94° to the Sun, to scan two of the instruments rapidly across the sky. A liquid helium cryostat like the IRAS cryostat will keep the two infrared instruments at 1.6 K for about 14 months, and a conical shield will protect the instruments from radiation from Sun and Earth.

2. DIFFERENTIAL MICROWAVE RADIOMETERS (DMRs)

The DMRs, a group of 6 Dicke switched dual-beam radiometers, will measure the anisotropy of the 3 K background on large angular scales, using three wavelengths (3.3, 5.7, and 9.6 mm) to discriminate the galactic electron emission by its spectral index. At the lowest frequency, electron emission dominates, and at the highest galactic dust is the limiting emission. The multiple wavelengths will also be used to show that any observed anisotropies have the proper spectral variation to be attributed to the cosmic background radiation. The sensitivity is about 0.2 mK (depending on frequency) per beamwidth after 1 year of observations, and about 10 μK per spherical harmonic. This sensitivity is sufficient to be resolve the galactic emission in much of the sky, reaching close to the ultimate sensitivity limit for this experiment. It would measure the dipole anisotropy to a noise level of 1 km/sec and an angular uncertainty of less than a degree, providing that systematic errors can be reduced to the receiver noise level.

The instrument is a conventional ferrite-switched radiometer, with mixers and IF amplifiers preceding square law detectors. For the two high frequency channels, the microwave parts are cooled to 140 K for improved sensitivity. The instrument beamwidth is 7°, set by the special low-sidelobe corrugated horn antennas. Calibration will be by noise source, with an accuracy of 1%, and will be confirmed by the Moon to 3%. The angle between the two horn antennas of each receiver is 60°, bisected by the spin axis. As the spacecraft spins, the instrument measures the difference between the brightnesses of all pairs of pixels separated by 60°. In the data processing, a least squares fit to the difference data will reconstruct a map of the sky and give estimates of stray light sources and those nonideal instrument behaviors which can be described mathematically.

3. FAR INFRARED ABSOLUTE SPECTROPHOTOMETER (FIRAS)

The FIRAS is a polarizing Michelson interferometric spectrometer with differential inputs. It will measure the spectrum of the cosmic background to unprecedented accuracy and sensitivity. It is calibrated against a full beam external blackbody to help detect small differences between the background spectrum and a blackbody spectrum. Such differences would be expected to reflect conditions from redshifts as high as 10^{6}, since electron scattering is slow to alter the spectrum and even slower to create or destroy photons to achieve equilibrium. A detection of such differences could show energy release from black holes, antimatter annihilation, galaxy formation, etc.

For λ from 0.5 to 5 mm, the sensitivity and accuracy after a year are 10^{-13} W/cm^{2}sr for λI_λ, for each 7° beamwidth and each spectral element, enough to give a signal-to-noise ratio of 1000 at the peak of the spectrum. The wavelength coverage is in two bands, 0.1 to 0.5 mm and 0.5 to 10 mm. Spectral resolution is <5%, but limited to 0.2 cm^{-1} (apodized). Some data will be lost because of the Moon, but will be reobserved 6 months later. The detectors are 4 composite bolometers, with an etendue of 1.5 cm^{2}sr and an NEP of about 10^{-14} W/Hz$^{1/2}$. The antenna is a parabolic Winston cone[2)] with a 7° field of view, and a flared aperture to suppress diffracted light. It looks along the spin axis of the spacecraft. The external calibrator, the source of the exceptional calibration accuracy of the experiment, is a re-entrant Eccosorb cone with an effective emissivity of .9999 and a temperature uncertainty of 1 mK at 2.7 K, capable of covering the entire aperture.

4. DIFFUSE INFRARED BACKGROUND EXPERIMENT (DIRBE)

The third instrument, the Diffuse Infrared Background Experiment (DIRBE), will measure the diffuse background from 1 to 300 μm. Such radiation is expected to come from the epoch of galaxy formation as well as from foreground sources such as interstellar and interplanetary dust, galactic starlight, IR galaxies, quasars, and galaxy clusters.

The DIRBE has 10 wavelength bands, at λ ($\Delta\lambda$) of: 1.3 (0.3), 2.2 (0.4), 3.5 (1.0), 4.8 (0.6), 11 (7), 22 (15), 60 (40), 100 (40), 160 (80), and 250 (100) microns (microns). The three short wavelength bands each have three detectors, a total power detector and two perpendicularly polarized detectors, to help recognize scattered light from interplanetary dust. As for FIRAS, the sensitivity per beamwidth and detector will be 10^{-13} W/cm^{2}sr in units of λI_λ. This is more than adequate to measure the foreground sources at all the DIRBE wavelengths, and sufficient to find small residual background radiation down to about 1% of the foreground brightness.

The instrument beamwidth is 0.7^o square. Calibration will be derived from known point sources (stars and planets) and confirmed by a commandable hot reference body. The telescope is an off-axis Gregorian for low stray light, and the beam is modulated against the dark instrument interior by a tuning fork chopper.

To measure the foreground interplanetary dust radiation, the instrument is mounted at 30° to the spacecraft spin axis, like the microwave radiometer horns, so that the line of sight sweeps from 64 to 124° away from the Sun. This pattern modulates both the ecliptic latitude and the Sun angle, which are the prime determinants of the interplanetary dust brightness. Simulations show that the dust radiation can be modeled and subtracted with an accuracy of a few percent, in a way similar to zenith angle scans for terrestrial photometry of stars. There are expected to be "windows" at 3 microns and 200 microns where the foreground emission is relatively faint, where quite small extragalactic backgrounds could be seen.

5. DATA PRODUCTS AND AVAILABILITY

The primary data reduction center will be at the GSFC. The data products will be made available to the public approximately three years after launch, following a careful calibration of the data. Data products will include the various time ordered, calibrated instrument data, calibration information, maps of the sky at all the wavelengths observed by COBE, spectra of individual pixels, and a variety of products giving the residual extragalactic background light after foreground models have been subtracted. The National Space Science Data Center at GSFC will welcome inquiries and will distribute data sets.

6. PROJECT STATUS

The three instruments have all been assembled and tested, and are now disassembled for refurbishment and cleaning prior to final test. Redesign of the spacecraft and the microwave radiometers to accomodate the Delta rocket has just begun (January 1987). Launch is expected in early 1989.

7. ACKNOWLEDGMENTS

We thank NASA Headquarters, the Goddard Space Flight Center and the COBE Project Office, and Nancy Boggess in particular, for their continued support. The COBE Science Team includes: C.L. Bennett, N. W. Boggess (Program Scientist), E.S. Cheng, E. Dwek (Deputy Project Scientist), S. Gulkis, M.G. Hauser (PI for DIRBE), M.A. Janssen, T. Kelsall, P.M. Lubin, J.C. Mather (PI for FIRAS and Project Scientist), S.S. Meyer, S.H. Moseley, T.L. Murdock, R.F. Silverberg, G.F. Smoot (PI for DMR), R. Weiss (Chairman), D.T. Wilkinson, and E.L. Wright (Data Team Leader).

8. REFERENCES

1. Mather, J.C., "The Cosmic Background Explorer (COBE)", Optical Engineering 21, 769-774, (1982).

2. Mather, J.C., Toral, M., and Hemmati, H., "Heat trap with flare as multimode antenna", Appl. Optics 25, 2826-2830 (1986).

LARGE SCALE ANISOTROPY OF THE INFRARED BACKGROUND: PRIMORDIAL FLUCTUATIONS VS. LOCAL EMISSION

de Bernardis, P., Masi, S., Melchiorri, F., Vittorio, N.
Istituto Astronomico, Università di Roma, La Sapienza

The observational limits on the cosmic microwave background (CMB) large scale anisotropy are especially powerful for constraining different galaxy formation scenarios. Infact, we can interpret them independently of any assumptions on the ionization history of the universe. Experiments for probing the angular distribution of the CMB are differential and pose upper limits on the first derivative of the CMB temperature field, Δ. From the theoretical point of view, it is possible to predict the rms value of Δ, expected in a given scenario, on a given angular scale. Usually, a model is accepted if the predicted value is consistent with the observed upper limits. This procedure seems to be a rather wasteful way of dealing with both theory and observations. In fact, knowing the theoretical probability distribution for the amplitude of density fluctuations, it is possible to calculate the entire pattern of the temperature fluctuations on the sky. For more details and references we refer to Vittorio and Juskiewicz (1987), who discussed the pattern of the microwave sky expected both on large and on small angular scale (see also Bond and Efstathiou, 1987). We want to report here few results on the large scale anisotropy, under the assumption that density fluctuations are initially Gaussian distributed. For an initially scale invariant density fluctuation spectrum and for a flat cosmological model, we expect on all the sky $N = 3800(1/\theta)^2$ maxima in the temperature field. Here θ is the antenna beam diameter. A fraction $f = 1.5\nu exp(-\nu^2/2)$ of these maxima has a temperature fluctuation, relative to the background, a factor ν higher than the rms value, σ. These regions appear as unresolved sources with their number increasing with improving resolution as θ^{-2}. With a given realization of the same ensamble of initial conditions, one can also generate a map of the sky (for details on the method see De Bernardis, et al. 1987). As an example, we show in Fig. 1 a map of the predicted microwave sky. The contour levels refer to the 0, 1, 2, and 3 σ (continuous lines) and to the -1,-2, and -3 σ (broken lines) temperature fluctuations. The geometry of the pattern is of course independent of the actual value of σ. We can test the capability of a given experiment of measuring temperature fluctuations at the expected level, by simulating the observations of our "theoretical" sky with different beam separations, integration times, and sky coverages. From our preliminary analysis, a sky coverage $\sim 30\%$, typical of a baloon flight, is sufficient to estimate the "true" theoretical value of the CMB quadrupole anisotropy. However, this constitutes only a lower limit to the observable quadrupole anisotropy since we neglected up to now both the instrumental noise and local effects. From this respect, the main contribution in the far infrared from our Galaxy is due to interstellar dust. Maps of the galactic emission at wavelengths near the CBR peak are up to now unavailable. Infrared dust emission at high galactic latitude has been detected for the first time by de Bernardis et al. ,1984 using two differential photometers operating at 220 μ m and 440 μ m. However, their analysis was based on a incomplete sky coverage and the North Galactic Pole was not observed directly. A complete absolute survey of the dust distribution in our Galaxy is provided by the IRAS All Sky Map at 100 μ m. In order to model the dust emissivity in the millimetric region (where the infrared CMB experiments are performed) we extrapolated the 100 μ m data as follows: $I_\lambda = \lambda^\alpha B_\lambda(T_d)$. Here B_λ is the black body emissivity at the dust temperature T_d=22 K. The spectral index α is usually taken between 1 and 2, as expected for silicon or carbonate dust grains (Aannestad, 1986). The extrapolated IRAS dust emission at the North Galactic Pole is consistent with an emission of $\sim 50\mu K$ at 3 mm (Lubin et al., 1985) for $\alpha \sim 1.5$. This value has to be considered only as a lower limit to α, since the galactic emission at the North Galactic Pole could be considerably smaller (Lubin et al., 1985). We produced maps of the dust emission at the millimiter wavelengths assuming $1.5 < \alpha < 2$. We simulated the flights of the Florence group (Fabbri et al., 1980),and of a future experiment of the Rome group (Melchiorri et al., 1987). We analyze the dust emissivity in spherical harmonics after subtracting from the full sky map the galactic plane ($\pm 10^o$ in galactic latitude). Preliminary results for the dipole and quadrupole component of the dust emission are given in Table 1. Again, these values have to be considered only as lower limits. A colder dust component could have escaped detection from IRAS (Pajot et al., 1986) and would of course be very important in the millimetric region. Work is in progress for a combined analysis of primordial anisotropy, dust emission, and instrumental noise.

Aannestad, P.A., Proc E.Fermi School, *"Evolution of Interstellar Dust and Related Topics"*, Varenna, 1986

Bond, J.R., and Efstathiou, G., 1987, preprint

de Bernardis, P. ,Masi, S., Melchiorri, F., Moreno, G., 1984, *Ap.J.*, **278**, 150

de Bernardis, P. ,Masi, S., Melchiorri, F., Vittorio, N., 1987, in preparation

Fabbri, R., Guidi I., Melchiorri, F., Natale, V. 1980, *Phys.Rev.Lett.*, **44**, 23

Lubin, P.M., Villela, T., Epstein,G., Smoot, G. 1983 *Phys.Rev.Lett.*, **50**, 616

Lubin, P.M., Villela, T., 1985, *The Cosmic Background Radiation and Fundamental Physics*, Melchiorri, F. Editor, Editrice Compositori, Bologna

Pajot, F., Boisse' P., Gispert, R., Lamarre, J.M. Puget, J.L., Serra, G., 1986 *Astron.Astrophys.*, **157**, 393

Vittorio, N., and Juskziewicz, R., 1987 *Ap.J.Lett.*, May 15

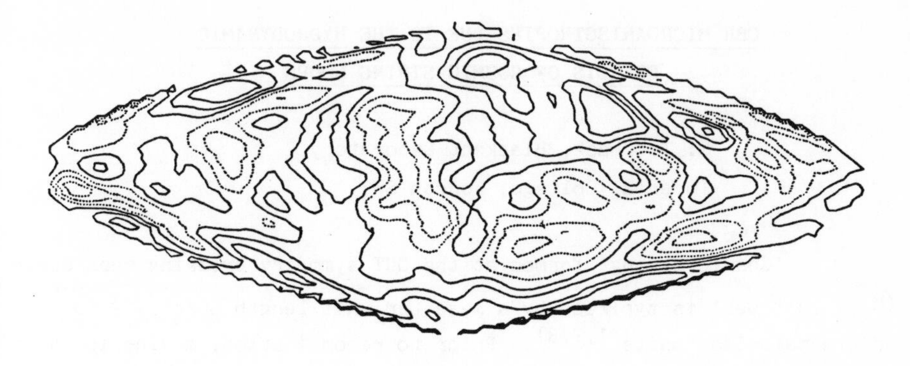

FIG.1

TABLE 1.DIPOLE AND QUADRUPOLE COMPONENTS FROM DUST EMISSION

	Q_d (mK)			D_d(mK)		
	$\alpha=1.5$	$\alpha=1.8$	$\alpha=2$	$\alpha=1.5$	$\alpha=1.8$	$\alpha=2$
FLORENCE FLIGHT	0.19 ± 0.03	0.09 ± 0.02	0.05 ± 0.01	0.25 ± 0.05	0.12 ± 0.03	0.07 ± 0.01
ROME FLIGHT	0.056 ± 0.004	0.022 ± 0.002	0.012 ± 0.001	0.039 ± 0.007	0.015 ± 0.004	0.008 ± 0.002

CBR MICROANISOTROPIES DUE TO THE HYDRODYNAMIC EFFECTS OF COSMIC STRING LOOPS

S. T. Chase, Blackett Laboratory,
Imperial College, London SW7.

'Cosmic string' formed at the GUT symmetry breaking mass scale ($M_X \approx 10^{15}$ GeV) is typified by a mass per unit length $\mu G/c^2 \approx 2\ 10^{-6}$ in dimensionless units [1,2,3]. Prior to recombination, moving loops of cosmic string interact hydrodynamically with the comic fluid via their gravitational potential[4]. At recombination this effect can produce a significant distortion of the CBR temperature for loops with radii greater than about 100 pc. These distortions occur on angular scales of a few arcminutes and may be as large as $\Delta T/T \approx 10^{-3}$ but the areal density of kpc-size loops is anticipated to be rather low, so that limited area searches for CBR microanisotropies[5,6] are unlikely to have found them yet.

The horizon scale at recombination ($\lambda_h \approx 100$-200 kpc), is much greater than the size of the loops considered here ($R \approx 0.5$-5 kpc), so it is a reasonable approximation to treat spacetime in the loop neighbourhood as being asymptotically Minkowskian. In general a loop will have a peculiar velocity V_p, and also support rapid oscillations[6] leading to a complicated and time-dependent metric. In the rest-frame of the loop centre-of-mass the time-dependent terms are essentially gravitational radiation, so that smoothed over times $t \approx R/c$ the metric will be approximately Schwarzschild in the region $r \geqq R$. The string mass is taken to be $M = \beta\mu R$, with $2\pi \leqq \beta \leqq 9$, in accordance with estimates based upon numerical simulations[8].

The cosmic fluid is described by an equation of state $p = p(\rho)$, where p and ρ are the comoving fluid pressure and mass-energy density respectively. Under the assumptions of ref.4) the Navier-Stokes equation can immediately be integrated to yield the Bernouilli equation,

$$1/2(V^2 - V_p^2) + \tilde{C}_s^2 \ln(\varepsilon/\varepsilon_0) + \ln\sqrt{-g_{00}} = 0 \qquad (1)$$

where $\varepsilon = \gamma^2(\rho + p)$, $\gamma^2 = (1-v_p{}^2)^{-1}$, and $\tilde{c}_s^2 = {}^{dp}/_{d\varepsilon}$ is an effective sound speed squared. V is the perturbed fluid velocity.

Eliminating ε from the continuity equation by substitution from the Bernouilli equation now gives

$$\nabla\cdot\{\mathbf{V}\exp[-(V^2-V_p^2+2\ln\sqrt{-g_{00}})/2\tilde{C}_s^2]\} = 0 \qquad (2)$$

It is sufficient for present purposes to take a 1-dimensional solution as illustrative of the more general solutions of equation 2). To second order in $(V - V_p)/V_p << 1$, and expanding $\ln\sqrt{-g_{00}} \approx \Phi_N$, the 'Newtonian' potential, this is

$$V = -V_p\exp\{-\Phi_N/(\tilde{C}_s^2-V_p^2)\} \qquad (3a)$$

and

$$\varepsilon = \varepsilon_0\exp\{\Phi_N/(\tilde{C}_s^2-V_p^2)\} \qquad (3b)$$

Before recombination the cosmic fluid pressure is dominated by the radiation pressure, so that $\varepsilon \approx \gamma^2(\rho_b+4\rho_\gamma/3)$, and $\tilde{C}_s^2\approx \{4\gamma^2(1+\frac{3}{4}d\rho_b/d\rho_\gamma)\}^{-1}$, (in the adiabatic case), where ρ_b is density due to non-relativistic interacting material including baryons, while ρ_γ represents the radiation energy density.

In the following numerical estimates, Hubble's constant is taken as $H_0 = 75h_0$ km/sec/Mpc, so that $z_{eq} \approx 2.2\ 10^4\Omega_b h_0^2$, where $\Omega_b = \rho_b(0)/\rho_{crit}$ is the 'baryonic' density parameter. Taking the epoch of recombination to occur at z = 1500, the effective sound-speed at recombination becomes

$$\tilde{C}_s(z_{rec}) \approx \{4\gamma^2(1+8.25\Omega_b h_0^2)\}^{-1/2} \qquad (4)$$

Under the foregoing assumptions, equation 3b) implies a local CBR temperature distortion in the hydrostatic limit of

$$\left.\frac{\Delta T}{T}\right|_{0,rec} = \frac{1}{3}\left.\frac{\Delta\varepsilon}{\varepsilon}\right|_{V_p=0} \approx 1.2\ 10^{-5}(1+8.25\Omega_b h_0^2)\mu_6$$

for $\beta \approx 9$ and $G\mu = 10^{-6}\mu_6$. For a loop with a peculiar velocity $V_{p,rec}$ at recombination, the above result is multiplied by a factor of

$$\{1-V_{p,rec}^2/\tilde{C}_s^2\}^{-1}.$$

The radiation suffers a gravitational redshift as it climbs out of the loop potential,

$$\Delta T/T\Big|_{gr} \approx -\phi_N = -10^{-6}\beta\mu_6,$$

so the final observable temperature distortion is

$$\frac{\Delta T}{T}\Big|_{V,obs} \approx \{1.2(\frac{1+8.25\Omega_b h_0^2}{1-V_p^2/\tilde{C}_s^2})-0.9\}\ 10^{-5}\ \mu_6$$

as shown for several cosmological models in Figure 1.

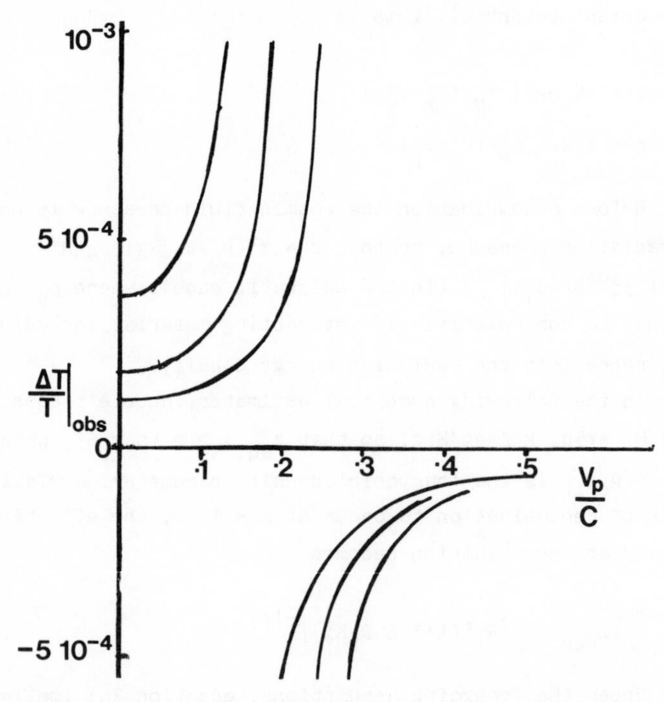

Figure 1.
The CBR microanisotropy due to a moving loop of cosmic string, as a function of the loop peculiar velocity, is shown here for three different cosmological models parameterized by the 'baryonic' density parameter Ω_b, and with $\mu = 2\ 10^{-6}$. a) $\Omega_b = 1$; b) $\Omega_b = 0.5$; c) $\Omega_b = 0.3$. The curves shown are discontinuous at $V_p = C_s$, whereas for 'real' loops the effect will be truncated at a finite value determined by the dynamical equilibration timescale.

The observable angular distribution and magnitude of CBR distortions caused by loops depends upon the number density and velocity distribution of the putative loop population, as well as on the effective sound-speed just prior to recombination, $\tilde{C}_{s,rec}$.

Loops are expected to form with peculiar velocities $V_{p,f}$ comparable to the speed of light[9,10], which are redshifted so that at a later time $t > t_f$, $V_p(t) = V_{p,f}z(t)/z_f$, where z_f is the redshift of formation. This implies that loops which are initially supersonic will become subsonic before recombination, so long as $z_f/z_{rec} > V_{p,f}/\tilde{C}_{s,rec}$.

During recombination the sound-speed drops rapidly to ≈ 100 km/sec. over a period of $\Delta t_{rec} \approx 5.8\ 10^4$ years ($\Delta z_{rec} \approx 400$) [11], so that essentially all loops will again become supersonic during recombination. The steady-state analysis leading to equations 3a) and b) remains approximately valid so long as the timescale for dynamical equilibration, which is roughly the sound-speed crossing time $t_{sc} \approx R/C_S$, is shorter than the timescale over which the sound-speed changes significantly. The resulting overdensity $\Delta\varepsilon/\varepsilon_0$ ceases to produce a corresponding CBR distortion when the photon mean-free path λ_{mf} becomes comparable to the loop radius. Just prior to recombination this is $\lambda_{mf} \approx 20(\Omega_b h_0^2)^{-1}$ pc, and varies approximately as the ionization fraction du recombination. For these reasons one expects loops of radius R ≦ a few kpc and with $V_{p,rec} \leq \tilde{C}_s$ to produce the largest distortions in the CBR.

The angular size of loops seen at recombination is about $0.5\Omega_0 h_0$ arcminutes/kpc, and the appropriate number density for loops with radii in the range $R \rightarrow R+\Delta R$ is[6,12] $n(R) = \nu R^{-2} t_{rec}^{-2}$ at recombination. Numerical estimates [6] put $\nu \approx 10^{-2}$, so taking a range of radii 0.5 kpc < R < 5 kpc, one expects about 50 to 70 loops in this range to be within the horizon volume at recombination, and about 10% of these to be contained within the surface of last scattering. Estimates of the areal density of loops for several cosmological models are presented in Table 1.

It is clear that a significant fraction of kiloparsec-sized loops might be expected to produce CBR anisotropies of order $\Delta T/T \approx 10^{-3} - 10^{-4}$ in the denser cosmologies, dropping to $\Delta T/T \geq 4\ 10^{-5}$ for $\Omega_b h_0^2 \leq 0.1$. While the magnitude of the anisotropies is well above the best current observational limits on arcminute scales, the areal density expacted in dense cosmologies ($\Omega_0 \geq 0.2$) is so low that detection by small-scale anisotropy stuεies[5,6] with small areal coverage is extremely unlikely. Present CBR microanisotropy limits on arcminute scales are based on the observation of typically less than 10 square arcminutes of sky, so that the probability of seeing a loop of angular size comparable to the beam area is about 10^{-2}-10^{-3}. Only for $\Omega_0 < 0.1$ is the areal density of the loops likely to be large enough that they should probably have been detected by now.

TABLE 1

Ω_0	Ω_M	Ang. size of horizon at t_{rec}.	Areal density of string loops.
1	1	1°.15	6.6 per sq. degree
1	0.1	2°	1.3 " " "
0.3	0.3	0°.5	27 " " "
0.1	0.1	0°.22	127 " " "

This work was supported by an S.E.R.C. Postdoctoral Research Fellowship.

REFERENCE LIST.

1). Kibble T.W.B. J. Phys. **A9** 1387-1398 (1976)

2). Vilenkin, A. Phys.Rev. **D24** 2082-2089, (1981)

3). Zeldovich Ya.B. M.N.R.A.S. **192** 663-667 (1980)

4). Chase S.T. Nature **323** 42 (1986)

5). Uson J.M., & Wilkinson D.T., Nature **312** 427-429 (1984)

6). Lasenby, A.N., & Davies R.D. M.N.R.A.S. **203** 1137-1170 (1983)

7). Brandenburger, R.H., & Turok, N., Institute for Theoretical Physics preprint NSF-ITP-85-88 (1985).

8). Albrecht A. & Turok N. Phys. Rev. Lett. **54** 1868- (1985)

9). Kibble T.W.B. & Turok N., Phys. Lett. **116B** 141 (1982)

10). Vilenkin A. & Shafi Q., Phys. Rev. Lett. **51** 1716-1719 (1983)

11). Peebles P.J.E.,'Physical Cosmoloogy' Princeton (1971)

12). Vilenkin, A., Phys. Rep. **121** 263-315 (1985).

SEARCH FOR DISTORTIONS IN THE SPECTRAL DISTRIBUTION OF THE COSMIC BACKGROUND RADIATION AT DECIMETRIC WAVELENGTHS

G.SIRONI(*), M.BERSANELLI(*)(#), G.BONELLI(+), G.CONTI(+),

G.MARCELLINO(*), M.LIMON(*), K.REIF(^)

(*) Dipartimento di Fisica dell' Universita' - Milano
(+) Istituto FCTR del CNR - Milano
(^) Radioastronomisches Institut der Universitat - Bonn

ABSTRACT

Preliminary results of measurements of the absolute temperature of the Cosmic Background Radiation (CBR) at λ=50 cm are presented

1. Introduction

Several of the processes which occurred in the early universe are capable of causing deviations from a pure planckian spectrum of the CBR especially in the Rayleigh -Jeans region[1]. However no definite distortion has been so far detected and recent measurements[2] set a 5% upper limit to the amplitude of possible distortions of the CBR spectrum in the wavelength range 12 cm - 3 mm. Above 12 cm, however, significant distortions may still be present. At decimetric wavelengths in fact the published data, obtained more than 15 years ago (for a review see ref. 3), display very large error bars which accomodate distortions as wide as 20 % or more. A joint program of new observations at decimetric wavelengths made in collaboration by our group and the group of George Smoot in Berkeley is currently underway.

Here we present preliminary results of absolute measurements of the CBR temperature at 50 cm (600 MHz) and 12 cm (2.5 GHz). 50 cm was chosen because at that wavelength the temperatures of the CBR and of the galactic background are comparable. The second wavelength, needed to separate the CBR from the galactic background, was chosen to get a link with our previous observations at 12 cm[4]. Results by Smoot and coworkers at 21 cm and 8.3 cm will appear soon[5].

2. The Experiment

The observations were made in 1985 - 86 from Alpe Gera (Italy) (Lat. 46.5N, Long. 10E, 2100 m a.s.l). The two radiometers were identical except for minor features. Both used Dicke switched superheterodyne receivers and scaled optimum rectangular horns with HPBW= 15°x 16°. The balance was given by the signal provided, for each system, by a second horn aimed at the North Celestial Pole. The main properties of the experiment were dictated by the 600 MHz system, i.e.:

(#) now at LBL Astrophysics Division - Berkeley

i)The physical dimensions of the horn (about 2 m x 2 m x 3 m at λ=50 cm) made it impossible to prepare a liquid helium reference source which completely filled the antenna aperture. The reference level was taken and the calibration made by substitution of the horn with a coax dummy load immersed in liquid helium; ii)The high level of electromagnetic pollution at VHF made it impossible to observe from an open site. We had to look for a remote site, naturally shielded by mountain ranges against TV station and radio link signals; iii)At 600 MHz the galactic background is always important and simple zenith scans are insufficient to work out the atmospheric contribution. The complete experiment included: i)Absolute measurements of the temperature of the sky at various right ascensions along declination +46.5; ii)Drift scans at dec. +46.5, +38, +30 and +11.5; iii)Zenith scans; iv)Measurements of the effective temperature of the cold load; v)Measurements of the elevation profile of the horizon and of the power pattern of the horns to be used for the evaluation of the ground contribution to the antenna temperature.

3. Data Reduction

The data reduction is underway. Preliminary values of the antenna temperature Ta at 600 MHz at a few positions along dec 46.5 N have been already obtained. By subtraction of Tgr, the ground contribution, we get Tvert, a mixture of the temperatures of the CBR (Tcbr), the galactic background (Tg) and the atmospheric noise (Tatm). At 600 MHz they cannot be readily separated. A modelling over the whole set of data at both frequencies and at various zenith angles is required to disentangle the three components. It will be made when all our data will have been reduced. At present we can only use values of Tg and Tatm obtained by scaling of values measured at other frequencies and/or under different conditions, by various authors [7-9]. We get in that way preliminary values of Tcbr and Tsky = Tg + Tcbr. They are shown in Table I and are in good agreement with the few data at 50 cm wavelength so far published [6,10]. The profile of Tsky along dec 46.5 N is plotted in figure 1.

Acknowledgments. This work was supported the Italian Ministry of Education (MPI 40%). Logistic support at the observing site was kindly provided by the Italian Electricity Board (ENEL)

REFERENCES

1. Danese, L. and De Zotti, G.F.: Riv. Nuovo Cimento 7, 277 (1977) and Astr. Ap. 84, 364, (1980)
2. Smoot, G. et al.: Ap. J. 291, L23, (1985)
3. Weiss, R.: Ann. Rev. Astron. Ap. 18, 489, (1980)
4. Sironi, G. and Bonelli, G. Ap. J. : 311, 400, (1986)
5. Smoot, G. et al : in preparation
6. Howell, T.F. and Shakeshaft, J.R.: Nature 216, 753, 1967
7. -----------: J. Atm. Terr. Phys. 29, 1159, (1967)
8. Partridge, R.B. et al.: Phys. Rev. D 29, 2683, (1984)
9. Haslam, C.G.T. et al.: Astr. Ap. Sup. 47, 1, (1982)
10. Stankevich, K.S. et al.: Aust. J. Phys. 23, 529, (1970)

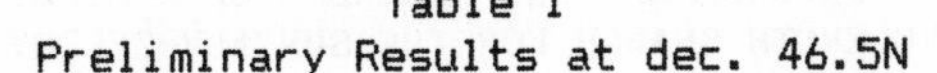

Table I
Preliminary Results at dec. 46.5N

r.a.	$14^h 34^m$	$16^h 27^m$
Tvert	17.1 + 0.4	17.2 + 0.4 K
Tgr	6.0 + 0.5	6.0 + 0.5 K
Tatm	1.3 + 0.2	1.3 + 0.2 K
Tsky	9.8 + 0.7	9.9 + 0.7 K
Tgal	7.0 + 0.2	7.8 + 0.2 K
Tcbr	2.8 + 0.7	2.1 + 0.7 K

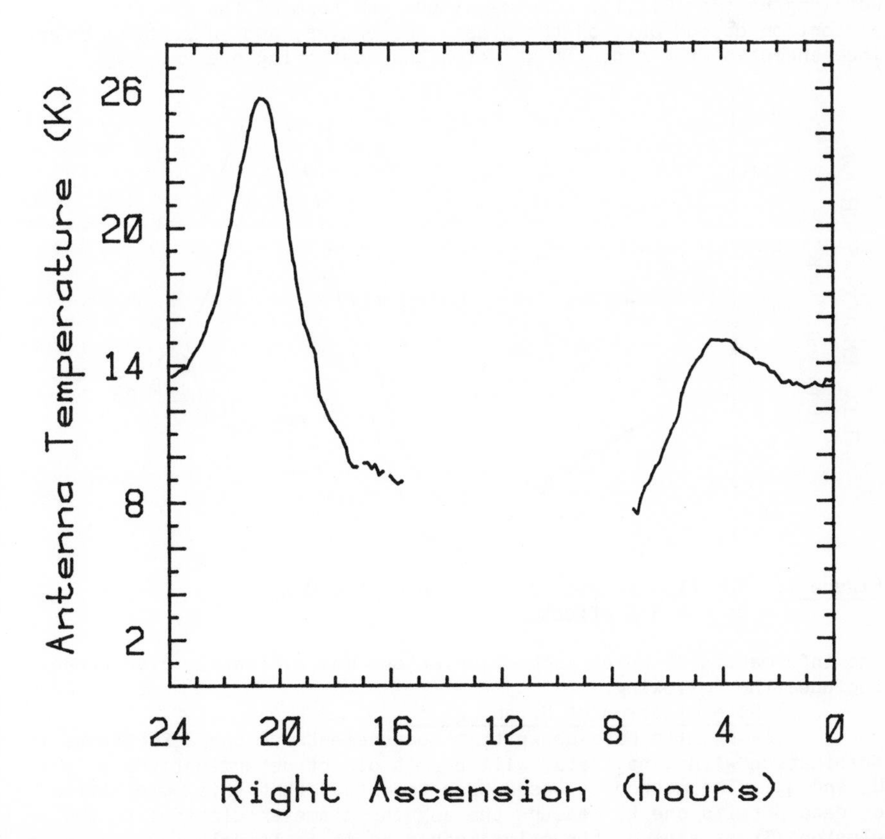

Figure 1 - Profile of the antenna temperature of the sky at 600 MHz measured along dec 46.5 N with an angular resolution of 15°

DEVELOPMENT AND PROGRESS OF A MILLIMETRE-WAVELENGTH SEARCH FOR THE SUNYAEV-ZELDOVICH EFFECT

S.T. Chase, R.D. Joseph, N.A. Robertson* & P.A.R.Ade†

Blackett Laboratory, Imperial College, London SW7;
*Dept. of Natural Philosophy, Glasgow University, Glasgow G12.
†Physics dept., Queen Mary College, London E4.

1. INTRODUCTION.

The Sunyaev-Zeldovich (S-Z) effect[1,2] is a spectral distortion of the Cosmic Background Radiation (CBR) induced by inverse Compton scattering of the microwave photons by energetic (≈10 keV) electrons in hot astrophysical plasmas. This process preserves photon numbers, but results in a net heating of the microwave photons, causing a flux decrement at long wavelengths and a corresponding increment at short wavelengths (see Fig.1). The magnitude and form of the spectral distortion depend only on the plasma properties, and in particular are independent of the redshift at which the scattering occurs.

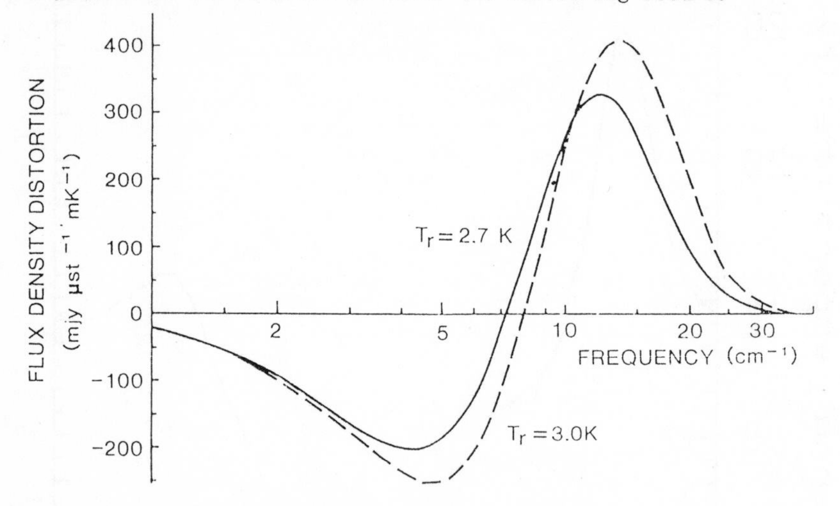

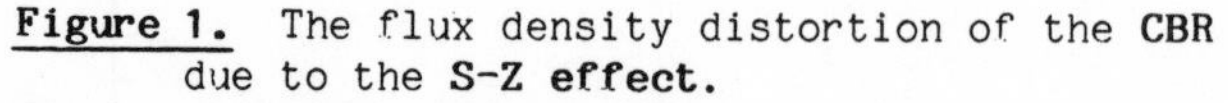

Figure 1. The flux density distortion of the **CBR** due to the **S-Z effect.**

The information to which such observations may ultimately give access includes the following.

1) Determination of H_0 and q_0[3].

Silk & White have shown that measurements of the S-Z effect, in combination with X-ray data, will permit direct determinations of both H_0 and q_0. The method is based upon the fact that this combination of data permits one to measure the angular-diameter-distance to the cluster. It is a very attractive method as it is largely independent of both the model and the evolution of the cluster atmosphere.

2) Probing the intra-cluster medium.

The x-ray luminosity of the intracluster medium is proportional to the volume integral of the electron number density squared, whilst the S-Z distortion is proportional to the line integral of the electron pressure. It is therefore possible to estimate the electron density, temperature, and the 'clumpiness' of the intra-cluster plasma by combining X-ray and S-Z data.

3) Cluster peculiar velocities.

If a cluster of galaxies has a radial component of peculiar velocity with respect to the local CBR frame of isotropy, the S-Z spectral distortion is Doppler-shifted. This shift may in principle be detected by multi-frequency observations at millimetre wavelengths.

4) Detection of possible cluster atmospheres around QSOs.

It is likely that at least some quasars occur within clusters of galaxies which will contain hot atmospheres capable of producing an S-Z distortion of the CBR. Multichannel millimetre-wave observations would allow one to separate out any intrinsic emission from the quasar itself. The redshift-independence of the S-Z effect means that it requires no greater sensitivity to detect at high redshift ($z \approx 3$ or 4).

5) Spectral distortions of the CBR.

Observations of the S-Z effect at several wavelengths would allow a differential measurement of the CBR temperature and estimates of possible deviations from a thermal spectrum [1,4]. These observations would best be performed at millimetre wavelengths.

On the basis of copious X-ray data [5] many clusters of galaxies are known to contain hot ($T \approx 10$ keV) intra-cluster plasmas. For the parameter range believed to be typical of such intra-cluster plasmas the magnitude of the distortion anticipated at radio wavelengths is in the range $\Delta J/J \approx \Delta T/T_r \approx 10^{-3}$ to 10^{-4}. Over the last 14 years there have been many attempts to measure the S-Z effect in a wide range of clusters, mainly conducted at centimetric wavelengths (9mm to 6cm) from large ground-based radiotelescopes, but also more recently at millimetre wavelengths. Despite the many thousands of telescope-hours devoted to radio measurements, the only significant claimed detections which remain uncontested were made only recently at $\lambda \approx 1.5$ cm [6]. There are a number of advantages in doing S-Z observations at millimetre wavelengths, using bolometric detection techniques.

a) The S-Z effect changes sign at 1.4 mm, so by measuring enhancements and decrements on the same cluster of galaxies one has unambiguous detection of the effect that is impossible to achieve by working only at radio wavelengths.

(b) The absolute magnitude of the flux distortion in the millimetre wavelength region is large compared to that at radio wavelengths (cf. Fig 1). This relatively large signal, combined with available bandwidths of at least 80 GHz, can provide detection sensitivities surpassing the best radio results [6,7] in smaller integration times.

c) Radio sources present in or near the cluster may produce flux contributions of either sign. Such sources are not always easy to identify [7]. Working at higher frequencies greatly reduces sensitivity to strong radio sources with their (generally) non-thermal spectra.

d) By virtue of the spectral form of the S-Z flux distortion, information on cluster peculiar velocities may in principle be obtained by multichannel observations in the 750 μm to 2.5mm spectral region [1,2].

2. GROUND-BASED OBSERVATIONS AT 1150 μm.

2.1 Technique.

The observations reported here were made at the United Kingdom Infra-Red Telescope in Hawaii (UKIRT), during the period 20th to 24th September 1983 using the QMC-Oregon millimeter-wave photometer[8], and in May 1986 with the common-user photometer UKT14. At 1150 μm wavelength the effective bandwidth was ≈ 80 GHz, and the Noise Equivalent Flux Density (NEFD) was typically ≈ 4 to 8 Jy√Hz. The beam profile was closely Gaussian with a full width at the 1/e-points of 135 arcsec. Signal modulation was achieved by secondary mirror chopping at 10 Hz frequency with chop and nod amplitudes of ≈ 3 arcmin. The reference beam positions were chosen to avoid contamination by known radio sources [5,9,10], and for both clusters this was possible with the chop oriented predominantly in Hour Angle.

During the 1983 run, data were recorded in blocks of typically 100 pairs of samples, with 10 seconds of integration per sample, and four samples (two pairs) per nod cycle. Each pair consists of one sample from each of the positive (+) and negative (-) nod positions. Each block took about 45 minutes of observing time (including dead time), and a calibration run was performed at the beginning of each block. Total integration times of 10 hours on the cluster 0016+16, and 6 hours on A478 were obtained over a four night period, out of total observing times of about 12.5 and 8 hours respectively. A further 20 hours of data on 0016+16 was obtained in 1986 using a similar but improved technique.

Jupiter was used as the primary calibration source during both the 1983 and 1986 observations. DR-21 and Orion were also used as secondary calibrators for the 1983 run. In summary, the calibration data shows that the system responsivity is constant to within ±5% on any one night, and there were slow variations in the atmospheric transmission at a marginally significant level of about ±4% during a whole night.

The clusters 0016+16 and A478 were chosen for observation because both have previously been extensively studied at radio wavelengths [5,11,12] and a detection of the S-Z effect in 0016+16 has been claimed [6,10]. These clusters are well placed for observation from UKIRT, and are also of small angular extent with centrally peaked X-ray emission [13] which makes them suitable for observation with our relatively small chop amplitude.

2.2 Environmentally induced effects in the data.

In the following, a 'raw data point' is a single sample from either the positive (+) or the negative (-) beam. The 'signal set' is formed from the raw data set by S = {(+)-(-)}, while the baseline set are b = 1/2{(+)+(-)}. If the baseline set for a whole night on one source are plotted against time, or Hour Angle, the most obvious feature is a marked change in the offset, as shown in Figure 2. This change in baseline offset is much too large to be accountable for solely by variations of the system responsivity. In fact it is not really a time-dependent effect, but depends instead on telescope attitude, as may be seen by a comparison of the calibration source and study-object baselines in Figure 2.

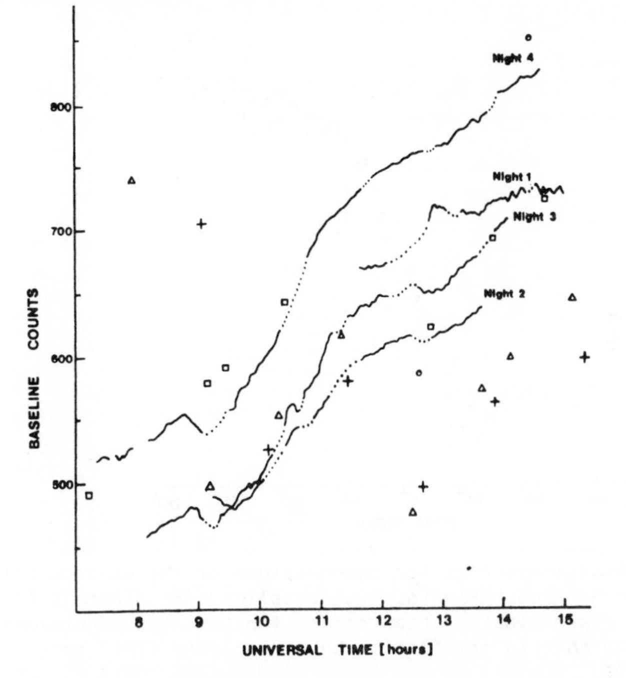

Figure 2. The solid line segments represent the data block baseline , while the dotted segments are interpolations over the calibration periods between the blocks. The symbols represent the calibration source baselines for DR-21 (early) and Orion (late) for nights 1 (o), 2 (+), 3 (Δ) and 4 (□) of the Sept. 1983 run.

If the changing offset were primarily due to time-dependent effects, the calibration source and study-object baselines should lie on the same curve, since thay are interleaved observations. Instead the baseline offset changes when the telescope is moved to a calibration source, and returns to an extrapolated point on the original baseline when the telescope is repointed to the study-object. This indicates that the changing baseline offset is due to gradients in the radiant background seen by the detector. The atmospheric emission, which is a function of airmass, is an obvious example of a radiant background which will produce such a changing offset. For the 1983 data the atmospheric contribution accounts for about 20 to 30% of the total change in offset. The baselines for 0016+16 on nights 2,3 & 4, are compared in Figure 3 after calibration and removal of the atmospheric component. This Figure shows vividly that the remaining change in offset reproduces from night to night. This is convincing evidence that the offset changes are due to flux gradients in the local environment, such as one expects to arise from the telescope and dome structure for instance. Flexure in the cryostat mounting is also a possible source of apparent flux gradients. Similar but smaller effects were also evident in the 1986 data.

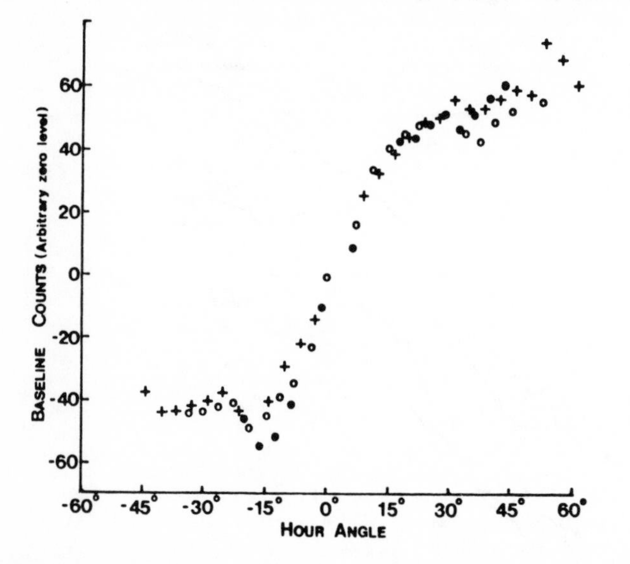

Figure 3. Baseline offset for observations of the cluster 0016+16 compared for different nights. The baselines for nights 2 (o), 3 (⊙) and 4 (+) are plotted after correction for atmospheric emission and the difference in calibration between nights (1983 data).

The critical importance of position-dependent offsets due to (real or effective) flux gradients is that the variation between nod positions of such a term will introduce a spurious systematic component into the S-Z signal dataset, whereas linear time-dependent terms are removed by our nodding and chopping procedure. Time-consuming reference-point observations are insufficient to properly identify such systematics. Special procedures for the identification and removal of position-dependent systematics are fully discussed elsewhere [14,15]. These procedures applied to the data discussed here result in residuals estimated at 50 to 100 μK.

2.3 Final data reduction and results.

After correcting the signal data sets for the systematics arising from changing offsets, it is convenient to convert the signal data to the corresponding flux equivalent Rayleigh-Jeans temperature decrement ΔT_{RJ}. This quantity is the most convenient single parameter by which to describe the size of the S-Z distortion, and is the accepted usage amongst radio astronomers. The signal power received is proportional to the integral of $\Delta J(\nu)$ over the instrumental passband, weighted for atmospheric absorption. The calibration factor (system responsivity) and the atmospheric weighting to be used in the conversion are determined from the calibration data.

For each data block, the mean, variance, standard deviation and standard error were calculated in the usual way, under the initial assumption of equal statistical weight for all points within the block, and subsequently with a spike rejection threshold set at approximately 5 times the initial block variance. A comparison of the block standard deviations obtained in this way shows that in general the noise increases with airmass, indicating a substantial 'sky noise' contribution of comparable magnitude to, or greater than, the intrinsic system noise. The 1986 data is not yet fully reduced, and the final corrected limits set on ΔT_{RJ} by the 1983 data are:

0016+16:
$\Delta T_{RJ} = -1.6\pm1.00$ mK, in 10 hours of integration.

A478:
$\Delta T_{RJ} = -0.20\pm1.00$ mK, in 6 hours of integration.

3. DISCUSSION.

3.1 Limits on the S-Z effect.

1). 0016+16. The corrected limit set on the observed tamperature decrement for this cluster is $\Delta T_{RJ} = -1.60\pm1.00$ mK in 10 hours of integration. This limit represents only a fraction of the total temperature decrement ΔT_0 at the cluster centre. The angular form of temperature decrement is proportional to the line of sight integral of the gas pressure through the cluster, and depends critically on the equation of state of the gas. The conversion from ΔT_{RJ} to ΔT_0 then depends on this form factor convolved with the beam profiles at the main and reference beam positions. The S-Z profile adopted by Birkinshaw et al.[6,9,11] is,

$$\Delta T(\theta) = \Delta T_0 \left(1+\theta^2/\theta_c^2\right)^{-\frac{1}{4}} \quad (1)$$

with $\theta_c \approx 30''$ for 0016+16. This corresponds to assuming a), that the X-ray surface brightness profile is given by

$$b_x(\theta) \propto \left(1+\theta^2/\theta_c^2\right)^{-\alpha} \quad (2)$$

with $\alpha = 1$, and b), that the gas is isothermal. Taking this form factor with the closely gaussian beam profile ($\theta_{fw} \approx 135''$) and chop amplitude ($\approx 3'$) used in the observations reported here, gives $\Delta T_{RJ} \approx 3\Delta T \approx -4.8\pm3.0$ mK. With our relatively small chop amplitude ($\approx 6\theta_c$) the conversion from ΔT_{RJ} to ΔT_0 is much more sensitive to the form factor adopted than in the case of Birkinshaw et al., who were able to employ a 7'.15 ($\approx 14\theta_c$) reference beam displacement. It should be noted that the observational uncertainty on the X-ray profile is large ($\theta_c \approx 30''\pm19''$) and that the exponent in Eqn.(2) is arrived at from the observed profiles of other, nearby clusters[16,17].

2): A478. The limit on this cluster is $\Delta T_{RJ} = -0.20\pm1.00$ mK in 6 hours of integration. This result on A478 is consistent with zaro, in agreement with the results of other workers (In view of the larger angular extent of this cluster ($\theta_c \approx 2'$ [8]), and the null results of previous workers, this result provides a valuable check that the correction procedures are reliable.

3.2 Limitations due to excess noise and systematics.

The reliability of the data reduction process depends crucially on the accuracy and confidence with which the systematic effects can be correctly identified and removed. The atmospheric correction amounts to approximately -0.40 mK when averaged over the four nights, while the correction for changing offset is -1.50 mK for 0016+16 and -0.90 mK for A478. The atmospheric flux depends approximately linearly on water vapour column density, for which the estimated error is 10%. Errors in the estimated atmospheric contribution are largely compensated for by the removal of the position-dependent offset term, of which the atmospheric contribution is really a special case.

If the chop throw has a declination component, the position dependent contribution to each point contains tha declination components of flux gradients. These cannot be determined from the baseline data set, because tha observing track is at constant declination, and must therefore be removed or at least estimated by other means.

The 1983 results show that a 1σ noise level of 1 mK was attained in 6 to 10 hours of integration, while the best 7 hours of data from 1986 yield a 1σ noise level of only 460μK. Fluctuations in the atmospheric emission ('sky noise') limit the attainable sensitivity at 1150 μm, and render the 800 μm window virtually unuseable from the ground. In addition, the confidence with which one can remove systematic effects is limited by the attainable signal-to-noise to which the baseline offsets can be determined. As sky noise increases with chop amplitude this problem becmes even more severe if one wishes to observe nearby clusters of large angular extent.

5. CONCLUSIONS.

The above results indicate a marginal detection of the S-Z effect in the distant cluster 0016+16, of $\Delta T = -1.6 \pm 1.0$ mK, uncorrected for the gas profile. No such effect was detected in the cluster A478, and this is consistent with previous work at radio wavelengths. These results comprise the first independent evidence in support of Birkinshaw et al.'s claimed detection of the S-Z effect in 0016+16 [6]. These results, obtained in only a few hours of integration with a general-purpose photometer, clearly demonstrate that bolometric millimetre wave techniques have a great potential in the field of S-Z and CBR micro-anisotropy investigations.

Position dependent offsets are the major source of systematic errors. Such effects are likely to have contributed to the many contradictory results obtained by radio observations of the S-Z effect. Errors arising from this source are removable by the procedures discussed in detail elsewhere [14,15], so long as the 'nod vector', which defines the data sampling pattern, is oriented predominantly in the Hour Angle direction. Although the traditional radio method of chopping in azimuth does eliminate the atmospheric contribution, other position dependent terms are neither removed nor identified by this procedure. Reliable removal of these effects depends upon the confidence with which the baseline offset function may be recovered from the data by polynomial fitting. Sky noise provides a significant contribution to the total NEFD, and in fact the overall system sensitivity was 'sky noise' limited at 1150 μm.

There is still considerable scope for improving ground based techniques for observing the S-Z effect at millimetre wavelengths, but very long integrations are required, and the ultimate sensitivity attainable is probably limited by sky noise and environmentally induced systematics. These problems are most severe when a large chop amplitude is used, and thus complicates the study of nearby clusters. In order to expliot the full potential of the S-Z effect as a cosmological and astrophysical tool it is clear that a space observatory equipped with broadband millimetre-wave detectors is demanded. The major technical requirement is that the beam diameter be of the order of 2 arcminutes, and the chopping secondary be capable of a 20 to 30 arcminute throw. The satisfaction of these requirements would also make such an instrument ideal for small angular scale studies of the Cosmic Background Radiation.

REFERENCES

1) Sunyaev R.A.,& Zeldovich Ya.B. Comm. Astro. & Sp. Sci. **4**,173 (1972).
2) Sunyaev R.A.,& Zeldovich Ya.B. MNRAS **190** 413 (1980)
3) White S.D.M & Silk J. Ap.J. **241**, 864 (1980)
4) Rephaeli Y., Ap.J. **241**, 858 (1980)
5) Johnson H.M. Ap.J. (Supp.) **47**, 235 (1981)
6) Birkinshaw M., Gull S.F., & Hardebeck H.; Nature **309**, 34 (1984)
7) Lazenby A.N., & Davies R.D. MNRAS **203**, 1137 (1983)
8) Ade P.A.R. et al. IR Physics **24** 46 (1984)
9) Birkinshaw M., Gull S.F. & Moffat A.T. Ap.J. **251**, L69 (1981)
10) Andernach H., et al. Astron. & Astrophys. **124**, 326 (1983)
11) Birkinshaw M., Gull S.F.,& Northover K.J.E.; MNRAS **185**, 245 (1978)
12) Gull S.F. & Northover K.J.E. Nature **263**, 572 (1976)
13) White S.D.M., Silk J. & Henry J.P.; Ap.J. **251**, L65 (1981)
14) Chase S.T., Joseph R.D., Robertson N.A. & Ade P.A.R. To appear in MNRAS (1987)
15) Chase S.T. Thesis, University of London, (1985)
16) Forman W. & Jones C.; Ann. Rev. Astron. & Astrophys, **20**, 547 (1982)
17) Cavaliere A. & Fusco-Femiano R.; Astron. & Astrophys. **49**, 137 (1976)
" " **100**, 194 (1981)

LARGE-SCALE STREAMING MOTIONS FROM FIRST-RANKED CLUSTER ELLIPTICALS

P. A. James, R. D. Joseph,
Blackett Laboratory, Imperial College, London SW7, England
C. A. Collins,
Astronomy Dept, University of Edinburgh, Edinburgh EH9 3HJ, Scotland

One of the most important constraints on cosmological models, and the cold dark matter model in particular, is the amplitude of the streaming motions of galaxies over scales of order 50-100 h^{-1} Mpc.[1,2] We have derived a new result in this field by using the V magnitudes and redshifts of first-ranked cluster ellipticals presented by Sandage and Hardy.[3] These galaxies are excellent standard candles, having a dispersion of only 0.3 in absolute magnitude. This dispersion permits us to infer galaxy distances to an accuracy of 14%, independently of their measured recession velocities. Thus they are excellent tracers of possible deviations from pure Hubble flow.

We have used this dataset to solve for that relative motion between the galaxy sample and the Local Group (LG) which minimises, in a least squares sense, the differences between distances inferred from apparent magnitudes, and those inferred from the redshifts. We restrict the solutions to one dipole, in the direction of the LG motion relative to the 2.8 K cosmic background radiation (CBR), since the sky coverage is incomplete in other directions. This is the direction in which an anisotropy in the Hubble flow about the LG would be expected to be most evident, were the galaxies at rest in the CBR frame. We have used the 60 galaxies in the Sandage & Hardy sample with redshifts $<15,000$ km/sec giving a mean effective redshift for the sample of 5400 km/sec.

We obtain a value for the velocity of the LG relative to the sample of -1 km/sec ± 105 km/sec. The error is the formal statistical error on the solutions, and was checked by doing a large number of resampling solutions with 25% of the points omitted at random from the minimisation procedure. Since the LG is moving relative to the CBR at 610 km/sec ± 50 km/sec, our solution provides evidence for a motion of the galaxy sample relative to the CBR with a component of magnitude 610 km/sec ± 116 km/sec along the direction $\ell = 272°$, $b = 30°$.

This result disagrees with that of Aaronson et al.[4], the only study of the large-scale velocity field to date which has found a galaxy sample to be essentially at rest in the CBR frame. However, these solutions using first-ranked cluster ellipticals are in good agreement with the conclusions of Collins et al.[1] and Dressler et al.[2], thus confirming the reality of streaming over scales of 50-100 h^{-1} Mpc.

REFERENCES

1. Collins, C., Joseph, R., and Robertson, N., Nature 320, 506 (1986).
2. Dressler, A., Faber, S., Burstein, D., Davies, R., Lynden-Bell, D., Terlevich, R., and Wegner, G. Preprint (1987).
3. Sandage, A., and Hardy, E., Ap.J. 183, 743 (1973).
4. Aaronson, M., Bothun, G., Mould, J., Huchra, J., Schommer, R. and Cornell, M., Ap.J. 302, 536 (1986).

CRYOGENIC DETECTORS FOR WEAKLY INTERACTING PARTICLES

B. Neuhauser, B. Cabrera. C.J. Martoff, and B.A. Young

Stanford University Physics Department
Stanford, CA 94305

ABSTRACT

We describe a new type of particle detector, called a silicon crystal acoustic detector (SiCAD), which senses ballistic phonons created by the collision of an incident particle with a nucleus or electron in a cube of single-crystal silicon cooled to a temperature below 1 K. A scattering event which deposits at least 1 keV within a 1 kg SiCAD would be measured with energy resolution better than100 eV and spatial resolution better than 1 mm^3.

1. INTRODUCTION

An ever more convincing body of astrophysical measurements suggests that the universe is permeated with dark matter which interacts gravitationally but eludes direct observation. One possible explanation for the invisible matter is that the various neutrinos have minute masses (≈10eV). Other arguments based upon the evolution of the early universe as well as our current understanding of elementary particle physics predict the existence of weakly interacting massive particles (WIMPs) as dark matter candidates with masses in the GeV/c^2 range. Identification of the dark matter, whether it be massive neutrinos or WIMPs, depends upon the development of particle detectors having lower thresholds and much larger cross sections than are offered by the best conventional detectors.

2. CRYSTAL ACOUSTIC DETECTORS

Motivated by these considerations, B. Cabrera, C. J. Martoff, and B. Neuhauser have proposed an entirely new type of particle detector called a silicon crystal acoustic detector (SiCAD)[1). The device consists of a 1 kg cube of ultracold ($T < 0.1K$) single-crystal silicon having arrays of phonon sensors on each face. When an incident particle collides with a nucleus or electron in the cube, the recoil kinetic energy creates electron-hole pairs which rapidly drop to the band edge by emitting very short wavelength phonons accounting for about 70% of the deposited energy. Because phonon decay rates are proportional to the fifth power of the frequency, within 10 nanoseconds the short wavelength phonons relax to a distribution of longer wavelength phonons which may be described crudely as a 10 Kelvin fireball inside of the ultracold crystal[2). These low energy phonons decay much more slowly. Furthermore, they propagate "ballistically" -- without scattering -- for distances of several centimeters because (1) the crystal is very pure and has few defects, and (2) there are very few thermally generated phonons. The long phonon mean free path is important because it allows the use of large detector volumes.

Figure 1(a) shows a cross section of the longitudinal and transverse phonon wavefronts a few microseconds after a scattering event within a crystal acoustic detector. These wavefronts are not spherical because the anisotropy of the crystal causes the group velocity and the wave vector for each mode to point in different directions. The wavefront distortion directs the acoustic energy into distinctive patterns determined by the orientation of the crystal faces. Figure 1(b) shows a Monte Carlo calculation of the phonon energy density resulting from a pointlike energy deposition in a cube with faces cut perpendicular to the [100] crystal directions. The projection of the position of the simulated scattering event is marked clearly on each face: the T1 (fast transverse) phonons produce a cross hair pattern, and the T2 (slow transverse) phonons create both an intense spot and a faint "X". Over 10% of the total phonon energy is focused into just 1% of the surface area. These patterns have been verified in a variety of heat pulse experiments. They can be used to locate the event to within 1 mm^3 inside of a 1 kg SiCAD cube, which has an edge length of 75 mm. As a result, it will be possible to reject spurious signals caused by radioactive surface contaminants.

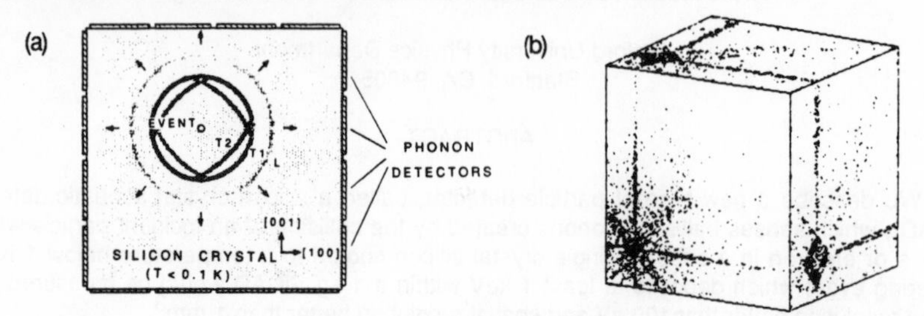

Figure 1: (a) SiCAD detection scheme; (b) Imaging of a phonon point source.

A likely phonon sensor configuration is diagrammed in Figure 2(a). Each cube face is covered by 64 parallel phonon sensing strips, and parallel faces have parallel strips. This arrangement allows each event coordinate to be determined with triple redundancy. For example, the y-coordinate can be found by comparing the arrival times of phonons in the peak channels on the pair of faces shown in Figure 2(b) or else by identifying the peak channel position on either face shown in Figure 2(c)

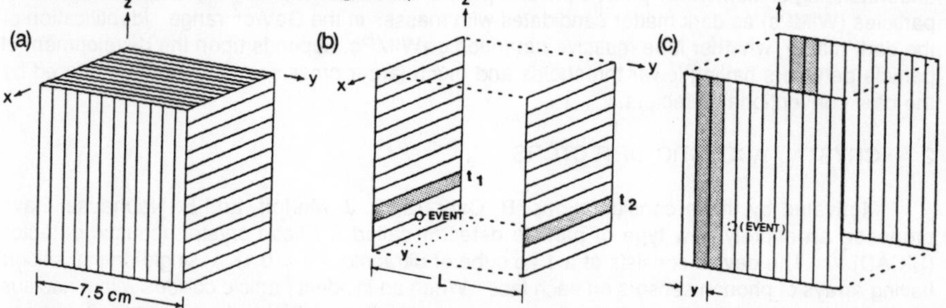

Figure 2: (a) Phonon sensors; (b) y-coord. from timing; (c) y-coord. from peak channel position.

The Monte Carlo simulation reveals that even in the most unfavorable case at least one sensor will receive 2% of the total recoil energy deposited in the crystal. Therefore detection of a threshold 1 keV recoil energy signal requires that each of the 384 sensors have an energy sensitivity of ≈ 20 eV. Note that for a 1 kg SiCAD the area of each sensor strip is 0.9 cm^2.

We have recently demonstrated particle detection mediated by ballistic phonons in a series of experiments utilizing superconducting transition-edge devices[3]. These consist of meander patterns made from 2 μm wide by 40 nm thick aluminum lines deposited on a 275 μm thick crystalline silicon substrate. When such a sensor is biased with a current of ≈1μA, which is slightly less than the critical current at the operating temperature of 1.3K, direct bombardment of the aluminum film with 5 MeV alpha particles produces self-terminating voltage pulses of amplitude ≈1 mV. These pulses occur because the alpha particles deposit enough energy in the film (≈100 keV) to initiate a superconducting-to-normal transition across the full width of a line[4]. Bombardment of the *back* side of the device substrate creates pulses of amplitude ≈0.1 mV. Since the range of a 5 MeV alpha particle in silicon is only about 20 μm, the sensor response must be due to phonons generated in the stopping process. This proves that the crystal acoustic detector concept is valid. However, the ballistic phonon energy arriving at the meander pattern is estimated to be on the order of a hundred keV, so it is clear that a much more sensitive type of phonon detector is needed.

3. SUPERCONDUCTING TUNNEL JUNCTIONS AS PHONON SENSORS

A superconductor-insulator-superconductor (S-I-S) tunnel junction is sufficiently sensitive to phonons to be used on a crystal acoustic detector because it exploits the extremely fragile nature of the superconducting state. When a metal is cooled below its superconducting transition temperature T_c, the conduction electrons begin to form pairs. Almost no unpaired electrons remain when the temperature has been lowered to 0.1 T_c. The energy required to break apart a pair typically is only a few hundred *micro* electron volts.

An S-I-S tunnel junction consists of two superconducting films, each a few hundred nanometers thick, separated by an insulating barrier that is only 1 or 2 nanometers thick. Current can flow between the superconducting layers only if charge carriers tunnel through the barrier. Two types of direct current tunneling are possible: if the voltage across the device is zero, electron pairs can tunnel as pairs; if the voltage is nonzero, quasiparticles can tunnel individually. Application of a magnetic field of the proper magnitude suppresses pair tunneling. Single electron tunneling is not affected by the magnetic field, but it too is negligible if the temperature is so low that there are essentially no unpaired electrons. Therefore if an S-I-S junction is cooled to $T<0.1T_c$ in an appropriate magnetic field, zero current flows through it. Suppose that an S-I-S junction is fabricated on the surface of an insulating crystal and biased with a voltage and magnetic field as discussed above so that the current is zero. If ballistic phonons are generated by a scattering event within the crystal, they will enter the adjacent superconducting film and break electron pairs. The liberated quasiparticles then will produce a pulse of tunneling current. Superconducting tunnel junctions have been used in this way for many years to study phonon focusing in crystals.

Recently two European research groups reported an intrinsic energy resolution of ≈50eV in the detection of 6 keV X-rays incident directly upon Sn-SnO_x-Sn superconducting tunnel junctions with areas in the range $(10)^{-3}$ to $(10)^{-4}$ square centimeters[5,6]. The high signal-to-noise ratio indicates that the threshold of sensitivity is on the order of 100 eV. For the purpose of SiCAD development, the challenge is to construct S-I-S junctions with similar sensitivity and resolution but with areas a thousand times greater.

4. ACKNOWLEDGMENTS

We want to thank Adrian Lee for participating in the transition-edge detector experiments. We are indebted to many staff members and students of the Stanford Integrated Circuit Laboratory: Jim McVittie, Peter Wright, Robin King, and Don Gardner were especially helpful.

This work has been funded in part by grants from Research Corporation and from Lockheed Corporation.

5. REFERENCES

1. B. Cabrera, C.J. Martoff, and B. Neuhauser, "Acoustic Detection of Single Particles", submitted to *Nuclear Instrumentation and Methods.*
2. R. Baumgartner, M. Engelhardt, and K.F. Renk, "Spectral Distribution of High-Frequency Phonons Generated by Nonradiative Transitions", *Physics Letters* 94, p. 55 (1983).
3. B. Neuhauser, B. Cabrera, C.J. Martoff, and B. A. Young, "Acoustic Detection of Single Particles for Neutrino Experiments and Dark Matter Searches", *Proceedings of the 1986 Applied Superconductivity Conference,* Baltimore, MD (in print).
4. D.E. Spiel, R.W. Boom, and E.C. Crittenden, Jr., "Thermal Spikes in Superconducting Thin Films of Sn and In", *Appl. Phys. Lett.* 7, p. 292 (1965).
5. H. Kraus, Th. Peterreins, F. Probst, F. v.Feilitzsch, R.L. Mossbauer, and E. Umlauf, "High-Resolution X-Ray Detection with Superconducting Tunnel Junctions", *Europhys. Lett.* 1, p. 161 (1986).
6. D. Twerenbold, "Giaever-Type Superconducting Tunnelling Junctions as High-Resolution X-Ray Detectors", *Europhys. Lett.* 1, p. 209 (1986).

PROSPECTS FOR DETECTING DARK MATTER PARTICLES BY ELASTIC SCATTERING

Bernard Sadoulet
Department of Physics
University of California
Berkeley ,CA 94720, USA

ABSTRACT

We discuss the rates,signatures and backgrounds likely to be encountered by experiments attempting to detect dark matter particles by elastic scattering.

1. GENERAL CONCEPTUAL FRAMEWORK.

In the following, we will assume that dark matter indeed exists, nearly closes the universe ($\Omega \approx 1$) and is non baryonic [1]. The hypothesis we would like to test, is that it is made out of heavy particles that we will call δ, not to call them heavy neutrinos[2], photinos[3], higgsinos[4], scalar neutrinos[5], cosmions[6], WIMPS[7] etc...., which would limit the generality of our remarks.

Can we say anything about the interaction rates of the δ's (which are presumably concentrated in the halo of our galaxy) with an ordinary matter target? It is clear that this is possible in specific models [3),4)]. We claim [8] that a general argument can be constructed in the case where the δ's have been in thermodynamical equilibrium in the early universe, with quarks and leptons, presumably through the annihilation channel

$$\delta \ \bar{\delta} \rightarrow q \bar{q}, e^+e^-, \nu \ \bar{\nu} \ ....$$

This excludes the axion from our considerations. Combining the Lee-Weinberg argument [2] relating the present density and the annihilation rate at freeze out, with crossing, it is possible to conclude in a rather model independent way that

$$\sigma_{el}(\delta N \rightarrow \delta N) \geq 10^{-38} cm^2 / (\Omega_\delta h^2) \text{ for } m_\delta \geq 1 GeV/c^2,$$

where Ω_δ is the current ratio of the δ average density to the critical density, h is the Hubble constant in units of 100km/s/Mpc and N stands for p or n. It can be also shown that accelerator data bound the δ mass from below around a few GeV, (although another allowable region exists around 1 electron-volt, that we will not discuss here).

2. RATES.

Let us first remark that although fairly model independent, the above arguments contain loopholes and it is possible to have a zero elastic cross section at low energy, for instance in the case of pure p-wave scattering (e.g. for AV or VA couplings), or of no coupling to the up and down quarks. This is however unlikely and does not occur in existing models.

Let us secondly assume that there is no initial asymmetry getween the number of δ and $\bar{\delta}$. In that case, $10^{-38} cm^2 / (\Omega_\delta h^2)$ is not a lower limit but the approximate value of the cross section on p or n.

For heavier targets, we should take into accounts coherence effects[9]. For certain couplings, (axial vector as for the photinos) the additive quantum number is the spin and the coherence factor is $\lambda^2 S(S+1)$, which at most of order unity (it is zero for a spinless target). For favorable targets (e.g. Boron), the event rate would be of the order given [8] by Fig 1_a. We assumed $\Omega_\delta h^2 = 1/4$, a mean root square velocity of 300km/s and a halo density of 0.7 10^{-24} g/cm^3.Note that because σ_{el} is roughly constant, the rate goes down as $1/ m_\delta$ if m_δ is comparable to the mass of the target.

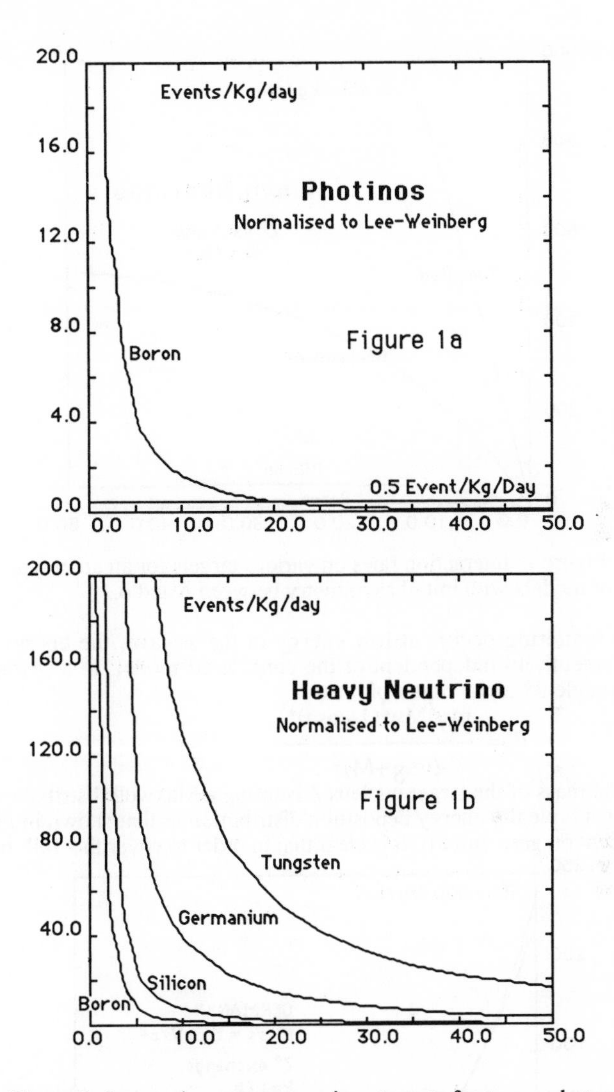

Figure 1: Interaction rates on various targets for two archetypes of dark matter models without initial asymmetry.

A third case, which is more favorable, is that of an additive quantum number proportional to A, or the number of protons or neutrons(as for heavy neutrinos). The rate is much bigger (Fig 1_b) but still decreasing as $1/m_\delta$.
As a last archetype of what could happen, we should consider an initial asymmetry between the number of δ and $\bar{\delta}$. The annihilation cross-section can be much bigger than the Lee-Weinberg limit. Fig 2 gives the example of a heavy neutrino coupling with the full Z^0 strength. The rates are much bigger and already double β experiments put limits on such models of dark matter[10].

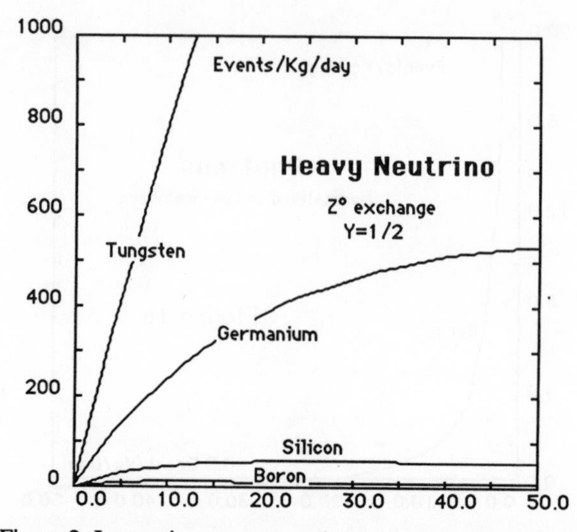

Figure 2: Interaction rates on various targets for an archetype of models with initial asymmetry between δ and δ.

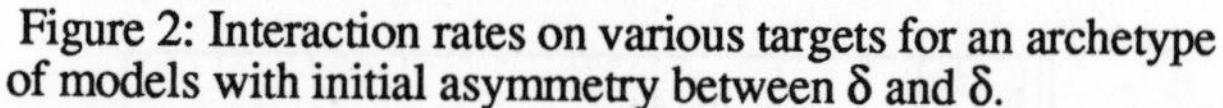

Because the scattering occurs at low energy in the s-wave, the energy deposition spectrum is essentially independent of the considered model. In a scattering of a δ particle at an angle θ* in the center of mass

$$E_d = \frac{m_\delta^2 M v^2 (1-\cos\theta^*)}{(m_\delta + M)^2}$$

where M is the mass of the target nucleus. Assuming a Maxwell distribution of the δ in the halo, we get typically energy deposition distribution as that shown in Fig 3 (m_δ=5 GeV/c^2 incident on germanium). It is clear that in order to cover the full mass range

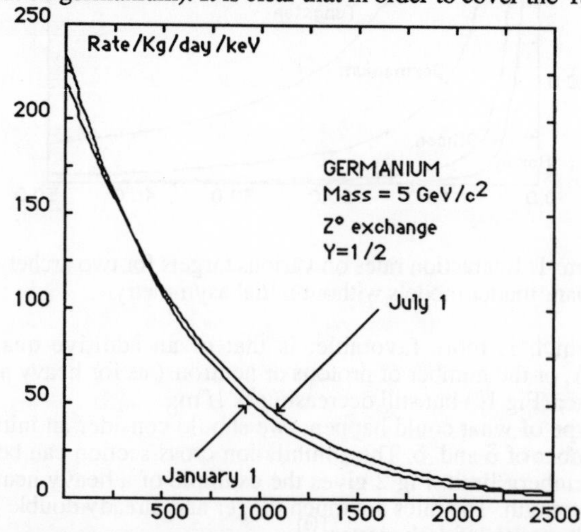

Figure 3: Example of energy deposition spectrum. The 2 curves refer to specific dates during the year.

not excluded by accelerators, threshold as low as 100 eV are needed. This is a strong motivation for the development of cryogenic detectors [11-13].

3. SIGNATURES.

Let us assume that a detector with such a low threshold indeed sees a potential signal at low energy. How could we be sure that it is due to dark matter particles from the halo of our galaxy?

The most convincing signature would be the annual modulation demonstrated in Fig 2. The halo has not collapsed significantly and is believed to have very small overall angular velocity. On the other hand, the sun goes around the galaxy and therefore through the halo at 200km/s and the earth is adding(substracting) half of its velocity to the sun velocity in the summer (winter). Both the mean energy deposition and the rate should vary by about $\pm 7\%$. In order to observe such an effect at 5σ, we would need about 5000 events. Therefore large mass detectors ($\approx$ 10kg) would have to be built!

Other important signatures include the shape of the energy distribution, which is expected to be different from X ray lines or Compton scattering (which is flat). The dependence of the energy deposition on the material used as a target might allow a measurement of the mass of the δ. The use of a mix of materials, may then allow to exclude the possibility that the incident particle is a γ or a neutron from the surrounding.

Finally, all the quantities used to discriminate against the radioactive background (e.g. distance from edges, ratio of heat to ionisation etc....),should have the distributions expected for dark matter. In principle, it is always possible to increase the radioactivity in the detector by a large amount, in order to show that the low-energy peak is not due to a feed-down from high energy contamination.

4. BACKGROUNDS.

The main background is from the residual radioactivity in the detector elements and in tits surrounding.

4.1 The **state of the art** is encouraging. The double β decay experiments of LBL/UCSB [12] and PNL/USC [10] get background rates of the order of 0.5 to 1 event/kg/keV/day at 20 keV. The rise observed below are related to specific instrumental problems which can be cured, and both groups are currently attempting to decrease their thresholds . This effort will bring new experimental information within the next year.

4.2 This level of radioactivity is quite compatible with attempts to **predict *a priori* background** [13].

a) Cosmic rays can be vetoed. Their indirect effects, spalliation and neutron production by muon capture, are more annoying and may require working underground.

b) Internal radioactivity of the detectors are expected to be negligible if crystals as Ge, Si, and may be B are used. On the other hand, spallation products such as ^{3}H [14] may quite disturbing, and it may be necessary to displace this tritium by hydrogen.

c) The main source of background will presumably come from the surrounding (refrigerator, dewar, shield).α's and β's can be eliminated completely in a position sensitive detector by imposing a fiducial region. Fast neutron from U and Th decays or μ capture, are potentially dangerous but can be thermalised easily with 40cm of water. Slow neutrons which may create γ rays can be absorbed by a borated shield.

By far the most difficult background to deal with are the γ rays from lines, nγ reactions and β bremstrahlung. Their main detrimental effect is a flat Compton background, which may be quite difficult to decrease appreciably with an active veto, because of the high energy of the parent γ. Typical computed levels [13] are compatible with the spectra observed by double β decay experiments.

4.3 **Scaling Laws.** If it is true that the observed background is flat in energy, experiments searching for dark matter particles, may run into signal to noise problems, if the interesting mass range is pushed up, for instance by new results of accelerators. If there is no initial asymmetry,the rate is proportional to

$$\frac{M}{(m_\delta + M)^2 m_\delta}$$

while the noise, if the background is flat, will be proportional to

$$\frac{m_\delta^2 \; M}{(m_\delta + M)^2}$$

Thus, the signal to noise goes down exactly as $1/m_\delta^3$ and the statistical accuracy roughly as $1/m_\delta^2$ at large masses!

A solution would be of course to improve dramatically the background rejection, by combining measurement of the heat deposited and of ionisation. The dark matter elastic scattering will give a nucleus recoil depositing comparitively little energy in ionisation, while ordinary radioactive background (except the neutrons which are easy to slow down) will deposit a much larger fraction in ionisation [15]. This method which has been suggested several times but not yet shown to be feasible, is presumably the only hope to detect dark matter if it is very heavy.

5. CONCLUSION

As a conclusion, we could list the necessary characteristics of a reasonnably complete detector for dark matter search, geared at the small mass region (≤ 20 GeV/c^2):10 kg detector,threshold 100-200 eV; measurement of the energy spectrum; mix of many materials, some with nuclear spins; localisation; very low radioactive background.

REFERENCES

1. Blumenthal, G.R. et al., Nature,311, 517(1984).
2. Lee, B.W. and Weinberg,S., Phys. Rev. Lett.,39, 165(1977).
3. Silk,J. and Srednicki,M., Phys. Rev. Lett.,53, 624(1984).
   Silk, J. and Olive, K. and Srednicki, M., Phys.Rev.Lett.,55,257(1985).
4. Ellis, J. et al., Nucl. Phys.,B238, 453(1984).
   Kane, G.L. and Kani,I., Nucl.Phys.,B277, 525(1986).
5. Hagelin,J.S. and Kane,G.L., Nucl.Phys.,B263, 399(1986).
6. Gelmini,G.B. and Hall,lL.J. and Lin,M.J., Nucl. Phys.,B281, 726(1987).
7. Press,W.H. and Spergel,D.N., Ap.J.,296, 679(1985).
   Faulkner,J. and Gough,D.O. and Vahia,M.N., Nature,321, 226(1986).
8. Greist,K. and Sadoulet,B., in preparation.
9. Goodman, M.W. and Witten, E., Phys.Rev.,D31, 3059(1985).
10. Ahlen,S. et al., Preprint Center for Astrophysics,(1986).
11. Niinikoski, T.O. and Udo , F., Cern Preprint,NP Report 74-6(1974).
    Cabrera, B. and Krauss, L.M. and Wilczek, F., Phys. Rev.Lett.,55, 25,(1985).
    Moseley, S.H. and Mather, J.C. and McCammon, D.,J.Appl.Phys.,56(5), 1257(1984).
    Drukier,A.K. and Stodolsky,L., Phys. Rev.,D30, 2295(1984).
12. Cabrera,B. , Caldwell,D. and Sadoulet,B., 1986 Summer Study on the Physics at SSC,Snowmass,(1986).
13. Sadoulet,B. et al., Lawrence Berkeley Laboratory Preprint,(1987).
14. Martoff, C.J., Dept. of Physics, Stanford note,(1985).

FIRST RESULTS FROM THE GALACTIC AXION SEARCH

S. De Panfilis, A. C. Melissinos, B. E. Moskowitz,
J. T. Rogers, Y. K. Semertzidis, and W. U. Wuensch
Department of Physics and Astronomy
University of Rochester, Rochester, NY 14627

H. J. Halama and A. G. Prodell
Brookhaven National Laboratory, Upton, NY 11973

W. B. Fowler and F. A. Nezrick
Fermi National Accelerator Laboratory, Batavia, IL 60510

ABSTRACT

Cold, light axions have been proposed as the main constituent for the halo of dark matter surrounding the galaxy. A microwave experiment designed to search for these galactic axions is described, and the first results presented. Future plans for the detector, and for a second generation detector, are also discussed.

1. INTRODUCTION

The axion was introduced[1] within the framework of quantum chromodynamics (QCD) as the Goldstone boson of the Peccei-Quinn symmetry which is invoked to explain the absence of CP violating effects in strong interactions[2]. One important parameter that is not predicted naturally by the theory is the mass of the axion, m_a, which is inversely proportional to the symmetry-breaking scale[3]. Upper limits on the axion mass come from accelerator searches[4], astrophysical arguments concerning the lifetimes of stars[5], and data from a low-background germanium detector[6]. Also, a cosmological lower limit on the mass arises since the matter density of axions can not exceed the critical density required to close the universe[7]. To summarize the best available limits:

$$10^{-3}\ \text{eV} \leq m_a \leq 10^{-5}\ \text{eV} \tag{1}$$

with the lower value of 10^{-5} eV preferred for closing the universe.

Axions are a leading candidate for the galactic dark matter since they should form a cold Bose gas which condenses into halos. Using a simple isothermal model by Turner[8], the calculated local density is 5×10^{-25} g/cm^3, and the r.m.s. velocity of the axions is of order 10^{-3} c, the galactic virial velocity. Thus, the energy spectrum should have a very narrow line:

$$Q_a = E_a/\Delta E_a = m_a c^2/(m_a v^2/2) = 10^6 - 10^7 \tag{2}$$

2. DETECTION

Although weakly interacting, axions should couple to charged fermions, and thus to two photons via a triangle diagram. Two-gamma decays are quite rare; however, by supplying a virtual photon from a static magnetic field, B_o, one may detect the decay of the axion into the second photon[9]. Since all of the energy of the axion converts into energy of the decay photon, the photon should have a frequency near 2 GHz for $m_a = 10^{-5}$ eV, conveniently in the mircrowave region. Since the interaction lagrangian is proportional to $(E \cdot B_o)\,\phi_a$, where E = electric field of outgoing photon and ϕ_a = psuedoscalar axion field, the conversion process may be enhanced by use of a resonant cavity mode chosen to maximize $E \cdot B_o$, such as the TM_{0n0} mode of a right cylindrical cavity. For such a mode, the expected cavity signal power is[10]

$$P_{sig} = (3.4 \times 10^{-24}\ W) \cdot \left[\frac{\langle\rho_a\rangle}{5\times10^{-25}\ g/cm^3}\right]\left[\frac{f_a}{1.2\ GHz}\right]\left[\frac{B_0}{6.6\ T}\right]^2\left[\frac{V}{1\times10^4\ cm^3}\right]\left[\frac{Q_L}{0.9\times10^5}\right]\left[\frac{4}{x_{0n}^2}\right] \quad (3)$$

where $\langle\rho_a\rangle$ = mean density of axions, Q_L = loaded Q of detector,

V = volume of cavity, and x_{0n} = n^{th} zero of Bessel fcn. J_0.

3. APPARATUS

The detector built by the Rochester-Brookhaven-Fermilab group is shown schematically in Fig. 1. Inside the 8-inch bore of a 6.6 T Nb-Ti superconducting solenoid is a OFHC copper cavity with a 40 cm length and 18 cm diameter. A typical measured value of the unloaded Q-factor for the TM_{010} mode is 2×10^5 ; the highest Q achieved is 3.6×10^5. Tuning of the resonant detector frequency is achieved by a single-crystal sapphire rod which can be raised or lowered into the cavity via a motor drive. The tuning range is 12%. The signal comes from an induction pickup in the cavity which is followed by a cryogenic GaAs FET amplifier, a two-stage mixer, and a 64-channel multiplexer which may be scanned and read by a computer.

4. PRELIMINARY RESULTS

A preliminary test run of the detector, covering the frequency range 1.09 - 1.22 GHz in 400 Hz bins, was completed in October 1986. A histogram of the number of standard deviations above mean for the power in each bin is shown in Fig. 2. Two false peaks from computer-generated r.f. stand above a Gaussian background. For bin width Γ_a, we observe the upper limit on the axion power at the 5 sigma confidence level to be:

$$P_a < 4 \times 10^{-21}\ W \cdot \frac{\Gamma_a}{400\ Hz} \qquad \text{for } \Gamma_a \geq 400\ Hz \quad (4)$$

5. FUTURE PLANS

As of this writing, two sweeps of the 1.09 - 1.22 GHz range have been completed with a 200 Hz multiplexer bin width at a sweep rate of approx. 200 Hz/sec. This is expected to yield an experimental limit on the cavity power which is 100 times greater than the simplest galactic halo model prediction. By using both the TM_{010} and TM_{020} modes, and by changing cavities and amplifiers, we plan to sweep the entire range 1 - 6 GHz in the next two years at the same sensitivity.

A planned second generation experiment featuring a 12 - 17 T magnet, a 1 - 2 m length cavity, an unloaded Q = 3.6×10^5 and an amplifier with a noise temperature near 4 K, offers the oppurtunity to truly test the axion galactic halo models.

REFERENCES

1) Weinberg, S., Phys. Rev. Lett. 40, 223 (1978).
   Wilczek, F., Phys. Rev. Lett. 40, 279 (1978).
2) Peccei, R., and Quinn, H., Phys. Rev. Lett. 38, 1440 (1977).
3) Dine, M., et. al., Phys. Lett. 104B, 199 (1981).
4) Zehnder, A., Phys. Lett. 104B, 494 (1981).
   Vuilleumier, L., et. al., Phys. Lett. 101B, 341 (1981).
   Edwards, C., et. al., Phys. Rev. Lett. 48, 903 (1982).

5) Fukugita, M., et. al., Phys. Rev. Lett. 48, 1522 (1982).
6) Aviognone, F., et. al., SLAC preprint no. SLAC-PUB-3872, (1986).
7) Preskill, J., et. al., Phys. Lett. 120B, 127 (1983).
   Abbott, L., and Sikivie, P., Phys. Lett. 120B, 133 (1983).
   Dine, M., and Fischler, W., Phys. Lett. 120B, 137 (1983).
8) Turner, M., Enrico Fermi Inst. preprint no. 85-67 (1985).
9) Sikivie, P., Phys. Rev. Lett., 51, 1415 (1983) and 52, 695 (1984).
10) Krauss, L., et. al., Phys. Rev. Lett. 55, 1797 (1985).

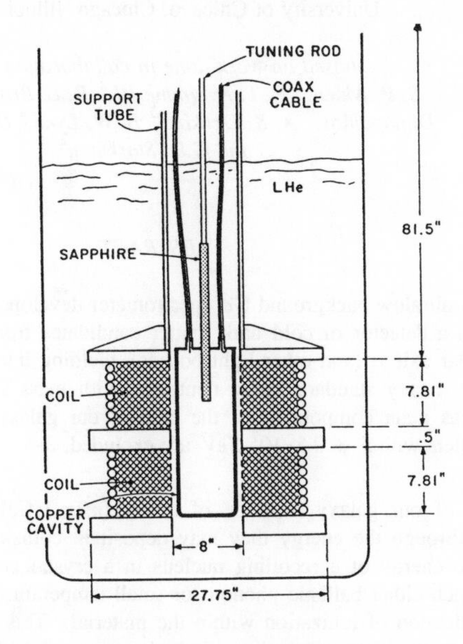

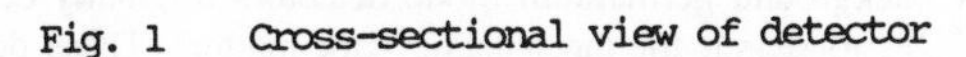

Fig. 1 Cross-sectional view of detector

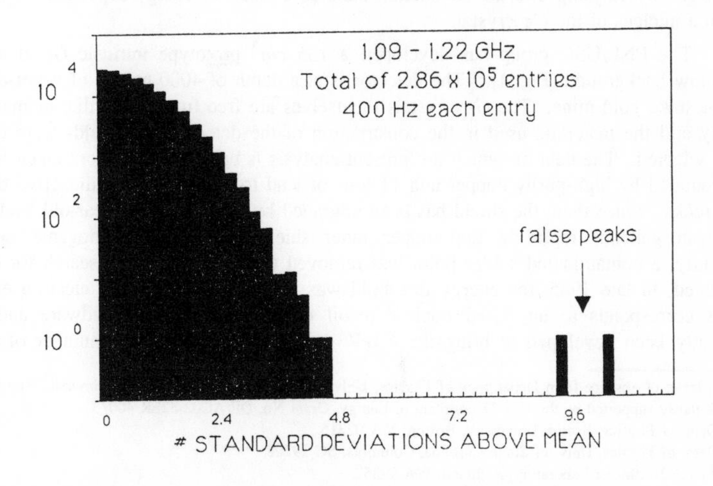

Fig. 2 Histogram of power in sweep bins

First Bounds from Direct Search of Galactic Cold Dark Matter and Solar Axions from a Ge-spectrometer

presented and written by
*G. B. Gelmini**

The Enrico Fermi Institute
University of Chicago, Chicago, Illinois 60637

based on work done in collaboration with
S. P. Ahlen,[1] *F. T. Avignone III,*[2] *R. L. Brodzinski,*[3] *S. Dimopoulos,*[4] *A. K. Drukier,*[5] *B. W. Lynn,*[4] *D.N. Spergel*[6] *and G.D. Starkman*[4]

ABSTRACT

The ultralow background Ge spectrometer developed by the USC/PNL group is used as a detector of cold dark matter candidates from the halo of our galaxy and of solar axions (and other light bosons), yielding interesting bounds. Some of them are: heavy standard Dirac neutrinos with mass $20\ GeV \leq m \leq 1\ TeV$ are excluded as main components of the halo of our galaxy; Dine-Fischler-Srednicki axion models with $F \leq 0.5 \times 10^7\ GeV$ are excluded.

If the halo of our galaxy is made of heavy ($m > 1\ GeV$) neutral particles they may be detected directly through the energy they may deposit in collisions with nuclei within a detector. Part of the kinetic energy of a recoiling nucleus in a crystal goes to the lattice (experiments are under study in which either ballistic phonons or small temperature increases are detected[1]) and part goes into the production of ionization within the material. This last mechanism allows the use of existing ultra low background germanium diode detectors originally developed to look for the double β decay of ^{76}Ge, to search for the galactic dark matter. These detectors count the number of electrons which jump into the conduction band as a result of energy deposited either in an electron or in a nucleus of the Ge crystal.

The PNL/USC group has developed a 135 cm^3 prototype intrinsic Ge detector having an ultralow background (Fig. 1).[2] The detector is at a depth of 4000 meters of water-equivalent in the Homestake gold mine. The Ge crystals themselves are free from primordial or man-made radioactivity and the materials used in the construction of the detector and shields were carefully studied and selected. The data on which the present analysis is based (Fig. 2) were taken with the detector surrounded by high-purity copper and 11 tons of lead to eliminate the radioactive background from the rocks. Since then, the shield has been upgraded by the use of 448 year-old lead (from a sunken Spanish galleon) replacing the copper inner shield which had cosmogenic radioactivity and, recently, a contaminated solder point was removed (Fig. 1). When the search for dark matter was initiated, in late 1985, the energy threshold was reduced to an incident electron energy of 4 keV. This corresponds to an initial nuclear recoil energy of 15 keV. Hardware and software have recently been developed to bring the 4 keV threshold to 1 keV. The number of electrons which

* On leave of absence from Department of Physics, University of Rome II, Via Orazio Raimondo, Rome, Italy, 00173.
Work partly supported by the U.S. Department of Energy, Grant No. DE AC02 82ER-40073.
1 Dept. of Physics, Boston University, Boston, MA 02215.
2 Dept. of Physics, Univ. of South Carolina, Columbia, SC 29208.
3 Pacific Northwest Laboratory, Richland, WA 99352.
4 Dept. of Physics, Stanford University, Stanford, CA 94305.
5 Applied Research Corp., 8201 Corporate Dr., Landover, MD 20785.
6 Institute for Advanced Studies, Princeton, NJ 08544.

jump into the conduction band when some energy is deposited on an electron is larger than the number when the same energy is deposited on a nucleus, by a factor that we call R.E.F., relative efficiency factor. The R.E.F. is an energy dependent quantity which was determined[3] (theoretically and through data from Chasman et al.[4]) to be $\simeq 3.75$ in the 10 to 100 keV region. The predicted rate R_P depends on the local density ρ_x of the dark matter particles, its mass m_x, its cross-section σ on Ge nuclei and the velocity, v, distribution in the halo, $R_P \sim (\rho_x)\sigma v/m_x$. A Gaussian v-distribution was used with $v_{r.m.s.} = 250\ km/\text{sec}$ and $v_{max} = 550\ km/sec$ and a local halo velocity of $80\ km/sec$ (which is the local bulge velocity). These are very conservative values. Integrating over velocities,an upper bound on $(\rho_x \sigma)$ is obtained by requiring $R_P < R_0$, the observed rate. To fix ideas the σ of standard heavy neutrinos was used. By fixing σ,an area in a (ρ_x, m_x) space is excluded (Fig. 3). By fixing $\rho_x = \rho_{halo}$, an area in a $(g/g_{weak} = \sqrt{\sigma/\sigma_{weak}}), m_x)$ space is excluded (Fig. 4).[3]

An interesting bound (similar to that obtained by requiring that the luminosity of the sun be mainly in photons) on very light scalars ϕ, such as axions, familons, Majorons, emitted by the sun was also obtained[5]: $g^{-1} \geq 0.5\ 10^7 GeV$, for a coupling with electrons $L = g m_e \bar{e} i \gamma_5 e \phi$ ($g^{-1} = F$, the Peccei-Quinn scale, in the case of Dine-Fischler-Srednicki axions).

1]. Smith, P.F., Rutherford Lab preprint RAL-886-029, 2nd ESO-CERN Symposium, Garching, March 1986. B. Sadouley in this proceedings.

2]. Brodzinski, R.L., Brown, D.P., Evans Jr., J.C., Hensley, W.K., Reeves, J. H., Wogman, N.A., Avignone III, F.T., and Miley, H.S., Nucl. Instr. and Meth. **A239**, 207 (1985).

3]. Ahlen, S.P., Avignone III, F.T., Brodzinski, R.L., Drukier, A.K., Gelmini, G., and Spergel, D.N., *Limits on Cold Dark Matter Candidates from the Ultralow germanium Spectrometer*, Harvard Center for Astrophysics, preprint No. 2292, 1986.

4]. Chasman, C., Jones, K.W., and Ristinen, R.A., Phys. Rev. Lett. **15**, 245 (1965).

5]. F. T. Avignone III, R.L. Brodzinski, Dimopoulos, S., Drukier, A.K., Gelmini, G., Lynn, B.W., Spergel, D.N., Starkman, G.D., *Laboratory Limits on Solar Axions from an Ultralow Background germanium Spectrometer*, SLAC preprint PUB-3872, 1986 to appear in Phys. Rev. D.

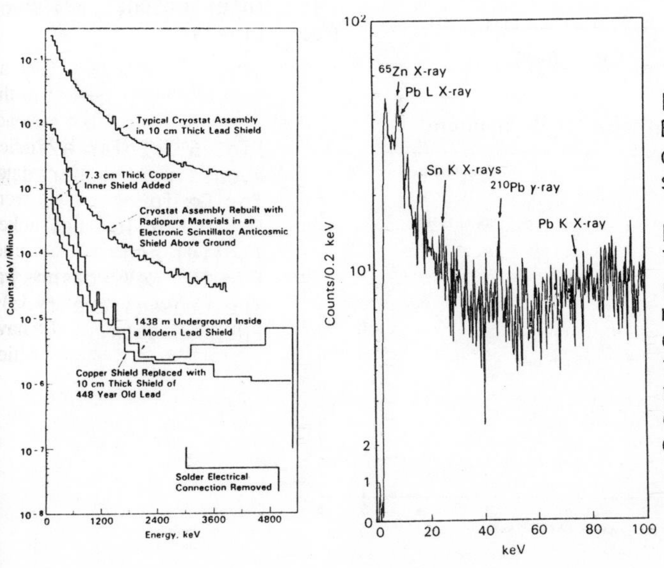

Fig. 1 (left) Background spectra of the PNL/USC Ge spectrometer.

Fig. 2 (right) 1000 hours of data (0.2 KeV per channel) taken with the copper shield. The identified peaks resulted from the decay products of one solder point (now removed).

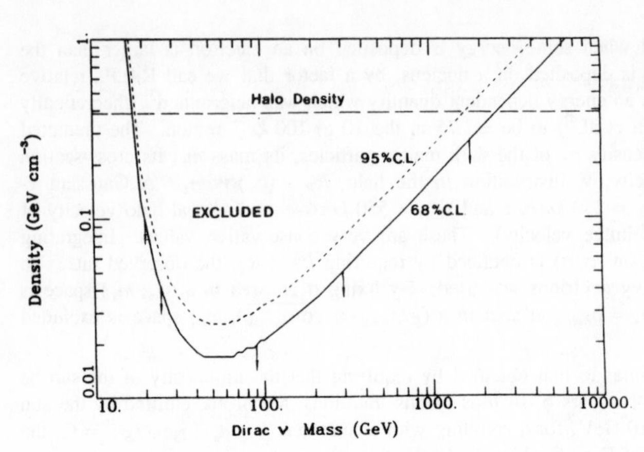

Fig. 3 (left). Maximum halo density of standard Dirac neutrinos consistent with the observed count rate at the 68% and 95% confidence levels.

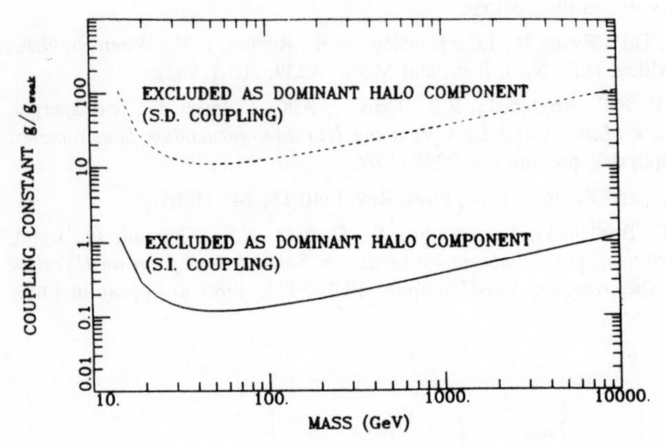

Fig. 4. Excluded regions in mass/cross section space $g/g_{weak}=\sqrt{\sigma/\sigma_{weak}}$) at the 68% C.L. for particles constituting the halo. σ_{weak} is the cross section of Dirac and Majorana neutrinos for spin independent and spin dependent interactions, respectively.

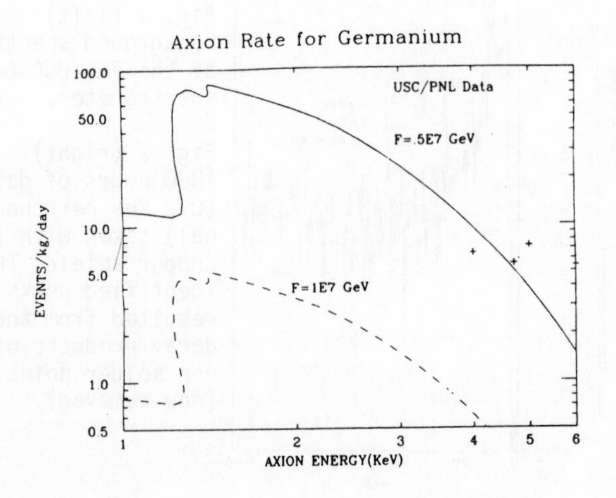

Fig. 5. Solar axion events per kg per day for Ge for $F = 0.5 \times 10^7$ GeV (solid) and $F = 10^7$ GeV (dashes). The crosses are from PNL/USC data.

GRAVITATIONAL LENSING OF SUPERNOVAE BY DARK MATTER CANDIDATES OF MASS $M \gtrsim 10^{-3}$ $M_\odot$

Robert V. Wagoner and Eric V. Linder

Dept. of Physics and Center for Space Science and Astrophysics
Stanford University, Stanford, CA 94305-4060

Because they are relatively well-understood, simple physical systems (compared to quasars, galaxies, etc.), supernovae are potentially important probes of the universe.[1] Here we shall summarize some results of recent studies[2,3] of their gravitational lensing by intervening condensed objects, which include dark matter candidates (such as dim stars and black holes). The expansion of the supernova beam within the lens produces characteristic time-dependent amplification and polarization, which depend upon the mass of the lens. Although we have considered the effects of the shearing of the beam due to surrounding masses (the parent galaxy), here we will confine our attention to isolated masses whose size is much less than that of the supernova ($\sim 10^{15}$ cm).

Consider a supernova of radius R, which is typically expanding with a velocity $v \cong 0.03$ c during the epoch of maximum luminosity. We take its surface brightness (in each polarization state) to be that produced by an electron-scattering dominated photosphere. The impact parameter L (distance from center of supernova to projection of lens mass into the source plane) will typically vary on a time scale long compared to that of R. The cross section $d\sigma/dA$ (referred to the source plane) for producing amplification A actually rises somewhat above the point-source result $2\pi\chi_o^2/A^3$ for $L \lesssim R$, but the maximum amplification (corresponding to L = 0) is given by

$$A \lesssim 2.2\ \chi_o/R. \tag{1}$$

The standard "lens radius"

$$\chi_o \cong (6 \times 10^3\ h^{-1}\ z_s\ M/M_\odot)^{1/2} \times 10^{15}\ \text{cm} \tag{2}$$

for supernova redshifts $z_s \cong 2z_\ell(\text{lens}) \ll 1$ and Hubble constant $H_o = 100$ h km s^{-1} Mpc^{-1}. It is then obvious that for $z_s \sim 0.2$, corresponding to proposed searches, masses $M \gtrsim 10^{-3}$ $M_\odot$ can produce significant amplification.

Figure 1 shows how the amplification A and polarization P change if the supernova beam expands enough to envelope the mass. As indicated by equation (1), the amplification decreases as R^{-1} for $R \gg L$, changing the shape of the supernova light curve. The polarization reaches its maximum at R = L, when the gradient in amplification across the beam produces the largest departure from spherical symmetry. These time dependences are a characteristic signature of such microlensing, but will be less likely the larger the mass of the lens. Other effects (such as image splitting and time delay) are unobservable for the masses $M \lesssim 10^6 M_\odot$ of interest.

Let us next consider the probability of observing these effects in a supernova search (such as the one begun by John McGraw at the Steward Observatory). For definiteness, we choose a large-scale Friedmann cosmology with $\Lambda = p_o = 0$ and $\Omega_o = 0.2$. The propagation of the supernova beam is described by the Dyer-Roeder model[4,5], with all the density in (conserved) objects of the same mass M. The supernova search is assumed unbiased, so we take the density of these lenses as uniform.

We consider supernovae of two classes, with all members of each class identical, and no evolution in their rate $[\alpha(1 + z_s)^3]$. The absolute blue magnitude and radius at maximum luminosity for Type I-a (standard) and Type II-P (plateau) are given by

$$\text{SNI-a:} \quad M_B = -18.2 + 5 \log h \; ; \; R_m = 1.2 \times 10^{15} \text{ cm} \tag{3a}$$

$$\text{SNII-P:} \quad M_B = -17.1 + 5 \log h \; ; \; R_m = 0.2 \times 10^{15} \text{ cm} \; . \tag{3b}$$

The spectrum for SNI-a is taken to be that of 1981b[6], and for SNII-P that of a 15,000 K blackbody with a Lyman limit cutoff.

We have computed the fraction $F(>A)$ of supernovae of a given apparent magnitude m_i which have been amplified by at least a factor A. In doing so, we have used the fact that the position of the supernova on the sphere at redshift z_s, not the observed direction of the supernova, is the correct random variable.[7] We have considered searches carried out using either blue (i = B) or red (i = R) bandwidths. Choosing $m_i = 22$, the results for masses $M = 10^{-2} M_\odot$ and $10^2 M_\odot$ are shown in Figures 2 and 3.

The most startling feature is the dominance of Type II-P supernovae, even though they are intrinsically fainter. This is explained by the severe ultraviolet deficiency of Type I supernovae, making them fainter than Type II at redshifts $z_s \gtrsim 0.25$ in the B band and $z_s \gtrsim 1.00$ in the R band. Since the amplified supernovae are at larger redshifts than the unamplified ones, the larger volume has a strong effect on the fraction amplified at fixed m_i. The other major feature (seen for the smaller mass) is the rapid falloff in the fraction with increasing amplification. This is a direct consequence of equation (1), which also explains why the cutoff amplification is larger for Type II supernovae (because R is smaller).

Candidates for lensing would be those supernovae at least as bright as their parent galaxy, or above the range of luminosities expected for their spectral class once their redshift was determined. Although a search for time dependence of the amplification and polarization would then be appropriate, we note from equation (1) that even a constant amplification would provide a lower limit on the mass of the lens.

This work is the result of a very pleasant and stimulating collaboration with Peter Schneider; and is being supported in part by National Science Foundation grant PHY 86-03273 at Stanford University, and by NASA grants NAGW-299 at Stanford and NSG-7123 at the Joint Institute of Laboratory Astrophysics.

REFERENCES

1] Wagoner, R.V., in Proceedings of the Vatican Observatory Conference on Theory and Observational Limits in Cosmology, in press (1987).
2] Schneider, P. and Wagoner, R.V., Astrophys. J., in press (1987).
3] Linder, E.V., Schneider, P. and Wagoner, R.V., in preparation (1987).
4] Dyer, C.C. and Roeder, R.C., Astrophys. J. **174**, L115 (1972).
5] Press, W.H. and Gunn, J.E., Astrophys. J. **185**, 397 (1973).
6] Panagia, N., in NATO Advanced Study Institute on High Energy Phenomena Around Collapsed Stars, to be published (1986).
7] Ehlers, J. and Schneider, P., Astron. Astrophys., in press (1986).

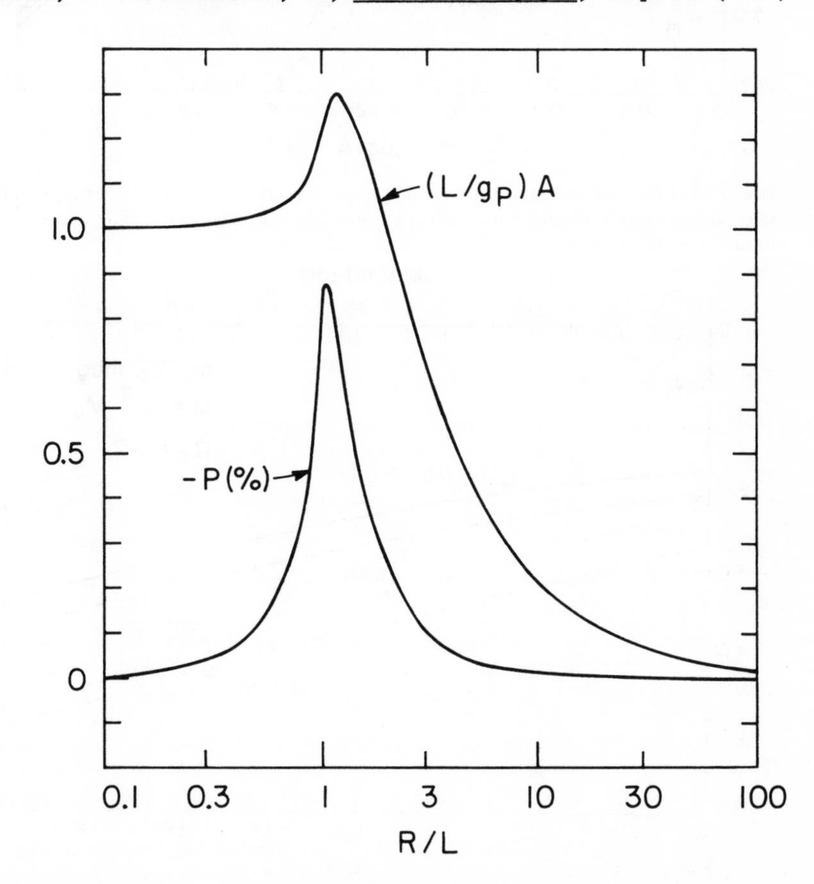

1. The dependence of the amplification and (percent) polarization on the supernova radius. For a point mass, the lens structure constant $g_P = \chi_($

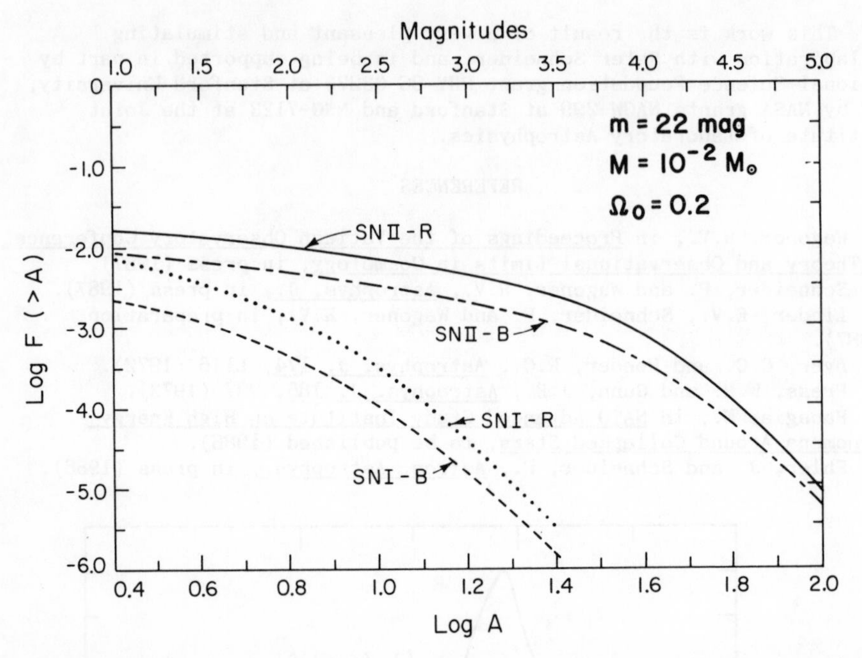

2. The fraction of supernovae amplified by at least A is presented for two wavelength bands and two classes of supernovae, for lenses of mass $M = 10^{-2}\ M_{\odot}$.

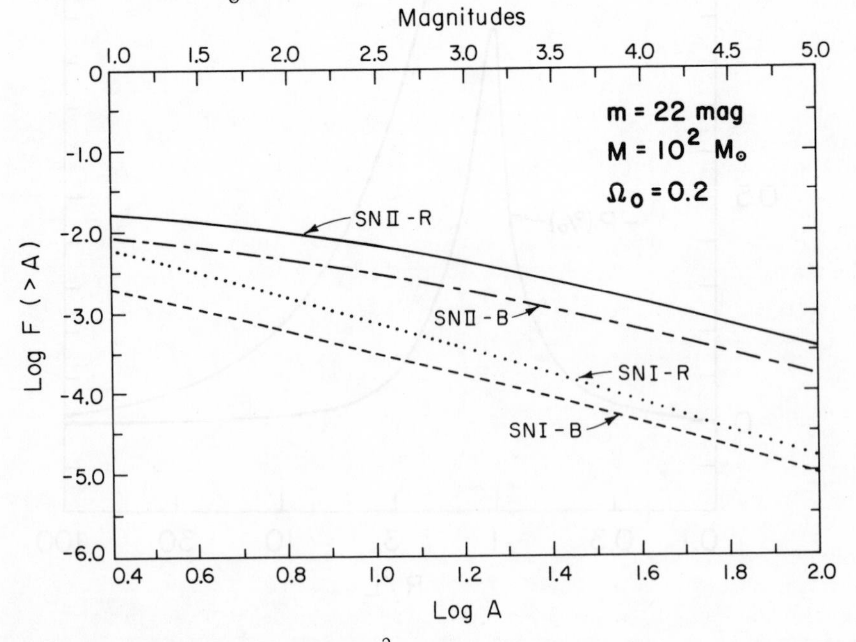

3. Same as Figure 2 for $M = 10^{2}\ M_{\odot}$.

REDSHIFT SURVEYS

JOHN P. HUCHRA AND MARGARET J. GELLER

Center for Astrophysics
60 Garden Street
Cambridge, Massachusetts 02138

1. INTRODUCTION

The last decade has seen a tremendous surge of efforts to measure redshifts of galaxies. The number of galaxies with redshifts has increased from less than 1000 in 1975 to well over 20,000 in 1986. This jump is the result of improvements in both optical and radio detectors.

Why are we measuring redshifts? Basically, we are trying to determine the distribution of mass-energy in the universe. Galaxies are (by definition) tracers of the luminous matter distribution and may also be tracers of dark matter. Measurements of the velocity distributions of aggregations of galaxies have so far been the only way in which dark matter has been detected on large scales.

The goals of the redshift surveyor are the determination of cosmological parameters (H_0 and q_0) and the determination of the topology of the matter distribution. This information is the key which we hope to use to unlock the secrets of galaxy formation, cluster formation as well as the creation of large voids and large scale motions.

2. TYPES OF SURVEYS

Redshift surveys can be classified in terms of their areal coverage and their completeness. Surveys can either be defined in terms of individual galaxies or in terms of galaxy custers. Table 1 lists many of the existing surveys plus several currently under way.

Large area surveys tend to be most useful for topological studies. Limits on the largest structures that can be seen are determined by the survey design, either the maximum angular extent on the sky or the maximum depth in redshift.

3. SURVEY REQUIRMENTS

In order to be of use for quantitative studies, redshift surveys must have well defined selection parameters. The edges of the sample must be well defined. The depth limit, usually given by a magnitude or diameter limit, must be well defined, and, for certain applications like the determination of the correlation function on large scales, must be *uniform* on the sky. A common problem in surveys is that the magnitude limit may vary with position on the sky. For example, the magnitudes in Volume 1 of the Zwicky Catalog are systematically off from those in the other volumes by 0.2 mag. Because $N \sim \Delta m^{0.6}$ in Euclidean space, the number counts in that volume are 30% off from the others. This introduces a rather large signal when computing the 2-point angular correlation function on scales larger than $5.^0$

TABLE 1
SOME REDSHIFT SURVEYS

Name	# of galaxies
A. LARGE AREA COMPLETE	
GALAXIES	
HMS (1956)[1]	500
RSA (ST 1981)[2]	1300
RC1, RC2 (dv 1964, 1976)[3]	2000
CFA1 (HDLT 1982)[4]	2400
UGC Galaxies (BBMH 1985)[5]	4000+
CFA Slice (dLGH 1986)[6]	1100
Southern Sky (DDTLHF 1987)[7]	1800
Nearby Gals (TF 1986)[8]	3000
CFA2 (HG 1987+)[9]	12,000+
ABELL CLUSTERS	
(HGT 1980)[10]	116
(PHGH 1985)[11]	561
B. LARGE AREA INCOMPLETE	
Markarian Gals (APDL+ 1970's)[12]	1300
Bootes (KOSS 1983, 1986+)[13]	400
ARECIBO (GH 1986+)[14]	2000+
IRAS Galaxies (DHTSY 1987+)[15]	2600
C. SMALL AREA (Usually Deep)	
GALAXIES	
AAT (EEP+ 1985)[16]	
KKS (1986+)[17]	
Century (GHKTW+ 1986+)[18]	2000+
ABELL CLUSTERS	
(CFH 1985)[19]	5
(PHG 1986) *Cor Bor*[20]	7
(HHP 1987)[21]	145

Other insidious selection effects that must be accounted for are surface brightness, confusion, mass-to-light rato, and morphological type. Surface brightness is the classic bugaboo because all surveys miss galaxies much fainter than the night sky and very few surveys are sensitive to compact galaxies. Confusion is a rampant problem in catalogs (like Zwicky or the UGC) where multiple faint galaxies are often listed as a single entry — another good way to mess up the correlation function. Mass-to-light ratio and morphological type most often plague 21-cm HI surveys; galaxies with very broad HI profiles are hard to detect; galaxies with little or no HI, eg. ellipticals and S0's, are impossible to detect.

Two other considerations are important for surveys. The sampling, or number of objects per unit volume, is very important for topological studies. The resolution of any map is a function of the number of points it contains. A good example of the effects of clustering can be seen in the comparison of the low velocity regions covered in the maps of the CfA slice.[6] The resolution of the first CfA survey with only 140 galaxies is quite poor compared to that of the deeper sample in the same volume of space. The survey dimensions – width, depth – are also important. To date, every survey made has detected a structure that is about as large as could have been seen. In general, the biggest thing that can be resolved is roughly 1/2 the maximum dimension of the sample.

4. THE LONG VIEW

What does the future hold? So far, almost all surveys have detected structures as large as they could. We have also nearly hit the limit of existing large area galaxy catalogs (CFA2 = Zwicky). We are almost there for cluster surveys, too, since the Abell catalog starts to be incomplete and biased at distance class 5.

There are two readily identifiable needs for further work: (1) a digital sky survey (multicolor/CCD) with $\sigma_M \lesssim 0.05$, (2) wide field survey telescopes ($\sim 1^0$) with multi-object spectrographs. Let's get going!

REFERENCES

1) Humason, m., Mayall, N. and Sandage, A. *A. J.* **61**, 97 (1956).
2) Sandage, A. and Tammann, G. *The Revised Shapley-Ames Catalog of Bright Galaxies*, (Washington: Carnegie Institution) (1981).
3) de Vaucouleurs, G., de Vaucouleurs, A. and Corwin, H. *Second Reference Catalog of Bright Galaxies*, (Austin: U. of Texas Press) (1976).
4) Huchra, J., Davis, M., Latham, D. and Tonry, J. *Ap. J. Suppl.* **52**, 89 (1983).
5) Bothun, G., Beers, T., Mould, J. and Huchra, J. *A. J.* **90**, 2487 (1985).
6) de Lapparent, V., Geller, M. and Huchra, J. *Ap. J.* **302**, L1 (1986).
7) daCosta, L., et al. in preparation (1987).
8) Tully, R. B. and Fisher, R. in preparation (1987).
9) Huchra, J. and Geller, M. in progress (1987).
10) Hoessel, J., Gunn, J. and Thuan, T. X. *Ap. J.* **241**, 486 (1980).
11) Postman, M., Huchra, J., Geller, M. and Henry, P. *A. J.* **90**, 1400 (1985).
12) Arakelian, M. et al. series of papers in Astrofizica (1971-1985).
13) Kirshner, R. Oemler, A., Schechter, P. and Shectman, S. in press (1987).
14) Giovanelli, R., Haynes, M., Meyers, S. and Roth, J. *A. J.* **92**, 250 (1986).
15) Davis, M., Huchra, J. Strauss, M and Yahil, A. in prep. (1987).
16) Peterson, B. et al. *M.N.R.A.S.* **221**, 233 (1986).
17) Kron, R., Koo, D., and Szalay, A. in preparation (1987).
18) Geller, et al. in progress (1987).
19) Ciardullio, R., Ford, H., and Harms, R. *Ap. J.* **293**, 69 (1985).
20) Postman, M., Huchra, J. and Geller, M. *A. J.* **92**, 1238 (1986).
21) Huchra, J., Henry, J. and Postman, M. in preparation (1987).

LARGE-SCALE STRUCTURE IN THE PISCES-PERSEUS SUPERCLUSTER

Martha P. Haynes

National Astronomy and Ionosphere Center and Astronomy
Department, Cornell University, Ithaca, NY 14853

ABSTRACT

The three-dimensional structure of the galaxy distribution in the Pisces-Perseus supercluster is complex and varied. While some galaxies lie on thin surfaces, the main ridge is more filamentary. Morphological segregation is seen over all density regimes, even among the spiral subclasses.

1. INTRODUCTION

One of the most prominent features of the galaxy distribution in the local universe is the Pisces-Perseus supercluster, noted as a possible "metagalactic" cloud even by Bernheimer[1)] in 1932. The main ridge of the supercluster can be traced over 90 degrees, outlining a continuous high-density filament in which the Pegasus and Pisces clusters, as well as A262, A347 and A426 are embedded. The ridge terminates abruptly in the zone of avoidance. Both Burns and Owen[2)] and Giovanelli and Haynes[3)] have suggested a physical connection between the Pisces-Perseus supercluster west of the zone of avoidance and clusters to the east. Batuski and Burns[4)] moreover propose a link in the third dimension to other clusters in the west.

The Pisces-Perseus supercluster stands out so prominently in the surface density distribution of nearby galaxies because at its characteristic distance of $50h^{-1}$ Mpc, member galaxies of even moderate luminosity are included in the galaxy catalogs; the luminosity function M^* corresponds to only m = +14.9. At such a distance, redshifts are easily measured in large quantity by both optical and radio 21 cm line techniques.

2. THE HI SURVEY OF THE PISCES-PERSEUS SUPERCLUSTER

Over the last several years, Riccardo Giovanelli and I and our co-workers[5,6,7,8)] have been undertaking a major survey of galaxies in the main region occupied by the Pisces-Perseus supercluster, primarily within the borders 22^h < R.A. < 4^h, 0° < Dec. < $+50^o$. 21 cm line redshifts have been measured for more than 3000 galaxies, bringing the number of measured redshifts in the area to over 4000. Because most of the redshifts have been obtained in HI-rich objects, the sample is not complete in the usual magnitude-limited sense[5)].

However, it should be noted that most of the galaxies outside cluster cores are typically HI-rich, and often of low optical surface brightness; it is often trivially easy to detect a galaxy of apparent magnitude +17 in the 21 cm line at 5000 kms^{-1}. Tracing the three-dimensional structure on large scales requires the study of objects of both high and low optical surface brightness[6].

The primary survey includes all galaxies in the Pisces-Perseus region bounded by $+3^o <$ Dec. $< +35^o$ that are included in the Zwicky Catalog[9] and have morphologies later than S0/a. This declination restriction is imposed by the tracking capabilities of the Arecibo 305 m telescope. The remainder of the region is being surveyed with lowered sensitivity because of vignetting at Arecibo and because of the transit nature of the 91 m telescope of the N.R.A.O. at Green Bank, used for galaxies north of declination $+38^o$.

The complexity of the Pisces-Perseus supercluster is visible even in the surface density maps. The ridge entends over some $45h^{-1}$ Mpc and is characterized by a large axial ratio. The width of the ridge, when averaged over its length, can be fit by a gaussian of full-width less than $10h^{-1}$ Mpc. From the redshift distribution, the depth of the ridge is also seen to be $10h^{-1}$ Mpc or less[6]. While de Lapparent et al.[10] have recently discovered clear evidence that galaxies lie on very thin surfaces, the high density ridge in Pisces-Persues is, in contrast, filamentary.

The concentration of galaxies to the narrow redshift window around 5000 kms^{-1} can be traced well outside the prominent ridge, to regions of significantly lower mean density. The structure in the declination zone south of $+20^o$[11] resembles that of the Coma region[10]. A prominent void covering at least 15° in right ascension, but seen over the entire 50° of declination, lies at a redshift of about 3200 $km s^{-1}$. Distinct, continuous structures outline this void, and connect the Local Supercluster to Pisces-Perseus[11].

3. LARGE-SCALE MORPHOLOGICAL SEGREGATION

The segregation of the major morphological classes - ellipticals, S0's, and spirals - in clusters of galaxies was first noticed over a half century ago and was quantified by Dressler[12]. The continuity of this morphological segregation over six orders of magnitude in space density to loose groups was shown by Postman and Geller[13]. The density regimes sampled by the Pisces-Perseus supercluster survey cover the whole range of densities occupied by galaxies. In the Pisces-Perseus region, the degree of morphological segregation is indeed gradual, extending beyond the volumes dominated by clusters, through the density domains represented by groups and filamentary enhancements to the general "field". Furthermore, the population fraction is seen to vary smoothly not only among the major morphological classes, but among the spiral types themselves. Sa galaxies cluster more so than do Sc and later types, and the gradient of this trend can be traced over nearly all density regimes. Because the peculiar velocities observed in the Pisces-Perseus region are small, usually on the order of 250 kms^{-1}, it is unlikely that the morphological segration could result from the diffusion of objects over a Hubble time. Furthermore, low optical surface brightness, HI-rich objects occupy a larger volume than do high optical surface brightness objects, but the former still do not fill the voids. Empty regions of characteristic dimension in excess of $20h^{-1}$ Mpc and containing no galaxies brighter than M = -18 exist.

Because of the growing degree of completeness of this redshift survey and the accumulation for large subsets of complementary data at optical and far infrared wavelengths, the variation with local space density of a number of fundamental galaxian properties, in addition to optical morphology, can be studied. The total mass of spiral galaxies can be estimated from the 21 cm line profile width. While the total masses of spirals show no trend, for the same galaxies, the optical luminosity increases significantly (by more than one magnitude over the range of density sampled) with increasing space density. The result is that the mass-to-light ratio decreases with increasing density, contrary to what would be expected purely as a result of morphological segregation[14]. This segregation of luminosities implies an unequal conversion of mass to light in locales characterized by different density enhancements at the epoch of galaxy formation.

The National Astronomy and Ionosphere Center is operated by Cornell University under contract with the National Science Foundation. The National Radio Astronomy Observatory is operated by Associated Universities, Inc. under contract with the National Science Foundation.

1) Bernheimer, W.E. 1932, Nature 130, 132.
2) Burns, J.O. and Owen, F.N., Astron. J. 84, 1474 (1979).
3) Giovanelli, R. and Haynes, M.P., Astron. J. 87, 1355 (1982).
4) Batuski, D.J. and Burns, J.O., Astrophys. J. 299, 5 (1986).
5) Giovanelli, R. and Haynes, M.P., Astron. J. 90, 2445 (1985).
6) Giovanelli, R., Haynes, M.P. and Chincarini, G.L., Astrophys. J. 300,77 (1986).
7) Giovanelli, R., Haynes, M.P., Myers, S.T. and Roth, J., Astron. J. 92,250 (1986).
8) Haynes, M.P., Giovanelli, R., Starosta, B. and Magri, C., preprint (1986).
9) Zwicky, F., Herzog, E., Wild, P. and Kowal, C.T., Catalog of Galaxies and Clusters of Galaxies, (Calif. Inst. Tech., Pasadena), six volumes (1961-68).
10) de Lapparent, V., Geller, M.J. and Huchra, J.P., Astrophys. J. (Lett.) 302, L1 (1986).
11) Haynes, M.P. and Giovanelli, R., Astrophys. J. (Lett). 306, L55 (1986).
12) Dressler, A., Astrophys. J. 236, 351 (1980).
13) Postman, M., Geller, M.J., Astrophys. J. 281, 95 (1984).
14) Giovanelli, R. and Haynes, M.P., in preparation (1987).

THE CENTER FOR ASTROPHYSICS REDSHIFT SURVEY: STATISTICAL MEASURES OF LARGE-SCALE CLUSTERING

VALÉRIE DE LAPPARENT, MARGARET J. GELLER, AND JOHN P. HUCHRA

Center for Astrophysics
60 Garden Street
Cambridge, Massachusetts 02138

1. THE CATALOG

The extension of the Center for Astrophysics Redshift Survey will eventually cover the northern sky with $|b^{\rm II}| \geq 40°$ and $m_{B(0)} \leq 15.5$. The complete portion of the survey contains 1770 galaxies in a 135° (r.a.) × 12° (dec.) strip centered at the north galactic pole.[1),2)] The distribution of galaxies contains large voids ($\leq 50h^{-1}$ Mpc) separated by thin sheets, suggesting a "bubble-like" topology: (1) most of the galaxies lie in portions of surfaces which extend naturally from the southern 6° of the slice into the northern half; (2) no corresponding filaments are present in projection on the sky.

2. LARGE-SCALE INHOMOGENEITIES AND THE MEAN GALAXY DENSITY

One of the most striking features of the new data is that the largest coherent structures ($\sim 100h^{-1}$ Mpc) are comparable with the extent of the survey. Thus, the region is *not* a fair sample of the distribution of galaxies in the universe, although it is one of the largest sample available.

As an example of the effect of large-scale inhomogeneities on the statistics of the galaxy distribution, we estimate the uncertainty in the mean density ρ of a slice-like sample as a function of its depth D. For distributions of voids in which the fractional r.m.s. fluctuation is determined by the number N of the largest structures in the sample, and where the fluctuations in N are Poisson:

$$\frac{\delta \rho}{\rho} \simeq \frac{F}{\sqrt{N}} \quad \text{with} \quad F \simeq \frac{N d^2}{D^2}; \tag{1}$$

F is the fraction of the survey volume occupied by the largest voids (diameter d). In a large solid angle survey, $F \propto d^3/D^3$.

In the completed 12° slice, $N = 1$, $d \simeq 50h^{-1}$ Mpc, and $D \simeq 100h^{-1}$ Mpc yield $F \simeq 0.25$, and $\delta(\rho)/\rho \simeq 0.25$. With $F = 0.25$, determination of the *universal* mean density to 10% requires a survey large enough to contain ~ 6 large voids. This requirement can be met by a wedge-shaped survey complete to a depth of $\sim 250h^{-1}$ Mpc ($m_{B(0)} \leq 17.5$, $\sim 3{,}000$ galaxies), or by an "all-sky" survey to a depth of $\sim 150h^{-1}$ Mpc ($m_{B(0)} \leq 16.5$, $\sim 40{,}000$ galaxies). The factor of ten in the number of galaxies illustrates the efficiency of probing the large scale structure with slices: in large solid angle surveys some of the information about the scale and frequency of the largest structures is redundant because of the large-scale coherence of the structures. If the structures are larger than $50h^{-1}$ Mpc and/or if they are more common than in the 12° slice (*i. e.* $F \geq 0.25$), even larger surveys are needed.

3. THE TWO-POINT CORRELATION FUNCTION

Because of its sensitivity to the mean density, the amplitude of the two-point correlation function at large scales is poorly constrained by this survey. Figure 1 shows the correlation function of the sample, $\xi(s)$ (s is the separation in redshift space), obtained with the luminosity function derived from the CfA Survey to 14.5 mag[3] and normalized internally to the 12° slice [□]. With the mean density for the CfA Survey to 14.5 mag[3] — 15% lower — the correlation function [○] has significantly more power between 20 and $100h^{-1}$ Mpc. The correlation functions in Figure 1 are *not* affected by removing the Coma cluster from the sample.

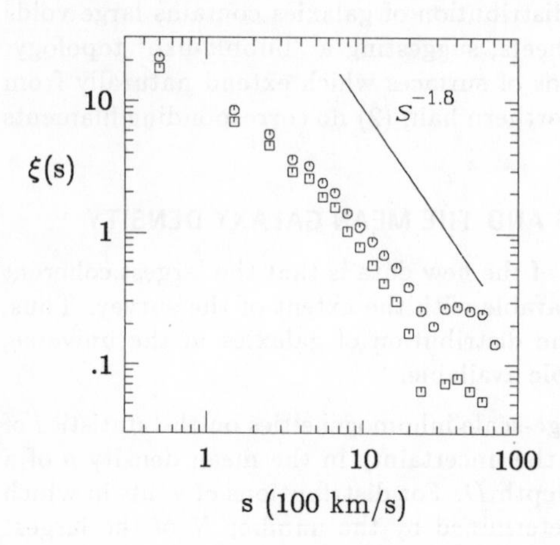

FIGURE 1. Correlation functions in the 12° slice with the internal mean [□] and the mean from the CfA Redshift Survey to 14.5 mag [○], as a function of separation in redshift space.

In the 3–$20h^{-1}$ Mpc range, the correlation functions in Figure 1 are flatter than $s^{-1.8}$: the slopes are respectively −1.5 [□] and −1.3 [○]. However, because of the large uncertainties in the mean density, these slopes are still consistent with −1.8. At these intermediate scales — and at large scales — the correlation function is determined by the geometry of the structures. Geometrical models[4] in which points are randomly distributed on the surface of close-packed bubbles have a correlation function on scales from 5 to $100h^{-1}$ Mpc which is indistinguishable in shape and amplitude from that for the data.

4. OTHER STATISTICAL MEASURES

The two-point correlation function may provide some information about the topology of the galaxy distribution. For example the correlation length — $\sim 8h^{-1}$ Mpc in Figure 1, in agreement with [5] — is a measure of the thickness of the structures. But average statistics, particularly on large scales, are compromised by the lack of a fair sample and by the lack of spherical symmetry in the observed large-scale structures. Statistical measures which depend upon local density contrast and/or on higher order moments of the distribution may be more effective for describing the large-scale features of the distribution.

Appropriate binning and smoothing of the distribution in the CfA Redshift Survey slice make the galaxies percolate, thereby defining the voids and ridges. From the smoothed distribution, we find that the galaxies occupy $\lesssim 20\%$ of the volume at the limiting magnitude of 15.5 mag. The surface density in the sheets is $\sim 0.3h^2$ galaxie per Mpc^2 at 5000 km/s, and they have a mean FWHM thickness of 500 km/s.[4] A general study of the network of voids and surfaces can provide additional measures of the topology of the distribution such as the spectrum of void radii and the degree of connectedness of the voids.

5. PROSPECTS

While only the hot dark matter models[6] can make $50h^{-1}$ Mpc voids, the cold dark matter models[7] give a better match to the data on intermediate scales ($\sim 10h^{-1}$ Mpc). But in both of these models, the coherence of the structures over scales of $\gtrsim 100h^{-1}$ Mpc is not quite reproduced. Statistical measures of the topology should quantify these differences and stimulate a further investigation of the models.

Larger redshift surveys are the next observational step in the quest for the maximum coherence scale of the galaxy distribution. Deeper samples will better constrain the upper end of the spectrum of void radii and provide better discrimination among the theoretical models. Deep photometric surveys are an important tool for measuring the mean density of galaxies and may obviate the need for complete deep ($\gtrsim 150h^{-1}$ Mpc) redshift surveys. A partial redshift survey of the region is still necessary to determine the size of the largest structures and to assure that the photometric sample is fair. As illustrated by the CfA Redshift Survey, such redshift survey samples must be of sufficiently high "signal-to-noise" to delineate the structures and to constrain the size of the largest ones.[8]

REFERENCES

1. de Lapparent, V., Geller, M. J., and Huchra, J. P. 1986, *Ap. J. (Letters)*, **302**, L1.
2. Geller, M. J., Huchra, J. P., and de Lapparent, V. 1987, *Observational Cosmology*, **IAU Symposium 124**.
3. Davis, M., and Huchra, J., 1982, *Ap. J.*, **254**, 437.
4. de Lapparent V. 1986, *Ph. D. Thesis*, Université de Paris 7.
5. Oemler, A. Jr., Schechter, P. L., Shectman, S. A., and Kirshner, R. P. 1987, in preparation.
6. Melott, A. L. 1987, preprint.
7. White, S. D. M. 1987, this volume.
8. White, S. D. M., Frenk, C. S., Davis, M., and Efstathiou, G., 1987, *Ap. J.*, in press.

DEEP REDSHIFT SURVEYS OF LARGE-SCALE STRUCTURES

D. C. Koo
Space Telescope Science Institute
R. G. Kron
Yerkes Observatory, University of Chicago
A. S. Szalay
Eötvos University and Johns Hopkins University

1. INTRODUCTION

In contrast to the wide-angle galaxy redshift surveys of 15th to 17th mag, we describe our field galaxy surveys that are 100 times fainter but which cover extremely tiny areas of the sky (0.3 deg^2). These skewer-like cones reach depths of a thousand Mpc or more, corresponding to look-back times great enough to directly measure, in principle, the evolution of clustering.

2. OBSERVATIONS

The selection and photometry of our redshift samples are based upon deep 4m prime-focus photographic plates that reach 24th mag, well beyond our spectroscopic limit of 22nd mag. The actual criterion for selecting galaxies is complex and controlled by the setup of our spectrograph, called the Cryogenic Camera, used on the 4m telescope at Kitt Peak National Observatory. The major advantage is that low resolution (15Å FWHM, 4Å per pixel) spectra from 4500Å to 7500Å can be acquired, usually in several hours, for 10 to 15 objects simultaneously over a 5 arcmin field.

3. DISCUSSION

Although the reduction of the redshift survey is still underway because of our desire to improve its accuracy, reliability, and completeness, we are able to show some preliminary results. In particular, two of our nine fields now have a total of 282 galaxies with crude ($\Delta z = 0.005$) redshifts, Selected Area 57 at 1305+30 near the North Galactic Pole and SA 68 at 0015+15 at galactic latitude $b = -45°$. The redshift distributions shown in Fig. 1 display strong overdensities (clusters) and underdensities ("voids") on scales of 100 Mpc h^{-1}.

An especially notable feature is the lack of galaxies between $z = 0.13$ and 0.17, where 15 galaxies are expected if the distribution of galaxies were smooth; one is observed. The now famous Bootes void was originally based upon better statistics of 25 expected with one observed[1)]. In an attempt to find if our apparent "void" extends to much larger volumes, we have examined the redshift catalog of Abell clusters[2] over an area extending from RA = 12^h to 14^h and DEC = 15° to 45°. Among 36 clusters with redshifts, only one at $z = 0.14$ fell between $z = 0.13$ and 0.16, corresponding to a volume of 3×10^6 Mpc or about 30 times the volume of the Bootes void!

Clearly a more detailed analysis of the clustering, and possibly its evolution with look-back time, is needed to assess if our observations can be reconciled with the predictions of various cosmological scenarios. In the meantime, our data can be compared qualitatively to the N-body simulations of a flat ($\Omega = 1$) cold-dark-matter universe as viewed in cones constructed to be rough estimates of our survey (Fig. 7a and 7b)[3].

4. REFERENCES

1) Kirshner, R. P., Oemler, A., Schechter, P., and Shectman, S., "A Million Cubic Megaparsec Void in Bootes?" Ap. J. (Letters), 248, L57 (1981).
2) Struble, M. F. and Rood, H. J., "A Compilation of Redshifts and Velocity Dispersions for Abell Clusters," Ap. J. Suppl., in press (1987).
3) White, S. D. M., Frenk, C. S., Davis, M., and Efstathiou, G., "Clusters, Filaments, and Voids in a Universe Dominated by Cold Dark Matter," Ap. J., in press (1987).

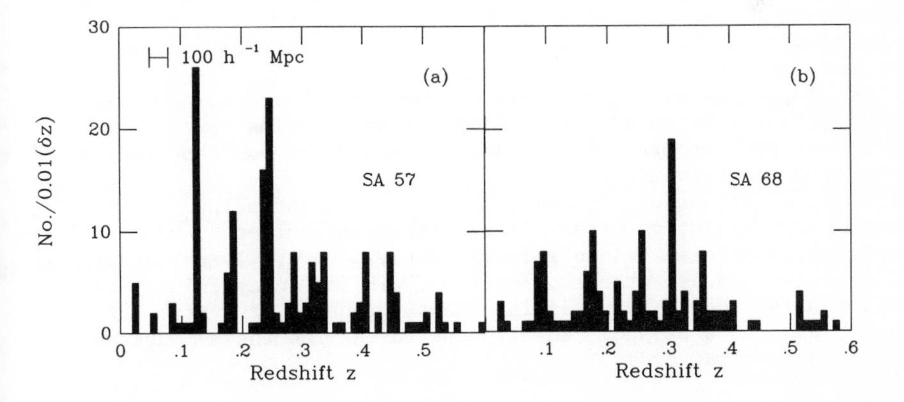

Figure 1 a) Histogram in bins of 0.01 in z of redshifts for 148 galaxies within a 0.3 deg^2 field (SA 57). b) Histogram of 134 redshifts in another field of 0.2 deg^2 (SA 68).

TRACING LARGE SCALE STRUCTURES WITH ABELL CLUSTERS

Jack O. Burns and J. Ward Moody
Institute for Astrophysics
The University of New Mexico
Albuquerque, NM 87131

David J. Batuski
Space Telescope Science Institute
Homewood Campus
Baltimore, MD 21218

ABSTRACT

A new, larger "complete" sample of Abell clusters is used to construct the two-point spatial correlation function at large separations. There are suggestions of "bumps" in the correlation function at separations out to 250 h_{75}^{-1} Mpc. None of the present density fluctuation scenarios are able to reproduce these large structures seen within the 2-point function and seen visually within the Abell cluster distribution. We also describe several large supercluster/void structures traced by Abell clusters.

1. INTRODUCTION

Although galaxy redshift surveys have greatly added to our knowledge of large scale structure, there remains several prominent questions concerning the largest scales. These questions include:

a) Have we surveyed a "fair sample", of the Universe? That is, have we sampled all scales and types of structures present in the Universe in correct proportions?

b) How large are the largest coherent structures? With each new expansion in observational volume, redshift surveys are finding structures on the scale size of the region sampled.

c) On what scale, if any, is the Cosmological Principle valid for the distribution of galaxies?

d) Do superclusters and voids connect? That is, does the Universe have a swiss-cheese topology in which voids are isolated entities, or a sponge-like topology in which the voids percolate throughout the Universe?

These questions are fundamental to our understanding of the origin and evolution of structure within the Universe. However, it will be some time before galaxy surveys cover large enough volumes to address these questions. In the intervening time, Abell clusters could serve as important tracers of very large structures (100's of Mpc) and allow us to examine these important questions to first order.

2. THE EXTENDED ABELL CLUSTER SAMPLE

Batuski et al. (1987) has recently defined a larger sample of Abell clusters that could be useful for statistical analysis. It is composed of 225 clusters of all richness classes with z < 0.085, excluding low galactic latitude and the north Galactic spur regions. It is complete in the sense that the density of clusters in the sample remains constant out to the redshift cutoff and the density is not a function of galactic latitude. However, one still must be cautious in using this sample because of the possible selection biases that may be inherent in the Abell catalog.

In recomputing the 2-point correlation function (see e.g., Bahcall and Soneira, 1983), we find that the larger sample allows us to probe out to larger separations. There are suggestions of "bumps" with positive amplitude (at the level of 0.1-0.2) in the 2-point function at separations out to 250 h_{75}^{-1} Mpc. Such large structure agrees with percolation analyses of the distribution of Abell clusters (Batuski and Burns, 1985a) where superclusters and voids with diameters of 200-300 Mpc are seen. This also appears to agree with new data on more distant clusters (Karachentsev, 1987) and deep redshift surveys (Koo, 1987) where there is a strong suggestion of galaxy sheets separated by 200 Mpc intervals.

Numerical simulations of the Abell cluster distribution fail to reproduce these very large structures (Batuski, Melott, and Burns, 1987). Isothermal, hot particle, and cold particle density fluctuation scenarios produce poor matches to the Abell cluster 2-point function at large separations (>40 h_{75}^{-1} Mpc), and produce too few large superclusters in comparison to those found in percolation studies for rich clusters. Something fundamental is missing from either the models or the cluster database.

3. STRUCTURES WITHIN THE DISTRIBUTION OF ABELL CLUSTERS

The percolation catalogs of candidate superclusters and voids by Bahcall and Soneira (1984) and Batuski and Burns (1985a) are useful for selecting regions of unusually high or low density for further study. These catalogs coupled with extensive three-dimensional models of the Abell cluster distribution have allowed us to form some general visual, qualitative impressions of second-order clusterings. First, it appears that the typical size of a supercluster composed of Abell clusters is about 200 Mpc. Second, the geometry of the superclusters is best described as irregular and nonspherical, with a range of structures such as filaments and sheets. Third, superclusters are generally separated by giant voids with diameters of 100-200 Mpc. Fourth, the voids are generally ellipsoidal in shape. Fifth, the voids in the Abell cluster distribution are generally larger than those in the galaxy distribution. Sixth, there are some suggestions that the voids may interconnect.

The rich clusters appear to be very good tracers of both galaxies in superclusters and the absence of galaxies in voids. Two particularly good examples of the rich cluster distribution near voids are in the Hercules-Bootes and Pisces-Cetus regions. In Figure 1, the Bootes void is very well traced by a ring of rich Abell clusters stretching from Hercules to the northeast to Corona Borealis to the southeast to Ursa Majoris in the west. The void of Abell clusters has dimensions of 400x150x100 h_{75}^{-3}, substantially larger than the galaxy distribution. The most recent galaxy data using a sparse sampling technique suggest that the center of the galaxy void agrees very well with the center of the cluster void (Kirshner et al., 1987).

In Figure 2, a wedge diagram near the heart of the Pisces-Cetus supercluster (Batuski and Burns, 1985b) also illustrates a void partially traced by 7 rich clusters. Both CfA galaxies and an incomplete deeper survey of the region at Lick (Brodie et al., 1987) show that a sphere of 50 h_{75}^{-1} Mpc is devoid of bright galaxies and clusters.

This work was supported by NSF grant AST-8520115 to J.O.B. We thank John Huchra for sending us the latest edition of the CfA catalog in advance of publication.

Bahcall, N. and Soneira, R. 1983, Ap.J., 270, 20.
Batuski, D. and Burns, J. 1985a, A.J., 90, 1413.
____. 1985b, Ap.J., 299, 5.
Batuski, D., Melott, A., Burns, J. 1987, Ap.J., submitted.
Batuski, D., Burns, J., Laubscher, B., and Elston, R. 1987, Ap.J., submitted.
Brodie, J., Burns, J., Willick, J., Batuski, D., Moody, J., and Bowyer, S. 1987, in preparation.
Karachentsev, I. D. 1987 in Observational Cosmology, IAU Symposium 124, ed. G. Burbidge and L. Fang (Reidel), in press.
Kirshner, R., Oemler, A., Schechter, P., and Shectman, S. 1987, Ap.J., in press.

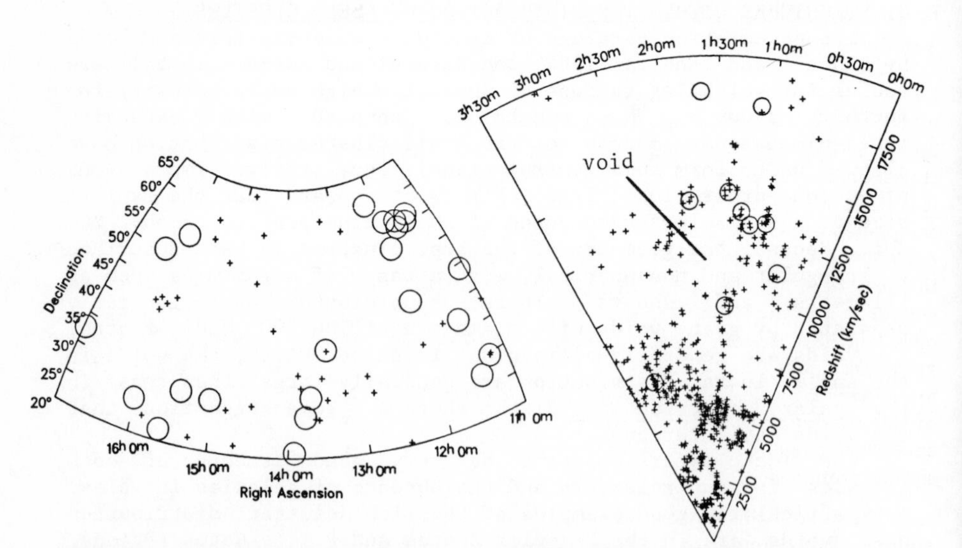

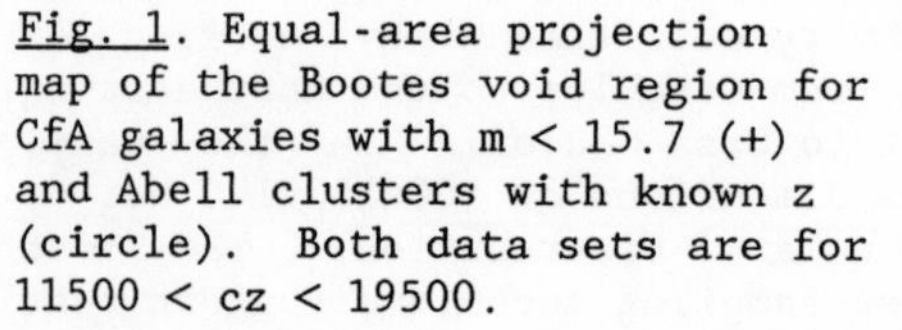

Fig. 1. Equal-area projection map of the Bootes void region for CfA galaxies with m < 15.7 (+) and Abell clusters with known z (circle). Both data sets are for 11500 < cz < 19500.

Fig. 2. Plot of cz vs. RA for a 20° strip centered on DEC = +5°. Symbols are the same as Fig. 1. The Pisces-Cetus void is indicated.

THE SOUTHERN SKY REDSHIFT SURVEY
and the
IRAS WHOLE SKY REDSHIFT SURVEY

Marc Davis
Department of Astronomy and Physics
University of California
Berkeley, California 94720

ABSTRACT

For the last 5 years a redshift survey of the Southern high latitude sky has been in progress. The sample consists of approximately 2000 galaxies selected from the ESO catalog with $\delta < -17.5°$ and $b < -30°$. The observations are nearly complete and a preliminary analysis has been undertaken. We find the usual filamentary, frothy appearance in the galaxy distribution, and an amazing reproducibility of the statistics of the clustering when compared to the Northern sky.

A new and different survey that should be completed within the year is drawn from the catalog of galaxies seen by the IRAS satellite. This sample is uniformly selected from the whole sky and offers a unique opportunity for a powerful cosmological test of the density parameter.

1. The Southern Sky Survey

This survey is a collaborative effort between groups at the Observatorio Nacional in Brazil, the South African Astronomical Observatory, and the Las Campanas Observatory. The principals involved are W. Sargent, J. Menzies, L. da Costa, J. Tonry, A. Meiksin, and myself. Our selection procedure was to take a diameter limited subset of the ESO galaxies; the primary list consists of 2028 galaxies with $log(D(0)) > 0.1$ in the region of sky above. Of this list all but approximately 250 low surface brightness dwarfs have measured redshift; the remaining objects are very difficult to study optically. Although this diameter limited catalog is very different from the usual magnitude limited catalog, it is still possible to define selection functions and to study the clustering statistics in the sample. Detailed results will be forthcoming soon; here I will simply give a representative example of our findings. Figure 1 shows the redshift wedge diagram for those galaxies in the zone $-30 < \delta < -17.5$, with the closed triangles denoting E/SO galaxies, the closed squares denoting Sa/Sb galaxies, the open squares denoting Sc galaxies, and the open triangles denoting Irregular galaxies. In this map there is one prominent foreground group (Eridanus) that is elongated in redshift space plus some elongated filaments. There are no very large voids in this slice but voids of size $30h^{-1}$ Mpc are evident in other slices of this catalog. In contrast to the Northern CfA catalog which is dominated by Virgo supercluster, and the Coma/A1367 superclusters, there are no clusters in this sample of similar prominence.

In spite of this obvious visual difference, the second order statistics of this sample are remarkably similar to the Northern CfA sample[1]. Shown in figure 2 is $\xi(r)$ as measured directly in redshift space. The closed squares are for the Northern CfA survey[1] while the open squares are for the new survey. In figure 3 is shown the relative pair velocity dispersion for the two samples as a function of projected separation. The two samples are identical in their correlation statistics within the error bars. Does this mean they are each fair samples of the Universe, or did we just get lucky?

2. IRAS Selected Redshift Survey

This is a progress report; we have been busy obtaining the redshift data and have not fully analyzed it and have not begun to do the relevant cosmological analysis. The IRAS survey is a collaborative effort between M. Strauss, J. Huchra, J. Tonry, A. Yahil, and myself. The survey consists of approximately 2700 galaxies with $f_{60\mu} > 1.92$ Jy, $f_{60\mu}/f_{12\mu} > 3$. The sample is selected with $|b| > 10°$, but some additional zones associated with higher latitude HII zones are excluded because of source confusion. In all, 76% of the sky is being surveyed. The map of the primary list of galaxies is shown in figure 4. These objects are mainly late–type galaxies with active star formation; almost all of the galaxies have optical emission lines so the redshifts can be obtained rapidly even on a modest sized telescope. At the present time we have nearly completed redshift measurements of the main list of galaxies, but we must continue to focus considerable attention on those sources that are partially resolved by the IRAS satellite, because for these sources the available IR flux is generally an underestimate which must be corrected by further, involved IRAS processing to insure that our survey list is complete and unbiased.

The major interest in this sample is that the angular dipole anisotropy of the IRAS selected catalog points within 30° of the microwave dipole anisotropy[2,3]. The microwave result is usually interpreted as a velocity effect and is a measure of our net motion relative to the comoving frame. The IRAS dipole anisotropy is a measure of the spatial inhomogeneity within approximately $100h^{-1}$ Mpc and so can be used to measure the gravitational acceleration of the local group or of other points within the survey volume. This peculiar acceleration is simply related to our peculiar velocity in linear perturbation theory and can be used as a powerful, relatively model free test of the cosmological density. We hope to report on this and other questions of cosmological interest within the next year.

REFERENCES

1. Davis, M. and Peebles, P.J.E., 1983, *Ap.J*, 267, 465.

2. Meiksin, A., and Davis, M. 1986, *A.J.*, 91, 191.

3. Yahil, A., Walker, D., and Rowan-Robinson, M. 1986, *Ap. J. (Lett.)* 301, L1.

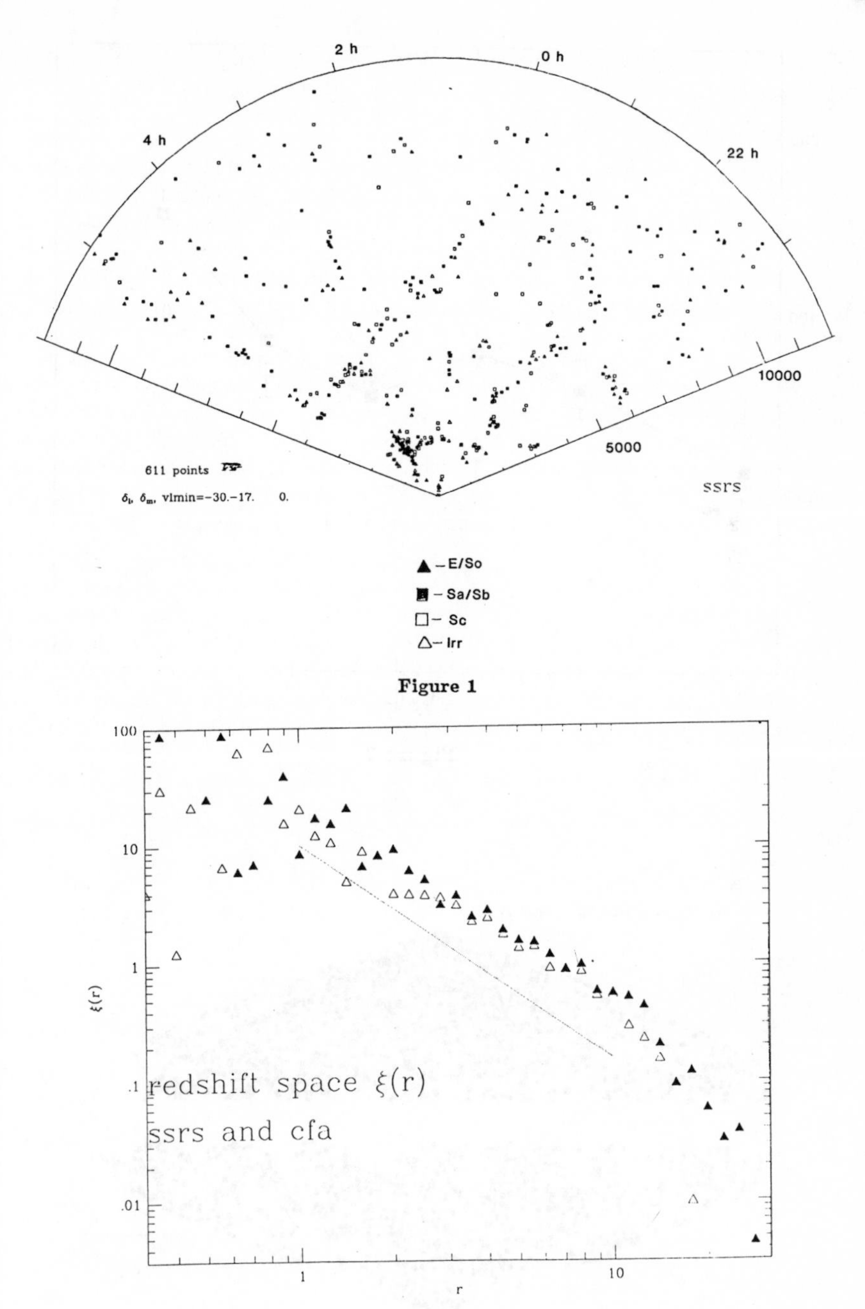

Figure 1

Figure 2

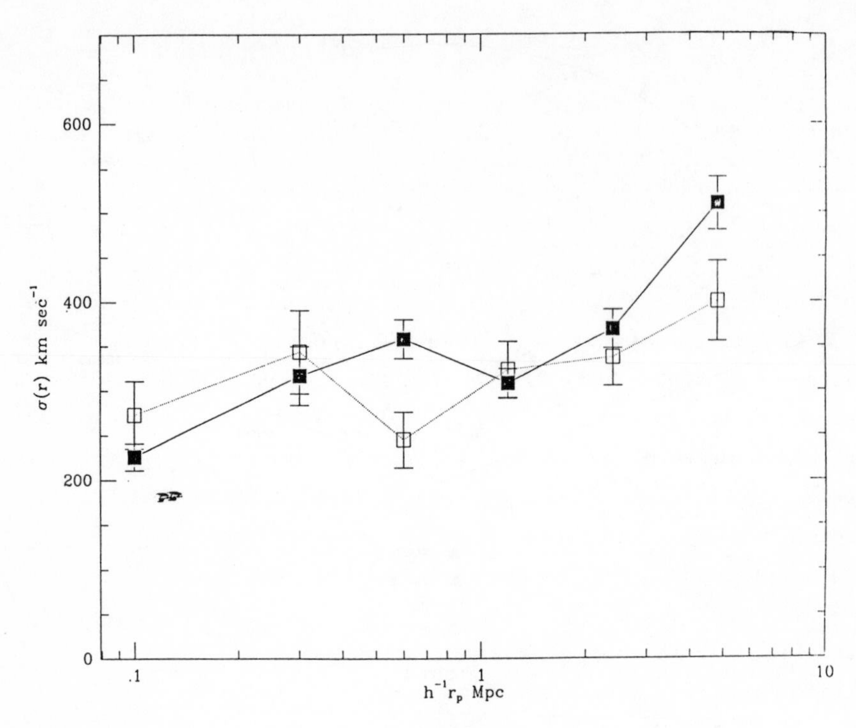

Figure 3

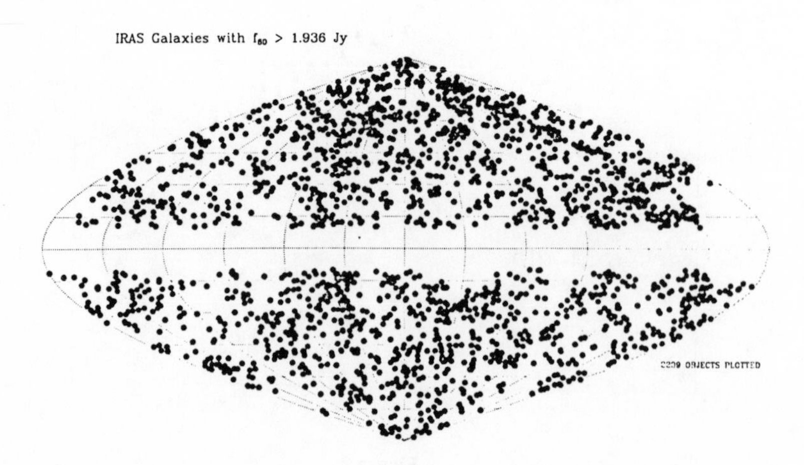

Figure 4

A LARGE-SCALE STREAMING MOTION IN THE LOCAL UNIVERSE

Alan Dressler (Mount Wilson and Las Campanas Observatories), S. M. Faber (UCSC-Lick Observatory), David Burstein (ASU), Roger L. Davies (KPNO, NOAO), Donald Lynden-Bell (IoA Cambridge), R. J. Terlevich (RGO), Gary Wegner (Dartmouth College)

ABSTRACT

The results of an all-sky study of the distances and velocities of ~400 elliptical galaxies were reviewed. The sample, typically at distances of 1000 - 6000 km s^{-1} has a net motion of ~600 km s^{-1} with respect to the frame of the cosmic microwave background. In direction and magnitude this motion is similar to the Local Group's motion with respect to the microwave background. This suggests that the Local Group's motion is characteristic of a large-scale amisotropy in the Hubble expansion field, rather than a fluctuation induced within r = 5000 km s^{-1}. Such a large-scale streaming motion appears incompatible with both cold and hot dark matter models of the formation of structure on very large scales (~100 Mpc).

Details are given in Dressler _et al_. February 15, 1987 _Astrophysical Journal Letters_, _313_, L37-L42.

GRAVITATIONAL LENSES AND LARGE SCALE STRUCTURE

Edwin L. Turner

Princeton University Observatory
Princeton, New Jersey 08544 USA

ABSTRACT

Four possible statistical tests of the large scale distribution of cosmic material are described. Each is based on gravitational lensing effects. The current observational status of these tests is also summarized.

INTRODUCTION

Because gravitational lensing effects depend on inhomogeneities in the distribution of material in and near the line of sight to high redshift objects, their study is closely linked to that of large scale structures (i.e., inhomogeneities). At present, the most urgent issue in large scale structure discussions is the question of whether or not galaxies "trace" (i.e., fairly sample) the total distribution of material in the universe. Are the voids/bubbles really empty or only dark? Is galaxy formation a biased process? These are the question phrased in observational and theoretical language, respectively. Four statistical approaches to the problem, each based on lensing effects, are described below.

DETERMINATION OF THE TOTAL DENSITY

Large scale structure studies are usually viewed as potentially illuminating the deeper issue of the universe's total mean density; however, this line of reasoning can be reversed. Knowing by some other means that $\Omega_o = 1$ or 0.1 would tell us that there must or could not, respectively, be dark matter distributed substantially differently from the galaxies by simple accounting arguments.

The best available lensing test for Ω_o is the variation of total lensing optical depth τ_{GL} with source redshift z_Q. As illustrated by the calculations of Turner, Ostriker, and Gott (1984) for instance, $\tau_{GL}(z_Q)$ is about twice as steep between z_Q=1 and z_Q=3 for Ω_o=0 as it is for Ω_o=1. The advantages of an Ω_o test based on this relation are that it does not depend on the absolute frequency of lensing, that it is nearly independent of the nature of the lensing objects, that it does not require detailed lens models of any particular system, and that it may employ biased samples of lenses so long as the bias is not dependent on (or may be corrected for) source redshift. This last condition is not too difficult to accomplish since all other lens properties are quite insensitive to source redshift for $z_Q \gtrsim 1$.

Currently available samples of lens candidates are too small, too heterogeneously sampled, and too uncertain to allow realization of this test. The most naive and uncritical evaluation of the present sample favors small Ω_o values weakly.

DISTANCE MEASURE FLUCTUATIONS

Classic analyses (Gunn 1967a, 1967b) and more recent numerical work (Alcock and Anderson 1985) show that fluctuations in the density of material in and near the line of sight to distant objects will produce fluctuations in various distance measures to the object for a given redshift. Thus, if one carried out H_0 determinations based on inter-image light curve time delays for several lens systems, the variations in the derived values of H_0 would carry information about total rms density fluctuations in the universe. Good agreement between the various values would indicate a relatively smooth universe. Variations in the values would indicate a lumpy universe (if they could be distinguished from errors due to inadequacies of the lens models on which such H_0 determinations must be based).

An entirely satisfactory theoretical treatment of this problem is not yet in hand; however, the order of magnitude of the effects are known. For instance, for a redshift unity lens and a redshift 3 source, the rms variations in derived H_0 values should be of order $0.5(\Delta z)^{0.5}$ where Δz is the largest scale of fluctuations with unit amplitude. This is $\lesssim$ 10% for even the most dramatic large scale inhomogeneities currently contemplated; thus, the test will be extremely difficult to realize. Only the absence of superior alternatives make it attractive.

No reliable lens based H_0 determinations have been achieved to date despite substantial efforts for one system (0957+561).

DARK LENSING OBJECTS

If dark matter occurs in sufficiently high density condensations, these can be detected directly by their multiple imaging effects. Moreover, even limits on the frequency of lensing in well defined samples of high redshift objects can be translated into limits on the abundance of specific hypothetical forms of dark matter. The difficulty with the former technique is that it is difficult in practise (and even in principle, logically) to exclude the possibility that a specific candidate dark lens system is not actually a pair (or set) of very similar but physically distinct objects. The latter method is quite powerful but depends on a careful analysis of a particular set of observations and their comparison to the predictions of a detailed dark matter model.

Several candidates for dark lenses exist (Turner 1987, Hewitt et. al. 1987) including one spectacular but controversial case (Turner et. al. 1986), but none are yet compelling.

Hewitt (1986 and elsewhere in this volume) has analyzed a set of careful VLA observations to set limits on the abundances of roughly galaxian mass point-like or spherical, isothermal objects. Her result, which is based on extremely conservative assumptions, indicates that such objects cannot contribute more than a few tenths to Ω_0.

MASSES OF GALAXIES

One of the most reliable (though somewhat redundant) uses of lens systems is the determination of mass distributions in and near (lensing) galaxies. This goal may be pursued by detailed modeling of specific multiple image lens systems; however, it may also be approached via statistical searches for the lens distortion of the images of background galaxies projected near closer intervening galaxies. The latter approach is observationally challenging but is conceptually simple and statistically unbiased.

Tyson (1987) has developed the idea for the statistical image distortion mass determinations and carried out a first observational realization. He obtains a null result which corresponds to a 2σ upper limit of 0.03 on the contribution of material within $80h^{-1}$ kpc of galaxies to Ω_0. This limit would allow only rather modest massive halos and appears to contradict the predictions of standard cold dark matter models. If it withstands the scrutiny it is sure to attract, Tyson's result is probably the most generally important product of lens studies to date.

DISCUSSION

Neither large scale structure nor gravitational lensing are well understood subjects at present. Both fields are presently very active and will probably continue to be for some time. Because they are connected to many of the same physical properties of the universe, it is likely that they will continue to be closely related and interacting disciplines. The four specific connections discussed above are by no means exhaustive of the possibilities.

S. White called my attention to the contradiction between Tyson's recent galaxy mass determinations and predictions of cold dark matter models. This work was carried out with the support of NASA grant NAGW-765.

REFERENCES

Alcock, C., and Anderson, N. 1985, *Ap. J. Lett., 291,* L29.

Gunn, J. E. 1967a, *Ap. J., 141,* 61.

Gunn, J. E. 1967b, *Ap. J., 150,* 736.

Hewitt, J. N. 1986, *MIT Ph. D. Thesis.*

Hewitt, J. N., Turner, E. L., Lawrence, C. R., Schneider, D. P., Gunn, J. E., Bennett, C. L., Burke, B. F., Mahoney, J. H., Langston, G. I., Oke, J. B., and Hoessel, J. 1987, *Ap. J.,* submitted.

Turner, E. L. 1987, in *Dark Matter in the Universe: Proceedings of IAU Symposium No. 117,* eds. J. Kormendy and G. Knapp (D. Reidel Publishing, Boston), 227.

Turner, E. L., Ostriker, J. P., and Gott, J. R. 1984, *Ap. J., 284,* 1.

Turner, E. L., Schneider, D. P., Burke, B. F., Hewitt, J. N., Langston, G. I., Gunn, J. E., Lawrence, C. R., and Schmidt, M. 1986, *Nature, 321,* 142.

Tyson, J. A. 1987, in *Theory and Observational Limits in Cosmology,* (Vatican Press, Vatican City), in press.

COMMENTS ON ASSOCIATED CIV ABSORPTION SYSTEMS AND BROAD ABSORPTION LINE SYSTEMS IN QSOS

by

Ray Weymann & Scott Anderson
Mount Wilson and Las Campanas Observatories
Carnegie Institution of Washington
and
Craig Foltz & Fred Chaffee
Multiple Mirror Telescope Observatory
University of Arizona

ABSTRACT

We present a summary of our recent and ongoing spectroscopic surveys of radio-loud QSOs for CIV absorption complexes. A small survey of twelve 3C QSOs has been completed and shows a high incidence (≈50%) of strong "associated" absorption ($z_{abs} \approx z_{em}$) compared with optically selected samples. Preliminary results are also presented from our ongoing survey of a larger sample of radio-loud QSOs (more varied in radio properties). This larger sample also shows a high incidence (≈25%) of strong associated absorption, but thus far, we are unable to determine which radio properties this incidence might depend upon. Our survey spectra also allow us to address the incidence of the broad absorption line (BAL) phenomenon in radio-loud QSOs. We find only one plausible BALQSO candidate out of a sample of 101 radio-loud QSOs, strengthening evidence for an anti-correlation between the BAL and the radio-loud phenomena.

I. INTRODUCTION

In 1979, Weymann et al.[1)] published the results of a photographic survey of intermediate redshift QSOs which appeared to show a clustering of CIV absorption systems near the CIV emission line redshift. A subsequent digital survey of an independent sample by Young et al.[2)] failed to confirm this result, and these latter authors suggested that the Weymann et al. result may have resulted in part from bias introduced by variations in the signal-to-noise from the emission line region to the continuum. Foltz et al.[3)] repeated the Weymann et al. sample digitally, and in an unbiased sample concluded that the original sample of Weymann et al. did show a statistically significant tendency for CIV absorption systems to occur near the emission line redshift at a level well above that expected on the basis of intervening matter and/or that expected on the basis of the simple galaxy-galaxy correlation function. Foltz et al. suggested that the difference between their results and those of Young et al. might have resulted from some real difference between the radio properties of the two samples.

IIa. THE 3C "MINI-SURVEY" AND ONGOING RADIO-LOUD SURVEY

This suggestion prompted us to undertake a small survey of a subclass of 3C QSOs (Anderson et al.[4]) as well as an ongoing survey of a larger sample of a variety (e.g., wide range in radio spectral index) of radio-loud QSOs. The 3C subclass sample shows a high incidence (≈50%) of strong associated CIV absorption. Lack of space prevents us from displaying these interesting 3C spectra here, but they are presented in Anderson et al. Results from the larger radio-loud survey reinforce our impression that the incidence of strong associated CIV absorption is significantly larger than in optically selected samples, but our survey has not yet revealed which radio properties (e.g., radio spectral index) this incidence depends upon. An example of a strong associated CIV absorption complex may be seen in Figure 1 (at ≈4610 Å), which displays our MMT spectrum of one of the QSOs, PKS 1157+014, from the larger radio-loud survey.

Recently, Barthel, Tytler, and Miley[5,6] have suggested that the presence of strong CIV absorption near the emission redshift might be expected to be found preferentially in radio sources showing resolveable structure that is "distorted" or "bent"; evidence for such a correlation appears to be strengthened by more recent preliminary results (Barthel[6]). These authors made the interesting suggestion that the same type of cloud that produces the absorption also produces the distortion in the radio plasma. If this is correct, then the characteristic distance of the absorbing clouds from the nucleus is on the same scale as that of the distortion, viz., a few to tens of kpcs. There are, however, a few discrepancies between these more recent preliminary results and those of Foltz et al., especially with regard to the incidence of strong associated absorption: Foltz et al., combining their results with the Young et al. sample, concluded that only a very small fraction of optically-selected objects exhibited <u>strong</u> associated CIV absorption, whereas the results reported by Barthel seem to show a higher fraction. It is important, in comparing various results, to define what is meant by "strong" and to be sure whether only CIV systems or systems appearing in Lyα and Mg II as well as CIV are being counted.

IIb. THE CASE OF 3CR 191, AND FURTHER HINTS ON THE DISTANCE OF THE ASSOCIATED SYSTEMS: "THE FORBIDDEN LINE REGION" CONJECTURE

3CR 191 provides an especially interesting case of strong associated absorption because of the presence of a low ionization system (in itself somewhat unusual) exhibiting the only well-documented case of Si II fine-structure in QSO spectra. Over ten years ago, Williams et al.[7] showed that the absorbing clouds were probably of order 10 kpcs from the ionizing source, consistent with the distance inferred from the Barthel et al. argument. The electron densities in the cloud, however (10^3 cm^{-3}), are higher than those of typical interstellar clouds in our galaxy--perhaps they are being confined by ram pressure connected somehow with the radio plasma. The analysis of Williams et al. assumes a photoionization model for the absorbing clouds, and as noted by these authors, there are some disturbing inconsistencies in such models.

Williams et al. also point out that with reasonable covering factors it is possible to explain the strength of the emission lines in 3CR 191. In this connection we draw attention to the strong narrow He II 1640 lines seen in several of the 3C objects, for example 3CR 181 and 3CR 268.4. These narrow lines in turn remind one of a very interesting correlation found by Boroson et al.[8] and also confirmed by Stockton and MacKenty[9]: In extended, steep-spectrum radio QSOs both resolvable and "nuclear" [OIII] emission tends to be much more prevalent than in the flat-spectrum or radio-quiet QSOs, the latter tending to show strong nuclear FeII emission instead. We speculate that a similar correlation may be found involving both the strong associated absorption and the HeII 1640 emission, these features being produced by the same clouds producing the [OIII].

An important bit of evidence towards understanding the nature of the associated CIV absorption and the interpretation of the apparent correlations with radio properties of the QSOs will surely come from the velocity distribution of the associated absorption. There is a conceptual problem involved in properly allowing for truly intervening absorbing complexes within 5000 km/sec of the QSO which is probably not of much practical importance. However, both the bias and dispersion introduced by the difference between the published emission line redshift and the "true" QSO redshift makes the interpretation of the present data ambiguous. The data to date from our ongoing sample yield a mean outflow velocity of 200 km/sec ±1600 km/sec --i.e. indistinguishable from zero. In addition, however, work by Junkkarinen[10] shows that there definitely appear to be some cases of "infall" involving velocities of several hundred km/sec. This must be accomodated in any model for the associated absorption.

III. IS THERE AN ANTI-CORRELATION BETWEEN BROAD ABSORPTION LINE QSOS AND RADIO LOUD QSOS; IF SO, WHAT IS THE EXPLANATION?

In Weymann et al.[11], a broad absorption line (BAL) QSO was defined to be one in which "contiguous absorption..." is "...present that is $\geq$2000 km/sec in breadth and whose blue edge is displaced at least 5000 km/sec blueward from the emission line center." This definition needs refining on several accounts. If taken literally, however, it is remarkable that out of the 101 objects in our on-going survey of radio-loud QSOs for which adequate spectroscopic data are available (88 based on our data and 13 others inspected in the literature), none are BALQSOs. This is a highly significant statistical result if the incidence of BALQSOs among optically selected QSOs is of order $\geq$5%. The radio source PKS 1157+014 does, however, appear to have a clump of "contiguous absorption" of order 1700 km/sec in width (after proper allowance for the CIV doublet spitting) with material displaced at up to about 6500 km/sec (see absorption at ≈4520 Å in Figure 1). This object has been discussed recently by Briggs et al.[12], and possibly represents a "transition" BALQSO. A method of analysis which allows for a continuum of absorption properties rather than one based merely upon a naive classification of object into either a BALQSO or a non-BALQSO should be developed. Pending this, suppose we nevertheless classify PKS 1157+014 as a full-fledged BALQSO. The probability of ≤ 1 QSO out of 101 showing this property is then 0.19 ("wishful thinking") if the actual incidence of BALQSOs is only 3%, 0.036 ("suggestive") if the incidence is 5%, and only 0.0003 ("conclusive") if the true BALQSO incidence rate (weighted to correspond to the redshift distribution of our radio-loud sample) is 10%. This sensitivity to the incidence rate points out the need to: (a) complete our spectroscopic survey of radio loud QSOs; and, (b) establish reliably the true incidence of BALQSOs as a function of redshift using strict, well-defined criteria.

What explanations might there be for an anti-correlation between radio-selected QSOs and BALQSOs, if this should prove to be the case. Explanations may simply involve geometrical orientation effects, or environmental effects (e.g., BALQSOs in spirals and extended radio sources in ellipticals). Another explanation is as follows: It seems to us that the characteristic time scale for the BALQSO phenomenon is likely to be a few thousand years or less. It is thus likely to be a quasi-continuous rather than one-shot affair, and very likely to have a time scale less than that involved in the production of the extended radio sources. Some phenomenon might then occur to shut off the BALQSO episode during the production of the extended radio structure. Suppose we adopt the following arbitrary, but not unreasonable assumption about BALQSOs: the absorbing clouds are shreds embedded in and confined by a sub-relativistically expanding semi-continuous wind. The mechanism for ionizing the clouds is photoionization with an ionization parameter obeying $N_e R_{psc}^2 \geq 10^9 L_{46}^{opt}$. Adopt a covering factor of 0.1 for the flow, T(cloud)=30,000K, and V(flow)=15,000km/sec. Then, the ratio of power in the flow to the optical power is of order 2%, independent of any specific assumption about the distance of the embedded clouds from the source. This ratio is not very different from that which must be channeled into the jets which power the large extended radio sources over several million years. Could it be therefore that in the process of developing extended radio structure, the power that was previously driving the BAL clouds is instead diverted into powering the radio jets?

IV. REFERENCES

1. Weymann, R.J., Williams, R.E., Peterson, B.M., and Turnshek, D.A., Ap. J., 218, 619 (1979).
2. Young, P., Sargent, W.L.W., and Boksenberg, A., Ap. J. Suppl., 48, 455 (1982).
3. Foltz, C.B., Weymann, R.J., Peterson, B.M., Sun, L., Malkan, M.A., and Chaffee, Jr., F.H., Ap. J., 307, 504 (1986).
4. Anderson, S.F., Chaffee, Jr., F.H., Foltz, C.B., and Weymann, R.J., to be submitted to A. J. (1987).
5. Miley, G.K., to be published in Proc. of the IAU Symp. 124,

"Observational Cosmology," (1986).
6. Barthel, P.D., priv. commun. (1986).
7. Williams, R.E., Strittmatter, P.A., Carswell, R.F., and Craine, E.R., Ap. J., 202, 296 (1975).
8. Boroson, T.A., Persson, S.E., and Oke, J.B., Ap. J., 293, 120 (1985).
9. Stockton, A., and MacKenty, J.W., to appear in Ap. J., May 15 (1987).
10. Junkarrinen, V.T., B.A.A.S., 18, 914 (1986).
11. Weymann, R.J., Carswell, R.F., and Smith, M.G., Ann. Rev. Astr. Ap., 19, 41 (1981).
12. Briggs, F.H., Turnshek, D.A., and Wolfe, A.M., 287, 549 (1984).

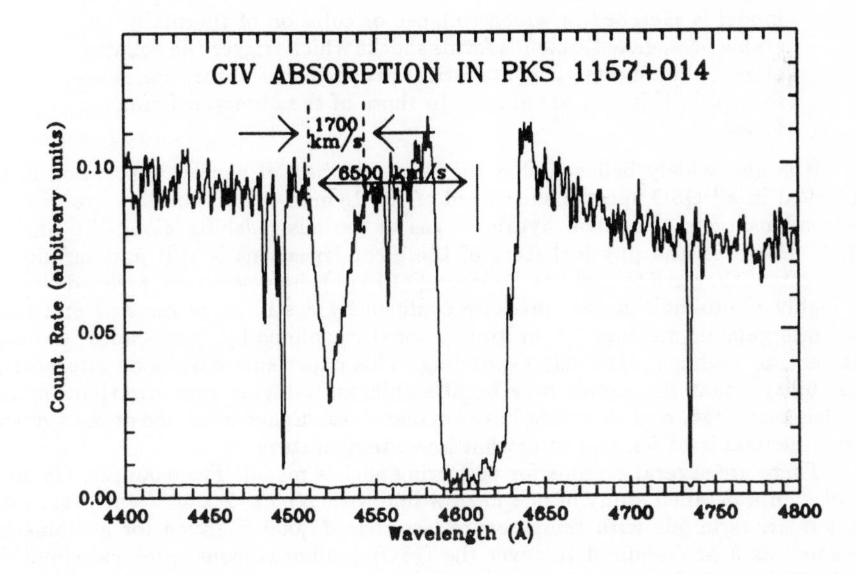

Figure 1. An MMT spectrum of PKS 1157+014, one of the objects in our spectroscopic survey of radio-loud QSOs. PKS 1157+014 has strong associated CIV absorption at 4610 Å, as well as a BAL-like feature at ≈4520 Å; it may be a "transition" BALQSO (Briggs et al.). The BAL-like trough is displaced up to 6500 km/sec from z_{em} but is too narrow (≈1700 km/sec) to strictly satisfy a defining criterion for the BAL class. This is the only BAL-like feature among the 101 radio-loud QSOs in our survey examined thus far.

QSO LYMAN-α ABSORPTION CLOUDS: PANCAKES IN PROTOGALAXIES?

Craig J. Hogan*

Steward Observatory, University of Arizona
Tucson, AZ 85721

ABSTRACT

Arguments are presented that QSO Lyman-α absorption lines might arise in denser environments than is usually thought. A model is sketched in which collapse or collision of thermally unstable protogalactic clouds forms shocks which trigger the collapse of cool thin sheets. The statistical properties of absorption lines formed in this way are similar to those of the observed Lyman-α forest.

It is now widely believed that the large numbers of narrow absorption lines observed in all QSO spectra shortward of the Lyman-α emission line are caused by condensations of neutral hydrogen gas at various redshifts along the line of sight. However, the physical state of these condensations is still undetermined. Most often, they are thought to be large ($\gtrsim$10 kpc), homogeneous, rarefied clouds of highly photo-ionized gas, pressure-confined by a still more rarefied and ionized intergalactic medium[1-4], or gravitationally confined by "mini-halos" of dark matter[5], or perhaps just freely expanding. This paper summarizes an alternative possibility[6], that the clouds may be at significantly higher pressure than in the models just cited, and therefore have smaller sizes, higher local densities, a much larger neutral fraction, and somewhat lower temperature.

There are several reasons for preferring such a model. For example, Chaffee *et al.* [7] in a detailed study of one narrow line with $N_{HI} \simeq 2 \times 10^{13}$ cm^{-2} showed that it arises in gas with temperature less than 17,000 K. Even for a diameter as small as 3 pc (required to cover the QSO) a homogeneous spherical cloud of such small neutral column density in the known ionizing background would have a temperature of at least 25,000 K. To bring T down to the preferred value of 12,000 K would require increasing the gas density, shortening the line-of-sight cloud dimension to $\simeq 10^{15}$ cm, the covering requirement being met by flattening clouds, or clustering them, or both. Tytler [8-9] brings up several other reasons for suspecting a larger-than-normal neutral fraction.

Suppose that thermally unstable protogalactic gas clouds are collapsing throughout the universe[10], producing condensations of cool dense phase confined by hot gas. If each of these clouds produces a Ly$-\alpha$ line over the range of epochs corresponding to observations of the Ly$-\alpha$ forest, there would be enough along a line of sight to explain the whole forest. From observations[11] of the lensed QSO 2345+007A,B, we know that the HI column density is highly correlated, and the

---

* Alfred. P. Sloan Research Fellow

dispersion in HI velocity must be less than 25 km/sec, for lines of sight separated by as much as 10 kpc; thus, the formation of Ly$-\alpha$ absorbers must be coherent in velocity on scales of at least 1/10 of the radius of the whole extended protogalaxy. This constraint would be satisfied if each protogalaxy is being assembled out of a small number of individually hydrostatic subunits, and Ly$-\alpha$ absorbers form in shocks as they collide. If the gas flow is coherent over scales $\simeq$ 10 kpc, so is the shock velocity. Shocks in these conditions would naturally lead to large, dense, coherent pancake-like structures with column density appropriate for the Lyman-α clouds.

It appears that a Lyman-α forest of some kind naturally arises as long as there is a substantial amount of gas present ($\Omega h^2 \sim 0.01 - 0.1$) at $z \sim 2$ to 3, and is being shocked in protogalaxies at velocities ~100–200 km/sec. Ly$-\alpha$ clouds formed in the collapse or collision of thermally unstable protogalaxies would have about the same column density and line density that observed QSO lines require. They would be large-scale versions of McCray *et al.*'s model[12] of sheets in the Vela supernova remnant.

I am grateful for useful conversations with J. Black, S. M. Fall, C. Foltz, J. Ostriker, M. Rees, W. Sargent, and R. Weymann. This work was supported by the Alfred P. Sloan Foundation and by NASA grant NAGW-763 at the University of Arizona.

REFERENCES

1) Sargent, W. L. W., Young, P., Boksenberg, A., and Tytler, D. (1980), *Ap. J. Suppl.*, **42**, 41.
2) Black, J. (1981), *M.N.R.A.S.*, **1971**, 553.
3) Ostriker, J. P., and Ikeuchi, S. (1983), *Ap. J. (Letters)*, **268**, L63.
4) Ikeuchi, S., and Ostriker, J. P. (1986), *Ap. J.*, **301**, 522.
5) Rees, M. J. (1986), *M.N.R.A.S.*, **218**, 25P.
6) Hogan, C. J. (1987), *Ap. J. (Letters)*, submitted.
7) Chaffee, R., Weymann, R. J., Latham, D. W., and Strittmatter, P. A. (1983), *Ap. J.*, **267**, 12.
8) Tytler, D. (1986a), *Ap. J.*, (submitted); Columbia Astrophysics Preprint 319.
9) Tytler, D. (1986b), *Ap. J.*, (submitted); Columbia Astrophysics Preprint 322.
10) Fall, S. M., and Rees, M. J. (1985), *Ap. J.*, **298**, 18.
11) Foltz, C. B., Weymann, R. J., Röser, H.-J., and Chaffee, F. H. (1984), *Ap. J. (Letters)*, **281**, L1.
12) McCray, R., Stein, R. K., and Kafatos, M. (1975), *Ap. J.*, **196**, 565.

THE REDSHIFT EVOLUTION OF QUASAR ABSORPTION SYSTEMS

DAVID TYTLER

Columbia Astrophysics Laboratory, Departments of Astronomy and Physics,
Columbia University, 538 West 120th Street, New York, N.Y. 10027

ABSTRACT

I introduce and summarize current understanding of the redshift distribution of the narrow absorption line systems in QSO spectra, emphasizing the overall scope of the subject and the future prospects.

1. INTRODUCTION

Paramount among current research questions are the understanding of the origin and the evolution of the absorbers. Although the detailed study of individual absorption systems is providing essential information, much emphasis is placed on studies of the absorption system population as a whole. There are at least three reasons why this is a productive approach.

1) Since absorption system will be produced by any and all large extended gaseous objects in the universe, we must expect all samples to be heterogeneous collections of absorbers. Observationally, this subject is inherently statistical.

2) Even systems of a given physical type will probably be found in a wide variety of complex environments. This variety will be confusing at first, but may ultimately help us isolate the physical fundamentals.

3) The evolution of the absorption systems is likely to occur on time scales much in excess of human lifetimes. By analogy with the utility of the Hertzsprung-Russel diagram, it is hoped that the study of large samples of absorption systems will provide clues about their evolution.

The redshift distribution of the systems is relevant to the above. On large scales comparable to the Hubble radius ($cH_0^{-1} \approx 3000h^{-1}$ Mpc) the redshift distribution yields information on (i) whether the absorption systems are ejected from QSOs or are an intervening population unconnected to the QSOs [1], (ii) their cosmological evolution, and (iii) H(z), the rate of expansion of the Universe as a function of epoch.

On small scales corresponding to velocity separations of less than about 5000 kms^{-1}, the absorption system distribution provides information on (iv) clustering [2] and (v) possible physical influences of the particular QSO under study [3,4].

2. EXPECTED REDSHIFT DISTRIBUTION

The simplest and standard assumptions [2] about the distribution of the systems are that all narrow lined systems are intervening and that they do not "evolve". This is taken to mean constant comoving number density, constant proper sizes and constant ionization, corresponding to a population of absorbers which are neither expanding or collapsing (ie. bound), with no net creation or destruction of clouds.

Using $H(z) = H_0\,(1+z)(1+2q_0z)^{0.5}$, these assumptions lead to an expected number density along a line of sight of $l(z) = l_0(1+z)(1+2q_0z)^{-0.5}$ systems per unit redshift, where $l_0 = cH_0^{-1}/($ mean free path $) = cH_0^{-1}\phi\sigma$, ϕ is the number of absorbers per volume, and σ is the mean cross-sectional area, all measured at the current epoch. Note that the latter two factors can not be determined separately using l(z) alone.

3. SYSTEM CHARACTERISTICS

Systems are currently classified in terms of their observed absorption line properties alone, with little or no reference to their likely physical properties. The 21 cm line of H I is very rarely detected, while molecular lines have yet to be found, even in systems with very large H I column densities. Thus the ultraviolet lines dominate. The strongest of these is Lyman-alpha ($L\alpha$) which is the only line seen in about 99 percent of systems. The remaining 1 percent of systems also show the strongest lines of the more abundant metals, including in particular, C IV and Mg II. For all systems with $N(H\ I) \geq 2 \times 10^{17} cm^{-2}$, the strongest spectral signature is Lyman continuum absorption. These Lyman Limit Systems or LLS are optically thick, unlike the $L\alpha$ systems. Table 1 contains the densities of these types of systems [3,5].

Table 1

Absorption System Densities

characteristic	MgII	LLS	CIV	$L\alpha$
W_r limit (Å)	0.6	$(\tau \geq 1)$	0.3	0.36
No. of QSOs	125	120	30	19
No. systems	36	41	26	475
system z range	0.1 – 2	0.2 – 3.8	1 – 2.5	1.6 – 3.8
$l_0(q_0 = 0)$	0.17±0.03	0.38±0.06	0.55±0.11	$\simeq 16$
R_* (kpc/h)	50	75	90	480
	Is evolution detected for $q_0 = 0$?			
	probably	probably	?	?
$\phi\sigma$ ratio	1.6	1.3	?	?
	Is evolution detected for $q_0 = 0.5$?			
	yes	yes	?	?
$\phi\sigma$ ratio	2.5	2.1	?	?

The data on the Mg II systems and the LLS has been obtained or compiled in collaboration with D. Kunth (Inst. d'Astrophysique) and T. Snijders (RGO). Although there exists an apparent abundance of information on the $L\alpha$ systems, recent analysis [3] reveals that they are very poorly represented by a monotonic functional from for l(z), and consequently these data can not be reliably interpreted in terms of evolution.

Of the four samples listed above, only the $L\alpha$ systems are an observationally distinct sample. Essentially all LLS are metal line systems, often with C IV or Mg II absorption. Although most Mg II systems are expected to be LLS, roughly 60 percent of even strong lined C IV systems do not have sufficient H I to qualify as LLS.

It should be stressed that the lowest redshift systems are of critical importance in these studies. It is at low redshifts that the redshift distribution is most sensitive to q_0 [2], and confusion arising from line blending and misidentification is minimal. If system evolution proceedes at a set rate per unit time, then it will be most pronounced at low redshifts since dt/dz is a largest at z=0. But of most interest is the possibility of directly comparing the lowest z systems with local objects which are comparatively well studied. It is only for low z systems that we stand a reasonable chance of being able to image the absorbing objects directly. There are now two definite cases in which ordinary Mg II absorbers have been identified with ordinary galaxies [6].

Aside from using l(z), there are also direct methods of measuring the evolution of the absorption system population. By analogy with the measurement of changes in the spectral energy distribution of emitting objects, we can measure the physical properties of individual systems and search for changes in the distribution of these properties with redshift. The properties include the column density [7], temperature, density, ionization, velocity dispersion, and metal abundances. In this way evolution can be detected without reference to the value of q_0. There will be major efforts and progress in this area in the next decade, and this is the key to a physical understanding of the l(z) distribution.

4. DO ABSORPTION SYSTEMS ARISE IN GALAXIES?

If we assume that some collection of high redshift absorption systems arise in galaxies, then a comparison of their observed number density with predictions derived from local galaxy observations will be sensitive to system and galaxy evolution [8,9]. The R_* values listed in Table 1 are the radii required of the disks of fiducial luminosity spiral galaxies assuming that the latter are to explain all absorption systems. Since $R_*^2 \propto l_0/(\phi\sigma)$, R_* values should be decrease by a factor of two if all galaxies absorb with spherical cross-sections, while it should be increased by two times if $h = 0.5$. In the cases where evolution is detected, it is in the sense of requiring the $\phi\sigma$ product to increase by about a factor of 1.3−2.5 as z increases across the middle two-thirds of the appropriate z range.

Recent comparisons of the observed density of systems at $z \simeq 2-3$ show that there are about 6 times more absorbers with $N(\text{H I}) \geq 3 \times 10^{18}\text{cm}^{-2}$ than would be produced by local galaxies [9,10]. It is not clear whether these comparisons are revealing the evolution of the size of galaxy disks, or simply the inadequacies in our understanding of local galaxies. They may alternatively show that most absorption systems, including most of those which show metal lines, do not arise in galaxies [10,11].

The values of R_* quoted in Table 1 above refer to relatively strong systems only. It is now well established that the number of absorption systems observed rises rapidly with decreasing rest frame equivalent width W_r. For example, for the Mg II systems the distribution function is approximately $g(W_r) \propto W_r^{-2}$ systems per Ångstrom, giving R_* values of about 80 kpc/h for $W_r \geq 0.25$Å and roughly 110 kpc/h by 0.15Å. For the Lα systems $R_* = 1$Mpc/h for $N(\text{HI}) \geq 10^{13}\text{cm}^{-2}$. These sizes are much larger than galaxies as we know them and consequently it is generally suspected that the Lα systems, at least, are intergalactic.

Note that there is every reason to expect that the number of Mg II systems observed will continue to rise steeply until most weak systems are on the linear portion of the curve of growth. This is because nearly all systems currently observed are on or near to the saturated portion, where slight changes in the sample minimum equivalent width limit allow the detection of systems with much lower column densities and gas velocity dispersions.

5. WHY DO ABSORPTION SYSTEMS EVOLVE?

Although questions on absorption system evolution can not be separated from those on the origin and nature of the absorbers, several general points can be made. Two potentially important effects have now been identified.

a) Ionization Changes

The first is the change in the intensity of the metagalactic ionizing radiation field. The *emitted* radiation is dominated by QSOs at low redshifts, and by either QSOs or young (especially elliptical) galaxies at redshifts above about 2. The *observed* flux at any z is the sum of all redshifted radiation emitted at earlier epochs, corrected for intervening absorption. Even at the current epoch, 90 percent of this might be radiation from high redshift young galaxies. We know the rough rate of evolution of the high luminosity QSOs, but not much about low luminosity QSOs or galaxies. Thus we have only lower limits on I_ν, the intensity of the radiation field.

A useful constraint on I_ν is that the integrated flux from all sources at various redshifts should not exceed the observed background intensity at $z = 0$. This limit is sufficiently large to permit all young galaxies to emit as indicated by the Bruzual models [12]. However this limit does not apply to short wavelength radiation emitted at redshifts in excess of about 3, because the distance from the emitter to our galaxy then exceeds the mean free path to absorption in LLS by a factor of a few.

At recent epochs from $z = 1$ to now, the flux should decline monotonically by about a factor of 3–10 to its current low value of about $I_\nu = 5 \times 10^{-22}$ erg cm^{-2}s^{-1}Hz^{-1}sr^{-1} (perhaps 3 times more if galaxies contribute a lot, or 3 times less if they are insignificant), simply because of the lack of new low redshift emitting sources. As z increases from 1–4, I_ν might vary by less than a factor of two if the increasing density of sources is compensated for by the lack of flux from much higher redshifts. In summary there is insufficient information to derive any accurate conclusions concerning the variation of I_ν with epoch.

The one percent of all absorption systems which are LLS are to a various extent self-shielded from the full intensity of the ionizing radiation. In contrast, systems that are optically thin in the Lyman continuum will be highly ionized, unless their densities are high enough that the recombination time is short compared to the photoionization time, that is $n_e \geq (I_\nu/10^{-22})\text{cm}^{-3}$ for temperatures of about 10^4 K. If ionization is a dominant effect on system evolution, we might expect that there would be close similarities in the l(z) for the optically thin systems (Lα and most C IV), and perhaps a different l(z) behaviour for the optically thick systems (the LLS, including most Mg II and a few C IV systems).

b) Size Changes

A second potential evolutionary effect is the propensity of systems to change size. Self-bound systems should collapse, and continue to collapse until they reach a state of (temporary) thermal equilibrium. This may occur when they emit heat to balance the input from I_ν, or to balance the heat from stars which may form inside them [13]. Dark matter might be important in determining the size of systems [14].

Note that if the systems were to expand freely with the Hubble flow, their number density would be $l(z) \propto (1+z)^{-1}(1+2q_0 z)^{-0.5}$ neglecting ionization, or $l(z) \propto (1+z)^{-(6+5S)}(1+2q_0 z)^{-0.5} I_\nu(z)^{1+S}$ if they are highly ionized and the column density distribution is $f(N\,H\,I) \propto N^S$. Although the former l(z) is incompatible with observations, the latter form may be acceptable [15] since $S \simeq -1.5$. However it is generally agreed that some form of binding, involving external pressure or gravity, is likely.

It is a major challenge to think of large structures which contain a lot of gas and which are stable over a Hubble time. Galaxies are an example. They remain of roughly constant sizes for billions of years because their mass is locked up in small dense objects (stars) which do not undergo significant two body interactions. They can also contain a lot of gas, in part because this is cycled through stars and also because of rotational support in disks. The complex velocity structure of many metal line systems suggests that they too consist of small dense clumps, while turbulent motions are inferred for many other systems.

The problem of making absorption systems is in many ways exactly the opposite of that of making galaxies. With galaxies, the difficulty is to get a sufficiently rapid collapse and density enhancement starting from low-contrast density perturbations at the epoch of the last scattering of the microwave background radiation. For the absorption systems one much try to stop this collape, especially at the late observable epochs. For many systems we must also allow for nearly solar metal abundances, either formed prior to the absorbing objects or *in situ*.

Several workers [16] have considered the possibility that the absorption systems can be explained within the context of galaxy formation hypotheses, which is both economical of effort and allows the observational constraints for galaxies to be considered together with those for absorption systems. In most of these discussions the absorption systems are identified with low mass "failed" or dwarf galaxies. Still, we are a long way from knowing whether absorption systems and galaxies can be treated together in the context of a single theory.

References

[1] Bahcall, J. N. and Peebles, P. J. E. 1969, Ap. J., **156**,L7; Tytler, D. 1982, Nature, **298**, 427.

[2] Sargent, W. L. W., Young, P., Boksenberg, A. and Tytler, D. , 1980, Ap. J. Suppl., **42**, 41.

[3] Tytler, D. 1986a, Ap. J., submitted.

[4] Foltz, C. B., Weymann, R. J., Peterson, B. M., Sun, L., Malkan, M. A. and Chaffee, Jr. F. H. 1986, Ap. J., **307**,504; Weymann, R. J., Anderson, S. F., Foltz, C. B. and Chaffee, Jr. F. H. this volume; and Bathel, P. D., Tytler, D. and Miley, G. K. in preparation.

[5] Young, P. J., Sargent, W. L. W. and Boksenberg, A. 1982, Ap. J. Suppl., **48**, 455; Bergeron, J. and Boissé, P., 1984, Astr. Ap., **133**, **374**.

[6] Bergeron, J. 1986, Astr. Ap., **155**, L8; Bergeron, J., Boulade, O., Kunth, D., Boksenberg, A., Vigroux, L. and Tytler, D. 1987, in preparation.

[7] Carswell, R. F., Webb, J. K., Baldwin, J. A. and Atwood, B. 1987, Ap. J. in press.

[8] Briggs. F. H., Wolfe, A. M., Krumm, N. and Salpeter, E. E. 1980, Ap. J., **238**, 510.

[9] Wolfe, A. M., Turnshek, D. A., Smith, H. E. and Cohen, R. D. 1986, Ap. J. Suppl., **61**, 249.

[10] Tytler, D. 1986b, Ap. J., submitted.

[11] Burbidge, G. R., O'Dell, S.L., Roberts, D.H. and Smith,H.E.1977, Ap. J., **218**, 33.

[12] Bruzual, A. G. 1981, Ph.D. Thesis, U. of California, Berkeley.

[13] Peebles, P. E. J. and Dicke, R. H. 1968, Ap. J., **154**, 891.

[14] Rees, M. 1986, M.N.R.A.S., **218**, 25p.

[15] Atwood, B., Baldwin, J. A. and Carswell, R. F. 1985, Ap. J., **292**, 58.

[16] Arons, J. 1972, Ap. J., **172**, 553; Fransson, C. and Epstein, R. 1982 M.N.R.A.S., **198**, 1127; Ikeuchi, S. and Ostriker, J. P. 1986, Ap. J., **301**, 552; and York, D. G., Dopita, M., Green, R. and Bechtold, J. 1986, preprint.

DAMPED LYMAN ALPHA ABSORPTION LINES IN QSOS

by

A.M.Wolfe
University of Pittsburgh

ABSTRACT

A survey for QSO absorption systems with damped Lα lines is described. A study of 68 QSOs with emission redshifts greater than 2.3 reveals the presence of 15 damped Lα absorbers with $N(HI) > 2 \times 10^{20}$ cm^{-2}. Besides filling 20% of the sky between z=2 and 3, the damped population of absorbers contributes a cosmological mass per unit comoving volume that dominates the baryon content of the universe in this redshift interval. Evidence is presented which suggests that the damped population is comprised of objects that resemble the gaseous disks of nearby spiral galaxies. A recent VLBI study of the gas causing 21cm absorption in one object shows it to extend ~10 kpc transverse to the line of sight . It is argued that the damped absorbers are at least three times larger than current stellar disks, and because of the coincidence between the density parameter of the damped population and of visible matter in galxies that these objects are the progenitors of the stellar disks of nearby spirals.

I. INTRODUCTION

Ever since Sargent et al. [1] published their comprehensive study of the Lα forest, research on QSO absorption lines has focused on the low column-density absorbers responsible for the weak Lα lines that dominate high-resolution spectra shortward of Lα emission. In this paper I discuss the results of a survey for absorbers with high column densities. Prior to our study [2,3], little was known about systems with HI column densities, $N(HI) > 10^{20}$ cm^{-2}. This is due to the fact that strong lines created by high column-density gas occur so infrequently that they rarely appear in the typical high-resolution study of a few QSOs.

As a result the first phase of our survey consisted of a

low-resolution study of large numbers of QSOs. Although there was interest in the relationship between absorbers with high column densities and the Lα-forest clouds, the primary objective was to discover the HI disks of high-z galaxies. It is argued below that the objects discovered do apprear to be the progenitors of present-day galactic disks;i.e., disks dominated by stars rather than gas.

II. THE SURVEY

The principal spectroscopic imprint of an HI disk is a Lα absorption line broadened by radiation damping. Surveys for 21 cm emission from spiral galaxies show that N(HI)>2×10^{20} cm^{-2} within $R=1.5\times R_{Ho}$ [4] (R_{Ho} is the Holmberg radius), and that the velocity dispersion $\sigma\sim10$ kms^{-1} across the entire gaseous disk [5]. This combination assures that Lα will be optically thick in the damping wings of the line profile. At the same time the lower column densities of abundant metals implies they they create saturated lines that are weak owing to the small σ. Furthermore the heavy-element transitions will be dominated by low-ions such as C^{+}, Si^{+}, and Fe^{+} rather than high ions like C^{+3} and Si^{+3} that dominate metal lines in most QSO spectra. Consequently a foreground HI disk imprints a unique signature on the spectrum of the background QSO, one in which the rest-frame equivalent width of Lα, W(Lα)>10 Å, while W(CII)>0.3 Å

An initial search for damped-Lα candidates was carried out with the 3.2m reflector at Lick Observatory [2]. Spectra of 68 QSOs with z>2.3 were acquired at a resolution of $\Delta\lambda=10$Å. Of the 476 absorption features identified as Lα, 47 qualified as candidates for damped absorption on the basis of W(Lα)>5 Å: while this criterion corresponds to N(HI)<2×10^{20} cm^{-2} , it was used to account for measurement errors in equivalent width. Subsequently accurate spectra were acquired at high resolution ($\Delta\lambda=1$-2 Å) at the MMT and CTIO 4m mirrors. Analysis of the spectra reveal that 30 of the 47 features fall into 2 categories. Based on accurate fits to Voigt damping profiles and the presence of narrow low-ion lines at the same redshift,16 features have been confirmed as damped Lα. The remaining 14 features appear to be multi-component systems that span velocity intervals of ~1000 kms^{-1}. While the damped systems dominate the equivalent-width distribution at W>10Å, the multi-component systems prevail near the W=5 Å threshold. Therefore the equivalent-width distribution at W<10 Å appears to be an ordered sequence in velocity width, and an ordered sequence in N(HI) at W>10 Å. The detection of damped Lα in 20% of the QSOs should be compared to the 4% predicted for intervening galaxies along the same redshift path. The radii of the damped systems $R>3\times R_{Ho}$, If their comoving density equals that of spiral galaxies.

III. EVIDENCE LINKING DAMPED Lα SYTEMS TO GALACTIC DISKS

There are various lines of evidence that support the disk hypothesis. First the HI column densities inferred from the damping profiles range from 2×10^{20} to 6×10^{21} cm^{-2}. Although the mean ,$\langle N(HI)\rangle=10^{21}cm^{-2}$, is typical of the inner regions of spirals, we recall that the average impact parameter $b>3\times R_{HO}$ where $N(HI)\sim10^{18}$ cm^{-2} in nearby spirals [6] . Thus the high-z damped systems are gas rich compared to their present-day counterparts, if the two groups are in a one-to-one correspondence. Secondly metal lines are detected in all 15 systems. Comparison of the CIV and CII equivalent widths shows that the damped systems are not at all like the usual QSO metal-line system in which CII is either weak or not detected. Rather in the damped systems CIV is either weak or not detected , while CII is detected in every case. As a result the damped systems more closely resemble the absorption spectra of the gas in the Galaxy [7]. Third 21cm absorption is frequently detected in the damped systems. We have detected it in 3 out of 5 tries. In comparison 21 cm absorption was detected in 2 out of 20 MgII absorption systems [8].

PKS 0458-02 is the only QSO in the sample with an absorption system that exhibits both damped Lα and a 21cm line.[9] The z=2.04 system exhibits CII,SiII,CIV,and SiIV lines that are produced by gas with $\sigma\sim35$ kms^{-1}. From the Lα profile we infer that $N(HI)=5\times10^{21}$ cm^{-2}. The 21 cm feature consists of two rather strong and narrow components ($\sigma\sim5$ kms^{-1}) separated by 15 kms^{-1}. Comparison between the Lα and 21cm features reveals that the spin temperature, $T_s<500K$. The absorber apparently consists of cold quiescent gas and a more turbulent component.

To determine whether the cold gas resides in a galactic disk, we [10] observed the 21 cm line with a VLB interferometer comprising Arecibo and the 140 ft. antenna at Green Bank. At the absorption frequency the source contains a compact core and an extended jet which emit 60% and 40% of the flux. If the foreground gas were a small cloud that just covered the compact core, the line depth of the VLB fringe amplitude would be 1.7 times that of the single-dish line depth. Instead we found that both the velocity structure and relative depth of the VLB and single-dish features were the same. As a result the absorbing gas must extend across a large fraction of the extended jet as well as the core. Because the jet is located between 0.5 and 2.0 arcsec from the core, this implies that the foreground gas has a transverse dimension greater than ~10 kpc. The similarity of the two profiles further indicates that the effective velocity dispersion is ~10 kms^{-1} across the same dimension. These facts imply a disk-like structure for the absorbing gas. If instead the gas were in a spherical configuration, extending ~10 kpc along the line of sight, σ would have to exceed ~70 kms^{-1} in order for the gas to be hydrostatically supported against self-gravity. Because the observed σ is so much smaller, the scale-height of the gas along the line of sight must be much smaller than 10 kpc. It is also reasonable to suppose that the gas is centrifugally supported by rotation in the transverse direction. All the data are best explained by a foreground rotating disk with R>>10 kpc.

The final link between the damped population and disk galaxies is the cosmological mass density. It is possible to estimate the density parameter Ω_{damp} from <N(HI)> and the 20% covering factor.[3] We find that Ω_{damp}~0.004. This is comparable to Ω_{vis}, the cosmological density contributed by visible matter,stars, in spiral galaxies. The equality of these factors is either a coincidence, or reflects the fact that we have detected the progenitors of disk galaxies at a time before most of the gas in the disk was converted into stars.

IV. IMPLICATIONS FOR GALAXY FORMATION

The results reported here suggest the detection of large galactic disks with z=2-3. These objects appear to belong to newly discovered phase of galactic evolution which occurs after the collapse of the protogalaxy, but before the formation of the current stellar disk. If this conjecture proves to be correct, questions such as how the disk contracts in its plane , and how the abundance of the elements evolve with time need to be answered.

V.REFERENCES

1. Sargent,W.L.W., Young,P.J., Boksenberg,A., and Tytler,D.D. 1980, Ap.J. Supp.,42,41.

2. Wolfe,A.M., Turnshek,D.A., Smith,H.E., and Cohen,R.D. 1986, Ap.J.Supp.61, 249.
3. Wolfe,A.M. 1986, Phil. Trans. Roy. Soc. 320,503.
4. Bosma,A. 1981, A.J., 86, 1825.
5. Lewis,B.M. 1984, Ap.J., 285, 453
6. Briggs,F.H., Wolfe,A.M., Krumm,N., and Salpeter,E.E. 1980, Ap.J., 218, 510.
7. Savage,B.D. and de Boer, K.S. 1981, Ap.J.,243, 460.
8. Briggs,F.H. and Wolfe,A.M.1983,Ap.J.268,76.
9. Wolfe,A.M., Briggs,F.H., Turnshek,D.A., Davis,M.M., Smith,H.E., and Cohen, R.D. 1985, Ap.J.Lett.294, L67.
10. Briggs,F.H., Wolfe,A.M., Davis,M.M., Liszt,H., and Turner,K. 1987, in preperation.

QSO-ABSORPTION SYSTEMS DUE TO FILAMENTARY SHELLS ?

Wolfgang Kundt

Institut für Astrophysik der Universität Bonn
5300 Bonn 1, F.R.G.

The IGM is populated by at least the following four systems of extended, non-galactic objects:

(1) Lyα forest: small-filling-factor, dense agglomerates of neutral hydrogen, of inferred temperature $T \gtrsim 10^4$ K, metallicity $Z \lesssim 10^{-3} Z_\odot$, column-density distribution $dN/dN_H \sim N_H^{-1.5}$ above $N_H := N(HI) \gtrsim 10^{14}$ cm^{-2}, and evolution with redshift $dN/dz = 3.4(1+z)^{2.3}$. Their inferred spatial extent D ranges from $D = N_H/n \gtrsim 10^{15}$ cm up to tens of kpc or even Mpc. No cosmic voids are detected [Wallace Sargent].

(2) Heavy-element absorption systems: small-filling-factor, dense agglomerates of hydrogen-depleted matter, of inferred temperature $T \gtrsim 10^4$ K, metallicity $Z \gtrsim 0.1\, Z_\odot$, equivalent hydrogen column densities going up to $N_H \lesssim 10^{20.5}$ cm^{-2}, velocity (fine) structure ranging from $\gtrsim 1500$ km s^{-1} down to $\lesssim 5$ km s^{-1} [Donald York], encountered $\lesssim 30$ times less often than the Lyα systems but with similar evolution [David Tytler]. Again, their inferred spatial extent ranges from $D \gtrsim 10^{15}$ cm up to kpc or larger. No cosmic voids are detected.

(3) Optical emission shells: faint systems of concentric partial shells, seen preferentially around isolated elliptical and SO galaxies, in opposing sectors, at separations between kpc and hundreds of kpc [9]. Mass estimates are of order 10^{10} $M_\odot$ per (large) shell. These shells have wedge-shaped surface brightnesses, with sharp outer edges, and BVR colours similar to starlight [2,10]. It is not clear whether the luminous arcs found by Petrosian & Lynds [8] are earlier versions of the same phenomenon.

(4) Massive thin shells in the halos of radio galaxies, of significant filling factor, inferred from optical maps (Cen A), radio maps (Cyg A, Her A), and linear polarization maps (3C277.3, Cyg A). They apparently interact with the beams [3,7], see figure 1. Mass estimates are of order 10^{10} $M_\odot$ per shell.

It is my impression that these four massive intergalactic systems are different aspects of the s a m e population, viz. massive filamentary shells ejected by the nuclei of galaxies during their active epoques [5]. They are equally numerous: several per massive galaxy (except for the - more compact - heavy-element systems), equally extended (large halo scale), with considerable fine structure in both size and velocity. Their (high) densities, sharp velocities and morphologies - similar to supernova shells - suggest ram pressure confinement. Hydrogen and heavy elements are diffusively separated but moving jointly. Their evolution reflects the fading of the AGN phenomenon. A power-law distribution in sizes, with a peak in number at the small end, is reminiscent of supernova shells [6].

This scheme rejects galaxy merging as an interpretation of phenomenon (3) [11] because of its high occurrence rate, sharp-edged morphology, anticorrelation with galaxy density and interaction with the beams in the case of Cen A. It also rejects ordered mass motions with filling factor unity [1,12] because such would be very different from what is encountered locally [4], and the flows would not fill the cosmic voids. On the other hand, systems (2) & (3) both anticorrelate with the host galaxy's (halo) gas density: radio-quiet (spiral) QSOs lack (intrinsic) high-velocity absorptions [Ray Weymann], and gas-rich (spiral) galaxies lack large emission shells. This common anticorrelation is a likely consequence of shell destruction by halo gas and supports their identity.

References

[1] Chernomordik, V.V., Ozernoy, L.M., Nature 303, 153 (1983).
[2] Fort, B.P., Prieur, Carter, Meatheringham, Ap. J. 306, 110 (1986).
[3] Gopal-Krishna, Saripalli, L., Astron. Astrophys. 141, 61 (1984).
[4] Kundt, W., Magnetic fields in SN and SN shells, Ringberg Workshop, eds. Beck & Gräve, Springer (1987).
[5] Kundt, W., Krause, M., Astron. Astrophys. 142, 150 (1985).
[6] Kundt, W., Krotscheck, E., Astron. Astrophys. 83, 1 (1980).
[7] Kundt, W., Saripalli, L., J. Astron. Astrophys., submitted (1987).
[8] Lemonick, M.D., Time, Jan. 19, 35 (1987).
[9] Malin, D.F., Carter, D., Astrophys. J. 274, 534 (1983).
[10] Pence, W.D., Ap. J. 310, 597 (1986).
[11] Quinn, P.J., Ap. J. 279, 596 (1984).
[12] Williams, R.E., Christiansen, W.A., Ap. J. 291, 80 (1985).

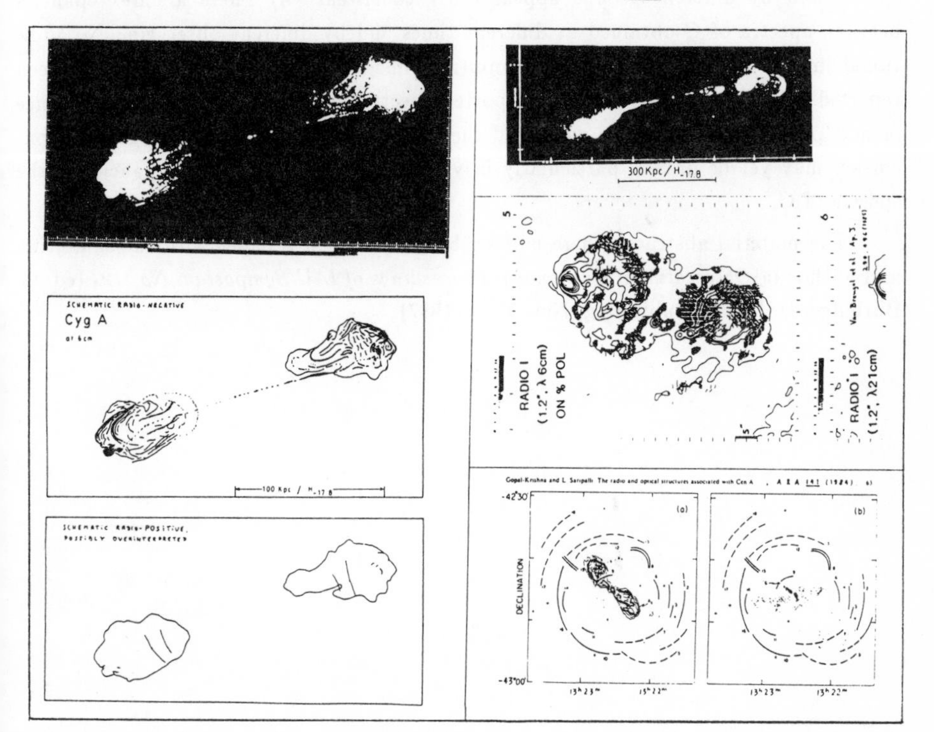

Figure 1.: Cyg A, Her A, 3C277.3 and Cen A at optical and/or radio frequencies, each showing signatures of beam interaction with thin, massive shells. For references see text.

1146+111B,C: A GIANT GRAVITATIONAL LENS?

Edwin L. Turner

Princeton University Observatory
Princeton, New Jersey 08544 USA

ABSTRACT

The currently available observations of the giant gravitational lens candidate 1146+111B,C are reviewed as of December 1986. Two attempts to detect microwave background distortions due to the putative lensing mass have yielded only upper limits (in the 0.1 to 1 mK range). Optical and near optical spectra have proliferated. The interpretation of these spectra taken as a whole is far more uncertain than when they are considered individually. Salient points include: 1) Some pairs of spectra show a striking similarity between B and C. 2) Some pairs show significant differences, particularly in Balmer line emission and C III] strength. 3) Spectra of B obtained at different epochs and by different groups appear fairly consistent. 4) There are discrepancies between spectra of C obtained at different times and by different observers. No additional images of B/C nor any other multiply imaged objects in the field have been reported. The negative evidence of reported spectroscopic differences and the absence of new positive evidence have weakened the case for lensing; however, no definite conclusion may yet be drawn, particularly in view of the unexplained discrepancies in the spectra of C.

The material abstracted here is described in detail in a paper with the same title and author in *Observational Cosmology: Proceedings of IAU Symposium No. 124* (ed. G. Burbidge, D. Reidell Publishing, New York, 1987).

A SEARCH FOR GRAVITATIONAL LENSES: SEARCH STRATEGIES AND A PRELIMINARY UPPER LIMIT ON THE DENSITY OF LENSES

J. N. Hewitt (MIT and Haystack Observatory), E. L. Turner (Princeton),
B. F. Burke (MIT), C. R. Lawrence (Caltech), C. L. Bennett (NASA/GSFC)

As of this writing, seven candidate gravitational lens systems have appeared in the literature: 0957+561[1)], 1115+080[2)], 2345+007[3)], 2016+112[4)], 1635+267[5)], 2237+031[6)], and 1146+111[7)]. The amount of evidence supporting the gravitational lens interpretation of these systems varies greatly, ranging from measurements at many wavelengths, including the detection of lensing galaxies (for example, 0957+561 and 2016+112), to just similar optical spectra at the same redshift (for example, 1146+111, 1635+267, and 2345+007). The complete lack of detection of a lensing galaxy in a few of these cases, and the difficulty encountered in modelling the lensing in some of the other cases, suggest that gravitational lensing may enable us to infer the presence of nonluminous matter. Other methods that have been used to detect dark matter, such as studies of galactic rotation curves and cluster dynamics, require luminous matter in the gravitational field that traces the dark matter distribution. This is not necessary in studies of gravitational lensing; the beacon we observe can be far (of order one thousand megaparsecs) from the lensing mass. In this paper we describe a radio-optical search for gravitational lens systems designed to produce a sample of gravitational lenses for studying the individual cases and the statistics of the sample. Statistical studies can be used to severely constrain the properties of lensing objects in the universe[8−10)]. We discuss here a preliminary result of just the radio data: we find that the density of matter in the form of a uniform, comoving density of 10^{11} to $10^{12} M_\odot$ compact objects, luminous or dark, must be substantially less than the critical density.[11)]

The procedure we have adopted for identifying gravitational lens candidates is as follows: (1) Map the arc second scale structure of a large sample of radio sources with the National Radio Astronomy Observatory (NRAO) Very Large Array (VLA). † The source list for this mapping effort is the MIT-Green Bank (MG) 5 GHz survey[12)]. (2) Select sources with multiple compact radio components and image them optically. (3) Further select sources with multiple radio-optical components and measure their optical spectra. Sources with multiple components at the same redshift with similar spectra are excellent gravitational lens candidates. In the past this was generally recognized to be sufficient evidence for classifying a source as lensed. However, the discovery of 1146+111 provides a candidate lens system which, because of the large angular separation of the "images", is relatively easily subjected to some other tests of gravitational lensing. Though the 1146+111 case is by no means closed, the failure of these tests to support the lensing interpretation[13−15)] alerts one to the possibility that there may exist pairs of physically distinct sources with very similar spectra that are not gravitationally lensed. Therefore, it may be necessary to subject the candidate lens systems to a battery of tests before the more skeptical will be convinced. Some possibilities that come to mind are: (1) VLBI — do the "images" have similar structure on scales much smaller than the image separation? (2) Flux monitoring — do the "images" have the same light curve, possibly offset by a time delay; and (3) Polarization measurements — are the polarization properties consistent with gravitational lensing?

Though positive proof of lensing may require measurements at many different wavelengths, many sources can be eliminated as lens candidates on the basis of their radio structure alone. In what follows we will restrict ourselves to a discussion of just some of the radio data collected during the course of the gravitational lens search. Considering what we *do not* see, i.e., the number of radio sources that are *not* lensed on scales detectable with the VLA, allows us to put limits on the number density of lenses in the universe. The sensitivity of the VLA to different populations of lenses can be calculated by converting the instrumental limits of the VLA into limits on the relative angular positions of the source and the lens that produce detectable pairs of images[16)]. The effect of the instrumental limits on the detectability of a population of lenses is a function of the lens mass model. Carrying out the calculation for a uniform, comoving density of point mass lenses, we find that the A array of the VLA is sensitive to point mass lenses in the mass range of 10^{11} to $10^{12} M_{\odot}$. We can compare these calculations to the

† The National Radio Astronomy Observatory is operated by Associated Universities, Inc., under contract with the National Science Foundation.

radio data and put an upper limit on Ω_L, the density of matter in the form of lenses, expressed as a fraction of the critical density. To do this, we will make three conservative assumptions. First, we ignore the effect of a flux intensification bias on the sample, that would tend to cause more lensed images to be included in the sample for a given population of lenses. Second, we assume the redshift distribution of the sources is the same as that measured for 30 optically bright sources (to the extent that faint sources have the same redshift distribution as bright sources, this assumption may not be conservative). Third, we make the ridiculously conservative assumption that *all* the structure seen in our VLA maps is due to gravitational lensing rather than to intrinsic source structure. Then, for a sample of 290 sources observed at a wavelength of 6 cm with the A array of the VLA, we count the number of sources that could be lensed. The maximum value of Ω_L consistent with the data and a background metric described by $\Omega = 1$ (again a conservative assumption in the sense that it gives a large value of Ω_L), for masses in the range of 10^{11} to $10^{12} M_\odot$, is 0.7. If we relax the third assumption by assuming that resolved, collinear components of a source represent intrinsic source structure and are not produced by lensing, the maximum consistent value of Ω_L is 0.4. These upper limits on Ω_L are independent of the value of Hubble's constant.

In summary, even at this very early stage in our gravitational lens survey, we can place limits on the density of compact objects that are completely independent of other methods. Future work will further constrain Ω_L as we continue to eliminate lens candidates and as we include the effects of the flux intensification bias in the calculations. Populations of other types of lenses can be constrained by inserting other mass distributions into the model calculations. Finally, the analysis can be extended to different mass ranges by using other instruments and techniques, such as the Space Telescope and VLBI.

This research was supported in part by grants AST-8415677 and AST-8512598 from the National Science Foundation.

REFERENCES

1. Walsh, D., Carswell, R. F., and Weymann, R. J. 1979, *Nature*, **279**, 381.

2. Weymann, R. J., Latham, D., Angel, J. R. P., Green, R. F., Liebert, J. W., Turnshek, D. A., Turnshek, D. E., and Tyson, J. A. 1980, *Nature*, **285**, 641.

3. Weedman, D. W., Weymann, R. J., Green, R. F., and Heckman, T. M. 1982, *Ap. J. (Lett.)*, **255**, L5.

4. Lawrence, C. R., Schneider, D. P., Schmidt, M., Bennett, C. L., Hewitt, J. N., Burke, B. F., Turner, E. L., and Gunn, J. E. 1984, *Science*, **223**, 46.

5. Djorgovski, S., and Spinrad, H. 1984, *Ap. J. (Lett.)*, **282**, L1.

6. Huchra, J., Gorenstein, M., Kent, S., Shapiro, I., Smith, G., Horine, E., and Perley, R. 1985, *A.J.*, **90**, 691.

7. Turner, E. L., Schneider, D. P., Burke, B. F., Hewitt, J. N., Langston, G. I., Gunn, J. E., Lawrence, C. R., and Schmidt, M. 1986, *Nature*, **321**, 142.

8. Press, W. H., and Gunn, J. E. 1973, *Ap. J.*, **185**, 397.

9. Canizares, C. R. 1981, *Nature*, **291**, 620 (*errata* **293**, 390).

10. Turner, E. L., Ostriker, J. P., and Gott, J. R., III. 1984, *Ap. J.*, **284**, 1.

11. Hewitt, J. N., Turner, E. L., Burke, B. F., Lawrence, C. R., Bennett, C. L., Langston, G. I., and Gunn, J. E. 1987, in *IAU Symposium 124, Observational Cosmology*, ed. G. Burbidge, A. Hewitt,and L.-Z. Fang (Dordrecht: Reidel), in press.

12. Bennett, C. L., Lawrence, C. R., Burke, B. F., Hewitt, J. N., and Mahoney, J. H. 1986, *Ap. J. (Supp.)*, **61**, 1.

13. Stark, A. A., Dragovan, M., Wilson, R. W., and Gott, J. R., III 1986, *Nature*, **322**, 805.

14. Lawrence, C. R., Readhead, A. C. S., and Moffet, A. T. 1986, *A.J.*, **92**, 1235.

15. Tyson, J. A., and Gullixson, C. A. 1986, *Science*, **233**, 1183.

16. Hewitt, J. N. 1986, Massachusetts Institute of Technology Ph. D. Thesis.

GRAVITATIONAL MICRO-LENSING AS A CLUE TO QUASAR STRUCTURE

S. Refsdal, B. Grieger, R. Kayser

Hamburger Sternwarte
Gojenbergsweg 112, D-2050 Hamburg 80
F.R. Germany

ABSTRACT

Micro-lensing due to stars in an intervening galaxy has many interesting astrophysical applications. We here discuss the possibility of determining the absolute size and luminosity profile of a lensed quasar from its light curve during a high amplification event (HAE) which occurs when the source crosses a critical curve (anticaustic).

1. INTRODUCTION

A general discussion of gravitational micro-lensing due to stars in an intervening galaxy, with special emphasis on astrophysical applications, has been given by Kayser, Refsdal and Stabell (1986))[1]. The "parallax effect" observable during a HAE by which the effective transverse velocity (and thereby the absolute size) of the quasar can be determined, was discussed in more detail by Grieger, Kayser and Refsdal (1986))[2]. We here discuss a method whereby the intrinsic luminosity profile of a lensed quasar can be determined from its light curve during a HAE, for more details see Grieger, Kayser and Refsdal (1987))[3].

2. HIGH AMPLIFICATION EVENTS

If a compact source crosses a critical curve, a typical asymmetric peak occurs in the observed light curve)[1]. The steep part of the light curve by such a HAE is due to the appearance (or disappearance) of a bright double image. It is an eclipse-like effect and by power expansion of the amplification factor close to the critical curve)[4,5] the flux F of this double image is given by

$$F(t)=F(\xi_c(t))=\int_{\eta_{min}}^{\eta_{max}}\int_{\xi_c}^{\xi_{max}} I(\xi,\eta)\frac{k}{\sqrt{\xi-\xi_c}}\,d\xi\,d\eta = k\int_{\xi_c}^{\xi_{max}}\frac{P(\xi)}{\sqrt{\xi-\xi_c}}\,d\xi \qquad (1)$$

compare Fig.1. Here $I(\xi,\eta)$ is the intrinsic surface brightness of the source and k is the "strength" of the critical curve, which is of the order 1 when normalized units[1] are used.

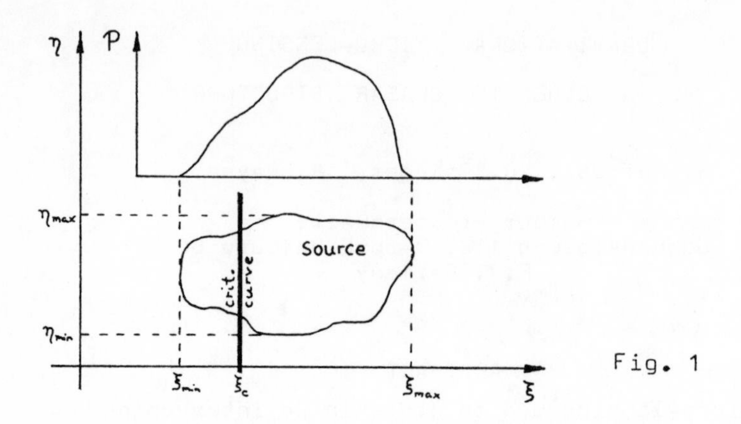

Fig. 1

If we assume the critical curve to move with constant velocity and that the other images have a constant brightness during the HAE, we can obtain the intrinsic luminosity profile $P(\xi)$ of the source perpendicular to the critical curve from the observed light curve.
We assume, that the steep part of the light curve appears at the left side of the peak (in the opposite case one makes simply a time inversion).
We denote the observed magnitudes by

$$m_i = m(t_i) \qquad ; t_{i+1} > t_i \quad ; i = 0, \ldots, n \tag{2}$$

The brightness increase due to the HAE should start after $t = t_0$. The observed intensity of the bright double image is then

$$F_i = F(t_i) = 10^{0.4(m_0 - m_i)} - 1 \tag{3}$$

3. DETERMINATION OF SOURCE PROFILE

We approximate the source profile by a polygon $P_i = P(\xi_i)$ and introduce $\Delta t_{ij} = t_i - t_j$. The first two points of the profile are given by

$$P_0 = 0 \qquad \text{and} \qquad P_1 = \frac{3}{4} \frac{F_1}{\Delta t_{1,0}^{1/2}} \tag{4a}$$

For $i > 1$ the value of P_i is given by

$$P_i = \frac{3}{4} \frac{F_i - \sum_{j=1}^{i-1} f_{ij}}{\Delta t_{i,i-1}^{1/2}} - \frac{1}{2} P_{i-1} \tag{4b}$$

where

$$f_{ij} = 2\,(p_{j-1} + s_j\,\Delta t_{i,j-1})(\Delta t^{1/2}_{i,j-1} - \Delta t^{1/2}_{i,j}) - \frac{2}{3}\,s_j\,(\Delta t^{3/2}_{i,j-1} - \Delta t^{3/2}_{i,j}) \tag{5}$$

and

$$s_j = \frac{p_j - p_{j-1}}{\Delta t_{j,j-1}} \tag{6}$$

5. COMMENTS

Numerical simulations with various source profiles have been presented by Grieger, Kayser and Refsdal (1987))[3]. For typical cosmological distances the source profile can usually be reproduced for source radii smaller then 10^{-3} pc$\cdot(M/M_\odot)^{1/2}$, where M is the typical star mass. This means, that the approximations we have made, including Eq.(1), are reasonable up to this source size.
The absolute length scale of the reproduced source profile is obviously proportional to the effective transverse velocity of the source. Possibilities to determine this velocity by the parallax effect have been discussed elsewhere)[2,3].

ACKNOWLEDGEMENT. This work is supported in part by the Deutsche Forschungsgemeinschaft under Az. Re 439/3.

REFERENCES

1. Kayser,R., Refsdal,S., Stabell,R. : 1986, Astron.Astrophys. 166, 36
2. Grieger,B., Kayser,R., Refsdal, S. : 1986, Nature 324, 126
3. Grieger,B., Kayser,R., Refsdal,S. : 1987, Astron.Astrophys., submitted
4. Chang,K., Refsdal,S. : 1979, Nature 282, 561
5. Chang,K. : 1981, Ph.D. Thesis, Hamburg University

COSMIC STRINGS IN THE REAL SKY

Craig J. Hogan*
Steward Observatory, University of Arizona
Tucson AZ 85721, USA

ABSTRACT

I discuss observational strategies for finding effects associated with the gravitational lensing of distant objects by strings. In particular, a proposed search program at Steward Observatory to find chains of galaxy image pairs is described.

Cosmic strings have emerged as possibly an important constituent of our universe, and may be the root cause of all galaxy formation and clustering.[1–7] Independent observational tests of the cosmic string scenario, such as discontinuities in microwave background temperature[8] or pulsar timing noise induced by strings through gravitational radiation[7], are precise enough to give it a level of respectability (as determined by falsifiability) on a par with other theories of galaxy formation. But the string scenario can be verified in principle at a much higher level of certainty by actually finding an actual string. Here I sketch one possibility of satisfying a skeptical astronomer who says "show me one." I will adopt the standard notation, where the mass per unit length of string is denoted by μ, and $\mu_{-6} \equiv G\mu/10^{-6}c^2$ is a convenient parameterization; for galaxy formation $\mu_{-6} \simeq 2$ is the currently preferred value[3], but this depends on uncertain factors such as the composition of the dark matter.

Optical searches can reveal a number of unique string-like effects. (1) Double images of QSO's can be formed by strings, with a string passing between the images.[9,10,11] This in itself is a stringy feature, as only a singular lens like a string can induce an even total number of images. In addition, velocities viewed on opposite sides of the string are shifted[8,11] by $\simeq 8\pi G\mu \simeq 8\mu_{-6}$ km/sec. Thus common Ly$-\alpha$ absorption lines in the two images should be systematically shifted in redshift by observable amounts, in some cases comparable to their width. For lines observed in the two images of QSO 2345+007A,B by Foltz *et al.* [12] the rms measuring error in comparing redshifts in the two images is about 25 km s^{-1}, so a test for a systematic shift should be possible with better data now being obtained. (2) Multiple (even) numbers of QSO images can be formed around compact string loops, whose rapid oscillatory behavior leads to rapid time variability. Rapid appearance and disappearance of images is also possible, and various features of the image appearance and brightness variations have a unique signature predicted from catastrophy theory.[13] (3) Long straight strings passing in front of a distant field of galaxies create a chain of galaxy pairs (with one image on each side of the string) whose angular separations are similar in magnitude ($\delta\phi \approx 4\pi G\mu$) and

* Alfred P. Sloan Research Fellow

whose position angles (as defined by the separation of the paired images) are nearly identical, although they are not in general perpendicular to the string.[11,14] (4) If the string itself actually passes in front of an extended resolved object of angular extent less than $4\pi G\mu$, it can produce an image with a sharp edge.[15]

The phenomenon of QSO lensing by strings has been extensively discussed elsewhere[9-11,13] and the search for stringlike effects is included in various programs studying and searching for lensed QSO's in general.[16] In this talk I will touch on optical search strategies for the effect (3) just listed. How it arises is illustrated in figure 1. Figure 1a shows a toy geometry with two distant galaxies, A and B, and a row of numbers floating in space. Light rays are shown bending around the string with angle $\theta_\mu \equiv 4\pi G\mu = 2.6\mu_{-6}$ arc sec, which is independent of impact parameter. Figure 1b depicts the appearance of the sky. The strip between numbers 2 and 4 can be seen around either side of the string. Galaxies such as A lying within this strip have duplicated images. Galaxies such as B which straddle the edge of this strip are duplicated also but one image is only a duplicate of the part which lies inside the strip, and hence has a sharp edge. For a string with typical velocity $\simeq c/2$, the position angles separating the images in each pair would not be perpendicular to the projected location of the string, although for a nearly straight string these position angles would be nearly parallel.[11]

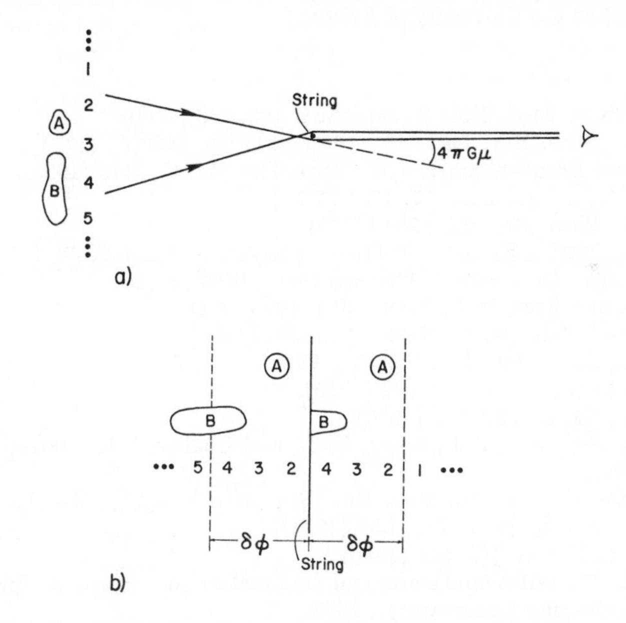

It should be mentioned that *any* field of the sky larger than a certain minimum size is likely to have at least one string loop in the field of view at $z \lesssim 1$. The total angular length of string ℓ at $z \lesssim 1$ in any survey field is proportional to the survey area A:

$$(\ell/rad) \simeq 0.38(A/sr)$$

This length is distributed evenly in equal octaves of ℓ which exceed the gravitational-wave-decay limit. Loops whose gravitational-decay timescale is comparable to H^{-1}, placed at a distance $ct_o/2$, have an angular circumference[13]

$\simeq 4\pi G\mu \simeq 2.6\mu_{-6}$ arc sec, and hence have a mean density of about one per $4\pi G\mu/.38 \simeq 3\times 10^{-5}\mu_{-6}$ sr $\simeq 400$ (arc min)$^2\mu_{-6}$. Any field larger than $20\mu_{-6}^{1/2}$ arc min on a side is thus almost certain to contain a loop within about 2000 h^{-1} Mpc. A square field $\theta_f \gtrsim 20$ arc min on a side in general contains a spectrum of loops, the largest of which however is still usually smaller ($\simeq 0.4\theta_f^2$ rad in circumference) than the field itself. A long strip of sky on the other hand in general intersects many loops larger than its thin dimension even though the length of string per area is the same as the square field. Exactly such a strip of deep digitized images is being created by Steward Observatory's CCD transit instrument (CTI).[17] The collection of data by this instrument is entirely automated, so the search for cosmic strings, like CTI searches for many other interesting phenomena, boils down to running various algorithms on an existing data base, with follow-up on ordinary telescopes. It is possible that CTI will collect enough images of distant galaxies by coadding data for a year or so that even with a non-detection of correlated image pairs to a specified level we will be able to put useful constraints on the string scenario. These constraints may be expected to be complementary to those already existing from microwave and gravitational-wave techniques.[18]

This work was supported by the Alfred P. Sloan Foundation and by NASA grant NAGW-763 at the University of Arizona.

REFERENCES

1. Zeldovich, Ya.B., Mon. Not. R. astr. Soc. 192, 663 (1980).
2. Vilenkin, A., Phys. Rev. Lett. 46, 1169 (1981); 46, 1496(E) (1981).
3. Turok, N. and Brandenberger, R.H., Phys. Rev. D. 33, 2175 (1986).
4. Turok, N., Phys. Rev. Lett. 55, 1801 (1985).
5. Vilenkin, A., Phys. Rep. 121, 263 (1985).
6. Vilenkin, A., "String Review", in *Inner Space/Outer Space*, Kolb, E.W. *et al.*, eds. (Chicago: Univeristy of Chicago Press, 1986), p. 269.
7. Hogan, C.J. and Rees, M.J., Nature 311, 109 (1984).
8. Kaiser, N. and Stebbins, A., Nature 310, 391 (1984).
9. Vilenkin, A., Phys. Rev. D. 24, 2082 (1981).
10. Gott, J.R., Astrophys. J. 288, 422 (1985).
11. Vilenkin, A., Nature 322, 613 (1986).
12. Foltz, C.B., Weymann, R.J., Röser, H.-J., and Chaffee, F.H., Astrophys. J. 281, L1 (1984).
13. Hogan, C. and Narayan, R., Mon. Not. Roy. astr. Soc. 211, 575 (1984).
14. Vilenkin, A., Astrophys. J. 282, L51 (1984).
15. Paczynski, B., Nature 319, 567 (1986).
16. Turner, E.L., "Gravitational lenses and dark matter: observations" (preprint, Princeton University Observatory), 1986.
17. McGraw, J.T., Stockman, H.S., Angel, J.R.P., Epps, H. and Williams, J.T., Opt. Eng. 23, 210 (1984).
18. This article was distilled from a contribution to the Proceedings of the "Quarks and Galaxies" workshop, Lawrence Berkeley Labs, August 1986.

MICRO-ARCSECOND RESOLUTION OF QUASAR STRUCTURE WITH ISOLATED MICROLENSES

Robert J. Nemiroff

Department of Astronomy and Astrophysics
University of Pennsylvania

ABSTRACT

Detection and analysis of lensing of quasars by isolated stars in nearby galaxies could tell us about internal structure of quasars, kinematics of lens motions, and the density of lenses.

When a star passes in front of a distant quasar, it does not block out an insignificant fraction of the quasar light. Quite the contrary: it can significantly amplify the light from the quasar. Some recent estimates show that such a situation is even a common occurrence[1-2]. If these lens induced light variations can be distinguished from intrinsic quasar light variations, the resulting analysis could provide much insight into quasar structure and lens motions[3].

The scenario assumed here is simplistic: a single star amplifies a blackbody continuum. These assumptions can be defended as reasonable approximations, although it is not the purpose of this paper to do so. Analysis of actual microlensing detection will also be a test of these assumptions.

To extract information, one should first note the time scale of the variation. Consider a lensing scenario wherein the background source is brightened about one magnitude by a single stellar lens, and assume the source is a featureless blackbody continuum. Note the two times on the light curves when the increased source flux is about 0.3 magnitudes brighter than the original baseline light level of the quasar. At this magnitude the stellar lens is, angularly, about one Einstein ring unit (ERU, the angular radius of the ring of light that would be seen by the observer at perfect observer-lens-source alignment) from the center of the source. Therefore the time between these two events is a measure of the relative velocity of the lens across the source, in ERU per time unit.

If the top of the light curve is flat, the lens is passing directly in front of the source and accurate measure of source size is possible. Locate the inflection points of the light curve. These give a measure of source size that is invariant to the maximum lensing amplification, the impact parameter of the lens to the center of the source, and the time base of the light curve. In fact, the inflection points are a direct measure of source size, but they are in time units. To convert source size to ERU, multiply by the velocity derived above. The source size measured by the inflection points is actually the length the lens moved across the source.

If the light curve is not flat on the top, the lens probably did not pass directly in front of the source. It is, then, usually a good approximation to consider the source to be a point. For this case, the maximum lensing amplification is a measure of the impact parameter: the distance between the lens and the source in ERU. In this way, by identifying the maximum lensing amplification of microlensing events, it is possible to estimate a lower limit on the number density of lensing stars along the line of sight to the quasar.

A continual tally of possible maximum lensing amplifications is itself useful, for these unobserved values can rule out stars passing closer than a given impact parameter[4]. In this way, non-detection of microlensing gives an upper limit on the number density of lensing stars along the light path to the quasar.

Information on quasar structure need not come only from photometric observations. Spectroscopic analysis can be quite useful as well. Specifically, the shapes of emission lines originating in the broad line emission region (BLR) can be changed by microlensing[4–5]. If there is a redshift gradient across the BLR, as is predicted from many hypothesizes of the kinematics of this region, microlensing can amplify specific redshift regions more dramatically than others. The result would be a change in the shape of the spectral lines as the lens moves. Such a line shift could provide information about both the structure and the kinematics of the inner quasar regions.

The angular size scale that microlensing brings to the observer is information at 10^{-6} arcseconds. This scale is independent of the wavelength of the radiation being analysed. No current telescope can approach this limit. The technique rests on the fundamental idea that photometric and spectroscopic analysis can give rise to increased effective angular resolution, and this idea is by no means new (as seen in close binary star analysis[6–7]. Microlensing can be a powerful tool to the informed observer.

References

1. Chang, K. and Refsdal, *Nature* **282**, 561 (1979).
2. Nemiroff, R. J., *Ap. Sp. Sci.* **123**, 381 (1986).
3. Refsdal, S., elsewhere in this issue (1987).
4. Canizares, C., *Ap. J.* **263**, 508 (1982).
5. Nemiroff, R. J., *Ph. D. Thesis, University of Pennsylvania* (1987).
6. Goodricke, J., *Phil. Trans. R. Soc. London* **73**, 474 (1783).
7. Russell, H. N., *Ap. J.*, **35** (1912).

ACTIVE GALACTIC NUCLEI

Martin J. Rees
Institute of Astronomy
Madingley Road
Cambridge CB3 OHA
England

ABSTRACT

Some aspects of the physics of jets and the central energy source are briefly reviewed; some comments are made on the evolution of AGNs, and the nature (and possible detectability) of the host galaxies of quasars with $z \gtrsim 2$.

1. INTRODUCTION

The implications of active galactic nuclei for theorists were becoming recognised at the time of the first Texas conference held in 1963. As Tommy Gold said in his banquet speech on that occasion, they implied that "relativists and their sophisticated work are not only magnificent cultural ornaments, but might actually be useful to science". In the time since then, an enormous body of data has accumulated, in all wavebands, about the various classes of objects that astronomers call quasars, radio galaxies, Seyfert galaxies, and so on. These phenomena seem related to one another, though it is not yet clear exactly how. They involve a range of physical processes in regions of dimensions extending up to the scale of the giant double radio galaxies, of order a megaparsec. The power derives from some central concentration of energy; and one belief on which the consensus has been steady since 1963 is that the energy is gravitational in origin. The prime reason for invoking collapsed objects is that any supermassive gravitationally-powered source which releases more than 1 percent of its rest mass energy contracts unstoppably; and a collapsed object, once formed, offers a more powerful and efficient power source than any precursor system. So if the power is gravitational in origin, it is inconsistent not to envisage that collapsed objects power the most powerful active nuclei— quasars and radio galaxies in particular. This paper is intended merely as an introduction to the more specialised contributions which follow it.

2. DOUBLE RADIO SOURCES AND JETS

I shall be discussing mainly the primary energy production, but let us start by considering the larger-scale phenomena – the double radio sources mapped so beautifully with the VLA. Extensive discussions available elsewhere [1] will not be repeated in the present written version of my talk. Over a hundred of these sources are found to display jets connecting them with the central nucleus. Computer simulations of jet propagation are proceeding apace, and some fascinating results have been reported at this meeting. It is now possible to include the effects of magnetic stresses, and to do genuinely three-dimensional simulations where axisymmetry is not artificially imposed. Such studies will help us to understand how stable the jets are, and whether they are confined by magnetic fields or by the pressure of an external cocoon.

Few entirely uncontroversial statements about jets can be made, other than simply stating that they somehow transport energy outwards. However, there is in my view an increasingly compelling case that in strong sources jets propagate with speeds of order c all the way out to the hot spots in the extended lobes. The propagation involves a high Mach number, permitting almost loss-free energy transport. The beams may consist of electron-positron plasma, and the apparent one-sidedness may arise from Doppler favouritism, rather than being intrinsic. Lower powered sources involve jets which are slower moving, having suffered more entrainment and dissipation.

Only a small fraction of AGNs display these jets. It could be that they are only generated under special circumstances. Alternatively, they may in many cases be rapidly stopped, failing to propagate effectively through the interstellar and intergalactic medium. There is evidence that large scale jets require a host galaxy whose potential well is pervaded by high-pressure hot gas. Big jets are primarily found in ellipticals or in interacting systems.

Even though they may be manifest in just a small subset of AGNs, jets pose a distinct problem: how can the power emerge primarily in this low-entropy form? The problem is posed especially by strong radio galaxies such as Cygnus-A, in which the kinetic energy transported by the jet exceeds the direct radiative luminosity of the galactic nucleus. The central mass must be at least $\sim 10^{8}\ M_{\odot}$, because of the huge overall energy content of extended sources. So the nuclear luminosity is far below the Eddington limit, suggesting a low accretion rate. How, then, can an intense relativistic outflow be generated, when $\dot{M}$ is low, especially in view of the fact that inflowing gas may then be unable to cool on the inflow timescale, so the efficiency of accretion may be low? The answer is that the jet production mechanism may be tapping the latent spin energy stored in the hole.

The possibility of tapping the rotational energy of a Kerr black hole was first discussed by Penrose [2]. Blandford and Znajek [3] showed how this could happen realistically, by exploiting the analogy between a Kerr black hole and a spinning conductor, and follow-up work [4,5] has led to a specific model for strong radio sources. This process requires a magnetic field threading the hole; a current system flowing through the hole; and, for maximal power dissipation in the external medium, a suitable impedance match between the resistivities of the hole and of its surroundings.

A small amount of plasma around the hole could carry currents sufficient to maintain a field pervading the ergosphere. Even though a low density plasma does not radiate very efficiently, it produces a few gamma-rays by bremsstrahlung and related processes. Collisions between such photons near the hole could create enough electron-positron pairs to supply a current flowing into the hole. (Indeed, the charge density would everywhere be high enough for relativistic MHD to apply, in the sense that charge neutrality is approximately preserved.) Electron-positron pairs moving with Lorentz factors of up to 100 would transport kinetic energy outwards, but most of the power outflow would initially be in the form of Poynting flux associated with the magnetic field coiled around the jet axis, and frozen in to the pair plasma. This Poynting flux may be converted into fast particles where the jet encounters ambient material, perhaps on the scale of the VLBI radio components. The expected magnetic field in the jet has just the kind of configuration that could cause magnetic confinement and collimation. The plasma around the hole that supplies the currents and anchors the field is just a catalyst. The power output of a source like Cygnus A could in principle be sustained with zero accretion rate if some of the hole's spin energy were channeled into the surrounding plasma to compensate for its small radiative losses.

Radio galaxies, according to this idea, harbour massive black holes which may, for most of their lifetime, be quiescent because they are not surrounded by plasma. If some event such as interaction with a companion were to trigger renewed infall (maybe at a low rate but sufficient to reactivate the nucleus by applying a magnetic field), the "clutch" is engaged, enabling the hole's spin energy to be converted into non-thermal directed outflow. The ejecta may plough their way out to scales 10^{10} times larger. If this is indeed what happens in Cygnus A and M87, then these very large scale manifestations of AGN activity may offer the most direct evidence for inherently relativistic effects.

3. GENERAL PHYSICAL CONSTRAINTS ON ELECTROMAGNETIC RADIATION FROM COMPACT OBJECTS

There are two quite distinct ways in which massive black holes can generate a high luminosity: straightforwardly by accretion, or via the electromagnetic process just described, where the power comes from the hole itself. The latter process tends to give purely non-thermal phenomena, whereas accretion yields an uncertain mixture of thermal and non-thermal power. The properties of an AGN must depend, among other things, on the relative contributions of these two mechanisms.

Irrespective of the details of the primary power production, there are simple thermodynamic constraints on the forms of radiation that could emerge directly from within a few gravitational radii. Figure 1 illustrates these for a typical quasar with $M = 10^8 M_\odot$, $L = 10^{46}$ ergs per second, and dimensions 10^{14} cms. If a luminosity L emerged with a black body spectrum, the temperature would be about 3×10^5 K, the flux therefore peaking in the far ultraviolet or soft X-ray region. A similar luminosity in harder X-rays could be generated via an optically thin thermal mechanism, e.g. Comptonized bremsstrahlung. Even were the radiation non-thermal, involving relativistic electrons with an effective kT up to $\sim 10^{12}$ K, synchrotron self-absorption produces a cutoff in the infrared, implying, of course, that no radio emission can come from dimensions 10^{14} cms unless a coherent mechanism is operative. The pair-production constraint, further discussed in Lightman's paper, prevents more than about 1 percent of L from emerging as γ-rays with energies above 1 Mev.

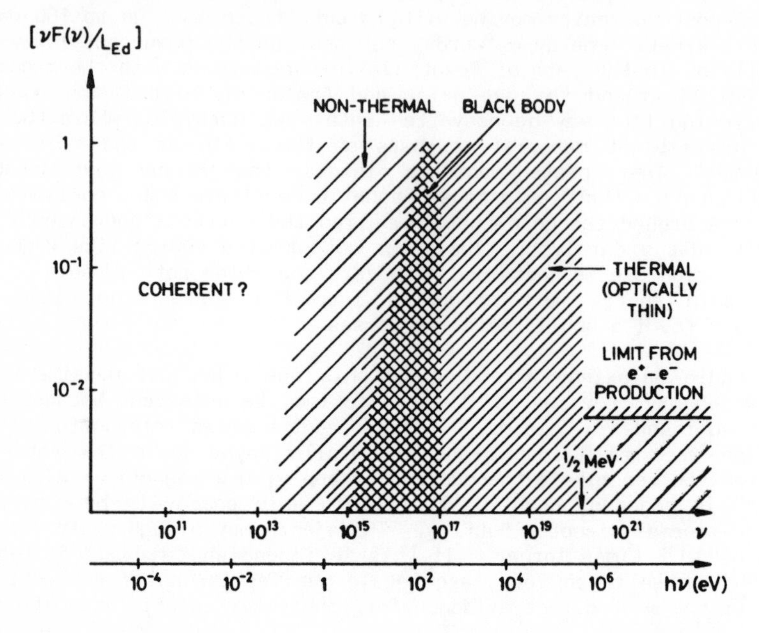

Figure 1. This diagram illustrates schematically the thermodynamic constraints on the radiation spectrum that could emerge directly from a source of luminosity $L_{Ed} \cong 10^{46}$ erg s^{-1} and dimensions $\sim 10^{14}$cm. Black body radiation with $T_{bb} \cong 3 \times 10^5$ K (heaviest shading) would peak in the UV. Optically thin thermal radiation from plasma with $T > T_{bb}$ could emerge in the X-ray band, but absorption would still prevent a high fraction of the radiation from emerging in the optical band even though the Rayleigh-Jeans Law would be higher than for the black body by a factor T/T_{bb} (intermediate shading). Synchrotron self-absorption yields a cut-off in the infra-red, so incoherent non-thermal radiation is restricted to the part of the diagram that is lightly shaded. (This is calculated assuming a magnetic field strength $\sim 10^4$ G, the equipartition value expected in radial accretion flow). Absorption of γ-rays via $\gamma + \gamma \to e^+ + e^-$ limits the luminosity above 1 Mev to $\lesssim 0.01\ L_{Ed}$. The main direct output from the central source would therefore be: (i) Thermal radiation in the UV or X-ray bands, (ii) non- thermal radiation, anywhere from the near infra-red to hard X-rays (but few γ-rays) and (iii) a relativistic $e^+ - e^-$ pair outflow.

The primary energy production must therefore be thermal radiation in the UV or X-rays, non-thermal radiation anywhere from the near-infrared up to hard X-rays (but rather few gamma-rays), and perhaps a relativistic electron-positron outflow. These simple considerations tell us immediately, even if we did not have other evidence, that much of the radiation we actually observe must be generated or reprocessed further out, in a manner depending on the environment within the host galaxy.

An AGN is too complicated, and involves far too many parameters, to be quantitatively modelled all in one go. However, most investigators agree on a general "scenario", which suggests what the essential ingredients are, and what physical processes merit detailed investigation by theorists. These certainly include photoionization equilibrium, relativistic plasmas, gas dynamics around Kerr holes, etc. The emission line region is perhaps the most extensively studied aspect of AGNs. The physics is standard and well-posed, but there is still uncertainty about the dynamics and the geometric configuration of the emitting gas. Time does not permit me to enter this subject here, but it is reviewed in the accompanying paper by Ferland.

In modelling the primary central source, one should ideally study the dynamics and radiative properties of a realistic fluid of relativistic plasma near a black hole. That is however too ambitious at the moment, and work has proceeded in a more piecemeal way, as reported in several contributions to the Workshop on AGNs. Unsteady flow patterns around black holes, the stability of accretion flows, etc, are being studied using the same programs already applied to jets. These do not yet include realistic dissipation or radiation processes, but offer clues to the likely flow patterns. In practice, the flow is likely to be unsteady on all timescales from the minimum dynamic timescale up to the overall lifetime of the system.

Complementary to this is the study of radiative processes in plasmas where many Mevs of energy are available per particle, and where electron-positron pairs are consequently important. Such work was reviewed by Lightman. In compact non-thermal sources, most gamma-rays will undergo pair production before escape, the resultant pair plasma having optical depth greater than unity (see Figure 2). The resultant spectrum is determined by two effects: firstly, the steepening of the slope of relativistic particles where the secondaries are taken into account; and, secondly, Comptonization of photons by subrelativistic pairs pervading the source. One would like to know to what extent this process can account for the observed X-ray spectral indices. Pairs also lower the effective Eddington limit, with the result that the relative importance of radiation pressure and gravity can be close to unity even for luminosities only 10 percent of the classical Eddington limit. A general result of these studies is to emphasize that any region containing a high concentration of non-thermally produced photons (for instance the magnetosphere of a spinning black hole generating a powerful jet) cannot be empty or transparent. Even if no ordinary matter is present, pair creation will be sufficiently prolific that the region "fogs up" and all emergent radiation is reprocessed.

Much of the photon emission from AGNs, when displayed in a $\nu F(\nu)$ plot, appears to be in the far-ultraviolet or X-ray part of the spectrum. This radiation could originate (either thermally or

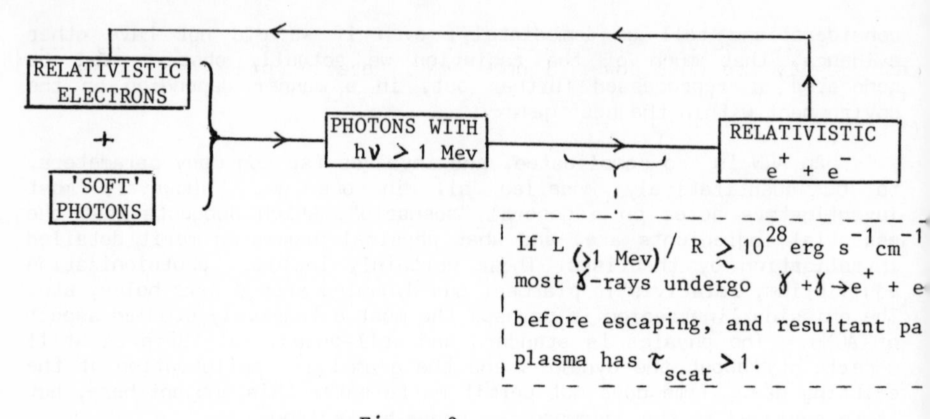

Figure 2

non-thermally) close to the hole. Alternatively, some could come from shocks associated with the line-emitting clouds, or even from non-thermal radiation located in the VLBI radio blobs. Increasing evidence of rapid variability implies that the bulk of the X-rays come from a compact region. There is evidence for both a power law hard X-ray component, and also a soft component (in some cases also variable) which could be the high frequency part of an optically thick XUV bump. The hard component is correlated with radio power, in the sense that it is stronger in radio AGNs than in radio quiet objects of the same optical power, suggesting that it is emitted non-thermally.

Conditions around black holes are extreme by astrophysical standards. The relevant basic physics, though, is well known — and certainly far from extreme by the standards of this conference. Moreover, the key problem is at least well-posed: plasma dynamics in a specified gravitational field, the aim being to calculate how much power is derived from accretion and how much is extracted from the hole's spin, and to find in what form these respective contributions emerge. Such calculations play the same role in the modelling of AGNs that nuclear physics does in theories of stellar structure and evolution. The confrontation of models with observations, indirect even for stars, is still more ambiguous for AGNs. This is because in stars the energy percolates to the observable surface in a relatively steady and well understood way. But in AGNs it is reprocessed into all parts of the electromagnetic spectrum on scales spanning many powers of ten, in a fashion depending on poorly-known environmental and geometrical factors within the host galaxy. But the generic black hole model is not infinitely flexible. It could be refuted in at least three ways. 1) By finding very regular periodicities, particularly on timescales below 1 hour; 2) By showing that the central masses were much less than 10^6 $M_\odot$ in Seyferts or much less than 10^8 $M_\odot$ in radio galaxies; or 3) By developing a theory of gravity more convincing than relativity which prohibits black holes.

Note, however, as a cautionary remark, that some AGNs, though not necessarily the most powerful ones, may represent precursor stages on the route to black holes, such as dense star clusters, "starbursts" or supermassive spinning stars.

Our theoretical scenario for the central energy source in AGNs suggests that there are a variety of channels for the power output. These can be illustrated in a flow diagram (Figure 3), which indicates that the relative importance of thermal and non-thermal processes depends on uncertain physics, but that the most purely non-thermal phenomena may be preferentially associated with starved holes. The diagram refers to reprocessing that happens close to the hole. Further reprocessing may occur on all scales from 10 up to as much as 10^{10} gravitational radii, in a manner sensitive to galactic environment.

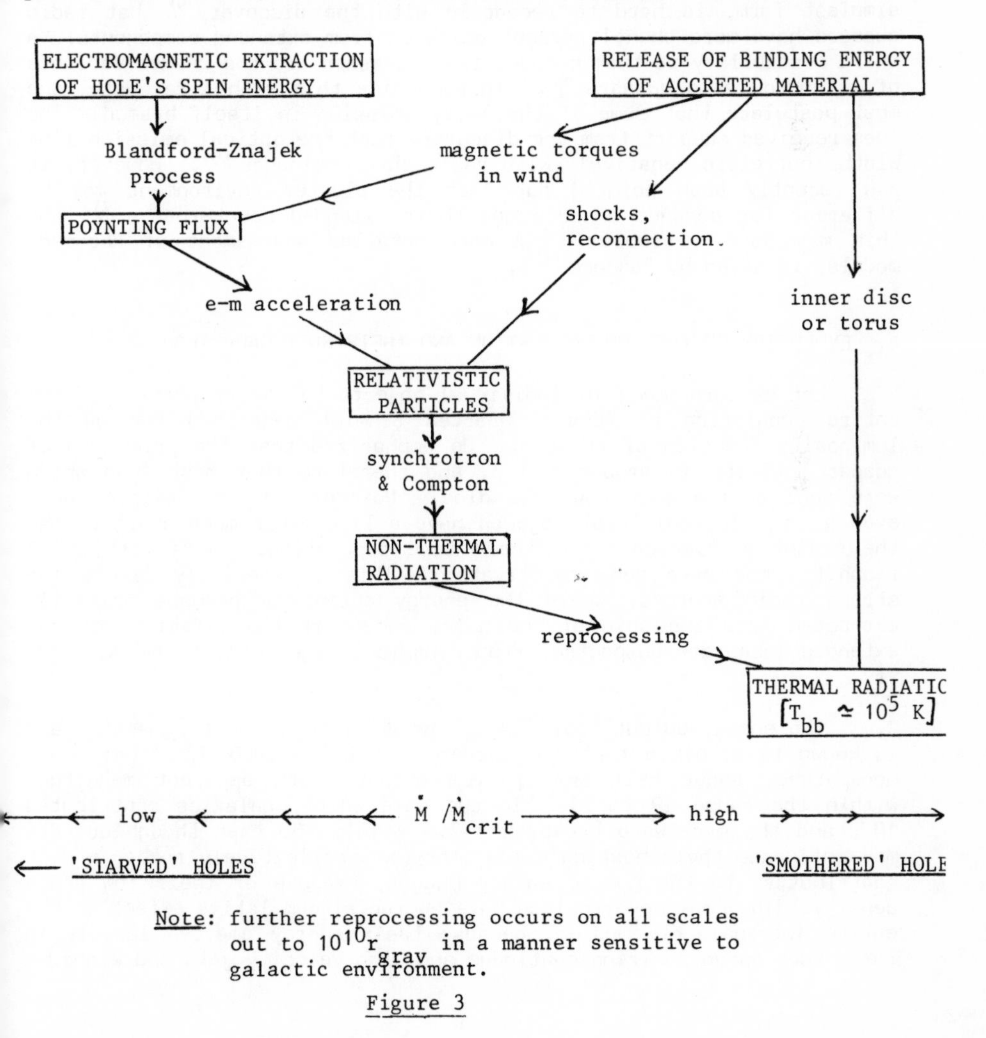

Figure 3

Apart from M, $\dot{M}$, and the nature of the host galaxy, many authors have speculated that the observed properties of an AGN may be sensitive to orientation. There is good evidence for relativistic beaming on scales of a few parsecs, from the superluminal motions revealed by VLBI. Moreover, even without relativistic outflow, non-spherical geometry could cause optially thick discs to appear brighter when observed face on, as well as absorption in larger scale discs. At least three types of partially unified models for AGNs have been proposed, involving beaming. The first such suggestion was that blazars represented extended radio galaxies with the jets pointing towards us [6]. This idea has received recent support from the low level activity observed in some nearby galaxies, e.g. Centaurus A. A more ambitious idea was that compact radio quasars differed from radio quiet quasars only in being beamed towards us [7,8]. This theory, in its simplest form, is hard to reconcile with the discovery [9] that radio quasars have more than 1 percent emission from extended components. A third possibility is that compact radio quasars are the beamed variants of extended radio quasars [10]. To reconcile this with X-ray data, one must postulate that some of the X-ray emission is itself beamed. The idea received support from the discovery that the optical emission line widths correlate negatively with radio core dominance [11]. However, it has recently been pointed out that the cluster environment may be different for compact sources and their extended counterparts [12], and this may pose a problem. (A more detailed assessment of "beamed" models, is given by Peacock [13]).

4. EVOLUTION OF THE AGN POPULATiON AND THEIR HOST GALAXIES

Let us turn now from individual objects to the properties of the entire population of AGNs. Maarten Schmidt described for us the luminosity function of quasars. He emphasized that the prime era of quasar activity is around z = 2, and therefore that most AGNs which were once active must now have died. However, it is possible that, even at z = 2, individual objects have a life cycle much shorter than the evolution timescale for the entire population. AGNs with small redshifts may have been reactivated. This is specially likely for strong radio sources, where the energy output is perhaps primarily extracted from the spin of the hole, and where the lifetimes of the extended lobes are suspected, from dynamical arguments, to be only $\sim 10^7$ yrs.

The energy output from AGNs is predominantly due to quasars, and is known to within a factor of order 2. It is $\sim 3{,}000\ M_\odot c^2$ per cubic megaparsec, about half coming from quasars with apparent magnitude within the range 19 to 21. In the same units, galaxies contributed 10^5, and the microwave background $\sim 7.5 \times 10^6$. So even though quasars may influence their host galaxies, they are collectively rather modest contributors to the cosmic energy budget, because of their low space density. (They may nevertheless have a crucial cumulative effect on the entire intergalactic medium, because their energy emerges largely in forms such as an ionizing continuum and high velocity jets and winds.)

At z = 2, the population decays on a timescale of roughly $t_{Evo} \simeq 2 \times 10^9$ years. But what is the lifetime t_Q of each object? The table below distinguishes two possibilities. In the first, one supposes that there is only one generation of quasars. As an alternative, one might imagine a shorter active lifetime, say 2 percent of the evolution timescale. The resultant masses are of course much greater in the first case, and this has a number of consequences for other aspects of the study of AGNs, such as the relative importance of radiation pressure and gravity, and the dynamics of the line-emitting clouds. Moreover, it has important implications for the kind of quasar remnants that might lurk in galaxies.

(i)	(ii)
$t_Q \cong t_{Evo}$	$t_Q \cong 4 \times 10^7$ yrs $\cong 0.02\ t_{Evo}$
$M = 2.5 \times 10^9\ \varepsilon_{0.1}\ M_\odot$	$M = 5 \times 10^7\ \varepsilon_{0.1}\ M_\odot$
$L << L_{Ed}$	$L \simeq L_{Ed}\ \varepsilon_{0.1}$
Broad-line regions gravitationally bound	Broad-line region <u>not</u> gravitationally bound
<u>Very</u> massive remnants in $\sim$ 2% of galaxies	$\sim 10^8\ M_\odot$ remnants in most bright galaxies

Evidence for a massive concentration of dark matter in the centre of M87 has been around for a number of years [14]. More recently [15,16], more modest central masses have been inferred from the study of stellar motions near the centre of M32 and NGC 3115. Even more interesting, the stars near the centre of M31 seem to be rotating around a mass concentration of about 2 x 10 M for which M/L must exceed 35 solar units [16]. The number and mas distribution of black holes must be determined primarily by phenomena at the epoch z = 2, where most energy was radiated. Although the M87 data suggests that monster black holes may grow within some galaxies (e.g. the central elliptical galaxies in clusters, perhaps as a result of cooling flows) a large proportion of AGNs may well involve lower-mass holes in a larger and more varied population of galaxies. The tentative indications of black holes in nearby galaxies would, moreover, suggest that our own galaxy would be underprivileged were it not endowed with a central mass of a few million solar masses.

What would the universe have been like at the epoch corresponding to z = 2, when it was perhaps 20 percent of its present age? There are growing reasons for believing that it would indeed have looked very different, and that galaxy formation was still going on at that time. This reappraisal comes from evidence that dark halos around galaxies extend out to 100 kpc. If galaxies extended only to 10 kpc, they could have formed on a timescale of $\sim 10^8$ years at $z \gtrsim 10$. But it is different if their halos indeed extend out beyond $\sim$100 kpc, as is suggested both by the dark matter evidence, and also by the idea that the discs of galaxies formed from tidally torqued material falling in conserving its angular momentum [17]. The infall time is of order 10^9 years, implying that infall from the edge of the halo, and disc formation, cannot be completed until the age of the universe is 2×10^9 years.

Are the AGNs different at z = 2, as well as merely more numerous? Only in the radio do we have real spatial information, and the evidence now suggests yes. Barthel and Miley[18,19] find that high redshift sources, whether associated with quasars or with AGNs, are more distorted than local sources of similarly high power. This suggests that the medium surrounding them is more disturbed. It might also mean that these sources involve interacting galaxies. Studies of smaller redshift objects suggest that the rarer renewed activity in nuclei at recent epochs is triggered by close encounters or mergers, though we do not fully understand why.

The considerations based on extensive massive halos do not rule out the possibility that galaxy formation started very early. All they imply is that it cannot be over and done with by z = 2. Any further statements become more model-dependent. However computations by Frenk et al.[20] based on the cold dark matter cosmogony suggest a drastically different cosmic scene at z = 2. Typical galaxies would then have been smaller, and most of the large present-day galaxies would have experienced a merger since that epoch. Moreover, most galaxies at z = 2 or 3 would be experiencing an interaction of some kind.

"Fuzz" has been seen around many quasars of modest redshift, and is interpreted as evidence of the host galaxy, or perhaps an associated cooling flow. At z = 3, physically similar fuzz would not be detectable, at least without ST resolution. However if an AGN were to light up within a galaxy where $\sim 10^{11}$ $M_{\odot}$ of gas were falling inward from a hundred kiloparsecs, then much of the quasar UV could be scattered or reprocessed in this material, yielding a detectable narrow emission line. Quantifying this depends on knowledge of uncertain geometry, covering factors, etc. However, even existing limits on fuzz may imply interesting constraints on galaxy formation.[21] It would be unfortunate if experimental searches for fuzz focussed only on low redshift objects, on the alleged grounds that there was only then a significant chance of a positive result.

Quite apart from the issue of whether the environment in a young galaxy may modify the appearance of a central AGN, observations confront us with the question of whether some objects interpreted as quasars could instead be young galaxies, in the sense that their luminous output is not dominated by a non-stellar nucleus, but comes from a population of young stars and gas spread through a region several kiloparsecs in size. This could be settled in favour of the quasar interpretation in individual instances if high polarization or rapid variability were discovered. But the typical quasar possesses neither of these attributes, and we cannot be sure that its optical spectra will be any different from that of the primordial galaxy unless and until we know what line widths to expect for the latter.

Maarten Schmidt mentioned phenomenological models for the evolution of the AGN population with redshift. One would like to interpret this evolution more physically, placing it within the context of ordinary galactic evolution. To do this we need to know the time dependence of at least four things. 1) The number of galaxies already formed. 2) The rate at which new galaxies form. 3) The merger rate, and 4) The gas supply, which is gradually depleted by star formation and stripping, but in some cases enhanced by cooling flows. We need also to understand how long it takes to accumulate a large compact mass at the centre of a galaxy.

The progress so far made with AGNs brings a new set of problems into sharper focus. By what route did the black holes first form? What is their relation to their host galaxies? All the issues of cosmogony and extragalactic astrophysics are so interdependent that we will make no progress with any unless we advance on a broad front.

REFERENCES

1) Begelman, M.C., Blandford, R.D. and Rees, M.J., Rev.Mod.Phys. 56, 255 (1984).
2) Penrose, R. Rivista Nuova Cim. 1, 252 (1969).
3) Blandford, R.D. and Znajek, R.L. MNRAS, 179, 433 (1977).
4) Rees, M.J., Begelman, M.C., Blandford, R.D. and Phinney, E.S. Nature 295, 17 (1982).
5) Phinney, E.S. Cambridge Ph.D. Thesis (1983).
6) Blandford, R.D. and Rees, M.J. on "Pittsburgh Conference on BL Lac Objects" ed. A.M. Wolfe p. 328 (1977).
7) Scheuer, P.A.G. and Readhead, A.C.S. Nature 277, 182 (1979).
8) Blandford, R.D. and Konigl, A. Astrophys.J. 232, 34 (1979).
9) Schilizzi, R.T. and de Bruyn, A.G. Nature, 303, 26 (1983).
10) Orr, M.J.L. and Browne, I.W.A. MNRAS 200, 1067 (1982).
11) Wills, B.J. and Browne, I.W.A. Astrophys.J. 302, 56 (1986).
12) Prestage, R.M. and Peacock, J.A. MNRAS (submitted).
13) Peacock, J.A. in "Neutron Stars, AGNs and Jets" ed. W. Kundt (Reidel, in press).
14) Sargent, W.L.W., Young, P.J., Boksenberg, A., Shortridge, K. Lynds, C.R. and Hartwick, F.D.A. Astrophys.J. 221, 731 (1978)
15) Tonry, J.L. Astrophys.J. (Lett) 283, L.27 (1984).
16) Kormendy, J. in "Structure of Elliptical Galaxies" ed. T. de Zeeuw (Reidel, in press).
17) Fall, S.M. and Efstathiou, G. MNRAS, 193, 189 (1980).
18) Barthel, P. Leiden Ph.D. Thesis (1984).
19) Barthel, P. and Miley, G.K. in preparation.
20) Frenk, C.S., White, S.D.M., Efstathiou, G. and Davis, M. Nature 317, 59 (1985).
21) Rees, M.J. MNRAS (submitted).

THE BROAD LINE REGION OF ACTIVE GALACTIC NUCLEI

G.J. Ferland
Astronomy Department
The Ohio State University

Properly understood, the emission line regions of AGNs could tell us of the luminosity of the central engine and chemical composition of the emitting gas. This understanding has been slow in coming because of the complexity of the microphysics governing these regions, and more importantly because of the very poor geometrical and spatial information on the BLR. In lieu of very high spatial resolution HST data, we must proceed by inference from self-consistent global models of the BLR.

The most popular model which has been examined in detail is one in which small BLR clouds (column density $N_H \sim 10^{23}$ cm^{-1}) are in pressure equilibrium with a surrounding medium at the Compton temperature of the AGN radiation field (see, Krolik McKee and Tarter, Ap. J. 249, 422). W.G. Mathews and I (Ap. J. submitted) have examined some consequences of this model, in view of the present understanding of the form of the radio - x-ray radiation field. We find that the Compton temperature of this field is low: Typical values are near 10^7K. This low temperature has several consequences, which together rule out Compton equilibrium as the agent governing the hot inter cloud medium HIM (if it exists at all). The problems are the following: The low temperature requires that the HIM have a high density, as a result the photoelectric opacity (especially the K- shell of Fe) is large. The low temperature also requires that the BLR clouds be entrained in the HIM (otherwise they

JETS IN LUMINOUS EXTRAGALACTIC RADIO SOURCES

Jack O. Burns and David A. Clarke
Institute for Astrophysics
The University of New Mexico
Albuquerque, NM 87131

Michael L. Norman
National Center for Supercomputer Applications
The University of Illinois
Champaign, IL 61820

ABSTRACT

Observations and numerical simulations of jets in classical doubles are briefly reviewed. Particular emphasis is placed on the solution of the jet confinement problem recently demonstrated in both fluid and MHD models.

1. INTRODUCTION

Although hundreds of radio jets have now been found in active galaxies and quasars, relatively few have been detected in the most luminous classical double sources (Bridle and Perley, 1984). This is due to an observational selection effect. Because of the high surface brightness of the hot spots in the radio lobes, high dynamic range is required on interferometer maps in order to detect even the most luminous jets. However, recent advances in computer post-processing (e.g., self-calibration) have allowed us to "dig out" such jets (e.g., Cyg A in Perley et al., 1984). There now exists a large enough (although incomplete) sample of luminous source jets from which we can glean some of their basic properties.

Similar recent advances in supercomputer numerical modeling of jets have allowed us to understand some of the basic physics that produces the wealth of jet morphologies. Such models have solved at least one outstanding problem, namely the confinement of "overpressured" jets. Our intention in this paper is to briefly review some of the recent exciting developments in both the observations and models of jets in classical doubles.

2. JET STATISTICS

2.1 Probability of Jet Detection

The integrated luminosity of a radio jet is apparently directly related to the strength of the radio core (Burns et al., 1984; Bridle, 1984). Some 50% to possibly 100% of classical doubles with strong cores (~ 1% of peak surface brightness of lobes) have radio jets (generally QSOs) . Whereas, jets in weak core 3CR galaxies have been detected only 5% of the time on VLA maps (Laing, 1985).

The correlation between core and jet luminosity is a surprising one. There are numerous physical processes that could alter the luminosity of a jet (e.g., entrainment, turbulence, particle acceleration) once it leaves the galaxy core and interacts with the surrounding medium. This correlation suggests that some preserved memory of the engine in the galaxy nucleus remains even at large scales in spite of the interaction. Thus, there is the hope that large scale jet structures could be used to constrain the nature of the active galaxy engine.

2.2 Jet Sidedness

Current observations indicate that nearly all jets in classical doubles are one-sided. The ratio of jet to the limits on counterjet brightness range from >2 to 100's to 1. This contrasts markedly with lower luminosity sources which typically have jet to counterjet ratios of 1-10. Recently, several possible examples of counterjets in classical doubles have emerged from high contrast VLA maps. The best example is 3C 219 which has a single, slightly resolved knot on the counterjet side (Bridle et al., 1986). Another possible example is 3C 334 which has a thin arc of emission emerging from the lobe and pointing back toward the nucleus (Bridle et al., 1987).

The origin of the jet one-sidedness remains controversial. Two models have been proposed. First, classical doubles may intrinsically possess two jets moving at bulk relativistic velocities inclined at some angle to the line of sight. The surface brightness of the jet pointing toward the observer will be Doppler boosted and the jet pointing away will be Doppler diminished. Some support for this model comes from the possible detection of counterjets noted above and the recent discovery of superluminal motion in the cores of 3 classical doubles (3C 179, 3C 245, 3C 263; Hough, 1986). On the other hand, the limits on the jet to counterjet ratios for samples of extended QSOs appear incompatible with differential Doppler boosting (Wardle and Potash, 1984). Second, the jets may be intrinsically one-sided with the plasma beams flipping from side to side on a time scale less than the age of the lobes (Rudnick and Edgar, 1984). Possible support for this model comes from the observed asymmetry of lobe hot spots where the lobe on the jet side generally appears more compact (Laing, 1985). One can only differentiate between these models with very deep VLA maps of a relatively large sample. Such observations are in progress (Bridle et al., 1987).

3. CONFINEMENT OF LUMINOUS JETS

The very small opening angles (several degrees) of jets in classical doubles strongly suggest that they are not freely expanding. However, these jets are overpressured (measured between knots) with respect to the external thermal pressure by a factor of 10 on average (Burns et al., 1984). How are these jets confined? Within the last year, two solutions to this problem have been explored.

The first is what we term a fluid solution since it involves only hydrodynamics (Norman and Burns, 1987; Begelman and Cioffi, 1987). The luminosity of a jet can be shown to be proportional to $M_j^3 \eta^{-1/2}$, where M_j is the Mach number of the jet measured relative to the external sound speed and η is the ratio of jet to external medium density. Luminous jets imply large M_j and small η. We performed a simulation with an initial $M_j = 100$ (which quickly dropped to 10 caused by shocking at the inlet) and $\eta = .001$ shown in Fig. 1. The jet propagates through an external medium with a z^{-2} decline in pressure. Because of the hypersonic motion, a bow shock with a small Mach cone angle is produced. Note that the cocoon and bow shock extend all the way back to the nozzel of the jet. Behind the strong bow shock, the cocoon and jet are overpressured with respect to the external medium by $\lesssim M_j^2$. This overpressure ratio can be as large as 10^4, easily satisfying the observational constraints.

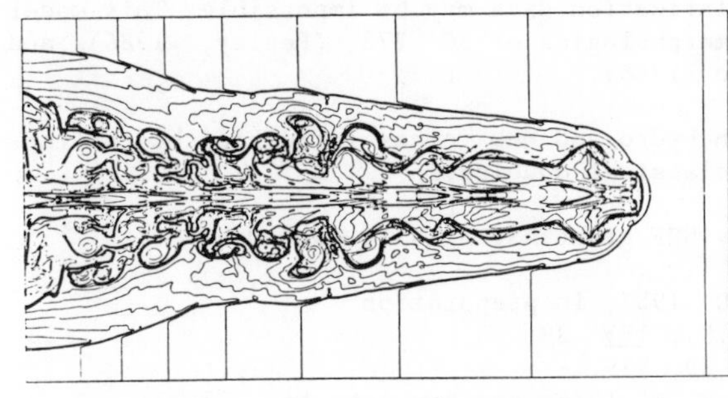

Fig. 1. Density isocontours of an axisymmetric hypersonic jet in a plane stratified atmosphere. The computational domain is 100 x 30 jet radii. Note the highly oblique bow shock which envelopes and confines the jet.

The second model is termed an MHD solution since it involves a jet with a dynamically dominant magnetic field (Clarke, Norman, and Burns, 1986). In analogy to laboratory charged particle beams, a current-carrying jet will generate a self-pinching azimuthal B-field (Benford, 1978). So far, there have been no direct detections of such a confining field in the polarization data of jet cocoons. But, as shown below, this may not be surprising. We performed a series of two-dimensional MHD simulations of jets in cylindrical symmetry with thermal to magnetic pressures of 0.2, $\eta = 0.1$, and $M_j = 6$ (with respect to jet sound speed). The magnetic field was entirely azimuthal throughout the simulation. As shown in Fig. 2, the following characteristics of the MHD model emerge:

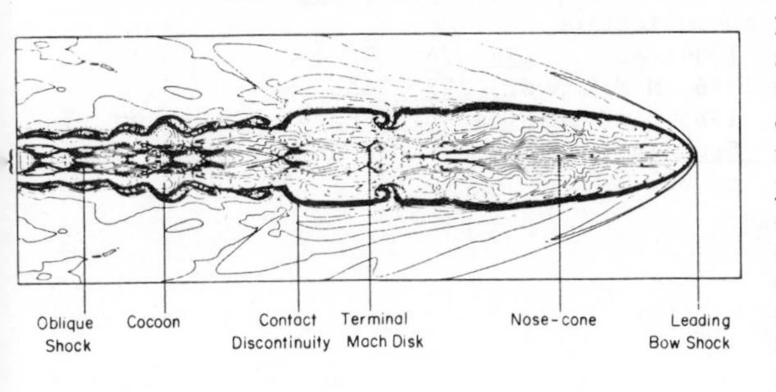

Fig. 2. This figure shows the density contours of a magnetically confined jet. The region shown is 20 by 6 jet radii. All along the axis of the nose-cone, and behind the criss-cross shocks, the jet is overpressured by a factor of 20 or more relative to the ambient pressure.

a) The self-pinching JxB force maintains a strongly collimated jet and produces reflection-mode shocks (X-shaped features in Fig. 2).
b) A Mach disk is formed at approximately the same location as in a purely hydro jet.
c) A new feature, that we call a nose-cone, is seen at the head of the jet. It is the product of the JxB Lorentz forces which confine the jet material between the Mach disk and the bow shock that would otherwise escape into the cocoon.

d) As a consequence of (c), there is no bulk return of jet material back down the jet in the form of a cocoon. What would have been cocoon plasma in the hydro simulation is nose-cone material in the MHD model. Therefore, the jet is quite "naked", and emission from the cocoon region is expected to be weak. Hence, detection of an azimuthal field from the polarization data may be impossible. This model appears to reproduce the morphologies of 3C 273 (Perley, 1986) and 0800+608 (Shone and Browne, 1986).

It is likely that both hydro and MHD forces influence the confinement of luminous jets in classical doubles.

This work was supported by NSF grant AST-8314558 to J.O.B.

Begelman, M. and Cioffi, D. 1987, in preparation.
Benford, G. 1978, M.N.R.A.S., 183, 29.
Bridle, A. H. 1984, A.J., 89, 979.
Bridle, A. H. and Perley, R. A. 1984, Ann.Rev.Astr.Ap., 22, 319.
Bridle, A., Perley, R., and Henriksen, R. 1986, A.J., 92, 534.
Bridle, A., Browne, I., Burns, J., Dreher, J., Hough, D., Laing, R., Owen, F., Readhead, A., Scheuer, P., and Wardle, J. 1987, in preparation.
Burns, J., Basart, J., De Young, D., and Ghiglia, D. 1984, Ap.J., 283, 515.
Clarke, D., Norman, M., and Burns, J. 1986, Ap.J. Lett., 311, L63.
Hough, D. 1986, Ph.D. Dissertation, Cal Tech.
Laing, R. 1985 in Physics of Energy Transport in Extragalactic Radio Sources, ed. A. Bridle and J. Eilek, (Proc. NRAO Workshop No. 9), p. 128.
Norman, M. and Burns, J. 1987, in preparation.
Perley, R., Dreher, J., and Cowan, J. 1984, Ap.J., 285, L35.
Perley, R. 1986, private communication.
Rudnick, L. and Edgar, B. 1984, Ap.J., 279, 74.
Shone, D. and Browne, I. 1986, M.N.R.A.S., 222, 365.
Wardle, J. and Potash, R. 1985 in Physics of Energy Transport in Extragalactic Radio Sources, ed. A. Bridle and J. Eilek (Proc. NRAO Workshop No. 9), p. 30.

SS433

J. I. Katz
Department of Physics and McDonnell Center for the Space Sciences,
Washington University, St. Louis, Missouri 63130

ABSTRACT

I discuss some controversies regarding SS433 and its relation to other astronomical jets, and offer some suggestions for further research.

On December 12-13, 1986 a conference "SS433 and Jets," complementing this workshop "AGN's and Jets," was held at Washington University. It is not possible in this space to summarize the 27 papers presented there, so I will only mention a few outstanding issues. There is no unanimity on any aspect of SS433; I will variously take the consensus, majority, or my personal view, with apologies to those who disagree or are ignored. Names in parentheses are citations of oral presentations.

X-ray observations (M. Watson) of the Ly α line of Fe XXVI establish that acceleration to the terminal velocity of $v/c = 0.26$ occurs very close to the central object, where even iron is nearly completely stripped. Direct geometrical arguments based on disc occultation limit the acceleration region to 10^{12} cm; comparison to observations of W50 implies that jet structure extends over a range of 10^8 in linear scale. Energetic considerations suggest the accelerating region may be as small as 10^7 cm. The X-ray continuum appears to be of thermal origin (disputed by D. Band).

Line-locking (M. Milgrom, P. Shapiro) is a very attractive explanation of the observed constancy of the jet speed, and of the observation that species from H I to Fe XXVI show the same speed. Line-locking requires strong clumping or a superabundance of the accelerating species (now believed to be Fe XXVI). It implies a very low mass efflux rate, corresponding to a jet power $\sim 10^{36}$ erg/sec.

The greatest quantitative uncertainty in SS433 is the mass efflux rate in its jets, which is conveniently expressed in terms of their power. Reported nuclear γ–ray lines led to extremely high values, but further observations (B. Geldzahler), laboratory experiments (E. Norman) and energetic arguments (J. Brown) discredit them. A naive interpretation (one recombination per proton) of the observed H lines leads to an estimate of 10^{42} erg/sec. A similarly naive interpretation (no reheating) of the X-ray emission leads to 10^{39} erg/sec, as does the assumption that the surrounding nebula W50 is powered by the jets. Line-locking models and radiation-hydrodynamic calculations of jet acceleration by supercritical accretion discs (G. Eggum) lead to 10^{36} erg/sec.

It may be possible to reconcile the lower estimates of jet mass efflux with the observed line strengths if the matter is sufficiently clumped. In the hydrogen line emission region clumping factors of 10^{8} may be required. Possible clumping mechanisms include thermal instability (G. Bodo) and radiation pressure driven instability (J. Katz), but it is unclear how well these work in practice.

The 164 day period is generally considered to represent jet and disc precession (disputed by W. Kundt), but its mechanism is controversial. R. Romani discussed the possibility that SS433 is triple, with an inner companion orbiting the jet source in a precessing orbit. This appears to predicts a stable 164 day period, but that period is known to be unstable (B. Margon). The more familiar suggestion that the disc is slaved to the binary companion, whose spin is misaligned and precessing, is inconsistent with calculations which suggest that the precession period is too long and its damping too rapid. The disc itself may undergo driven precession, but this must be reconciled with its apparent rapid transport of angular momentum. Precession self-excited by oblique shadowing of the companion has been suggested, and is consistent with the ubiquity of disc precession.

The discovery of additional objects like SS433 will greatly aid our understanding. If they all have $v/c = 0.26$ then line-locking will be inescapable, and if their jets all precess some universally excited mechanism must be found. Galactic interstellar extinction makes it more promising to search for objects in nearby external galaxies. These may have $m_v \sim 18$ and a soft X-ray flux $\sim 10^{-11}$ erg/cm^2/sec, if a high X-ray luminosity is broadly directed along the jets (possibly explaining W50) (J. Katz).

Is SS433 related to bipolar nebulae (C. Lada) or to extragalactic radio jets and double-lobed radio sources? The bipolars and SS433 are both thermal phenomena, and may have similar hydrodynamics. The radio sources are nonthermal, and may exceed in power any thermal phenomena in their parent galaxies. In contrast, evidence suggests that the jets carry only a minor part of the luminosity of SS433. The radio sources may be a very different phenomenon, with only the jet geometry, determined by an angular momentum axis, in common with SS433.

We thank the NSF and the McDonnell Center for the Space Sciences for support of the meeting "SS433 and Jets."

Effects of Active Galactic Nuclei on the nature of host galaxies

Smita Shanbhag and Ajit Kembhavi

Tata Institute of Fundamental Research
Homi Bhabha Road, Bombay 400 005, India

It has recently been suggested by Begelman[1] that high energy radiation from an active galactic nucleus can, through compton heating, evacuate the hot interstellar medium (ISM) in the host galaxy. We examine the evacuation scenario in the light of realistic initial mass functions (IMF), star formation histories and recent results on quasar X-ray spectra.

Begelman[1] has estimated the flux of the ISM out of the host galaxy, due to compton heating by high energy photons, using the formalism of Begelman, Mckee and Shields[2]. He compares this with the mass input into the galaxy by stars by virtue of stellar winds, and finds that the ISM can be evaluated to a distance of $\sim$ several KPC. The mass injection is estimated using a formula due to Tinsley[3], which is applicable to the situation where stars in the galaxy are at the turnoff point in the HR diagram. However, Seyfert galaxies, as well as the hosts of the majority of quasars are believed to be spirals. The stellar population in the disks of these galaxies is very different from the evolved population considered by Tinsley. Massive stars are continuously formed in disks, and the mass input due to these can be quite different from that due low mass stars. Moreover the star formation rate in the host galaxies could have been higher in the past, leading to a time dependence in the evacuation radius. These factors warrant a more thorough consideration of Begelman's scenario. We report some preliminary steps in this direction, concentrating on the mass injection into the galaxy, while accepting Begelman's[1] results for mass outflow due to heating.

We have used the spiral galaxy initial mass function as described by Miller and Scals[4], together with the main sequence corrections given by Scalo[5]. Values for stellar mass loss rates are taken from various sources[6,7,8]. For the time variation of the IMF we have used the maximally decreasing star formation rate of Miller and Scalo[4]. It should be noted that this history is consistent with the SFR in the Solar neighbourhood at the present time. The histories relevant to active galaxies could be quite different, involving a much higher SFR in the past. Using these data, and assuming an exponential stellar density profile for the disks of spirals, we have obtained the mass injection rates as a function of the distance from the galactic centre. The results are shown in Fig.1 for various values of galaxy redshift (which determines the age of the galaxy, assuming all galaxies were formed 12 Gyr ago). Also shown in the diagram is the mass loss rate, due to comptor heating, obtained using Begelman's[1] equation 25. Using these to curves (or solving the equation "mass injection = mass outflow" numerically), we obtain the evacuation radius as a function of redshift. This is shown in Fig.2. Recent results on quasar X-ray spectra[9], indicate that quasar X-ray luminosities could be very much smaller than the value $Lx \sim 10^{46}$ $ergsec^{-1}$ used in deriving the mass outflow rate. The result of using $Lx = 10^{44}$ $ergsec^{-1}$ is shown in both figs 1&2. The value for the evacuation radius we obtain for Z=0 is a factor 2 less that of Begelman for $Lx=10^{46}$ $ergsec^{-1}$. For highe redshift and lower X-ray luminosity the evacuation radius is substantially smaller.

References

1. Begelman, M.C., 1985, Ap.J. 297, 492
2. Begelman, M.C., Mckee, C.F., and Schields, G.A., 1983, Ap.J.271,7
3. Tinsley, B.M., 1980, Fund. Cosmic Phys. 5, 287.
4. Miller, G.E., and Scalo, J.M., 1979, Ap.J.Supp.41,513
5. Scalo, J.M., 1986, Fund. Cosmic Phys. 11, 1
6. Proceedings ofIAU 66, 1973, Ed. Tayler, R.J.
7. Proceeding of IAU Collq.59, 1980, Ed. Chiosi, C., and Stalio, R.
8. Proceedings of IAU 83, 1978, Ed. Conti, P.J., and DeLoose, W.H.
9. Elvis, M. et.al., 1986, Center for Astrophysics Preprints No. 2311, 2342.

Fig 1

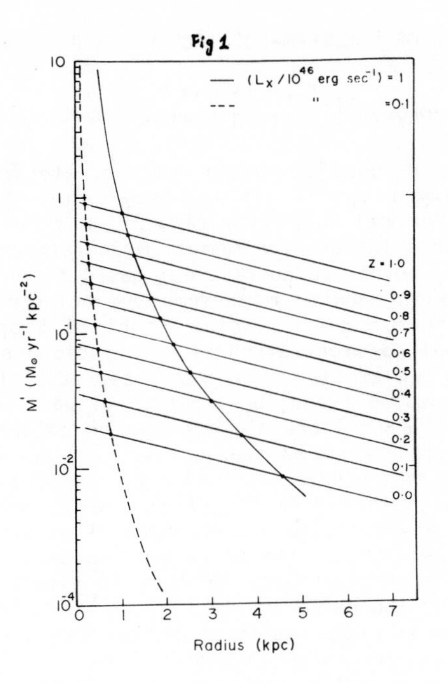

— (L_X/10^{46} erg sec^{-1}) = 1
--- " =0·1
Z = 1·0
0·9
0·8
0·7
0·6
0·5
0·4
0·3
0·2
0·1
0·0
10
1
10^{-1}
10^{-2}
10^{-4}
M' (M☉ yr^{-1} kpc^{-2})
0 1 2 3 4 5 6 7
Radius (kpc)

Fig. 2

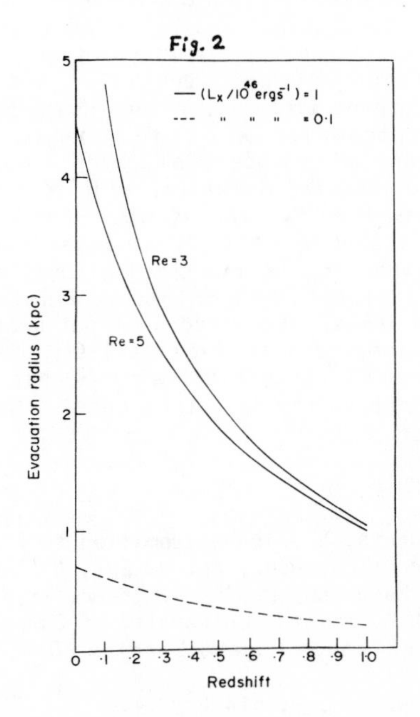

— (L_X/10^{46} ergs^{-1}) = 1
--- " " " = 0·1
Re = 3
Re = 5
5
4
3
2
1
Evacuation radius (kpc)
0 ·1 ·2 ·3 ·4 ·5 ·6 ·7 ·8 ·9 1·0
Redshift

EFFECTS OF ELECTRON-POSITRON PAIRS ON ACCRETION FLOWS

S. Tsuruta and B. Tritz
Department of Physics, Montana State University, Bozeman, Montana

We investigated possible importance of electron-positron (e^-e^+) pairs on accretion flows[1]. It was shown that the pair effects are negligible in spherical accretion flows[2], but the situation could be quite different when the angular momentum is introduced[3][4].

As a starting point, we carried out basic structure calculations of accretion flows with angular momentum, by including the e^-e^+ pairs in the two-temperature accretion disk model of Shapiro, Lightman and Eardley[5]. To treat comptonization of soft photons and calculate the e^-e^+ pair densities, we adopted the general approach of Zdziaski, with the best available cross sections for various relevant processes involved[6]. The work of Takahara and Tsuruta[7] was used to calculate the soft photon source. Our results are summarized below.

Our most important finding is that a two-temperature torus collapses to a (geometrically) thin disk (or ring), inside a critical radius r_{cr}. This is caused by the catastrophic enhancement of proton-electron energy transfer rates within r_{cr} when pairs are included. This happens when the rates of energy transer from protons to electrons through Coulomb interactions exceed the rates at which the gravitational energy released through accretion is supplied to protons. The size of the collapsed thin disk is larger for larger accretion rates $\dot{M}$. It may be noted that by applying a more analytic approach to quasi-spherical inflows with angular momentum, Begelman, Sikora and Rees[8] also reached a similar conclusion.

The effects of pairs on the structure of accretion flows, e.g., the distribution of proton and pair densities and temperatures, are investigated. They depend strongly on the input parameters, such as $\dot{M}$ and the viscosity parameter α. In general, the effects of e^-e^+ pairs are significant with moderately high $\dot{M}$ and in the presence of substantial amount of angular momentum, with $\alpha < \sim 0.5$. For a typical example of $M = 10^7 M_\odot$, $\dot{M} = 1 M_\odot$/year, $\alpha = 0.1$, radiation efficiency $\epsilon_\gamma = 0.1$ and thermal efficiency $\epsilon_T = 0.25$, we found that the pairs dominate inside $r_p \sim 13r_g$ (where r_g is the gravitational radius), and the two-temperature torus collapses to a thin disk within $r_{cr} \sim 10r_g$. Within the two-temperature torus, the effects of pairs on the distribution of electron and proton temperatures are relatively small, and the electron temperature stays near 10^9 K within the torus, but its inward decline is slightly steeper (than in the pair free case). More complete reports will be found in Ref. 1).

REFERENCES

1) Tritz, B. and Tsuruta, S., to be submitted to Ap. J. (1987).
2) Schultz, A.L. and Price, R.H., Ap. J. 291, 1 (1985).
3) Takahara, F. and Kusunose, M., Prog. Theor. Phys. 73, 1390 (1985).
4) Phinney, E.S., Ph.D. thesis, University of Cambridge (1983).
5) Shapiro, S.L., Lightman, A.P. and Eardley, D.M., Ap. J. 204, 187 (1976).
6) Zdziaski, A.A., Ap. J. 289, 514 (1985).
7) Takahara, F. and Tsuruta, S., Prog. Theor. Phys., 67, 485 (1982).
8) Begelman, M.C., Sikora, M. and Rees, M.J., submitted to Ap.J. (1986).

NUCLEOSYNTHESIS IN THE THICK ACCRETION DISKS

Sandip K. Chakrabarti

Theoretical Astrophysics,
California Inst. Of Technology, Pasadena.

ABSTRACT

We briefly examine the possibilities of the nucleosynthesis in thick radiation supported disks formed around black holes.

Recently, through detailed numerical simulations Chakrabarti, Jin and Arnett(1987); and Jin, Arnett and Chakrabarti (1987) have suggested that in an wide range of the parameter space as spanned by the accretion rate, mass of the hole and the viscosity parameter, it is possible to have a considerable amount of nucleosynthesis within the matter accreting into the hole. These calculations are carried out using an approximate model of the thick disk (Chakrabarti, 1985) surrounding a black hole. Figure 1 shows the thermodynamic quantities such as the density and temperature at the center of a typical disk for various values of β, the ratio of gas pressure to radiation pressure. Also shown is the variation of the accretion rate (measured in the unit of the critical accretion rate) when α is kept fixed at $\alpha = 10^{-6}$ at the center of the disk. The α parameter prescription is adopted, though α is not constant throughout the disk; it rises outward very rapidly consistent with the assumption of a constant accretion rate and the isentropic disk condition. The diagram clearly shows that for a given viscous mechanism and accretion rate, higher temperature is attainable at the center of the disk for a lighter mass hole.

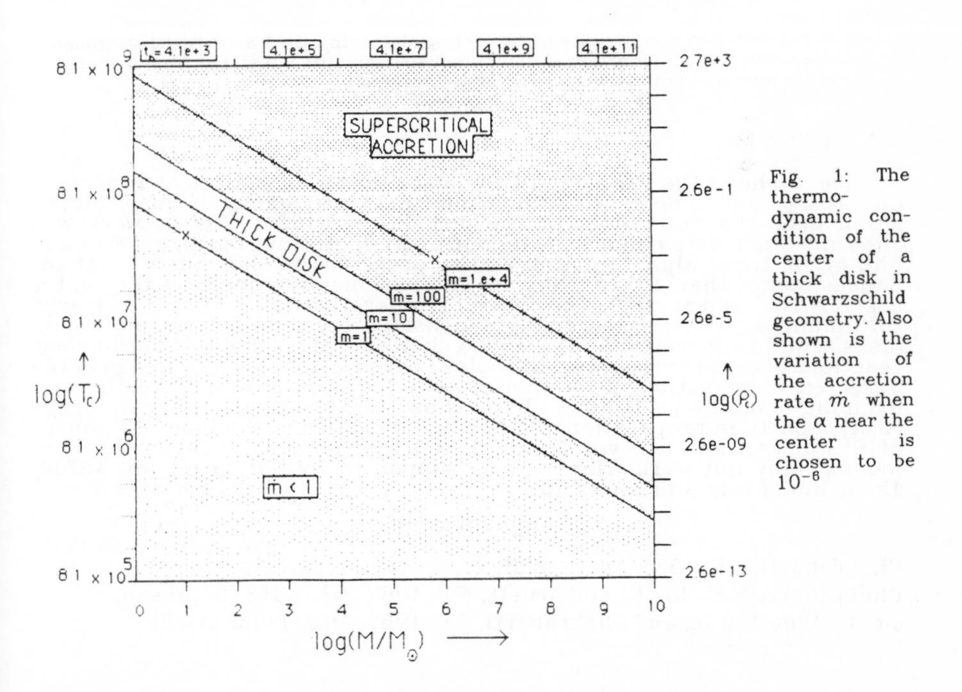

Fig. 1: The thermodynamic condition of the center of a thick disk in Schwarzschild geometry. Also shown is the variation of the accretion rate $\dot{m}$ when the α near the center is chosen to be 10^{-6}

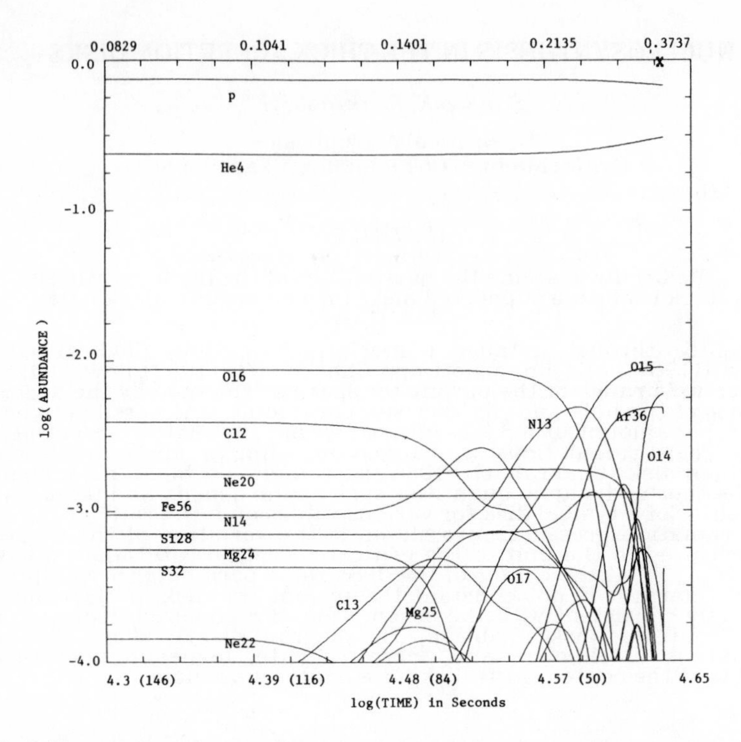

Fig. 2: The variation of composition of accreting matter in a thick accretion disk around a $10M_{\odot}$ black hole as a function of time (bracketed quantity is distance from the hole). The α parameter near the center is 10^{-5} and the accretion rate $\dot{m} = 10$.

Fig. 2 shows the variation of the composition of matter as it accrete on a $10M_{\odot}$ central hole. The simulation starts with matter of solar composition near the outer edge of the disk (at $r = 1000$). The network was chosen to include elements participating in deuterium burning, PP chain, CNO cycle, triple alpha reaction and rp process. The elements with abundances higher than 10^{-4} are drawn. The parameters chosen at the center are: $\alpha = 10^{-5}$ and $\dot{m} = 10$. The central temperature is $T_9 = 0.3737$. Clearly, the abundance of matter changes considerably. Similar calculation in disks around supermassive holes also shows considerable thermo-nuclear reactions provided viscosity parameter is even lower. The knowledge of the observed composition of jet matter may help pin-pointing the thermodynamic condition inside a disk. Thus, for example, one can argue (Chakrabarti, in preparation) that Fe line observed in the blue component of SS433 can not be overabundant and mechanisms such as the line-locking may not work. This work is supported by NSF grant No. AST84-15355 and a Tolman Fellowship.

REFERENCES

Chakrabarti, S.K. 1985, *Ap. J.*, **288**, 1.

Chakrabarti, S.K., Jin, L., and Arnett, W.D. 1987, *AP.J.*, **313** (In press).

Jin, L., Arnett, W.D., and Chakrabarti, S.K. 1987, *Ap.J.* (submitted).

KELVIN-HELMHOLTZ INSTABILITY OF MAGNETIZED ROTATING JETS

G. Bodo[1], R. Rosner[2], A. Ferrari[3], E. Knobloch[4]

[1]*Osservatorio Astronomico di Torino, Italy*
[2]*Harvard-Smithsonian Center for Astrophysics, Cambridge, MA*
[3]*Istituto di Fisica Generale, Università di Torino, Italy*
[4]*Physics Department, UCB, Berkeley, CA*

DISCUSSION

We consider a rotating cylindrical jet with a longitudinal uniform magnetic field separated by a vortex sheet from an unmagnetized external medium at rest. From the linearized equations we can obtain the following dispersion relation for the axisymmetric modes[1—3]:

$$\nu\tau_i \frac{\phi_+^2 - M_A^2}{\phi_+^2/\alpha^2 R^2 - \left(\phi_+^2 - M_A^2\right)^2} \frac{J_0'(\tau_i)}{J_0(\tau_i)} = -\frac{\tau_e}{\phi^2} \frac{H_0^{(1)\prime}(\tau_e)}{H_0^{(1)}(\tau_e)} \quad ,$$

where

$$\tau_i^2 \equiv \alpha^2 \left[\frac{\left(\phi_+^2 - M_A^2\right)^2 - \phi_+^2/(\alpha^2 R^2)}{\phi_+^2(1 + M_A^2) - M_A^2} \right] \left[\frac{\phi_+^2 - 1}{\phi_+^2 - M_A^2} \right] \quad ,$$

$$\tau_e^2 \equiv \frac{\alpha^2}{\eta^2} [\phi^2 - \eta^2] \quad ,$$

are the non-dimensional radial wave numbers respectively in the internal and external regions, and $\alpha \equiv ka$ is the wave number along the beam axis, $M \equiv U/c_{si}$ is the Mach number, $M_A \equiv v_A/c_{si}$ is the Alfvenic Mach number, $1/R \equiv 2\Omega a/c_{si}$ is the nondimensional rotation frequency, $\eta \equiv c_{se}/c_{si}$ is the ratio of the external to the internal sound speeds, $\phi \equiv \omega/kc_{si}$ is the nondimensionalized phase velocity measured in the reference system of the external fluid, and $\phi_+ \equiv \phi - M$ (related to ϕ by a simple Doppler shift formula) is the corresponding phase velocity in the frame of the internal fluid.

Dispersion relation of this type are known to lead to a rich variety of unstable modes, ranging from the surface (ordinary) modes to the body (reflected) modes. A deeper insight into its mode structure can be obtained considering the energetics of the perturbed configuration. Analytical considerations and numerical analyses show that the time-averaged energy associated with a perturbation tends to be negative when the phase velocity $Re(\phi_+)$ in the proper frame is negative, that is for $\phi < M$[1]. By negative energy we mean that the total energy (zeroth order plus perturbation) is lowered by the presence of the perturbation. A global mode will be formed by the coupling of an internal to an external perturbation (with the appropriate matching conditions) and, since the frequencies seen in the two respective rest frames are related by a Doppler shift, the internal perturbation will couple, at different Mach numbers, to different types of external perturbations: namely, for $\phi < \eta$ it will couple to a surface wave, while for $\phi > \eta$ it will couple to an outward propagating body wave. In the first case we have a stable mode, in the second case, since energy can be extracted from a negative energy system, we have an unstable situation.

For each type of waves that the medium inside the jet can support we will therefore have one series of negative energy modes, in each series the different modes can be distinguished by the number of nodes that the perturbation has in the interior of the jet. These modes will become unstable at a critical Mach number, when they can couple to external propagating waves. More precisely, in the case of a non rotating magnetized jet, we can have slow and fast magnetosonic waves and therefore we have two series of unstable modes that we call slow and fast reflected modes[1]. The series of the slow reflected modes was not found in previous studies of the Kelvin-Helmholtz instability, in general the growth rate of these modes is much lower than that of the other modes, however there is a range of values of the Alfvenic Mach number for which it cannot be neglected.

The inclusion of rotation introduces a new type of waves that the medium can support and, therefore, a new series of unstable modes. In addition we note that the terms containing the rotation effect in the dispersion relation can be comparable to the other terms only for large wavelengths, and for these wavelengths the growth rate of the ordinary mode is lowered by the presence of rotation.

CONCLUSIONS

We have discussed the stability of a rotating magnetized jet. We have found that the mode structure is very complicated, and this has to be taken into account when considering shear profiles or when interpreting the results of numerical simulations. We have interpreted the reflected modes in terms of negative energy modes and we have found that when they are destabilized there is a transfer of energy from the jet to the external medium, this an be of importance in the interpretation of the numerical simulations[4], in which a large region around the jet is perturbed, with the formation of a cocoon. As a last point, we have found that rotation introduces a new series of unstable modes and has a stabilizing effect for the ordinary mode, at least in the axisymmetric case.

REFERENCES

[1]Bodo, G., Rosner, R., Ferrari, A., Knobloch, E., *M.N.R.A.S.*, submitted, (1987).
[2]Ferrari, A., Trussoni, E. and Zaninetti, L., *M.N.R.A.S.*, **196**, 1051, (1981).
[3]Payne, D.G., and Cohn, H., *Ap. J.*, **291**, 655, (1985).
[4]Norman, M.L., Smarr, L., and Winkler, K.-H.A., in *Numerical Astrophysics: A Meeting in Honor of James Wilson*, eds. J. Centrella, R. Bowers, and J. LeBlanc (Portola Valley: Jones and Bartlet), (1984).

JETS, GALACTIC HALOS, AND THE LINEAR SIZE-DISTANCE EFFECT IN RADIO GALAXIES

Paul J. Wiita

Department of Physics and Astronomy, Georgia State University
Atlanta, GA 30303-3083, USA

Gopal-Krishna

Radio Astronomy Group, Tata Institute of Fundamental Research
Post Box 1234, Indian Institute of Science, Bangalore 560 012, India

ABSTRACT

In our model a radio beam first propagates through the essentially power-law gaseous halo of a parent galaxy and then enters a much hotter and less dense intergalactic medium (IGM). These two media are taken to be pressure matched at their interface, so that the boundary moves closer to the center of the galaxy with increasing redshift because the IGM pressure rises much faster with z than should the galactic pressure. We adopt from the literature "best values" for jet parameters for a typical strong radio source, its associated X-ray emitting halo, and the IGM. Our model can explain both the observed linear sizes of nearby sources as well as their cosmological evolution. The assumed IGM is cosmologically important, with $\Omega_b \approx 0.25$.

1. Introduction

Most powerful (FR Type II) radio sources consist of a pair of radio lobes on either side of a massive elliptical galaxy or quasar, which are presumably energized by jets. The separation between the hot spots defines the overall linear extent of the radio galaxy, which for nearby powerful sources ($\langle z \rangle = 0.15$) has a median value of about 350-400 kpc if $H_0 = 50$ km/s/Mpc[1,2]. Recent studies have also shown that the linear size for samples of similar intrinsic radio luminosities decreases rapidly with cosmological redshift[1-3].

Meanwhile, *Einstein* X-ray observations have shown that hot gaseous halos or coronae are associated with the majority of massive early type galaxies, including those lying near the periphery of the Virgo cluster and relatively isolated large elliptical galaxies[4]. These types of galaxies typically spawn powerful double radio sources.

Further, a reanalysis of the 5-200 keV X-ray background emission favors the hypothesis that a hot IGM with density and temperature at the current epoch of $\sim 7 \cdot 10^{-7}$ cm^{-3} and ~ 18 keV, respectively, pervades the universe[5]. In this paper we argue that both the present epoch median linear size and its cosmological evolution can be understood in terms of a simple but reasonable model for jet propagation.

2. The Model

A jet with a relativistic equation of state and power Q is assumed to propagate at constant opening angle θ through a halo gas of declining density: $n(D) = n_o[1+(D/a)^2]^{-\delta}$, with the core radius $a \approx 2$ kpc, $n_o \approx 0.01 cm^{-3}$ and $\delta = 0.75$[4]. Our key assumption is that these galactic parameters do not change much for $z \lesssim 1$. When it reaches the pressure matched boundary with the IGM at $R_h(z)$ the jet is expected to flare[6] as it enters a hotter, but less dense, medium. Because the IGM density and temperature both rise rapidly with redshift we find $R_h(z) \approx 170(1+z)^{-5/2\delta}$kpc. Generalizing simple analytical models[7] we find that $D(t) = [(2-\delta)V_* D_*^{1-\delta} t]^{1/(2-\delta)}$ for $D < R_h$ while for $D > R_h$ $D(t) = \sqrt{2}\{V_* D_*^{1-\delta} R_h^{\delta}(z)t + R_h^2(z)[\frac{1}{2}-(2-\delta)^{-1}]\}^{1/2}$, where $D_* = 10$ kpc and V_*, the average advance rate of the head of the beam, is ~0.03c. All other quantities are related by: $a^{-\delta}(4KQ/\pi m\theta^2 c n_o)^{1/2} = V_* D_*^{1-\delta}$, with K a constant between 1 and 2 and m the mass of a proton.

After penetrating the IGM the forward progress of the jet slows because it flares; also, since the sound speed in the IGM is high [and goes as (1+z)] the head of the jet soon advances subsonically. Subsequently the jet is expected to lose its identity and the source should fade rapidly, even if nuclear activity continues.

3. Results

Using the above parameters we find that at z = .15, R_h= 107 kpc, and the median value of the lobe separation, ⟨2D⟩ = 400 kpc, while for z = 1.0, R_h = 17 kpc and ⟨2D⟩ = 60 kpc. These values for ⟨D⟩ are in excellent accord with observations[1-3], while alternative assumptions such as: (1) a continuing halo and no IGM, or (2) no flaring at the halo/IGM interface, both imply too large values of linear-size at all epochs[6]. The ratio of the sizes of the largest sources at both epochs is also in accord with our scenario. Further, it seems that at $z \approx 2$ most radio sources are very small[8]. This is expected from our model, although extrapolating the halo parameters to $z \approx 2$ is uncertain.

Since the IGM required in our picture is cosmologically interesting, with $0.22 < \Omega_{IGM} < 0.32$, this scenario clearly deserves further study. Better statistics on the distribution of sizes of radio sources at different epochs should be compared with more sophisticated models. *AXAF* measurements showing that X-ray halos around more distant galaxies were smaller would support our view.

References

1] Kapahi, V.K., Mon. Not. Roy. Astr. Soc. [MNRAS], **214**, 19P (1985).
2] Gopal-Krishna, Saripalli, L., Saikia, D.J. and Sramek, R.A. in *Quasars, IAU Symp. No.* **119**, eds. G. Swarup and V.K. Kapahi (Dordrecht: Reidel), p.193 (1986).
3] Eales, S.A., MNRAS, **217**, 179 (1985).
4] Forman, W., Jones, C. & Tucker, W., Astrophys. J., **293**, 102 (1985).
5] Guilbert, P.W. and Fabian, A.C., MNRAS, **220**, 439 (1986).
6] Gopal-Krishna and Wiita, P.J., MNRAS, in press (1987).
7] Scheuer, P.A.G., MNRAS, **166**, 513 (1974).
8] Barthel, P.D., *Quasars, IAU Symp. No.* **119**, *op cit.*, p.181 (1986).

Sco X-1
A Unique Laboratory for Testing the Dual-Beam Model of Double Radio Sources

by

Barry Geldzahler
Applied Research Corporation/Naval Research Laboratory

and

Ed Fomalont
National Radio Astronomy Observatory

Our studies of Sco X-1 have been motivated by a desire to gain insight into the physics of quasars and the interaction of beams with the ambient medium. Scaling laws have been proposed (e.g. Begelman, Blandford, and Rees 1986) that relate the physical phenomena in QSO's to stellar scale objects.

Sco X-1 has been considered as a "microquasar" because it is about 10**-6 in extent as extragalactic double radio sources, yet it has morphological (see Figure 1), spectral, and polarization characteristics similar to those seen in quasars. In addition, Sco X-1, too, is overly luminous when one looks at the statistics of X-ray sources. Because of the close proximity of Sco X-1 to the Earth, ~500 pc, the structural and intensity variability time scales of the extended features are much faster han in the extragalactic sources. This allows for a unique opportunity to study the effects of beams emanating from the core and their interactions with the surrounding medium.

The results from six years of VLA observations are presented in Geldzahler and Fomalont (1986). They fall into two catagories which we briefly summarize.

Proper Motions

1. All components of Sco X-1 have an absolute proper motion consistent with that of the optical star, 0.03 "/yr, pa 180 deg.
2. The NE lobe is moving radially away from the core at 0.015 "/yr (35 km/sec projected velocity). See Figure 2.
3. The relative proper motion of the SW hot spot at the outer edge of the SW lobe has larger errors because of its intrinsic faintness (~0.5mJy). Its motion with respect to the core is < 0.030 "/yr. This is the first evidence for motion of a hot spot relative to the core. See Figure 2.
4. The assumption of linear motion of all three components is consistent with the post-fit residuals. Any evidence of nonlinear motion is too sparse at present to make a definitive statement.

Flux Density Variations

1. The core can vary by an order of magnitude in a period in an hour.
2. The NE lobe varies by about 10% over a year or less.
3. The SW hot spot does not appear to vary within the errors, but percentage variations comparable to those in the NE lobe could be present and may be reflected in the actual values of the flux density.
4. The integrated flux density of the SW lobe varies by 30% over a year or less, and the variation is not localized to a hot spot.
5. The variations of the NE and SW lobes may be correlated.

These data are summarized in Table I. Note that this is the first evidence for flux density variation of extended radio lobes.

The new observational result we present at this meeting is a VLBI observation of the NE radio lobe. MkIII VLBI observations were carried out in April 1985 at 18 cm using 4 US antennas. The measured size of the NE lobe was about 7 mas, or 5 x 10**13 cm for an assumed distance of 500 pc.

The results in Geldzahler and Fomalont (1986) argue strongly against the model of Sco X-1 proposed by Achterberg et al. (1983) which involves magnetic focussing of the beams. The results of Geldzahler and Hertz (1987) suggest that the model of Sco X-1 being inside a supernova remnant proposed by Kundt and Gopal-Krishna (1984) is also incorrect. The only model offered so far which can accomodate all the aspects of the radio data is that of Blandford and Rees (1974): the dual-opposing beam model. To test this model, we have used equations that balance kinetic energy flow in the beam and the radiative energy loss from the hot spots and that balance the momentum of the flow with that of the working surface.

After some manipulation of the equations, described in Geldzahler and Fomalont (1986), we find that the density external to each radio lobe necessary to satisfy the initial conditions is N =30/Eb atoms/cm for the NE lobe and 0.03/Eb atoms/cm for the SW hot spot.

There are some problems, though, with the outcome of the calculations. The hot X-ray source should heat the interstellar medium so that at the distance of the lobes the density is about 0.01. This is consistent with the results for the SW lobe but not the NE. Also, the NE lobe still appears to be unresolved, and a smaller size will force N still farther out of line with expectations. The quantity E, the factor giving the efficiency of conversion of kinetic to radiant energy, may not be 100%, and the factor b=v/c may be << 1. Both of these factors could cause increases in the values of the ambient density. The distance to

Sco X-1 could be in error by a factor of 2. Finally, not all the kinetic energy may be transformed into radiant energy so that the hot spot does not reflect the site where the entire beam is stopped. This is analogous to a rock in the midst of a flowing river.

Conclusion

While other models have been proposed to explain the radio properties of Sco X-1, the dual-beam model is the most successful to date. The fact that the current form of the model leads to some uncomfortable numbers means that continuing studies on a variet of fronts will encourage the development of a more refined model.

References

Achterberg, A., Blandford, R.D., and Goldreich, P. 1983, Nature 304, 607.

Blandford, R.D., and Rees, M.J. 1974, MNRAS 169, 395.

Kundt, W. and Gopal-Krishna 1984, Astron. Astrophys. 136, 167.

Geldzahler, B.J. and Fomalont, E.B. 1986, Ap.J. 311, 806.

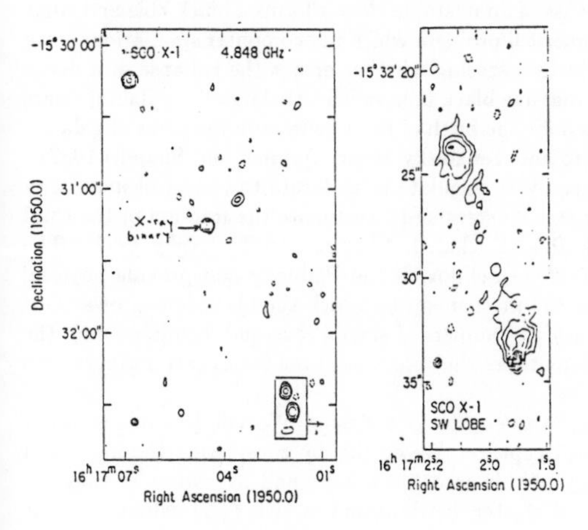

Table I

6cm Flux Densities of Sco X-1 Components (mJy)

Epoch	Integrated Core		Integrated NE Lobe		Peak SW Hot Spot		Integrated SW Lobe	
1980.108	16.3 ±	0.5	9.2 ±	0.4 ↑	--		↑ 7.9 ±	1.5
1981.098	10.21	0.15	8.74	0.28 ↓	0.45 +	0.07	↓ 4.56	0.21
1982.212	10.90	0.24	10.06	0.20 ↑	0.55	0.07	↑ 6.61	0.38
1983.910	1.05	0.09	9.38	0.12 ↓	0.40	0.04	↓ 6.40	0.15
1984.712	2.05	0.18	10.15	0.16 ↑	--		↑ 7.56	0.21
1984.964	21.32	0.44	10.47	0.22	0.53	0.04	4.72	0.20
1985.183	3.26	0.18	9.74	0.15 ↑	0.47	0.04	↑ 6.80	0.44

Figure 1. VLA maps of Sco X-1

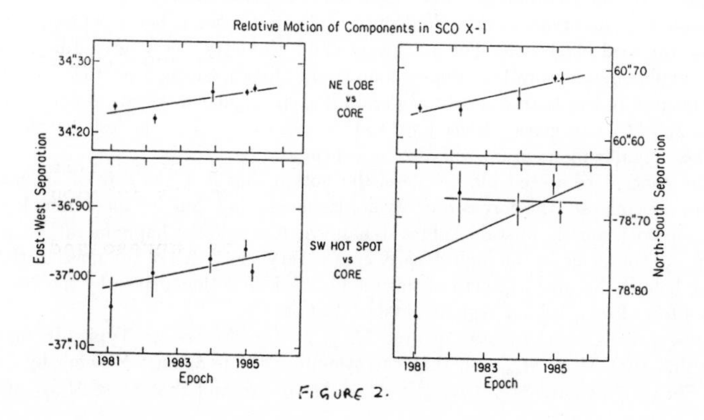

FIGURE 2.

The Size of Supermassive Black Holes Formed From Star Cluster Collapse

Christopher S. Kochanek

Theoretical Astrophysics, California Institute of Technology

Stuart L. Shapiro and Saul A. Teukolsky

Center for Radiophysics and Space Research, Cornell University

It has long been conjectured that an equilibrium sequence of spherical relativistic star clusters, parametrized by central redshift, becomes unstable to catastrophic collapse to a black hole beyond the first binding energy maximum. A sufficient condition was found by Ipser (1980), and perturbation calculations by Ipser, Thorne, and Fackerell (1968,1969ab,1970) found unstable modes past the maximum. Only recently have relativistic Vlasov simulations by Shapiro and Teukolsky (1985abc, 1986) determined both the stability of equilibrium clusters and, for those that are unstable, their final state. These calculations provide numerical support for the binding energy conjecture. In general, the final state of an unstable cluster just beyond the maximum consists of a central, massive black hole surrounded by a halo of orbiting stars.

Determining what fraction of the total mass of an unstable cluster forms a black hole and what fraction remains outside is a very difficult numerical problem which the computer simulations solve only for selected cases. The answer is crucial to assessing whether or not the collapse of a dense cluster of compact stars can produce a supermassive black hole sufficiently large to explain quasars and AGNs. Such a scenario has been proposed for the birth of these sources in the cores of galactic nuclei (Zel'dovich and Podurets 1965; Shapiro and Teukolsky 1985c; Quinlan and Shapiro 1987).

We have performed approximate, semi-analytic calculations to locate the onset of instability along an equilibrium sequence of relativistic star clusters and to estimate the fraction of the total cluster mass which goes into the final black hole (Kochanek, Shapiro and Teukolsky 1987). The calculations are based on the "avalanche effect" of Zel'dovich and Podurets and provide physical insight into the nature of the instability and the mechanism by which the black hole grows. Our analytic estimates are in good agreement with the numerical simulations and "explain" why the final black hole mass may be considerable (many times the core mass) even for an extreme core-halo cluster.

Consider a seed black hole of mass $M_{\rm seed}$ at the center of a cluster. We will determine a *lower limit* to the mass that the seed black hole will capture. We do this by considering the motion of particles in the *fixed* Schwarzschild geometry of this seed black hole, and by assuming that the ambient stars remain described by the original cluster distribution function $f(E)$, where E is the energy of a particle. An exact analysis would treat the full dynamical geometry of the evolving cluster, but our stationary approximations lead to the required lower limit.

We first estimate the minimum angular momentum per unit mass $\hat{L}$ below which a particle will be captured by the seed black hole. For a Schwarzschild geometry characterized by mass M, there is a critical angular momentum that depends on energy for capturing a particle. *All* bound particles will be captured in less than one orbital period if their angular momentum per unit mass is less than $\hat{L}_{crit} = 2\sqrt{3}M$. This gives a *lower limit* to the mass captured by the hole because some particles with higher angular momentum will also be captured.

Underlying the concept of a seed black hole is the notion that it is the *core* of an unstable cluster that triggers the collapse: the core is roughly homogeneous and thus collapses homologously when the cluster is gravitationally unstable. This dynamical core collapse happens rather quickly because the dynamical timescale in the high-density core is very short. All of the mass in the core collapses to a black hole. Subsequent growth of this seed black hole is then driven by the avalanche capture of particles from lower density regions outside the core.

For a *given* cluster, consider the mass captured M_{capt} by a seed mass M. Typically, for small $M < M_{thresh}$ we find that $M > M_{capt}$ so that the system is stable against growing by further capture of stars. For intermediate M, $M_{thresh} < M < M_{satr}$, we find that $M < M_{capt}$ and the hole is unstable to growing by capturing stars. For sufficiently large $M > M_{satr}$, once again the system is stable. Thus, along a *sequence* of equilibrium clusters, the onset of dynamical instability occurs when the core mass M_{core} equals M_{thresh}. Moreover, the black hole in unstable clusters will grow beyond the core up to a mass $M = M_{satr}$.

The mass M_{satr} is a lower bound on the final mass of the black hole in unstable clusters, and is in good agreement with the results of the numerical simulations. We find that unstable clusters that have nearly homogeneous profiles undergo total collapse, while those with extremely centrally condensed profiles form black holes with masses much less than the total cluster mass. However, even in the case of these extreme core-halo configurations the resulting black hole contains approximately ten times the core mass of the initial cluster.

References

Fackerell, E.D., 1970, *Ap. J.,* **160**, 859.

Ipser, J.R., 1980, *Ap. J.,* **238**, 1101.

————. , 1969a, *Ap. J.,* **156**, 509.

————. , 1969b, *Ap. J.,* **158**, 17.

Ipser, J.R., and Thorne, K.S., 1968, *Ap. J.,* **154**, 251.

Kochanek, C.S., Shapiro, S.L., and Teukolsky, S.A., 1987, *Ap. J., in press.*

Quinlan, G.D., and Shapiro, S.L., 1987, *Ap. J., in press.*

Shapiro, S.L., and Teukolsky, S.A., 1985a, *Ap. J.,* **298**, 34.

————. , 1985b, *Ap. J.,* **298**, 58.

————. , 1985c, *Ap. J. (Letters),* **292**, L41.

————. , 1986, *Ap. J.,* **307**, 575.

Zel'dovich, Ya. B., and Podurets, M.A., 1965, *Astron. Zh.,* **42**, 963. [English translation in *Sov. Astron.-A.J.*, **9**, 742.]

THE UV EXCESS OF QUASARS: LUMINOSITY DEPENDENCE

Amri Wandel

Center for Space Science and Astrophysics
Stanford University, ERL, Stanford, California 94305

ABSTRACT

The IR-to-near-UV continuum of quasars can be described roughly as a combination of a power-law and a "bump" or flatening at the near UV. Studying a complete set of ~100 quasars we find that the flattening, measured by the IR-to-blue spectral index (in the quasar's restframe) is strongly correlated with luminosity and redshift. This result is explained by a simple model consisting of an optically thick accretion-disc spectrum imposed on a nonthermal power-law. If quasars have characteristic efficiency ϵ and Eddington ratio L/L_E, then we find that the ratio between the thermal disc spectrum and the nonthermal power-law scales as $L^{1/3}$, and the data are fit by $L/L_E \sim 0.003 - 0.01(\epsilon/0.1)^{-1}$.

It is well established that the continuum spectrum of quasars and active galactic nuclei has approximatly a power-law form, $f(\nu) \propto \nu^{-\alpha}$, with a slope $\alpha \sim 1 - 1.2$ extending from the infrared to the UV (Neugebauer *et. al.* 1979, Malakan and Filippenko 1983). In the blue and near UV, however, there is an excess over the power law, which causes a flattening of the spectrum. This feature, which is sometimes refered to as the "Bump", is present in almost all quasar and AGN spectra. Blackbody (Shields 1978, Malkan and Sargent 1982), and accretion disc (Malkan 1983) spectra have been suggested to fit the bump.

We use the complete set of quasars from the Palomar Green survey, observed by Neugebauer *et. al.* (1986), and monitor the continuum flux at two points: blue (4200Å or log $\nu = 14.84$), and red (7500Å or log $\nu = 14.6$), which we denote by subindices b and r respectively. Define the r-b slope (or spectral index)

$$\alpha_{rb} = -\log[f(\nu_r)/f(\nu_b)]/\log(\nu_r/\nu_b), \tag{1}$$

where $f(\nu)$ is the flux at (rest frame) frequency ν.

Fig. 1 shows the r-b spectral index versus continuum luminosity, $L_b = \nu f(\nu)$ at 4200Å($H_o = 50$ and $q = 0$). Although the distribution is somewhat scattered, it is clear that the spectral index is decreasing with luminosity. In other words, the slope of the energy spectrum from the near IR to the blue becomes flatter with increasing luminosity.

We model the monotonous increase in the flattening of the spectrum with luminosity by a combination of a power law spectrum and a standard geometrically thin, optically thick accretion-disc. The spectrum of an optically thick accretion disc (Sakura and Sunyaev, 1973) has a characteristic power-law spectrum $f(\nu) \propto \nu^{1/3}$, with a cutoff at $\nu_{max} \sim kT_{max}/h$,

$$T_{max} = \left(\frac{3GM\dot{M}}{8\pi\sigma R_{min}^3}\right)^{1/4} = (1.5 \times 10^5 \,^\circ\mathrm{K}) M_8^{-1/4} \dot{m}^{1/4}, \tag{2}$$

where $M_8 = M/10^8 M_\odot$, $\dot{m} = \dot{M}c^2/L_E = 4.4(\dot{M}/M_\odot \mathrm{y}^{-1})M_8^{-1}$, $L_E = 1.3 \times 10^{46} M_8 \mathrm{erg\ s}^{-1}$, and σ is the Stefan-Boltzmann constant. The inner radius of the disc, R_{min}, has been taken as $6GM/c^2$, the innermost stable orbit around a Schwartzshild black hole. For a maximally rotating Kerr black hole, $R_{min} = 1.23GM/c^2$, hence T_{max} is larger by a factor of 3.3. The total luminosity of the disc is given by

$$L_d = \epsilon_d GM\dot{M}/R_{min} = 0.057\dot{M}c^2 \tag{3}$$

where ϵ_d is a factor of order unity (=1/2 in the Newtonian approximation, and less when general relativity is included). The coefficient on the right hand side is for a Schwartzshild black hole. From eqs. (2)-(3) we find

$$f_d(\nu) = (1.9 \times 10^{29} \mathrm{erg\ s^{-1} Hz^{-1}}) M_8^{4/3} \dot{m}^{2/3} (\nu/\nu_b)^{1/3}. \tag{4}$$

Note that the dependence on R_{min} cancels out, so that to first order eq. (4) holds for a rotating black hole as well.

The differential spectrum of a nonthermal component of total luminosity L_{nt} and power-law index α is

$$f_{nt}(\nu) = (1.4 \times 10^{31} \mathrm{erg\ s^{-1} Hz^{-1}}) \phi^{-1} L_{nt,46} \left(\frac{\nu}{\nu_b}\right)^{-\alpha}, \tag{5}$$

where ϕ is a numerical factor of order unity, depending on α and on the lower and upper boundaries. The flux ratio of the disc-to-power-law components is given by

$$R(\nu) \equiv \left(\frac{f_d}{f_{nt}}\right) = 0.013\phi L_{b,46}^{-1} M_8^{4/3} \dot{m}^{2/3} \left(\frac{\nu}{\nu_b}\right)^{\alpha+\frac{1}{3}} = 0.010 \left(\frac{\epsilon}{\phi}\right)^{-\frac{2}{3}} \left(\frac{L_b}{L_E}\right)^{-\frac{2}{3}} L_{b,46}^{-1/3} \left(\frac{\nu}{\nu_b}\right)^{\alpha+\frac{1}{3}}, \tag{6}$$

where $L_{nt} = \epsilon \dot{M} c^2$, so that $\dot{m} = 0.8\phi\epsilon^{-1} L_{b,46}/M_8$, and we have used the Eddington ratio in order to express M in terms of the luminosity. The spectral index of the combined spectrum is given by eqs. (1) and (8),

$$\alpha_{rb} = \log\left(\frac{\delta_{rb}^{\alpha} + R_b}{1 + R_b \delta_{rb}^{-1/3}}\right) \Big/ \log(\delta_{rb}), \tag{7}$$

where $R_b = R(\nu_b)$ and $\delta_{rb} = \nu_b/\nu_r = 1.78$.

For the the spectral index of the infrared-to-UV power law we adopt a characteristic value of ~1.2 (Malkan and Filippenko 1983). We also assume that the power-law componenet extends over three orders of magnitude (e.g. $\log\nu = 13 - 16$), corresponding to $\phi \approx 7$. For the efficinecy of producing the non thermal continuum we take ~ 0.1 (Begelman 1984).

The solid curves in Fig. 1 show the theoretical dependence of the spectral index on luminosity, given by eqs. (6) and (7) with $\alpha = 1.2$, $\epsilon/\phi = 0.1/7 = 0.014$, for $L_b/L_E = 0.003$, 0.03 and 0.3, marked a,b and c respectively. Note that the data are consistent with $\log(L_b/L_E) \sim -2 \pm 0.5$ (Cf. Wandel and Yahil 1985).

We note that low luminosity ($\log L_b < 45$) objects have steeper spectra than predicted by the model, some even have $\alpha_{rb} \gg 1$. This is probably due to contamination by starlight from the host galaxy. The dashed curves in Fig. 1. show the efffect of a stellar luminosity of $5 \times 10^{10} L_\odot$ (assumed to be a blackbody with $T = 3000°$K).

Additional effects which would cause a deviation of the observed spectral index from the model prediction are inclination of the disc to the line of sight (cf. Netzer 1986), and the actual disc spectrum differing from the idealized $\nu^{1/3}$ power law; in the region $h\nu_b \approx kT_{max}$ the disc spectrum has a wide plateau, so the disc contribution to the spectral index α_{rb} will have a slope of $\alpha \sim 0$ rater than -1/3. This effect can explain the objects at the high luminosity end of Fig. 1 with a relatively steep spectral index.

From IUE data on the near UV spectrum of a number of quasars and AGN it appears that the average blackbody temperature that would best fit the bump is 27000±5000°K (Malkan 1983; O'brian, Gondhalekar and Wilson 1986). For a Kerr hole eq. (2) gives

$$T_{max} = 33{,}000°\mathrm{K} \left(\frac{\epsilon/\phi}{0.014}\right)^{1/4} \left(\frac{L_b/L_E}{0.01}\right)^{1/2} L_{b,46}^{-1/4}, \tag{8}$$

in agreement with the value of L_b/L_E independently derived from the $\alpha_{rb} - L$ relation.

REFERENCES

Begelman, M.C. 1985, in *Astrophysics of Active Galaxies and Quasi-Stellar Objects*, ed. J.S. Miller, University Science Books.

Malkan, M.A. 1983, *Ap. J.*, **268**, 582.

Malkan, M.A., and Sargent, W.L.W. 1982, *Ap. J.*, **254**, 22.

Malkan, M.A., and Filippenko, A.V. 1983, *Ap. J.*, **275**, 477.

Netzer, H. 1986, *M.N.R.A.S.*, **216**, 63.

Neugebauer, G., *et. al.* 1986, *Ap. J. Sup.*, ,

Neugebauer, G., Oke, J.B., Becklin, E.E., and Mathews, K. 1979, *Ap. J.*, **230**, 79.

O'brian, P.T., Gondhalekar, P.M., and Wilson, R. 1986, in *New insights in Astrophysics*, Proc. Joint NASA/ESA/SERC Conference, University College London, p. 601.

Sakura, N.I. and Sunyaev, R.A. 1973, *Astr. Ap.*, **24**, 337.

Shields, G. 1978 *Nature*, **272**, 706.

Wandel, A. 1987, *Ap. J. (Letters)*, , Lsubmitted.

Wandel, A., and Yahil, A. 1985, *Ap. J. (Letters)*, **295**, L1.

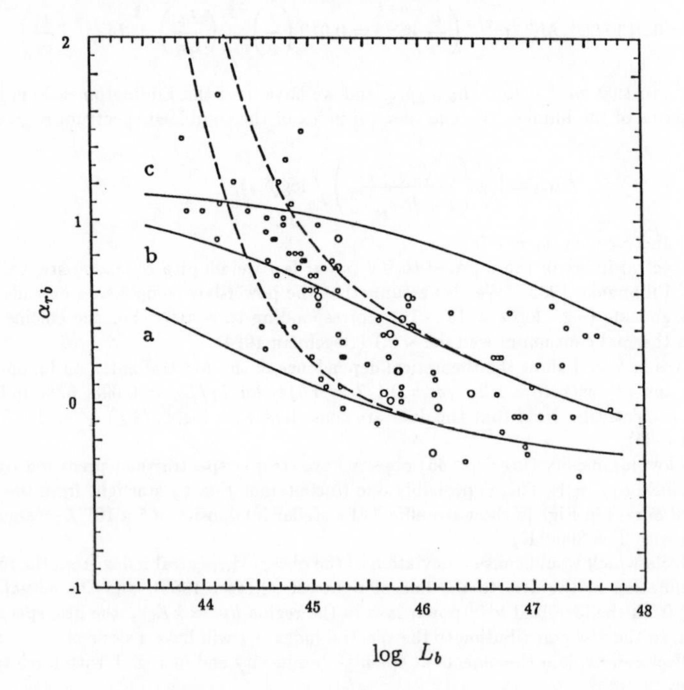

Figure 1. The spectral index between 4200Å and 7500Å (rest frame) vs. luminosity ($L_b = \nu f_\nu(4200\text{Å})$). Data taken from Neugebauer *et. al.* (1986). Large circles denote radio loud quasars. Solid curves correspond to the model prediction (eqs. [6]-[7]), with $\alpha = 1.2$, $\epsilon/\phi = 0.014$ and $L_b/L_E = 0.003$, 0.03 and 0.3 (marked a, b and c respectively). The dashed curves demonstrate the effect of adding a galactic (stellar) luminosity of $5 \times 10^{10} L_\odot$.

SMALL BROAD-LINE EMITTING REGIONS IN ACTIVE GALACTIC NUCLEI AND EVIDENCE FOR SUPERMASSIVE BLACK HOLES

Bradley M. Peterson
Department of Astronomy
The Ohio State University, Columbus, Ohio 43210

Spectroscopic monitoring of a few bright Seyfert galaxies over the past several years has led to important results that strengthen the argument that the dynamics of the broad-line region (BLR) are dominated by a supermassive black hole.

First, cross-correlation analyses of the continuum and broad Balmer-line light curves reveal that the delay between continuum and line changes is less than ~1 month;[1,2,3] this means that the BLR is about an order of magnitude smaller than previous estimates based on ionization equilibrium arguments, probably because of underestimated electron densities. Small BLRs can be virialized by reasonable central masses (~10^8 $M_\odot$), and BLR emission is expected to be gravitationally redshifted by an observable amount. Both of these predictions are consistent with the observations.

Second, an enormous increase in the strength of He II λ4686 during a 1984 continuum outburst in NGC 5548 suggests the appearance of new material in the vicinity of the nucleus[4]. Peterson and Ferland have interpreted the appearance of a new, very broad component of He II λ4686 as the signature of an accretion event in this galaxy.

Third, it has been shown that the Balmer lines in NGC 5548 arise in two physically distinct regions[3,5]. Peterson, Korista, and Cota[5] suggest that the two regions represent two independent BLRs surrounding two separate supermassive black holes. If these form a binary system, estimates of the component masses can be made, and these prove to be consistent with other estimates.

References

1. Peterson, B.M., Meyers, K.A., Capriotti, E.R., Foltz, C.B., Wilkes, B.J., and Miller, H.R., Astrophys. J., 292, 164 - 171 (1985).
2. Gaskell. C.M., and Sparke, L.S. 1986, Astrophys. J., 305, 175 - 186.
3. Peterson, B.M., Astrophys. J., 312, 79 - 90 (1987).
4. Peterson, B.M., and Ferland, G.J., Nature, 324, 345 - 347 (1986).
5. Peterson, B.M., Korista, K.T., and Cota, S.A., Astrophys. J. (Letters), 312, L1 - L4 (1987).

Interstellar Material Near the Galactic Center

D. T. Jaffe

Department of Astronomy, University of Texas
Austin, Texas 78712

ABSTRACT

The interstellar medium in the inner 10 pc of our Galaxy is unique. At r = 2-10 pc, the interstellar material is in the form of a rotating disk. This 3 x 10^4 $M_\odot$ disk is mostly molecular and neutral atomic gas with a density of 10^4-10^5 cm^{-3} and a temperature of ~ 400 K. The disk is at once highly clumped and very turbulent. At r = 2 pc there is a ring of ionized gas at the inner edge of the disk. Two streamers of ionized material extend into a cavity at r < 2 pc, joining a region of higher velocity ionized gas in the inner 0.5 pc of the Galaxy. Observations of infrared continuum emission from dust imply a luminosity ~ 2 x 10^7 $L_\odot$ for the exciting source in the galactic center. The surface temperature of this source is ~ 35,000 K.

1. Introduction

Interstellar material in the inner 10 pc of the Galaxy has a major effect on the appearance of this region at all wavelengths from the microwave through the near-infrared, and dominates the emergent spectral energy distribution. This material serves as a telltale for the dynamics in the inner 10 pc. The interstellar matter also is important indirectly as a generator of energy - either through accretion onto a massive collapsed object or from periodic formation into young UV and luminosity producing stars. Figure 1 [1)] shows a composite spectrum of the inner 4 pc of the galactic center. Thermal bremsstrahlung from ionized gas dominates the continuum emission in this region from greater than 6 cm to around 1 mm. From 1 mm to 3 μm dust (T_{dust} ranging from ~ 30 K to 200K) provides most of the continuum flux. CO and H_2 lines are responsible for most of the emergent intensity from molecular gas. Infrared fine structure lines of C^+ and O^0 are the major radiative losses for the neutral atomic gas. Lines from Ne II, O III and He I as well as hydrogen recombination lines cool the ionized gas.

Morphologically and physically, the interstellar medium of the inner 10 pc of our Galaxy divides into two regions. (1) A "disk" at 2-10 pc which consists mostly of dense clumps of warm molecular and neutral atomic gas; (2) A central cavity at r < 2 pc consisting of: (a) a region which contains long filaments or "arms" of ionized gas at 0.5 pc < r < 2 pc; (b) a "bar" at r < 0.5 pc made up of ionized material and a population of relativistic electrons; (c) a "high velocity" region less than 0.1 pc from the radio point source Sgr A* where ionized material has velocities up to ± 650 km s^{-1}. We will review the physical characteristics and excitation of the disk, summarize the state of the

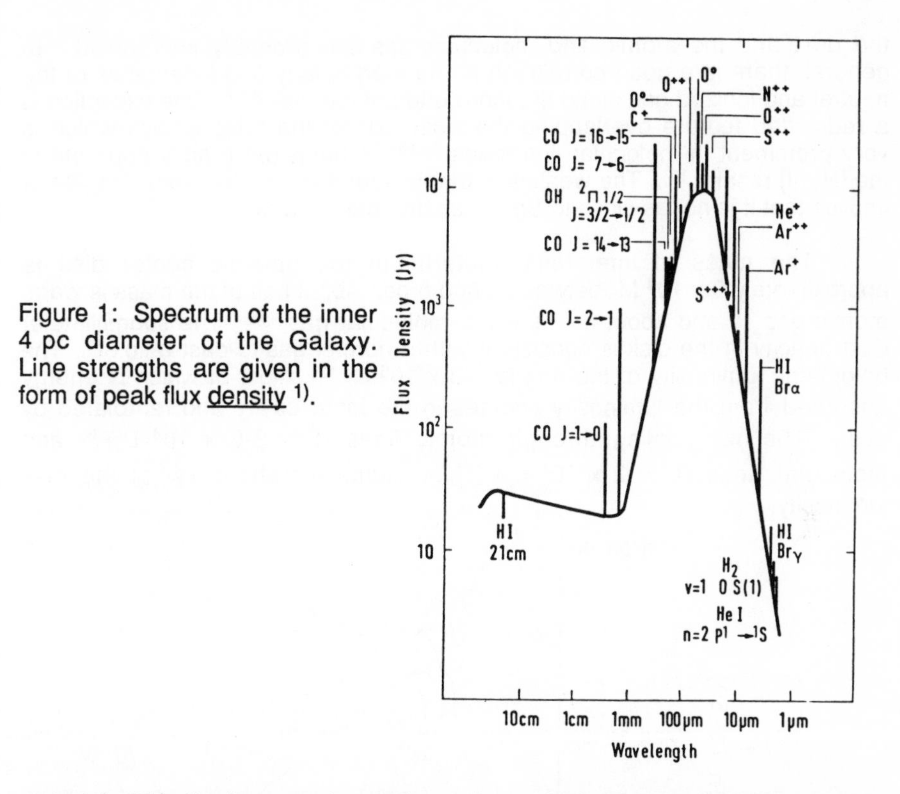

Figure 1: Spectrum of the inner 4 pc diameter of the Galaxy. Line strengths are given in the form of peak flux density [1].

material within the inner cavity, and use the excitation and energetics of this gas to infer the luminosity and temperature of the central energy source. Recent reviews and the proceedings of a recent symposium can provide more additional information about the interstellar material in the galactic center [1-2].

2.1 Morphology and Physical Conditions in the 2-10 pc Disk

In the region from 2 to 10 pc from the ultra-compact radio source Sgr A*, the interstellar material lies in a thick disk. This disk is turbulent and perturbed, but its primary motion is rotation about the galactic center at 80-120 km s^{-1}. It lies almost in the galactic plane, and is therefore highly inclined to the line-of-sight (i ~ 60° - 70°). Most of the ionized material associated with the disk lies in an apparently incomplete thin ring at its inner edge [3-4], although there is [Ne II] in the disk up to 4 pc from Sgr A* [5]. Predominantly neutral atomic material exists throughout the disk. The [O I] and [C II] fine-structure line emission from this material peaks sharply, however, at the inner edge [6-8]. The molecular material forms an azimuthally complete, well-defined disk running from 2 to > 5 pc [9-10]. Figure 2, a 4 arc second resolution map of the HCN 1 → 0 transition toward Sgr A [11] shows the bright molecular ring at the inner edge of the disk and constrains its thickness (< 5" = 0.25 pc at 2 pc radius). The HCN observations show that the disk is clumpy on scale sizes of 4" (= 0.2 pc). The distribution of dust column density [12-13] implies that the dust and the atomic and molecular gas are probably well mixed. In general, there is a good correlation in the morphology and kinematics of the neutral and ionized gas along the inner edge of the disk [8-9]. One exception is a redshifted feature overlapping the west side of the rotating disk which is very prominent in molecular line maps [9, 14-15], but is not readily apparent in the [Ne II] results [16]. The weakness of this strong molecular feature in [Ne II] implies that it is not as close to Sgr A* as the rotating disk.

The mass of interstellar material in the galactic center disk is approximately 3×10^4 $M_\odot$ between 2 and 5 pc. About half of the mass is warm atomic gas [7], and about half is warm molecular gas [17]. The submillimeter dust opacity in the disk is consistent with a normal gas-to-dust ratio [18]. The bolometric luminosity of the ring is ~ 3×10^6 $L_\odot$ [12], most of which is energy absorbed from the luminosity sources in the inner cavity and reradiated by dust. The gas cooling through atomic lines ($L \simeq 3\text{-}6 \times 10^4$ $L_\odot$) [6] and molecular lines ($L \simeq 2 \times 10^4$ $L_\odot$) [17] accounts for about 3% of the disk luminosity.

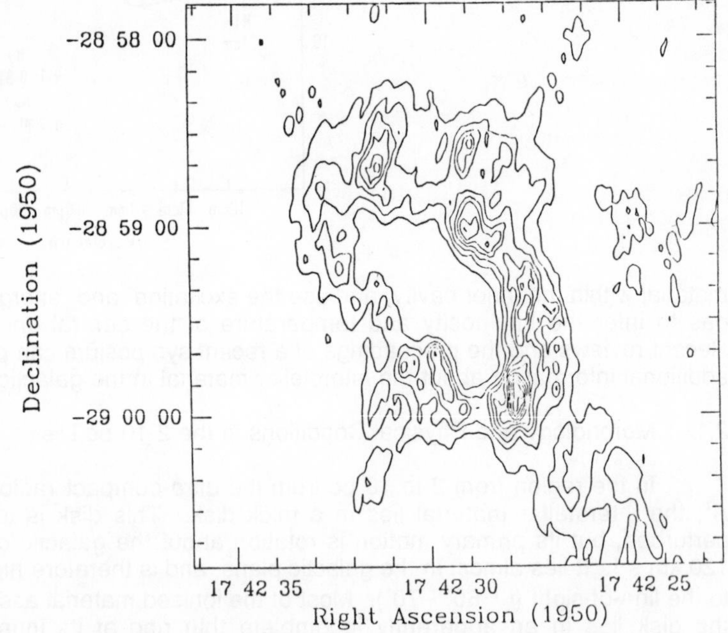

Figure 2: HCN 1 → 0 integrated intensity map of the galactic center disk at 4" resolution [11].

The physical properties of the atomic and molecular gas in the galactic center disk are very similar. Multi-transition CO studies place the kinetic temperature of the molecular disk at 200 - 600 K and its density at 10^4 - 10^5 cm^{-3} [17,19]. Excitation calculations based on observations of CS and NH_3 confirm these high densities and temperatures [20-21]. The molecular hydrogen emission requires higher temperatures (≈ 2000 K) if collisions excite the H_2 [21,13]. The molecular gas is highly clumped (volume filling factor ~ 0.1). Observations of far-IR lines of [O I] and [C II] yield a temperature of ~ 400 K, a density at $10^{4.5}$, and a volume filling factor ~ 0.1 for the neutral atomic gas [7]. The dust color temperature in the galactic center disk is 50-80 K [11].

The brightness distributions of all the atomic and high excitation molecular lines peak at the inner edge of the galactic center disk and drop with increasing radius. The pressure in the neutral atomic gas (nT) drops by a factor of about 2 between 2 and 4.5 pc [7]. A multi-transition study of CO line intensity study implies a pressure drop proportional to R^{-1} or R^{-2} in the molecular gas between 2 and 5 pc[20]. The far-IR dust temperature drops from 80 to 60 K over the same distance [12,24].

2.2 Excitation of Hot Gas in the Disk

Either shocks, UV radiation from the central source(s) or a combination of the two provide the excitation for the material in the galactic center disk. Conventional heating by gas-grain collisions will not work since the dust temperature is less than the gas temperature. The shock heating would be a result of dissipation of the observed internal turbulence [7,9] or possibly of a fast wind from a central source impacting the inner edge of the disk [22,12]. The 4" resolution HCN results show that local supersonic turbulence dominates the line widths in the disk. Even on scales of 5" (0.25 pc), the lines are 40 km s^{-1} wide [11]. The H_2/CO luminosity ratio in the galactic center is about 100 times lower than in shock excited regions like Orion/KL [17]. The lower ratio is only possible if the galactic center shocks raise the gas temperatures much less than the shocks in star formation regions. The far-IR fine structure line ratios in the galactic center agree poorly with the predictions of Orion shock models [23-24].

The photoelectric heating mechanism proposed for H II region/molecular cloud interface regions may be important in heating the neutral gas in the galactic center [25]. The mechanism relies on electrons ejected from dust grains by UV photons to heat the gas. The low UV opacity of the dust within the inner cavity, the volume filling factor of the warm molecular and atomic gas and the warping of the disk all make it possible to have photoelectrically heated material throughout the disk. The mechanism is only effective, however, at heating the outer few 10^{22} cm^{-2} of hydrogen to temperatures significantly above the dust temperature. This creates problems for heating CO which may not survive directly behind the H II/H I interface. Self-shielding of CO possibly could allow enough warm CO to survive in a high UV environment to explain the observed temperatures [17].

The gas along the inner edge of the disk is most likely photoionized by UV emission from the center. The electron temperature of this gas, derived from radio recombination line measurements is low - 5000-8000 K [27-28]. The line and continuum emission is strong along the western side of the central cavity. The radio continuum emission is weak or absent along much of the corresponding eastern edge [3], leading to the suggestion that ionizing radiation does not reach this part of the ring [16]. There is, however, weak [Ne

II] emission along the eastern edge of the cavity at velocities appropriate to the inner edge of the disk. A somewhat lower density would result in a greatly lowered [Ne II] surface brightness and could also explain the absence of a sharp feature in the radio interferometer data.

3.1 Physical Conditions in the "Cavity" at $r \leq 2$ pc.

All material at $r < 2$ pc appears ionized. The brightest features in the outer cavity at 0.5 pc < 4 < 2 pc are the two long arms stretching north and east from the center, seen principally in the thermal radio continuum [3-4] and in the 12.8 μm line of [Ne II] [17]. The electron temperature in the northern arm is similar to Te along the outer edge of the cavity [27]. This material is apparently ionized by the central ionization source. Both the northern arm and the eastern arm have broad (60 - 80 km s^{-1}) wide lines [16] reflecting a combination of large systematic velocity gradients with turbulent motions. Polarization of the infrared continuum emission from the northern arm implies magnetic field strengths of > 10 mG [29].

At distances less than 0.5 pc from Sgr A*, in the "bar", there is a considerable amount of material seen in [Ne II] at very high velocities (± 200 km s^{-1}) [30]. The absence of radio recombination lines from this region has been interpreted as a result of high local electron temperature [27]. Recent millimeter continuum measurements indicate that about 30 percent of the 15 GHz continuum from the bar is non-thermal [31].

At $r < 0.1$ pc from Sgr A*, toward the source IRS 16, there is emission in infrared recombination lines of hydrogen and helium at velocities up to 650 km s^{-1} from line center [32-33]. This high velocity gas accounts for only a small fraction of the total number of ionizing photons in the cavity. This extreme velocity material appears to extend over 0.1 → 0.2 pc [33].

4.1 Global Properties of the Central Source

The total energy absorbed and re-emitted by the dust in the Galactic Center disk is 3-6 x 10^6 $L_\odot$ [12]. Given the apparent solid angle occupied by the disk as observed from Sgr A* [11], we derive a total luminosity for the exciting shource of ~ 2 x 10^7 $L_\odot$. This estimate is in good agreement with earlier values [12]. The thermal radio continuum flux together with the fluxes in various near infrared fine structure lines provides an estimate for the surface temperature of the ionizing source. This temperature is about 35,000K for a source with a spectrum similar to a stellar atmosphere [12,33].

References

1 Genzel, R., in "The Galaxy", (D. Reidel, Dordrecht), (1986).
2 "Symposium on the Galactic Center", D. Backer ed., (Amer. Inst. Phys.), (1986).
3 Lo, K. Y., and Claussen, M. J., *Nature*, **306**, 647 (1983).

[4] Ekers, R. D., van Gorkom, J. H., Schwarz, U. J., and Goss, W. M., *Astron. Astrophys.*, **122**, 143 (1983).
[5] Serabyn, E., Ph.D. Thesis, University of California, Berkeley (1984).
[6] Genzel, R., Watson, D. M., Townes, C. H., Dinerstein, H. L., Hollenbach, D., Lester, D. F., Werner, M. W., and Storey, J. W. V., *Ap. J.*, **276**, 551 (1984).
[7] Genzel, R., Watson, D. M., Crawford, M. K., and Townes, C. H., *Ap. J.*, **297**, 766 (1985).
[8] Lugten, J. B., Genzel, R., Crawford, M. K., and Townes, C. H., *Ap. J.*, **306**, 691 (1986).
[9] Güsten, R., Genzel, R., Wright, M. C. H., Jaffe, D. T., Stutzki, J., and Harris, A. I., *Ap. J.*, in press (1987).
[10] Kaifu, N., Hayashi, M., and Gatley, I., "Symposium on the Galactic Center", D. Backer ed., (Amer. Inst. Phys.) (1986).
[11] Güsten, R., Genzel, R., Wright, M. C. H., Jaffe, D. T., Stutzki, J., and Harris, A. I., "Symposium on the Galactic Center", D. Backer ed., (Amer. Inst. Phys.) (1986).
[12] Becklin, E. E., Gatley, I., and Werner, M. W., *Ap. J.*, **258**, 134.
[13] Lester, D. F., Joy, M., Harvey, P. M., and Ellis, H. B., "Symposium on the Galactic Center", D. Backer ed., (Amer. Inst. Phys.) (1986).
[14] Gatley, I., Jones, T. J., Hyland, A. R., Wade, R., Geballe, T. R., and Krisciunas, K., *M.N.R.A.S.*, **222**, 299 (1986).
[15] Serabyn, E., Güsten, R., Walmsley, C. M., Wink, J. E., and Zylka, R., *Astron. and Astrophys.*, **169**, 85 (1986)
[16] Serabyn, E., and Lacy, J. H., *Ap. J.*, **293**, 445 (1985).
[17] Harris, A. I., Jaffe, D. T., Silber, M., and Genzel, R., *Ap. J. (Letters)*, **294**, L93 (1985).
[18] Novak, G., personal communication (1986).
[19] Lugten, J. B., Harris, A. I., Stacey, G. J., and Genzel, R., *Ap. J.*, in press (1987).
[20] Evans, N. J. II, "Symposium on the Galactic Center", D. Backer ed., (Amer. Inst. Phys.) (1986).
[21] Serabyn, E., and Güsten, R., *Astron. Astrophys.*, **161**, 334 (1986).
[22] Gatley, I., Jones, T. J., Hyland, A. R., Beattie, D. H., and Lee, T. J., *M.N.R.A.S.*, **210**, 565 (1984).
[23] Chernoff, D. F., Hollenbach, D. J., and McKee, C. F., *Ap. J. (Letters)*, **259** L97 (1982).
[24] Draine, B. T., and Roberge, W. G., *Ap. J. (Letters)*, **259**, L91 (1982).
[25] Tielens, A. G. G. M., and Hollenbach, D. J. *Ap. J.*, **291**, 711 (1985).
[26] van Dishoek, E. F., and Black, J. H., in "Physical Processes in Interstellar Clouds", M. Scholer ed. (1987).
[27] van Gorkom, J. H., Schwarz, U. J., and Bregman, J. D., in "The Milky Way Galaxy", eds. van Woerden *et al.*, (Reidel: Dordrecht), (1985).
[28] Mezger, P. G., and Wink, J. E., *Astron. Astrophys.*, **157**, 252 (1986)
[29] Aitken, D. K., Roche, P. F., Bailey, J. A., Briggs, G. P., Hough, J. H., and Thomas, J. A., *M.N.R.A.S.*, **218**, 363 (1986).
[30] Lacy, J. H., Townes, C. H., Geballe, T. R., and Hollenbach, D. J., *Ap. J.*, **241**, 132 (1980).
[31] Wright, M. C. H., Genzel, R., Güsten, R., and Jaffe, D. T., "Symposium on the Galactic Center", D. Backer ed., (Amer. Inst. Phys.) (1986).

[32] Hall, D. N. B., Kleinmann, S. G., and Scoville, N. Z., *Ap. J. (Letters)*, **262**, L53 (1982).
[33] Geballe, T. R., Wade, R., Krisciunas, K., Gatley, I., and Bird, M. C., preprint (1986).

THE GALACTIC CENTER COMPACT NONTHERMAL RADIO SOURCE

K. Y. Lo

Astronomy Department
University of Illinois, Urbana-Champaign
and
California Institute of Technology

ABSTRACT

The current observational status of Sgr A*, the compact nonthermal radio source at the galactic center, is reviewed. It is a unique radio source at a unique position of the Galaxy. It is unlike any compact radio sources associated with known stellar objects, but it is similar to extragalactic nuclear compact radio sources. The positional offset between Sgr A* and IRS16 places little constraint on the nature of the underlying energy source, since the nature of IRS16 itself is not well understood and may not be the core of the central star cluster. With its unique properties in the Galaxy and being the only unusual object at the center with dimensions approaching the gravitational radius of a $\sim 10^6$ $M_\odot$ black-hole, Sgr A* is still the best candidate for marking the location of a massive collapsed object.

1. INTRODUCTION

At the center of Sgr A, the powerful radio source at the center of the Galaxy, is an extremely compact nonthermal radio source.[1,2] Such a radio source was anticipated as a possible signature of a massive black-hole that was proposed to be the energy source for the luminosities observed towards the center.[3]

Efforts to determine the structure of this compact radio source have gone on for many years,[4-6] but because of the practical difficulties there is still no brightness distribution map of the source. The principal limitation, given the existing radio telescopes and the angular scale of the source, is that there are very few sufficiently short baselines on which the source is not completely resolved.

We review here the current status of the properties of the radio source and the constraints one can place on the nature of the underlying energy source. We also discuss briefly the relevance of various other constraints that have been placed on the source.

2. THE COMPACT NONTHERMAL RADIO SOURCE

The properties of the radio source are summarized in Table 1.

Table 1 Observed properties of the galactic centre compact radio source

Source size	< 17 AU ($\sim 2 \times 10^{14}$ cm)
Wavelength dependence of size	$\lambda^{2.0}$ ($\lambda \geq 1.35$ cm)
Position angle of elongation	$98 \pm 15°$ ($\lambda = 3.6$ cm)
Axial ratio	0.55 ± 0.25 ($\lambda = 3.6$ cm)
Upper limit to source expansion	< 13 km s^{-1}
Flux density variabiality ($\Delta S/2S$)*	$\lesssim 0.2$ (1975-86)
Spectral index	~ 0.25
Turnover frequency	$\gtrsim 90$ GHz
Radio luminosity	$\sim 1.3 \times 10^{34}$ erg s^{-1}
Brightness temperature	$> 7 \times 10^8$ K (λ 1.35 cm)

*ΔS, peak-to-peak variation; S, mean flux density; ref. 15 and J. van Gorkom, K. Y. Lo and M. Claussen, unpublished results.

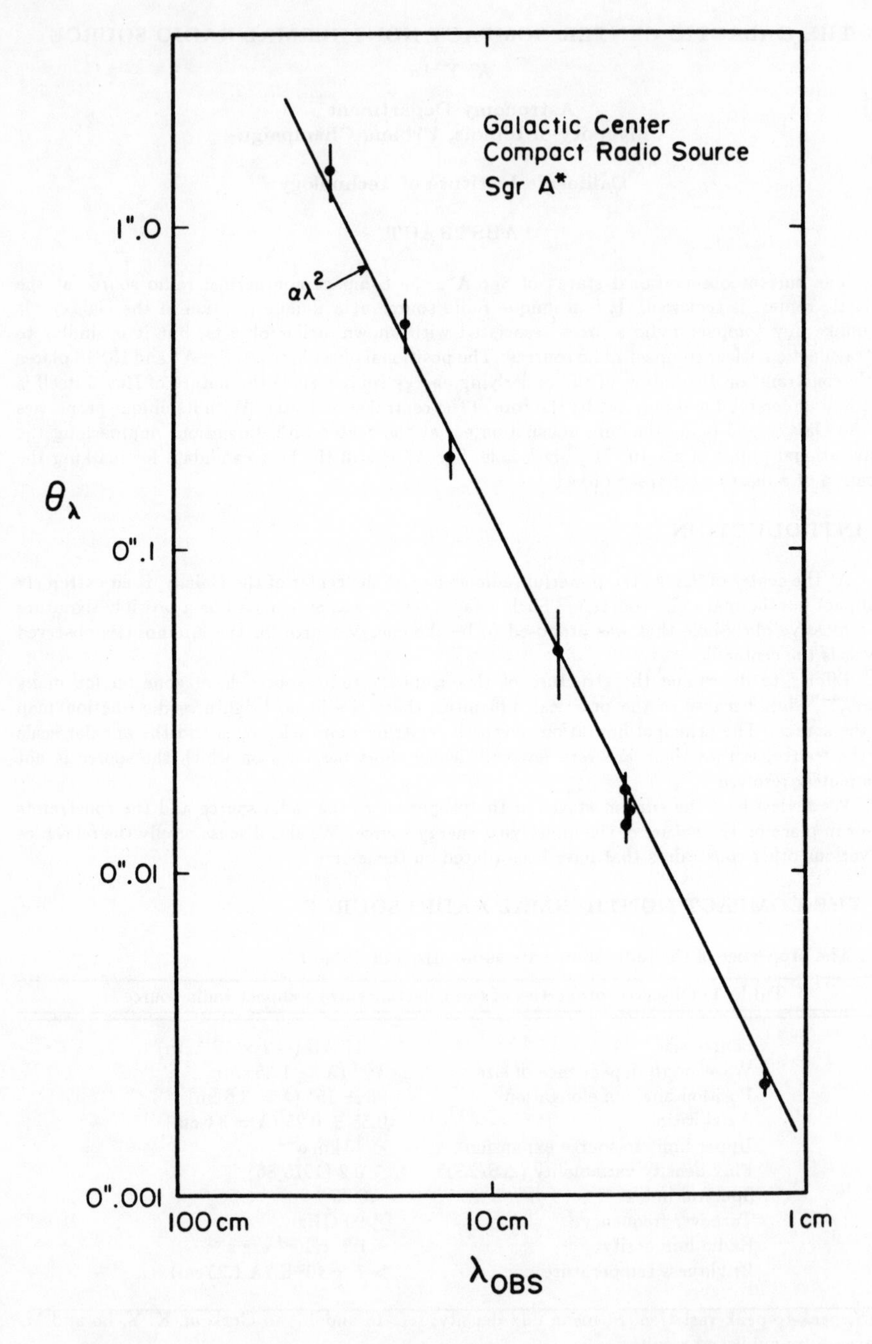

Figure 1. A plot of measured source diameter versus observing wavelength.[4-8]

2.1 Scale Size

Interferometric and VLBI observations have yielded source diameters (FWHM of a Gaussian brightness distribution, the simplest source model) of Sgr A* at various wavelengths.[2,6,7,8] The results are summarized in figure 1. The smallest measured size is 0.002" ± 0.0003", corresponding to 17 AU at a distance of 8.5 kpc. The diameters are seen to vary very nearly as λ^{-2}. The formal value of the index derived from the measurements as summarized in figure 1 is 2.08 ± 0.05. Given the systematic errors involved and the lack of proper maps, the index cannot be distinquished from 2.

Davies *et al*[7] first suggested that the apparent source size of Sgr A* could be affected by ***interstellar scattering*** of the radio wave, due to the irregular distribution of interstellar electrons along the line of sight. In this case, the observed size is an upper limit to the intrinsic size of the source. However, the amount of scattering is among the highest observed and the amount of electron density fluctuations required depends on where the bulk of scattering occurs.[9]

There are indications that sources within 100 pc of the center have large observed sizes implying a source of strong scattering near the center,[10] but Cordes *et al*[11] suggest that the observed scattering in Sgr A* is not an unrealistic extrapolation from pulsar results which can be explained by a two-component model of general interstellar electrons. The reasons for where to place the blame of scattering are twofold – one has to do with whether to believe scattering is affecting Sgr A* because a λ^{-2} dependence of the source size could be due to a 1/r distribution of thermal electrons within the source,[12] and two is the possible variability time scales involved, due to scintillation and refractive scattering.[13] In any case, the current measurements are only upper limits to the source size.

2.2 Source Shape

Recent VLBI observations have revealed structure more complicated than the simplest model of a symmetrical Gaussian distribution.[2] At 3.6 cm, the observations are best fit by an elliptical Gaussian, with an axial ratio of 0.55 ± 0.25. The position angle of the major axis is 98 ± 15 degrees, whereas the minor axis of the Galaxy is at 122 degrees. Due to the limited amount of data, the observed elongation in the source structure should be verified with more extensive observations. Clearly, brightness distribution maps of Sgr A* would help to determine the nature of the radio source. The best way to obtain them is at high frequency where both propagation effects and the opacity due to thermal electrons would be minimized.

Many issues involved in the effects of propagation through the interstellar electrons on the observations of Sgr A* are not definitively resolved, and more observations are needed. For example, if the radio waves from Sgr A* are strongly scattered, the observed image is unlikely to be full of details. Refractive scattering may also shift the positions of the radio source with time as well as with wavelength.[14] The presence of intensity scintillation in Sgr A*, if observed, would place a very strong constraint on the source size.[9]

2.3 Variability

Since Sgr A* was identified in 1974, there have been many measurements of its flux density by various observers. The observed flux density was often different but never varying by more than ~ 20 percent ($\Delta S/2S$). The only published study of the variability of Sgr A* was that of Brown and Lo[15] between December 1976 and May 1978. They found a few instances of low level variation on hourly time scale as well as a secular increase at 11 cm. The secular increase at 11 cm has not continued indefinitely since later measurements at nearby wavelengths have not shown an unusually large value. The variability properties of Sgr A* are not very well defined, except that it is clear that large outburst is not common and has not been observed.

A multi-wavelength study of variability at different time scales may reveal intrinsic source variability, or changes due to scintillation[9] and refractive scattering effects,[13] all of which could help to resolve the important issues of source sizes, propagation effects and radiation mechanisms. This kind of study requires dedicated, carefully calibrated and sensitive interferometer observations.

2.4 Source Spectrum

The spectrum of a radio source may also provide information on the source structure.[16] Figure 2 shows a "mean" spectrum of Sgr A* along with the spectra of the compact radio sources

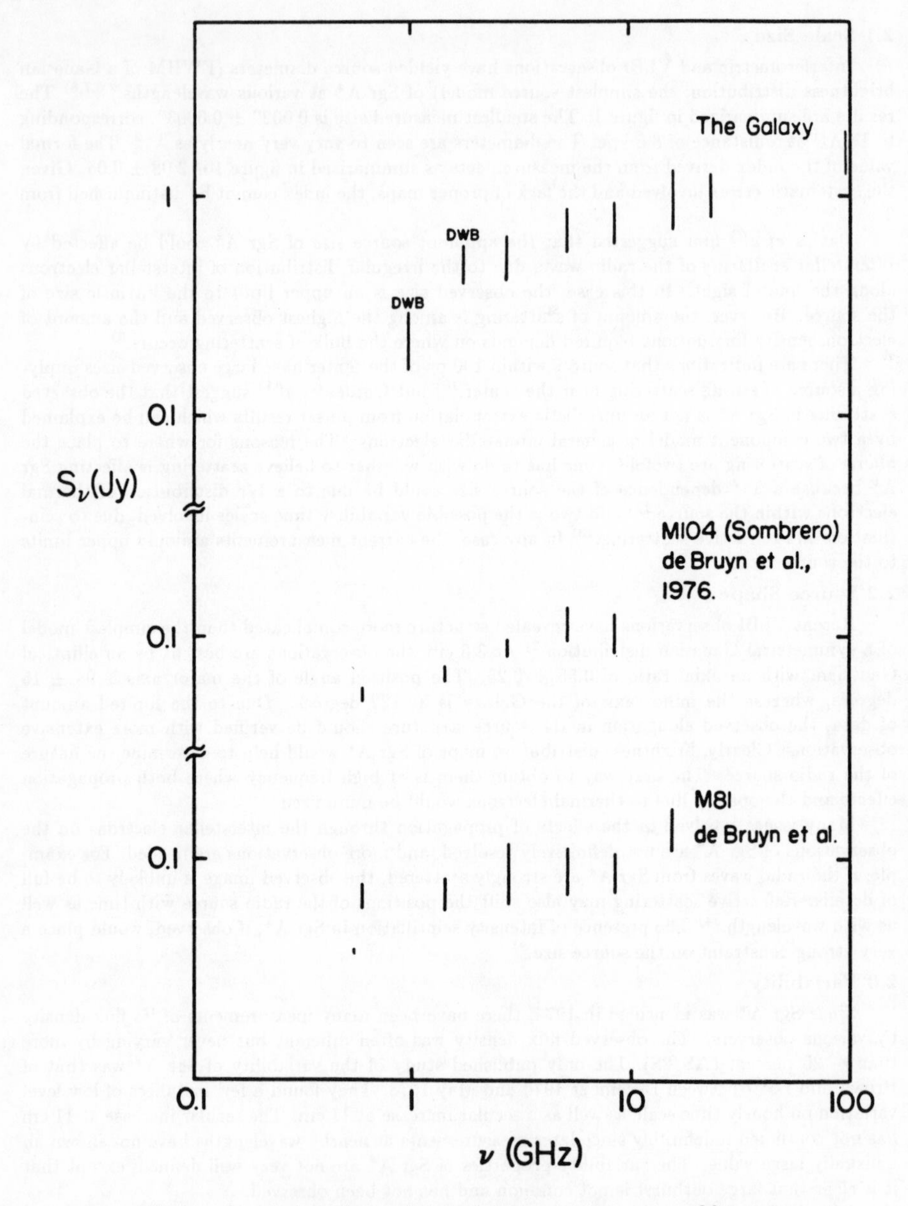

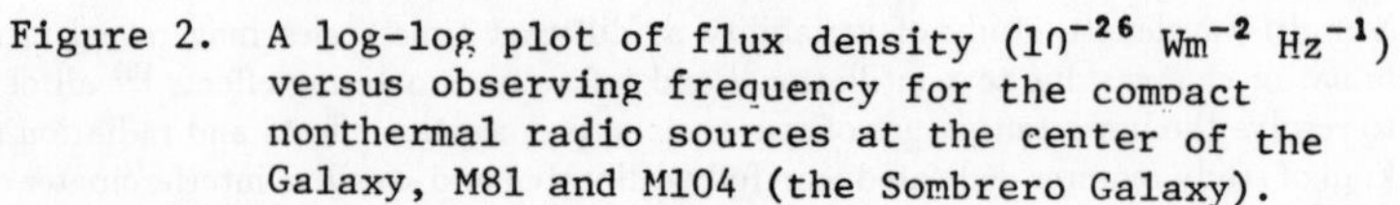

Figure 2. A log-log plot of flux density (10^{-26} Wm^{-2} Hz^{-1}) versus observing frequency for the compact nonthermal radio sources at the center of the Galaxy, M81 and M104 (the Sombrero Galaxy).

in the nuclei of M81 and M104.[17] The slowly rising spectra are most simply interpreted as that of a self-absorbed incoherent synchrotron source. The highest frequency measurement is at ~ 90 GHz where the flux density is 1.05 ± 0.15 Jy.[18] The turn-over frequency, above which the source becomes optically thin, is thus > 90 GHz, implying a very compact source. An important empirical point is that the source spectrum of Sgr A* is completely unlike that of pulsars.

3. UNDERLYING ENERGY SOURCE

Even if the brightness distribution of Sgr A* is as yet unknown, its very small scale size poses a very strong constraint on the underlying energy source and requires an object with stellar dimensions. In the Galaxy, there are very compact radio sources associated with known stellar objects - pulsars, binary stellar radio sources, young supernovae or supernova remnants. However, empirical arguments can already exclude all of these as possible energy source.

Ordinary pulsars are ruled out because the spectral index of pulsar radiation, α, generally lies in the range $-3 < \alpha < -1$, whereas that of Sgr A* is ~0.2 and the most luminous pulsars have an average radio luminoisty of about 10^4 times less. Furthermore, pulsars are very common, whereas Sgr A* has unique properties in the Galaxy. Sgr A* is unlikely to be a binary stellar radio source because such sources are characterized by frequent, large outbursts in which $\Delta S/2S$ could be 10 to 1000, while their steady radio luminosity is typically 10^5 times smaller that of Sgr A*. The ratio of their X-ray to radio luminosities is least 30 times larger than that of Sgr A*.

Recently, there have been detections of very bright compact nonthermal radio sources in external galaxies that are most likely young supernovae and supernova remnants.[19] They are different from Sgr A* in that they have been observed to expand or their flux densities are declining steadily on the time scale of a few years. For more than 12 years, Sgr A* have been essentially steady in its intensity and the expansion velocity is constrained to be <13 km/s averaged over 8 years.[2]

While one cannot explain Sgr A* in terms of known stellar radio sources, the radio luminosity of Sgr A* is only ~ 10^{34} erg s^{-1} or ~ 10 $L_\odot$. As up to 10^{38} erg s^{-1} may be available from a rotating neutron star, at least in principle, Sgr A* can be powered by a rotating neutron star. A model of pulsar-driven wind has been proposed for Sgr A*,[16] but the small scale size and the lack of dense gas clouds at the position of Sgr A* (which is needed to pressure confine the source) have ruled out such a model.

The mass of Sgr A* is thus a critical parameter needed to determine the nature of the energy source. There is an indirect way of constraining the mass – its proper motion. If Sgr A* is a stellar mass object near the bottom of the potential well at the center, very likely it will possess a large velocity. Hence, a measurable proper motion would rule out the possibility that it is supermassive. The latest proper motion result shows that the motion of the sun is observed relatively to Sgr A* and not relative to the neighbouring reference sources and that the residual proper motion is less than 40 km s^{-1}.[20] Thus, Sgr A* is at the galactic center, and the low upper limit to the proper motion, while not a proof, is consistent with Sgr A* being supermassive.

4. SIMILARITY TO EXTRAGALACTIC COMPACT RADIO SOURCES

While Sgr A* is a unique radio source in a unique position in the Galaxy, it has properties very similar to those of extragalactic compact radio sources. In particular, in the nucleus of M81, a nearby bright spiral (Sb) galaxy, there exists such a radio source which has been quite well observed.[21] It is compact (1000 AU × 4000 AU), with the major axis essentially aligned with the minor axis of the galaxy. It has a rising spectrum (cf. figure 2) and is mildly variable[17]. It is also a unique radio source at the center of a bright galaxy.[22] However, in this case, the radio luminosity (~10^{37} erg s^{-1}) is high enough that powering by a rotating neutron star is unlikely. Rees *et al* have proposed that extragalactic radio sources that are not luminous in the optical are explainable by massive black holes with a low accretion rate.[23] In particular, the model has been applied to Sgr A*.[24]

5. IDENTIFICATION OF SGR A*

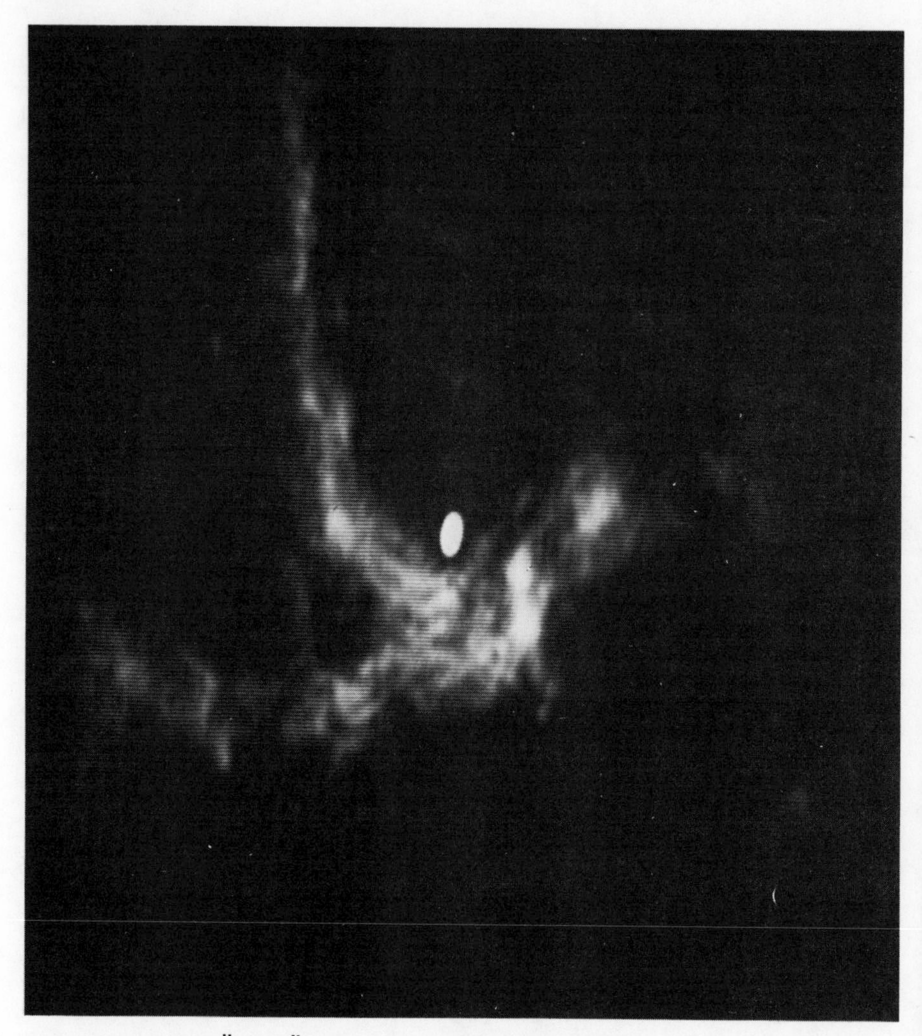

Figure 3. A $0\overset{''}{.}3 \times 0\overset{''}{.}6$ resolution λ6cm radio continuum VLA map of the central 45" of SgrA West.[45] The ellipse is the "saturated image" of SgrA* and is much bigger than the actual size. Note the filamentary structure in the arms and suggestive braided structure in the filaments along the northern arm.

The relatively low luminosity of Sgr A* at radio wavelenghts ($\sim 10^{34}$ erg s^{-1}) makes it still possible for a rotating neutron star to power it. Identification of Sgr A* with objects in the infrared is thus important for determining its bolometric luminosity. For example, a total luminosity of 10^7 L$_\odot$ is inferred to originate from the central pc of the Galaxy.[25] Whether all this luminosity is produced in a single object or a cluster of young stars has not been resolved.[26,27] A large bolometric luminosity would place a further important empirical constraint on the underlying energy source. Identification would also be helpful for identifying the location of Sgr A* relative to the central star cluster and for dynamical probes of the mass of the radio source.

The identification process has been beset by systematic errors involved in determining absolute positions of the near-infrared sources.[26,28] In the radio, the best absolute position of Sgr A* has been determined with the VLA,[29] but the atmospheric modelling sets the limiting systematic errors of ±0.2". Ekers was the first to suggest a method which bypassed many of the systematic errors involved in comparing the absolute positions of the radio source and the near-IR sources.[30] He compared the Brγ (2.16 μm) map to a 6 cm radio map. By aligning the ionized gas features observable at both 2 μm and 6 cm, he concluded that Sgr A* is offset from IRS16, a 2 μm continuum source. The best result using this method is that of Forrest *et al* who obtained a Brα (4.05 μm) image with an 1 - 5 μm array camera.[31]

Thus, the evidence seems to point to a positional offset between Sgr A* and the various components of IRS16.[30,28,31] However, since the nature of the IRS16 components is uncertain,[32] the implication of this "non-identification" is not clear. The non-association of an obvious near-infrared source with Sgr A* does not preclude the possibility of a massive black-hole, since in the case of low accretion rate onto a massive black-hole, which could explain Sgr A*, little luminosity is expected from the ion-supported accretion disk.[23] In addition, the actual accretion process is highly uncertain, Sgr A* may be in a "starved" phase.[24]

If Sgr A* were a $10^6 M_\odot$ massive black-hole, a cusp of stars is expected to form around it with an accretion radius of 20".[33,34] The 2 μm light distribution within the central few arc-seconds is complicated by a few bright discrete sources, so that the determination of the peak of the light distribution is not straight-forward.[35] High resolution measurements of stellar velocity dispersion may be the most reliable way to define the dynamical center of the Galaxy.[36,37] Even then, uncertainties in the intrinsic velocity distributions may still lead to ambiguous results.[38]

6. UPPER LIMITS TO MASS OF SGR A* ?

There have been various discussions on the mass of the possible black hole at the galactic center based on the separation of Sgr A* from IRS16,[39,34] the broad He line[40,41] and positron production[42,41]. It has been argued that the mass of the possible black hole cannot be large.[41,34] But, given the uncertainties in the nature of the components of IRS16, the location of the dynamic center, the ill-defined location of the 0.5 Mev e^+e^- annihilation line and alternate ways to produce positrons,[23,43] the arguments are not definitive.

The most direct observational indication of the possible presence of a concentrated mass at the center is still given by the large velocities of the ionized gas.[44] Given the complexities of the ionized gas distribution (figure 3) and the uncertain origin of its motion, the gas velocities may not accurately probe the gravitational potential. The origin and motion of the ionized gas will be much better constrained if the three dimensional velocity field of the gas is known. We plan to measure the proper motion of the ionized gas by obtaining another high resolution map with the VLA in 1987, for comparison with the map in figure 3 obtained in 1983. A transverse motion of 100 km s^{-1} corresponds to a proper motion of 0.01" in this period. This proper motion measurement can only improve with time as the time base increases.

7. CONCLUSION

Accretion onto black-holes was proposed to account for the inordinately large luminosity observed from a very compact volume of space in quasars and radio galaxies. Observationally, there is almost no hope of spatially resolving such collapsed objects in such distant objects. In the center of our Galaxy, the proximity permits probing within the interesting scales approaching the

Schwarzschild radius of a 10^6 $M_\odot$ black-hole, as in the case of Sgr A*. Ironically, while the unique properties of Sgr A* preclude models involving known stellar systems, the radio luminosity is too low to demand the mechanism of accretion onto a massive black hole. The exact nature of Sgr A* can only be resolved with more observations.

8. REFERENCES

1. Balick, B., and Brown, R. L., *Astrophys.J.*, **194**, 265-270 (1974).
2. Lo, K. Y., *et al, Nature*, **315**, 124-126 (1985).
3. Lynden-Bell, D., and Rees, M. J., *M.N.R.A.S.*, **152**, 461-475 (1971).
4. Lo, K.Y., Schilizzi, R. T., Cohen, M. H., Ross, H. N., *Astrophys. J.*, **218**, 668-670 (1977).
5. Kellermann, K. I., Shaffer, D. B., Clark, B. G., and Geldzahler, B. G., *Astrophys.J.(Lett.)*, **214**, L61-L62 (1977).
6. Lo, K. Y., Cohen, M. H., Readhead, A. C. S., and Backer, D. C., *Astrophys.J.*, **249**, 504-512 (1981).
7. Davies, R. D., Walsh, D., and Booth, R., *Mon. Not. R. astr. Soc.*, **177**, 319-333 (1976).
8. Marcaide, J. M., *et al.*, in *Conference on Active Galactic Nuclei* (ed. Dyson, J. E.) (Manchester University Press) 50-53 (1985).
9. Ozernoi, L. M., Shishov, V. I., *Sov. Astron. (Lett.)*, **3**, 233-235 (1977).
10. Backer, D. C., *Astrophys.J.(Lett.)*, **222**, L9-L12 (1978).
11. Cordes, J. M., Ananthakrishnan, S., and Dennison, B., *Nature*, **309**, 689-690 (1984).
12. de Bruyn, A. G., *et al, Astron. & Astrophys.*, **46**, 243 (1976).
13. Rickett, B. J., *Astrophys.J.*, **307**, 564-574 (1986).
14. Romani, R. W., Narayan, R., Blandford, R. B., *Astrophys. J.*, in press.
15. Brown, R. L., Lo, K. Y., *Astrophys.J.*, **253**, 108-114 (1982).
16. Reynolds, S. P., and McKee, C. F., *Astrophys.J.*, **239**, 893-897 (1980).
17. de Bruyn, A. G., *Astron. & Astrophys.*, **52**, 439 (1976).
18. Wright, M. C. H., this volume.
19. Kronberg, P. P., Sramek, R. A., *Science*, **227**, 28 (1985).
20. Backer, D. C., Sramek, R. A., this volume.
21. Bartel, N., *et al. Astrophys.J.*, **262**, 556-563 (1982).
22. Bash, F., Kaufman, M., *Astrophys.J.*, **310**, 621-636 (1986).
23. Rees, M., *et al Nature*, **295**, 17-20 (1982).
24. Rees, M., *AIP Conf. Proc.*, **83**, 166-176 (1982).
25. Becklin, E., Gatley, I., and Werner, M., *Astrophys.J.*, **258**, 135-142 (1982).
26. Henry, J. P., Depoy, D. L., and Becklin, E. E., *Astrophys.J.(Lett.)*, **285**, L27-L30 (1984).
27. Rieke, G. H. and Lebofsky, M. J., *AIP Conf. Proc.*, **83**, 194-203 (1982).
28. Storey, J. and Allen, D. A., *M.N.R.A.S.*, **204**, 1153 (1983).
29. Brown, R. L., Johnston, K. J., and Lo, K. Y., *Astrophys.J.*, **150**, 155 (1978).
30. R. D. Ekers, presented at the Galactic Center Workshop at Caltech, Jan. 1984.
31. Forrest, W., *et al*, this volume.
32. Allen, D. A., this volume.
33. Frank, J., *M.N.R.A.S.*, **187**, 833 (1979).
34. Allen, D. A. and Sanders, R. H., *Nature*, **319**, 191 (1986).
35. Rieke, G. H., this volume.
36. Sellgren, K. *et al*, this volume.
37. McGinn, M. *et al*, this volume.
38. Sargent, W. L. W., this volume.
39. Gurzadyan, V. G., Ozernoi, L. M., *Astron. & Astrophys.*, **86**, 315 (1980).
40. Hall, D. N. B., Kleinman, S. G., and Scoville, N. Z., *Astrophys.J.(Lett.)*, **262**, L53-L58 (1982).
41. Ozernoi, L. M., this volume.
42. Ramaty, R., and Lingenfelter, R. E., *Highlights of Astronomy*, **6**, 525-529 (1983).
43. Blandford, R. D., *AIP Conf. Proc.*, **83**, 177-179 (1982).

44. Lacy, J. H., *et al*, this volume.
45. Lo, K. Y. and Claussen, M. J., in preparation.

X-RAY AND GAMMA-RAY OBSERVATIONS OF THE GALACTIC CENTER

M. Leventhal

AT&T Bell Laboratories
Murray Hill, New Jersey 07974

ABSTRACT

X-ray (XR) and gamma-ray (GR) continuum observations and electron-positron (e^+e^-) annihilation line measurements of the Galactic Center (GC) are reviewed indicating the presence of a powerful and unique compact source thought to be a black hole but not necessarily located at Sgr A West.

XR and GR observations of the GC began in the early 1970's and continue through the present, with the peak in activity occurring in the late 1970's when the HEA0 satellites were launched. While balloon observations have played an important role (particularly in discovering the e^+e^- annihilation line source at 511 keV and tracking its temporal variability) in general the highest quality data have been produced by the HEA0 1-3, satellite experiments. Hence these experiments will be emphasized in the brief review that follows along with two recent space shuttle XR observations and some new developments concerning the annihilation line.

XR emission from the GC was originally discovered by the Uhuru satellite in 1971. A source region 1 ° in extent called GCX, was indicated by scans with a 0.5 ° collimator. In 1979 the Imaging Proportional Counter on HEAO 2 (Einstein Observatory) mapped GCX with 1′ resolution over the energy range 0.5-4.0 keV. GCX is revealed to be a complex of ~12 weak point sources embedded in diffuse emission stretched out along the galactic plane (see Fig. 1).[1] One of the point sources, no. 3, labeled 1E 1742.5-2859 contains Sgr A West and is thought to be associated with it. It is the most luminous source within the square box with $L = 1.5 \times 10^{35}$ ergs^{-1} and was constant to ~2 percent in two observations separated by six months. The luminosity is what one might expect for an OB star and falls well short of the brightest galactic XR sources ($L \sim 10^{38}$ ergs^{-1}).

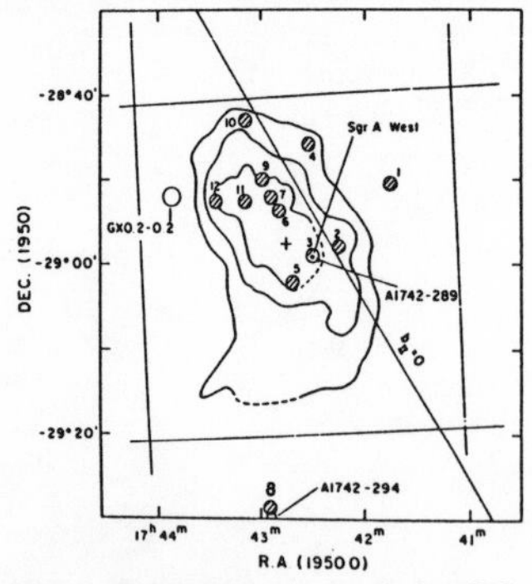

Fig. 1. Schematic map of the GC XR radiation as observed by the IPC on HEA0 2. The circles represent the point sources and the heavy contours represent the diffuse radiation centered on the cross. At 10kpc 30′ ~ 100 pc.

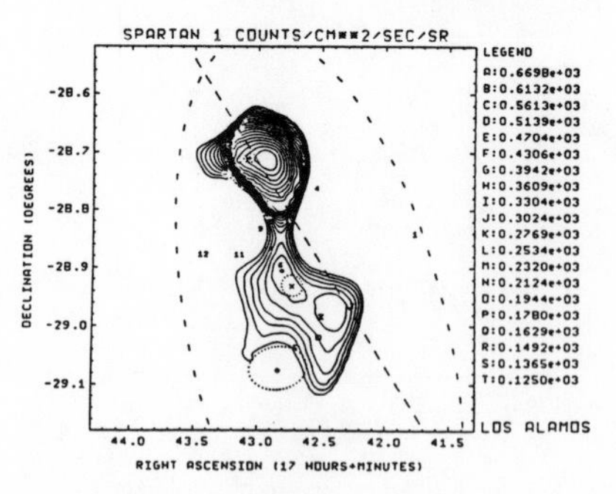

Fig. 2. Spartan 1 image of the diffuse emission surrounding the position of Sgr A West indicated by an hourglass. Note how the diffuse emission is connected to Einstein source no. 10 to the NE.

The picture changes significantly as one moves to hard XR's and low energy GR's. In this energy regime imaging instruments are only now being developed. A 6 °×6 ° square region centered on Sgr A West was imaged over the energy range 2.5-32 keV using dual coded mask and proportional counter telescopes flown by the University of Birmingham group on spacelab 2 during July 1985. Within ~1 ° of the GC only four Einstein point sources were visible, Sgr A West, no. 10 (1E 1743.1-2843), no. 8 (1E 1742.9-2928) and 1E1740.7-2942 all embedded in an extended source.[20] Sgr A West was the weakest of these sources and appears as an extension of source no. 10. In an image achieved with 3′ resolution Sgr A West is marginally resolved. These workers also generated a series of images in successive energy bands. The most notable feature here is that the source 1E1740.7-2942, ~1 ° to the SW of Sgr A West, increases in intensity with increasing energy until in the top band it becomes the brightest source in the region.

A 2.5 °×2.5 ° region centered on Sgr A West was studied over the energy range 1-15 keV with a collimated (5′×3 °) proportional counter system that flew on spartan 1 during June 1985. Multiple scans taken across the region from different directions allowed images of the region to be constructed by the Los Alamos, NRL and NW group.[21] In general the results are consistent with those of spacelab 2 except for the fact that Sgr A West is *not* detected as a point source by spartan 1. This is seen clearly in Fig. 2. Only diffuse emission ~ 100 pc in extent is seen at the GC. Perhaps the spacelab 2 detection represents a local enhancement in the diffuse source. In any event it seems clear that if the Sgr A West XR source has not in fact turned "off" between 1979 and 1985 it has probably turned down considerably. This may be related to the behavior of the annihilation line source discussed below.

All of the measurements reported so far at energies $\gtrsim$ 30 keV have been achieved with detectors and either active or passive mechanical collimators. The resolution is observed to degrade with increasing energy because of the increasing penetrating power of the radiation. Fig. 3 shows scans made across the GC region with the low energy detector on HEAO 1 which had a 1.6 ° FWHM collimator.[2] In the 13 to 50 keV range the brightness of GCX is typical of the other sources in the region. However in the 50 keV to 100 keV range GCX is the brightest source and accounts for ~1/3 of the net 100 keV flux from the region ±15 ° from the GC.

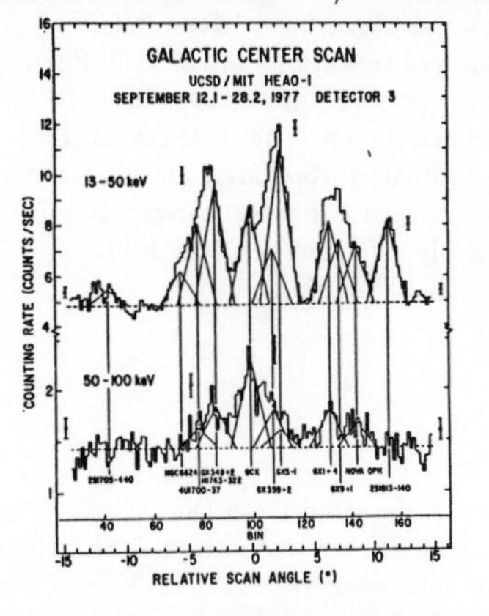

Fig. 3. HEAO 1 LED scans of the GC. The data are fitted by a background, indicated by a dashed line, and the flux from discrete sources. The triangular shape for the sources is due to the transmission collimator.

Numerous measurements of the GC spectrum have been made in the energy region between 100 keV and 1 MeV with typical resolution of ~15 °. A sample of these spectra are indicated in Fig. 4.[2] Two important conclusions can be drawn from these observations. Every group that has looked at the GC in this energy range has detected an intense source of inverse power law continuum radiation with spectral index ~−2 and maximum luminosity $\sim 3\times 10^{38}$ $ergs^{-1}$. Every observing group that has come back to look at the source for a second time has found it to be highly variable on a time frame of months. Perhaps the most striking example of the variability is indicated by the HEA0 3 observations (8 and 9 of Fig. 4) which seem to have caught the source in its extreme high state in the fall of 1979 and its extreme low state in the spring of 1980 (a factor of 8 decrease at 300 keV in 6 months). There is a trend for the spectrum to soften as the intensity decreases. The sparse data which exists between 1 and 10 MeV (obtained by the high energy detector on HEA0 1 and a Rice University balloon flight) may indicate a softening of the spectrum at energies above ~ 2 MeV.[2] Indeed the spectrum above 10 MeV must necessarily soften not to exceed the known >100 MeV flux.

The spark chamber that flew on COS B mapped our Galaxy in the energy range 70 MeV to 5 GeV producing beautiful contour maps of the GR luminosity. The Galaxy is seen to be a continuous source of diffuse emission probably due to the interactions between the cosmic-rays and the interstellar gas. However embedded in the diffuse emission are 25 point like sources.[3] Only 4 of them have been identified with known objects. One of the 21 remaining sources, 2CG 359-00, contains Sgr A West within its 1 ° error radius. The luminosity of the source is $\sim 8 \times 10^{36}$ erg s^{-1} if it is at a distance of 10 kpc. Little more is known about the source and indeed considerable controversy exists as to whether or not the unidentified COS B sources are truly point sources or represent local enhancements in the cosmic-ray or interstellar gas density.

In summary, the picture of the GC continuum source that emerges is dominated by a bright, hard and variable low energy GR power law source which in its high state is the most luminous in the Galaxy and has the hardest spectral index of any known galactic source. While the source is present in soft XR's and high energy GR's it does not appear particularly distinguished at these energies. The scanning measurements have located the source to within a few degrees of Sgr A West and because of its unusual nature most observers have tended to associate it with the galactic nucleus itself. The new spacelab 2 results challenge that assumption and indicate that the source may be located ~ 1 ° to the SW of the GC. However it is clear that the XR source at Sgr A West was much brighter in the past suggesting the possibility that the recent spacelab 2 result may be misleading us.

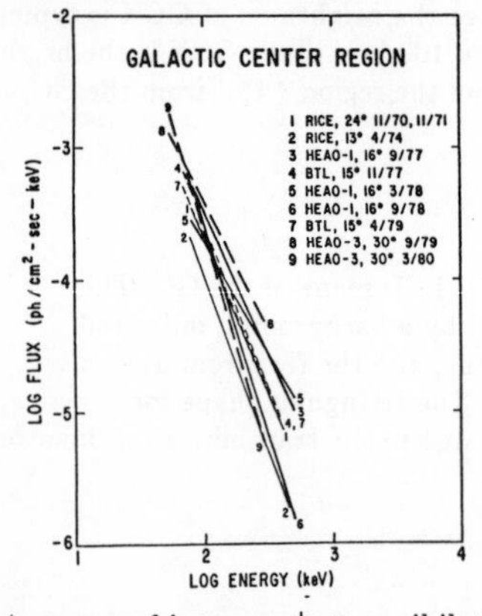

Fig. 4. A sample of hard XR/soft GR continuum observations of the GC. For simplicity, only the power law fits to the data are shown. The instrumental FOV is indicated. Repeated measurements by the same group have consistently yielded significantly different results. BTL indicates Bell/Sandia.

A source of intense e^+e^- annihilation line radiation was detected from the GC early in the 1970's during a series of Rice University balloon flights with low resolution NaI detectors. However it was not until 1977 that the annihilation line energy of 511 keV was clearly identified with high-resolution Ge detectors flown by the Bell/Sandia balloon group. A total of 17 different balloon and satellite observations (including the recent SMM results) involving seven different scientific groups has now clearly established the existence of a powerful, compact and variable source of sharp annihilation line radiation within ± 4 ° of the GC.[4] The compact source appears to be embedded in an extended annihilation line source associated with the galactic plane.[5] Surprisingly, the compact source was observed to "turn off" rather abruptly at the beginning of this decade and has since remained in a low state. A summary of the line flux measurements is given in Fig. 5. Again the observations are being made with wide field of view instruments and it is the unusual nature of the source which suggests an association with the galactic nucleus itself.

The highest quality data was obtained with the Ge spectrometer that flew on HEA0 3.[6] The galactic plane was scanned for 2 weeks in the fall of 1979 and again in the spring of 1980. In Fig. 6 a net GC spectrum near 511 keV is shown for the full 2-week scan in the fall of 1979. The line is narrow ($\lesssim$ 2.5 keV FWHM) with an intensity of $(1.85 \pm 0.21)\times 10^{-3}$ photons cm^{-2} s^{-1}. A striking decrease in source intensity by a factor of 3 to $(0.65 \pm 0.27)\times 10^{-3}$ photons cm^{-2} s^{-1} was detected when the source was reobserved 6 months later. This decrease, which has been confirmed by repeated balloon observations,[7] implies that the extent of the source and annihilation region are less than the light-travel distance of 10^{18}cm. The variability also requires the gas density to be $\gtrsim 10^5$ cm^{-3} so that the positrons can slow down and annihilate in less than six months. The gas temperature must be $<5\times 10^4$ K so as to keep the Doppler broadening of the line consistent with observations.

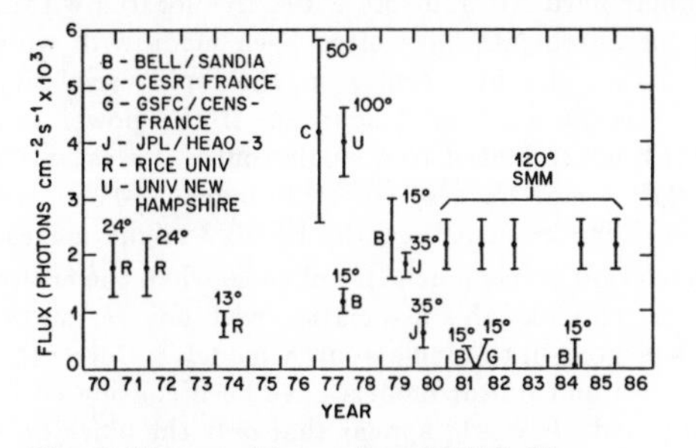

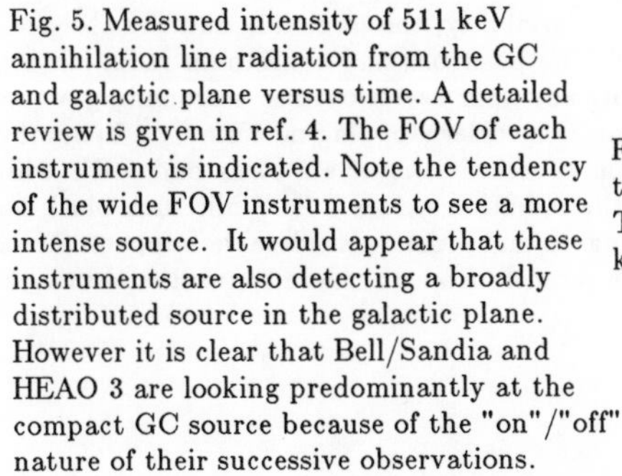
Fig. 5. Measured intensity of 511 keV annihilation line radiation from the GC and galactic plane versus time. A detailed review is given in ref. 4. The FOV of each instrument is indicated. Note the tendency of the wide FOV instruments to see a more intense source. It would appear that these instruments are also detecting a broadly distributed source in the galactic plane. However it is clear that Bell/Sandia and HEAO 3 are looking predominantly at the compact GC source because of the "on"/"off" nature of their successive observations.

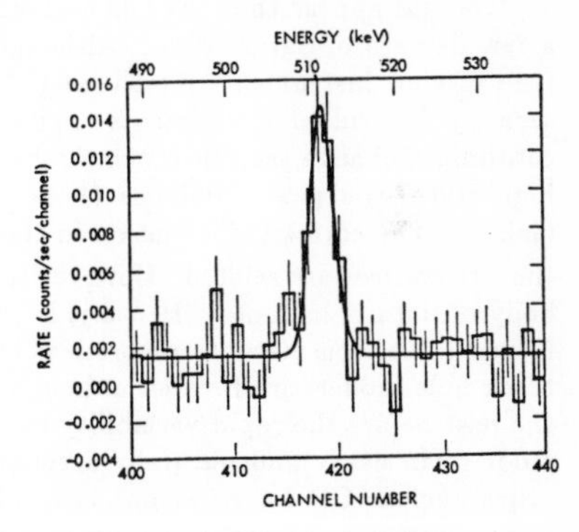

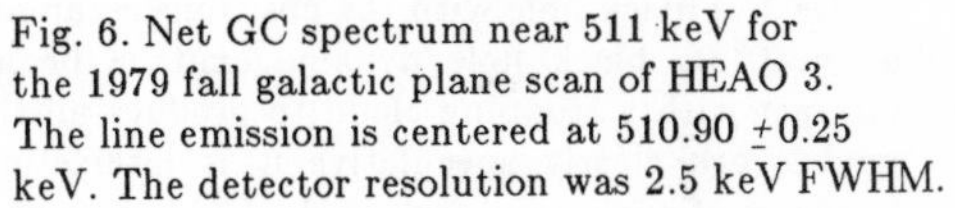
Fig. 6. Net GC spectrum near 511 keV for the 1979 fall galactic plane scan of HEAO 3. The line emission is centered at 510.90 $\pm$0.25 keV. The detector resolution was 2.5 keV FWHM.

Strong evidence for the presence of a three-photon triplet positronium (Ps) continuum is seen in the HEA0 3 spectrum shown in Fig. 7.[8] While in principle, positrons can annihilate from either the free state or bound Ps state, detailed Monte Carlo calculations[9] and laboratory simulation experiments show that for most astrophysical situations Ps annihilation should dominate, occurring $\sim 90\%$ of the time in a cold neutral medium.[10] The telltale sign of Ps annihilation is a low energy tail on the annihilation line decreasing monotonically with energy. The presence of such a continuum seems evident in Fig. 7 along with the power law continuum discussed earlier. It is possible to deduce a Ps fraction f (defined to be the fraction of incident positrons annihilating from the bound Ps state) from the HEA0 3 data and some of the balloon data by making some simple assumptions about the overall spectrum. This fraction, shown in Fig. 8, is found to be satisfyingly large. Some confusion exists in the literature on this point because of inconsistent definitions of f.[11,12] Taking f to be ~ 0.9 requires an overall annihilation rate $\sim 4\times 10^{43}$ positrons^{-1} at a distance of 10 kpc corresponding to a luminosity of $\sim 6\times 10^{37}$ ergs^{-1} in two and three photon radiation. Hence the power law and annihilation luminosities are comparable.

In principle a detailed study of the lineshape, linewidth and Ps fraction can greatly constrain the parameters of the annihilation medium. Based on the Monte Carlo[9] calculations it was originally thought that the medium must be at least partially ionized with $n_e \gtrsim 0.1$ and $T \lesssim 5\times10^4$ K. A cold neutral medium was ruled out because it was felt that the annihilation linewidth would be dominated by a severely Doppler broadened (6.5 keV FWHM) singlet Ps line from inflight Ps formation before thermalization. However recent simulation experiments have shown this not to be the case.[13] As seen in Fig. 9 for H_2, the composite linewidth is dominated by the very narrow direct annihilation line from the 10.3% of the positrons that survive below the positronium formation threshold (8.6 eV) in this gas. The composite lines from neutral H_2 and H (FWHM $\sim$ 2.2 keV) have been shown consistent with the GC observations[14]. Hence little can be said at the present moment about the ionization state of the annihilation medium. It has also been suggested that dust grains may play a significant role if the medium is hot and partially ionized.[15]

It would appear that the GC harbors a unique hard XR and soft GR source located within a few degrees of Sgr A West. Although the measurements have often been made with wide field of view instruments introducing the possibility of source confusion, the rapid variability seen by individual experiments argues for an origin in 1 or 2 sources. If the power law continuum change seen in 6 months by HEAO 3 is attributed to a single source it was, in its high state, the most luminous ($\sim 4\times10^{38}$ $ergs^{-1}$) and hardest (spectral index $\sim$-1) in the Galaxy. The correlated variation in the 511 keV line intensity seen by HEA0 3 might suggest the two sources are related. However this correlation seems much less obvious when the entire body of data is included. Hence it is premature to make this association with any certainty. Nevertheless considerable theoretical effort has gone into single source models. Currently, black hole models are favored although supernova and pulsar models have been considered in the past before the rapid variability was established. It would appear that only the black hole models can easily and naturally account for the variability, the large flux and the absence of other nuclear GR lines. In one class of models the positrons and continuum radiation arise from thermal processes occurring in a hot, optically thin accretion disc surrounding a small ($\sim 100M_o$) black hole with the positrons escaping to a cooler stopping medium.[16,17] In another, an $\sim 10^6 M_o$ black hole dynamo emits a beam of relativistic e^+e^- pairs and GR's which interact with a stopping cloud to produce all the radiation we see.[18,19] While such black hole models are clearly speculative it is interesting to note that considerable support for the presence of such an object at the GC has come recently from the more conventional branches of astronomy.

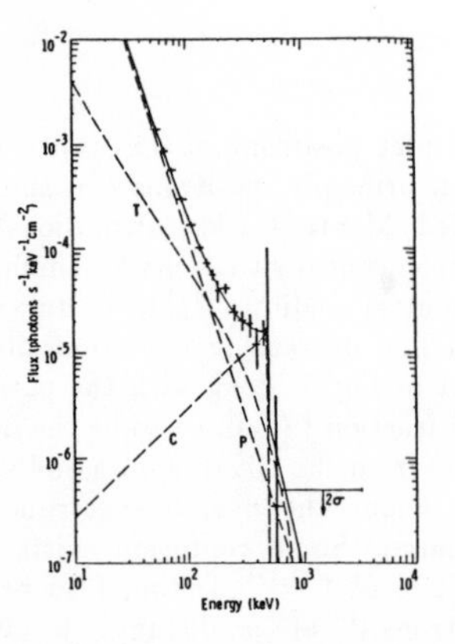

Fig. 7. Net GC continuum spectrum for the spring 1980 galactic plane scan of HEAO 3. The data have been fit to the sum of a power law (P), a Ps continuum (C) and a thermal spectrum (T). Little evidence for emission above 511 keV is seen.

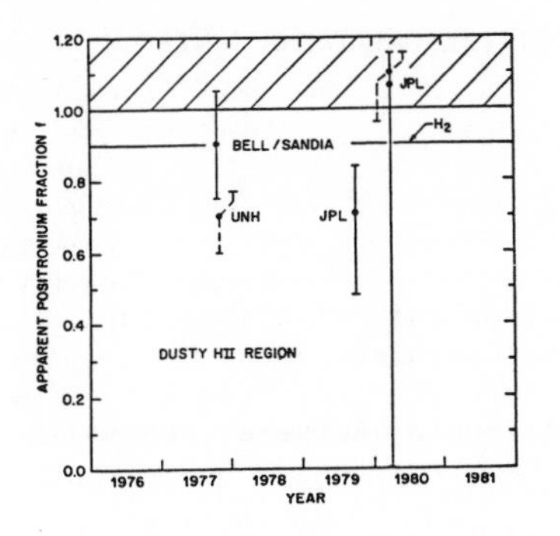

Fig. 8. The apparent Ps fraction f is shown for four different GC observations. Values greater than f=1 are non-physical.[12]

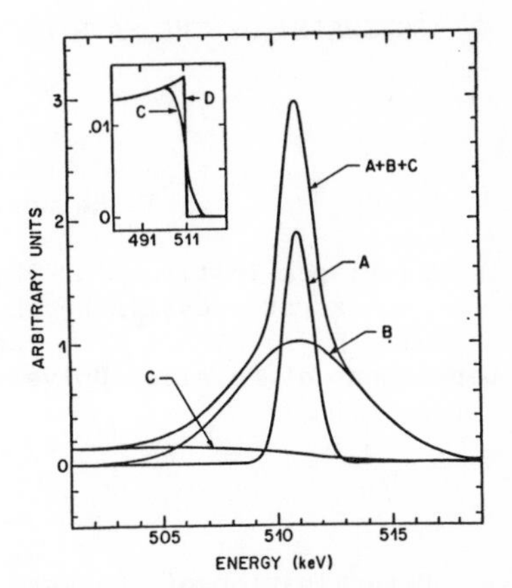

Fig. 9. Composite spectrum for positron annihilation in a neutral H_2 medium.[13]

Finally we would like again to emphasize that a definite connection has not yet been made between the GR source and Sgr A West. While the roughly concurrent decrease in intensity of the Sgr A West XR source and 511 keV line intensity seem suggestive, the new spacelab 2 images indicate alternative hard XR sites. If however this does turn out to be the case, the GR source still only accounts for about 0.1% of the total bolometric luminosity inferred for the central pc of the Galaxy.

REFERENCES

1. Watson, M. G, Willingale, R., Grindlay, J. E. and Hertz, P., Ap.J., *250,* 142 (1981).
2. Matteson, J. L., in "The Galactic Center", edited by G. R. Riegler and R. D. Blandford, AIP Conf. Proc. No. 83, p.109 (1982).
3. Swanenburg, B. N., et al., Ap.J. (Letters), *243,* L69 (1981).
4. MacCallum, C. J. and Leventhal, M., in "Positron-Electron Pairs in Astrophysics", ed. by M. L. Burns, A. K. Harding and R. Ramaty, AIP Conf. Proc. No. 101, p. 211 (1983).
5. Share, G., private communication.
6. Riegler, G. R., et al., Ap.J. (Letters), *248,* L13 (1981).
7. Leventhal, M., et al., Ap.J., *302,* 459 (1986).
8. Riegler, G. R., et al., Ap.J. (Letters), *294,* L13 (1985).
9. Bussard, R. W., Ramaty, R., and Drachman, R. J., Ap.J., *228,* 928 (1979).
10. Brown, B. L., Leventhal, M. and Mills, A. P., Jr., Phys. Rev. A, *33,* 228 (1986).
11. Riegler, G. R., et al., Ap.J. (Letters), *305,* L33 (1986).
12. Brown, B. L. and Leventhal, M., Ap.J. submitted (1987).
13. Brown, B. L., and Leventhal, M., Phys. Rev. Lett., *57,* 1651 (1986).
14. Brown, B. L., Ap.J. (Letters), *292,* L67 (1985).
15. Zurek, W. H., Ap.J., *289,* 603 (1985).
16. McKinley, J. M., in "Positron (Electron)-Gas Scattering", ed. by W. E. Kauppila, T. S. Stein and J. M. Wadehra, World Scientific, p. 152 (1986).
17. Lingenfelter, R. E., and Ramaty, R., ref. 4, p. 267.
18. Burns, M. L., ref. 4, p. 281.
19. Kardashev, N. S., et al., ref. 4, p. 253.
20. Willmore, P., and Skinner, G., private communication.
21. Kawai, N., Fenimore, E. E., Middleditch, J., Cruddace, R., Fritz, G., Snyder, W., and Ulmer, M. preprint.

ESTIMATES OF THE MASS DISTRIBUTION IN THE GALACTIC CENTER

Reinhard Genzel *

Max-Planck-Institut fur Physik und Astrophysik, Institut fur extraterrestrische Physik, D-8046 Garching, FRG
and
Department of Physics, University of California, Berkeley, CA 94720

Determination of the mass distribution in the center of the Galaxy is of fundamental importance, especially as a direct test of whether or not there is a massive central object. I am going to consider three approaches to this important problem. First, the stellar surface brightness distribution can be converted to a mass distribution if an approximately constant mass to luminosity ratio can be adopted. Second, the velocities of interstellar gas clouds can probe the mass distribution if the gas clouds rotate about the dynamical center, or if the velocities can be interpreted in terms of other well defined orbits in the center's gravitational field, or else if the virial theorem can be applied in a statistical sense. Third, the mass distribution can be determined from the statistics of stellar velocities provided that the stellar distribution has a relatively simple symmetry, such as spherical or spheroidal. We now discuss these three categories in turn.

1. Stellar Surface Brightness Distribution

Because of strong foreground extinction by interstellar dust ($A_V \approx 30$), [6], the galactic center stellar cluster can only be observed at $\lambda \geq 1$ μm. 2 μm observations by Becklin and Neugebauer [6),7)], Allen et al. [1)] and Matsumoto et al. [28)] have shown the presence of an extensive stellar cluster with a brightness distribution scaling with about $R^{-0.75}$ from $\approx 1°$ to $\approx 1''$. The stellar light is peaked within a few arcsec of a complex of infrared sources called IRS 16 which is, therefore, commonly interpreted to be the center of the galaxy. Beyond $R \approx 6°$, the surface brightness distribution steepens and scales approximately like R^{-2} [28)]. Most of these stars probably represent M and K giants of surface temperature ≈ 4000 K. Allen et al.[1)] have interpreted the central cusp of the brightness distribution as an indication of a small core radius of the cluster ($R_{core} \leq 0.05$ pc). However, the central parsec of the Galaxy happens to lie in a minimum of line of sight extinction [36)]. Furthermore, a number of the

* Review given at 13th Texas Symposium on Relativistic Astrophysics, Chicago, 1986

2 μm sources, including the bright supergiant IRS 7 and the "blue" IRS 16 complex itself, are not part of the general red giant population, thus artificially decreasing the observed core radius. With a correction for these effects, current estimates for the cluster's core radius range between ≈ 0.1 pc [2] and 1 pc [36].

The near-infrared stellar luminosity of the central 2 pc radius region, by extrapolation of the measured near-infrared flux with T ≈ 4000 K and allowance for reddening due to intervening dust, is $L_{IR} \propto 6.5 \times 10^6 L_{\odot} R_o{}^2(10)$ * [6].

Assuming a constant mass to light ratio, the 2 μm light distribution then suggests that the mass of stars emitting in the near-infrared increases somewhat steeper than linear with radius ($M(R) \propto R^{1.2}$) between about 1 and a few 10^2 pc [6),1),28]. Inside of the cluster's core radius, mass increases much faster ($M_*(R) \approx R^2$ to R^3). Beyond about one kpc, the mass inferred from the light distribution increases much slower than linear with radius [28]. To assign to the so derived distribution an absolute scale, Becklin and Neugebauer [6] adopted a mass to total near-infrared luminosity ratio equal to that of the nucleus of M 31 ($M/L \approx 3 M_{\odot}/L_{\odot}$) with the reasoning that their 2 μm distributions and total near-infrared luminosities are quite similar. On the basis of Becklin and Neugebauer's data, Sanders and Lowinger [38] and Oort [30] constructed more refined models, taking into account deviations from spherical symmetry as suggested by the flattening of the cluster's light distribution.

2. Interstellar Gas Dynamics

2.1 100 to 1000 pc

Historically, the rotating "nuclear disk" of neutral atomic hydrogen on a scale of a few hundred pc to about one kpc was the other early indicator of the mass distribution near the center. The mass distribution derived from the rotation curve of the atomic gas along the galactic plane [37),30),44] with the assumption of spherical symmetry is shown in Fig. 1. From the mass derived from the 21 cm HI rotation curve, an improved M/L ratio for the stellar cluster can be derived [30]. In the best fit of their 2 μm data to the HI rotation curve, Matsumoto et al. [28] find $M/L_{IR} = 2 M_{\odot}/L_{\odot}$. A further correction comes from the finding [9),25] that the "nuclear disk" is actually tilted with respect to the galactic plane and has significant noncircular motions. The observed terminal velocities along the plane thus are strict upper limits to the equilibrium rotation velocities. Burton and Liszt's [9] analysis suggests that radial and tangential velocities in the "nuclear disk" or "nuclear bar" are about comparable. With $v_{rot} \approx v_{tan} = 180$ km s^{-1}, the mass to luminosity ratio calibration at a galactocentric radius of a few hundred pc gives $M/L_{IR} \approx 1 M_{\odot}/L_{\odot} R_o{}^{-1}(10)$. The medium thin line in Fig. 1 marks the mass distribution derived from the 2 μm light distribution with this calibration of M/L_{IR} and under the assumption that M/L_{IR} does not change

* $R_o(10)$ is the Earth-Galactic Center distance in units of 10 kpc. The most likely value of R_o is between 7 and 8 kpc.

with radius.

2.2. 10 to a few 10^2pc

Between about 10 and a few hundred pc radius, there is little certain information on the mass distribution from interstellar gas dynamics, although there are many massive molecular clouds in this region. However, these clouds do not show a strong signature of rotation and have large (50 to 100 km s^{-1}) non circular velocities. In their recent analysis of extensive mapping of ^{13}CO J=1→0 line emission in the central few degrees [4], Bally et al. [5] deduce an approximately flat rotation curve for clouds between 10 and 10^2 pc from the center.

2.3. 1.7 to 10 pc

As has been discussed by Jaffe earlier in this volume (see also [14]), high spatial resolution infrared and millimeter observations demonstrate that there is an essentially complete 2 pc radius ring of gas which forms the inner edge of a thin, disk-like structure extending to at least 8 pc from the center [27),41),11),21),15], in agreement with the earlier inference by Becklin, Gatley and Werner [8]. The ring/disk structure is inclined by 60° to 70° with respect to the line of sight. As an example, Fig. 2 shows a velocity integrated map of the 3 mm HCN J=1→0 transition, obtained with the Hat Creek interferometer at 10" resolution, together with typical spectra.

The available infrared and microwave spectroscopic measurements give consistent and detailed information on the neutral gas dynamics in the disk. The dominant large scale motion of the majority of the neutral gas is rotation about the center. Several investigations find that systematic radial motion of the disk must be less than about 30 km s^{-1} while a few clouds do show strong deviations from pure circular motion emission. Gatley et al. [11], and Kaifu et al. [21], on the other hand, conclude that systematic radial motions as large as 50 km s^{-1} are characteristic of the ring as a whole [27),41),15]. The derived rotation velocity as a function of distance from the center, corrected for inclination ($i \approx 70°$), is approximately constant between 2 and 4 pc ($v_{rot} = 110 \pm 5$ km s^{-1}).

Beyond 4 pc the rotation curve may be more complex. The millimeter lines in the southern part of the disk show a splitting into two velocity components (-110 and -80 km s^{-1}, Gusten, priv. comm. [41]), and the velocity centroids of the high excitation C^+ and CO=7→6 lines decrease from -100 to about -70 km s^{-1} between 4 and 6 pc [19),27),15]. It is presently not clear which (if any) of the two components are representative of the equilibrium rotation velocity.

While a circulation disk is a good first order description of the neutral gas dynamics, the rotation pattern is strongly perturbed. Large line widths <u>everywhere</u> in the disk ($\Delta v_{FWHM} \approx$ 50 to 60 km s^{-1} in a 5 to 10" beam) indicate large <u>local</u> non circular motions. A bright cloud near the minor axis at $R \approx 2$ pc west of the nucleus, prominent in H_2, CO and HCN emission, does not fit the general rotation pattern, and is clearly an example where the radial velocity of a particular cloud may be comparable to its circular velocity. Furthermore, the spatial distribution and velocity field of the HCN gas at the inner edge of the disk indicate that not all the material is in one plane, but that the inner disk is warped or kinked. Finally, the disk's position angle on the sky changes with distance from the center. The plane of rotation is within 5° of the galactic plane at the inner edge of the disk, but is tilted by about 20 to 25° relative to the plane at radii of 3 to 5 pc.

The corresponding inferred mass distribution ($M(R) \approx 2.8 \pm 0.3 \times 10^6 R(pc) M_\odot$ between 2 and 4 pc, heavily hatched area in Fig. 1) is in very good agreement among several different data sets, including the "western arc" of ionized gas at $R \approx 1.7$ pc $R_o(10)$ [40]. The distribution and velocities of ionized gas clouds in the western arc suggest that this streamer is rotating about the center with a velocity of 110 ± 10 km s^{-1} and probably is the inner ionized surface of the circum-nuclear ring [40),45),46)] (heavy black dot in Fig. 1). The inferred mass distribution also agrees with the stellar light distribution and M/L_{IR}=0.8 to 1.0 $M_\odot/L_\odot$ $R_o^{-1}(10)$. As discussed above, the velocity field of the neutral ring at $R > 4$ pc is more complex and shows a velocity component $|100$ km $s^{-1}|$. If this dropoff is representative of the equilibrium rotation curve, the mass distribution outside 4 pc has to be more concentrated than that of the 2 μm light distribution or of an isothermal cluster ($\rho_*(R \propto R^{-\alpha}$ with $\alpha > 2.5$).

2.4. <u>0.1 to 1.7 pc</u>

Fig. 3 shows the velocities of ionized gas clouds within 1.7 pc of the center, as derived from 12.8 μm [NeII] and radio recombination line spectroscopy [22),40),45),46)], superposed on the 5 GHz radio continuum map of Lo and Claussen [26)]. An inspection of the figure clearly shows that inside of about 1.7 pc, the average velocities increase toward the center, from about 100 km s^{-1} at R=1.7 pc $R_o(10)$ to 150 km s^{-1} at 0.7 pc, 260 km s^{-1} at 0.3 pc and 700 km s^{-1} at 0.1 pc. If that increase can be interpreted in terms of the virial theorem, a mass of 1 to 4 x 10^6 $M_\odot$ $R_o(10)$ has to be concentrated within 0.1 pc of IRS 16/Sgr A*, [10),40),29)]. This conclusion confirms the earlier results of Lacy et al. [22)]. However, since the velocity field is dominated by a few coherent gas streamers, an orbital analysis rather than a statistical approach may be more appropriate.

An interpretation of the velocity field of the ionized streamers other than the western arc discussed above in terms of pure circular motion is not appropriate since the streamers show strong noncircular velocities. As a demonstration, Fig. 4 gives the velocity centroids of the 12.8 μm[NeII] lines along the northern arm, plotted against declination offset from the bright cloud IRS 1 (from [40)]). The mean velocity gradient of the northern arm is about 4 times larger than that of the western arc, indicating that it comes much closer to the center ($R_{min} \approx 0.5$ pc). There are also clear deviations from a mean linear slope, indicating that the gas is moving radially. Serabyn and Lacy [40)] have demonstrated that the spatial distribution and velocity field of the northern arm can be well fitted by <u>simple</u>, single particle orbits (ellipse, parabolas) in gravitational fields of various forms.

The gas dynamics of the northern arm is well represented by a parabolic or elliptical orbit in the field of a 2 to 4 x 10^6 $M_\odot$ $R_o(10)$ point mass (Fig. 4a). More specifically, a mass of at least 3 x 10^6 $M_\odot$ $R_o(10)$ has to be contained within ≈ 0.5 pc, the pericenter distance of the northern arm (upper limit in Fig. 1). Stellar clusters with an isothermal ($\rho_* \propto R^{-2}$) or shallower distribution do not fit the data since they require ≥ 12 x 10^6 $M_\odot$ to be contained within 1.7 pc, in order to provide adequate acceleration at the radius of the northern arm. That large a mass is inconsistent with the mass estimated from the velocity field of the western arc ($M(R \leq 1.7$ pc$) = 4.8 \pm 1 \times 10^6$ $M_\odot$, [40]). To fit the data without a central point mass, any stellar cluster model has to have a density distribution $\rho_*(R) \propto R^{-\alpha}$ with $\alpha >$ 2.7. In a similar analysis for the eastern arm and central edge, Serabyn, Lacy and Townes [42] find that at least 2 to 3 x 10^6 $M_\odot$ $R_o(10)$ have to be contained within ≤ 0.2 to 0.3 pc from the center (upper limit in Fig. 1). Again, isothermal or shallower cluster models do not fit the measurements.

Because of the small scale size, an orbital analysis for the innermost [NeII] gas clouds and the HeI/HI broad line zone is not available yet, but there is some evidence for short streamers even in the innermost region of the bar [42),24]. Nevertheless, an estimate of the mass can be made in this case assuming, for example, that the clouds have fallen in to their current radius R_1 from radius R_o with initial radial velocity $v_o \ll v_{rot}$ through gravitational fields of various forms. For a point mass M, v_1, the velocity reached at R_1 is, of course, proportional to $(M/R_1)^{1/2}$. For an isothermal cluster $v_1 = v_{rot}\sqrt{2 \ln (R_o/R_1)}$. From this simple, conservative model (it does not take into account an upward correction of the velocities for projection effects), the ± 260 km s^{-1} clouds (Fig. 3) can just barely be explained without a massive central object if these clouds have fallen in from the radius of the neutral ring. The ± 700 km s^{-1} velocity range of the HeI/HI lines within 1" of IRS 16, [20),12] and recently discovered [NeII] clouds with velocities up to ± 400 km s^{-1}, and possibly 600 km s^{-1}, [24] cannot be explained by an isothermal or shallower stellar distribution, or by the stellar mass distribution derived from the 2 μm surface brightness distribution unless unreasonably large initial radii are assumed.

A likely physical interpretation of the noncircular motions of the ionized streamers may be that gas clouds in the neutral disk collide with each other, in part lose angular momentum, and fall toward the center [26]. It should be noted that the above arguments on the mass distribution also hold at least approximately if the ionized streamers represent <u>outward</u> and not inward motion. The main reason is that the <u>change</u> of velocity along a given path is used to measure the shape of the gravitational field. This change, of course, is independent of the direction of motion, as long as gravitation is the dominant force and single particle orbits (one set of initial conditions) are a viable approximation to the streamers' motions.

3. Stellar Dynamics

Measurements of stellar velocities in the galactic center come from a sample of about 55 OH-IR stars in the central ± 200 pc studied by Habing et al. [17] and Winnberg et al. [47], and from a sample of 5 M giants and the supergiant IRS 7 in the central 2 pc of the Galaxy [43]. The OH-IR stars probably sample mass-losing asymptotic giant branch stars of mass $\geq 1\ M_{\odot}$ and age 10^8 to 10^9 years [18]. There is some evidence that the OH-IR stars at distances ≥ 50 pc from the center (the "Habing et al. 1983 sample") represent somewhat older stars in the bulge while the OH-IR stars at 1 to 50 pc galactocentric radius in the sample presented by Winnberg (which are more concentrated toward the galactic center and the galactic plane) are younger and somewhat more massive. The enclosed masses at different R derived from these observations from a calculation of the velocity dispersion and a derivation of the mass from the virial theorem ($M(R) = \beta \sigma_v^2\ R/G$, $\beta = 2$ to 3, assuming a spherical distribution) are shown in Fig. 1 as large open quadrangles. The surrounding shaded boxes give the uncertainties, as estimated from the statistical error bars and the application of different mass estimators discussed by Sellgren et al. [43]. While the stars within a few pc on either side of the center show no rotation and a line of sight velocity dispersion of about 80 km s^{-1}, the stars beyond about R = 3 pc show a radially increasing component of rotation in the same sense as galactic rotation. Gwinn et al. [16] find a rotation velocity of about 50 to 90 km s^{-1} for the M stars at ≈ 4 pc. For the OH-IR stars Winnberg et al. [47] and Habing et al. [17] derive a rotation velocity of $v_{rot} \approx (270 \pm 77) l$ km s^{-1}, where l is galactic longitude in degrees.

4. Discussion of the Analysis of the Mass Distribution

The different estimators of mass shown in Fig. 1 appear to give a very consistent determination of the mass distribution in the central few hundred pc of the Galaxy. While individual stars do not have the same velocities as do gas clouds along the same line of sight, the statistical distribution of velocities agree very well within the uncertainties with the velocities derived from the gas dynamics.

Overall, the enclosed mass decreases approximately linearly or steeper with decreasing galactocentric radius from a few hundred pc to a few pc, representing a stellar cluster with $M/L_{IR} \approx 1\ M_{\odot}/L_{\odot}\ R_o^{-1}(10)$. There is some - although not conclusive - indication from low gas and stellar velocities that the mass distribution decreases less steeply than linear between about 30 and 50 pc, and more steeply between a few hundred and 30 pc, possibly suggesting different populations in the central region (galactic center cluster vs. bulge?: Winnberg et al. [47]). Inside of about 2 pc, the enclosed mass decreases much slower with decreasing radius and may approach a constant value. The measurements are best consistent with a central point mass of 1 to $3 \times 10^6\ M_{\odot}\ R_o(10)$ in combination with a near isothermal cluster of mass 1.5 to $2.5 \times 10^6\ R(pc)\ M_{\odot}$ outside of its core radius ($0.1 \leq R_c \leq 1$ pc). Two possible models combining a central point mass and a stellar cluster of two different core radii (0.1 and 1 pc) are shown as thin dotted curves in Fig. 1. The gas and stellar dynamics measurements are also consistent with only a stellar cluster and no central point source if the stellar density increases with increasing radius as fast or faster than $R^{-2.7}$ inside of ≈ 2 pc and about R^{-2} between 2 and 4 pc. The 2 μm stellar surface brightness distribution, however, does not show such a distribution, making it unlikely that a cluster with such a density distribution dominates the mass.

Note that the mass distribution of the stellar cluster may have to be scaled (down) for deviations from spherical symmetry. Evidence for such deviations comes from the flattened distribution of OH-IR stars (minor to major axis ratio $c/a \approx 0.7$ at $R \leq 50$ pc [47]), and $c/a \approx 0.6$ for the bulge at R a few 10^2 pc [18]), and possibly from the 2 μm stellar surface brightness distribution (see section 1.). Further evidence comes from the fact that the cluster appears to have a significant amount of angular momentum [47),16)]. For $v_{rot}/\sigma \approx 0.8$ at 4 pc [16]), the calculations by Peebles and Ostriker [33]) give $c/a \approx 0.7$ to 0.8. The corresponding correction factor for the mass enclosed within a sphere of a given radius compared to the spherical case is about 0.9. This factor is derived from the relationship between rotation velocity and mass in non-homogeneous spheroids given by Schmidt [39]).

Therefore, if the 2.2 μm light distribution is a measure of the total mass distribution of all stars in the center, the central mass concentration could be in form of a supermassive star (should such objects be stable) or more likely, in form of a massive black hole. The presence of $\approx 10^6$ $M_\odot$ black hole is by far sufficient, but not necessary to explain the current UV luminosity of the galactic center. The required mass accretion rate for a massive black hole is about 10^{-5} $M_\odot$ y^{-1} $R_o{}^2(10)$, small compared to the estimates of current gaseous accretion in the central pc (10^{-3} to 10^{-2} $M_\odot$ y^{-1} $R_o{}^2(10)$), and probably, [31),23)], but not necessarily, [35)], small compared to the estimated average tidal disruption rate of stars.

The evidence for an unusual mass distribution in the galactic canter could be misleading, however. Each estimator of mass requires certain assumptions which may be challenged. The conversion of surface brightness to mass not only requires calibration of the M/L ratio by other methods (note that the M/L_{IR} ratio discussed above is strictly an upper limit to M/L, because of additional radiation in the visible, UV or far-infrared regions), but also relies on a reasonably constant M/L ratio. This assumption may be hazardous, especially in the very center of the Galaxy, where there appear to be sources of 2 μm radiation which are not part of the M/K giant population.

In the case of measurements of the dynamics, it is essential that gravitation is the dominant force. While this assumption is more likely to hold for the stars than the gas, a straightforward analysis of the stellar velocities requires that the stellar distribution function is not explicitly a function of angular momentum (as would be the case for triaxial distribution, for example). For the gas dynamics, equilibrium may not be a good assumption and forces other than gravitation may be important. This concern is especially pertinent for the motions of the ionized gas streamers and the HeI/HI broad line region which are the most important for the black hole question. While the broad line zone near IRS 16 would be the most definite indication of a massive point source if the motions were either due to rotation or inflow of gas, the current data are equally well [13]) or better consistent with outflow [12),31)]. In this case the HeI/HI data do not constrain the central mass. Outflow also has to be considered as a possibility for the dynamics of the ionized gas streamers, decreasing their value as a mass probe. For the [NeII] clouds at R < 1 pc magnetic forces or friction with hot, distributed gas [34)] could strongly affect or invalidate the "single orbit" analysis of Serabyn and Lacy [40)]. Such forces would, however, not qualitatively affect the evidence for a massive central object from the mere presence of high velocities, assuming that the gas is falling in. In the case of significant dissipation of kinetic energy by friction, the evidence for a massive object would even be strengthened.

In summary on the mass distribution, it has become clear that definite statements about the stellar mass distribution and the possible presence of a massive black hole can now be made with greater certainty than 5 years ago, but only under certain assumptions. The large number of new, high quality measurements on the gas dynamics have allowed to tighten the arguments in some areas (infalling streamers, rotating disk), but also have opened up new complications (deviations from equilibrium, warping). The evidence for a massive black hole in the galactic center from the gas and stellar dynamics is very suggestive, but not fully convincing. Better measurements of the ionized gas dynamics in the central 5" and measurements of the dynamics of the central stellar cluster are highly desirable.

References

1) Allen, D.A., Hyland, A.R. and Jones, T.J. 1983, MNRAS 204, 1145
2) Allen, D.A. 1987, in Proceedings of "Townes"-Symposium (see Backer 1987
3) Backer, D. 1987, Proceedings of the Symposium on "The Galactic Center" in honor of Charles H. Townes, American Institute of Physics, in press
4) Bally, J., Stark, A.A., Wilson, R.W. and Henkel, C. 1986a, Ap.J. (Suppl.), in press
5) Bally, J., Stark, A.A., Wilson, R.W. and Henkel, C. 1986b, Ap.J., in press
6) Becklin, E.E. and Neugebauer, G. 1968, Ap.J. 151, 145
7) Becklin, E.E. and Neugebauer, G. 1975, Ap.J. (Letters) 200, L71
8) Becklin, E.E., Gatley, I. and Werner, M.W. 1982, Ap.J. 258, 134
9) Burton, W.B. and Liszt, H.S. 1978, Ap.J. 225, 815
10) Crawford, M.K., Genzel, R., Harris, A.I., Jaffe, D.T., Lacy, J.H., Lugten, J.B., Serabyn, E. and Townes, C.H. 1985, Nature 315, 467
11) Gatley, I., Jones, T.J., Hyland, A.R., Wade, R., Geballe, T.R. and Krisciunas, K.L. 1986, MNRAS 222, 299
12) Geballe, T.R., Krisciunas, K.L., Lee, T.J., Gatley, I., Wade, R., Duncan, W.D., Garden, R. and Becklin, E.E. 1984, Ap.J. 284, 118
13) Geballe, T.R., Krisciunas, K.L., Gatley, I. and Bird, M.C. 1987, in Proceedings of "Townes"-Symposium (see Backer 1987)
14) Genzel, R. 1987, Proceedings of NATO Summer School on "The Galaxy", eds. G. Gilmore and R.S. Carswell, Reidel (Dordrecht), in press
15) Gusten, R., Genzel, R., Wright, M.C.H., Jaffe, D.T., Stutzki, J. and Harris, A.I. 1987, Ap.J., in press
16) Gwinn, M.T., Sellgren, K., Becklin, E.E. and Hall, D.N.B. 1987, in Proceedings of the "Townes"-Symposium (see Backer 1987)
17) Habing, H.J., Olnon, F.M., Winnberg, A., Matthews, H.E. and Band, B. 1983, Astr.Ap. 128, 230
18) Habing, H.J. 1987, in Proceedings of the NATO Summer School on "The Galaxy", eds. G. Gilmore and R.S. Carswell, Reidel (Dordrecht), in press
19) Harris, A.I., Jaffe, D.T., Silber, M. and Genzel, R. 1985, Ap.J. (Letters) 294, L93
20) Hall, D.N.B., Kleinmann, S.G. and Scoville, N.Z. 1982, Ap.J. (Letters) 262, L53
21) Kaifu, N., Hayashi, M., Inatani, J. and Gatley, I. 1987, in Proceedings of "Townes"-Symposium (see Backer 1987)
22) Lacy, J.H., Townes, C.H., Geballe, T.R. and Hollenbach, D.J. 1980, Ap.J. 241, 132
23) Lacy, J.H., Townes, C.H. and Hollenbach, D.J. 1982, Ap.J. 262, 120
24) Lacy, J.H., Lester, D. and Arens, J.F. 1987, in Proceedings of the "Townes"-Symposium (see Backer)
25) Liszt, H.S. and Burton, W.B. 1980, Ap.J. 236, 779
26) Lo, K.Y. and Claussen, M.J. 1983, Nature 306, 647
27) Lugten, J.B., Genzel, R., Crawford, M.K. and Townes, C.H. 1986, Ap.J. 306, 691
28) Matsumoto, T., Hayakawa, S., Koizumi, H., Murakami, H., Uyama,

K., Yamagami, T. and Thomas, J.A. 1982, in "The Galactic Center", eds. G.R. Riegler and R.D. Blandford, American Institute of Physics Conference Proceedings No. 83

29) Mezger, P.G. and Wink, J.E. 1986, Astr.Ap. 157, 252
30) Oort, J.H. 1977, Ann.Rev.Astr.Ap. 15, 295
31) Ozernoy, L.M. 1984, Astron.Zirk. 1342, 1
32) Ozernoy, L.M. 1987, this volume
33) Peebles, P.J. and Ostriker, J.P. 1973, Ap.J. 186, 467
34) Quinn, P.J. and Sussman, G.J. 1985, Ap.J. 288, 377
35) Rees, M. 1982, in "The Galactic Center", eds. G.R. Riegler and R.D. Blandford, American Institute of Physics Conference Proceedings No. 83
36) Rieke, G. and Lebofsky, M.J. 1987, in Proceedings of "Townes"-Symposium (see Backer 1987)
37) Rougoor, G.W. and Oort, J.H. 1960, Proc.Natl.Acad.Sci.USA, 46, 1
38) Sanders, R.H. and Lowinger, T. 1972, A.J. 77, 292
39) Schmidt, M. 1965, in "Galactic Structure", eds. A. Blaauw and M. Schmidt, The University of Chicago Press, pp. 513-529
40) Serabyn, E. and Lacy, J.H. 1985, Ap.J. 293,445
41) Serabyn, E., Gusten, R., Wink, J.E., Walmsley, C.M. and Zylka, R. 1986, Astr.Ap. 169, 85
42) Serabyn, E., Lacy, J.H. and Townes, C.H. 1987, in prep.
43) Sellgren, K., Hall, D.N.B., Kleinmann, S.G. and Scoville, N.Z. 1987, Ap.J., in prep.
44) Sinha, R.P. 1978, Astr.Ap. 69, 227
45) van Gorkom, J.H., Schwarz, U.J. and Bregman, J.D. 1985, in IAU Symposium "The Milky Way Galaxy", eds. van Woerden et al., Reidel (Dordrecht)
46) van Gorkom, J.H., Schwarz, U.J. and Bregman, J.D. 1987, in Proceedings of the "Townes"-Symposium (see Backer 1987)
47) Winnberg, A., Baud, B., Matthews, H.E., Habing, H.J. and Olnon, F.M. 1985, Astr.Ap. (Letters) 291, L45

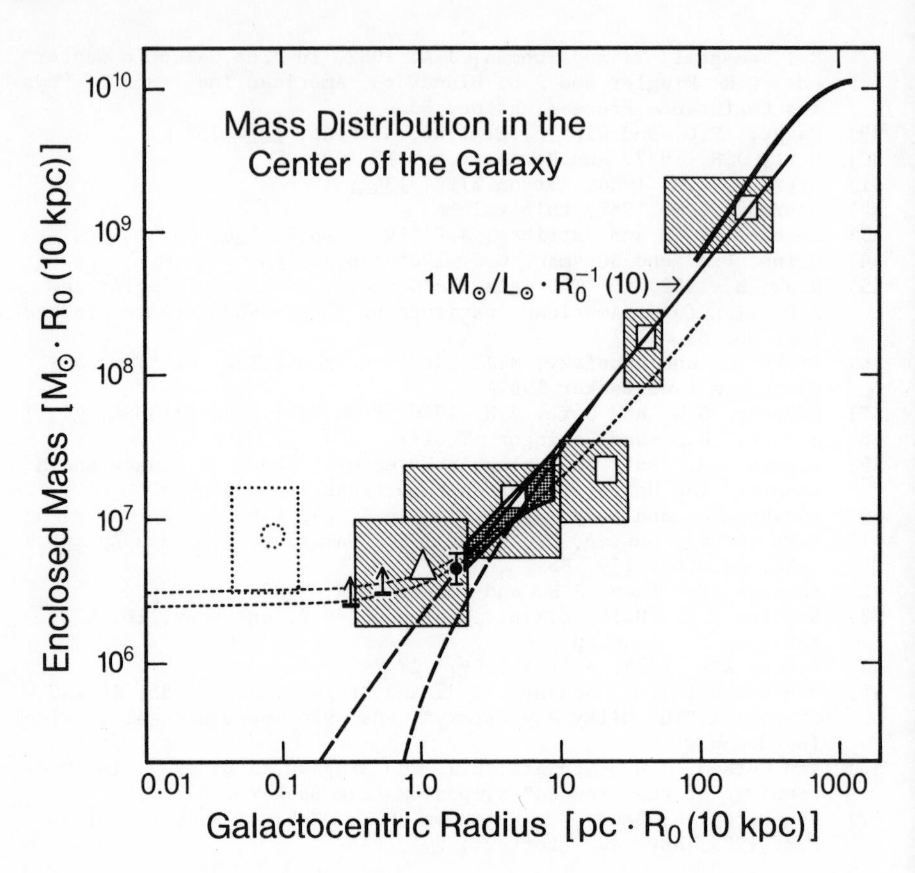

Fig. 1 Composite Mass Distribution in the Galactic Center. The heavy curve is the observed HI rotation curve along the galactic plane [37),30),44)]. The heavily shaded region is the rotation curve in the neutral circum-nuclear ring. The black dot and lower limits represent mass estimates from the ionized streamers at $R \leq 1.7$ pc, [40)]. The medium thin line is the mass distribution derived from the 2 μm surface brightness distribution, [6),28)], with M/L ratio of 1 $M_\odot/L_\odot$ $R_0^{-1}(10)$, calibrated on the mass at $\approx$ 300 pc derived from the HI rotation (v_{rot} = 180 km s $^{-1}$, corrected for non-circular motions, [9)]. The dashed continuations of that curve at R < 3 pc are for core radii of 1 and 0.1 pc. The large shaded boxes with central quadrangles represent mass estimates from the velocity dispersion of OH-IR stars, [47),17)]. The large shaded box with central triangle represents the mass derived from the velocity dispersion of 6 stars emitting at 2 μm, [43)]. The thin dotted box with central circle is a mass estimate derived from the HeI/HI "broad line" region, [20),12),13)], assuming virial equilibrium. The thin dotted curve which is lower on the left represents a model with a 2.5 x 10^6 $M_\odot$ point mass, a core radius of 1 pc, and mass distribution derived from 2 Mm observations with M/L = 1 $M_\odot/L_\odot$. The thin dotted curve which is uppermost on the left represents a mass model with 3.0 x 10^6 $M_\odot$ in a central point mass and a stellar cluster of core radius 0.1 pc and mass distribution derived from the 2 μm surface brightness distribution and M/L = 0.6 $M_\odot/L_\odot$.

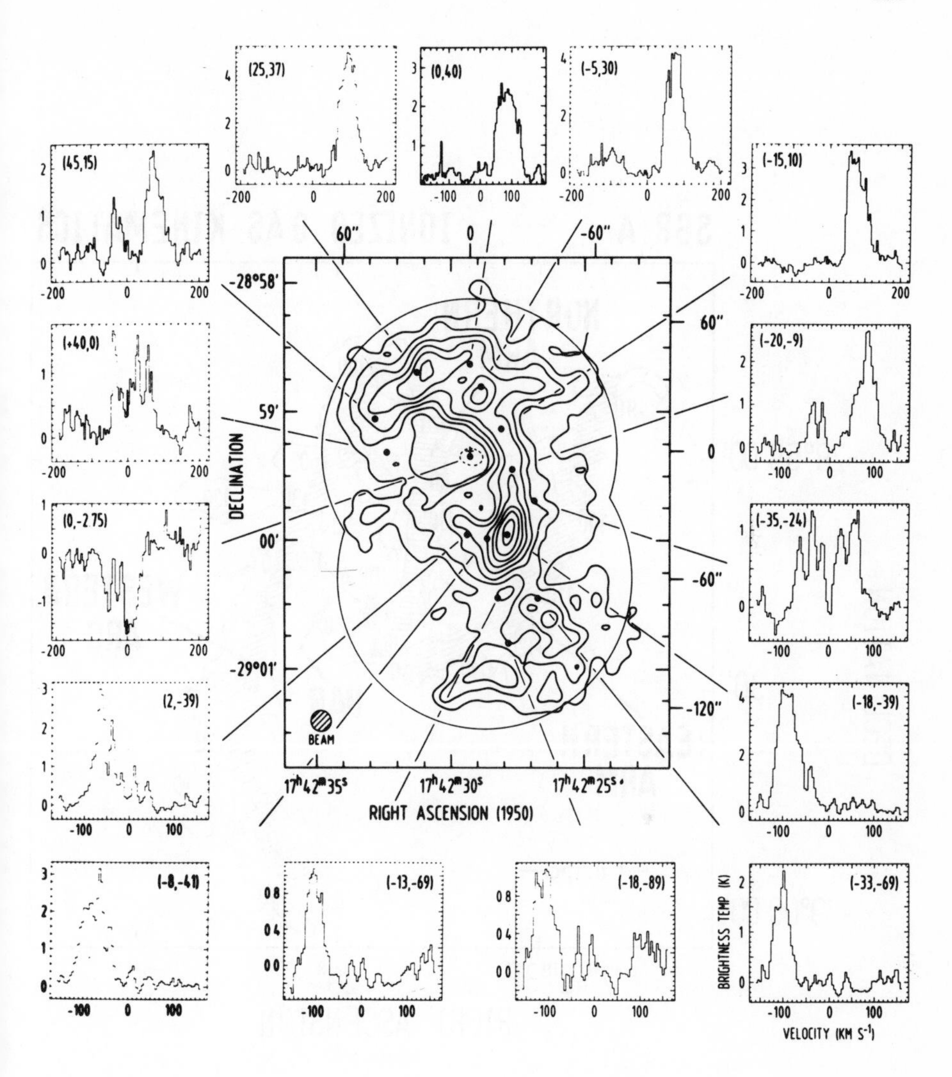

Fig. 2 10" resolution cleaned map of the 88 GHz HCN J=1→0 transition toward Sgr A, together with typical spectra, [15].

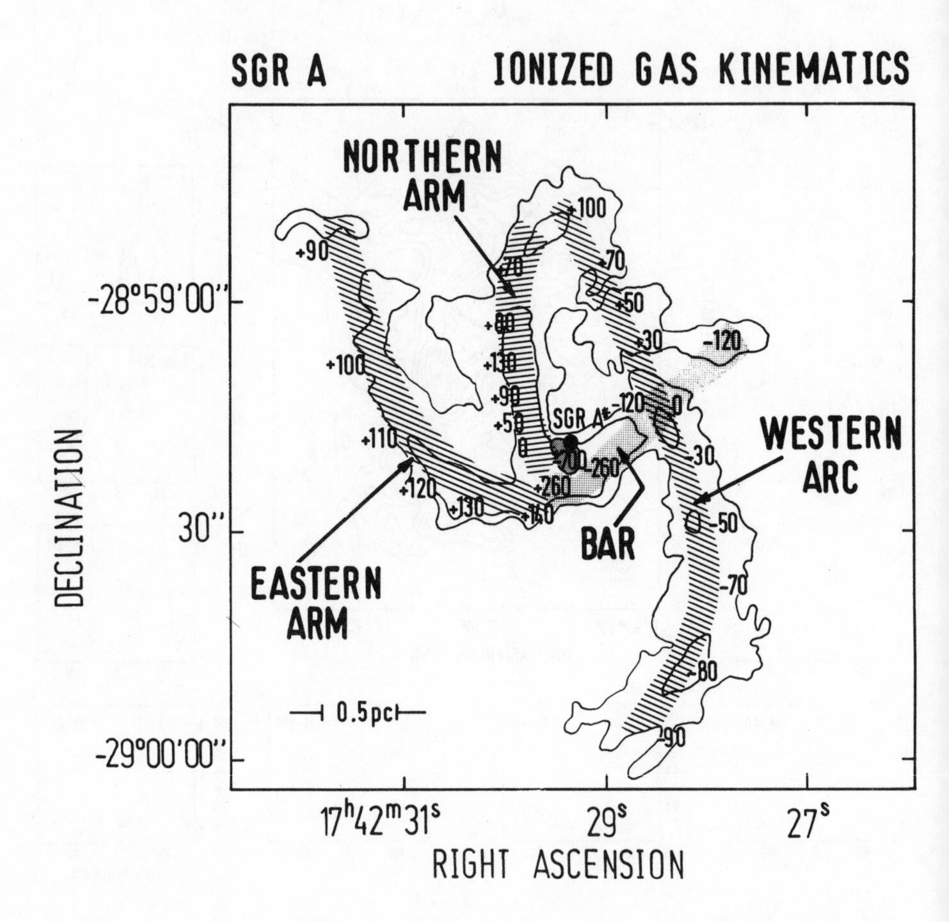

Fig. 3 Overview of the ionized gas velocities superposed on the Lo and Claussen [26] 5 GHz map. The measurements in the western arc, bar, northern and eastern arms are from 12.8 μm [NeII] spectroscopy by Lacy et al. [22] and Serabyn and Lacy [40]. The ± 700 km s $^{-1}$ broad line emission region is from 2 μm HeI/HI spectroscopy by Geballe et al. [12].

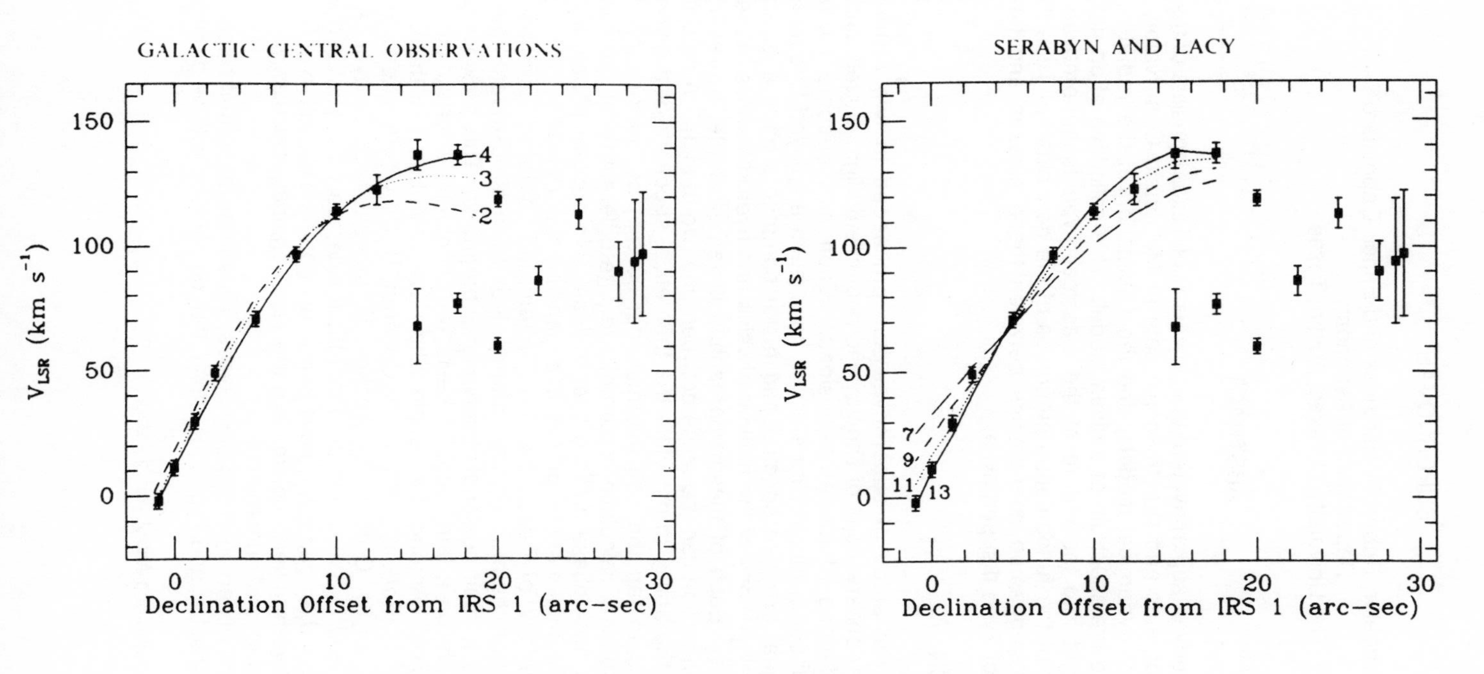

Fig. 4 [NeII] velocities in the northern arm (filled quadrangles) and best fit parabolic orbits about a point mass (models marked in units of 10^6 $M_\odot$) centered 2.7" W and 0.7 N of IRS 1 (a: left side) and best fit orbits in a R^{-2} (isothermal) stellar cluster (b: right side). Models are labeled in terms or mass in units of 10^6 $M_\odot$ enclosed within 1.7 pc of the center, [40].

SUPERNOVAE

J. Craig Wheeler, Robert P. Harkness and Enrico Cappellaro*
Department of Astronomy
The University of Texas, Austin, Texas

ABSTRACT

The program of study of supernovae at the University of Texas consists of the collection of photometric and spectroscopic data at McDonald Observatory, the construction of dynamical models, and the calculation of the radiative transfer and hence the spectrum of a given model. This collective effort has brought new insight into the nature of the classical Type Ia supernovae, established many of the fundamental properties of the new class of Type Ib supernovae, and suggested connections between these classes and the different varieties of Type II supernovae.

1. INTRODUCTION

Historically, the study of supernovae has proceeded in two rather independent ways--observations of spectra and photometric light curves, and theoretical calculations of model explosions. Each has become more sophisticated as modern electronic detectors have been employed to obtain digitized high signal to noise spectra, and faster computers have allowed more elaborate calculations of the nuclear physics and hydrodynamics of the theoretical models. Each of these approaches is rather sterile, however, unless a direct connection can be made between the two of them. A crucial recent adjunct is the ability to calculate the theoretical spectrum of a given model, and hence to tap the rich informational content of the observed spectra. These spectral calculations currently require different techniques in the early optically thick phase and the later optically thin nebular phase, but give two powerful complementary probes of expanding supernovae.

At the University of Texas we have attempted to bring all these techniques to bear on the problem. An informal federation of observers have added supernovae to their personal observational programs. This provides a supply of good data on a fairly regular basis with minimum interruption of telescope schedules. Among the people who have contributed in the last several years are: E. Barker, J. Brown, A. Cochran, H. Corwin, A. Crotts, H. Dinerstein, M. Frueh, D. Garnett, M. Gaskell, R. Levreault, S. Sawyer, A. Shafter, B. Wills, and D. Wills. On the theoretical side, we have explored a variety of models, and, in particular, have begun a vigorous exploration of the supernovae radiative transfer problem. We are also beginning calculations of the later nebular stages of supernovae.

Supernova are observed in two basic types, Type II for those that display hydrogen in their spectra, and Type I for those that do not. As will be argued

* current address: Osservatorio Astrofisico di Asiago

below, Type I supernovae are now understood to occur in two spectroscopically and physically distinct subclasses, Type Ia and Type Ib. There are also subclasses of Type II supernovae, based on their light curve morphology, those which decline linearly in magnitude from maximum, Type II-L and those that have a pronounced plateau, Type II-P. These may represent extremes of a continuum, but there is a possibility that they, too, represent different physical origins. Figure 1 shows characteristic spectra of

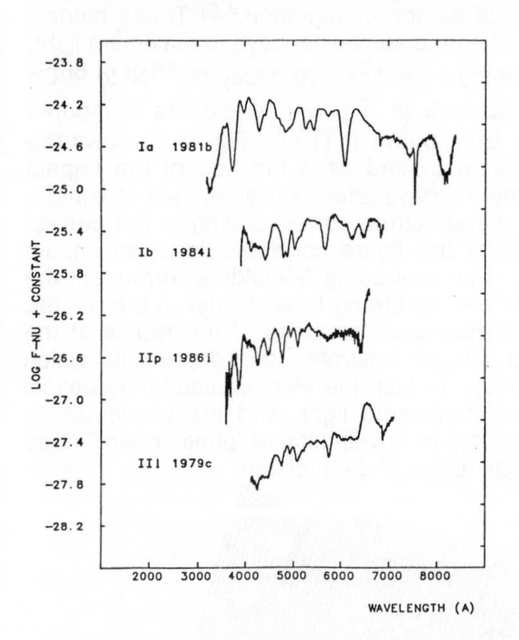

Figure 1. McDonald Observatory spectra of four types of supernovae: SN Ia 1981B in NGC4536 near maximum light, SN Ib 1984L in NGC 991 near maximum light, SN II-P 1986I in M99 about 3 weeks after maximum, and SN II-L 1979 in M100 about 6 weeks after maximum.

the four subclasses of supernovae. Zwicky[1] identified Types III, IV, and V as well (see Doggett and Branch[2] for a recent discussion). These classes are generally represented by only one reasonably well observed example, and so most supernovae fall into the two main categories. A possible recent exception is SN 1986J which has been suggested as a Type V candidate on the basis of its rather narrow lines.[3] Most of the recent work on spectral analysis has been done on the Type I events. Note, however, that there are interesting differences between the spectra of the Type II events in Figure 1, for instance, the strength of the absorption dip at 5700 Å. It will be interesting to investigate whether these differences represent some fundamental property, or are just part of the natural dispersion of properties among Type II events. In this review, we will concentrate on recent developments on Type Ia and Ib supernovae, and present some results pertaining to the origin of Type II-L events.

2. TYPE Ia SUPERNOVAE

2.1 Atmosphere Calculations of the Carbon Deflagration Model

The evolutionary origin of Type Ia supernovae is still uncertain, although the suspicion remains strong that mass transfer onto a white dwarf in a binary system is involved. Supernova atmosphere calculations for Type Ia events are based on the detailed models of carbon deflagration.[4,5] These models are expanded from a few seconds of age to somewhat beyond maximum light, accounting for the deposition of energy by radioactive decay of ^{56}Ni to ^{56}Co and then to ^{56}Fe. Theoretical spectra at a given phase are computed assuming Local Thermodynamic Equilibrium (LTE).[6] Figure 2 shows the emergent flux in the observer frame plotted as a function of the impact parameter for the deflagration model 14 days after explosion, near maximum light. At the rear of the figure is the spectrum corresponding to the central impact parameter, while the front of the figure corresponds to an impact parameter passing near the limb. The protruding "shoulders" represent the redward emission of radiation scattered by strong lines deeper in the matter. The length of the protrusions is a qualitative measure of the radius of the ejecta at a given frequency, and this figure illustrates the degree to which the "radius" is a function of frequency. In fact, the electron scattering optical depths of the ejecta are <20 near maximum light, and the whole ejecta constitute the atmosphere. The concept of a well-defined "photosphere" does not correspond well with the complex reality of the problem.

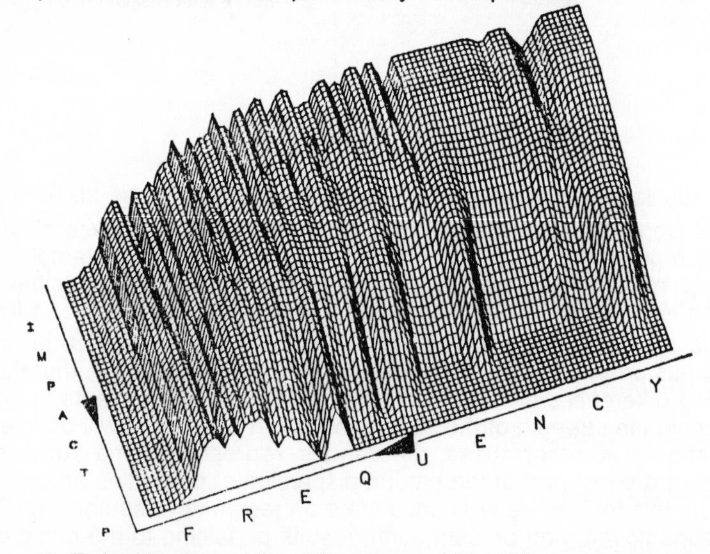

Figure 2. Emitted flux from carbon deflagration model for SN Ia as a function of impact parameter through the ejecta.

The result of these calculations is a theoretical spectrum that agrees rather well with the observed spectrum (see Figure 2 of Harkness, this volume). The ability to reproduce the ultraviolet spectrum, in particular, is a major accomplishment of the atmosphere calculations. In the current model, the ultraviolet spectrum is due to the combined effect of several hundred resonance lines, mainly of Fe II.

2.2 Detonation versus Deflagration

The layers of partially burned matter and overall kinematics that naturally arise in the deflagration model thus give a remarkably good agreement with the spectra of SN Ia. A fundamental conflict is shaping up, however, concerning the nature of degenerate carbon ignition in such an environment. In their arguments in favor of the formation of deflagration, Mazurek, Meier, and Wheeler[7] neglected the cumulative effects of the burning-induced shocks to amplify the shock from previously burned matter. Blinnikov and Kholkov[8] pointed out that when care is taken to examine the process of carbon ignition and runaway at very fine mass scales, less than 10^{-5} $M_{\odot}$, one finds that a region of spontaneous burning exists in which the timescale for adjacent regions to burn to nuclear statistical equilibrium is less than the sound crossing time. A small, but finite amount of matter thus burns at speeds for which the phase velocity of the "front" (which depends on the initial temperature distribution) is supersonic. Blinnikov and Kholkov concluded that this finite ignition region would lead to the formation of a detonation. Similar conclusions have been reached by Woosley and Weaver.[9] Barkat, Swartz and Wheeler[10] find that a Chapman-Jouguet detonation already is formed within the region of spontaneous burning.

This conclusion seems to be independent of whether the ignition is central or somewhat off-center (in the context of spherical models), and applies to all ignition temperatures in excess of 5×10^8 K. Whether there is an ignition temperature below which a detonation will not form is under investigation. This question is complicated by the increased length scale required to form a detonation as the ignition temperature is decreased.

If these conclusions are accepted at face value, then a dilemma exists. The deflagration model matches the observations fairly well, but is not self-consistent physically. Detonation may be more physically correct but the simplest detonation models lead to virtually complete incineration of the white dwarf, and hence do not correspond to observations. A careful investigation must be made of ways to damp a pre-existing detonation in a degenerate carbon/oxygen white dwarf.

3. TYPE Ib SUPERNOVAE

3.1 The Early Phase

Twenty-five years ago, Bertola identified what he called "peculiar" Type I supernovae.[11,12] The two events he studied, SN 1962L in NGC 1073 and SN 1964L in NGC 3968, were characterized by lack of evidence for hydrogen, placing them in the spectral category of Type I, but showed no evidence for the strong absorption at 6150 Å, attributed to Si II λ6355, which characterizes the classical Type I events. The discovery by Evans and subsequent study of SN 1983N in M83[13)] and SN 1984L in NGC 991[14)] have led to a rush of developments and the conviction that the "peculiar" Type I events represent a distinct class of supernovae. They have been formally identified as Type Ib supernovae to distinguish them from the classical events, now know as Type Ia. The SN Ib tend to occur in or near H II regions,[14-17)] they appear to be dimmer than SN Ia by about a factor of three or four,[14-18)] some emit radio radiation characteristic of the collision of the supernova shock with a wind,[19-21)] and they may have characteristic infrared light curves that are identifiably different from SN Ia.[22)] A current list of candidates for SN Ib and some of their properties is given in Table 1.

Table 1
SN Ib CANDIDATES

SN #	HOST	LOCATION	EARLY PHASE HE	SUPERNEBULAR PHASE
1954A	NGC 4214	?	high?	?
1962L	NGC 1073	H II region	low	?
1964L	NGC 3968	Spiral arm	low	?
1975B	anonymous	?	low?	?
1982R	NGC 1187	?	low?	?
1983I	NGC 4051	?	low	?
1983N	M83	H II region	high	O rich
1983V	NGC 1365	H II region	low	?
1984L	NGC 991	H II region	high	O rich
1985F	NGC 4618	H II region	?	O rich
1986M	NGC 7499	?	low?	?

An important step was the identification of a series of features attributable to He I[23,24)] in a fine series of spectral data obtained over the first two months subsequent to maximum of SN 1984L and the construction of atmosphere models which reproduce the spectra near maximum light.[17)] The development of He I λ5876 is particularly distinct. It appears with a P-Cygni line profile with a Doppler shifted absorption minimum at about 5700 Å. Figure 2 shows the spectrum of the SN Ia 1981B in comparison with the maximum light spectrum of SN Ib 1984L. Note the presence of the 6150 Å feature in the Ia and its absence in the Ib, but the deep minimum at 5700 Å in the Ib which is not present in the Ia.

Harkness et al.[17)] construct power law density profiles of varying slopes and uniform compositions rich in helium, carbon, and oxygen. Atmosphere calculations reveal that even with a large helium abundance (~90 percent) the helium lines are not as strong as observed. Harkness et al. then introduce a "departure coefficient" b by which the Boltzmann populations of He I are multiplied. A theoretical spectrum corresponding to a model with a density profile with $r \propto r^{-7}$, a composition of about 90 percent He, 1 percent C, 9 percent O, and the remaining heavy elements in solar proportions, all by mass, and a departure coefficient of b = 100 is given in Figure 3 along with maximum light spectrum of SN 1983N. The simple model spectrum reproduces the observations remarkably well, confirming the identification of He, but raising the challenge of understanding the physical origin of the implied strong departure of He from LTE in the SN Ib.

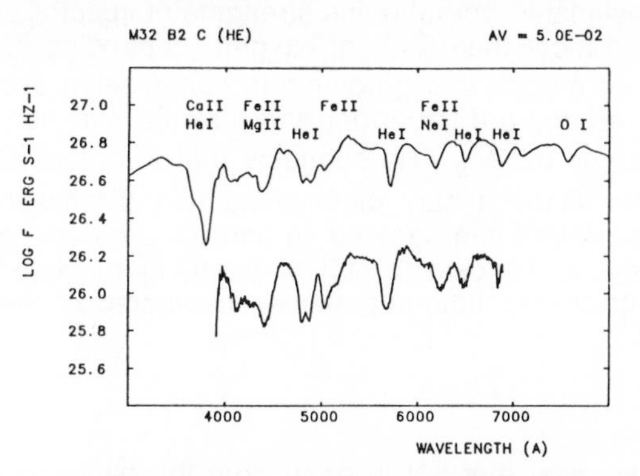

Figure 3. The observed spectrum of SN Ib 1984L near maximum light is compared to a theoretical spectrum of a model with a power-law density profile and excited states of He I enhanced by a factor of 10^2 (see text).

The success in reproducing the spectra of SN 1983N and 1984L led to a re-examination of spectra of other supernovae. Wheeler et al.[16] present theoretical spectra of SN 1983I and 1983V, based on the same simple power law profiles with homogeneous composition. They suggest that SN 1983 I and V are also members of the class SN Ib, but with a smaller helium abundance. The helium abundance can not be specified without a better understanding of the departures from LTE but the calculation suggests that events such as SN 1983 I and V are relatively helium deficient compared to SN 1983N and 1984L. Table 1 suggests that the events with large helium abundance may be in the minority, but the statistics are still poor.

3.2 The Supernebular Phase

About a month after maximum light, supernovae evolve from the early optically thick phase dominated by permitted lines to an optically thin nebular phase dominated by forbidden lines. Wheeler et al.[24] coined the phrase "supernebular" to describe this important later phase. A spectrum of SN 1983N about eight months after maximum light revealed that SN Ib evolve to a phase dominated by forbidden emission lines which is very different than observed for SN Ia or SN II.[25] The supernebular phase of SN II is dominated by a strong H alpha emission line (see the SN II spectra of Figure 1), and SN Ia by forbidden lines of Fe II and III and Co II and III.[26,27] The supernebular phase of SN Ib is dominated by the strong blended emission lines of [O I] λ 6300, 6364. This phase has also been observed in SN 1984L[28] and in SN 1985F, which was discovered only many months after maximum, and identified as a SN Ib on the basis of its supernebular spectrum.[25,29,30] The

6300 Å emission can be seen about two months after maximum in SN 1984L,[17] and also shows up in the supernebular phase of some SN II.[31]

Efforts are now underway by several groups to analyze the supernebular phase of SN Ib. Study of this phase represents a crucial complement to attempts to reproduce the maximum light spectra because the nebula is virtually optically thin so that all the ejecta can be sampled at once and because the detected lines form in, and hence sample, the ejecta in different ways than the permitted lines that characterize the earlier phase.

Begelman and Sarazin[32] estimated from the line strengths of spectra of SN 1985F that that event ejected more than 5 $M_{\odot}$ of oxygen (depending on the distance). Pinto (this volume) argues that gradients in composition and deposition of radioactive decay energy are important and that the minimum emitting mass could be as small as 0.2 $M_{\odot}$. He suggests that a feature at 5900 Å may be Co, rather than Na as preliminary suggestions held. Fransson (this volume) shows that the line profiles can be used, in principal, to deduce the density distribution of the ejecta, and stresses that [O I] and [O II] may arise in different spatial locations, introducing inhomogeneities neglected in the simplest analyses.

3.3 The origin of SN Ib

Several lines of evidence suggest that SN Ib arise from the hydrogen-stripped cores of moderately massive stars. The association with H II regions implies that the progenitors are moderately young. The atmosphere models suggest that several solar masses of helium, carbon, and oxygen enriched matter are ejected. The analyses of the supernebular phase are consistent with this. The width of the light curve peak, which is comparable to that of SN Ia, seems to require more than 1 $M_{\odot}$ but less than about 6 $M_{\odot}$.[33-35] The radio emission is consistent with an origin in the shocked wind from a massive core, although the wind could have other sources. Even with the relatively poor statistics, the rate of explosion of SN Ib events seems to be comparable to SN Ia and SN II.[18,36] Since the bulk of SN II are estimated to come from stars of less than about 25 $M_{\odot}$, the same is probably true of SN Ib.

A model that is consistent with all the constraints to first order is one in which an SN Ib is essentially identical to an SN II-P, but the SN II-P events occur in single stars, and the SN Ib events occur in close binaries. The single star would form a core of helium and heavier elements, evolve through core collapse, and explode, still surrounded by a red supergiant envelope. The result would be an SN II-P and the neutron star would plausibly be left magnetized and spinning to form a pulsar. A star of the same mass in an appropriate binary system would form a core of the same mass, but it would be laid bare after mass transfer to the companion or mass loss from the system. When the core exploded, still retaining an appreciable abundance of helium, but no hydrogen, the result would be an SN Ib. In this case, less than half the total mass of the system is likely to be ejected and the neutron star would be left in an elliptical orbit in a close binary system. Eventually, such a system should evolve to become an X-ray binary. Such binary X-ray sources are well known, and have long been surmised to form by the collapse of a stellar core in a binary system. The SN Ib may provide a key missing link in the evolutionary origin of such binary X-ray systems.

4. TYPE II-L SUPERNOVAE

The light curves of SN II-P are consistent with explosions in the red supergiant envelopes of moderately massive stars.[37,38] Such stars presumably explode by core collapse, as just discussed. The origin of the SN II-L events is less clear. Litvinova and Nadozhin discuss the possibility that they are closely related to SN II-P, and that a SN II-L occurs in the limit that the envelope mass is small. Doggett and Branch[2] have, however, raised an interesting alternative possibility. They note that the linear light curves resemble those of SN I as much or more as they do SN II-P, and that some SN II-L, e.g., SN 1979C,[39] are nearly as bright as SN Ia. Doggett and Branch hypothesize that while SN II-P arise in core collapse, SN II-L arise in the same kind of thermonuclear explosion that is thought to form SN Ia events.

A number of models have been run to check the hypothesis of Doggett and Branch. Carbon deflagration models have been calculated in which the inner degenerate core is surrounded by a constant density envelope of mass ranging between zero and several solar masses. The corresponding light curves have been calculated in the approximation of a simple constant opacity for the core and Rosseland opacity for the hydrogen envelope. For masses in excess of about 2 $M_\odot$ the light curves are too broad to match those of SN II-L. Figure 4 presents light curves for a model with an envelope of 0.7 $M_\odot$ and three observed B magnitude light curves, for SN 1959D, SN 1979C, and SN 1980K, normalized to the same maximum luminosity. The explosion is assumed to produce 1 $M_\odot$ of radioactive ^{56}Ni. Shown are the theoretical

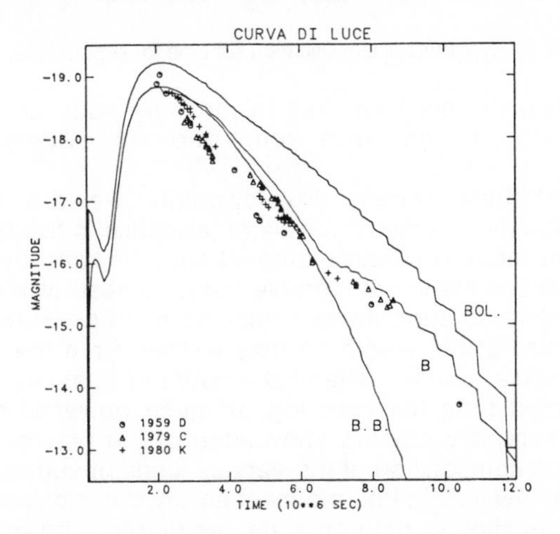

Figure 4. Blue magnitude light curves for SN II-L events 1959D in NGC 7331, 1979C in M100 and 1980K in NGC 6946 compared with theoretical light curves based on a deflagrated white dwarf within a hydrogen envelope of mass 0.7 $M_\odot$ (see text).

curves for the bolometric luminosity, the magnitude in the B band if one assumes a black body emissivity at the temperature corresponding to the computed luminosity and the radius corresponding to optical depth 2/3, and the magnitude in the B band if one adopts the temperature evolution implied by observed B-V color. The latter provides a remarkably good agreement with the observations. These calculations do not prove that SN II-L arise in thermonuclear explosions, but they do give encouragement to continue to investigate this possibility.

5. CONCLUSIONS

Figure 5 presents a scheme which is not meant to be a solution, but a framework in which to pose further questions about the nature of the different types of supernovae. If the SN II-L are thermonuclear explosions like SN Ia, they should all be quite bright, but the evolutionary origin of either type remains a difficult problem. If the SN Ib are the explosions of the bare cores of massive stars, they have a significance far beyond the establishment of a new spectral category of supernovae. They give us for the first time the ability to see directly into the heart of an exploding massive star. This will give the opportunity to confront directly with observations all the many calculations of massive star evolution and explosion (see papers by Bruenn and by Arnett in

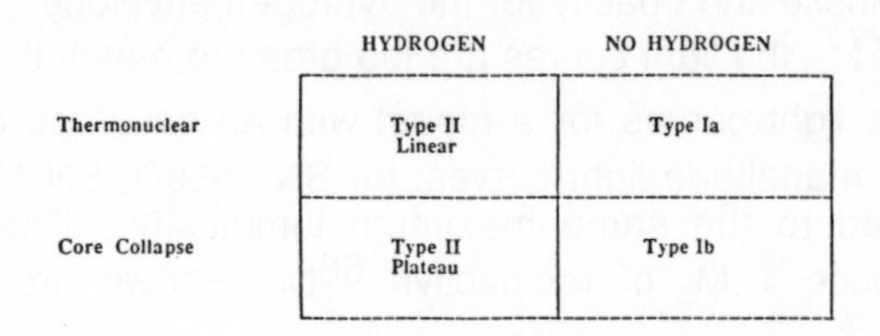

	HYDROGEN	NO HYDROGEN
Thermonuclear	Type II Linear	Type Ia
Core Collapse	Type II Plateau	Type Ib

Figure 5. Hypothetical classification scheme for supernovae

this volume) which until now have had to measure success by the simple criterion of explosion or no, with some indirect evidence based on nucleosynthesis.

In the rush of these recent developments we are witnessing a renaissance in supernovae research. This renaissance is being fueled by the excellent data from modern electronic spectrographs, and by the ability to calculate theoretical spectra in comparable detail. These twin developments give us the capacity to do supernovae tomography, to peer ever deeper into the complex structure of the ejecta as they expand from the early optically thick to the later nebular phase. The vast amount of quantitative information that can be extracted from the coupling of these powerful new elements means that we are rapidly gaining knowledge of the precise nature of the mass, structure, and composition of the various kinds of supernovae. This in turn should give us new insight into the mechanism of explosion and into such questions as the formation of neutron stars versus black holes which are the most fundamental goals of supernova research and which have so very dearly needed more direct information. Finally, with greater physical insight we will surely be able to make more firm statements about the absolute luminosity of supernovae, and hence, by means of their distances, to address even grander questions such as the age, shape, and structure of the Universe.

This research is supported in part by NSF grant 8413301 and by the Italian Ministry of Education. We are grateful to the many colleagues who have shared their insights into the developments summarized here.

REFERENCES

1. Zwicky, F., in *Stars and Stellar Systems*, University of Chicago Press, eds. L. H. Aller and D. B. McLaughlin, VIII, 367 (1965).
2. Doggett, J. B. and Branch, D., *A.J.*, 90, 2303 (1985).
3. Rupen, M. P., van Gorkom, J. H., Knapp, G. R., and Gunn, J. E., preprint (1987).
4. Nomoto, K., Thielemann, F.-K., and Yokoi, K., *Ap. J.*, 286, 644 (1984).
5. Thielemann, F.-K., Nomoto, K, and Yokoi, K., *Astron. and Ap.*, 158, 17 (1986).
6. Harkness, R. P., in *Supernovae as Distance Indicators,*, ed. N. Bartel (Berlin: Springer-Verlag), 183 (1985); Harkness, R. P., in *Radiation Hydrodyanamics in Stars and Compact Objects*, eds. D. Mihalas and K.-H. A. Winkler (Berlin: Springer-Verlag), 166 (1986)); Harkness, R. P., this volume.
7. Mazurek, T. J., Meier, D. L., and Wheeler, J. C., *Ap. J.*, 215, 518 (1977).
8. Blinnikov, S. I.. and Khokhlov, A. M., preprint (1986).
9. Woosley, S. E. and Weaver, T. A., in *Radiation Hydrodynamics in Stars and Compact Objects*, eds. D. Mihalas and K.-H. A. Winkler (Berlin: Springer-Verlag), 91 (1986) and this volume.
10. Barkat, Z., Swartz, D., and Wheeler, J. C., in preparation (1987).
11. Bertola, F., *Annals d'Astrophysique*, 27, 319 (1964).
12. Bertola, F., Mammano, A., and Perinotto, M., Contributions of the Asiago Observatory, No. 174 (1965).
13. Panagia et al., preprint (1987).
14. Wheeler, J. C. and Levreault, R., *Ap. J. (Letters)*, 294, L17 (1985).
15. Uomoto, A. and Kirshner, R. P., *Astron. Ap.*, 149, LL7 (1985).
16. Wheeler, J.. C., Harkness, R. P., Barker, E. S., Cochran, A. L., and Wills, D., *Ap. J. (Letters)*, 313, XXX (1987).
17. Harkness, R. P., Wheeler, J. C., Margon, B., Downes, R. A., Kirshner, R. P., Uomoto, A., Barker, E. S., Cochran, A. L., Dinerstein, H. L., Garnett, D. R., and Levreault, R. M., *Ap. J.*, 317, XXX (1987).
18. Branch, D., *Ap. J. (Letters)*, 300, L51 (1986).
19. Sramek, R. A., Panagia, N., and Weiler, K. W., *Ap. J. (Letters*, 285, L63 (1984).
20. Panagia, N., Sramek, R. A., and Weiler, K. W., *Ap. J. (Letters)*, 285, L63 (1986).
21. Chevalier, R. A., *Ap. J. (Letters)*, 285, L63 (1984).
22. Elias, J. H., Matthews, K., Neugebauer, G., and Persson, S. E., *Ap. J.*, 296, 379 (1985)

23. Wheeler, J. C. and Harkness, R. P., *Proceedings of the NATO Advanced Study Workshop on Distances of Galaxies and Deviations from the Hubble Flow*, eds. B. M. Madore and R. B. Tully (Dordrecht: Reidel), 45 (1986).
24. Wheeler, J. C., Harkness, R. P., Barkat, Z., and Swartz, D., *P.A.S.P.*, 98, 1018 (1986).
25. Gaskell, C. M., Cappellaro, E., Dinerstein, H. L., Garnett, D. R., Harkness, R. P. and Wheeler, J. C., *Ap. J. (Letters)*, 306, L77 (1986).
26. Axelrod, T. S., Ph.D. Thesis, University of California, Santa Cruz (1980); in *Type I Supernovae*, ed. J. C. Wheeler (Austin: University of Texas), 80 (1980).
27. Woosley, S. E., Axelrod, T. S., and Weaver, T. A., in *Stellar Nucleosynthesis*, eds. C. Chiosi and A. Renzini (Dordrecht: Reidel), p. 263 (1984).
28. Kirshner, R. P., private communication (1986).
29. Filippenko, A. V. and Sargent, W.L.W., *Nature*, 316, 407 (1985).
30. Filippenko, A. V. and Sargent, W.L.W., *Ap. J.*, 91, 691 (1986).
31. Branch, D., Falk, S. W., McCall, M. L., Rybski, P., Uomoto, A. K., and Wills, B. J., *AP. J.*, 244, 780 (1981).
32. Begelman, M. C. and Sarazin, C. L., *Ap. J. (Letters)*, 302, L59 (1986).
33. Schaeffer, R., Cassé, M., and Cahen, S., *Ap. J. (Letters)*, in press (1987).
34. Ensman, L. and Woosley, S. E., in preparation (1987).
35. Sutherland, P. G., Wheeler, J. C., Harkness, R. P., Cappellaro, E., and Swartz, D., in preparation (1987).
36. van den Bergh, S., McClure, R. D., and Evans, R., preprint (1987).
37. Falk, S. W. and Arnett, W. D., *Ap. J. Suppl.*, 33, 515 (1977).
38. Litvinova, I. Yu. and Nadyoozhin, D. K., *Ap. and Space Sci.*, 89, 89 (1983).
39. de Vaucouleurs, G., de Vaucouleurs, A., Buta, R., Ables, H. D., and Hewitt, A. V., *P.A.S.P.*, 93, 36 (1981).

EARLY TIME SPECTRA OF TYPE Ia SUPERNOVAE

Robert Harkness

Astronomy Department, University of Texas at Austin

Until recently, it was a widely held belief that the near-maximum light spectra of all Type I supernovae were identical. For many years[1)] there was a suggestion that some objects departed from the norm, but maximum light observations of supernova 1983N in M83 (NGC 5236) really demonstrated that there were at least two distinct subclasses of explosions that had the light curve signature of a "Type I." The "classical" Type I features a strong Si II absorption line at around 6150 Å, which is not seen in events similar to 1983N. Furthermore, 1983N and more recent examples of its subtype show a strong absorption feature at about 5700 Å, not seen in cases that feature the Si II line. Supernovae with the Si line are thus designated "Type Ia" and those without, "Type Ib." Recent progress in analyzing the Ib case suggests that the two subtypes may have a radically different origin (Wheeler et al., this volume). Comparison of late-time spectra in the "supernebular" phase demonstrates dramatically the disparity of Ia and Ib. In SN Ia the spectrum is formed by [Fe II], in SN Ib the most prominent feature is due to [O I].

While the spectroscopic differences are highlighted at late stages, the early time spectra give important clues in understanding the basic explosive mechanisms involved. This is particularly true for SN Ia, because the spectral features undergo major changes shortly after maximum light, and because of the relative richness of the maximum light spectrum. Figure 1 shows the spectral development of the bright Type Ia supernova 1981B in NGC 4536. The data were obtained by several workers at McDonald Observatory over a period of several months. Following Branch[2,3)] and others, the major spectral lines are identified as being due to the intermediate-mass elements O, Mg, Si, S and Ca, together with some contribution from Fe. One month after maximum light only Ca and Fe lines are apparent. The spectrum changes gradually thereafter as we see essentially a pure iron spectrum emerge.

The most successful theoretical model for SN Ia involves the thermonuclear incineration of a carbon/oxygen white dwarf as it approaches the Chandrasekhar limit due to mass accretion from a companion. Deflagration models, in which the combustion occurs as a subsonic burning front propagates toward the surface, can synthesize the necessary intermediate mass elements as a mantle surrounding a core of radioactive Ni^{56}. This nickel isotope originates in the matter burned to nuclear statistical equilibrium (NSE); progressively lighter elements are synthesized as the burning front expands the matter ahead of it, and the peak temperature falls. The energy from the nickel decay powers the supernova through maximum light, and the subsequent, less energetic cobalt decay accounts for the energy output during the exponential decline phase of the light curve. Recent finely zoned calculations indicate that central ignition will lead to a detonation and essentially the entire white dwarf will be burned to Ni^{56}, with no intermediate

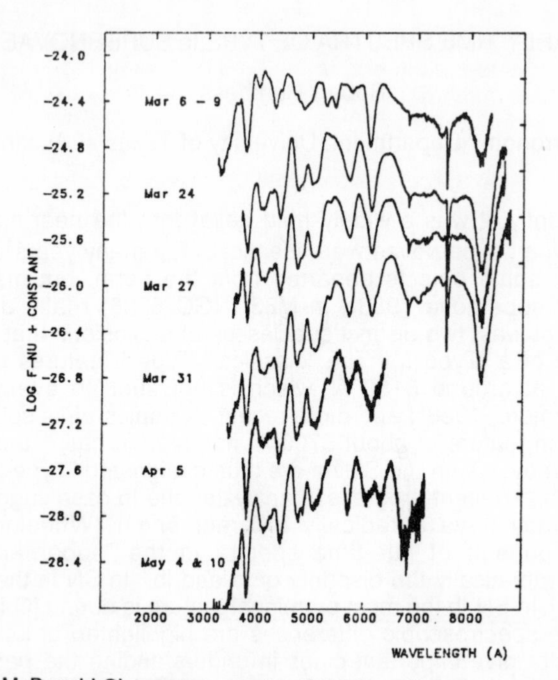

Figure 1. McDonald Observatory spectra of the Type Ia Supernova 1981B in NGC 4536. The maximum light spectrum is formed by intermediate mass elements. Two weeks later, the spectrum is totally dominated by Fe II.

mass elements required to explain the maximum light spectrum. This apparent clash with observation may be resolved if the detonation falters in some way, but for the time being the deflagration model has the clear advantage.

The early spectral evolution of the Nomoto et al. "W7" deflagration[4)] has been computed in some detail. The properties of this particular model illustrate the most interesting features of supernova "atmospheres." Radiative transfer in a supernova atmosphere is, of course, influenced by the large range in expansion velocity. For Type Ia supernovae, however, the most interesting problems arise due to the unique, radially dependent composition which is apparently devoid of hydrogen and helium, and the deposition of radioactive decay energy throughout the ejecta. Sufficient modelling has been done so that we can be fairly certain that all the major spectral features can be accounted for, at least qualitatively. The calculated spectra of the W7 model are in good agreement with the observed optical spectrum at maximum light.

IUE observations of SN 1981B near maximum light[5)] showed that Type Ia supernovae have an 'ultraviolet deficiency' compared to the ultraviolet flux that would be expected from a blackbody with a temperature corresponding to the (optical) color temperature. In the context of the W7 model, there could be several contributing factors which account for this. The continuous opacity at optical wavelengths is mainly due to electron scattering throughout the entire atmosphere. The ratio of continuous absorption to electron scattering is typically 1/50 and the color temperature reflects the local temperature at the thermalisation depth. If the only source of opacity were Thompson scattering, UV photons generated at that depth would escape after several scatterings. In this instance we would have to conclude that the central temepratures were much less than expected. This could arise if the radioactive matter surrounded a large inert core. In the W7 model there is a small central region of stable iron isotopes, but the effect is not large enough to account for the deficit. A more straightforward explanation is the effective scattering opacity due to many resonance lines occurring in the ultraviolet. The line opacities are so great that they can indeed produce a deficit in practical calculations. There remains a problem of completeness: one can make the deficit using a few hundred of the strongest lines, but many thousands of other lines could make an appreciable difference to the magnitude of this effect. Figure 2 shows the result of including about 1000 of the strongest resonance lines; for this calculation the contribution from lines of Ti II, Cr II and Mn II was found to

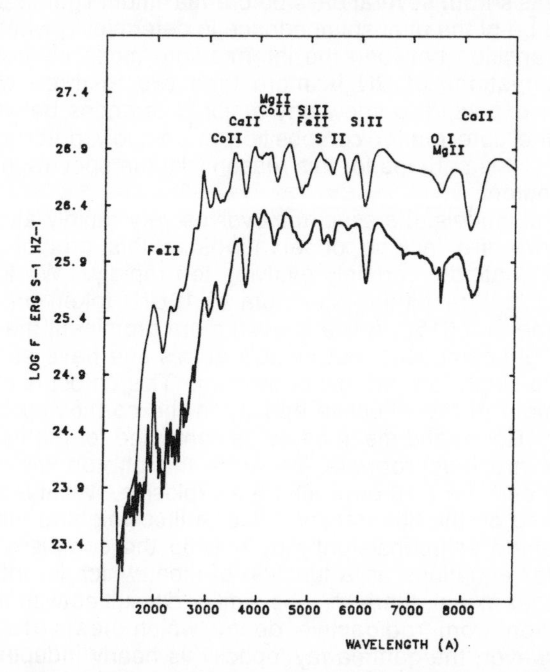

Figure 2. The theoretical spectrum of a carbon deflagration supernova at 14 days (upper) compared to a composite of observations of the Type Ia supernova 1981B in NGC 4536, at approximately the same epoch. Ions responsible for the major spectral features of the model are indicated. The large 'ultraviolet deficit' results from the combination of several hundred lines of all of the elements up to nickel, but the most important ion appears to be Fe II. The composition has been artificially mixed for the matter with expansion velocity greater than 10,000 km/s.

make an appreciable difference although these elements make up less than 0.1% of the total mass. The calculations also indicate that the UV deficit is quite sensitive to the amount of high velocity matter. While the UV resonance line description of the deficit arises in a natural way, it is possible that there is no deficit in reality. If the optical spectrum reflects non-thermal excitation of the iron core in the same manner that the gamma radiation excites the nebula at later stages, there may not be a source of UV radiation to begin with. It would be difficult, however, to extrapolate backwards to maximum light conditions where the densities are still large enough to ensure frequent collisions. If we take the W7 model at face value, and evolve the model to maximum light conditions, assuming only that the matter is fully mixed at velocities greater than 10,000 km/s, we find an excellent agreement with observation from 1500 Å to 10000 Å. If the matter is left unmixed the spectral lines tend to have abrupt edges not apparent in the observations and also appear to be too narrow because the ions responsible are confined to a narrow range of velocity. While the burning is expected to be extremely turbulent and even fragmented (see Woosley, this volume) the gross observable consequences are unknown at present. Certainly, observations taken on a daily basis from several days before maximum light to perhaps two weeks later would be of the utmost importance in determining whether there is indeed a sharp transition between the intermediate mass elements and the NSE core. Observations of SN Ia more than two to three weeks after maximum are not expected to show any major differences between events, simply because the core matter composition is uniquely determined by the conditions of NSE. It is perhaps for this reason that the spectral homogeneity of this class is apparent.

In theoretical models, the spectrum evolves very rapidly after maximum light. Although we are lacking observations in this crucial period, the theoretical model is almost certainly evolving too rapidly. We know from a narrow waveband high-resolution spectrum of 1981B taken on March 13th (not shown) that the Si II 6150 Å line is even more prominent than it appears in the maximum light composite, but models at this age have already lost all features not due to singly ionized iron or calcium. This defect is probably due to the underestimate of the effective opacity in the co-moving frame of the expanding matter. Rosseland mean opacities computed for the hydrodynamic expansion phase are inappropriate, because the photon mean free path becomes large as early as 7-10 days after the explosion. What is required is a mean opacity based on the flux transport, i..e. a flux weighted mean, but this can only be obtained self-consistently by solving the complete frequency-dependent transfer equations as a function of time, which is not practical at present. To a lesser extent, there may be similar difficulties with the transport of gamma radiation from radioactive decay which heats the expanding atmosphere. However, the gamma ray 'opacity' is nearly independent of the composition and is also larger than the optical opacity in the NSE core at maximum light, so thermal diffusion is the primary energy transport mechanism.

Accurate modeling of the total energy transport is of importance for the distance determination methods based on supernovae.[6,7] The models evolved with a Rosseland opacity reach maximum light too early and are also too bright. Future calculations should be able to resolve these remaining difficulties.

REFERENCES

1. Bertola, F., *Annals d'Astrophysique*,, 27, 319 (1964).
2. Branch, D., in *Supernovae Spectra*, eds. R. Meyerott and G. H. Gillespie (New York: American Institute of Physics), 39 (1980).
3. Branch, D., in *Proceedings of the Eleventh Texas Symposium on Relativistic Astrophysics*, ed. D. S. Evans (New York: N.Y. Acad. of Science), 186 (1984).
4. Nomoto, K., Thielemann, F.-K., and Yokoi, K., *Ap. J.*, 286, 644 (1984).
5. Benvenuti, P. Sanz Fernandez de Cordoba, L., Wamsteker, W., Macchetto, F., Palumbo, G. C., and Panagia, N., *An Atlas of UV Spectra of Supernovae* (Paris: European Space Agency ESA SP-1046) (1982).
6. Harkness, R. P., *Supernovae as Distance Indicators*, ed. N. Bartel (Berlin: Springer-Verlag), 183 (1985).
7. Wheeler, J. C. and Harkness, R. P., in *Proceedings of the NATO Advanced Study Workshop on Distances of Galaxies and Deviations from the Hubble Flow*, eds. B. M. Madore and R. B. Tully (Dordrecht: Reidel), 45 (1986).

AGES FROM NUCLEAR COSMOCHRONOLOGY

F.-K. Thielemann[1], J. J. Cowan[2], and J. W. Truran[3]

ABSTRACT

The nuclear production ratios $^{232}Th/^{238}U$, $^{235}U/^{238}U$, and $^{244}Pu/^{238}U$, when calculated in the context of a model for r-process nucleosynthesis which predicts an overall abundance pattern in good agreement with solar system matter, are found to be $^{232}Th/^{238}U = 1.60$, $^{235}U/^{238}U = 1.16$, and $^{244}Pu/^{238}U = 0.40$. These production ratios, when interpreted within a simple exponential model of galactic evolution allowing values of initial disk enrichment ranging from approximately 10 to 30 percent, lead to galactic ages ranging from 12.4 to 14.7 billion years. Uncertainties in the nuclear data introduce an additional spread of approximately 2×10^9 y. More complex galactic evolution models with infall and without instantaneous initial metal enrichment *could* increase the age of the Galaxy by as much as 2-3 billion years.

[1]Department of Astronomy, Harvard University
[2]Department of Physics and Astronomy, University of Oklahoma
[3]Department of Astronomy, University of Illinois and Max-Planck- Institut für Physik und Astrophysik, Garching b. München

1. INTRODUCTION

The determination of the age of the universe presents a persistent and challenging problem in modern astrophysics and cosmology. One classical approach to this problem is that involving the determination of the age of matter in our Galaxy from studies of nucleocosmochronology. It is appropriate here to make use particularly of the long-lived radioactive nuclei: ^{187}Re ($\tau_{1/2} = 4.5 \times 10^{10}$y), ^{232}Th ($\tau_{1/2} = 1.405 \times 10^{10}$y),^{238}U ($\tau_{1/2} = 4.46 \times 10^9$y), and ^{235}U ($\tau_{1/2} = 7.038 \times 10^8$y). In the absence of a detailed knowledge, it has usually been assumed that the time dependence of the nucleosynthesis production follows an exponential behavior[12]. An alternative "model independent" approach [25] allows the determination of the mean age, but ultimately itself becomes somewhat model dependent in its determination of a total galactic age [21,34].

Age determinations resulting from these parametrized galactic production models are found to be sensitive to several critical physical and astrophysical boundary conditions: (1) the production ratios for the radioactive isotopes of interest in their appropriate nucleosynthesis sites; (2) the time dependence of the nucleosynthesis rate over the course of galactic evolution (including specifically the level of initial enrichment of the gas from pre-disk stellar populations); and (3) the isotopic ratios in the matter which condensed to form meteorites at the time of formation of the solar system, 4.6 billion years ago. It is important to note that the cosmochronologically important radioactive isotopes ^{187}Re, ^{232}Th, ^{235}U, ^{238}U, and ^{244}Pu are formed in a single astrophysical process: the rapid neutron capture process (r-process) which synthesizes heavy elements from the mass region A$\sim$70 through the actinides.Constraints upon predictions of nucleocosmochronology and galactic chemical evolution are imposed by the fact that the astrophysical site of r-process nucleosynthesis has not yet been unambiguously determined [7], for a review see Truran [35].

2. ISOTOPIC PRODUCTION RATIOS

The theoretical calculations which have been performed predict an abundance pattern which reproduces the gross features of the solar system r-process curve; this indicates that the r-process capture path is in approximately the right position in the N-Z plane and that the different mass numbers are produced in roughly the correct proportions. Using these results should give reliable predictions for the production ratios, independent of the actual r-process site.

Most early estimates of production ratios from r-process calculations ignored the effects of beta-delayed fission [9,26–27]. Preliminary attempts to include the effects of fission [18,20,39] ignored the details of the abundance distribution along the capture path by assuming a constancy of nuclear abundances as a function of mass number. Thielemann, Metzinger and Klapdor [30–31] assumed an $(n,\gamma) \rightleftarrows (\gamma,n)$ equilibrium for each isotopic chain in the r-process calculation which, depending on the environment, is not necessarily realized. These calculations nevertheless made it clear that the inclusion of beta-delayed fission has a profound influence on the production ratios.

For a detailed description of the present calculations see Cowan, Thielemann, and Truran[8]. Here we want to concentrate on the changes to previous results. The percentages of fission and neutron emission after β-decay (β-delayed fission and neutron emission) for nuclei in the mass range $75 \leq Z \leq 100$ were calculated by Thielemann, Metzinger, and Klapdor [30], with the appropriate β-strength functions[17], fission barrier heights [16] and the nuclear masses from three different mass formulae [14,16,38]. The recent analysis of Hoff [15], who studied the production yields required in the uranium isotope chain to reproduce the observed abundance yields in thermonuclear explosions, indicated that the best agreement is obtained when the Howard and Möller mass formula[16] is used. This means that earlier calculations [31] with the Hilf *et al.* mass formula[14] overestimated the effect of beta delayed fission.

The r-process calculations were performed similar to previous calculations[6], but with the explicit inclusion of beta delayed fission. The ratios of the nuclear chronometers are $^{232}Th/^{238}U = 1.60$, $^{235}U/^{238}U = 1.16$ and $^{244}Pu/^{238}U = 0.40$. Further calculations were made to assess the sensitivity of the production ratios to the nuclear input physics and the agreement between the theoretical and the solar system r-process abundance curve. These turned out to give errors of 0.10 in the production ratios of all chronometric pairs.

3. CHEMICAL EVOLUTION AND AGE DETERMINATIONS

Age determinations within the framework of cosmochronology, utilizing the r-process chronometer pairs $^{232}Th/^{238}U$, $^{235}U/^{238}U$, and $^{244}Pu/^{238}U$ have been performed by many authors (refs. [5,9–13,21,23,25,28–29,31–33,36]) Other age determinations have been made using the pairs $^{187}Re/^{187}Os$, $^{87}Rb/^{87}Sr$, and $^{207}Pb/^{235}U$ rather than the actinide chronometers (refs. [2–4,40–41]).

All of these calculations start from the basic equation for the gas mass in the galactic disk

$$\frac{dM_G}{dt} = -\Psi + R\Psi + f - o, \tag{1}$$

where M_G denotes the mass of gas in the galactic disk, Ψ is the star formation rate (the mass of gas turned into stars per unit time), R is the fraction of mass ejected by stars during and at the end of their evolution (stellar winds, supernovae etc.), f describes the infalling gas into the disk from the galactic halo, and o denotes a possible outflow into the galactic halo. The form used above already includes the instantaneous recycling approximation. The elements and isotopes which are of concern here are produced in the r-process which is believed to be associated with Type II supernovae and to involve massive stars with short evolutionary time scales, where the instantaneous recycling approximation is well founded.

Based on Eq.(1), one can write an equation for the rate of change of the number of nuclei of species A which reads

$$\frac{dN_A}{dt} = P_A\Psi - \frac{(1-R)}{M_G}\Psi N_A + \frac{f}{M_G}\frac{Z_f}{Z}N_A - \frac{o}{M_G}N_A - \lambda_A N_A. \tag{2}$$

The term including $(1-R)\Psi$ is the same as in Eq.(1). The infall term has to be modified here because the contribution of the infalling gas to the abundance of a heavy nucleus A is reduced according to its lower metallicity Z_f, in comparison to the disk metallicity Z; this is not the case for outflow. Besides the pure reejection of unprocessed gas in the outer envelope of stars R, there is a term (denoted by $P_A\Psi$) which describes the number of nuclei A being newly synthesized in material processed during stellar evolution and explosive processing. λ_A is the decay rate of radioactive nucleus A. A general solution to this equation, when abbreviated as $\dot{N}_A = P_A\Psi - \omega N_A - \lambda_A N_A$, has been given by Tinsley[32] with $\nu(t) = \int_0^t \omega(t')dt'$ as

$$N_A(t) = exp - [\lambda_A t + \nu(t)] \int_0^t P_A\Psi(t') exp[\lambda_A t' + \nu(t')]dt' \tag{3}$$

which depends on the so-called "effective nucleosynthesis rate" $\Psi e^{\nu(t)}$. Eq.(1), together with the definition of $\alpha = (1-R)\Psi/M_G$, leads to solutions for $M_G(t)$ and $\Psi(t)$. Using this result for $\Psi(t)$ and the definition of $\nu(t)$, the effective nucleosynthesis rate Ψe^{ν} reads

$$\Psi e^{\nu} = \Psi_0 exp(-\int_0^t [\frac{f}{M_G}(\frac{Z_f}{Z}-1) - \frac{1}{\alpha}\frac{d\alpha}{dt'}]dt') \equiv \Psi_0 exp(-\int_0^t \lambda_R dt'). \quad (4)$$

Thus the general result for the abundance of nuclei of species A becomes

$$N_A(t) = N_A(0) exp - [\lambda_A t + \nu(t)] + exp - [\lambda_A t + \nu(t)] \int_0^t P_A \Psi_0 exp[-\int_0^{t'} \lambda_R d\tau] exp(\lambda_A t')dt'. \quad (5)$$

We want to describe the initial enrichment $N_A(0)$ in such a way that its contribution to the total abundance of a stable nucleus A at time $t = \Delta$ is equal to a given value $S_0(< 1)$. This leads to

$$N_A(0) = \frac{S_0}{1-S_0} P_A \Psi_0 \int_0^{\Delta} exp[-\int_0^{t'} \lambda_R d\tau]dt'. \quad (6)$$

For the ratio of the abundances of two nuclei A and B at the time of the formation of the solar system, i.e. after a nucleosynthesis epoch of duration $t = \Delta$, we find from Eqs. (5) and (6),

$$\frac{N_A(\Delta)}{N_B(\Delta)} = \frac{P_A}{P_B} f(\lambda_A, \lambda_B, S_0, \lambda_R, \Delta). \quad (7)$$

The ratio of the two radioactive nuclei N_A/N_B is then a function of the unknowns λ_R, S_0, and Δ. Given the knowledge of the production ratios P_A/P_B in a nucleosynthesis process and the abundance ratios N_A/N_B at the time of the formation of meteorites in the solar system, i.e. $4.6 \times 10^9 y$ prior to today, these three unknowns can be determined; if for three chronometric pairs A/B, the production ratios and meteoritic ratios are known. This is, however, only the case if we treat λ_R to first approximation as a constant. Such an approximation corresponds to the exponential model for Ψe^{ν}, which was initially introduced by Fowler [9].

In the previous section we presented the ratios of the three pairs $^{232}Th/^{238}U$, $^{235}U/^{238}U$, and $^{244}Pu/^{238}U$. Unfortunately the very short half-life of ^{244}Pu, which is only sensitive to the last few $10^8 y$ before the formation of the solar system, makes the ratio of $^{244}Pu/^{238}U$ relatively useless for age determinations. Therefore we have used only two equations and must use observational knowledge to constrain one of the three free parameters, preferably S_0.

The most important shortcoming of the simple exponential model - forcing f/M_G to be constant in time. This assumption is true for a closed model with $f = 0$[24]. In order to fulfill observational constraints for the metallicity distribution of stars [22] and the age metallicity relation[37] when $f = 0$, it is necessary to assume an initial enrichment in metals at the beginning of the galactic disk evolution of the order $Z_0/Z_{sol} = 0.1 - 0.3$ [19,34,41]. This would be the same value as S_0 if the initial r-process enrichment is the same as the initial metal enrichment.

In infall models f/M_G might be approximated by a constant (λ_R) during the declining phase of f and M_G with comparable time scales. The first rise and burst of star formation can be put in a separate term in Eq.(5) by performing the integral over two intervals in time. Technically, such a term corresponds to an initial enrichment. One should be aware of the fact that the results give lower limits to the age of the Galaxy, because the duration of the initial burst of star formation is not included. While this duration time might be as short as $\sim 2 \times 10^8$ years, it could be as long as $(2-3) \times 10^9 y$[5].

4. RESULTS

Eq.(7) was utilized for the two chronometric pairs $^{232}Th/^{238}U$ and $^{235}U/^{238}U$. with the production ratios 1.60 and 1.16 and the meteoritic abundance ratios 0.317 for $^{235}U/^{238}U$ and 2.32 for $^{232}Th/^{238}U$[1]. When varying S_0 within the limits 0.1 and 0.3, as discussed above, results range from $7.8 \times 10^9 y$ to $10.1 \times 10^9 y$ for Δ which translates into ages of the Galaxy from $12.4 \times 10^9 y$ to $14.7 \times 10^9 y$. Nuclear uncertainties add errors of $2 \times 10^9 y$ to the allowed ranges. These ages do not include the duration of galactic evolution which produced the initial enrichment which could be as large as $(2-3) \times 10^9 y$.

REFERENCES

(1) Anders, E. and Ebihara, M., *Geochim. Cosmochim. Acta* **46**, 2363 (1982).
(2) Arnould, M., Takahashi, K., and Yokoi, K., *Astron. Astrophys.* **137**, 51 (1984).
(3) Beer, H., and Macklin, R.L., *Phys. Rev.* **C32**, 738 (1985).
(4) Beer, H., and Walter, G., *Ap. Space Sci.* **100**, 243 (1984).
(5) Clayton, D.D., *Ap. J.* **285**, 411 (1985).
(6) Cowan, J.J., Cameron, A.G.W., Truran, J.W., *Ap.J.* **294**, 656 (1985).
(7) Cowan, J.J., Cameron, A.G.W., Truran, J.W., and Sneden, C., in *Nuclear Astrophysics*, ed. J. Audouze (in press) (1986).
(8) Cowan, J.J., Thielemann, F.-K., Truran, J.W., submitted to *Ap. J.* (1987).
(9) Fowler, W.A., in *Cosmology, Fusion and Other Matters* ed. F. Reines (Boulder:Colorado Associated University Press), p. 67 (1972).
(10) ——., *Proceedings of the Welch Foundation Conferences on Chemical Research* **XXI**, ed. W.D. Milligan (Houston:Robert A. Welch Foundation) p. 61 (1978).
(11) ——., submitted to *Q.J.R.A.S.* (1987).
(12) Fowler, W.A. and Hoyle, F., *Ann. Phys.* **10**, 280 (1960).
(13) Hainebach, K.L., and Schramm, D.N., *Ap. J.* **212**, 347 (1977).
(14) Hilf, E.R., von Groote, H. and Takahashi, K., in *Proc. Third International Conference on Nuclei Far From Stability* (Geneva: CERN 76-13), p. 142 (1976).
(15) Hoff, R., in *Weak and Electromagnetic Interactions in Nuclei*, ed. H.V. Klapdor, in press (1986).
(16) Howard, W.M. and Möller, P., *At. Data Nucl. Data Tables* **25**, 219 (1980).
(17) Klapdor, H.V., Oda, T., Metzinger, J., Hillebrandt, W., and Thielemann, F.-K., *Z. Phys. A* **299**, 213 (1981).
(18) Krumlinde, J. Möller, P., Wene, C.O., and Howard, W.M., *Proc. 4th Intl. Conference on Nuclei far from Stability* **CERN 81-09**, p. 260 (1981).
(19) Lynden-Bell, D., *Vis. in Astron.* **19**, 299 (1975).
(20) Meyer, B.S., Howard, W.M., Mathews, G.I., Möller, P. and Takahashi, K., in *Proc. 85 Meeting of the American Chemical Society*, Chicago (1985).
(21) Meyer, B.S. and Schramm, D.N., *Ap. J.* **311**, 406 (1986).
(22) Pagel, B.E.J. and Patchett, B.E., *M.N.R.A.S.*, **172**, 13 (1975).
(23) Reeves, H., *Ap. J.* **237**, 229 (1979).
(24) Schmidt, M., *Ap. J.* **137**, 758 (1963).
(25) Schramm, D.N. and Wasserburg, G.T., *Ap. J.* **162**, 57 (1970).
(26) Seeger, P.A., Fowler, W.A., and Clayton, D.D., *Ap.J. Suppl.* **11**, 121 (1965).
(27) Seeger, P.A., and Schramm, D.N., *Ap. J. Letters* **160**, L157 (1970).
(28) Symbalisty, E.M.D., and Schramm, D.N., *Rep. Prog. Phys.* **44**, 293 (1981).
(29) Thielemann, F.-K., and Truran, J.W., in *Galaxy Distances and Deviations from the Universal Expansion* eds. B.F. Mador and R.B. Tully (Dordrecht:Reidel), p.185 (1986).
(30) Thielemann, F.-K., Metzinger, J., and Klapdor, H.V., *Z. Phys.* **A309**, 301 (1983).
(31) ——. , *Astron. Astrophys.* **123**, 162 (1983).
(32) Tinsley, B.M., *Ap. J.* **198**, 145 (1975).
(33) ——. , *Ap. J.* **216**, 548 (1977).
(34) ——. , *Fund. Cosmic Phys.*, **5**, 287 (1980).
(35) Truran, J.W., *Ann. Rev. Nucl. Part. Sci.*, **34**, 53 (1984).
(36) Truran, J.W., and Cameron, A.G.W., *Astrophys. Space Sci.* **14**, 179 (1971).
(37) Twarog, B.A., *Ap. J.* **242**, 242 (1980).
(38) von Groote, H., Hilf, E.R., and Takahashi, K., *At. Data Nucl. Data Tables* **17**, 418 (1976).
(39) Wene, C.O., and Johansson, S.A.E., *Proc. 3rd Intl. Conference on Nuclei far from Stability* **CERN 76-13**, p. 584 (1976).
(40) Woosley, S.E., and Fowler, W.A., *Ap. J.* **233**, 411 (1979).
(41) Yokoi, K., Takahashi, K., and Arnould, M. *Astron. Astrophys.* **117**, 65 (1983).

THE STELLAR EVOLUTION TIME SCALE

Robert T. Rood
Astronomy Department, University of Virginia
and National Radio Astronomy Observatory,* Charlottesville, VA 22903

ABSTRACT

If stellar models are properly normalized to fit the sun, errors in input physics do not introduce large errors in ages determined from the luminosity of stellar cluster turnoff points. Much larger errors arise from the uncertainty in composition, distance modulus, and reddening. For globular clusters it is possible to estimate the age in several ways. The virtues and deficiencies of each are mentioned. The ages of the globular clusters are estimated to be 15^{+6}_{-4} Gyr. It seems unlikely that cluster ages can be consistent with $H_\circ$ much larger than 50 km s^{-1} Mpc^{-1}.

1. The Ages of Solar-type Stars

Generalizations of simple dimensional arguments such as those found in Chapter 2 of Schwarzschild's famous book can give astonishingly accurate estimates for the luminosity and its initial rate of increase with time for main sequence stars.

$$L \sim \frac{M^{5.5}\mu^{7.5}}{\kappa_\circ} \quad \text{and} \quad \frac{1}{L}\frac{dL}{dt} \sim \frac{35\mu L}{4\epsilon_H M}$$

where the opacity has been parametrized as $\kappa = \kappa_\circ \frac{\rho}{T^{3.5}}$, ϵ_H is the energy produced by the fusion of a gram of hydrogen into helium, and μ is the mean molecular weight. Since to lowest order $lifetime \sim \frac{M}{L}$, the only major uncertainty introduced in stellar lifetimes by uncertainties in the input physics is through the opacity. Even that is a fairly small effect. The shape of evolutionary tracks changes only slowly as opacity is changed. Thus, while the mass of a star at given isochrone turnoff luminosity changes with opacity, the age varies only slowly.

Since cluster ages should not be dramatically affected by input physics, one might be puzzled by the large variation in the estimates of the age of the old disk cluster NGC 188 obtained by Iben in 1967[1] (11 Gyr), Demarque and McClure in 1977[2] (5 Gyr), and VandenBerg[3] in 1983 (10 Gyr). These large variations arise because stellar evolutionists (including me) have not routinely adjusted μ (or the

---

* The National Radio Astronomy Observatory is operated by Associated Universities Inc. under contract with the National Science Foundation.

Change	δ age/Gyr
CHANGES IN INPUT PHYSICS	
Use opacity for Z=0.032 rather than 0.02[a]	0.89
3σ decrease in p-p rate	0.79
3σ decrease in ^{3}He-^{4}He rate	-0.34
3σ decrease in ^{14}N-p rate	0.05
3σ decrease in ^{16}O-p rate	0.00
Assume $Z_\odot = 0.01$ rather than 0.02	-2.17
Use density rather than pressure scale height	-0.05
CHANGES IN COMPOSITION	
$Z = 0.01$ and Y constant	-1.65
$Z = 0.01$ and $dY/dZ = 3$	-2.11
$\delta Y = -0.03$ and Z constant	-0.24

[a] This is roughly the equivalent to the change in one generation of Los Alamos opacities.

helium abundance Y) and the mixing length parameter l_{mix} to fit the sun at the proper age (except when considering the solar neutrino problem). Basically it is not fair to change input physics at constant Y and l_{mix}. Input physics should be changed at constant *sun*. This is not to argue that l_{mix} and Y are constants over all of astrophysics, but that they require some adjustment as the physics is changed. Some of the confusion arose because various formulations of the mixing length theory differ by factors of 2 and $\sqrt{2}$.

Indeed Adler and Rood[4)] have found that for models calibrated to the sun, only opacity or opacity related errors in the input physics are very important. For 9 Gyr isochrones of solar metalicity ($Z_\odot = 0.02$) they find age variations given in the table. Consideration of the size of likely errors shows that errors in input physics will produce an error in the age of an old disk cluster like NGC 188 of about 1 Gyr. The error in metalicity (perhaps a factor of 2) leads to twice as large an error.

2. Globular Cluster Ages

The ages of globular clusters are of particular interest to this meeting because they are the oldest objects in the universe and thus can place an upper limit on H_o. The physics of globular cluster stars is quite similar to that of solar-type stars so the arguments above should apply here as well. There is some uncertainty in propagating the solar helium abundance to these low metalicity systems but this is at least partially offset by the fact that opacity errors should be smaller in the metal poor gas. It is probable, then, that errors in input physics introduce errors of ~1 Gyr in globular cluster turnoff errors.

There are a number of ways of obtaining globular cluster ages. These have been reviewed at length by Iben and Renzini.[5] Buonanno, *et al.*[6] give a particularly nice discussion of the errors of the specific case of NGC 6752. Further references may be found in those papers. I can only discuss the three methods very briefly here.

2.1 *Turnoff Luminosity*

The luminosity of the turnoff point is the physical quantity related to age which is least affected by uncertainties in the models. Observationally it can be determined either from the color-magnitude (c-m) diagram or from luminosity functions.[7] Uncertainties in the mixing length or in the color-temperature relation play a minor role. Bolometric corrections should be no problem. It is assumed that the distance modulus $(m - M)$ is determined in some way independent of the age determination, *e.g.*, from empirically determined RR Lyrae absolute magnitudes.

The largest errors arise from $(m - M)$ and composition uncertainties.[5]

$$\frac{d \log age}{d \log L_{turnoff}} \sim -1 \quad \text{or} \quad \frac{d \log age}{d\,(m-M)} \sim 0.4$$

From the difference in the assorted estimates of RR Lyrae absolute magnitudes we can infer that the uncertaintity in $(m - M)$ is at least a few $\times$ 0.1. This will lead to an error in age of about 10% or 1.5 Gyr. The composition leads to errors

$$\frac{d \log age}{d\,Y} \sim 0.5 \quad \text{and} \quad \frac{d \log age}{d \log Z} \sim 0.12$$

Typical errors might be 0.3 in log Z and 0.03 in Y, again resulting in a 10% or more error in age.

Coupled with a 1 Gyr error from input physics, and an error of at least 0.1 mag in turnoff, one might expect a final error of 4–5 Gyr.

2.2 *Main Sequence Fitting*

The age can be determined by fitting theoretical isochrones to the observed c-m diagram. In essence, this is equivalent to determining $(m - M)$ from fitting to the lower main sequence. It has become the most popular technique perhaps because of the seductively good fits obtained to CCD data (*e.g.*, ref [6]). There are a number of problems. The fits depend critically on the color-temperature relation, the mixing length theory, and reddening. The reader is urged to trace an example isochrone[4] which is shifted in color (*i.e.*, reddening error) or slightly rotated (error in color-temperature relation, convection theory, or photographic photometry) and to then try to fit this isochrone to the sets of same or different composition. A relatively trivial error of 0.03 mag in redding can lead to a 3 Gyr error in age.

The composition induced errors are the same as above. So even though main sequence fitting can produce small formal errors, it still leads to an absolute error of order 4–5 Gyr.

2.3 *The* $\Delta M^{RR}_{turnoff}$ *Method*

This method makes use of the observation that the difference in magnitude between the horizontal branch and main sequence turnoff, $\Delta M^{RR}_{turnoff}$, is normally 3.4 mag. It is more-or-less equivalent to taking $(m - M)$ from the theoretical luminosity of RR Lyrae variables. It has the advantage of not depending on colors. In addition some errors in input physics (opacities) and metalicity tend to cancel. *E.g.*,

$$\frac{d \log age}{d \log Z} \sim 0.06 \quad \text{but unfortunately} \quad \frac{d \log age}{d\,Y} \sim 2$$

Because the error associated with the helium abundance increases, the total error associated with composition uncertainty is still 10% or larger. In addition, errors in input physics can be far more important because evolutionary stages through core helium burning enter into this method. *E.g.,* adding neutral currents to neutrino losses during red giant evolution decreases the age by 0.75 Gyr.

In short, the $\Delta M^{RR}_{turnoff}$=3.4 mag method replaces one set of errors with another. While there may be some aesthetic reason to prefer the model errors associated with this method, at this point the results are about the same, 4–5 Gyr. It is important to note that using this method one might achieve very small errors (perhaps $\lesssim$1 Gyr) for the age of one cluster relative to another. To do this requires a solution to the "Sandage Effect" problem with the RR Lyrae, which I do not have space to discuss.

My bottom line subjective estimate for globular cluster ages is 15^{+6}_{-4} Gyr. If $\Omega_{\circ} = 1$ and clusters form at Z (redshift) = 4, then $H_{\circ} = 38^{+10}_{-15}$ km s^{-1} Mpc^{-1}.

REFERENCES

) Iben, I., Jr., *Astrophys. J.* **147**, 624 (1967).

) Demarque, P., and McClure, R.D., in *The Evolution of Galaxies and Stellar Populations,* eds. B.M. Tinsley, R.B. Larson (New Haven: Yale U. Obs.), 199 (1977).

) VandenBerg, D.A., *Astrophys. J. Suppl.* **51**, 29 (1983).

) Adler, D.S., and Rood, R.T., *Bul. Am. Astron. Soc.* **17**, 593 (1985).

) Iben, I., Jr., and Renzini, A., *Phys. Reports* **105**, 331 (1984).

) Buonanno, R., Corsi, C.E., Iannicola, G., and Fusi Pecci, F., *Astron. Astrophys.* **159**, 189 (1986).

) Paczynski, B., *Astrophys. J.* **284**, 670 (1984).

STELLAR POPULATION PROBLEMS

Richard B. Larson

Yale Astronomy Department
Box 6666, New Haven, Connecticut 06511

A "stellar population" may be regarded as a group of stars that have a common origin and thereby share some common properties. It is then important to know what populations or population sequences can be distinguished in galaxies, and to understand what underlying astrophysical parameters determine their properties. The ultimate goal of stellar population studies is to understand the origin of the observed stellar populations and hence the origin of the galaxies that they constitute.

Classically, two major components of our Galaxy have been distinguished: a metal-poor halo containing globular clusters (population II), and a metal-rich disk containing mostly younger stars (population I). However, it is now recognized that this classification is too simple to encompass all of the properties of stellar populations in our Galaxy and others. For example, the globular clusters themselves exhibit both halo and disk sub-populations separated by metallicity at [Fe/H] $\sim$ -0.8 (1); some of the disk globular clusters are in fact quite metal-rich. Moreover, the halo or spheroid components of galaxies are not always metal-poor, nor are disks necessarily metal-rich; elliptical galaxies and spiral bulges may contain old yet metal-rich stars, while galaxies like the Magellanic Clouds have relatively metal-poor disks. Thus the traditional distinction between a metal-poor halo and a metal-rich disk may just reflect an accident of how things happened in our own Galaxy.

More fundamentally, the distinction between halos and disks may itself originate just as a result of dynamical accident. Most star formation, whether of metal-poor or metal-rich objects, may actually occur in disks or flattened systems (2), and some disks may later be disrupted by collisions that leave the stellar debris in a rubble heap that is more nearly round than flat and would be called an elliptical galaxy or spheroid (3). Even globular clusters may mostly form in disks by gravitational instabilities characteristic of flattened systems (4). If galaxies experience many collisions and mergers during their early history, all of the stars in a galaxy that formed prior to the last major collision would end up in a nearly spherical distribution or halo, while those that formed after the last major collision would mostly remain in a disk. The main astrophysical distinction between the different populations in a galaxy would then just be in their time of formation, which controls the metallicity through a steady increase of metallicity with time in the residual gas.

Many properties of stellar populations depend on the initial mass function (IMF) with which the stars are formed, so a key question is whether the IMF varies with time or location in galaxies. A possible explanation of the weak age-dependence of stellar metal abundances and the paucity of metal-poor stars in the solar neighborhood (the "G-dwarf problem") is that the IMF contained a higher proportion of massive stars at earlier times (5). The higher oxygen-to-iron ratio in the oldest stars of our Galaxy could also be due to an IMF that was enriched in massive stars at early times. Several pieces of evidence suggest that the present-day IMF is not universal but varies with location, the formation of massive stars being favored in regions where the star formation rate is higher (6, 7). It is therefore not implausible that the IMF has varied with time such that the formation of massive stars was more important at early times when the star formation rate was higher than it is at present (7).

If this is the case, then the remnants of the intermediate- and high-mass stars that formed in abundance at early times become an important contributor to the total mass of a stellar system; for example, they may account for all of the unseen mass in the solar neighborhood (8). Indeed, a model with a time-varying IMF can solve both the G-dwarf problem and the local unseen mass problem. Other explanations are of course possible; for example, the G-dwarf problem may be solved by gas infall, although there is not much evidence for this, or by including the thick disk and halo components of our Galaxy in a chemical evolution model (9). Also, the unseen mass could be in very low-mass stars or "brown dwarfs", although no confirmed brown dwarfs have yet been found. If the IMF does vary with time in the above manner, a general consequence is that older stellar populations should have larger mass-to-light ratios because of the larger contribution of stellar remnants to their masses.

Some fundamental clues to stellar population problems such as those discussed above may be provided by the properties of the smallest galaxies, especially the dwarf spheroidal galaxies in the Local Group (10, 11, 12). The dwarf galaxies are very metal-poor and are generally regarded as very primitive objects; indeed, they may resemble the building blocks from which larger galaxies formed, so they may tell us something about how star formation occurred during early galactic evolution. Among the dwarf spheroidals, both the mass-to-light ratio and the metallicity appear to be closely correlated with total luminosity; the mass-to-light ratio varies by a large factor from $M/L \sim 2$ for Fornax ($MV = -12.3$, $[Fe/H] = -1.4$) to $M/L \sim 100$ for Ursa Minor ($MV = -8.8$, $[Fe/H] \sim -2.2$). The fact that both [Fe/H] and M/L vary systematically with luminosity suggests that a variation of the IMF may be responsible; certainly if it is assumed that the mass in these systems is in stars or stellar remnants, the only way to account for the variation in M/L is to postulate a variation of the IMF. The dwarf spheroidal galaxies are apparently not unique in showing an increase in M/L with decreasing luminosity, since a similar trend is also seen in recent results for dwarf irregular and spiral galaxies (13, 14).

One possibility is that both the high mass-to-light ratios and the low metallicities of the faintest dwarfs reflect a greatly increased proportion of low-mass stars in the IMF. However, a simple closed-box model of chemical evolution based on this assumption would predict $Z \propto (M/L)^{-1}$, in disagreement with the data which show a weaker variation of Z with M/L. Alternatively, if the IMF in the faintest dwarfs was dominated by massive stars, as might be the case if they formed early, the high mass-to-light ratios of these systems might be due to a large content of stellar remnants. The metallicity variation among the dwarf spheroidals would then have to be explained in another way, but this is almost certainly required anyway because these systems are so weakly bound that they probably did not retain the heavy elements produced within them by supernovae. Most, if not all, of their nucleosynthesis products were probably ejected in a hot metal-enriched wind (15, 16) into a hot intergalactic medium. If so, the very low metallicities of systems like Ursa Minor might reflect a very low degree of retention or recapture of hot enriched gas, or very inefficient mixing of the hot gas into the dense cool clouds in which the stars formed.

A possible interpretation of the properties of the dwarf spheroidal galaxies is then that they formed at early times and with a predominance of massive stars, so that stellar remnants now contribute importantly to their masses. The chemical enrichment of the star-forming gas was probably very inefficient, and it may have taken considerable time for mixing processes to enrich the gas. If the IMF evolved toward a more normal form as the metallicity rose (such a dependence of the IMF on metallicity is suggested by some of the evidence (17)), the proportion of massive stars formed would decrease with time and the correlation between present M/L and metallicity would be qualitatively explained. It is consistent with this picture that Ursa Minor, the dwarf spheroidal with the lowest Z and highest M/L, is also the only one with a purely very old stellar content, while the spread in stellar ages toward younger values generally increases with increasing Z and decreasing M/L (12). Thus the properties of the stellar populations in the dwarf galaxies may be determined basically by the time at which most of the presently observed stars formed, coupled with a general increase in metallicity and an evolution of the IMF with time. Although the mass of a dwarf galaxy could also be an important parameter controlling chemical evolution and the IMF, it is curious that current data indicate very similar masses for all of the local dwarfs.

If large galaxies form by a progressive merging of smaller stellar aggregates into larger ones (18), then the dwarf galaxies that we presently see may represent building materials left over from the formation of larger galaxies like our own. If the first stellar systems formed with a predominance of massive stars, stellar mass loss would have tended to unbind them and most of them would probably have been disrupted by gas loss or collisions and merged into larger systems; the present dwarfs may then be just the lucky survivors of a once much larger population. The systems formed by the mergers of such dwarfs would be metal-poor and roughly spherical, and would have large mass-to-light ratios; thus they would have many of the properties of present galactic halos. The halo globular clusters may have originated in some of the larger pregalactic dwarfs, since the Fornax dwarf contains four apparently typical globular clusters. Thus an understanding of the properties of the dwarf galaxies may help us to understand the origin of at least the older stellar populations in larger galaxies like our own.

REFERENCES

1. Zinn, R. J., Astrophys. J. 293, 424 (1985).
2. Larson, R. B., in Nearly Normal Galaxies from the Planck Time to the Present, ed. S. M. Faber, in press. Springer-Verlag (1987).
3. Toomre, A., in The Evolution of Galaxies and Stellar Populations, ed. B. M. Tinsley and R. B. Larson, p. 401. Yale University Observatory (1977).
4. Larson, R. B., in Globular Cluster Systems in Galaxies, IAU Symposium No. 126, eds. A. G. D. Philip and J. E. Grindlay, in press. Reidel (1987).
5. Schmidt, M., Astrophys. J. 137, 758 (1963).
6. Scalo, J. M., Fundam. Cosmic Phys. 11, 1 (1986).
7. Larson, R. B., in Stellar Populations, eds. C. A. Norman, A. Renzini, and M. Tosi, in press. Cambridge Univ. Press (1987).
8. Larson, R. B., Mon. Not. Roy. Astron. Soc. 218, 409 (1986).
9. Gilmore, G. and Wyse, R. F. G., Nature 322, 806 (1986).
10. Zinn, R. J., Mem. Soc. Astron. Ital. 56, 223 (1985).
11. Aaronson, M., in Stellar Populations, eds. C. A. Norman, A. Renzini, and M. Tosi, in press. Cambridge Univ. Press (1987).
12. Da Costa, G. S., in Globular Cluster Systems in Galaxies, IAU Symposium No. 126, eds. A. G. D. Philip and J. E. Grindlay, in press. Reidel (1987).
13. Sargent, W. L. W. and Lo, K.-Y., in Star Forming Dwarf Galaxies, eds. D. Kunth, T. X. Thuan, and J. Tran Thanh Van, p. 253. Editions Frontieres (1986).
14. Freeman, K. C., in Nearly Normal Galaxies from the Planck Time to the Present, ed. S. M. Faber, in press. Springer-Verlag (1987).
15. Vader, J. P., Astrophys. J. 305, 669. (1986).
16. Vader, J. P., Astrophys. J., in press (1987).
17. Campbell, A. W., Terlevich, R. J., and Melnick, J., Mon. Not. Roy. Astron. Soc., in press (1987).
18. Tinsley, B. M. and Larson, R. B., Mon. Not. Roy. Astron. Soc., 186, 503 (1979).

NUCLEOSYNTHESIS YIELDS AND PRODUCTION TIMESCALES

J. W. Truran
Max-Planck-Institut für Astrophysik
Garching b. München, FRG

Department of Astronomy
University of Illinois, USA

ABSTRACT

An overview is presented of nucleosynthesis as a function of stellar mass and the implied timescale for the return of these nucleosynthesis products to the interstellar medium. We identify, specifically, the mass ranges of stars which are expected to produce significant quantities of heavy elements, either during the normal course of their hydrostatic evolution or in ensuing supernova events. We note the particular nucleosynthesis products to be expected and the relevant stellar evolutionary timescales. We then briefly survey existing observations of the abundances of these interesting elements in the stellar populations of our galaxy, emphasizing particularly the extremely metal deficient field halo stars and globular cluster stars. We conclude with a brief discussion of possible implications of these combined observational and theoretical studies for models of the early history and chemical evolution of our galaxy.

1. INTRODUCTION

Studies of galactic chemical evolution generally seek to account for the observed distributions of abundances of the elements in the stellar populations and interstellar gas in galaxies. Critical input to such calculations includes a knowledge of: (1) the rate of star formation as a function of time; (2) the stellar initial mass function as a function of time; (3) the fraction of gas processed through earlier stellar generations; (4) the evolutionary characteristics of stars as a function of their initial composition; and (5) stellar and supernova nucleosynthesis. These diverse factors can strongly influence the relative abundances of heavy elements produced in generations of stars and thus the abundance patterns observed in stars associated with the distinct stellar populations identified in our galaxy.

It should be apparent that models of galactic chemical evolution can rapidly become extremely detailed[1]. In an attempt to simplify at least one aspect of this problem, our aim in this paper is to extract the essential features of nucleosynthesis studies and to establish a quite straightforward prescription for stellar and supernova nucleosynthesis as a function of stellar mass (and associated evolutionary timescale) which may be utilized as a probe of the gross features of galactic chemical and dynamic evolution. A discussion of these nucleosynthesis considerations is presented in section 2. A survey of elemental abundance patterns observed in stars as a function of their total metallicities is presented in section 3. Significant abundance patterns and trends are then discussed and some possible implications for the early history of stellar activity in our galaxy are noted[2-4].

2. NUCLEOSYNTHESIS AS A FUNCTION OF STELLAR MASS

Studies of galactic chemical evolution require as input a variety of types of information concerning the consequences of stellar evolution and nucleosynthesis. Significant progress in nucleosynthesis theory has occurred over the past decade[5]. Detailed predictions of nucleosynthesis associated both with relatively stable phases of stellar evolution and with the ejection of matter in supernova events are now available. For present purposes, we confine our attention to the following interesting elements or classes of nucleosynthesis products: carbon, nitrogen, oxygen, the elements from neon to calcium, the iron-peak nuclei, the s-process heavy elements, and the r-process heavy elements. We also identify the following critical ranges of stellar mass for which the nucleosynthesis contributions are to be considered: (1) the range of massive stars $M \gtrsim 10\text{–}12\ M_\odot$ which are believed to represent the progenitors of Type II supernovae; (2) the range of intermediate mass stars $1 \lesssim M \lesssim 10\ M_\odot$ which are important sites of nucleosynthesis during the asymptotic giant branch phase of their evolution; and (3) Type I supernovae, which are now thought to arise as a consequence of the evolution of stars of intermediate mass in binary systems. We note that there exists a critical distinction in the production timescales for these nucleosynthesis sources as defined by their stellar lifetimes. Intermediate mass stars typically evolve on timescales $\tau > 10^8\text{–}10^9$ years, while massive stars $M \gtrsim 10\ M_\odot$ evolve on timescales $\tau < 10^8$ years compatible with a halo collapse timescale. One might therefore anticipate that extreme halo population stars and globular cluster stars can be characterized by "anomalous" abundance patterns, relative to solar abundances, which reflect nucleosynthesis in massive stars. We now summarize nucleosynthesis occurring in each of the environments indicated above.

2.1 Massive Stars and Type II Supernovae

Nucleosynthesis occurring in the cores of massive stars during the late stages of presupernova evolution and in the ensuing supernova events themselves produces nuclei from carbon to nickel[6-9]. It appears, however, that both carbon and iron-peak elements are underproduced in such events relative to oxygen and the neon-to-calcium elements. While the mass of iron ejected is quite uncertain, due to uncertainties associated with the position of the mass cut and the expected remnant mass, the tendency for iron to be underproduced seems clear. We conclude that massive stars $M > 10\ M_{\odot}$ are the primary source of oxygen and the intermediate mass elements from neon to calcium.

We also assume that these massive stars provide the site of production of the r-process heavy nuclei. While the r-process environment remains somewhat uncertain, we believe that models involving the ejection of highly neutronized matter from the vicinity of the mass cut in supernovae which leave neutron star remnants are most promising[5,10,11]. The production history of the interesting nuclear chronometers ^{232}Th, ^{235}U, ^{238}U, and ^{244}Pu, formed in the r-process, is thereby tied to the history of Type II supernova activity.

2.2 Intermediate Mass Stars

Stars in the mass range $1 \lesssim M \lesssim 10\ M_{\odot}$ provide important contributions to the abundances of the heavy elements in the galaxy as a consequence both of the occurrence of thermal pulses in their helium burning shells on the asymptotic giant branch and of the subsequent evolution of their cores to Type I supernovae in binary systems. Estimates of nucleosynthesis yields from asymptotic giant branch stars[12,13] indicate that significant production of carbon, nitrogen, and the s-process elements can be achieved in this environment. It should be recalled that the s-process elements are secondary nucleosynthesis products, in that their formation demands the presence of seed iron-peak nuclei from the ashes of prior stellar generations.

Theoretical studies currently suggest that nucleosynthesis in Type I supernovae nicely compliments that of Type II supernovae by forming predominantly iron and iron-peak nuclei. Calculations of explosive nucleosynthesis associated with carbon deflagration models of Type I supernovae[14] predict that sufficient iron-peak nuclei are formed to explain both the powering of the light curves of Type I supernovae by the radiative decay of ^{56}Ni and the observed mass fraction of iron-group nuclei in galactic matter. The predicted yields of the intermediate mass nuclei neon-to-calcium are consistent with observations of Type I spectra, but insufficient to contribute significantly to the abundances of these nuclei in the galaxy. Serious questions remain, however, concerning the nature and evolutionary histories of the binary progenitors of Type I supernovae and their frequencies of occurrence in different stellar populations. For the purpose of the present discussion, we will assume that such binary models for Type I supernovae[15,16] produce iron and iron-peak nuclei on timescales compatible with the lifetimes of the intermediate mass stars which are assumed to characterize these systems.

2.3 Nucleosynthesis Summary

Massive stars ($M \gtrsim 10\ M_{\odot}$), evolving on timescales $\tau < 10^8$ years to Type II supernovae, are the major galactic sources of oxygen, the intermediate mass elements from neon to calcium, and the r-process elements.

Intermediate mass stars ($1 \lesssim M \lesssim 10\ M_\odot$), evolving on timescales $\tau \gtrsim 10^8$–10^9 years, are the main galactic sources of carbon, nitrogen, and the s-process elements. Type I supernovae produce the iron-peak nuclei on comparable timescales.

3. COMPOSITION TRENDS IN METAL DEFICIENT STARS

We now briefly survey significant trends in galactic abundances which reflect the abundance history of the average matter in our galaxy over the course of its evolution. We adopt solar system abundances as the standard against which others are to be compared. Particular attention is given to observations of the abundance patterns characterizing the most metal deficient stellar components of the galaxy.

3.1 Metal Deficient Field Halo Stars

Observations of the abundances in the most metal deficient stars in our galaxy reveal the presence of interesting deviations from solar abundances which we seek to explain in the context of theories of nucleosynthesis and galactic chemical evolution[2,3,17]. In the following discussion, we draw heavily on the nice review of the composition of field halo stars by Spite and Spite[18].

High oxygen to iron ratios [O/Fe] $\approx$ +0.5 are found to characterize halo stars. Similarly, the elements Mg, Si, and Ca are found to be enriched relative to Fe by approximately 0.5 dex[19,20]. In contrast, both [C/Fe] and [N/Fe] are found to be compatible with solar and approximately constant[21] for a sample of disk and halo stars which span a range in [Fe/H] of −2.45 to +0.5.

Further interesting trends appear in the heavy element region. The data clearly establishes the existence of depletions in the abundances of the s-process elements Sr and Ba, relative to iron, in stars of low Fe/H[18,19]. Moreover, both theory[22] and observations[23] now strongly suggest that the heavy element abundance patterns in the extreme metal deficient stars are dominated by the r-process contributions. This is again compatible with our assumption that massive stars represent a site of r-process nucleosynthesis.

3.2 Globular Cluster Stars

Observational studies now indicate that globular cluster stars exhibit abundance patterns similar to those of the metal deficient field halo stars[24,25]. These include specifically the enrichments of the intermediate mass elements Mg, Si, and Ca relative to Fe by ~0.5 decs and the consistency of the heavy element abundance patterns with an r-process origin. High O/Fe ratios are also encountered in a substantial fraction of the studied clusters.

3.3 Summary of Observed Trends

Extreme metal deficient field halo stars and globular cluster stars exhibit anomalous abundance patterns relative to solar abundances which are consistent with the ejecta of Type II supernovae arising from the evolution of massive stars. These patterns persist through the growth of the metallicity to at least [Fe/H] ≈ −1, presumably indicating the point at which the ejecta of intermediate mass stars of longer lifetimes introduces carbon, nitrogen, iron, and s-process elements into the interstellar medium.

4. DISCUSSION

A number of useful inferences and conclusions can be drawn from the combined theoretical and observational results we have considered.

(1) The trends in elemental abundance patterns in the different stellar populations in the galaxy are consistent with the general predictions of nucleosynthesis theory: oxygen, the elements from neon to calcium, and the r-process heavy elements are products of the evolution of massive stars of lifetimes $\lesssim 10^8$ years, while carbon, nitrogen, the iron-peak nuclei, and the s-process heavy elements are produced in stars of lower masses and longer lifetimes.

(2) The realization of abundance patterns in our galaxy or other galaxies which are consistent with that of solar system matter requires a timescale $\gtrsim 10^9$ years, compatible with the evolutionary timescales of the low and intermediate mass stars which contribute significantly to nucleosynthesis.

(3) The anomalous abundance patterns which have been found to characterize field halo stars of Population II are quite consistent with the contamination of the gas from which they were formed by the ejecta of normal stars of masses $\gtrsim 10\ M_\odot$, and associated Type II supernovae, on a halo collapse timescale. One cannot strictly rule out contributions from more exotic sources (eg. supermassive stars or a Population III) to the abundances of the most extreme metal deficient halo stars, but neither is there any compelling evidence to suggest that such sources may have contributed significantly.

(4) The identification of the site of the r-process with massive stars allows one better to define the production history of the important nuclear chronometers ^{232}Th, ^{235}U, ^{238}U, and ^{244}Pu. The implied level of initial disk enrichment, depending on ones choice of an age-metallicity relation[26,27], can yield lower estimates of the galactic age from cosmochronology[28].

(5) The change in abundance patterns at [Fe/H] $\gtrsim$ −1 occurs approximately at the metallicity of the most metal deficient disk stars. This would appear to be associated with the termination of the halo phase on a timescale of order 10^9 years.

(6) The similarities in the abundance patterns characterizing globular cluster stars and extreme halo population field stars are suggestive of a similar nucleosynthesis origin. It is of interest to determine whether they were independently contaminated by the ejecta of

massive stars or, rather, if they were formed from the debris of a common earlier stellar generation. In fact, an interesting possible test of whether a self-enrichment model is appropriate to globular clusters might be provided by a careful study of the abundance patterns in the most metal-rich clusters of $Z \gtrsim 0.1\ Z_{\odot}$. Field stars of such metallicity exhibit patterns of abundances relative to iron which are consistent with those of solar system matter, since the nucleosynthesis contributions from the intermediate mass stars of relatively longer lifetimes have by this time enriched the interstellar medium in carbon, nitrogen, s-process elements, and iron-peak nuclei. Self-enrichment of a globular cluster must necessarily be realized on a much shorter timescale, hence the abundance patterns of stars in even the more metal-rich clusters should reflect only the contributions from the more massive stars. If self-enrichment indeed occurred for the globular clusters, one might then expect even the metal-rich clusters to exhibit, to some degree, both the high O/Fe, Mg/Fe, Si/Fe, and Ca/Fe ratios and the r-process heavy element abundance pattern which are found to characterize the most metal-deficient field halo stars.

5. ACKNOWLEDGEMENTS

The author wishes to express his thanks to the Alexander von Humboldt Foundation for support by a U.S. Senior Scientist Award and to Professor R. Kippenhahn for the hospitality of the Max-Planck-Institut für Astrophysik, Garching bei München.

6. REFERENCES

1. Tinsley, B.M., Fund. Cosmic Phys. 5, 287-388 (1980).
2. Truran, J.W., in J. Audouze and J. Tran Thanh (Eds.), "Formation and Evolution of Galaxies and Large Structures in the Universe", D. Reidel, Dordrecht, 391-399 (1984).
3. Truran, J.W. and Thielemann, F.-K., in C.A. Norman, A. Renzini and M. Tosi (Eds.), "Stellar Populations", Cambridge University Press, Cambridge, in press (1987).
4. Matteucci, F., Mem.S.A.It., in press (1987).
5. Truran, J.W., Ann. Rev. Nucl. Part. Phys. 34, 53-97 (1984).
6. Arnett, W.D., Astrophys. J. 219, 1008-1016 (1978).
7. Weaver, T.A., Zimmerman, G.B., and Woosley, S.E., Astrophys. J. 225, 1021-1029 (1978).
8. Thielemann, F.-K. and Arnett, W.D., Astrophys. J. 295, 604-619 (1985).
9. Woosley, S.E. and Weaver, T.A., in D. Mihalas and K.H. Winkler (Eds.), "Radiation Hydrodynamics in Stars and Compact Objects", D. Reidel, Dordrecht, in press (1987).
10. Hillebrandt, W., Space Sci. Rev. 21, 639-702 (1978).
11. Truran, J.W., Cowan, J.J., and Cameron, A.G.W., in W. Hillebrandt (Ed.), "Nuclear Astrophysics", Max-Planck Publication MPA 199, Munich (1985).
12. Iben, I., Jr. and Truran, J.W., Astrophys. J. 220, 980-995 (1978).
13. Renzini, A. and Voli, M., Astr. Astrophys. 94, 175-194 (1981).

14. Thielemann, F.-K., Nomoto, K., and Yokoi, K., Astr. Astrophys. 158, 17-33 (1986).
15. Nomoto, K., Ann. New York Acad. Sci. 470, 294-310 (1986).
16. Woosley, S.E. and Weaver, T.A., Ann. Rev. Astr. Astrophys. 24, 205-253 (1986).
17. Tinsley, B.M., Astrophys. J. 229, 1046-1056 (1979).
18. Spite, M. and Spite, F., Ann. Rev. Astr. Astrophys. 23, 225-238 (1985).
19. Luck, R.E. and Bond, H.E., Astrophys. J. 292, 559-577 (1985).
20. Gratton, R.G. and Sneden, C., preprint (1986).
21. Laird, J.B., Astrophys. J. 289, 556-569 (1985).
22. Truran, J.W., Astr. Astrophys. 97, 391-393 (1981).
23. Sneden, C and Pilachowski, C.A., Astrophys. J. Letters 288, L55-L58 (1985).
24. Pilachowski, C.A., Sneden, C., and Wallerstein, G., Astrophys. J. Suppl. 52, 241-287 (1983).
25. Gratton, R.G., Quarta, M.L., and Ortolani, S., preprint (1986).
26. Carlberg, R.G., Dawson, P.C., Hsu, T., and Vandenberg, D.A., Astrophys. J. 294, 674-681 (1985).
27. Twarog, B.A., Astrophys. J. 242, 242-259 (1980).
28. Thielemann, F.-K. and Truran, J.W., in B.F. Madore and R.B. Tully (Eds.), "Galactic Distances and Deviations from Universal Expansion", D. Reidel, Dordrecht, 185-195 (1986).

THE CHEMICAL EVOLUTION OF THE GALAXY'S HALO

Bruce W. Carney

Department of Physics and Astronomy
University of North Carolina
Chapel Hill, NC 27514, U. S. A.

1. INTRODUCTION

Our Galaxy's juvenile state differed greatly in both form and content from the mature state in which we find it. The majority of its stars and gas clouds are now distributed in a rotating disk, and a few percent by mass of its mass is composed of elements heavier than Li, elements which were synthesized within stars following the Galaxy's birth. All these elements are referred to as "metals" by astronomers, for the joint sakes of brevity and obstinance, and this convention will be followed here. This paper reviews our understanding of the earliest stages of the chemical evolution of our Galaxy, during which time it was a contracting spheroid of gas and young stars, based upon studies of the lower mass stars whose lifetimes are long enough that they still shine.

2. PRIMORDIAL ABUNDANCES

The proto-Galaxy inherited from the Big Bang H, He, and Li. Abundances of the latter two with respect to the former in the Galaxy's oldest stars are thus of great interest, although they are very difficult to measure. Little has changed since a recent review by Boesgaard and Steigman[1]. Briefly, the best-determined ratio is Li/H, found by the Spites[2] to be so small it implies $\Omega << 1$ for standard hot Big Bang nucleosynthesis models. Helium has not been detected in any halo objects except hot, core-helium burning stars whose atmospheres are subject to gravitational diffusion processes that seriously alter the original He/H ratio, and in planetary nebulae[3,4,5], where it is not clear how to distinguish between the inherited and the newly-synthesized helium. Only indirect methods may be used for the less-evolved halo stars[6,7,8], but they agree with abundances derived from extragalactic metal-poor newly-formed emission nebulae[9], with a helium mass fraction of $Y \sim 0.23$.

3. TIMESCALE

The transition from the low-metallicity spheroidal phase to the higher-metallicity disk was brief compared to the Galaxy's age, but whether it was as short as a rotation or free-fall timescale ($\sim 10^8$ years) or ≥ 10 times longer remains disputed. One direct and two indirect methods have been applied to the question of the timescale.

In principle, a direct measure is available by age-dating halo globular star clusters of diverse mean metallicities. New, accurate cluster data and theoretical models of stellar evolution and spectra are in excellent agreement. The few metal-poor and the metal-rich clusters so studied to date differ by no more (and possiby less than) about 2 x 10^9 years[10]. Inclusion of differing CNO/Fe ratios (§5) may further decrease the differences[11]. However, the clusters' helium abundances and their distances remain free parameters in the fits, and there are suggestions that the resultant distance scale is incorrect[12], so that age differences of a few billion years may be indicated.

If the halo's contraction lasted much longer than the nucleosynthesis timescale, simple models predict the development of a metallicity gradient, with lower metallicities and smaller dispersions in the mean metallicity occuring at larger Galactocentric radii. Measurement of the halo's metallicity gradient is complicated by the presence of a second population of globular clusters[13] which are metal-rich compared to the usual globulars, and belong to a rotating disk population within the inner Galaxy ($R_{GC} \lesssim 8$ kpc). Eliminating these clusters shows the halo's metallicity gradient to be absent[13] or weak[14,15]. In any case, the metallicity dispersion neither decreases with metallicity nor with R_{GC}. However, this does not necesarily rule out the slow halo contraction model. If the proto-Galaxy was not a closed system, but rather a slow merger of independent fragments, such a global metallicity distribution would ensue[16]. Further, an age spread of a few billion years could explain the unusual distributions of the surface temperatures of the core-helium burning stars within the outer halo globular clusters (the "second parameter problem").

The discovery[17,18] of a correlation between stars' metallicities and their eccentricities of their Galactic orbits (projected onto the Galactic plane) suggests the halo's chemical enrichment proceeded on a dissipationless (*i.e.*, brief) timescale. However, a similarly extensive and kinematically-biased survey[19] and another very large and kinematically-unbiased survey[20] have not confirmed such conclusions.

In summary, the timescale for the evolution of the metal-poor spheroid into the metal-rich disk remains uncertain, with a few x 10^8 years to a few x 10^9 years as the likely boundaries.

4. THE HALO'S METALLICITY DISTRIBUTION

Following Hartwick[21], if one assumes the halo began with zero metals, evolved as a closed system with a fixed initial stellar mass function, and that instantaneous recycling applies (*i.e.*, that the nucleosynthesis enrichment time is shorter than the contraction time), the cumulative number of stars, S, with metallicities less than Z, relative to the total number of stars, S_f, with limiting metallicity Z_f, is

$$\frac{S}{S_f} = \frac{1 - \exp(-Z/y)}{1 - \exp(-Z_f/y)} \qquad (1)$$

where y is the yield (the stellar ejecta mass of the metals, Z, divided by the total mass trapped in stellar remnants). For astrophysically plausible values of y, the above expression does not match the observed distributions of field star and globular cluster metallicities[21], but can be made to do so if the effective yield is reduced by a factor of ten to twenty, due perhaps to mass loss from the halo to the proto-disk[22,23].

A possible discrepancy at the lowest metallicities has been suggested by Bond[24], who failed to find the expected numbers of stars deficient in Fe/H by factors of a thousand or more compared to the Sun, despite an exhaustive search. Bond even suggested his observed metallicity distribution function is better explained if the halo had a non-zero initial metallicity, due to a generation of pre-Galactic stars or even unpredicted traces of Big Bang nucleosynthesis. However, recent efforts[23,25] have begun to uncover the lowest metallicity stars in their more-or-less expected low proportions.

5. EVOLUTION OF THE HALO'S ELEMENTAL ABUNDANCE PATTERN

Here we address the evolution of the element-to-iron ratios as functions of Fe/H, which we take to be a monotonic (albeit imprecise) function of time. Sneden[26] reviewed the trends of C,N, and O *vs*. Fe, Spite and Spite[27] have summarized the behavior of as many elements as had been studied through 1984, and Lambert[28] has more recently reviewed the lighter elements Na through Ca.

Element/Fe *vs*. Fe/H should vary for at least three reasons. First, nucleosynthesis probably occurs in several different types of environments, each of which should produce different element-to-iron ratios. For example, massive stars' supernovae are thought to produce larger O/Fe than those of intermediate-mass stars. The observed high O/Fe ratio in low Fe/H stars suggests nucleosynthesis at the earliest times in the Galaxy was due primarily to massive, shorter-lived stars. In principle, if we can more conclusively identify such sites' production ratios and their progenitors' lifetimes, we will be able to use such element/Fe *vs*. Fe/H data to infer the timescale of the halo's chemical enrichment. Second, the lack of complete mixing within the extensive halo of the supernovae ejecta will result in variations in Fe/H in subsequent generations of stars. Scatter in element/Fe *vs*. Fe/H diagrams may be caused in part by such dilution-patchiness effects. Since many chemical evolution models assume uniform mixing, it is important to measure the intrinsic scatter in such diagrams to determine how well such an assumption is satisfied. Third, some elements are "secondary" in that they require seed nuclei for their manufacture. The s-process elements Sr, Y, and Ba, and the r-process element Eu all require the presence of Fe. Unless the Fe is manufactured by the same star prior to the later synthesis of s-process or r-process elements, we expect s-process/Fe and r-process/Fe to depend linearly on Fe/H.

In brief, both O and the other "α-nuclei" (*e.g.*, Mg, Si, Ca, Ti) increase in abundance relative to iron as Fe/H drops to about 0.05 times solar, below which O/Fe and "α"/Fe level off at about 5 and 3 times solar, respectively. C/Fe is solar until Fe/H drops to about 0.02 times solar, then it appears to rise. These breaks at 0.05 and 0.02 times solar Fe/H are probably due to the appearance of new nucleosynthesis sources, perhaps the type II (massive stars) SNe being joined by the longer-lived progenitors' of SNe types Ia and Ib. If this can be demonstrated, a knowledge of progenitor lifetimes leads to much-improved estimates of the halo's chemical enrichment timescale. It must be pointed out, however, that all the cited trends show considerable scatter. At the lowest metallicities (less than 0.003 solar Fe/H), only a very few stars have been studied. Another possibility, at least for the break at 0.05 solar Fe/H, is that the stellar initial mass function began to shift toward lower-mass stars due to star-formation processes involving metallicity-induced cooling of protostellar clouds[29], or dynamical reasons having to do with the disk's formation. Precise studies of element/Fe *vs.* Fe/H and *vs.* kinematics will prove valuable. Finally, the r-process element Eu/Fe ratio remains roughly solar to very low Fe/H (although with much scatter again). The nominal s-process elements such as Sr, Y, and Ba remain solar with respect to Fe doen to roughly 0.03 solar Fe/H, below which they decline. Only Ba/Fe, however, then declines linearly with Fe/H, as expected for a secondary element. Interpretation of these results remains complicated by uncertain r-process contributions to their abundances.

6. REFERENCES

1. Boesgaard, A. M. and Steigman, G., Ann. Rev. Astron. Astrophys., 23, 319 (1985).
2. Spite, F. and Spite, M., Astron. Astrophys., 163, 140 (1986).
3. Hawley, S. A. Miller, J. S., Astrophys. J., 220, 609 (1978).
4. Adams, S., Seaton, M. J., Howarth, I. D., Auriere, M. and Walsh, J. R., Mon. Not. Royal Astron. Soc., 207, 471 (1984).
5. Barker, T. and Cudworth, K. M., Astrophys. J., 278, 610 (1984).
6. Carney, B. W., Astrophys. J., 233, 877 (1979).
7. Lacy, C. M., Astrophys. J., 218, 444, 1977.
8. Paczyński, B. and Sienkiewicz, R., Astrophys. J., 286, 332 (1984).
9. Kunth, D. and Sargent, W. L. W., Astrophys. J., 273, 81, (1983).
10. VandenBerg, D. A., and Bell, R. A., Astrophys. J. Suppl., 58, 561 (1985).
11. VandenBerg, D. A., Preprint.
12. Jones, R. V., Carney, B. W., Latham, D. W. and Kurucz, R. L., Astrophys. J., in press.
13. Zinn, R., Astrophys J., 293, 424, (1985).
14. Pilachowski, C. A., Astrophys. J., 281, 614 (1984).
15. Carney, B. W., Publ. Astron. Soc. Pacific, 96, 841, (1984).
16. Searle, L. and Zinn, R., Astrophys. J., 225, 357 (1978).
17. Eggen, O. J., Lynden-Bell, D. and Sandage, A., Astrophys. J., 136, 748 (1962).
18. Sandage, A. and Fouts, G., Astron. J., 93, 74, (1987).
19. Carney, B. W. and Latham, D. W., in preparation.
20. Norris, J., Astrophys J. Suppl., 61, 667 (1986).
21. Hartwick, F. D. A., Astrophys. J., 209, 418, (1976).
22. Hartwick, F. D. A., Mem. Soc. Astron. Ital., 54, 51 (1983).
23. Laird, J. B., Carney, B. W. and Latham D. W., in preparation.
24. Bond, H. E., Astrophys. J., 248, 606 (1981).
25. Beers, T. C., Preston, G. W. and Shectman, S., Bull, Amer. Astron. Soc., 17, 803 (1986).
26. Sneden, C., in Proc. ESO Workshop on Production and Distribution of C,N,O Elements, p. 1.
27. Spite, M. and Spite, F., Ann. Rev. Astron. Astrophys., 23, 225 (1985).
28. Lambert, D. L., Preprint.
29. Fall, S. M. and Rees, M., Astrophys J., 298, 18 (1985).

THE CHEMICAL EVOLUTION OF THE GALACTIC DISK

Rosemary F.G. Wyse* and Gerard Gilmore**

*Astronomy Department, UC Berkeley, CA 94720
**Institute of Astronomy, Cambridge, England, CB3 0HA

ABSTRACT

The distribution of enriched material in the stars and gas of our Galaxy contains information pertaining to the chemical evolution of the Milky Way from its formation epoch to the present day, and provides general constraints on theories of galaxy formation. The separate stellar components of the Galaxy cannot readily be understood if treated in isolation, but a reasonably self-consistent model for Galactic chemical evolution may be found if one considers together the chemical properties of the extreme spheroid, thick disk and thin disk populations of the Galaxy. The three major stellar components of the Galaxy are characterised by their distinct spatial distributions, metallicity structure, and kinematics. The thick disk provides a straightforward physical mechanism for the desired pre-enrichment to resolve the thin disk 'G dwarf problem'. Breaks in the relations between element ratios and overall enrichment occur at the characteristic metallicities of each of these three stellar components, *i.e.*, where a change of stellar kinematics occurs, and by inference, star formation and gaseous dissipation rates also change. The breaks are usually explained as being a reflection of different stellar masses contributing to the overall yield of a given element, but the coincidence of the break metallicities with those metallicities where a change of stellar kinematics also occurs suggests a more compelling explanation, as discussed here.

1. CHEMICAL EVOLUTION WITH A THICK DISK

The starting point for most theories of chemical evolution is the 'simple, closed box model', where one considers a cloud of gas to form stars and self-enrich under the following assumptions : There are no gas flows, either into or out of the system; the gas is homogeneous and well-mixed chemically at all times; the nucleosynthetic yield per generation of stars (and the stellar initial mass function or IMF) is time-independent; the enriched gas ejected from stars is 'instantaneously recycled' *i.e.*, stars that contribute to the chemical evolution live for an infinitesimal time; and the system starts from zero metallicity. The evolution of this model may be described by straightforward analytic equations, which have the additional advantage that the relation between metal enrichment and gas fraction is independent of the assumed star formation rate. However, as has been known for 25 years, application of this model to the Galactic thin disk fails because the predicted number of long-lived low metallicity stars greatly exceeds those observed[1,2] – the 'G dwarf problem'. At least one of the assumptions of the above model must be wrong, and the most appealing resolution is that the thin disk formed from gas that was not metal-free, perhaps as a result of enriched inflow[3], further meaning that the thin disk did not evolve as a closed system. One could also allow the stellar IMF to vary with time, allowing more massive stars at early times[4,5]. The chemical evolution of the solar neighbourhood tightly constrains the possible variations, and though not required by the observational data for our Galaxy, such a model is viable[5]. Here we retain the more conservative assumption of an invariant IMF.

The original motivation for applying the closed box model to the Galactic disk came from the proposal[6] that the formation of the spheroid of the Galaxy, and the collapse of residual gas to a thin disk, was extremely rapid, essentially being completed on a freefall time. However, recent evidence from *in situ* observations for a stellar component of the Galaxy with kinematics and metallicity distribution intermediate between those of the extreme spheroid and thin disk[7,8,9,10,11] has shown that this picture of Galaxy formation and evolution is oversimplified. In any case stellar evolution in the extreme spheroid cannot produce sufficient metals to provide the pre-enrichment for the thin disk[12], even though one *can* understand the chemical structure of the spheroid by an adaptation of the 'simple model' to include outflow of the stellar ejecta[12,13]. Thus with a two-component Galaxy one can understand the spheroid if one allows for outflow of enriched material, but the sink of this gas is unknown, and one can understand the thin disk if one allows pre-enrichment by inflow prior to the bulk of star formation there, but the source of the enriched material is unknown.

This situation is rectified if one investigates[14] the metal enrichment history of a three component Galaxy with an evolutionary sequence whereby a short-lived phase of rapid star formation forms the extended, metal poor, pressure supported extreme spheroid[6], with the thick disk forming next, before a final equilibrium state is attained[15] and the centrifugally supported thin disk forming over a long timescale once the Galaxy potential is unperturbed and steady[15,16]. The adopted model[14] starts with a cloud of zero-metal gas, and evolves by the modified 'simple model' with gas outflow at a rate proportional to the star formation rate[12], the constant of proportionality being fixed by the observations of the metallicity structure of the extreme spheroid. The gas need not actually flow away from the Galaxy, but simply be (temporarily) removed from the star-forming process by, for example, heat input due to supernovae from young, massive stars. Observational constraints set the outflow rate to be ~ 10 times the star formation rate, which results in a predicted mean enrichment of the outflow to be $[Fe/H] \simeq -1.5$ dex, which is 7σ below the observed mean of the thin disk[3] – confirming the inability of spheroid stellar evolution alone to account for the thin disk G dwarf problem – but which is only 3σ below the thick disk mean metallicity[8]. The process of simple model plus outflow is then started again, with initial metallicity of -1.5 dex, until the observed thick disk metallicity distribution is obtained. The outflow here is constrained to be at a rate about equal to the star formation rate, with an average enrichment in the ejecta of -0.6 dex, which is $\sim 3\sigma$ below the thin disk mean metallicity. This latter level of pre-enrichment allows the G-dwarf metallicity distribution in the solar neighbourhood to be fit[3], with an uncontrived physical mechanism for the pre-enrichment (c.f. ref 17 for similar, independent conclusions).

This model also allows one to estimate the mass ratios in the three components, modulo any gas flows to or from the Galactic system itself. Predicted values for the ratios extreme spheroid : thick disk : thin disk are approximately 1:3:12, in reasonable agreement with the observations, assuming that the local solar neighbourhood thick disk normalisation and scaleheight[7,10] are applicable globally (we are currently deriving global values from a large *in situ* survey in various lines of sight round the Galaxy). It should also be remembered that even the luminosity ratio of such well accepted components of the Galaxy as the extreme spheroid and thin disk is uncertain from the observations and quoted values range from[18] 1:3.6 to[19] 1:12.6.

2. CORRELATIONS WITH ELEMENT RATIOS

The chronological formation sequence of the three stellar components of the Galaxy proposed here may be tested by study of their detailed elemental and isotopic abundances. The formation of the thick and thin disks is likely to occur on timescales long compared to the lifetimes of the massive stars which explode as Type II supernovae, but of order the lifetimes of the stars which produce carbon, nitrogen and Type I supernovae. The available data on the variation of oxygen to iron as a function of metallicity in stars over a range of metallicities spanning extreme spheroid to thin disk show a well-defined change of slope near [Fe/H] ~ -1[20,21], intermediate between the extreme spheroid and thick disk. The relative amounts of oxygen and iron is constant, independent of [Fe/H], in stars of the lowest metallicities, as expected if oxygen and iron are produced in stars of the same (high) mass, but [O/Fe] decreases, with a slope of -0.5, in stars of metallicity [Fe/H] $\gtrsim -1$. This latter behaviour is expected if the major contributors of iron are stars of lower mass than those producing the oxygen so that the iron behaves as a secondary element. There are indications that the slope flattens again at near solar metallicity[22]. Very similar behaviour is seen in the relation between the α-nuclei (Mg, Si and Ca) and [Fe/H], although here there is stronger evidence for the flattening at high metallicities, the change of slope occuring at -0.3 dex[22,23], intermediate between the thick and thin disks. Iron and the α-nuclei are believed to have origins in stars of the same wide range of masses, so that about half the solar system abundances of these elements could come from Type I supernovae[24]. This makes the similarity between [O/Fe] *versus* [Fe/H] and [α/Fe] *versus* [Fe/H] rather difficult to understand if the changes in slope are attributed to differences in the lifetimes of the stars producing iron and the other elements.

It is interesting, and probably important, to realise that the breaks in these relations occur at metallicities where the kinematics of the stars also change, the stellar populations becoming less pressure supported and more rotation dominated. Assuming that the relative amount of pressure support in a stellar component is related to the relative rapidity with which gas was transformed into stars during formation of that component, the positions of the breaks should contain information about the temporal evolution of the star formation rate of the

Galaxy. If one knew in detail the nucleosynthetic yields for different elements as a function of stellar mass and could thus deconvolve the effects that differing progenitor stellar masses, and hence lifetimes, have on the slope of such relations as [O/Fe] *versus* [Fe/H] then one would gain new insight into the interplay between star formation rate, chemical enrichment and gaseous dissipation. The ultimate goal of understanding the luminosity evolution and rate of star formation in the Galaxy would then be one step closer.

We thank NATO for a research grant (490/84) to aid our collaboration.

REFERENCES

1. van den Bergh, S., *Astr. J.*, **67**, 486-490, (1962).
2. Schmidt, M., *Astrophys. J.*, **137**, 758-769, (1963).
3. Pagel, B.E.J. and Patchett, B.E., *M. N. R. A. S.*, **172**, 13-40, (1975).
4. Larson, R.B., *M. N. R. A. S.*, **218**, 409-428, (1986).
5. Wyse, R.F.G. and Silk, J., *Astrophys. J. Lett.*, in press (1987).
6. Eggen, O.J., Lynden-Bell, D. and Sandage, A.R., *Astrophys. J.*, **136**, 748-766, (1962).
7. Gilmore, G. and Reid, I.N., *M. N. R. A. S.*, **202**, 1025-1047, (1983).
8. Gilmore, G. and Wyse, R.F.G., *Astr. J.*, **90**, 2015-2026, (1985).
9. Zinn, R., *Astrophys. J.*, **293**, 424-444, (1985).
10. Sandage, A. R. *Stellar Populations,* eds. A. Renzini, M. Tosi and C. Norman (CUP, Cambridge) in press (1987).
11. Norris, J., Bessel, M.S. and Pickles, A.J., *Astrophys. J. Suppl.*, **58**, 463-492, (1985).
12. Hartwick, F.D.A., *Astrophys. J.*, **209**, 418-423, (1976).
13. Fall, S.M. in *The Milky Way Galaxy,* eds. H. van Woerden, R.J. Allen and W. Butler 603-610 (Dordrecht, Reidel, 1985).
14. Gilmore, G. and Wyse, R.F.G., *Nature*, **322**, 806-807, (1986).
15. Jones, B.J.T. and Wyse, R.F.G., *Astr. Astrophys.*, **120**, 165-180, (1983).
16. Gilmore, G., *M. N. R. A. S.*, **207**, 223-240, (1984).
17. Pagel, B.E.J., in *The Galaxy,* eds. G. Gilmore and R.F. Carswell (Reidel, Dordrecht), in press (1987).
18. de Vaucouleurs, G. and Pence, W., *Astron. J.*, **83**, 1163-1173, (1978).
19. Van der Kruit, P.C., *Astr. Astrophys.*, **140**, 470-475, (1984).
20. Sneden, C., Lambert, D.L. and Whitaker, R.W., *Astrophys. J.*, **234**, 964-972, (1979).
21. Clegg, R.E.S., Lambert, D.L. and Tomkin, J., *Astrophys. J.*, **250**, 262-275, (1981).
22. Lambert, D. L., preprint *Chemical Evolution of the Galaxy : Abundances of the Light Elements Na to Ca,* (1986).
23. Nissen, P.E., Edvardsson, B., Gustafsson, B., in *Production and Distribution of C,N,O Elements,* eds. I.J. Danziger, F. Matteucci and K. Kjär, (Garching:ESO) p131-149, (1986).
24. Woosley, S.E and Weaver, T.A., *Ann. Rev. Astr. Ap*, **24**, 205-254, (1986).

THE FORMATION OF GALACTIC SPHEROIDS

S. Michael Fall
Space Telescope Science Institute
3700 San Martin Drive, Baltimore, MD 21218

and

Department of Physics and Astronomy
The Johns Hopkins University
Homewood Campus, Baltimore, MD 21218

The purpose of this article is to discuss some physical processes that may have been important during the formation of galactic spheroids. Special emphasis is given to the dual roles played by heavy elements as coolants and as the products of chemical enrichment in a collapsing proto-galaxy. The following arguments were developed in collaboration with Martin Rees in an attempt to understand the origin of globular clusters. A technical description of our theory is given in reference 1 and a general review, which contains all the material in this article, is given in reference 2. Our starting point is the widely held view that fragmentation and star formation should occur in a proto-galaxy when it can cool in a free-fall time[3,4,5)]. This condition picks out a mass of order $10^{12}\ M_\odot$ and a radius of order 10^2 kpc. The cooling arguments have been extended to a picture in which the luminous components of galaxies form by the dissipative collapse of gas in dark halos that cluster hierarchically from small perturbations in the early universe[6)]. In the latest version of this story, the halos are assumed to consist of "cold dark matter"[7)]. Our theory is motivated in part by these ideas but the general features should apply in a much wider range of cosmological pictures.

For a proto-galaxy to collapse in free fall, the radiative cooling must remain at least as efficient as the gravitational heating. If the gas is lumpy, as expected in any realistic proto-galaxy, the overdense regions will cool more rapidly than the underdense regions. This process—a thermal instability—will produce a two-phase medium, i.e., cold dense clouds embedded in and confined by hot diffuse gas. Now there are two characteristic temperatures in the problem. One, the temperature of the hot gas, can be expressed as $T_h \approx (\mu_h/3k)\ V_{gal}^2$, where V_{gal} is a typical velocity for large-scale, gravitationally-induced motions and $\mu_h \approx 0.6\ m_p$ is the mean mass per particle of ionized gas. The other characteristic temperature, $T_c \approx 10^4$ K, is where hydrogen recombines and the cooling rate drops precipitously. We assume for the moment that the clouds do not cool to lower temperatures and justify this later. The densities of the two phases, once they reach pressure balance, are related by $\rho_c/\rho_h = (\mu_c/\mu_h)(T_h/T_c)$, where $\mu_c \approx 1.2\ m_p$ is the mean mass per particle of neutral gas. For $V_{gal} = 300$ km s^{-1}, a value appropriate to the Milky Way, we find $T_h \approx 2 \times 10^6$ K and therefore $\rho_c/\rho_h \approx 400$. Our detailed calculations show that this state is reached during the collapse of a proto-galaxy if the initial amplitudes of the perturbations giving rise to the clouds are of order 10%. Perturbations with larger amplitudes grow even more rapidly.

Any clouds with masses greater than some critical value will be gravitationally unstable and will collapse. The standard formula for an isothermal sphere confined by an external pressure ρ_h is

$$M_{crit} = 1.2(kT_c/\mu_c)^2 G^{-3/2} p_h^{-1/2}. \tag{1}$$

This can be simplified by noting that the hot gas as a whole remains near the threshold for gravitational instability while it collapses. Combining eqn (1) with a similar expression for a proto-galaxy of mass M_{gal} then gives

$$M_{crit} \approx \frac{1}{4}(T_c/T_h)^2 f_h^{-1/2} M_{gal}, \qquad (2)$$

where f_h is the fraction of the mass in the hot phase. In general, we expect f_h to be near unity when the first clouds form and to decrease thereafter. The exact value is not crucial, however, because f_h enters eqn (2) only through a square root. Another way to estimate p_h and hence M_{crit} is by assuming that the cooling time of the hot gas is comparable to the free-fall time, as in the arguments that lead to a preferred scale for proto-galaxies. (This is what was done in ref 1.) For $T_h \approx 2 \times 10^6$ K and $M_{gal} \approx 3 \times 10^{11}$ $M_{\odot}$, we find $M_{crit} \approx 2 \times 10^6 M_{\odot}$. This is somewhat higher than but reasonably close to the masses of globular clusters. Since T_h scales roughly as $M_{gal}^{1/2}$ (the Faber-Jackson and Tully-Fisher relations), the critical mass, as given by eqn (2), should have little variation from one galaxy to another.

The clouds produced by a thermal instability can have a wide range of masses. In the absence of magnetic fields, thermal conduction sets a lower limit, which, unless the clouds are highly flattened or filamentary, is well below M_{crit}. If there is a tangled magnetic field with a strengh of only 10^{-15} G, conduction is suppressed and the lower limit on the masses is even smaller. Most of the clouds will therefore be gravitationally stable. They will persist in pressure balance with the hot gas until collisions produce agglomerations that are massive enough to collapse. We therefore expect the proto-clusters to have a narrow range of masses near M_{crit}. The value $M_{crit} \sim 10^6$ $M_{\odot}$ is, however, special only if the temperature of the cold gas "hangs up" at 10^4 K. A necessary condition for this to occur is that the cooling times of the clouds be comparable to or longer than their internal free-fall times so that they contract quasi-statically. If this condition were not satisfied, the gas would cool rapidly through 10^4 K and M_{crit} would be drastically reduced. Some of the smaller clouds might eventually reach temperatures low enough to become gravitationally unstable, but if the cooling time is large in comparison with the free-fall time, a feature near 10^6 $M_{\odot}$ will still be imprinted in the mass spectrum.

In gas with a primordial composition, the only significant cooling at temperatures just below 10^4 K is caused by molecular hydrogen. This would spoil our theory were it not that H_2 can be destroyed by radiation just longward of the Lyman limit[8)]. Even the hot gas in a proto-galaxy emits enough ultraviolet photons to keep molecular cooling at modest levels and this could be reduced further by radiation from massive stars or an active galactic nucleus. Once heavy elements are produced and dispersed within a proto-galaxy, they provide another source of cooling. In an idealized model with no heat input, we find that the temperatures of the clouds would remain near 10^4 K as long as the metallicity is less than or of order 10^{-2} $Z_{\odot}$. This estimate is in reasonable agreement with the abundance of heavy elements in many globular clusters. A completely realistic treatment would include heating mechanisms and might therefore be compatible with the higher metallicities of some clusters. There are several possibilities: (a) heating by supernovae, stellar winds, etc. within the proto-clusters, (b) photo-ionization by massive stars elsewhere in the proto-galaxy, (c) heating by cosmic rays, (d) photo-ionization by an active galactic nucleus. Any of these effects could raise the metallicity at which cooling becomes important but none of them can be calculated without additional assumptions.

A consequence of the previous arguments is that the first generation of stars would form in clouds with masses of order $10^6\ M_\odot$. As the proto-galaxy is progressively enriched in heavy elements, cooling becomes more important and clouds with smaller masses can collapse. These objects would be more susceptible to disruption and the stars that formed in them would be released into a field population. The metallicity at which the transition occurs is a little vague because of the uncertainties in the heating mechanisms and the possibility of some self-enrichment in the proto-clusters. Moreover, some of the field stars with very low metallicities may have formed in globular clusters that were later disrupted. Nevertheless, we do expect the field stars, on average, to be slightly younger and to have higher metallicities than the globular clusters. The field stars, by forming later in the collapse of a proto-galaxy, should also have a space distribution more centrally concentrated than that of the globular clusters. As the result of various selection biases, these suggestions are not easy to test for the Milky Way but they are consistent with the available data for other galaxies[9,10,11,12].

REFERENCES

1. Fall, S. M., and Rees, M. J., *Ap. J.*, **298**, 18 (1985).
2. Fall, S. M., and Rees, M. J., in *IAU Symposium 126, Globular Cluster Systems in Galaxies*, eds. J. E. Grindlay and A. G. D. Philip (Dordrecht: Reidel), in press.
3. Binney, J., *Ap. J.*, **215**, 483 (1977).
4. Rees, M. J., and Ostriker, J. P., *M.N.R.A.S.*, **179**, 541 (1977).
5. Silk, J., *Ap. J.*, **211**, 638 (1977).
6. White, S. D. M., and Rees, M. J., *M.N.R.A.S.*, **183**, 341 (1978).
7. Blumenthal, G. R., Faber, S. M., Primack, J. R., and Rees, M. J., *Nature*, **311**, 517 (1984).
8. Stecher, T. P., and Williams, D. A., *Ap. J. (Letters)*, **149**, L29 (1967).
9. Forte, J. C., Strom, S. E., and Strom, K. M. *Ap. J. (Letters)*, **245**, L9 (1981).
10. Harris, W. E., *A. J.*, **91**, 822 (1986).
11. Mould, J., in *Stellar Populations*, eds. C. A. Norman, A. Renzini and M. Tosi (Cambridge: Cambridge University Press), in press.
12. Mould, J. R., Oke, J. B., and Nemec, J. M., *A. J.*, **93**, 53 (1987).

GAMMA RAYS FROM SOLITARY NEUTRON STARS

M. Ruderman*
Department of Physics
Columbia University

1. Introduction There are several families of observed γ-ray sources which may consist of isolated neutron stars: six claimed 10^{12} eV sources which are rapidly spinning radiopulsars[1,2]; almost 20 otherwise unidentified 10^2-10^3 MeV Cos B sources; at least 10^3 transient 10-10^5 keV "γ-ray burst" sources[3] which include the brightest γ-ray sources in the sky for durations of up to 10^2 s. Most is known about the energetic radiation (optical to at least 10^{12} eV) of the Crab and Vela pulsars, and we shall consider how efforts to understand its origin (Sections 2 and 3) may possibly also suggest the nature of other sources. In particular there can exist in the Galaxy a large population of rapidly rotating magnetized neutron stars (Sections 4 and 5) which do not radiate like the Crab or Vela pulsars except when temporarily ignited by transient events such as those mentioned in Section 6. During such events the expected energetic emission should usually resemble the steady emission from the Crab pulsar in the γ-ray regime but be relatively much weaker in X-rays for the same reasons that this is true for the Vela pulsar. Possible connections to other families of transient and steady γ-ray sources are also summarized in Section 7.

2. Outermagnetosphere Plasma Gaps and Crab Pulsar Radiation

About 10^{-3} of the spindown energy of the Crab pulsar is radiated into two subpulses whose arrival times are energy independent over a photon energy range extending from optical to above 10^{12} eV.[8] The rest seems to be emitted in an $e^{\pm}$ wind of intensity ~ 10^{39} s^{-1} which sustains the surrounding nebula's emission and accelerated expansion.[4,2]. Since $e^{\pm}$ pairs are produced by γ-rays such as those observed it seems simplest to suppose that there is a particular region near the neutron star which is the source of both. Because that region must allow the escape of 10^{12} eV γ-rays the local magnetic field (B) cannot greatly exceed 10^6 G. The region cannot be too far from the star or the γ-rays from it will not find sufficient target photons or B on which to convert to $e^{\pm}$ pairs. The needed combination suggests a strong particle accelerating region near the "light cylinder" at $r \sim cP(2\pi)^{-1} \sim 10^8$ cm in which there is a large electric field ($\vec{E}$) component along $\vec{B}$ to accelerate e^-/e^+ to energies enough to account for the pulsar's double pulsed energetic photon spectrum and $e^{\pm}$ production.

A region within a pulsar's light cylinder with an $\vec{E} \cdot \hat{B}$ which

---

Work supported by NSF Grant PHY-86-02831.

accelerates e^-/e^+ to ultrarelativistic energies can exist only where there is no magnetospheric plasma instead of the one of net charge density $\vec{\Omega}\cdot\vec{B}/2\pi c$ needed to achieve $\vec{E}\cdot\vec{B} \sim 0$ in the rest of the magnetosphere. Such a charge deficient region ("gap") in the outer-magetosphere is a common feature of a variety of models[2,5,6]: an outergap can occur on the "open" magnetic field lines, extending from the stellar surface polar cap out through the light cylinder, which connect the star to its distant environment. Where these open field lines do not have a copious source of plasma with both signs of charge, current flow can generate a growing gap especially beyond the "null surface," $\vec{\Omega}\cdot\vec{B} = 0$. When such a gap begins to grow the accelerating $\vec{E}\cdot\vec{B}$ in it also increases until the resulting e^-/e^+ accelerator becomes strong enough to support so much $e^\pm$ production that the supply of pair plasma quenches further gap growth. For the Crab pulsar this is the result of a series of processes[7]:

a) An $e^\pm$ pair produced in the gap is instantly separated by the large $\vec{E}\cdot\vec{B}$ there which accelerates the e^- and e^+ in opposite directions. Because of magnetic field line curvature each lepton radiates multi-GeV curvature γ-rays.

b) These are converted into $e^\pm$ pairs in collisions with keV x-rays [from d) below]. Pairs created in the gap repeat process a).

c) Pairs created beyond the gap boundary lose their energy to synchrotron radiation (optical to MeV) and to higher energy γ-rays from Compton scattering on the same X-ray flux responsible for the pair creation.

d) The X-ray flux from the synchrotron radiation of c) is that which initially caused the curvature radiation γ-rays of a) to materialize and the inverse Compton scattering of the pairs in e). Since the entire series of processes is powered by the extremely energetic e^- and e^+ of a) moving in opposite directions within the gap, all of the resulting fluxes of photons and $e^\pm$ pairs are also oppositely directed. The pairs and γ-rays moving in one direction then interact mainly with the X-rays moving oppositely.

e) A third generation of $e^\pm$ pairs comes from partial materialization of the crossed γ-ray beams from c).

The above processes a) - d) "bootstrap" the creation of an $e^\pm$ plasma until enough is produced to form a gap boundary layer which quenches further growth. A calculated spectrum from 1 - 10^9 eV which results from them is compared to observations in Fig. 1. Some 10^{13} eV emission (not shown) is expected from inverse Compton scattering of gap e^-/e^+ on optical photons in the gap from process c) and from the synchrotron radiation by the $e^\pm$ pairs these ultra high energy γ-rays make beyond gap boundary. Radiation reaction limits gap e^-/e^+ to about 10^{13} eV. (Higher energies[9] would be given to protons which traverse the entire 10^{15} Volt gap potential drop. These 10^{15} eV protons can make π°-mesons in collisions with the soft X-ray photons from the Crab pulsar,[2,7] but 10^{15} eV γ-rays from π°-decay could escape without conversion to $e^{\pm}$ by the magnetospheric B only if the π° is made well beyond the pusar light cylinder.)

In the specific model which gave the spectrum of Fig. 1 an outergap which is a source of pairs and energetic photons always has a mirror gap on the opposite side of the magnetosphere. An immediate

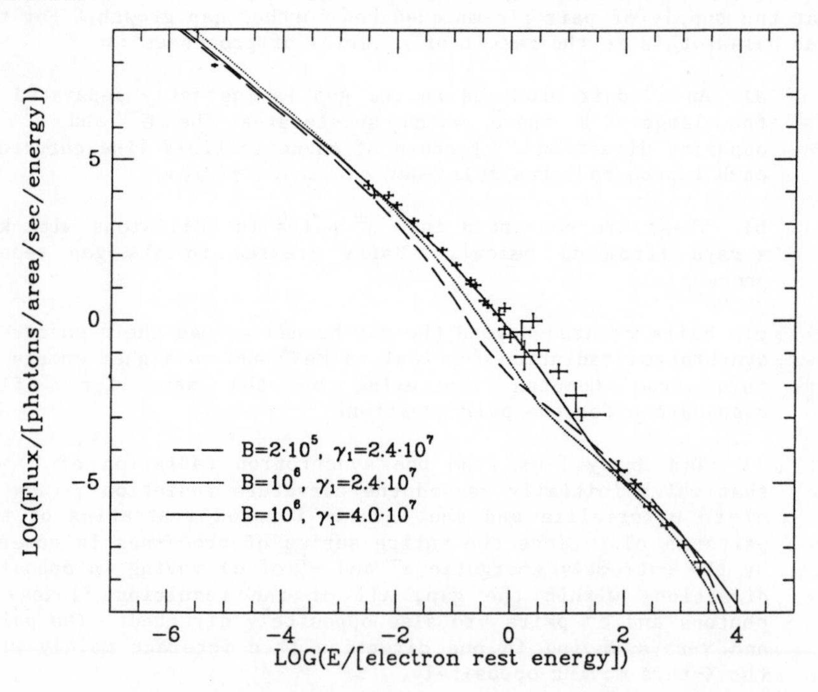

Fig. 1. Calculated and observed spectra for the Crab pulsar.[7] The parameters are for different possible local magentic fields and characteristic gap e^-/e^+ energies (in units of electron rest energy) at that part of the gap where the magnetic field is oriented along the line of sight to the observer.

consequence is a pair of beams for each stellar rotation with a large observed subpulse separation because of different beam travel times across the magnetosphere and aberration from the near relativistic rotation of the gaps.

3. High Energy Radiation from the Vela Pulsar

The series of mechanisms which limit the growth of an outermagnetosphere gap in the Crab pulsar are not nearly so efficient in the Vela pulsar. The synchrotron X-rays of 2c), necessary for the $e^{\pm}$ conversion of curvature γ-rays from the gap, are greatly suppressed in Vela. This is because the time it takes for a relativistic e^{-}/e^{+} to synchrotron radiate down to a particular characteristic frequency is proportional to B^{-3}. Since Vela has about the same dipole moment as the Crab, but spins 3 times less rapidly, the local outer-magnetosphere B is smaller by 3^3. This is enough to suppress strong X-ray synchrotron radiation since the radiating pairs will leave the magnetosphere before such radiation is dominant. Vela seems to limit its outergap growth by mechanisms rather different from those of the Crab, and the gap grows to almost 1/3 the available outermagnetosphere volume before this is accomplished. Calculated and observed spectra are compared in Fig. 2.

4. Strongly Magnetized Neutron Stars as Latent γ-ray Sources

After Vela spins down significantly, it will no longer be expected to be able to maintain copious outermagnetosphere pair production and thus to remain a source of the energetic radiation associated with such activity. However, a rapidly spinning ($P \sim 10^{-1}$ s), strongly magnetized (surface dipole $B_s \sim 10^{12}$ G) neutron star with great charge depletion (and thus a large $\vec{E}\cdot\hat{B}$) on its open field lines far from the star could again be made to radiate like the Crab or Vela pulsar if a sufficient X-ray flux is caused to pass through the charge deficient region. The number of such potentially "inflammable" objects in our Galaxy waiting to be temporarily ignited by an X-ray "match" is unknown because it is not yet clear what happens to such a star's magnetosphere and polar cap currents: a) such stars may pull charge in from beyond their light cylinders in amounts sufficient to keep about the same net electric current flow in their magnetospheres that they sustained before outergap $e^{\pm}$ creation failed. b) Alternatively, magnetosphere and polar cap currents may be greatly diminished or may cease. (A possible solution of this sort has been demonstrated only for the case of an aligned charged spinning magnetic dipole[10a,b].) If case b) holds for a non-aligned pulsar, further spindown would come only from the Maxwell stresses on the rotating dipole moment. A case b) star would align its dipole along its spin if its shape were not significantly affected (over a 10^4-10^5 year span) by its solid crust or its magnetic field. In this event, the aligned neutron star would spin down at a greatly reduced rate. This uncertainty about

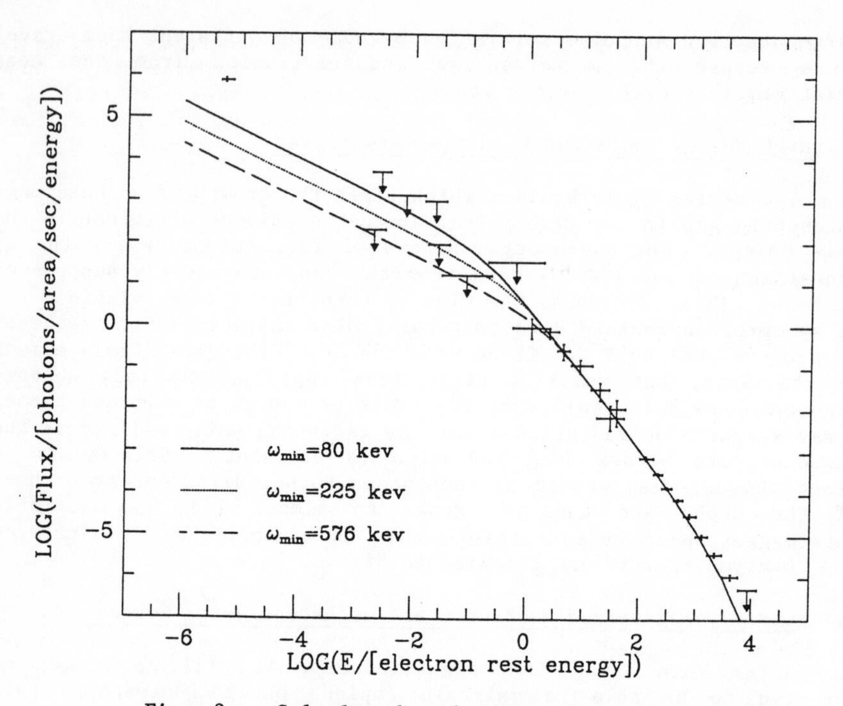

Fig. 2. Calculated and observed spectra of the pulsed electromagnetic radiation from the Vela pulsar[7]. (All of the calculated spectra are normalized to fit the integrated observed intensity. The spectral breaks at various energies reflect an uncertainty in local outergap magnetic field which affects the residence time needed before strong synchrotron radiation reaches the X-ray regime.)

further spindown leads to several possibilities for the galactic population of such latent Crab/Vela type pulsars. For case a), where canonical spin-down continues, the number would not be expected to exceed 10^3, an estimate based upon that for Vela/Crab-like radiopulsars in the Galaxy. But because it is not understood just what leads to observable coherent radioemission from neutron stars such a number may be a large underestimate. If in case b) spin-down ceases and the strong dipole field decays after 10^7 years, the number grows to over 10^5. Finally, if the magnetic fields of isolated neutron stars survive to a much greater age[11,12] the number could appraoch 10^8.

In case b) and possible even in case a) radiopulsar emission would cease. But typical radiopulsars with P ~ 1 sec, or even more, almost certainly do have polar cap current flow even though they are unable to generate the very large potential drops needed for outermagnetosphere $e^{\pm}$ production. Thus such outergap $e^{\pm}$ creation is not necessary to maintain the needed magnetosphere currents on open field lines, perhaps as in case a). However it may also reflect a possible consequence of certain field line geometries in surface $e^{\pm}$ production from the conversion of curvature γ-rays on the 10^{12} G local field near the star (which takes less than 10^{-2} the potential drop needed for outergap $e^{\pm}$ creation). If an adequate $e^{\pm}$ plasma is created there on all open polar cap field lines magnetosphere current flow will not result in a charge depleted region in the outermagnetosphere.[13a,b)]

5. Millisecond Pulsars as Active and Latent γ-ray Sources

Taylor et al.[14)] estimate that 10^{-1} of all presently observable radiopulsars are members of the millisecond radiopulsar family: weak dipole ($5 \times 10^8 \lesssim B_s \lesssim 10^{10}$ G) short period ($P \lesssim 10$ ms) old neutron stars, commonly assumed to have been spun-up by accretion from a companion. This fraction would imply about 10^4 such radiopulsars in the Galaxy. In several ways the magnetospheres of this neutron star family closely resemble the younger, high field, longer period family of latent Vela-like pulsars of Section 4. For example, the fastest observed radiopulsar in this family, PSR 1937+21, has a surface dipole $B_s \sim 5 \times 10^8$ G and $P = 1.5 \times 10^{-3}$ s. But the magnetosphere potential drop and current flow which it can sustain are about the same as those of the Vela pulsar; its near light cylinder magnetic field is about twice that of the Crab. Among the somewhat slower members of the family of accretion spun-up neutron stars (most with larger dipole moments), one might then expect a substantial subset of latent γ-ray pulsars--neutron stars which, as in section 4, cannot sustain strong outermagnetosphere $e^{\pm}$ production and γ-ray emission unless some "match" is lit to supply the soft x-rays needed for bootstrapping the series of processes of section 3. In addition, there may be an even larger family of weak dipole neutron stars which are formed initially as very rapid rotators but are not yet observed at all in radiopulsar surveys. Many of these may also be candidates for latent γ-ray pulsars.

There is again great uncertainty about the number of neutron stars in this family of latent pulsars since the fraction of such rapidly spinning weak dipole stars which are observable in a flux limited radiopulsar search is unknown. The estimate of 10^4 for the number of observable millisecond radiopulsars may suggest an at least comparable number for latent γ-ray pulsars with similar properties.

So far in this section, as in the previous one, we have considered only a potential for converting into γ-ray emission some of the electric field energy of outermagnetosphere open field line charge depletion in certain neutron star families--the same process which seems to be the self-sustaining source of γ-rays from the Crab and Vela pulsars. The loss of plasma and the consequent charge depletion on open field lines appears as a consequence of magnetospheric current flow through the light cylinder. There are also processes especially in the weak dipole stars, which can cause the loss of much of the closed field line magnetospheric charge which otherwise keeps $\vec{E}\,\vec{B} = 0$ on these field ines (or, equivalently, keeps $\vec{E} = 0$ in the corotating frame). A significant flow of e^-/e^+ charge across field lines occurs in the presence of a polarized Poynting flux as long as the flux frequency does not greatly exceed the local e^-/e^+ cyclotron frequency, such as that of the very soft X-rays from the surface of an old cooling spinning down neutron star which has an internal energy source from superfluid vortex line unpinning and repinning.[15] (As long as the surface kT does not greatly exceed $e\hbar B/mc$ there, this flux is expected to be polarized.)

Discharging a gap by $e^\pm$ production and γ-ray emission would have different time histories for the case when a match ignites the outermagnetosphere's open field lines in an otherwise latent γ-ray pulsar and that when the closed field line plasma drift supplies the discharge energy (and perhaps also its own match). In the former case the γ-ray production can last as long as the match remains sufficiently strongly lit, during which the maximum energetic radiation power from the neutron star is

$$\dot{E} \sim B_s^{\,2} R^3 \, (\Omega R/c)^3 \Omega. \quad (1)$$

For Vela-like parameters this is about 3×10^{36} erg s^{-1}. In the latter case, a sudden filling of part of a closed magnetosphere by $e^\pm$ production, there would be a single brief pulse of emitted energetic radiation with maximum total energy

$$E_m \sim B_s^{\,2} R^3 (\Omega R/c)^2 \sim 10^{35} B_9^{\,2} P_{-3}^{\,2} \text{ ergs}, \quad (2)$$

where the numerical estimate is appropriate for a latent millisecond pulsar family. If the magnetosphere is empty only for radii exceeding r, the released energy $E \sim (R/r)E_m$.

6. Matches: Ignition of Latent γ-ray stars

After the first observations of transient γ-ray bursts (Section 1) a very large number of proposals were made for the sources of these events, many of which involved neutron star. Among such accretion disk instabilities, cometary or asteroidal impacts, internal readjustments, and nuclear explosions of accreted material are still popular candidates for burst energy sources.[3] Any of these might possibly instead be just a match which supplies the soft X-rays (or perhaps just injects some plasma into a plasma starved gap) to temporarily ignite $e^\pm$ production and γ-ray emission from the partly empty magnetosphere of an otherwise relatively dormant neutron star. Since the needed match energy can be very much less than the total γ-ray energy ultimately released from a partially empty stellar magnetosphere, versions of a γ-ray burst source model too mild to give required burst energies may still be suitable as matches. Moreover, the emission spectrum will not be restricted to the same soft X-ray regime as that of the match.

A plausible match for an isolated latent γ-ray pulsar is the nuclear explosion whch would ultimately follow continuing slow accretion of interstellar medium matter onto a cool neutron star polar cap. When the integrated layer of such accretion on the cap surface reaches about 3×10^8 g cm^{-2} explosive He burning would occur. For Vela-like surface magnetic fields ($B_s \sim 4 \times 10^{12}$ G), magnetic rigidity is strong enough to hold this surface density to a 10^9 cm^2 polar cap area. An Eddington limit explosive X-ray flux would then be a 10^{34} erg s^{-1} X-ray match. The match could, however, sustain 10^{36} erg s^{-1} of magnetospheric γ-rays (and hard X-rays) for most of the duration of the surface explosion. For observed X-ray burst sources this duration is typically about 10 sec but it can be as long as 10^2 sec (both of which times are also characteristic of many observed γ-ray burst sources). Interstellar medium accretion rates onto such a polar cap have been estimated to be of order 10^{-1} g cm^{-2} s^{-1} for a neutral flow,[16] about the same as that for the maximum charge separated plasma flow that can be sucked onto a polar cap by open field line electric fields. The match ignition repetition rate would then be about once per 10^2 yrs. Such an interstellar accretion, less than 10^{-17} $M_\odot$ yr^{-1}, would give an $L_x \lesssim 10^{29}$ erg s^{-1}, so that, while accreting, the neutron star would not be an observed X-ray source.

Low B_s neutron stars may shift enough closed field line plasma for growing closed field line gaps to become essentially self-igniting. They would need only some small initiating seed to discharge the energy of Eq (2) into a single brief pulse of energetic radiation. The time needed to empty all of the vulnerable part of the closed field line magnetosphere beyond r is $\tau \sim c^3/3\varepsilon\Omega^4 r^3$, with ε is the ratio of internal heating to the total spin-down energy loss rate. For heat generation from superfluid vortex line detachment of slowing down neutron stars, $\varepsilon \sim 10^{-4} - 10^{-5}$ is a canonical estimate.[17,18] Then with $r \sim R$, and $P = 6 \times 10^{-3}$ s we would have for the recurrence time of such bursts $\tau \sim$ 1-10 days, intervals of the same order as that observed for the only confirmed transient γ-ray burst repeaters, 5 March 1979 (3 days), 24 March 1979, and 7 January 1979 (< 1 day), all of which are characterized by a dominant very brief pulse.[21] However, the implication that in such a repeating γ-ray burst source a larger interval between bursts (i.e. smaller r) should precede stronger bursts is not supported by observations of the rapid 7 January 1979 repeater.[22]

7. γ-ray Burst Sources

Are the very many observed isotropically distributed transient γ-ray burst sources (GRB's) a population of solitary neutron stars[25] within a few $\times 10^3$ light years of the sun? If so, could they be some of the neutron star populations of Sections 4 and 5, temporarily turned into γ-ray sources by the matches of Section 6? The answers seem to be "perhaps." The observed spectra, isotropy,[23] and intensities can be interpreted as supportive, but the required population size may exceed that available.

a) Spectra: Various occasional features such as intensity variations on a time scale of less than 10^{-2} s and a possible appropriately gravitationally redshifted $e^- - e^+$ annihilation γ-ray feature seem to implicate neutron stars.[3] At energies above about one MeV typical GRB spectra follow about the same power law as that of the Crab and Vela pulsars; at lower energies they lie somewhat above the calculated Vela spectrum of Fig. 2 and well below that observed for the Crab shown in Fig. 1. Such a spectrum is not implausible when Crab/Vela type pulsars pass over into those of Section 4.

b) Intensities: The time integrated flux (fluence) of GRB's is shown as a function of burst duration in Fig. 3. The data are not inconsistent with a relatively fixed luminosity candle which remains lit for times between about 10^{-1} and 10^2 seconds: superimposed upon the expected fluence variation with source distance there is a proportionality of fluence to pulse duration. Sources with luminosity 10^{36} erg s^{-1} at a distance of 10^3 light years would lie on the upper diagonal of Fig. 3; at a 10^2 light year distance they would lie on the lower diagonal. The presumed luminosity is near that of Eq. (2) for Vela-like parameters and for those of millisecond pulsars, but whether such neutron stars would mobilize almost that maximum energy in an $e^{\pm}$ discharge while temporarily filling their open field line outermagnetospheres is not known.

c. Population: If the isotropic distribution of localizable GRB sources is interpreted as that of a disk distribution of neutron stars less than the disk thickness away, then the number of such sources in the whole Galaxy should exceed 10^5. As indicated in Sections 4 and 5, this probably exceeds the Galactic population of accretion spun-up weak dipole neutron stars (but perhaps not that of such stars formed in other ways); the population of latent Crab/Vela type stars may reach or exceed 10^5 if stellar spin-down is supressed by alignment after outergap extinction. The much smaller observed population (3, so far) of rapidly repeating GRB sources is not incompatible with the candidates for them of Section 6.

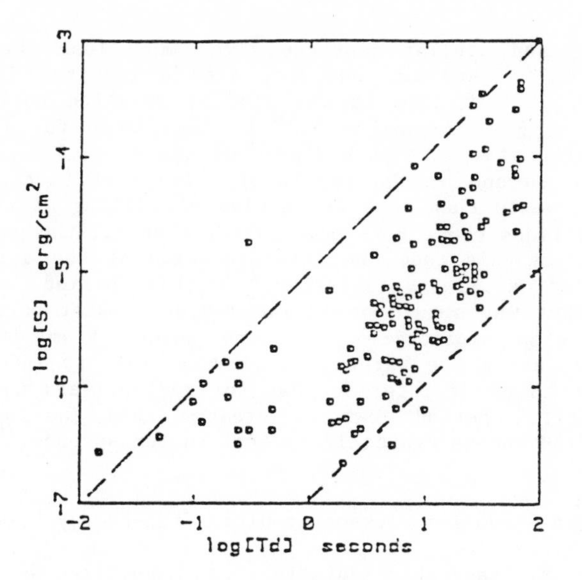

Fig. 3. Burst fluence S as a function of burst duration T_D, using the data from the KONUS experiment.[21,3] The diagonals give the fluence of 10^{36} erg s^{-1} sources at 10^3 and 10^2 light year distances.

d) Other γ-ray Sources

Could many of the unidentified Cos B sources be similar to the neutron stars of Sections 4 and 5 which have $L \sim 10^{36}$ erg s^{-1}, except that instead of being marginally unable to self sustain needed outergap $e^{\pm}$ producton (unless lit by a Section 5 match) they might be barely able to self-sustain such production at the cost of mobilizing most of their spindown power to accomplish it? For Vela-like parameters the required luminosity would be supportive of this possibility. With the relatively weak X-ray spectrum of Fig. 2, such sources at 10^4 light year distances would not yet have been observed as X-ray sources.

e. The GRB of March 5, 1979[21]

This most famous of all GRBs comes from a repeating source. Its brief but extraordinarily intense inital pulse has a very long but weaker (10^{-5} erg cm^{-1} s^{-1}) tail whch is strongly modulated with an 8 s period. Like the two other repeaters, its spectrum is much softer than and it may be very different in nature from that of most GRB sources. It has for example been interpreted as a distant Crab-like pulsar but spinning much more rapidly. If the observed 8 s period is the rotational one of neutron star then Eq. (1) with plausible B_s and stellar distance gives much less than 10^{-3} the observed flux. There is, however, also a reported 23×10^{-3} s quasi-period.[24] If this is the stellar rotation period and the dipole part of B_s = several $\times 10^{10}$ G (reasonable for an accretion spun-up neutron star with such a period) the March 5 event might be interpretable as one of the family of pulsars of Section 5. (The inital pulse would then have to involve $e^{\pm}$ filling of near surface closed field lines whose B is much larger than B_s, the purely dipole component.) In this case the interpretation of the 8 s period is obscure. As a free precession,[25] this period may not be implausible but why should an isolated neutron star sustain such a large precession amplitude? And the torque from magnetosphere current flow during the burst is much too small to induce it. It may turn out to be of interest that for radiopulsars the so-called "subpulse drift" period (P_3) is greater than the neutron star rotation period and is typically several to ten seconds.[26,27]

References:

1. 13th Texas Symposium Workshops: Ultra High-Energy Cosmic Rays.

2. Ruderman, M. "Energetic Radiation from Magnetized Neutron Stars," 1985 NATO Advanced Study Institute--Lecutres on High Energy Phenomena Around Collapsed Stars, F. Pacini, ed. (in press) and references therein.

3. Liang, E. and Petrosian, V., eds. Gamma Ray Bursts AIP Conference Proceedings 141 (1986).

4. Kennel, C. and Coroniti, F., Ap.J., **283**, 710 (1984).

5a. Mestel, L. 13th Texas Symposium Workshop: Neutron Stars.

5b. Arons, J., IAU Symposium 74: Origin of Cosmic Rays 175 (1981).

6. Cheng, K., Ho, C., and Ruderman, M., Ap.J., **300**, 500 (1986).

7. Cheng, K., Ho, C., and Ruderman, M., Ap.J., **300**, 522 (1986).

8. Dowthwaite, J., et al., Ap.J. (Letters), **286** (L35 (1984).

9. Wdowczyk, J., Rencontre de Moriond: High Energy Astrophysics, J. Tran Thanh. Van, ed. (1984) and references therein.

10. Krause-Polstorff, J.and Michel, F.C., M.N.R.A.S., **213**, 43 (1985).

11. Blair, D. and Candy, B. 13th Texas Symposium Workshop: Neutron Stars.

12. Kundt, W. in the Evolution of Galactic X-ray Binaries, J.

Truemper, et al., eds. 263 (1986).

13a. Jones, P., M.N.R.A.S., in press.

13b. Beskin, V., Guravich, A., and Istomin, Y., Astrophys. and Space Sci., **102**, 301 (1984).

14. Taylor, J., "Pulsars," 13th Texas Symposium on Relativistic Astrophysics, Proceedings.

15. Pines, D. "Neutron Stars: Interior," 13th Texas Symposium, Proceedings.

16. Taam, R. 13th Texas Symposium Workshop: Gamm-Ray Bursts.

17. Shaham, R. private communication.

18. Lamb, F., and Shibazaki, N., 13th Texas Symposium Workshop: Neutron Stars.

19. Hurley, K., Laros, J., 13th Texas Symposium Workshop: Gamma Ray Bursts.

20. Matz, S., Los Alamos Workshop on Gamma-Ray Stars, July 20-25, 1986.

21. Mazets, E., and Golenetskii, S. Astrophys. and Space Phys. Rev. **1**, 205, (1981).

22. Laros, J., 13th Texas Symposium Workshop: Gamma-Ray Burst Workshop.

23. Hurley, K., 13th Texas Symposium Workshop: Gamma-Ray Bursts.

24. Hurley, K., IAU Symposium 115: High Energy Transients in Astrophysics, S. Woosley, ed. (1984).

25. Brecher, K., IAU Symposium 115, loc. cit.

26. Manchester, R. and Taylor, J., Pulsars (1978) (W.H. Freeman & Co., San Francisco).

27. Rankin, J., Ap.J. **301**, 901 (1986).

NEUTRON STARS AS COSMIC NEUTRON MATTER LABORATORIES

David Pines

Center for Materials Science
Los Alamos National Laboratory, Los Alamos, NM 87545
and
Department of Physics, UIUC
1110 W. Green St., Urbana, IL 61801[+]

ABSTRACT

Recent developments which have radically changed our understanding of the dynamics of neutron star superfluids and the free precession of neutron stars are summarized, and the extent to which neutron stars are cosmic neutron matter laboratories is discussed.

1. Introduction

Remarkable progress has been made in the past few years in both theory and observation relevant to understanding the structure of neutron stars and the behavior of the neutron superfluids that make up the greater part of these stars. Indeed, it may be argued that it is now possible to use neutron stars as cosmic neutron matter laboratories, in that by combining theory and observation one can obtain detailed information (unobtainable in any other way) on the nature of the neutron superfluids in the stellar interior, as well as on the equation of state of neutron matter at densities equal to or somewhat greater than that found in laboratory nuclei. Space does not permit me to describe this progress in any detail in these proceedings; I shall therefore simply summarize briefly some major developments and refer the reader interested in more detail to recent

[+]Permanent address.

reviews,[1,2] and accompanying articles in this volume.[3,4]

2. Neutron Superfluids

Two distinct neutron superfluids are expected in neutron stars:

- in the inner part of the crust, at densities, 4×10^{11} gm $cm^{-3} \lesssim \rho \lesssim \rho_o$, a ^1S_o paired superfluid coexists with a lattice of crustal nuclei
- in the liquid core ($\rho \gtrsim \rho_o$), a $^3P_2$ paired superfluid coexists with a comparatively small fraction (~5%) of superconducting protons and normal relativistic electrons.

Our current understanding of these superfluids is that:[2,3]

- the core superfluid rotates rigidly with the crust
- because of the pinning of vortices in the crustal superfluid to crustal nuclei, the crustal superfluid, which contributes $\lesssim$ 10% of the stellar inertial moment, can be out of equilibrium with the crust for timescales of weeks to years and is responsible both for pulsar glitches and post-glitch relaxation
- vortex creep theory, which describes the motion of pinned vortex lines in the crustal superfluid, provides an excellent fit to the Vela pulsar timing observations of Downs,[5] which span the decade 1969-79 and include four giant glitches, as well as to the Crab timing observations of Lohsen,[6] which span the decade 1969-79, and include two large glitches, and the observation by Downs[7] of a glitch and its consequences in PSR 0525+21.[8]
- the observed post-glitch relaxation times in the Vela and Crab pulsars, and PSR 0525+21 provide a means of estimating the interior temperature of these pulsars
- interior temperatures of the Crab and Vela pulsars obtained in this way, while consistent with observations, are not consistent with a standard cooling scenario[4]
- the dominant source of heat in pulsars old enough to have

radiated away their original heat content is vortex creep

- o strong pinning of vortices to crustal nuclei is ruled out by an upper limit on the rate of dissipation of energy by vortex creep found through EXOSAT observations of PSR 1929+10.[9]

3. Neutron Star Precession and the Neutron Matter Equation of State

Trumper et al.,[10] on the basis of their EXOSAT observations of Her X-1, have concluded that the clock mechanism for its 35-day cycle is the free precession of the neutron star. Since the existence of a stable free precession mode requires a solid crust astronomical object, it is intuitively obvious that the wobble frequency will depend upon the extent of the neutron star crust. Calculations based on various models for the equation of state of neutron matter show that the crustal extent is determined primarily by the stiffness of the equation of state for neutron matter at densities $\rho \gtrsim \rho_o$. If one makes the "Occam's razor" assumption that all neutron stars possess nearly the same mass ($\sim 1.4 M_\odot$) and that the initial spin period, P_o, of Her X-1 is not far from that inferred for the Crab pulsar ($P_o \sim 15$ms), the identification of the 35^d period as stellar wobble leads to the conclusion[1] that the equation state of neutron matter is comparatively stiff, somewhat stiffer indeed than one would conclude from the microscopic calculations of Friedman and Pandharipande.[11]

Alpar and Ögelman have shown how vortex creep enables the rotation vector of the pinned superfluid to follow the precession of the crust.[12] As Alpar discusses elsewhere in these Proceedings,[3] further accretion torques on the crust of the star are likely large enough to balance the internal torques applied by the creeping crustal superfluid. A further requirement is that these need to act in phase with the wobble in order to pump it to the magnitude required by observation; thus one needs to posit a wobble-operated accretion gate.[13]

4. Conclusion

In the Table, taken from Ref. 1, the predictions of neutron

matter theory for various density regions in the star are summarized, as are the observational consequences of those predictions. Pulsar timing irregularities provide the principle probe of the superfluid neutron stellar interior. In view of the excellent agreement between vortex creep theory and the observed behavior of the Vela and Crab pulsars, and of PSR 0525-21, following giant glitches, it could be argued that timing irregularities provide us with more detailed imformation about the interior of a neutron star than, to date, observations of "regular" pulsar behavior provide us with a detailed picture of the particles and fields outside the star responsible for characteristic pulsar radiation.

There remain a number of challenging problems concerning the structure and physical behavior of neutron stars. For example, while vortex creep theory has been extremely successful in describing the postglitch behavior of the Vela pulsar, following its giant glitches, as well as the Crab pulsar and PSR 0525+21 following their large glitches, the postglitch relaxation observed following the recent glitch from PSR 0355+54 is different from that observed for the other pulsars.[14] This is perhaps not surprising, since the relative magnitude of the glitch, $|\Delta P/P| = 4.4\times10^{-6}$, is the largest observed to date, as is the relative magnitude of the jump in the period derivative $|\Delta\dot{P}/\dot{P}| = 0.10 \pm 0.04$. It is quite possible that under these circumstances, in which 10 ± 4% of the moment of inertia of the neutron star is involved in the glitch, different, and possibly new regimes of vortex creep will play a role.

Such a large inertial moment for the pinned neutron superfluid is easily accomodated in neutron star structures calculated from an equation of state sufficiently stiff to yield the 35^d wobble period of Her X-1. However, such pinned superfluid inertial moments would not exist for stellar models of $1.4M_\odot$ neutron stars with an equation of state which is appreciably softer. Given the fact that constraints on the mass-radius relation for neutron stars,[15,16] which are obtained by combining theory with observation for X-ray bursts assuming this mass value, favor softer equations of state, it would seem that if both the identification of the 35^d periodicity in Her X-1 and the current canonical description of X-ray bursts are to be correct, then the neutron stars being observed in these two cases must have rather different masses.

Another challenging problem is the calculation of the pairing energy gaps for neutron matter at densities between 10^{12} gm cm^{-3} and nuclear matter density. Current theoretical calculations yield widely disparate values; it appears likely that the constraints placed on these energy gaps by post-glitch behavior and vortex creep heating of old neutron stars will be decisive in sorting out the correct theoretical description. Indeed, it is increasingly clear that neutron stars provide the <u>only</u> observational constraints on the description of neutron matter.

Finally, there is the issue of the nature of neutron star matter at densities greater than nuclear matter density, ρ_o. While strange matter, because it leads to strange stars with no internal structural differentiation, can be ruled out by the observation of glitches,[17] the appearance of a pion condensate in this region continues to be a distinct possibility; and might be required to explain the cooling curves of young neutron stars.[4]

This work has been supported in part by the Department of Energy and by NSF Grant PHY 86-00377 and NASA Grant NSG 7653.

Table - Neutron Stars as Cosmic Hadron Matter Laboratories

Density range of Hadron Matter	Theoretical Predictions	Observational Consequences
$\frac{\rho_o}{500} \lesssim \rho \lesssim \rho_o$	o Superfluid crustal neutrons o Pinning of superfluid vortices to crustal nuclei o Vortex creep theory of postglitch relaxation o Internal temperature of glitching pulsars o Vortex creep heating of old neutron stars o Wobble period fluctuations	o Origin of macroglitches in the Vela and other pulsars o Postglitch behavior and surface temperatures of Vela and PSR 0525+21 o Surface temperatures of old neutron stars o Changes in wobble period
$\rho \gtrsim \rho_o$	o Equation of state of neutron matter can be soft or stiff o Crustal extent, wobble period, M-R relation, M_{max}, and starquake frequency are sensitive e.o.s. barometers	o 35^d periodicity observed in Her X-1 o Initial spins of Her X-1 like stars o Starquakes in Crab and other very young pulsars
$\rho \gtrsim 2\rho_o$	o Possible exotic new forms of matter, such as pion condensates, quark liquids, which lead to novel cooling mechanisms	o Reduced surface temperatures of young neutron stars

References

1. Pines, D., "Neutron Stars as Cosmic Hadron Physics Laboratories," Proc. of NATO Institute, "High Energy Phenomena and Collapsed Stars," D. Riedel, in the press (1987).
2. Pines, D. and Alpar, M. A., Nature 316, 27 (1985).
3. Alpar, M. A., these proceedings.
4. Tsuruta, S., these proceedings.
5. Downs, G. S., Ap. J., 249, 687 (1981).
6. Lohsen, E., Astron, Astrophys. Suppl., 44, 1 (1981).
7. Downs, G. S., Ap. J. Lett., 257, L67 (1982).
8. Alpar, M. A., Nandkumar, R., and Pines, D., Ap. J., 288, 191 (1985).
9. Alpar, M. A., Brinkmann, W., Kiziloglu, V̈., Ögelman, H., and Pines, D., Astronomy and Astrophysics, in the press, 1987.
10. Trumper, J., Kahabka, P., Ögelman, H., Pietsch, W., and Voges, W., Ap. J., (1986).
11. Friedman, B., and Pandharipande, V. R., Nucl. Phys. A, 361, 502 (1981).
12. Alpar, M. A. and Ogelman, H., preprint.
13. Lamb, D. Q., Lamb, F. K., Pines, D. and Shaham, J., Ap. J., 198, L21 (1975).
14. Lyne, A. G., Nature, in the press, (1987).
15. Ebisuzaki, T. and Nomoto, K., Ap. J., 305, L67 (1986).
16. Sztajno, M., Fujimoto, M. Y., van Paradijs, J., Vacca, W. D., Levin, W. H., Pennix, W., and Trumper, J., Mon. Not. Roy. Astr. Soc., in the press, (1987).
17. Alpar, M. A., preprint.

Pulsars: An Overview of Recent Developments

J. H. Taylor
Joseph Henry Laboratories and Physics Department
Princeton University, Princeton, NJ 08544, USA

Pulsars have always been prime material for discussion at these Texas Symposia, because in a number of ways they are inherently relativistic objects. As neutron stars, their internal structure is determined under conditions in which gravitational forces dominate, or at least are competitive with, all other forces. Pulsar magnetospheres, which after nearly two decades of study are still not particularly well understood, are at least known to be places where relativistic electrodynamics is crucially important. And because pulsars make such good natural clocks, they have turned out to be exceedingly useful in experiments that evoke memories of the "rigid rods" and "rapidly moving ideal clocks" which most of us encountered in textbooks when we first studied relativity.

Much of what has been happening in the pulsar world in the past few years is related to two small and overlapping subclasses, the binary and millisecond pulsars. Only seven binary pulsars are presently known, and only three millisecond pulsars, two of which are also among the binaries. Nevertheless, this group of eight objects has taught us much about the present characteristics and evolutionary histories of the population of pulsars as a whole, and an overall picture of the neutron-star population in the Galaxy is becoming more clear. Moreover, the accuracies achievable in timing observations of the binary and millisecond pulsars has made them into exquisite tools for determining neutron star masses, for probing fundamental subtleties of the basic gravitational interaction, and for placing limits on the energy density of a cosmic background of gravitational radiation.

I will begin this talk with some introductory comments on the statistics of the entire pulsar sample population, and will then concentrate most of my remarks on the binary and millisecond pulsars. In doing so, I hope to bring you up to date on where recent observations and theoretical developments seem to be leading us.

1 The Pulsar Population

New pulsars have been discovered at a mean rate of just over 20 per year since late 1967. The cumulative total, now some 440 objects, probably amounts to somewhat less than one percent of the active pulsars in the Galaxy. This is a small enough fraction that many observational surprises probably still await us; at the same

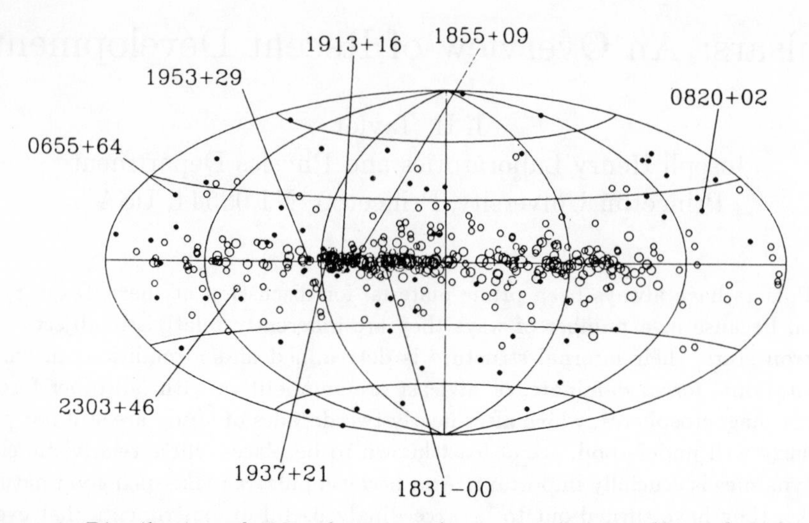

Figure 1: Distribution of 430 pulsars in galactic coordinates. Sizes of circles correspond to dispersion measures in the ranges < 30, 30 to 100, 100 to 300, and > 300 cm^{-3} pc. Binary and millisecond pulsars are identified by name.

time, the sample is large enough that the general characteristics of the galactic pulsar population are becoming reasonably well established.

Figure 1 shows the distribution of the known sample of pulsars on the sky, plotted in galactic coordinates. In this figure the sizes of circles give a rough indication of each pulsar's dispersion measure and hence its distance. One can see that the closest pulsars to the sun have a nearly isotropic distribution, while the more distant ones are concentrated toward the galactic equator. The majority of known pulsars are thought to lie within about 3 kpc of the sun, although the most luminous ones can be seen at distances of 10 kpc or more [9].

The eight pulsars identified by name in Figure 1 make up the subclass of known binary and millisecond pulsars. The fact that five of them are located in the narrow sector of the galactic disk between longitudes 30° and 70° is a selection effect reflecting the high sensitivity of pulsar searches carried out with the Arecibo telescope, which has limited declination coverage. The other three binary pulsars lie farther from the galactic equator and in other regions of the sky. As a group the binary and millisecond pulsars seem to have somewhat lower absolute luminosities and smaller distances from the galactic plane than are typical of pulsars as a whole [18].

In all cases where timing observations of sufficient accuracy have been made, pulsar periods have been found to be gradually lengthening. A scatter plot of period derivative against period is shown for 361 pulsars in Figure 2. In this figure the large black circles identify the seven known binary pulsars; the three known millisecond pulsars (i.e., those with period $P < 10$ ms) are located near the lower left corner. I have also identified the three youngest pulsars known, the Crab

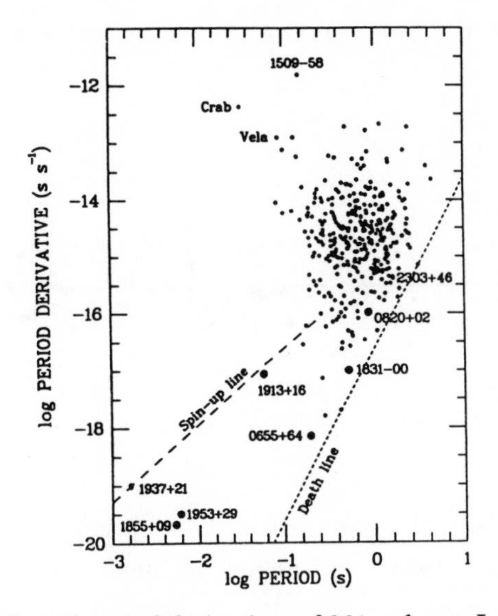

Figure 2: Periods and period derivatives of 361 pulsars. Large filled circles identify the seven known binary pulsars.

and Vela pulsars and PSR 1509-58, at the top left of the diagram. These objects are still surrounded by visible supernova remnants, tell-tale signs of the violent circumstances of their birth.

Current understanding of pulsar evolution [9,10] suggests that most pulsars first appear in the upper portion of the P, $\dot{P}$ diagram, with periods in the range ~ 0.01 to 0.5 s. Dipole magnetic fields of order 10^{12} G at the neutron-star surfaces cause large braking torques, so that within a million years or so a young pulsar moves downward and to the right until it joins the dense clump of points in Figure 2. Further evolution, including decay of the braking torque, will carry a pulsar along a steepening path across Figure 2's "death line," beyond which it ceases to generate sufficient electric potential in the polar cap region to allow it to function as an active pulsar. Most pulsars probably die within 10^7 years or so, with terminal periods of ~ 0.3 to 3 s.

Pulsars lucky enough to have a gravitationally bound companion star get a chance at "life after death" [1]. When the companion evolves off the main sequence and becomes a red giant, material lost from its outer layers is accreted by the neutron star, bringing with it a large amount of angular momentum. In this way the pulsar can spin up until it reaches an equilibrium period determined by its magnetic field, which by this time is perhaps no stronger than 10^8–10^{10} G. The parameters of all seven of the presently known binary pulsars are consistent with this scenario: in particular, they all lie between the spin-up line and the death line in Figure 2. It is likely that even the millisecond pulsar PSR 1937+21, which is not a member of a binary system, evolved in this way—eventually coalescing with

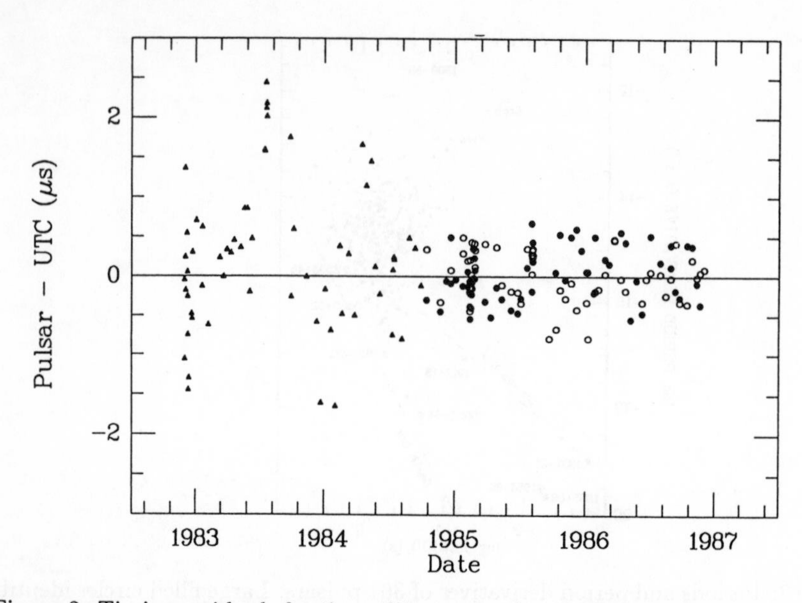

Figure 3: Timing residuals for the millisecond pulsar PSR 1937+21, obtained over 4.2 years.

its companion at the end of the process [24].

2 Millisecond Pulsar Timing

The narrow, periodic pulse waveforms of most pulsars make it possible to conduct timing observations with accuracies of order 10^{-3} periods. For millisecond pulsars, therefore, microsecond timing accuracies should be possible. Indeed, as shown in Figure 3, recent measurements of PSR 1937+21 by Rawley *et al.* [11] have achieved root mean square timing residuals (differences between observed and expected pulse arrival times) $\delta t < 1$ μs over a time span exceeding 4 years, and $\delta t \approx 0.3$ μs over the past 2 years.

It is interesting to note that between the first and last measurements plotted in Figure 3, exactly 81,143,313,239 pulse periods had elapsed. There are no pulse numbering ambiguities, even over gaps in the data as large as 10^9 periods. The implied rotational stability of the pulsar is impressive: a crude estimator is the r.m.s. residual divided by the data span, $\delta t/T \approx 3 \times 10^{-15}$—which makes the pulsar stability comparable with that of the very best atomic clocks.

With this kind of stability and achievable measurement accuracy, millisecond pulsars become useful tools for relativity experiments. For example, Figures 4 and 5 show what happens when each of two important terms are omitted in the general relativistic transformation from proper atomic time to coordinate time at the solar system barycenter. Figure 4 shows the timing residuals obtained for

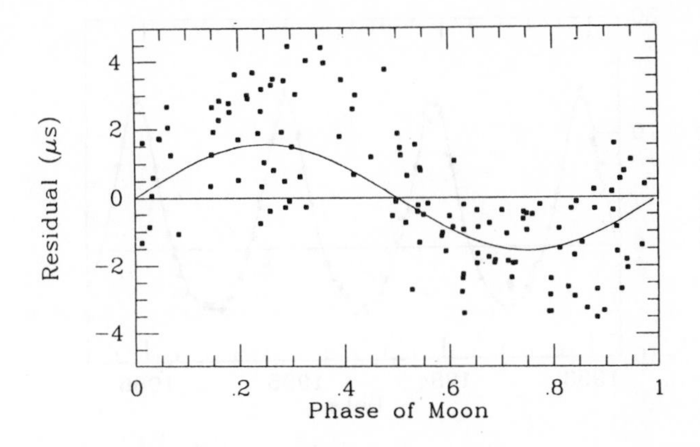

Figure 4: Timing residuals for PSR 1937+21, with the monthly term in the correction from proper atomic time to coordinate time intentionally omitted. The sinusoidal curve represents the omitted term.

PSR 1937+21 since October 1984 plotted as a function of lunar phase, with the monthly variation in solar gravitational potential at Earth intentionally omitted. Here is direct proof, based on a clock some 15,000 light years from the solar system, that clocks on Earth run more slowly near the phase of full moon — because at this time of the month we are closer to the Sun and deeper in its gravitational potential.

In a similar way, Figure 5 illustrates the annual variation of the Shapiro time delay [15] as the signals from PSR 1937+21 propagate through the solar system toward Earth. We can, in fact, carry out the Shapiro experiment with these data; the result is $(1+\gamma)/2 = 1.015 \pm 0.025$, where γ is the parametrized post-Newtonian parameter which determines the amount of spatial curvature produced per unit mass, and which has the value 1.0 in general relativity. For this purpose the pulsar data are not really competitive with those obtained from the Viking Lander on the surface of Mars [12], but the result is an interesting and independent confirmation nevertheless, and a test of general relativity at the 3% level.

Detweiler [4] first pointed out that pulsar timing residuals can yield an upper limit to the energy density contained in a cosmic background of gravitational radiation. Results based on slowly rotating pulsars observed for more than a decade [7,13] have shown that at wavelengths of a few light years the energy density in gravitational waves is less than 5×10^{-4} of that required to close the universe. Various authors have pointed out [8,26] that upper limits much below this level will begin to place interesting constraints on possible conditions in the early universe.

The discovery of millisecond pulsars made this method several orders of mag-

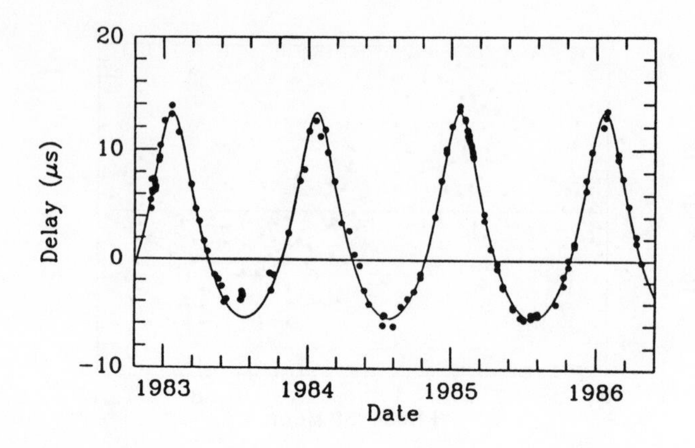

Figure 5: Timing residuals for PSR 1937+21 with the Shapiro delay in the solar system intentionally omitted. The smooth curve represents the omitted correction.

nitude more sensitive overnight. Consequently, based on just 2 years of data on PSR 1937+21, Davis *et al.* [3] were able to obtain an upper limit $\Omega_g < 5 \times 10^{-4}$ for gravitational wave frequencies in the range 1 to 3 y^{-1}. Today, with almost twice as much data and much of it of higher quality, the limits are $\Omega_g < 5 \times 10^{-5}$ for $f \sim 1$ y^{-1} and $\Omega_g < 1.5 \times 10^{-6}$ for $f \sim 0.3$ y^{-1}. It is worth noting that any claims to have *detected* a cosmic background of gravitational radiation, as opposed to specifying upper limits, will probably have to wait until a number of millisecond pulsars are available so that inter-comparisons can be made among them [11].

3 Binary Pulsars

Timing observations of binary pulsars yield the same kind of Doppler information as that obtainable for single-line spectroscopic binary stars. Three examples of binary pulsar velocity curves, corresponding to the three most recently discovered members of the class, are shown in Figure 6. The curve for PSR 2303+46 is decidedly non-sinusoidal, from which it follows that its orbit is highly eccentric. The other two pulsars have nearly circular orbits. PSR 1855+09, in addition to being in a binary system, is the third millisecond pulsar discovered. It thus joins PSR 1953+29 in the subclass of pulsars that are both binary and millisecond pulsars.

High quality timing data are now available for all eight binary pulsars, and Table 1 summarizes some of their astrophysically interesting parameters. Orbital periods of the binaries range from less than 8 hours to more than three years. Five of the orbits are essentially circular, with eccentricities $e < 0.01$. Four of these also have very small mass functions, $f_1(m_1, m_2) = (m_2 \sin i)^3/(m_1 + m_2)^2$,

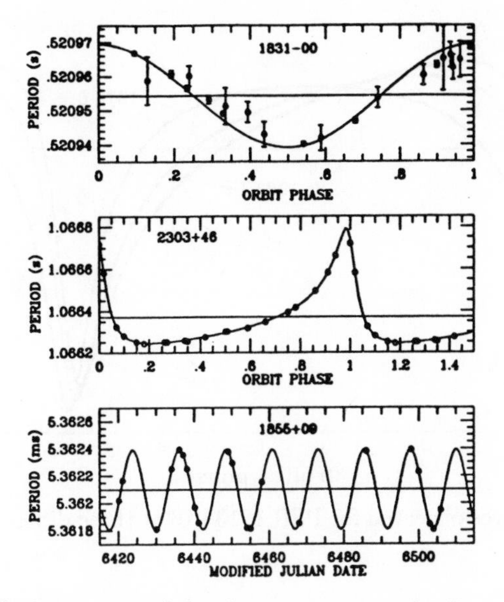

Figure 6: Velocity curves of the three most recently discovered binary pulsars.

implying probable companion star masses no more than a few tenths of a solar mass. (Here m_1 and m_2 are the masses of the pulsar and companion, respectively, and i is the inclination between the plane of the orbit and the plane of the sky.) The remaining two systems, in contrast, have orbits with $e > 0.6$ and much larger mass functions. Table 1 also lists the approximate separations a between the pulsars and their companion stars, computed (except for PSR 1913+16) under the assumptions $m_1 = 1.4\ M_\odot$ and $\cos i = 0.5$. The estimated separations range from 2.8 to 500 solar radii.

PSR 1913+16, the best studied of all binary pulsars, has been observed at more-or-less regular intervals [20,21,22,25] since 1974. Figure 7 shows examples of the orbital velocity curves observed at three different epochs spaced over 10

Table 1: Parameters of binary and millisecond pulsars.

PSR	P (ms)	$\log \dot{P}$	$\log B$ (G)	$\lvert z \rvert$ (pc)	a ($R_\odot$)	P_b (days)	e	$f(m_1, m_2)$ ($M_\odot$)	Likely m_2 ($M_\odot$)
1937+21	1.6	−19.0	8.6	20	—	—	—	—	—
1855+09	5.4	−19.7	8.5	20	23	12.33	0.00002	0.0052	0.2 - 0.4
1953+29	6.1	−19.5	8.6	20	100	117.35	0.0003	0.0027	0.2 - 0.4
0655+64	195.6	−18.2	10.0	120	5	1.03	< 0.00005	0.0712	0.7 - 0.8
1913+16	59.0	−17.1	10.3	190	2.8	0.32	0.6171	0.1322	1.38
1831−00	520.9	−17.0	10.9	190	6	1.81	<0.005	0.00012	0.06 - 0.13
0820+02	864.9	−16.0	11.0	280	500	1232.40	0.0119	0.0030	0.2 - 0.4
2303+46	1066.4	−15.4	11.8	480	29	12.34	0.6584	0.2463	1.2 - 1.8

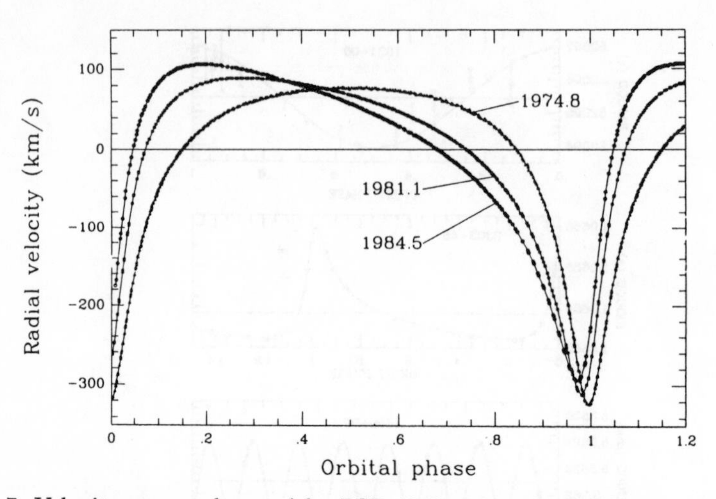

Figure 7: Velocity curves observed for PSR 1913+16 at three different epochs over 10 years.

years. Relativistic periastron advance causes the shape of the curve to change rather rapidly. The effect is detectable over time spans as short as ten days, and has now been measured to an accuracy of a few parts in 10^5.

Because the PSR 1913+16 system has the ideal combination of short orbital period, short pulsar period, large component masses, and large eccentricity which combine to permit the detection of higher-order secular and periodic relativistic effects, it is possible to determine the component masses quite accurately. In addition to the rate of precession of the orbit, one can measure the variable part of second-order Doppler shift and gravitational redshift, γ; the rate of decay of the orbit due to gravitational radiation, $\dot{P}_b$; and, with marginally significant accuracy, the orbital inclination, $\sin i$. The most recent analysis [19,25] yields for the masses $m_1 = 1.451 \pm 0.007$ and $m_2 = 1.378 \pm 0.007$. These results are important in that they provide a firm experimental foundation on which to build models of neutron stars and their formation [2].

These mass determinations, as well as evolutionary arguments [2,16,23], suggest strongly that the companion of PSR 1913+16 must be another neutron star, even though it is not observed as a pulsar. In that case the well-measured parameters of the system provide a clean prediction (from the general relativistic "quadrupole formula") for the rate at which the orbital period should decay because of energy losses to gravitational waves. The measured value is in excellent agreement with this prediction: the latest data yield the ratio

$$\frac{\text{Observed } \dot{P}_b}{\text{Predicted } \dot{P}_b} = 1.00 \pm 0.04 \quad .$$

For those who like to see and follow the progress of the orbital decay with time, Figure 8 shows that the pulsar is now passing periastron more than 5 s earlier than

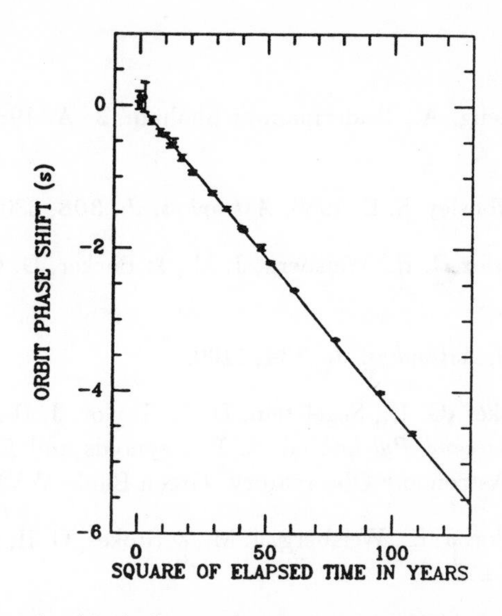

Figure 8: Measured orbital phase advance in the PSR 1913+16 system, plotted as a function of the square of the time elapsed in years since 1974.8. The sloping line gives the expected rate of orbital decay from gravitational radiation.

it would have been if the orbital period had remained fixed at its value in late 1974. It also shows that the orbital phase is, as predicted, advancing quadratically with time.

4 Conclusion

Although less than 2% of the 430 presently known pulsars are in binary systems, the fraction increases to 13% (4 of 30) in the period range $P < 200$ ms, and to 60% (3 of 5) of those with $P < 60$ ms. Six percent of the 50 pulsars with the smallest dispersion measures, and therefore presumably those closest to the sun, are binaries. Until very recently all major pulsar surveys were relatively insensitive for periods less than about 200 ms [5], and only one high-sensitivity survey has been made looking explicitly for low-luminosity pulsars in the solar vicinity [6]. It therefore seems likely that as more extensive surveys for fast and nearby pulsars are undertaken, many more of both the millisecond and binary pulsars will be found. Experience gained in Phase I of the Princeton-Arecibo pulsar survey [14,17] suggests that at least 10% of the pulsars detected in a survey with good sensitivity down to $P \approx 5$ ms will be millisecond pulsars, and probably most of these will be binaries. Future fulfillment of these expectations will be a great boon for understanding of the physics of neutron stars and their evolution, and perhaps for gravitation physics and cosmology as well.

References

[1] Alpar, M. A., Cheng, A., Ruderman, & Shaham, J. A. 1982, *Nature*, **300**, 728.

[2] Burrows, A., & Woosley, S. E. 1986, *Astrophys. J.*, **308**, 680.

[3] Davis, M. M., Taylor, J. H., Weisberg, J. M., & Backer, D. C. 1985, *Nature*, **315**, 547.

[4] Detweiler, S. 1979, *Astrophys. J.*, **234**, 1100.

[5] Dewey, R. J., Stokes, G. H., Segelstein, D. J., Taylor, J. H., & Weisberg, J. M. 1984, in *Millisecond Pulsars*, ed. S. P. Reynolds and D. R. Stinebring, (National Radio Astronomy Observatory: Green Bank, WV), p. 234.

[6] Dewey, R. J., Taylor, J. H., Weisberg, J. M., & Stokes, G. H. 1985, *Astrophys. J. (Letters)*, **294**, L25.

[7] Hellings, R. W., & Downs, G. S. 1983, *Astrophys. J. (Letters)*, **265**, L39.

[8] Hogan, C. J., & Rees, M. J. 1984, *Nature*, **311**, 109.

[9] Lyne, A. G., Manchester, R. N., & Taylor, J. H. 1985, *M.N.R.A.S.*, **213**, 613.

[10] Narayan, R. A. 1987, *Astrophys. J.*, in press.

[11] Rawley, L. A., Taylor, J. H., Davis, M. M., & Allen, D. W. 1987, to be published.

[12] Reasenberg, R. D., Shapiro, I. I., MacNeil, P. E., Goldstein, R. B., Breidenthal, J. C., Brenkle, J. P., Cain, D. L., Kaufman, T. M., Komarek, T. A., & Zygielbaum, A. I. 1979, *Astrophys. J. (Letters)*, **234**, L219.

[13] Romani, R. W., & Taylor, J. H. 1983, *Astrophys. J. (Letters)*, **265**, L37.

[14] Segelstein, D. J., Rawley, L. A., Stinebring, D. R., Fruchter, A. S., & Taylor, J. H. 1986, *Nature*, **322**, 714.

[15] Shapiro, I. I. 1964, *Phys. Rev. Lett.*, **13**, 789.

[16] Smarr, L. L., & Blandford, R. D. 1976, *Astrophys. J.*, **207**, 574.

[17] Stokes, G. H., Segelstein, D. J., Taylor, J. H., & Dewey, R. J. 1986, *Astrophys. J.*, **311**, 694.

[18] Taylor, J. H. 1987, in *Neutron Stars*, I. A. U. Symposium No. 125, ed. D. J. Helfand, (Dordrecht: Reidel), in press.

[19] Taylor, J. H. 1987, in *Proc. 11th International Conference on General Relativity and Gravitation*, ed. M.A.H. MacCallum, (London: Cambridge University Press), in press.

[20] Taylor, J. H., Fowler, L. A., & McCulloch, P. M. 1979, *Nature*, **277**, 437.

[21] Taylor, J. H., Hulse, R. A., Fowler, L. A., Gullahorn, G. E., & Rankin, J. M. 1976, *Astrophys. J. (Letters),* **206**, L53.

[22] Taylor, J. H., & Weisberg, J. M. 1982, *Astrophys. J.,* **253,**, 908.

[23] van den Heuvel, E. P. J. 1984, in *Millisecond Pulsars*, ed. S. P. Reynolds and D. R. Stinebring, (National Radio Astronomy Observatory: Green Bank, WV), p. 86.

[24] van den Heuvel, E. P. J., & Bonsema, P. T. J. 1985, *Astron. Astrophys.*, **139**, L16.

[25] Weisberg, J. M., & Taylor, J. H. 1984, em Phys. Rev. Letters, **52**, 1348.

[26] Witten, E. 1984, *Phys. Rev.*, **D30**, 272.

TOWARDS A MODEL FOR QPOs*)

Jacob Shaham
Physics Department and Columbia Astrophysics Laboratory
Columbia University
New York, New York 10027

1. INTRODUCTORY REMARKS

We have to date reports of Quasi-Periodic-Oscillation (QPO) observations in some thirteen X-ray sources, of which at least seven are low mass X-ray binaries (van der Klis 1986). They constitute a formidable zoo of phenomenae with so much variety that they, at times, do not at all even seem amenable to a single model. Several models have been constructed to explain the QPO phenomenon, which divide into two categories--mgnetospheric and configurational. As my time is very short I have to refer to excellent recent reviews of the various models, e.g., by Lewin (1986), and will concentrate here on the Beat Frequency Model which, it seems to me, is by far the prime model for at least some of the QPOs.

The Beat-Frequency-Model (BFM) was constructed in order to account for what is, perhaps, the simplest of the QPOs, GX5-1. As you know, by some good fortune (or bad?) GX5-1 happened to have also been discovered first. I would therefore like to open my talk by first describing the GX5-1 story.

To remind everyone of the observations: van der Klis et al. (1985a,b) reported the discovery of quasi-periods between about 25 and 50 msec in GX5-1 with a Q value ($\equiv f/\Delta f$, where f is the corresponding frequency) of at most 4. The quasi periods were correlated with the overall source's 1-18 keV count rate I, such that $\Lambda \equiv d\log f/d\log I$ was about 2.5, while the size of the modulation seemed to decrease steadily with count rate - from a value of 8% for $I = 2.4\times10^3$ cps to zero at $I = 3.1\times10^3$ cps. In addition, low frequency noise (LFN) was observed in GX5-1. Its intensity correlated well with the size of the QPOs modulation.

*) Based on a talk given in the IAU Symposium #125, Nanjing, China (1987, in press).

A periodicity - albeit a quasi-periodicity - so short, has triggered in Ali Alpar and me (and, presumably, in many others) the following immediate reaction: In our previous scenario for the formation of millisecond pulsars (Alpar et al. 1982) we have argued, that they were weak-field neutron stars spun up by accretion from a companion in a galactic-bulge-type binary system. Millisecond periods were therefore expected in these binaries while they were still X-ray-active, and yet none has yet been observed. The GX5-1 observation was the first such kind of period reported. Yet, The large variability of the period length ruled out stellar rotation because of the needed large variability in angular momentum, so that another - external - location for the origin of the oscillations had be be considered. Alpar and I concluded it had to be the magnetospheric boundary: We recalled that accepted models for the region just outside the surface of an accreting neutron star (see, e.g. Ghosh and Lamb 1979) talk about a magnetic-field dominated regime of plasma motion (for surface fields $B_s \gtrsim 10^6$ G) which turns over into a gravity-dominated one (a "Keplerian disk") outside of some boundary region at a radius R_o. R_o can be roughly determined by equating the ram pressure of the Keplerian plasma, ρv_K^2, to the magnetic pressure $B^2/8\pi$; it is given, more or less, by

$$R_0 \sim 10^6 B_8^{4/7} R_6^{12/7} \dot{M}_{17}^{-2/7} (M/M_\odot)^{-1/7} \text{ cm} \tag{1}$$

where in (1), B_8 is the surface magnetic dipolar field in units of 10^8G, R_6 is the stellar radius R_S in units of 10^6 cm, M is the neutron stellar mass, $M_\odot$ is the solar mass and $\dot{M}_{17}$ is the total mass accretion rate in units of 10^{17} g/sec. The Keplerian angular frequency at R_0, Ω_0, is

$$\Omega_0 \equiv \left(\frac{GM}{R_0^3}\right)^{1/2} \sim 10^4 B_8^{-6/7} R_6^{-18/7} \dot{M}_{17}^{3/7} (M/M_\odot)^{5/7} \text{ rad/sec} \tag{2}$$

Ω_0 is therefore a prime candidate for variability because of its dependence on $\dot{M}$ hence on the count rate I; however, it only varies weakly with $\dot{M}$ to account for a Λ of 2.5. So, Alpar and I (1985) postulated that the QPO frequency, $\Omega_{QPO} \equiv f/2\pi$, is to be identified with Ω_B, the beat frequency,

$$\Omega_B = \Omega_0 - \Omega_S \equiv A\dot{M}^{3/7} - \Omega_S \tag{3}$$

where A is some constant and Ω_S is the stellar spin. This, provided the stellar rotation axis is perpendicular to the disk. We shall make this assumption throughout this talk, and only modify it towards the end (p. 12).

While we are at it, let us recall that, even though it was hard to see how this might be the source of the GX5-1 period, there are two additional points of interest outside of a rotating, accreting neutron star:

The corotation radius R_c is that radius for which the Keplerian angular velocity equals the stellar spin angular frequency Ω_S:

$$\left(\frac{GM}{R_c}\right)^{3/2} = \Omega_S$$

hence

$$R_c \sim 1.5\times10^6 \ (M/M_\odot)^{1/3} \ (P_S/1 \text{ msec})^{2/3} \text{ cm} \tag{4}$$

where P_S ($\equiv 2\pi/\Omega_S$) is the neutron stellar period. In a rotating star with a magnetic field structure attached, R_c plays a very important role: If any field lines are threading the accretion disk inside of R_c, they develop an azimuthal component B_ϕ downstream which tends to slow down the plasma - hence spin-up the star; the reverse holds outside of R_c. Thus, spin-up and spin-down torques act simultaneously on the star; for some value of R_c their resultant should vanish - that determines the equilibrium period P_{eq} of the accreting star. It is roughly given by (Alpar _et al._ 1982)

$$P_{eq} \sim (0.33 \text{ msec}) B_8^{6/7} (M/M_\odot)^{-5/7} (\dot{M}/10^{-8} M_\odot \text{yr}^{-1})^{-3/7} R_6^{15/7} \tag{5}$$

Secondly, the velocity-of-light radius R_L is the distance for which corotation with the star would mean moving at the speed of light,

$$R_L \sim 5\times10^6 \text{ cm } (P_S/1 \text{ msec}) \tag{6}$$

For millisecond pulsars R_0, R_c and R_L close in on each other, still with $R_0 < R_L$; and, as long as the neutron stars are still spinning up, then, on the average, $R_0 < R_c < R_L$. If the mass accretion rate fluctuates, R_0 fluctuates too, but R_c and R_L do not. Now, in (3), A is some constant which depends on B_S, the surface magnetic dipolar field. Comparing (3) with the correlation data of Ω_{QPO} and the count rate I yielded $2\pi/\Omega_S \sim 7$ msec, $B_S \sim 5\times10^9$ G which were amazingly close to what one may have expected from the above millisecond pulsar model. Also, (3) gives

$$\mathrm{dlogf/dlogI} = (3/7)\left(1-\Omega_S/(\Omega_S+\Omega_{QPO})\right) \simeq 2.7 \text{ average},$$

in close agreement with the GX5-1 observations.

What was the physical mechanism, giving rise to the beat frequency (3), going to be be? We had originally felt that a variety of physical phenomenae at R_0 would cause a modulation of flow onto the star at the beat frequency (BF) (one is, of course, immediately reminded of Saturn's spokes, see e.g. Porco and Danielson 1982). To be specific, we suggested one class of phenomenae - a 'magnetic gate' process: Assume that plasma coming down to R_0 is magnetized (for example, after having been temporarily magnetized by the stationary stellar field at R_c or by some random magnetization process). Then, it will cross R_0 preferentially depending on the stellar field at R_0 (e.g., where the stellar field is opposite in direction to the plasma magnetization so that field reconnections could occur); that will give a modulation of the crossing rate at Ω_B.

Naturally, any BF process can only come about if the plasma does not flow into R_0 hydrodynamically. Instead, every region in the plamsa must carry its own magnetization along and remain pretty localized until crossing R_0. Fred Lamb has suggested to us that perhaps matter actually comes down in blobs - evidence for the blobby nature of accretion flow onto neutron stars in some compact X-ray sources has been around for a while. As we shall see below, blobs do make a fine physical scenario for realizing the Ω_{QPO}; however, as we shall also see below, they are not the only one.

It is also important to realize, that the magnetic axes of the neutron stars in QPO sources may well be at an angle to their rotation axes, perhaps even perpendicular to the latter. Therefore, the magnetospheric boundary may not be a circle in the accretion plane at all, but only some closed curve with mirror symmetry through the center, even if the <u>average</u> curve for large scale phenomenae <u>is</u> circular. This, of course, can provide for an Ω_B modulation without the incoming plasma being previously magnetized, and even in the hydrodynamic flow limit there may be an Ω_B component. These things are very hard to discuss in any great detail because of their great mathematical complexity, and we are (nevertheless) presently involved in having a crack at them. In the meantime, however, I shall assume that some kind of function of stellar and material field strengths and directions determines an Ω_B modulation such that any group of particles at R_0 will produce a luminosity signal modulated by the Ω_B clock.

2. THE BFM

There are, essentially, four basic assumptions that go into the BFM:

(1) That the QPO phenomenon, being essentially universal (from

having now observed several sources), may actually be independent of the observer's relative geometry and can therefore depend, at most, on internal stellar geometry;

(2) that some special radius exists outside the star, that is the source of the oscillations, possibly R_0;

(3) that Ω_{QPO} is related to the Keplerian frequency Ω_0 - and with the given variety of observed Ω_{QPO} values possibly related to $\Omega_B \equiv \Omega_0 - \Omega_S$; and

(4) that any observed intensity-correlated variability of Ω_{QPO} is related to variability in mass accretion rate $\dot{M}$, even if intensity changes do not necessarily relate to changes in $\dot{M}$ in any simple way (Priedhorsky 1986).

Given the above, we now describe the QPO modulation as follows:

Let us view everything in a coordinate system corotating with the matter at R_0 and have a density delta function disturbance $\delta(\phi-\phi_i, t-t_i)$ appear on the R_0 circle at azimuth ϕ_i and time t_i. As that disturbance goes down into the neutron star, let it produce a luminosity signal $\ell(\Omega_B t_i - \phi_i,\ t-t_i)$ which depends on the starting position relative to the $\Omega_B \equiv \Omega_0 - \Omega_s$ clock. Note, that an extra phase parameter should come into ℓ due to, say, a random direction of matter field $\Omega_B t_i - \phi_i \rightarrow \Omega_B t_i - \phi_i - \alpha_i$; but we shall, for simplicity, absorb α_i into ϕ_i. Assume now that the luminosity arising from a superposition of many delta-function density disturbances at R_0 is the sum of the individual signals - i.e., no non-linear effects on the way down to the star. Then, if the linear density profile at R_0 is

$$g(\phi,t) = \int d\phi_i dt_i\, g(\phi_i,t_i)\, \delta(\phi-\phi_i, t-t_i) \tag{7}$$

we have

$$L(t) \equiv \int d\phi_i dt_i\, g(\phi_i,t_i)\, \ell(\Omega_B t_i - \phi_i,\ t-t_i) \tag{8}$$

Let us go over to Fourier components,

$$g(\phi,t) = \sum_{\nu} \int d\omega\, e^{i\nu\phi} e^{i\omega t}\, \tilde{g}(\nu,\omega) \tag{9}$$

$$\ell(\psi,\tau) = \sum_{\bar{\nu}} \int d\bar{\omega}\, e^{i\bar{\nu}\psi}\, e^{i\bar{\omega}\tau}\, \tilde{f}(\bar{\nu},\bar{\omega}) \tag{10}$$

Hence

$$L(\omega) = \sum_{\nu} \tilde{\tilde{g}}(\nu, \omega - \nu\Omega_B)\, \tilde{\tilde{f}}(\nu,\omega) \tag{11}$$

The simplest BFM (Lamb et al. 1985) utilizes a simple form for $\ell(\psi,\tau)$,

$$\ell(\psi,\tau) = \theta(\tau)\, e^{-\gamma\tau} \left[1 + \beta \cos(\Omega_B \tau + \psi)\right] \tag{12}$$

where $\theta(\tau)$ is the step function: 1 for $\tau > 0$, 0 for $\tau < 0$. (8) and (12) describe a rather subtle process. One firstly needs, of course, the linearity, namely that various regimes on the R_0 circle do not influence each other's dynamics as they go down the magnetospheric boundary. Secondlay (but also related to the first point), one requires that the specific R_0 bunch of particles retains it coherence during the process; otherwise, if the coherence lifetime is shorter than γ^{-1}, we may need to replace (12) by

$$\ell(\psi,\tau) = e^{-\gamma\tau} \left[1 + \beta e^{-\Gamma\tau} \cos(\Omega_B \tau + \psi)\right]$$

and, for $\Gamma^{-1} << \gamma^{-1}$, the BF essentially disappears from the spectrum. We ask for the process to be, in that sense, not a topologically simple or a hydrodynamical flow process between R_0 and the luminosity-generating location. We do not know yet whether the inner disk dynamics really allows for these physical assumptions to be valid-disk dynamics is still too hard to handle down to such detail. We shall, however, assume that they are valid. Work on testing that is in progress, and it seems that an embedded magnetic field has much to do in securing a small Γ.

Before we proceed, we ought to comment on the possible values for β, since some data analysis suggests that $\beta > 1$ while a negative ℓ, which could result from such β, is certainly nonphysical.

We note, that $\ell(\psi,\tau)$ could, in principle, also include higher Ω_B harmonics, as long as their magnitude is consistent with the data. It is well known, that cosine or sine Fourier transforms $\tilde{f}(\nu)$ of positive functions $f(\tau)$ have the following property:

$$\left|\tilde{f}(\nu)\right| \equiv \frac{1}{\pi} \left|\int f(t) {\cos \nu t \atop \sin \nu t} dt\right| \leq \frac{1}{\pi} \int \left|f(t)\right| \left|{\cos \nu t \atop \sin \nu t}\right| dt \leq$$

$$\leq \frac{1}{\pi} \int f(\tau)\, \alpha t \equiv 2\tilde{f}(0)$$

Hence, in (12), the only strict requirement on the coefficient of $\cos(\Omega_B t + \psi)$ is really $|\beta| \leq 2$, provided of course, higher harmonics are present in $\ell(\psi,\tau)$ or else ℓ will indeed not be always positive. For example, notice that

$$0 < \sin^{2k}(\theta/2) = 2^{-2k}\Big[\binom{2k}{k} + 2\sum_{p=0}^{k-1}(-)^{k-p}\binom{2k}{p}\cos[(k-p)\theta]\Big]$$

$$\equiv 2^{-2k}\binom{2k}{k}\Big\{1+2\sum_{p=0}^{k-1}(-)^{k-p}\frac{\binom{2k}{p}}{\binom{2k}{k}}\cos[(k-p)\theta]\Big\}$$

$$[\theta \equiv \Omega_B t + \psi]$$

and the effective β value (i.e., the coefficient of cos θ) here is

$$\beta = \binom{2k}{k-1}\binom{2k}{k}^{-1} \equiv \frac{2k}{k+1}\ (\rightarrow 2 \text{ for large } k)$$

So, β can be above 1 provided the observational upper bound on the relative power at higher harmonics is not below the corresponding value of $\binom{2k}{p}^2\binom{2k}{k}^{-2}$. If $\beta = 1.35$ is found (Elsner et al. 1986), then we can take k=2, and expect $\ell(\psi,\tau) = e^{-\gamma\tau}(1-\frac{4}{3}\cos\theta + \frac{1}{3}\cos 2\theta)$ hence a relative power of 2nd to 1st harmonic 1.4%!

After this detour - back to the calculation of $L(\omega)$. Given the expression for ℓ, we find

$$\tilde{\tilde{\ell}}(0,\omega) = (\gamma + i\omega)^{-1}$$

$$\tilde{\tilde{\ell}}(\pm 1,\omega) = \frac{1}{2}\beta[\gamma + i(\omega \mp \Omega_B)]^{-1} \tag{13}$$

hence

$$L(\omega) = \tilde{\tilde{g}}(0,\omega)\frac{1}{\gamma+i\omega}$$

$$+ \frac{1}{2}\beta\tilde{\tilde{g}}(1,\omega-\Omega_B)\frac{1}{\gamma+i(\omega-\Omega_B)}$$

$$+ \frac{1}{2}\beta\tilde{\tilde{g}}(-1,\omega+\Omega_B)\frac{1}{\gamma+i(\omega+\Omega_B)} \tag{14}$$

(14) is the fundamental equation of the BFM. The fundamental conclusion we draw from it is that to observe the QPO line, g must have power at $e^{\pm i\phi}$; to observe the LFN, g must have a ϕ-independent (i.e. isotropic) component. These are necessary, not sufficient requirements.

Anyone who can imagine processes involving various contributions from an isotropic and/or a $\cos\phi$ term in the density distribution on the R_0 circle can explain any corresponding ratio of QPO to LFN intensities. I have heard it said that the BFM cannot cope with a finite-QPO no -LFN situation. This, however, is not the case, as the following example would show:

Consider

$$g(\phi,t) = A + B(t)\cos(\phi + \alpha) \qquad (15)$$

where α is a fixed phase. What (15) describes may be a toy model which works as follows: Imagine that at R_c, plasma does not only get partly magnetized with azimuthal dependence but also acquires - again, due to the magnetic field - an azimuthal density variation by magnetostriction. On the way down to R_0 that density variation, like the magnetization, will be partially preserved; that will lead to (15). (15) may actually be a good approximation for any reasonably smooth flow at the boundary region, outside of R_0. We should keep in mind that the non-steady-state apsect of the BFM is reflected in (12) and is not necessarily needed for (8).

(15) could also describe the following: Remember that we have absorbed the magnetic field angle into ϕ (p. 5). Suppose the azimuth angle is, in fact, random, but that the probability for a chunk of matter to begin entry into the magnetosphere depends on its relative magnetization angle. A situation like that may arise when the process is sufficiently non-linear. This will bring about a probability function of type (15), which can therefore be treated as an effective R_0 density.

We now have

$$\tilde{\tilde{g}}(0,\omega) \propto A\delta(\omega)$$

$$\tilde{\tilde{g}}(\pm 1,\omega) \propto \frac{1}{2}\,\tilde{B}\,(\omega)^{\pm i\alpha}$$

hence

$$L(\omega) \propto c\delta(\omega) + \frac{1}{2}\,\tilde{B}(\omega-\Omega_B)\;\beta\,\frac{e^{i\alpha}}{\gamma+i(\omega-\Omega_B)} + \text{c.c.} \qquad (16)$$

where c is some constant. The $c\delta(\omega)$ represents a dc term in the spectrum, not LFN; hence only the QPO lines exist, shaped according to $\tilde{B}(\omega)$ (or rather $|\tilde{B}(\omega)|^2$). It is easy to see how the dc term appears: If the density distribution at R_0 is isotropic, interference from all ϕ regimes will destroy the Ω_B modulation. If, furthermore, the flow has no temporal structure, no frequency except for the zero one will appear. Also note, that B should only have power around $\omega = 0$, but can be a random process with the $\cos\phi$ term turning randomly on and off. If (15) represents the magnetic field angle, then in (16) $\tilde{B}(\omega-\Omega_B)$ should be replaced by $\tilde{B}(\omega)$, so that B should have power around Ω_B.

Naturally, if A were time dependent, LFN would arise, shaped by $\tilde{A}(\omega)$. But, in general, different LFN and QPO lineshapes can be easily understood via different time dependence of A(t) and B(t).

The easiest way of envisaging power at $e^{\pm i\phi}$ is, of course, in a process which has power at <u>every</u> ϕ harmonic (including $\nu = 0$). A fully blobby flow can accomplish that (Lamb <u>et al.</u> 1985), with

$$g(\phi,t) = \sum_j a_j \delta(\phi-\phi_j)\, \delta(t-t_j) \tag{17}$$

where a_j, ϕ_j and t_j are all random variables. Since independent evidence for the blobby nature of flow in some X-ray sources does exist, this is a natural process to examine. Then

$$\tilde{\tilde{g}}(\nu,\omega) \propto \sum_j a_j e^{-i\omega t_j} e^{-i\nu\phi_j}$$

and the corresponding LFN and QPO powers are

$$\mathrm{LFN} \rightarrow \frac{1}{\gamma^2+\omega^2} \sum_{p,q} \bar{a}_p a_q\, e^{i\omega(t_p-t_q)} \equiv \frac{1}{\gamma^2+\omega^2} F(\omega) \tag{18}$$

$$\mathrm{QPO} \rightarrow \frac{(\frac{1}{4})\,\beta^2}{\gamma^2+(\omega-\Omega_B)^2} \sum_{p,q} \bar{a}_p a_q\, e^{i(\phi_p-\phi_q)}\, e^{i(\omega-\Omega_B)(t_p-t_q)}$$

$$\equiv \frac{(\frac{1}{4})\,\beta^2}{\gamma^2+(\omega-\Omega_B)^2}\, G(\omega-\Omega_B) \tag{19}$$

For uncorrelated a_j, ϕ_j and t_j,

$$F(\omega) = G(\omega-\Omega_B) \equiv \Sigma|a_p|^2$$

hence the LFN and QPO powers are always correlated here, much as they are found to be in GX5-1 (van der Klis et al. 1985) with power ratio of $(1/2)\ \beta^2$; as we pointed out earlier, under some circumstances that ratio may be close to 2. Processes in which some cross correlations between a_j, ϕ_j and t_j can be imagined, which therefore have various degrees of coherence, can change the LFN/QPO power. For example, Lamb (1987) shows specific numerical results for blob clusterings in which the LFN/QPO ratio is substantially reduced, in a fashion intermediate between (17) and (15). Numerical simulations by Lum (1986) show a similar trend.

3. PROBLEMS

The preceding discussion represents, probably, all the distance in mathematics that one could dare to cover before more detailed physical computations will be at hand. Let me, therefore, address myself to some of the obvious problems the BFM may have - which may all be attributed to our ignorance, not necessarily to the inadequacy of the BFM.

The most important problems are:

(i) Underlying all is the assumption that Ω_S exists, i.e. that the neutron star rotates at angular frequency Ω_S which, in the case of GX5-1, say, is of order 10 msec. Yet, Ω_S is yet to be detected in the X-ray luminosity of any of the LMXBs, let alone the QPO ones. This, while the very assumption of the existence of $R_0 > R_S$ means that if the magnetic field is dipolar one should see an X-ray modulation across the stellar surface. Some ways of understanding that have been discussed in Lamb et al. (1985). Effects like the low efficiency of channeling of plasma onto polar caps due to the relative closeness of R_0 and R_S, gravitational lensing (but see Chen and Shaham 1986), optical scattering in the disk corona, preferential viewing along the direction perpendicular to the disk (hence the magnetic field?) because the disk is optically thick - all of these may combine to quench the Ω_S modulation substantially, but possibly not below some of the observational upper bounds. A promising possibility is, however, the following:

Eichler and Wang (1986) have shown that as the magnetic field in neutron stars decays, higher multipoles become relatively stronger on the surface. The resultant multi-peaked surface field will cause a very different X-ray pulsation pattern than one finds in simple dipole fields: With sufficiently large "cap" areas, already an octupolar field component may smear the Ω_S pulsations substantially; higher mutlipoles will produce low level modulation at much higher - and at a wide range - of frequencies. We note, that this complex surface field need not be relevant to R_0 (hence the accretion process) or R_L (hence the pulsar action after the X-ray source became a pulsar): Even for a 1.5 msec pulsar, $(R_L/R_S)^2 \sim 50$, which is roughly by how much the dipole will be amplified over any other multipole at R_L over their surface ratio. The amplification at R_0 will be [see (5)] ~50 for a 10 msec neutron star.

(ii) The simple f vs. I relation in GX5-1 becomes inverted and even bimodal or indefinite in ScoX-1 (Priedhorsky et al. 1986). How to interpret that? We note, that the BFM predicts an f vs. $\dot{M}$ correlation, not directly an f vs. I. I itself is really not defined until one specifies the frequency range, $\Delta\nu_i$. Now, since torques (magnetic? material?) on the disk vary with $\Omega_S/\Omega_0 \equiv (1 + \Omega_B/\Omega)^{-1}$, and since these fine-tune the amount of total enery available (accretion energy onto the neutron stellar surface ± $\text{spin}^{\text{down}}_{\text{up}}$ energy of the neutron star), a marked difference may develop between dlog Ω_0/dlog $\dot{M}$ and the various dlog Ω_0/dlogI($\Delta\nu_i$), as was first pointed out by Priedhorsky (1986; see also Shaham and Tavani 1986). In fact, Priedhorsky writes

$$I(\text{hard X-rays}) = \dot{M}\left[\frac{GM}{R_S}\left\{1 - \frac{1}{2}\left(\frac{R_S}{R_c}\right)^3\right\} - f(\dot{M}, \Omega_S)\right]$$

$$I(\text{soft X-rays}) = \dot{M}\, f\, (\dot{M}, \Omega_S) - N\Omega_S$$

where N is the total torque on the star and f is some function. Priedhorsky suggests

$$f = \frac{GM}{2R_0} + \Omega_S(R_0^2\ \Omega_0 - R_S^2\Omega_S)$$

while Shaham and Tavani (1986) investigated the possibility

$$f = \frac{GM}{2R_0}$$

We find, that for a variety of choices of f, dlog I(hard)/dlog$\dot{M}$ indeed changes sign for some value of $\dot{M}$. Furthermore, an extended region with dlogI(hard)/dlog$\dot{M}$ >> 1 can exist, where f is essentially constant with I. A well defined I(hard) vs. $\dot{M}$ curve can be constructed to correspond to the full extent of the Ω_{QPO} vs. I ScoX-1 observations. Thus, even ScoX-1 can be interpreted on the basis of the BFM.

There is certainly a lot more that we do not know about the magnetospheric boundary, and hence about the detailed physics of the BFM. One may want to see the observations in order to either probe the magnetospheric boundary with the aid of the BFM or come up with a clearly better model for QPOs. A group of us is actively following the first option, mostly because the BFM looks, to date, more promising than other suggested models (see W.H.G. Lewin 1986).

One important physical method of probing an unknown region is to have it subject to some controlled external perturbation. Within the BFM, one important perturbation of the magnetospheric boundary can be neutron stellar free precession (as in Her X-1, see Trumper <u>et al.</u> 1985). Free precession will periodically change the opening angle of the cone, on the surface of which any particular magnetic field line segment rotates, so that the actual beat frequency may change (for details see Shaham 1986).

To fix our ideas, assume that the magnetic gate operates via $(\vec{B}\cdot\vec{r})^n$ where $\vec{r}$ is the radius vector of material orbiting at R_0, $\vec{B}$ is the local field at R_0 and n is some integer. Then $\vec{B}\cdot\vec{r} \propto \vec{\mu}\cdot\vec{r}$ where $\vec{\mu}$ is the stellar dipole. An extreme case is when the stellar rotation axis is in the disk plane with $\vec{\mu}$ perpendicular to it. Then $\vec{\mu}$ crosses the disk twice each $2\pi/\Omega_S$ period, and for n = 1 this gate gives both $\Omega_0 \pm \Omega_S$ freqeuncies at equal powers. Another extreme case for n = 1 is with $\vec{\mu}$ aligned with the rotation axis which is, again, in the plane of the disk. Then only Ω_0 appears, i.e. the BF is replaced by the magnetospheric boundary Keplerian frequency. In the most general orientation for n = 1, one obtains the $\Omega_0 \pm \Omega_S$ and Ω_0 frequencies at different powers, generally modulated by the free precession frequency. For larger n values more frequenices show up, among them Ω_S, $2\Omega_S$ and $2\Omega_0$, and I find it interesting that in CygX-2 there is some evidence for Ω_0 showing up (Hasinger 1986). In CygX-2 these things also correlate with hardness ratio (Hasinger 1986). The relation between the above mentioned cone angle and the hardness ratio is to date a complete unknown, but it seems quite worthwhile to look into free precession of QPOs with the Cyg X-2 observations in mind.

4. CONCLUSION

The QPO phenomenon has grown rapidly from a clean, simple one when only GX5-1 was known to a complex, puzzling one as other sources were discovered. Does it mean that there are many ways to produce QPOs - or that the geometries and environments change from source to source while the fundamental process remains the same? Scientific debate is always the hottest when very little is known. So, one should keep looking to present and future observations for guidance and also try to get a better theoretical handle on the magnetospheric boundary.

This work was supported under NAGW-567 and NAG8-497. This is contribution number 332 of the Columbia Astrophysics Laboratory.

4. REFERENCES

Alpar, M.A. and Shaham, J. 1985, Nature, **316**, 239.
Alpar, M.A., Cheng, A.F., Ruderman, M.A., and Shaham, J. 1982, Nature, **300**, 728.
Chen, K.Y and Shaham, J. 1986, Proc. IAU Symp. #125.
Eichler, D.S. and Wang, Z. 1986, Proc. IAU Symp. #125.
Ghosh, P. and Lamb, F.K. 1979, Ap.J., **234**, 296.
Hasinger, G. 1986, Proc. IAU Symp. #125.
Lamb, F.K., Shibazaki, N., Alpar, M.A., and Shaham, J. 1985, Nature, **317**, 681.
Lamb, F.K. 1987, this volume.
Lewin, W.H.G. 1986, this volume.
Lum, K.S. 1986 (private communication).
Porco, C.C. and Danielson, G.E. 1982, Ast. J., **87**, (826).
Priedhorsky, W. 1986, Ap.J. (Letters), **306**, L97.
Priedhorsky, W., Hasinger, G., Lewin, W.H.G., Middleditch, J., Parmar, A., Stella, L., and White, N. 1986, Ap.J. (Lett.), **306**, L91.
Shaham, J. and Tavani, M. 1986, Proc. IAU Symp. #125.
Shaham, J. 1986, Ap.J..
Trumper, J., Kahabka, P., Ogelman, H., Pietsch, W. and Voges, W. 1985, preprint.
van der Klis, M. 1986, Proc. IAU Symp. #125.
van der Klis, M., Jansen, F., Van Paradijs, J., Lewin, W.H.G., Trumper, J. and Sztajno, M. 1985a, IAU circular 4043.
van der Klis, M., Jansen, F., Van Paradijs, J., Lewin, W.H.G., Trumper, J. and Sztajno, M. 1985b, Nature, **316**, 225.

DYNAMICS OF THE NEUTRON STAR INTERIOR

M.Ali Alpar

Scientific and Technical Research Council of Turkey
Research Institute for Basic Sciences
P.K.74, Gebze, Kocaeli 41470, Turkey

ABSTRACT

Current models for the superfluid interior of neutron stars are described. Applications to neutron star timing observations are discussed.

I. INTRODUCTION

The dynamics of the neutron star interior is of interest in connection with timing observations of pulsars. Among the most spectacular findings of pulsar timing observations are sudden jumps (glitches) in the pulsar rotation frequency (for references to the observations and theoretical work see Ref.1.The two most recent glitches are reported in Ref.2.).The electromagnetic properties of the radio pulses do not exhibit any systematic changes accompanying the glitches. Hence, these events must have their cause in the dynamics of the neutron star interior rather than in external (magnetospheric) torques. Timing noise from radio pulsars (e.g.Ref.3 and 4) is a related phenomenon which may be influenced by the internal structure of neutron stars. Among the accretion powered X-ray sources, the recent observational evidence[5] that the 35-d cycle of Her X-1 is due to precession of the neutron star brings up questions concerning the dynamics of the interior.

The interiors of neutron stars are believed to be mostly superfluid. The theoretical argument for superfluidity in neutron stars rests on a general theorem that the ground state of a collection of fermions with attractive interactions must be superfluid. The bulk of a neutron star consists of neutrons in extended states, and the nuclear interactions between these neutrons are indeed attractive. The protons in the star's core are likewise expected to be superconducting.

The observational evidence for superfluidity in neutron stars comes from the timescales of postglitch relaxation. The initial jumps in frequency, and in spindown rate, which typically have a magnitude $\Delta\dot{\Omega}/\dot{\Omega} \sim 10^{-3}-10^{-2}$, gradually relax towards an extrapolation of preglitch timing behaviour. A succession of relaxation times, ranging from a few days to several years are observed. As these timescales are long compared to viscous relaxation times of normal (non-superfluid) matter, a superfluid neutron star interior is indicated.

Two different regimes of neutron superfluidity can be identified in models of the neutron star interior. The crust superfluid coexists with a crystal lattice, at densities 4×10^{11} gm-cm^{-3}-2×10^{14} gm-cm^{-3} in the inner crust of the neutron star. At larger densities there is no crystal lattice, and we have the core superfluid, which is a homogeneous mixture of superfluid neutrons, superconducting protons, and normal relativistic electrons.

The purpose of this paper is to outline the body of recent work on the dynamical properties of both types of superfluid and the applications of these models to observations. In the next section the basic properties of a rotating superfluid are reviewed. This is followed by discussions of the core superfluid and the crust superfluid respectively. The behavior of both superfluids in a precessing star will be discussed briefly in a separate section.

2. BASIC PROPERTIES OF A ROTATING SUPERFLUID

The minimum free energy state of a system with constant angular momentum is the state of rigid rotation at the appropriate uniform angular frequency. For rigid body rotation at Ω, the curl of the fluid's velocity (i.e., the vorticity) is simply 2Ω everywhere. A superfluid cannot achieve exact rigid body rotation because its velocity is a gradient-the gradient of the phase of a quantum-mechanical wavefunction. Such a velocity field is curl-free unless there are singularities in the phase. Physically, such singularities are quantized vortices. A superfluid can mimic rigid body rotation quite closely by forming an array of quantized vortices. The quantum of vorticity is $K \equiv h/2m$ where h is Planck constant and m the neutron mass. To mimic rigid rotation at frequency Ω, a vortex density $n=2\Omega/K$ is set up.

To spin up or spin-down a superfluid, the density of vortices must change: a radial vortex flow is needed. The equation of motion for the superfluid is:

$$\dot{\Omega} = -\frac{nK}{r} V_r \tag{1}$$

In this equation r is the distance from the rotation axis and V_r is the mean radial velocity of the vortex lines. Each physical model for the spindown or spinup of a superfluid is a model for this radial velocity in response to the driving forces of the system.

The motion of individual vortex lines is determined by the Magnus equation,

$$\vec{f} = \rho\vec{K} \times (\vec{V} - \vec{V}_L) \tag{2}$$

where $\vec{f}$ is the force applied per unit length of vortex lines, ρ is the superfluid's density, $\vec{K}$ is the vorticity directed along the line, $\vec{V}_L$ is the velocity of the line and $\vec{V}$ the velocity of the ambient superfluid. The force $\vec{f}$ arises from the interactions between normal matter, for example electrons or the crystal lattice, and the "core" of the quantized vortex line, which is made of neutrons in the normal (non-superfluid) state.

This equation of motion will set up the appropriate radial migration of vortex lines to change the superfluid's rotation rate. For example, when the star's outer crust is slowing down due to external torques, electrons in the interior will also slow down since they are coupled viscously to the outer crust. Their relative motion with respect to the vortex core will lead to an azimuthal drag force on the vortex core. Eq.(2) shows that the vortex will then move radially outward. As vortices move outward, the superfluid spins down.

3. THE CORE SUPERFLUID

The core superfluid is a homogeneous mixture of superfluid neutrons, superconducting protons and normal, relativistic electrons. In most neutron star models this core contains almost the entire moment of inertia of the star. The first link in the coupling of the core to the outer crust is the viscous coupling of the electrons. The coupling times are of the order of minutes or less[6]. The protons also follow the motion of the crust on these timescales, as any relative motion of protons and electrons would be damped immediately by the large ensuing electromagnetic fields. As the charged particles couple to the other crust, the neutron superfluid will follow through their interaction with the neutron vortices. The most effective interaction is due to magnetization of the neutron vortex lines.

The source of this magnetization is the interaction between the neutron and the proton superfluids. In a mixture of two superfluids consisting of interacting species of particles, there will be drag effects so that a current in one superfluid will drag some particles of the other species whose contribution on the flow depends on the strength of the microscopic interaction. The effect leads to a drastic enhancement of the interaction between the vortices and the charged particles. The microscopic superfluid neutron velocity around each vortex drags a solenoidal proton current which depends on the neutron-proton interaction and can be expressed in terms of the effective nucleon masses in the medium. This proton current sets up a magnetic field of $\sim 10^{15}$ G within a London length ~ 50 fm of the neutron vortex. Thus these lines, which carry the rotation of the neutron superfluid, have a very efficient electromagnetic coupling to the charged particles. The relaxation time τ_v for any relative velocity between the vortex line and the system of charged particles is calculated to be[7]

$$\tau_v \cong 5\left(\frac{m_p}{\delta m_p}\right)^2 P(s) \qquad (3)$$

where P is the pulsar's rotation period, m_p the bare proton mass and δm_p the difference between the bare mass and the effective mass.
Given this timescale for the interaction of a single vortex, the

dynamical coupling time τ_d of the neutron superfluid can be found .

$$\tau_d \cong \frac{\tau_v}{x} \cong 100 \left(\frac{m_p}{\delta m_p}\right)^2 P(s). \tag{4}$$

This is the timescale on which differential rotation between the neutron superfluid and the charged particles relaxes. On this timescale vortices move radially in response to forces arising from the relative velocities. x is the ratio of the number of electrons (and protons) to the number of neutrons, which is typically about 5%. $\tau_d \cong 400$ P(s) has occasionally been quoted as a typical value, using $\delta m_p/m_p = 0.5$. As the nucleon effective masses in neutron matter are not very well known, τ_d could be somewhat longer; for example $\tau_d \cong 10000$ P(s) for $\delta m/m \cong 0.1$. On timescales longer than a few hours the core superfluid is rigidly coupled to the neutron star's crust. This is much shorter than the timescales of the previously considered coupling mechnisms for the core superfluid and does not have temperature dependence.

Such tight coupling has important astrophysical consequences. Since the core superfluid is rigidly coupled to the crust on timescales longer than τ_d, it cannot be responsible for the observed postglitch relaxation of pulsars, which continues for months and years. This is consistent with the fact that the glitches involve fractional charges $\Delta\dot{\Omega}/\dot{\Omega} \cong 10^{-3}$ - 10^{-2} in the spindown rate. This can be interpreted as the fractional moment of inertia of the crust superfluid which must then be responsible for the glitches and the postglitch relaxation. Thus, the core superfluid is already coupled to the observed outer crust on a timescale shorter than the resolution of the actual glitch time, and behaves effectively as a part of the outer crust during the subsequent relaxation.

Another important consequence concerns the precession of the neutron star. Recent observations of the 35-d cycle of Her X-1 strongly suggest precession. If the neutron star were freely precessing, the coupling to the core superfluid would damp the precession on a timescale

$$\tau_{pr} = \tau_d P_{pr}/P \cong 100 \left(\frac{m}{\delta m_p}\right)^2 . \ 35 \text{ days} \tag{5}$$

Thus the observed precession of Her X-1 must be continuously excited by some external torque against damping by the core superfluid.

4. THE CRUST SUPERFLUID

In the inner crust of a neutron star, at densities 4×10^{11} gm-cm^{-3} $\lesssim \rho \lesssim 2\times10^{14}$ gm-cm^{-3}, the neutron superfluid coexists with a crystal lattice. Neutron rich nuclei containing the protons and some of the neutrons in bound states occupy the lattice sites. The rest of the neutrons, as well as the electrons, are in extended Bloch states. These continuum neutrons form a superfluid. There is a finite local density of the neutron superfluid inside the nuclei, different from the density in interstitial regions. This is an extremely inhomogeneous medium for the motion of vortex lines. The typical lattice spacing is 50 fm, the radius of the nuclei 7 fm , and the radius ξ of the vortex line cores is 10 fm. This latter lengthscale is the coherence length of the neutron superfluid. The vortex line core contains neutrons in the normal rather than the superfluid state. The energy cost per particle of setting up this normal core is $3\Delta^2/8\varepsilon_F$ where Δ is the superfluid energy gap and ε_F the Fermi energy of the neutrons. These quantities, particulary Δ, depend on the local density of superfluid neutrons. Thus the energy of the vortex line has different values depending on whether it coincides with a nucleus or an interstitial site. At 10^{13} gm-cm^{-3} $\lesssim \rho \lesssim 2\times10^{14}$ gm-cm^{-3} a vortex core coinciding with nuclei is energetically favored and one can estimate a pinning energy E_p per each nucleus in the vortex core. E_p depends sensitively on the gap $\Delta(\rho)$. There are uncertainties in calculations of the gap, but recent results indicate that $E \sim 0.1$-1 MeV.

When vortices are thus pinned to the lattice, the superfluid cannot spin down freely. As the lattice is spun down by the pulsar torques, the pinned vortex lines, which move with the lattice, are moving with respect to the superfluid. The required force (see Eq.(2)) is supplied by pinning. The force per unit length on each pinned vortex is radially inward, directed towards the rotation axis of the star. The associated energy is :

$$\Delta E = \rho KR(\Omega-\Omega_c)\, b\xi \equiv \rho KR\omega b\xi. \tag{6}$$

Here R is the distance to the rotation axis, Ω and Ω_c the rotation rates of the pinned superfluid and the crust lattice respectively, and $\omega \equiv \Omega-\Omega_c$ is the lag between these rotation rates. b is the distance between successive pinning sites along a given vortex line. The coherence length ξ represents the distance scale of the pinning force. The available pinning energy, E_p at each site, defines a maximum (critical) lag $\omega_{cr} \equiv E_p/(\rho Krb\xi)$ that pinning can sustain; if $\omega \geqq \omega_{cr}$, vortex lines will become unpinned. A statistical theory of the motion of vortex lines in this pinning medium has been formulated[8]. Describing the medium as a random potential with typical barrier height E_p and distance scale b, one notes that since a radially inward pinning force is required at a finite lag ω, the energy potential is biased in the radially outward direction. Random motions of the vortex lines in the radially outward direction will face barrier heights $E_p-\Delta E$, while for radially inward motions, the barrier height is $E_p+\Delta E$. At a temperature T and microscopic vortex line velocity V_o, the thermal expectation value for the radial velocity V_r of the vortex lines, the rate of "vortex creep", is

$$V_r = 2V_o \exp -\frac{E_p}{kT} \cdot \sinh \frac{E_p}{kT}\frac{\omega}{\omega_{cr}} . \tag{7}$$

Using this expression in Eq.(11), and writing the equation of motion

$$I_c \dot{\Omega}_c = N_{ext} + N_{int} = N_{ext} - \int dI_p \dot{\Omega} \tag{8}$$

for the observed rotation rate $\dot{\Omega}_c$ of the outer crust, one obtains a dynamical model for the coupling of the pinned crust superfluid by vortex creep. In Eq.(8) N_{ext} is the external (pulsar) torque, and N_{int} the internal torque exerted on the crust by the pinned superfluid, with moment of inertia dI_p in each pinning region characterized by a particular pinning regime. I_c, the crust moment of inertia, effectively includes the entire star other than the pinned inner crust superfluid, since, as discussed above, the core superfluid is coupled rigidly to the crust-normal matter system on the timescales of interest. This vortex creep model has been applied successfully and consistently to postglitch relaxation data showing the following salient features:

1) The jumps in $\dot{\Omega}$,and the subsequent relaxation involve the decoupling and recoupling of vortex creep in regions of fractional moment of inertia $I_p/I_c \cong 10^{-3}$-10^{-2}, in agreement with (in fact, historically, in anticipation of) the tight coupling of the core superfluid to the observed outer crust. Thus, the pinned crust superfluid seems solely responsible for the glitches and the postglitch relaxation phenomena.

2) The solutions for the relaxation of vortex creep after a glitch involve relaxation times of the form:

$$\tau = \frac{kT}{E_p} \frac{\omega_{cr}}{|\dot{\Omega}_c|} \tag{9}$$

The fits to the data produce a hierarchy of relaxation times. Evaluating these with estimates of the pinning energy, one can arrive at estimates of the temperature in the innercrust(which, at least for the older pulsars, is the isothermal core temperature). Although there are factor 2 uncertainties in these estimates, the temperature estimate for the Vela pulsar (internal temperature 1.5×10^7 K, surface temperature 3×10^5 K) like the observational upper limit to its surface temperature, falls significantly below the temperatures predicted by standard cooling theories[9].

3) Vortex creep dissipates energy at the rate $\dot{E}_{diss} = I_p\, \omega_{cr}\, |\dot{\Omega}_c|$. For old pulsars, which have already radiated away their initial heat content, this rate should determine the surface luminosity and effective temperature. For the old pulsar PSR 0525+21, estimates of the surface temperature from $\dot{E}_{diss}$ are in agreement with estimates from the observed postglitch relaxation times, using Eq.(9)[10].

The recent large glitch from PSR 0355+54 poses a challenge to vortex creep theory. Its postglitch relaxation is different from the postglitch behavior in other pulsars, suggesting possibly a new regime of vortex creep. It must also be noted that according to the present theoretical framework, the available pulsar timing data contain direct evidence only for the crust superfluid, while the core superfluid, although theoretically believed to be there,can't supply signature of its presence on the observed timescales. Evidence for its presence may, perhaps, be supplied by a novel interpretation of timing noise[11].

5. PRECESSION

Before concluding,let me allude briefly to recent work[12] that applies these models to the case of a precessing neutron star. The chief conclusions of the work are twofold. First, it is argued that there is a mode of thermal vortex creep that will allow the rotation velocity of the pinned superfluid to precess. This alleviates constraints on the rate o precession which would follow if the vortex lines were absolutely pinned, so that they did not move at all in the star's reference frame.Secondly, the above dynamical models for the core and the crust superfluids can be used to obtain steady state precessional motion of these fluids from the Euler equations. A crucial question is whether the torques that the superfluids apply on the outer crust (the observed precession) are small enough to be balanced by the driving external torques available in the system. Such a balance is needed since Her X-1 seems to be executing almost free precession. The internal torques implied by the above models indeed do not exceed estimates of the available external torques in magnitude. Whether the time dependence of the external torques can appropriately "lock in" to the precession is an open question at present.

6. CONCLUSION

A comprehensive dynamical model of the neutron star interior has been emerging in response to the accumulation of pulsar timing observations with the many surprises they offered. The models are of a rather simple and general type, and have succeeded in providing a consistent interpretation of the data so far. Whether this will continue to be so will be seen as we face new challenges from the observations, particularly more glitches.

I thank the Physics Department of the University of Illinois at Urbana-Champaign for hospitality, and K.-S.Cheng, Hakkı Ögelman, and David Pines for discussions on these topics.

REFERENCES

1. Pines,D. and Alpar,M.A., Nature, 316,27(1985).
2. Lyne,A.G., submitted to Nature (1987); Lyne,A.G.et.al., to be submitted (1987).
3. Cordes,J. and Downs,G.S., Astrophys,J.Suppl. 59,343 (1985).
4. Boynton,P.E. and Deeter,J.E.,Astrophys.J., in press (1986).
5. Trűmper,J.,Kahabka,P.,Ögelman,H.,Pietsch,W.& Voges,W.,Astrophys.J. Lett.300,L63(1986).
6. Easson,I.,Astrophys.J.,228,257 (1979).
7. Alpar,M.A.,Langer,S. and Sauls,J.A,Astrophys.J.,282,533(1986).
8. Alpar,M.A.,Anderson,P.W.,Pines,D. & Shaham.J.,Astrophys.J.,276,325 (1984).
9. Nomoto,K. and Tsuruta,S.,Astrophys.J.Lett., 305,L19(1986).
10. Alpar,M.A.,Nandkumar,R. and Pines,D.,Astrophys.J.,288,191 (1985).
11. Cheng,K.S.,to be submitted (1987).
12. Alpar,M.A. and Ögelman,H.B., submitted to Astr.Astrophys, (1987).

NEUTRON STAR COOLING

Sachiko Tsuruta
Department of Physics, Montana State University, Bozeman, Montana

1. INTRODUCTION

The studies of thermal evolution of neutron stars are important in connection with various temperature related problems[1]. To be complete, evolution calculations should include all possible heating and cooling mechanisms, both internal and external. The previous investigations show, however, that cooling determines the thermal history of an isolated neutron star during the earlier period, of less than about 10^4 years[2]. In the following review, we shall confine our attention to isolated neutron stars, in the sense that the external effects such as mass accretion are negligible, and we shall devote our effort mostly on the earlier stages where cooling dominates over heating.

A neutron star cools mainly through neutrinos escaping directly from the central core and photon radiation from the surface[3]. In the "standard" scenario, the dominant neutrino mechanisms are the modified URCA and neutrino bremsstrahlung involving nucleons in the core, the crust neutrino bremsstrahlung involving heavy nuclei, and plasmon processes[4]. However, a neutron star could cool much faster through various "exotic" means, such as the neutrino processes involving pions and quarks, or the emission of "exotic" particles such as axions. The latter possibilities are, for convenience, called "non-standard" cooling[3]. In the following, we shall give a brief review of the most recent work and the work in progress, on both standard and non-standard cooling. Earlier more complete reviews are found in Refs. 2) and 3).

2. RESULTS

2.1. Standard Cooling

Cooling curves are obtained with two methods, the "exact" method in the sense that the basic stellar structure-evolution equations are solved simultaneously without approximations, and the "isothermal" method where the hydrostatic and thermal parts of these equations are decoupled through the "isothermal" assumption. Both results should converge when the isothermal state is reached. The conclusion of our most recent calculations[5] is that for nuclear models with stiff equations of state, it takes $\sim 10^5$ years before the isothermal state is reached. All observable objects, e.g. supernova remnants (SNRs), which are suitable for comparison with the cooling theory, are younger than this[3]. Therefore, the "exact" method should be used when we are to compare the theoretical cooling results with the substantially improved observational data, from the Einstein Observatory and those expected in the future. For typical models of intermediate softness, this time scale is 500 to 10^5 years, much longer than predicted earlier. The major cause for these relatively long time scales we found for thermal relaxation is that in these most recent calculations we adopted more accurate opacities[5].

Our representative results are shown in Fig. 1, for the FP model of an intermediate softness[6], with $M_A = 1.4 M_\odot$, R = 10.9 km, and $\rho_c = 1.17 \times 10^{15}$ gm/cm^3 (where M_A, R and ρ_c are the baryon mass, radius and the central energy density of the star). Here, the fully general relativistic equations were solved, with the best physical input currently available[5].

The curves (1) correspond to the "standard" cooling. The solid curve is obtained with the "exact" method while the "isothermal" method was used for the dashed curve. In the dot-dashed curve the effect of superfluidity is maximized.

2.2. Non-Standard Cooling

There may be various possibilities for the "non-standard" scenario including emission of as yet undiscovered particles[7a](see also Iwamoto in this volume). Here we shall discuss the preliminary results of our work[7b] in progress, on the pion, quark, and axion cooling where sufficient physical information is already available. In order to see a measure of seriousness of non-standard cooling, cooling curves are obtained first with the isothermal method, for a star with a pion and quark core. In both cases the core starts at $\rho = 2\rho_N$ (where ρ_N is the nuclear density). Otherwise, the input physics is the same as in the standard case and, as before, $M_A = 1.4M_\odot$. In Fig. 1, the curves (2) and (3) show, respectively, pion and quark cooling. In the solid curve (2) we adopted the most realistic emissivity for charged pion (π^c) condensates derived by Tatsumi[8], while Iwamoto's emissivity[9a] of quark matter was used in the curve (3). The Kyoto group[8],[10] have shown that the earlier simple emissivity for π^c condensates by Maxwell et al.[11] was seriously overestimated. The dashed curve (2) shows the older π^c cooling[11]. It is very close to the quark curve. An important conclusion is that when the emissivity is treated more realistically, the π^c curve lies substantially higher than the older simple model. They also found that when neutral (π^o) and charged (π^c) pions coexist (a realistic possibility), the cooling rates will be further reduced, with about the same amount[10][12]. That is, the curve for the π^o- π^c mixture should lie about half way between the standard and the new π^c curve, the solid curve (2)[7b].

A neutron star can cool very efficiently through the emission of axions if the symmetry breaking parameter F is sufficiently small[13]. F is inversely related to the axion mass, and we can set the upper limit to the axion mass from cooling theory[15]. We have, therefore, calculated axion cooling with different values of F[7b][15], using the emissivites derived by Iwamoto[13] and Itoh[14].

2.3. Vortex Creep Heating

It has been pointed out that the vortex creep dissipation can be an important mechanism for heating a neutron star[16],[17] (see Pines and Aplar in this volume). Evolutionary calculations were carried out[16] for the vortex creep heating (see Shibazaki and Lamb in this volume). The details are found in these other articles in this volume. In Fig. 1 are shown our results for vortex creep heating (dotted curves)[7b], which were obtained with essentially the same heating equation as in Ref. 16) but treating the evolution more realistically (e.g. including all important neutrino cooling mechanisms). We note that the effect of the vortex creep heating is larger when the critical frequency $\bar{\omega}_{cr}$ is large. The observation agrees with small $\bar{\omega}_{cr}$ (~ 0.01). In that case, the effect of heating is negligible before the age of $t \sim 10^6$ years. This conclusion is probably true qualitatively for other heating mechanisms, too. Therefore, it confirms the earlier estimate that heating is negligible for younger stars such as the Vela and Crab pulsars.

3. DISCUSSION AND CONCLUSION

3.1. Comparison with Observation and Vortex Creep Theory

The latest theoretical cooling curves are compared with the data from the Einstein obsevatory in Refs. 3) and (5a). The most interesting points are summarized in Fig. 1. We note:

a) The observed upper limit for the Vela is below the standard cooling curves while it is above the non-standard curves with both charged pion and quark cores. Since a pulsar was observed in Vela, this is a strongest indication for the presence of non-standard cooling. This statement applies for other nuclear models with stiffer and softer equations of state.

b) The data points for the Crab, RCW 103, and 3C 58 are all consistent with the standard cooling theory. Therefore, if some of these points, e.g. RCW 103, are shown to be the actual temperature measurement, at least in these stars, there should be no "exotic" particles such as charged pion condensates or quarks.

c) The temperature upper limits for Cas A, Tycho and SN 1006 are all below the standard cooling curves. Since no pulsars were discovered nor any central activities detected within these SNRs, the indication may be that there are no neutron stars in these remnants. This view agrees with a Type I supernova scenario[18)].

d) In the presence of superfluidity (the dot-dashed curve (1)), heating is required in order to be consistent with the data point for PSR 1929+10. The vortex creep heating agrees with this interpretation.

We may point out here that magnetic fields are not included in the theoretical results shown in Fig. 1. It has been pointed out that the magnetic effects on neutron star cooling are minor[19)]. However, we shall reserve our final words until after these effects are carefully investigated. Also, the observed points will be lowered further if the stellar surface contains elements as light as hydrogen and helium (Romano in this volume). In that case, our conclusions a) and c) will be further enhanced.

The surface temperature of a neutron star in the Crab and Vela were estimated independently from the vortex creep theory and observed glitch data[17)] (see Alpar, Pines in this volume). These points are shown by a symbol ⊗. For the Crab, it is consistant with the standard cooling, while the Vela point is consistent with the non-standard cooling with a π^c core. Therefore, the vortex creep theory strongly supports the possibility that the Crab neutron star has no pion or quark core while the Vela contains a core of charged pion condensates. That is a theoretically reasonable possibility if the Vela neutron star is a little more massive than the Crab[3)].

3.2. Constraints on the Axion Mass

If some of the data points are the actual temperature detection, we can impose a strong constraint on the axion mass. For instance, there is a strong possibility that this applies to RCW 103[20)]. If so, from the constraint that the axion cooling should not be too fast to bring the cooling curves below the observed point for this source, we can place the upper limit to the axion mass, as $m_a < 0.001$ eV, implying that the cosmological constant $\Omega > 0.01$[15)]. That is, the axion contribution should be as significant as baryons. This will be important in the sense that no stronger constraints have been given by any other astrophysical means.

3.3. Conclusion

Theories of neutron star cooling and heating should remain as exciting subjects. Some of the reasons are summarized below.

a) The satellite programs planned in the near future, such as ROSAT, AXAF, LXAO and XMM, could potentially confirm the actual temperature measurement (not just upper limits) for sources such as RCW 103, and detect even some more new sources like RCW 103. That may very well lead to more conclusive and exciting results when compared with cooling theory.

b) Neutron star temperatures could be estimated independently with some other means, such as the vortex creep theory.

c) Heating theories could be tested by the observation of older pulsars, through e.g. EXOSAT, Space Telescope, ROSAT and XAO.

d) Stability of old neutron stars depends critically on their temperatures[21].

Therefore, it will be doubtlessly invaluable to continue further careful work in this area, including studies of i) effects of magnetic fields on cooling, temperature modulation over the surface and the URCA process in the interior, ii) other possibilities for non-standard cooling, and (iii) various other possible heating mechanisms.

We wish to thank many of our colleagues, especially Drs. K. Nomoto, N. Iwamoto, D. Pines, A. Alpar, M.J. Rees, G.G. Raffelt, R. Tamagaki, T. Takatsuka, T. Tatsumi, N. Shibazaki, and M.A. Ruderman, for valuable discussions. This work was supported in part by the NSF under grant AST-8602087 and by MONTS under grant 16404216.

REFERENCES

1) Tsuruta, S. and Nomoto, K., to appear in Phys. Reports (1987).
2) Tsuruta, S., Phys. Reports 56, 237 (1979).
3) Tsuruta, S., Comments on Ap. 11, 151 (1986).
4) See e.g. Maxwell, O.V., Ap. J. 231, 201 (1979).
5a) Nomoto, K. and Tsuruta, S., Ap. J. Lett. 250, L19 (1986).
5b) Nomoto, K. and Tsuruta, S., Ap. J. 312, Jan. 15 (1987).
6) Friedman, B. and Pandharipande, V.R., Nuc. Phys. A361, 502 (1981).
7a) Iwamoto, N., Nomoto, K. and Tsuruta, S., in preperation (1987).
7b) Tsuruta, S. and Nomoto, K., in preperation (1987).
8) Tatsumi, T., Prog. Theor. Phys. 69, 1137 (1983).
9a) Iwamoto, N., Phys. Rev. Lett. 44, 1637 (1980).
9b) Kiguchi, M. and Sato, K., Prog. Theor. Phys. 66, 725 (1981).
10) Takatsuka, T. and Tamagaki, R., Proceedings of the International Workshop on Condensed Matter Theories, San Francisco (1985).
11) Maxwell, O., Brown, G.E., Campbell, D.K., Dashen, R.F. and Manassah, J.T., Ap. J. 216, 77 (1977).
12) Muto, K. and Tatsumi, T., Neutron Stars, Kyoto, in press (1987).
13) Iwamoto, N., Phys Rev. Lett. 53, 1198 (1984).
14) Itoh, N., private communication (1986).
15) Tsuruta, S. and Nomoto, K., Observational Cosmology, in press (1987).
16) Shibazaki, N. and Lamb, F.K., preprint, 1986.
17) see e.g. Pines, D. and Alpar, A., Nature 316, 27 (1985).
18) See e.g. Nomoto, K., Ann. N.Y. Acad. Sci. 470, 294 (1986).
19) Hernquist, L., M.N.R.A.S. 213, 313 (1985).
20) Tuohy, R., Garmire, G.P., Manchester, R.N. and Dopita. M.A., Ap. J. 268, 778 (1983).
21) L. Lindblom, Ap. J., in press (1987).

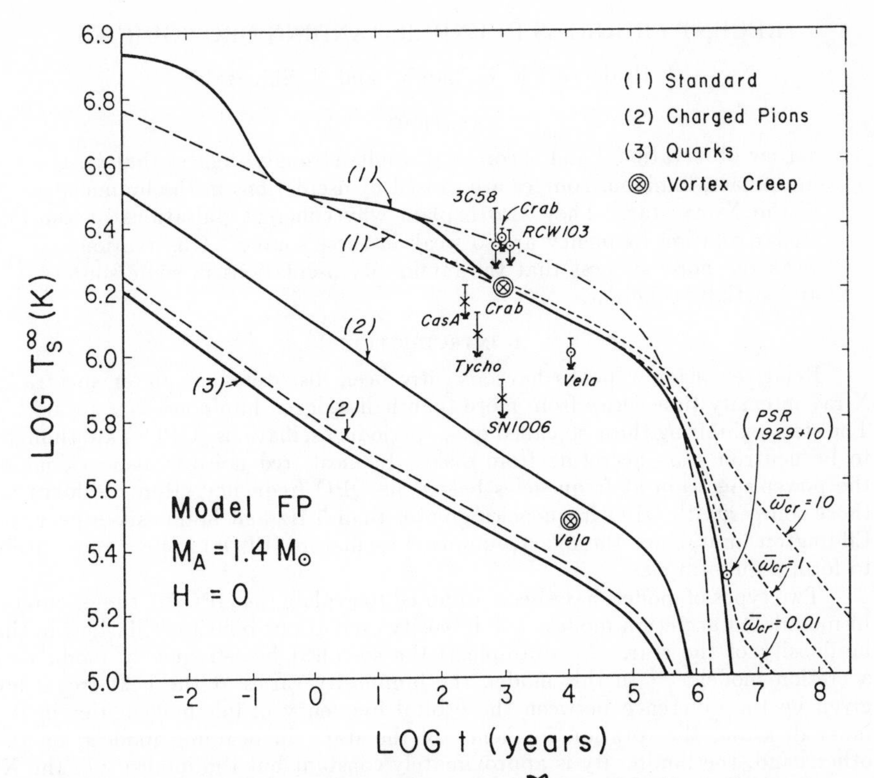

Fig. 1 - The observed surface temperature T_s^∞ as a function of age t. Standard (1) and non-standard (2)(3) cooling curves are compared with the oservational data and temperatures from the vortex creep theory. The circles refer to the observed temperature upper limits where point sources are detected, while crosses indicate true upper limits with no detected radiation. These points are accompanied by the error bars which are mostly due to the uncertainty in interstellar absorption. See the text for further details.

RECENT PROGRESS IN UNDERSTANDING QPO SOURCES

J. Brainerd,* F. K. Lamb,† and N. Shibazaki‡

ABSTRACT

New observational and theoretical results strongly suggest that QPOs in the fast, luminous sources are caused by oscillations in the luminosity of the X-ray star. They also explain why coherent pulsations at the stellar rotation frequency are so weak in these sources. Observations of weak red noise suggest that the luminosity oscillations in some sources are partially coherent.

1. INTRODUCTION

Relatively narrow peaks have recently been discovered in power spectra of X-ray intensity time series from more than half a dozen luminous X-ray stars.[1,2] The stars exhibiting these so-called quasi-periodic oscillations (QPOs) are thought to be neutron stars accreting from disks. In most, red noise is also evident in the power spectrum at frequencies below the QPO frequency. Here we focus on those sources with QPO frequencies greater than 5 Hz and luminosities near the Eddington limit, since these fast, luminous oscillators (FLOs) appear most likely to form a uniform class.

Two types of models have been proposed to explain the QPOs in these sources. In modulated accretion models, the intensity oscillations reflect oscillations in the luminosity of the star. An example is the so-called beat-frequency modulated accretion model.[3–6] In this model, the luminosity varies at the beat frequency given by the difference between the orbital frequency of inhomogeneities in the inner disk and the rotation frequency of the star. In beaming models, on the other hand, the luminosity is approximately constant but the intensity of the X-rays emerging in our direction varies. Examples of this type include beaming of radiation by luminous matter orbiting in the boundary layer at the surface of a nonmagnetic neutron star[7] and beaming by shock waves revolving at the magnetospheric boundary.[8] It has also been suggested that the QPOs are due to beaming of X-rays by disk oscillations[9,10] or self-luminous[9], scattering,[11] or obscuring[9] matter orbiting in the inner disk. We describe recent observational and theoretical results on central coronae which strongly suggest that the oscillations observed in the FLOs, at least, reflect variations in the accretion rate. We also discuss the implications of other recent observations which indicate that in some sources the total power in the oscillations greatly exceeds that in the red noise.

2. CENTRAL CORONAE

A recent cross-correlation analysis[12] of EXOSAT data has provided strong evidence for the presence of a central corona in the luminous QPO source Cyg X-2.

---

* Center for Space Science and Astrophysics, Stanford University, Stanford, CA 94305 and Lawrence Livermore National Laboratory.

† Departments of Physics and Astronomy, University of Illinois at Urbana-Champaign, Urbana, IL 61801.

‡ Space Science Division, NASA Marshall Space Flight Center, Huntsville, AL 35812.

This analysis shows that the QPOs observed in higher energy X-rays lag the QPOs in lower energy X-rays by a few milliseconds. When combined with modeling of the X-ray spectrum, this result indicates that the oscillating X-rays are Comptonized by passage through a scattering cloud of radius $\sim$ 200 km and electron scattering optical depth $\sim$ 3–10. This conclusion is consistent with earlier observational results[5] which indicated that central coronae with properties like these are present in luminous low-mass X-ray binaries, and with theoretical arguments[5] which suggested that heating of the surface of the inner disk should create such coronae. A somewhat smaller lag of the QPOs in higher energy X-rays has now also been reported in GX 5-1.[2] These results have profound implications for QPO models.

First, if the lag of the oscillations at higher energies is indeed caused by Comptonization, the QPOs must be imposed on the intensity well inside the scattering cloud, *i.e.,* at radii much smaller than $\sim$ 200 km. Otherwise the oscillations at higher energies would not lag those at lower energies. This argues against models in which the QPO frequency reflects the orbital frequency of self-luminous, obscuring, or scattering matter in the accretion disk, since the radius at which such matter must be located is $\sim$ 200 km for a QPO frequency of 20 Hz, as observed in Cyg X-2 and GX 5-1, or $\sim$ 600 km for a QPO frequency of 5 Hz, as observed in Sco X-1.

Second, a central corona with the properties indicated by these observations is very effective in reducing the amplitude of oscillations in the X-ray intensity produced by beaming. Recent analytical and Monte Carlo results[13] show that for a rotating δ-function beam at the center of a spherical cloud of scattering optical depth τ_c, the fractional rms variation seen by a distant observer is $\gamma_{rot} \approx 2^{1/2}/(1+\tau_c)$. For a more realistic axisymmetric beam with a polar intensity pattern of the form $I_0(\theta) \propto 1 + \cos n\theta$, the pulsation amplitude falls off with τ_c even more rapidly, as one might expect. For cloud optical depths $\tau_c \geq 5$, analytic and Monte Carlo results yield

$$\gamma_{rot} \approx 2^{1/2} \left| \frac{1}{4-n^2} \right| \left(\frac{1}{1+\tau_c} \right),$$

for n odd and $\gamma_{rot} \approx 0$ for n even. The total power at the rotation frequency is of course $\propto \gamma_{rot}^2$. These calculations assume that photons escape from the scattering cloud in a time short compared to the time required for the beam to rotate through an appreciable angle. If this is not the case, the power at the rotation frequency is further reduced by photon trapping.[14]

These results show that models[7–11] in which the luminosity of the source is approximately constant and the QPOs are produced by a rotating pattern of luminous, obscuring, or scattering matter face serious difficulties. The reason is that such beaming models have difficulty accounting for the relatively large observed amplitudes of the oscillations, which are typically $\sim$ 5–10% but sometimes reach 20% of the total flux,[2] if the beaming takes place in the interior of a central corona of optical depth $\sim$ 3–10 as indicated by the observational evidence discussed earlier.

The presence of a moderately dense central corona is *not* a difficulty for models — such as the beat-frequency modulated accretion model[4–6] — in which the

QPOs reflect oscillations in the *luminosity* of the star. In such models, the amplitude of the oscillations is unaffected by scattering as long as the time for photons to escape is small compared to the oscillation period.[14,15] For central coronae with the properties discussed above, this is the case.

These recent observational and theoretical results also suggest why coherent pulsations at the rotation frequency of the neutron star have proved to be so weak in the FLOs. (Current upper limits on rms pulsation amplitudes range from 10–30% in many low-mass X-ray binaries[16,17] to 0.3–0.8% in GX 5-1,[9] Sco X-1,[18] and Cyg X-2.[19]) For the luminosities observed and the field strengths $\sim 10^9$ G expected in these neutron stars, the magnetosphere is small, the radiation pressure is very large, and channeling of the accretion flow by the stellar magnetic field is likely to be only partially effective.[4,5] As a result of weak channeling and scattering of photons by the relatively dense plasma within the magnetosphere, the X-radiation emerges from the magnetosphere much more isotropically in these stars than in canonical accretion-powered pulsars. Also, the thick plasma torus expected around the small magnetosphere restricts the path for unimpeded propagation of X-rays to directions near the rotation axis, where modulation of the intensity by beaming and rotation is inherently weak. If the amplitude of the rotational modulation just outside the magnetosphere is no more than $\sim 20\%$ and the beams have, as expected, multiple lobes, scattering in a central corona of optical depth ~ 5–10 will reduce the amplitude seen by a distant observer to a few tenths of 1%. If the stellar rotation frequency is ~ 100 Hz or greater, photon trapping will reduce the observed amplitude still further.[14]

3. WEAK RED NOISE

In the FLOs so far discovered, the total observed power P_{QPO} in the oscillations is typically comparable to or greater than the total observed power P_{RN} in the red noise.[1,2] Recently, power spectral analyses have been extended to much lower frequencies than before. These analyses show that in some sources, such as Sco X-1[10,18,20,21] and the Rapid Burster,[2] the power ratio P_{QPO}/P_{RN} is at times ~ 10 or larger. The relative strength of the QPOs in these sources is a potential difficulty for any physical model, since if individual X-ray wave trains (pulses) last only a finite time, are uncorrelated, are positive at all times, and vary sinusoidally at the QPO frequency, P_{QPO}/P_{RN} is necessarily less than 0.5[5].

There are several ways in which this limit can be exceeded:[5,6] (1) the pulses may persist for a long time, shifting the power density in the red noise to frequencies below the observational bandpass without changing the actual power ratio; (2) the pulses may be highly nonsinusoidal; (3) some mechanism may act to suppress the mean value of the pulses; or (4) the pulses may be partially coherent. The last three possibilities actually enhance the total power in the quasi-periodic oscillation relative to the power in the red noise.

It is not possible to account for power ratios as large as 10 solely by a nonsinusoidal waveform, since to achieve such a ratio would require a waveform that has many very strong overtones, contrary to observation. Power ratios this large could be explained if most pulses last as long as 60 s. However, such long-lived pulses appear unlikely in an environment where most time scales are measured in milliseconds. Power ratios of 10 or more could also be explained if some mechanism 'fine tunes' the waveform to reduce its mean value by a factor of 3–10. This is not a possibility for models in which the oscillations are produced by self-luminous clumps that are occulted by the star or disk.[7,9] For oscillating disk,[9,10] occulting clump,[9] or scattering[11] models, the required 'fine-tuning' of the pulse waveforms places severe restrictions on the angles of inclination of these systems,[5] on the dynamics of the inner disk, and on the disk emissivity as a function of radius.

Perhaps a more promising possibility is that the pulses are partially coherent, in the sense that the oscillating components are partially correlated from one pulse to another.[6,22] For example, the QPO/RN power ratio would be ~ 10 if pulses occur at random times but the oscillating components of some seven groups of pulses each containing $\sim 7\%$ of the total number of pulses are correlated at any one time.[6] Figure 1 shows the spectrum in this case, which agrees qualitatively with the power spectrum observed in Sco X-1.

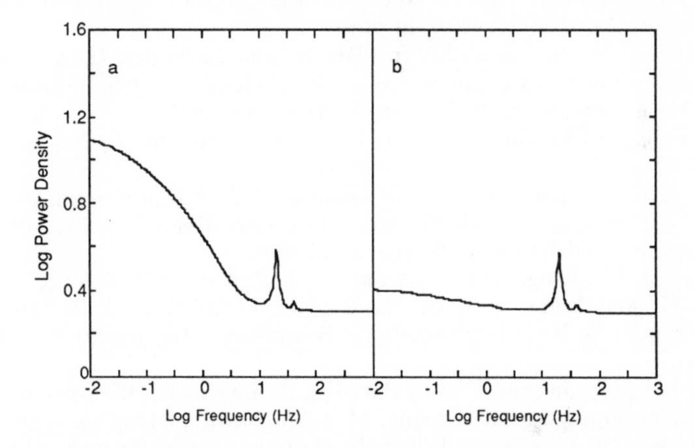

FIG. 1.—Model power spectra showing the effects of pulse-to-pulse correlations in the phases of the oscillating components of pulses. (a) No correlations. (b) Degree of correlation described in the text. Note the relative weakness of the red noise in (b). From ref. 6.

In the context of the beat-frequency modulated accretion model, this type of correlation might occur in several ways:[6] as the result of fragmentation of density or magnetic field fluctuations just outside the magnetospheric boundary layer into many smaller fluctuations that drift into the boundary layer, as the result of persistence of randomly occuring local excesses of plasma within the boundary layer, or interaction of the magnetosphere with density patterns in the inner disk that persist.[23]

4. CONCLUSIONS

Recent observations have provided strong evidence that the QPOs in the fast, luminous sources are produced inside moderately dense central coronae. These observations appear to pose serious difficulties for models in which the QPO frequency reflects the orbital frequency of matter in the inner disk, since the observed frequencies correspond to radii comparable to or larger than the observed dimensions of the coronae. Moreover, recent theoretical work has shown that such coronae strongly suppresses QPOs produced by luminous, obscuring, or scattering matter orbiting inside. These results taken together strongly suggest that the QPOs in these sources reflect oscillations in the luminosity of the X-ray star. They also explain why coherent pulsations at the stellar rotation frequency are so weak in these sources. Observations of weak red noise suggest that the luminosity oscillations in some sources are partially coherent.

It is a pleasure to thank W. C. Priedhorsky for numerous helpful discussions. F.K.L. also thanks the John Simon Guggenheim Memorial Foundation for generous support while part of this work was completed. This research was also supported in part by NASA grant NSG 7653 and NSF grant PHY 86-0037 at Illinois.

5. REFERENCES

1. van der Klis, M., in *Accretion onto Compact Objects,* Proceedings of the Tenerife Workshop, April 21-25, 1986, in press.
2. van der Klis, M., in *Variability in Galactic and Extragalactic X-Ray Sources,* Proceedings of the Meeting in Como, Italy, October, 1986, in press.
3. Alpar, M. A., and Shaham, J., *Nature,* **316**, 239 (1985).
4. Lamb, F. K., Shibazaki, N., Alpar, M. A., and Shaham, J., *Nature,* **317**, 681 (1985).
5. Lamb, F. K., in *The Evolution of Galactic X-Ray Binaries,* ed. J. Trümper, W.H.G. Lewin, and W. Brinkmann (Dordrecht: Reidel), p. 151 (1986).
6. Shibazaki, N., and Lamb, F. K. *Ap. J.,* in press.
7. Hameury, J. M., King, A. R., and Lasota, J. P. *Nature,* **317**, 597 (1985).
8. Morfill, G., and Trümper, J., in *The Evolution of Galactic X-Ray Binaries,* ed. J. Trümper, W.H.G. Lewin, and W. Brinkmann (Dordrecht: Reidel), p. 173 (1986).
9. van der Klis, M., Jansen, F., van Paradijs, J., Lewin, W.H.G., van den Heuvel, E.P.J., Trümper, J., and Sztajno, M. 1985, *Nature,* **316**, 225.
10. van der Klis, M., Stella, L., White, N., Jansen, F., and Parmar, A. N., *Ap. J.,* in press.
11. Boyle, C. B., Fabian, A. C., and Guilbert, P. W., *Nature,* **319**, 648 (1986).
12. Hasinger, G., in *IAU Symposium 125, The Origin and Evolution of Neutron Stars,* ed. D. J. Helfand and J. H. Huang (Dordrecht: Reidel), in press.
13. Brainerd, J., and Lamb, F. K., *Ap. J. (Letters),* submitted.
14. Kylafis, N. D., and Klimis, G., in preparation.
15. Chang, K. M., and Kylafis, N. D., *Ap. J.,* **265**, 1005 (1983).
16. Leahy, D. A., Darbro, W., Elsner, R. F., Weisskopf, M. C., Sutherland, P. G., Kahn, S., and Grindlay, J. E., *Ap. J.,* **266,** 160 (1983).
17. Mereghetti, S., and Grindlay, J. E., *Ap. J.,* in press.
18. Middleditch, J., and Priedhorsky, W. C., *Ap. J.,* in press.
19. Norris, J. P., and Wood, K. S., *Ap. J.,* submitted.
20. van der Klis, M., and Jansen, F. in *The Evolution of Galactic X-Ray Binaries,* ed. J. Trümper, W.H.G. Lewin, and W. Brinkmann (Dordrecht: Reidel), p. 129 (1986).
21. Priedhorsky, W., Hasinger, G., Lewin, W.H.G., Middleditch, J., Parmar, A., Stella, L., and White, N., *Ap. J. (Letters),* in press.
22. Elsner, R. F., Shibazaki, N., and Weisskopf, M. C., *Ap. J.,* submitted.
23. Zurek, W. H., and Benz, W., *Ap. J.,* **308**, 123 (1986).

New Developments in Studies of Compact X-ray Binaries

Jonathan E. Grindlay
Harvard Smithsonian Center for Astrophysics
60 Garden St., Cambridge, MA 02138

ABSTRACT

Several recent developments, both observational and theoretical, on the study of x-ray binaries and the compact objects they contain are discussed. The recent discovery of the first binary periods for the globular cluster x-ray sources has stimulated a new model for their origin. As a variant of the "standard" tidal capture origin model, this predicts an enhanced number of neutron stars in globular clusters. Long term timing studies of x-ray binaries may be consistent with many of these systems, primarily x-ray burst sources, being in fact hierarchical triple systems. Finally, the radio studies of Cyg X-3 and other x-ray binaries suggest that non-thermal processes are as important, energetically, as accretion processes in these systems.

1. INTRODUCTION

With the recent EXOSAT observations of compact x-ray sources on timescales much longer than possible previously and with more sensitive and sustained observations of compact x-ray sources at radio through optical wavelengths, a number of new discoveries have been made. We shall concentrate on only a few which further elucidate the nature and origin of the so-called low mass x-ray binaries (that is, compact x-ray binaries in which the donor star mass is less than that of the neutron star) as well as summarize some recent work which indicates that non-thermal processes may also be important in these systems.

We begin with an update on the process of tidal capture for the formation of x-ray binaries in globulars since several recent calculations have indicated this is more complex than previously recognized. We then turn to the recent discoveries of the binary periods for the first two globular cluster x-ray sources: those in NGC 6624 and in M15. These apparently disparate systems, with periods of 11.4 minutes and 8.5 hours (respectively), may in fact be linked by a common evolutionary scenario which we outline. The tidal capture of a triple companion by a compact binary is then briefly discussed. It is possible that a significant number of x-ray binaries both in and out of globular clusters may be hierarchical triples which, if confirmed, would allow new constraints on the origin and nature of these systems (possibly arising within globulars since disrupted) to be derived. One possible triple is the spectacular non-thermal radio emitting x-ray binary Cyg X-3. Other examles are also mentioned in which it seems that the energy released into non-thermal particles must be comparable with the x-ray (thermal) luminosity. Our conclusions and directions for future work follow.

2. NEW RESULTS AND MODELS FOR GLOBULAR CLUSTER X-RAY SOURCES

Globular clusters contain both high ($\geq 10^{36}$ erg/sec) and low ($\leq 10^{34.5}$ erg/sec) x-ray sources which are probably due to the tidal capture of white dwarfs and neutron stars, respectively, by normal main sequence stars in the dense cores of clusters (cf. Grindlay 1985, 1986 for reviews). The process of tidal capture (Fabian et al 1975, Press and Teukolsky 1977), whereby a binary can be formed by the dissipation of orbital energy of the pair with the raising of a tide on one or both of the stars during a close encounter, results in capture into a compact binary system if the two stars encounter each other within approximately 3 stellar radii. Recently, these calculations have been extended, and calculations by McMillan, McDermott, and Taam (1987) have also included a more realistic treatment of the effects on the structure of the interacting stars by the tidal dissipation of energy in their encounter. In spite of these improvements, the maximum tidal capture impact parameter is still found to be approximately 3.3 stellar radii (McMillan et al 1987), so that the various inferences drawn for the number of neutron stars required in globular clusters to produce the numbers of sources observed (e.g. Lightman and Grindlay 1982, Grindlay 1987) are essentially unaffected by the new capture cross section results, though they may be enhanced by other considerations (cf. section 2.3 below).

The study of x-ray binaries in globular clusters has taken a great quantum leap forward with the recent discoveries of the first binary periods for members of this class of objects. An x-ray periodicity of 11.4 minutes was found (Priedhorsky, Stella and White 1986) for the archetype globular cluster source 4U1820-30 in NGC 6624 using the EXOSAT x-ray observatory. Within two months, an 8.5 hour period was announced from the optical photometry (Ilovaisky et al 1986) and absorption line spectrum (Naylor et al 1986) of the proposed (Auriere et al 1984) optical counterpart, AC211, for the bright x-ray source 4U2127+12 in M15. The period and the optical counterpart, the first for any globular cluster source, was confirmed by a detection of the periodicity in archived x-ray data for this source (Hertz 1986). We have incorporated these discoveries into a model (Bailyn and Grindlay 1987a; hereafter referred to as BG) for the formation and evolution of a class of tidal capture x-ray binaries in globulars which is summarized below (section 2.3).

2.1 Shortest Binary Period (11 min) Known: Discovered in a Globular Cluster

The x-ray source 4U1820-30 in the globular cluster NGC 6624 was among the first to be found in a globular cluster and was the first identified source of x-ray bursts (Grindlay et al 1976); it has now been found to be the shortest period binary system known (cf. Stella et al 1987). As the best determined period (685.011836(32) sec; cf. Morgan and Remillard 1987) for any suspected binary in a globular cluster, it is certainly also the most convincing detection of a binary of any sort in a globular cluster. The a posteriori detections of the period in 10 years of historical x-ray data on the source (cf. Morgan and Remillard 1987) showed that the period remains constant to within ~0.1 msec per year and greatly strengthened the original claim of Stella et al that the period must be orbital, not rotational (which, given the large mass transfer rate inferred from the x- ray luminosity of ~5 x 10^{37}erg/sec, would lead to a much larger rate of change of period as the neutron star spins up). The ultra-short binary period requires the mass- losing star be a degenerate dwarf to fit in so small an orbit. A helium white dwarf of mass 0.06 $M_\odot$ orbiting a 1.4 $M_\odot$ neutron star could supply the necessary mass transfer by Roche lobe overflow at a rate of 1 x $10^{-8}M_\odot$/yr (as required by the x-ray luminosity) by virtue of the angular momentum losses expected from the gravitational radiation from such a system (Stella et al 1987). The orbital period should increase at a rate only a factor of ~2 below the current upper limits (Morgan et al 1987) for such a system. This should be easily testable with the recently launched (February 1987) Japanese X-ray Satellite ASTRO-C.

It has been proposed (Verbunt 1987) that 4U1820-30 is the result of a collision of a red giant (containing a white dwarf core) and a neutron star in the cluster, but BG show that this is relatively less likely than an alternative mechanism (described below) and in any case subject to a possible mass transfer instability which would cause the white dwarf and neutron star to coalesce. The system parameters imply that the x-ray lifetime for the system in its current (second) mass transfer phase (cf. section 2.3 below) has been relatively short ($\lesssim 10^7$ years). This suggests that the total number of such systems in a dormant (i.e. not transferring mass) state must be large and that the corresponding number of capture binaries involving neutron stars is also large.

2.2 The Peculiarly Variable, UV-Bright 8.5 hour X-ray Binary in M15

The x-ray source 4U2127+12 in the globular cluster M15 is peculiar in having a relatively luminous optical counterpart with absolute magnitude M_v = ~ 0.5 (uncertain because of variability and spectrum, not distance) which is uv-bright and variable (Auriere et al 1984). The optical variations are apparantly modulated at the 8.5 hour period (Ilovaisky et al 1986), whereas the x-ray flux was apparantly similarly modulated in the 1977 HEAO-1 data (Hertz 1986) but not the 1984 EXOSAT observations (Callanan et al 1986).

It is likely (Grindlay 1985, 1986) that the M15 source is much more luminous than suggested by its relatively modest x-ray flux (~5-10 UFU), which would indicate a luminosity of ~$10^{36.4}$erg/sec. Rather, its luminosity is probably in excess of 10^{38}erg/sec and the source is surrounded by a so-called accretion disk corona (ADC) which prevents us from seeing the compact object directly. Fabian et al (1987) have also recently invoked an ADC to explain the relatively brighter (than in other low mass x-ray binaries) optical to x-ray flux ratio as due to excess optical luminosity generated in an ADC that is physically larger because of the decreased ionization (and hence larger effective ionization radius) in the low metallicity of the binary in a metal poor cluster such as M15. Callanan et al (1986) also argue that the lack of bursts is due to low metallicity, and Fabian et al (1987) suggest that the system inclination is relatively low because of the lack of eclipses in the EXOSAT data. However, the large optical modulation would suggest that the system is at relatively large inclination if the modulation is due to the changing aspect of an ADC which is limited in size and shape by the Roche geometry of the system. In this case, a high luminosity x-ray source in which much of the intrinsic luminosity is scattered out of our line of sight (Grindlay 1985) is indicated.

A high luminosity is also suggested for 4U2127+12 in M15 by the apparent lack of burst activity for this source, by the anomalously low x-ray to optical flux ratio (only ~20 vs. ~10^3 for typical low mass x-ray binaries), and by the unusual variable low energy x-ray variability of the source which suggests variable absorption around the system. The latter variability was discovered for the M15 source by Hertz and Grindlay (1983) (cf. Figure 3 in their paper) but has been found to be generally present in a re- analysis of the bulk of the Einstein observations of M15. In Figure 1 below, taken from Garcia and Grindlay (1987), we show the general correlation evident between the low energy absorption (expressed as the equivalent neutral column density N_H) and the hard x-ray flux from the source which measures the overall x-ray luminosity: when the luminosity increases, the apparent low energy absorption decreases.

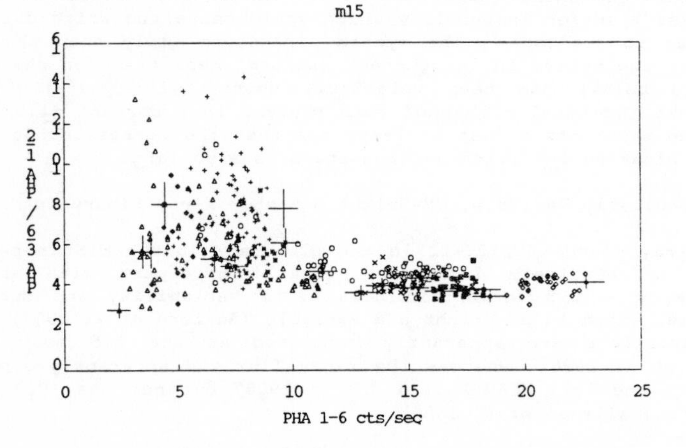

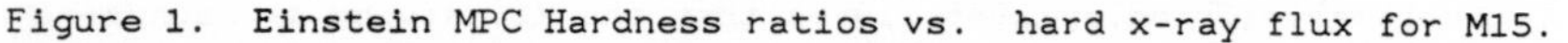

Figure 1. Einstein MPC Hardness ratios vs. hard x-ray flux for M15.

The apparantly variable low energy absorption indicates that there is substantial material surrounding the binary system which can be partially ionized, and made more transparent, when the x-ray flux increases. This interpretation, rather than partial eclipses (Grindlay 1985), is more likely in view of the totality of the x-ray and optical data now available. We note that variable absorption is also indicated by the EXOSAT results of Callanan et al (1986) and that, despite their claim that low energy absorption increases with flux (which their observations seem to show on long time scales--i.e between widely separated observations), their data appear to show the same anti-correlation evident in the Einstein data on the shorter timescales comparable with the binary period.

If the system is accreting at or near the Eddington limit to produce the inferred high luminosity, and is surrounded by the inferred circumsource absorbing material, then mass loss from the system in an outflowing wind is expected. This is the likely explanation for the systemic velocity of ~- 150 km/sec in the λ4471 He I absorption line reported by Naylor et al (1986), rather than their suggestion of ejection of the binary from the cluster core since this would occur in only $\sim 10^3$ years. A marginal confirmation of the blueshifted helium line velocity was obtained by Grindlay and Huchra (1987) in an average of three short exposure MMT spectra obtained on three successive nights in September 1986. It is encouraging that the measured velocity is close to that expected from typical wind velocities in giants and to what might be expected from around the Roche surface of the binary. The velocity variation of ~45 km/sec at the 8.5 hour orbital period (Naylor et al 1986) would then be due to the variable potential gradient, and thus velocity of the wind, around the Roche surface as it rotates at the binary period relative to our line of sight. A detailed treatment of the mass loss expected and its expected velocity variation is needed.

2.3 Dormant Tidal Capture Model

Formation of compact binaries in globular clusters by tidal capture does not necessarily mean production at the outset of a mass transfer binary. Instead, if the binary is formed by an encounter with closest approach less than the maximum ($\sim 3.3R_*$) for capture but greater than the maximum for Roche lobe overflow of the secondary (with radius R_*), then a "dormant" x-ray binary will be formed. Mass transfer and x-ray emission will only begin when the captured star evolves off the main sequence and expands to fill its Roche lobe. This is the key idea in the evolutionary scenario of BG for the 11.4 minute binary system and possibly also the source in M15. In this case, a $0.8M_\odot$ main sequence star is captured into an orbit which circularizes at a separation $\geq 3R_{TAMS}$ (i.e. it encountered the neutron star with a separation of half this value), where R_{TAMS} is the radius of the star when it (later) has evolved just off the terminal age main sequence and is perhaps ~1.5-2 times the radius R_* of the star on the main sequence. Hence the required range of closest approach distances is $\sim 2.3\text{-}3.3R_*$, or for approximately one-third of tidal capture encounters (since the cross section scales linearly with closest approach distance due to gravitational focusing).

We have shown (BG) that when mass transfer finally begins, it will be unstable if it proceeds via an accretion disk and if the mass ratio $q = m_2/m_1 \geq 0.67$, where the subscripts 1 and 2 denote the primary and secondary masses. This critical mass ratio would be exceeded for stars more massive than the present $0.8M_\odot$ turnoff mass which are captured by $1.2\ M_\odot$ neutron stars, where such a neutron star mass is expected if the neutron stars in globular clusters are predominately the result of the accretion induced collapse of (massive) white dwarfs (Taam and Van den Heuval 1986, Grindlay 1987). The unstable mass transfer will result in accretion onto the neutron star at approximately the Eddington limit as well as accumulation of the excess mass (which cannot accrete) in the disk and in a surrounding circumstellar shell. This will lead to the formation of a common envelope and an evolution of the binary which is very difficult to predict in detail or in timescale. It is likely, however, that ultimately the circumstellar envelope will cause enough frictional drag on the secondary that it will spiral into the primary (probably finally on a rapid timescale) to a separation where the orbital energy of the binary exceeds the binding energy of the common envelope and atmosphere of the secondary exterior to the remaining core of the secondary. Since the secondary star is assumed to have (just) evolved off the main sequence, it will have developed a small ($\sim$0.1 $M_{\odot}$) white dwarf core. Thus the common envelope phase is expected to end with a detached white dwarf- neutron star binary in a $\sim$31 minute orbit (BG). Gravitational radiation will then spiral the detached low mass white dwarf towards the neutron star primary on a timescale of 10^7 years after which it will begin a second stage of mass transfer onto the primary and a luminous ($\sim10^{38}$ erg/sec) x-ray source will again be present. This is the stage at which the 11.4 minute binary is now found; the 8.5 hour binary in M15 may well be the first phase (i.e. post main sequence star secondary, entering a common envelope stage) of this generic tidal capture evolution.

3. X-RAY TRIPLES ?

A compact binary in a globular cluster should be an effective target for the tidal capture of a third star, as suggested by Grindlay (1985, 1986). This process has now been partially calculated by McMillan (1986) and is being studied in greater detail by Bailyn, Grindlay and McMillan (1987). Preliminary results suggest that at least 3% of the collisions of a third star with a compact binary at a minimum separation of at least three times the binary semi-major axis result in a stable hierarchical triple system. Such systems must have a ratio of inner to outer semi-major axes of at least $\sim$3 for dynamical stability. In this case, and in fact out to separation ratios of $^{3}0$, the outer binary is still "hard" in the frame of the globular cluster (i.e. its orbital velocity exceeds the central velocity dispersion of the cluster) and it will survive.

A number of low mass x-ray binaries have been found to have long-term variations with apparent periods in the vicinity of $\sim$200 days (cf. Priedhorsky and Holt 1987 for a recent review). The discovery now of an 11 minute binary period for one of these (the NGC 6624 source), for which there was a previously discovered (Priedhorsky and Terrell 1984) long- term period of 176 days again suggests the system may have a hierarchical triple companion (cf. Grindlay 1986). If the 176 day period is due to precession of the 11.4 minute orbit, an outer orbit with a period of $\sim$15 hours is required (cf. Bailyn and Grindlay 1987b). However, as discussed in the evolutionary scenario of BG, the binary period for 4U1820-30 may have been originally $\sim$9 hours in which case a $\sim$15 hour triple period would not be stable; the resolution may be (Bailyn 1987) that the longer (i.e. $\geq$40 hour) periods required in for stability of the hierarchical triple companion to a $\sim$9 hour binary could still give rise to the 176 day precession period when the binary has shrunk to 11.4 minutes if the system is not co-planar (as was assumed in Bailyn and Grindlay 1987b). Indeed, non-coplanarity would be expected for a tidally captured triple companion.

We note that the suggestion by Priedhorsky and Holt (1987) that the long term periods may arise from a disk feedback mechanism would appear difficult given the wide range of disk sizes and physical conditions in the systems showing long-term periods: from the ultra compact system in NGC 6624 to Aql X-1 with an orbital period of 1.3 days and a long period of 1.19 years. If the periods are precession of hierarchical triples formed by tidal capture, however, then a natural outer period is given by the period appropriate to the capture of a giant at ~3 stellar radii, or of order 1 day. The binary period, on the other hand, is that appropriate to the capture of a main sequence star at ~3 of its radii (a giant-neutron star binary could not subsequently capture another giant or a main sequence star and remain stable). Thus if the long term periods are due to precession of a triple, where the precession period is then $P_{outer} \equiv (P_3)^2\}/P_2$ and P_3 is the period of the third star and P_2 is the binary period, then the precession period is expected to be "naturally" a factor $[R_3 / R_2]^3$ times the orbital period, where the R values are the radii of the outer and inner stars. This ratio is then 10^{3-4}, or approximately that observed for the bursters summarized by Priedhorsky and Holt(1987), for giant and main sequence radii in ratios of ~10-30 as expected in globular clusters.

Perhaps the best bet for a hierarchical triple is the 4.8 hour binary Cyg X-3, which continues to show a 4.95 hour period in its low-amplitude radio flares (Molnar et al 1987). The radio and x-ray properties of Cyg X-3 are reviewed by Molnar (1986), in which a discussion of our recent determination of the expanding-source nature of the Cyg X-3 (mini-) flares and the VLBI detection of the expanding jet in Cyg X-3 is also given. Preliminary evidence for this discrepant radio period was presented by Molnar, Reid and Grindlay (1984, 1985) but this has been strengthened now by the addition of considerably more observations. In addition, two possible examples have been found of the phase "glitches" expected (Bailyn 1987) in a triple whereby the phase of the beat period (4.95 hours) "catches up" twice per outer orbit (4.2 days) with the orbital period (4.8 hours). If further examples of the phase glitches can be detected (two are expected each outer orbit of ~4.2 days), the triple nature of Cyg X-3 would be confirmed. In this case, the mass ratios and relative inclinations of the system can be determined from a careful analysis of the relative amplitudes and glitch timings of the radio light curve.

4. NON-THERMAL PROCESSES IN X-RAY BINARIES 4.1 Cyg X-3

4.1 Cyg X-3

It is clear that the quiescent radio emission from Cyg X-3 is due to a succession of "mini" radio flare events, each with the characteristic frequency vs. time behaviour of an expanding synchrotron source. The radio data, including our VLBI observations of Cyg X-3 which resolve the extended and probable jet structure of the low-level flare emission regions, are summarized by Molnar (1986). The energy input per flare into relativistic electrons is model dependent but is of order 10^{42} ergs. Time averaged, this then represents an energy input rate into non-thermal particles of some 10^{38} erg/sec, or comparable to the Eddington limit if the system contains a neutron star and in any case comparable to the total x-ray luminosity. Thus the production of non-thermal particles is potentially of major importance in the overall energetics of this high luminosity x-ray binary.

The giant radio flares, which recur approximately periodically once per year, are the extreme manifestations of the non-thermal energy release in the system. The period of the flares is again commensurate with being a precession period of a hierarchical triple system, with outer period ~4.2 days to give the 4.95 hour beat period (radio mini-flares) with the 4.79 hour orbital period (x-ray emission) as mentioned above. In this case, the giant flares may be triggered by the periodic change of inner orbital eccentricity, and thus mass transfer, due to the perturbing effects of the outer companion. This could lead to build up of a large amount of material in the disk (which is unable to accrete since the rate is already Eddington) that is released in a violent instability into the jets to provide the particle injection for the giant radio outbursts.

4.2 GX13+1, etc.

Cyg X-3 is not alone as a variable radio emitter among the low mass x-ray binaries. Sco X-1 is of course a classic double-lobed radio source, with variable emission and expansion observed in the lobes (Fomalont et al 1986), and a significant number of other bright low mass x-ray binaries have been detected as variable point radio sources.

In a recent survey of a number of low mass x-ray binaries in and out of globular clusters with the VLA (Grindlay and Seaquist 1986), we recently found that the bright galactic bulge source GX13+1 is a low-level radio emitter, like a number of other of the "GX sources". In a follow-up detailed radio vs. x-ray study, Garcia et al (1987) conclude that the radio and x-ray emission are not correlated on short timescales at least but that the radio emission may be consistent with a model proposed by Priedhorsky (1986) whereby the radio arises from spin-down torques exerted by the rapidly rotating neutron star on the surrounding accretion disk during periods of relatively lower accretion. In fact, Garcia (1987) has found that there appears to be a significant correlation between those x-ray binaries showing the relationship between their hardness ratio and x-ray flux (cf. Figure 1 above) that is exhibited by Sco X-1, where the relationship defines a double valued curve (as opposed to the single valued behavior for the M15 source shown in Figure 1), and their detectability as variable radio emitters. In the case of GX13+1, the variable radio emission again implies that the luminosity in non-thermal particles may again be comparable to the total x-ray luminosity. This would be consistent with the Priedhorsky (1986) model for the overall energetics, as mentioned above. It is not clear, however, what is the particle acceleration mechanism or site (e.g. an accretion disk corona ?).

5. CONCLUSIONS

We have summarized a number of recent developments for the origin and nature of compact x-ray binary systems. The globular cluster sources offer particularly attractive possibilities for probing these questions since their luminosities and progenitor systems are particularly well constrained. The burgeoning field of studying the other sources of emission in these systems, such as their radio and non-thermal particle production, deserves considerably more study. It appears that the energy released in high energy particles is an appreciable fraction of the total luminosity (in x-rays) and that high energy particles may therefore be dynamically important for the structure of winds and coronae around compact x-ray binaries. In certain systems, such as Cyg X-3 and Sco X-1, the particle production appears to be confined to jets and may therefore resemble the processes occuring in the much more luminous active galactic nuclei. The detailed study of both spectra and temporal variability may allow the trigger and injection mechanisms to be specified; it is likely that in some of these objects at least the perturbative effects of a third companion star are important.

This work was supported in part by NSF grant AST-84-17846 and NASA grants NAGW-624 and NAS 8-30751.

REFERENCES

Auriere, M., Le Fevre, O., and Cordoni, J. 1984, Astron. Astrophys., **138**, 415.
Bailyn, C. and Grindlay, J., 1987a, Ap.J. (Letters), in press.
Bailyn, C. and Grindlay, J., 1987b, Ap. J., **312**, 748.
Bailyn, C., Grindlay, J., and McMillan, S. 1987, in preparation.
Bailyn, C. 1987 in preparation.
Callanan, P. et al 1986, MNRAS, in press.
Charles, P., Jones, D. and Naylor, T. 1986 Nature, **323**, 426.
Fabian, A., Pringle, J. and Rees, M., 1975, MNRAS, **172**, 15P.
Garcia, M. and Grindlay, J., 1987, in preparation
Fabian, A., Guilbert, P. and Callanan, P. 1987, preprint.
Grindlay, J. E., 1984, Adv.Sp.Res., **63**, No. 10, 19.
Grindlay, J. E., 1985, in Proc. US-Japan Seminar on Galactic and Extragalactic Compact X-ray Sources, Y. Tanaka and W. Lewin, eds., ISAS, Tokyo, 215.
Grindlay, J. E., 1986, in The Evolution of Galactic X-ray Binaries, J. Trumper, W. Lewin and W. Brinkman, eds., NATO ASI Series, **167**, 25.
Grindlay, J. E., 1987, in IAU Symposium 125. Origin and Evolution of Neutron Stars, D. Helfand and J. Huang, eds., Reidel, Dordrect, in press.
Grindlay, J. E. et al., 1976, Ap.J. (Letters), **205**, L127.
Hertz, P., 1986, IAU Circular No. 4272.
Hertz, P. and Grindlay, J., 1983, Ap.J., **275**, 105.
Ilovaisky, S. Chevalier, C., Auriere, M. Angebault, P., 1986, IAU Circular No. 4263.
Lightman, A. P. and Grindlay, J. E., 1982, Ap.J., **262**, 145.
McMillan, S., 1986, Ap.J., **306**, 552.
McMillan, S., McDermott, P. and Taam, R., 1987, preprint.
Molnar, L. 1986, in The Physics of Compact Objects, Proc. Tenerife Mtg., (K. Mason, M. Watson, and N. White, eds.), Springer-Verlag, p. 313.
Molnar, L., Reid, M., and Grindlay, J. 1984, Nature, **310**, 662.
Morgan, E. and Remillard, R., 1987, preprint.
Priedhorsky, W. 1986, Ap. J., in press.
Priedhorsky, W. and Terrell, J., 1984, Ap.J. (Letters), **284**, L17.
Priedhorsky, W. and Holt, S. 1987, preprint.
Press, W. and Teukolsky, S., 1977, Ap.J., **213**, 183.

Statler, T., Ostriker, J. and Cohn, H., 1987, Ap.J., in press.
Stella, L., Priedhorsky, W. and White, N., 1987, Ap.J. (Letters), **312**, L17
Taam, R. and Van den Heuvel, E., 1986, Ap.J., **305**, 235
Verbunt, F., 1987, Ap.J. (Letters), **312**, L23.

ACCRETION-INDUCED COLLAPSE OF WHITE DWARFS

Ken'ichi Nomoto

Department of Earth Science and Astronomy
College of Arts and Sciences
University of Tokyo
Meguro-ku, Tokyo 153, Japan

ABSTRACT

The origin of neutron stars in several low mass X-ray binaries and binary radio pulsars has been suggested to be accretion-induced collapse of white dwarfs. To examine this scenario, we first discuss under what conditions white dwarfs will collapse to form neutron stars rather than explode. The outcome of the evolution of accreting white dwarfs is summarized as functions of accretion rate and the initial mass of the white dwarf. We then discusss some possible evolutionary scenarios that lead to accretion-induced collapse of white dwarfs and apply them to radio binary pulsars.

1. INTRODUCTION

The final fate of accreting white dwarfs will be either thermonuclear explosion or collapse, if the white dwarf mass grows to the Chandrasekhar mass. Recent developments have shown a very good agreement between the exploding white dwarf models (the carbon deflagration model where the white dwarf is disrupted completely with no neutron star remnant left behind) and the observed features of Type Ia supernovae (Nomoto 1986a,b; Woosley and Weaver 1986b). Though the exact precusor systems are not known yet, this success indicates that some accreting white dwarfs actually evolve to the Chandrasekhar mass.

On the other hand, there are indications that neutron stars in some low mass X-ray binaries and binary radio pulsars were formed by accretion-induced collapse of white dwarfs (van den Heuvel 1984). The strong argument for the accretion-induced collapse model is as follows (Taam and van den Heuvel 1986):

Some X-ray pulsars (Her X-1, 4U1626-67, 1E2259+59, GX1+4) are supposed to have magnetic field larger than 10^{11} G to account for X-ray pulsations. The presence of such a strong magnetic field implies that the age of the neutron star is shorter than 2×10^7 yr. On the contrary, the age of the binary system as inferred from the evolutionary timescale of the low mass companion is longer than $10^8 - 10^9$ yr. This implies that the neutron star must have formed from accretion-induced collapse of a white dwarf rather recently. This is also the case for the binary radio pulsar PSR 0820+02.

Some low mass X-ray binaries and QPOs are thought to have magnetic field less than 10^{10} G. However, the presence of the relatively weak magnetic field in old systems may not be an argument for the accretion-induced collapse model. This is because recent discoveries of a very old white dwarf companion in the radio binary pulsars PSR 0655+64 (Kulkarni 1986) and 1855+09 (Wright and Loh 1986) indicate that the magnetic field of neutron stars decays very slowly after it becomes weaker than $\sim 10^{10}$ G.

The questions adressed in this review are:

1) Under what condition will white dwarfs collapse to form neutron stars rather than explode?
2) What kind of binary systems can actually undergo accretion-induced collapse? Is it consistent with the binary parameters of radio binary pulsars?

2. THE FATE OF WHITE DWARFS AS A FUNCTION OF MASS AND ACCRETION RATE

Possible models for the accretion-induced collapse so far presented involve solid CO white dwarfs, in which carbon and oxygen may or may not have chemically separated (Canal et al. 1980; Isern et al. 1983) and ONeMg white dwarfs (Nomoto et al. 1979). The CO white dwarfs could either explode or collapse, depending on the conditions of the white dwarfs and binary systems in which they are formed. Chemical separation between carbon and oxygen may occur but not necessarily lead to a complete differentiation. A siginificant fraction of carbon (> 10 %) is expected to coexist with oxygen in the the central region (Iyetomi and Ichimaru 1987). If the carbon abundance is larger 10 %, carbon deflagration will be induced. Therefore, it is worth determinig the critical condition for which a carbon deflagration induces collapse rather than explosion. In the following we focus on the CO white dwarfs in which carbon is uniformly distributed with a mass fraction of 0.5 irrespective of solid or liquid.

2.1 Effects of Accretion and Carbon Ignition Density

Isolated white dwarfs are simply cooling stars. In close binary systems they evolve differently because mass accretion from their companion provides gravitational energy that rejuvenates them. The gravitational energy released at the accretion shock near the stellar surface is radiated away and does not heat the white dwarf interior. However, the compression of the interior by the accreted matter release additional gravitational energy. Some of this energy goes into thermal energy (compressional heating) and the rest is transported to the surface and radiated away (radiative cooling). Therefore, the interior temperature is determined by the balance between heating and cooling and, thus, strongly depends on the mass accretion rate, $\dot{M}$.

Compression first heats up a layer near the surface because of the small pressure scale height there. Later, heat diffuses inward (Nomoto 1982a). The diffusion timescale depends on $\dot{M}$ and is small for larger $\dot{M}$ because of the large heat flux and steep temperature gradient generated by rapid accretion. For example, the time it takes for the heat wave to reach the central region is about 2×10^5 yr for $\dot{M} \sim 10^{-6}$ $M_\odot$ yr^{-1} (Nomoto and Iben 1985) and 5×10^6 yr for $\dot{M} \sim 4 \times 10^{-8}$ $M_\odot$ yr^{-1} (Nomoto et al. 1984). Therefore, if the initial mass of the white dwarf, M_{CO}, is smaller than 1.2 $M_\odot$, the entropy in the center increases substantially due to the heat inflow and carbon ignites at relatively low central density ($\rho_c \sim 3 \times 10^9$ g cm^{-3}). On the other hand, if the white dwarf is sufficiently massive and cold at the onset of accretion, the central region is compressed only adiabatically and thus is cold when carbon is ignited in the center. In the latter case, the ignition density is as high as 10^{10} g cm^{-3} (e.g., Isern et al. 1983).

2.2 Collapse of C+O White Dwarfs Induced by Cabon Deflagration at High Density

Nomoto (1986b) has examined the critical condition for which a carbon deflagration leads to collapse of the white dwarf. If a carbon deflagration is initiated in the center of the white dwarf when $\rho_c \sim 10^{10}$ g cm^{-3} and if the propagation velocity of the deflagration wave, v_{def}, is slower than a certain critical speed, v_{crit}, the outcome is collapse, not explosion, as seen in Figures 1 and 2 (see also Isern et al. 1985). For $v_{def} > v_{crit}$, complete disruption results (and the ejecta contain too much neutron-rich matter). The value of v_{crit} depends on ρ_c at carbon ignition. For $\rho_c \sim 1 \times 10^{10}$ g cm^{-3}, $v_{crit} \sim 0.15\ v_s$ (v_s denotes sound speed). A lower ρ_c implies a lower v_{crit}. In our case of $\rho_c \sim 1 \times 10^{10}$ g cm^{-3}, $v_{def} < v_{crit}$ for both conductive and convective deflagrations and, therefore, collapse will result.

Accordingly, the ultimate fate of accreting CO white dwarfs depends on $\dot{M}$ and the initial mass of the white dwarf, M_{CO}, as summarized in Figure 3. Here $\dot{M}$ denotes the growth rate of the CO white dwarf mass irrespective of the composition of the accreting matter. There are three regimes in the parameter space where neutron star formation from accretion-induced collapse occurs:

1) The accretion with $\dot{M} > 2.7 \times 10^{-6}\ M_\odot\ yr^{-1}$ occurs for merging white dwarfs as will be discussed in §3.3.
2) For $2.7 \times 10^{-6}\ M_\odot\ yr^{-1} > \dot{M} > \dot{M}_{det}$ and $M_{CO} > 1.2\ M_\odot$, a central density as high as 10^{10} g cm^{-3} is reached by adiabatic compression if the white dwarf is sufficiently cold at the onset of accretion.
3) For $\dot{M} < 10^{-9}\ M_\odot\ yr^{-1}$ and $M_{CO} > 1.13\ M_\odot$, the white dwarf is too cold to initiate a helium detonation. Eventually pycnonuclear carbon burning starts in the center when ρ_c reaches $\sim 10^{10}$ g cm^{-3}.

For $10^{-9}\ M_\odot\ yr^{-1} < \dot{M} < \dot{M}_{det}$, a helium flash grows into a helium detonation that ejects at least the overlying helium layer. (This explosion might explain some features of Type Ib supernovae; Branch and Nomoto 1986). Thus the white dwarf cannot evolve to the Chandrasekhar mass. The value of $\dot{M}_{det}$ depends on the Population of the mass transferring star. For X(CNO) > 0.005, the $^{14}N(e^-,\nu)^{14}C(\alpha,\gamma)^{18}O$ (NCO) reaction ignites helium flashes that are too weak to initiate a detonation (Hashimoto et al. 1986). Thus $\dot{M}_{det} \sim 1 \times 10^{-8}\ M_\odot\ yr^{-1}$. For smaller X(CNO), $\dot{M}_{det} \sim 4 \times 10^{-8}\ M_\odot\ yr^{-1}$ (Nomoto 1982a, b)

The frequency of systems above cases 2 and 3 may not be large. First, if hydrogen-rich matter accretes at $\dot{M} < 10^{-9}\ M_\odot\ yr^{-1}$, nova-like explosions will decrease the growth rate of the white dwarf mass and might even prevent the mass from growing (see, however, §3.1). Secondly, massive CO white dwarfs (> 1.2 $M_\odot$) may be rare. The formation of such white dwarfs depends on when the precursor star loses its

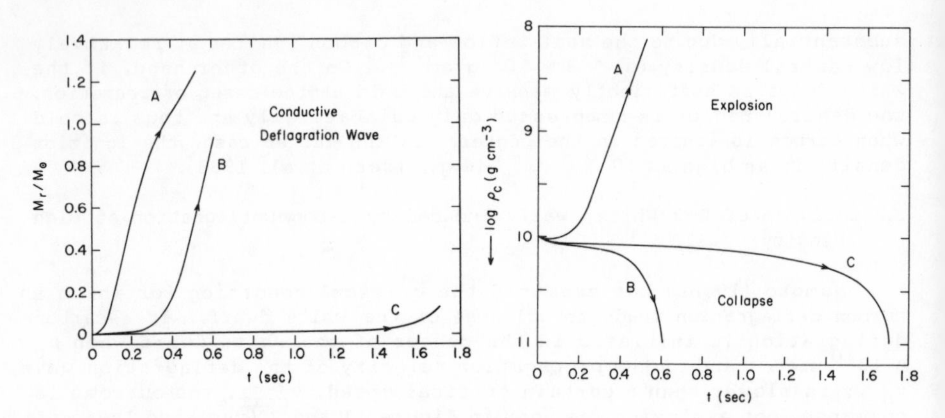

Figure 1 (left): Propagation of the deflagration wave in the C+O white dwarf. The location (M_r) of the deflagration front is shown as a function of time, t, for three cases (A, B, C) of propagation velocity.

Figure 2 (right): Changes in the central density of the C+O white dwarf associated with the propagation of the deflagration wave. Relatively Slow propagation in cases B and C leads to the increase in the central density, i.e., collapse of the white dwarf. On the other hand, faster propagation in case A induces the explosion of the white dwarf.

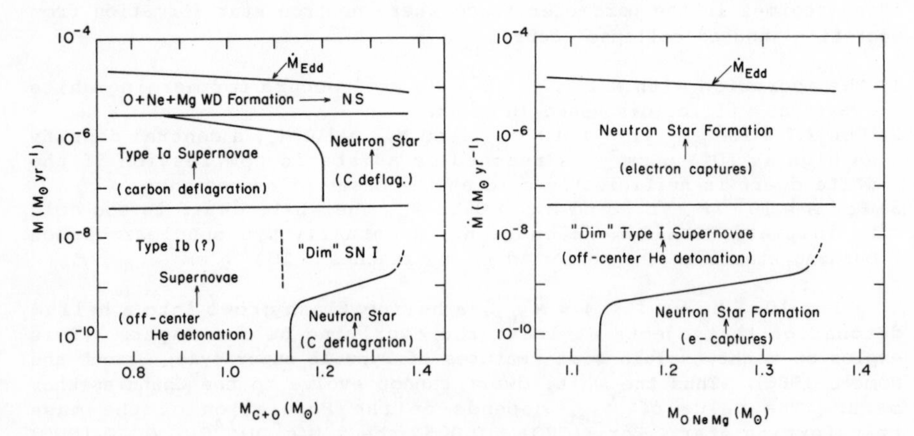

Figure 3 (left): The final fate of accreting C+O white dwarfs expected for their initial mass M_{CO} and accretion rate M. For two regions indicated by "Neutron Star", carbon deflagration is ignited in the center when the central density is as high as 10^{10} g cm^{-3}. Propagation of the deflagration wave will induce collapse to form a neutron star rather than explosion.

Figure 4 (right): Same as Figure 3 but for O+Ne+Mg white dwarfs. Collapse is triggered by electron capture on ^{24}Mg and ^{20}Ne.

hydrogen-rich envelope by either a stellar wind or Roche lobe overflow.

2.3 Collapse of O+Ne+Mg White Dwarfs

A similar diagram for the ONeMg white dwarf is shown in Figure 4. For a wide range of mass accretion rates and initial white dwarf masses, the ONeMg white dwarfs collapse due to electron capture on ^{24}Mg and ^{20}Ne (Nomoto et al. 1979; Miyaji et al. 1980). The initial mass of the ONeMg white dwarfs, M_{ONM}, is larger than $\sim$ 1.1 $M_\odot$ (Nomoto 1984). In many cases, M_{ONM} is very close to the Chandrasekhar mass, so that only a small increase in mass is enough to trigger collapse. However, ONeMg white dwarfs form from 8 - 10 $M_\odot$ stars (Nomoto 1984). The number of such systems may be siginificantly smaller than the number of systems that contain CO white dwarf whose precusors are 1 - 8 $M_\odot$ stars, perhaps, by four order of magnitude (Iben and Tutukov 1984).

3. EVOLUTIONARY ORIGIN OF BINARY RADIO PULSARS AND LOW MASS X-RAY BINARIES

Several evolutionary scenarios for low mass X-ray binaries and binary radio pulsars have been proposed (see van den Heuvel 1984 for a review). The neutron stars in these systems may or may not be formed by accretion-induced collapse of white dwarfs. So it is interesting to see if these proposed scenarios are consistent with the conditions for the neutron star formation as summarized in Figures 3 and 4.

3.1 Subgiant - White Dwarf Systems and Binary Radio Pulsars

The evolution of close binary systems consisting of a subgiant and a more massive neutron star has been examined to explain the origin and properties of low mass X-ray binaries and binary radio pulsars (Webbink et al. 1983; Taam 1983; Joss and Rappaport 1983; Paczynski 1983; Savonije 1983; Helfand et al. 1983). The subgiant of mass $\sim$ 0.8 $M_\odot$ filles its Roche lobe and transfers hydrogen-rich matter over to the neutron star because its envelope is expanding as the helium core grows. Since mass losing star is smaller than the neutron star, the mass transfer is stable and proceeds on a nuclear timescale until the hydrogen-rich envelope of the subgiant is exhaused by mass transfer and hydrogen shell burning. Because of the increasing mass ratio, the separation of the system is getting larger ("spiral out"). After the termination of the mass transfer, the system now consists of a helium white dwarf (smaller than $\sim$ 0.5 $M_\odot$) and a neutron star, which would be a binary radio pulsar.

The average mass transfer rate from the subgiant is

$$\langle \dot{M} \rangle \sim 1 \times 10^{-8} \ (P_0 \ / \ 20 \ d) \ M_\odot \ yr^{-1}, \qquad (1)$$

where P_0 denotes the orbital period of the system at the beginng of mass transfer (Webbink et al. 1983). For longer P_0, the helium core mass is larger so that the luminosity, radius of the subgiant, and $\langle \dot{M} \rangle$ are larger. Savonije (1987) and Joss et al. (1987) estimated the mass of the white dwarf companion to be 0.45, 0.30, 0.22, and 0.16 $M_\odot$ for PSR 0820+02, 1953+29, 1855+09, and 1831-00, respectively. The accretion rates just before their termination are $\dot{M}$ / $M_\odot$ yr^{-1} $\sim$ 9 x 10^{-8} (PSR 0820+02), 8 x 10^{-9} (1953+29), 7 x 10^{-10} (1855+09), and 4 x 10^{-11} (1831-00) (Savonije 1987).

Among these binary radio pulsars, magnetic field of PSR 0820+02 is 3×10^{11} G (Dewey et al. 1986) so that the neutron star is much younger than the companion star. PSR 1953+29 and 1855+09 have a weak magnetic field ($< 10^9$ G), i.e., the neutron stars are old. However, they lie at 20 pc above the galactic plane, which implies a very small space velocity (< 10 km s^{-1}) given to the system at the neutron star formation (Helfand et al. 1983). Also these systems have quite small eccentricity. These features favor the accretion-induced collapse origin.

The question is whether the accretion-induced collapse model is consistent with our constraints on the mass accretion rate. Let's start from the system consisting of a subgiant and a white dwarf. The white dwarf accretes hydrogen-rich matter at a rate $\langle \dot{M} \rangle$ given by equation (1). Then a hydrogen flash is ignited. Let's assume that a helium layer is gradually built up at $\langle \dot{M} \rangle$ (even if some accreted matter is ejected, compressional heating of the helium layer is determined by $\langle \dot{M} \rangle$; Nomoto and Hashimoto 1987). In order for the white dwarf mass to reach the Chandrasekhar mass, $\langle \dot{M} \rangle$ must be higher than $\dot{M}_{det}$ or lower than 1×10^{-9} $M_\odot$ yr^{-1} (and $M_{CO} > 1.13\ M_\odot$) to avoid a helium detonation (Fig. 3).

$\langle \dot{M} \rangle$ for PSR 0820+02 is consistent with these conditions; i.e., the white dwarf can grow to the Chandrasekhar mass by recurring weak hydrogen and helium shell flashes. PSR 1953+29 had $\langle \dot{M} \rangle$ near the borderline of 1×10^{-8} $M_\odot$ yr^{-1} if we assume that the $^{14}N(e^-,\nu)^{14}C(\alpha,\gamma)^{18}O$ (NCO) reaction ignites a mild helium flash; this case is marginally consistent with the accretion-induced collapse model.

PSR 1855+09 and 1831-00 correspond to $\langle \dot{M} \rangle < 10^{-9}$ $M_\odot$ yr^{-1}, so that the white dwarf can avoid a helium detonation if the initial white dwarf mass is larger than 1.13 $M_\odot$. However, at such low $\dot{M}$, a hydrogen shell flash may be as strong as nova explosions and thus eject a significant fraction of the accreted material. If we argue that it is hard to account for the nature of PSR 1855+09 by the spontaneous collapse of iron core in massive helium stars (see below), even a nova-like explosion should be able to increase the white dwarf mass. If this is the case, the white dwarf mass should initially be close to the Chandrasekhar mass because of low efficiency in mass increase. Then the ONeMg white dwarf is favored because its mass is as large as 1.35 $M_\odot$.

However, if we argue that nova-like explosions cannot increase the white dwarf mass or even decrease it by mixing, the neutron stars in PSR 1855+09 and 1831-00 system should come from spontaneous collapse of iron cores in helium stars more massive than 2.5 $M_\odot$. (Smaller mass helium stars leave ONeMg or CO white dwarfs; Nomoto 1984.) If the neutron star mass is assumed to be $\sim 1.4\ M_\odot$, the helium star should eject at least 1.1 $M_\odot$ at the explosion. Then the system would obtain a substantial recoil velocity. The spontaneous collapse scenario needs to find a way to reconcile this with the small distance of the current system from the galactic plane (Helfand et al. 1983). This scenario also requires the efficient circularization due to tidal interaction (Dewey et al. 1986).

If the mass of the companion of PSR 1831-00 is such small as 0.06 - 0.13 $M_\odot$ (Dewey et al. 1986), it is too small to be a remnant of a helium core of subgiant stars. We need a more complicated scenario probably invoking spiral-in (Dewey et al. 1986).

To summarize, the accretion-induced model for the binary radio pulsars with the current orbital period longer than $\sim$ 110 d (PSR 0820+02 and 1953+29) is consistent with the current view of the evolution of accreting white dwarfs. However, the accretion-induced collapse model for PSR 1855+09 and 1831-00 with the orbital period shorter than $\sim$ 15 d would require that the white dwarf mass is increased by a nova-like explosion. If so, accretion-induced collapse model predicts the gap between $\sim$ 15 d and $\sim$ 110 d which corresponds to a helium detonation that prevents the white dwarf mass from growing. If not, the latter two binary radio pulsars belong to a different class that originated from spontaneous collapse of iron cores.

3.2 Helium Star - White Dwarf System

Recent studies of intermediate mass close binary stars (Tornambè and Matteucci 1986; Iben and Tutukov 1987; Iben et al. 1987) have shown that, for a relatively wide range of initial parameters, systems may evolve through common envelope phase into a configuration consisting of a more massive CO (or ONeMg) white dwarf and a less massive helium star. If they are close enough, mass transfer from the helium star to the white dwarf is driven by gravitational wave radiation. Such a system may be called as "helium star cataclysmics". If gravitational wave radiation is the only source of angular momentum loss, accretion rate of helium is $\dot{M}_{He}$ = 2 - 3 x 10^{-8} $M_\odot$ yr^{-1} for the helium star with M > 0.3 $M_\odot$. When the helium star becomes smaller than 0.3 $M_\odot$, the star can no longer burn helium and will attempt to evolve into a degenerate dwarf. Because of the decrease in the separation, the mass transfer rate is enhanced to 4 - 9 x 10^{-8} $M_\odot$ yr^{-1} (Savonije et al. 1986; Nomoto and Hashimoto 1987). When the helium star becomes sufficiently degenerate, its radius starts to increase as its mass decreases. Then the mass accretion rate declines monotonically as the timescale of gravitational wave radiation increases.

Since the system evolves at $\dot{M}_{He}$ = 2 - 3 x 10^{-8} $M_\odot$ yr^{-1} for a fairly long period, it is crucial whether $\dot{M}_{det}$ is 4 x 10^{-8} $M_\odot$ yr^{-1} (without the NCO reaction) or 1 x 10^{-8} $M_\odot$ yr^{-1} (with the NCO reaction). If the NCO reaction is not effective, the helium star cataclysmic will eventually give rise to a helium detonation. On the other hand, if the NCO reaction works or some additional angular momentum loss mechanism operates, $\dot{M}_{He}$ > $\dot{M}_{det}$ as far as M > 0.1 $M_\odot$. Then the white dwarf will evolve to the Chandrasekhar mass.

If the latter is the case and if a CO white dwarf is a companion of the helium star, a Type Ia supernova explosion or collapse would result depending on whether M_{CO} > 1.2 $M_\odot$. If an ONeMg white dwarf is the companion, the white dwarf will collapse for a wide range of parameters and the system will become a configuration consisting of a neutron star and a helium star. It will be observed as an ultrashort period X-ray binary system such as X-ray pulsars 4U 1626-67 and 1E 2259+586 whose orbital periods are 41 min and 38 min, respectively (Savonije et al. 1986; Taam and van den Heuvel 1986). The orbital period decreases to its minimum of 11 min due to the decreasing radius of the helium star and then starts to increase as the mass ratio increases. Eventually, the helium star becomes a dwarf-like star. If the mass transferred from the helium star to the white dwarf is not enough to trigger collapse, the system would become a binary like a helium variable PG 1346+082 which has an orbital period of 1490 min (Wood et al. 1987).

The 685 s orbital period of an interesting ultrashort period X-ray binary 4U 1820-30 (Stella et al. 1987) might be accounted for by the helium burning star - neutron star system because its orbital period could be as short as 11 min. One of the arguments against this model is that this X-ray source is located in the globular cluster NGC 6624 so that a helium star, which is supposed to form from an intermediate mass star, would be too young to exist in the globular cluster (Stella et al. 1987). However, there is another way of forming a 0.5 - 0.6 $M_{\odot}$ helium star in globular clusters, i.e., merging of two helium white dwarfs after complicated binary evolution (Saio and Nomoto 1987). Another stronger argument against the helium burning star model is that mass accretion rate onto the neutron star at the orbital period of 11 min is as high as 10^{-7} $M_{\odot}$ yr^{-1} (Savonije et al. 1986) which is almost ten times higher than the observed value inferred from X-ray luminosity (Stella et al. 1987). Therefore, it is more likely that 4U 1820-30 originated from tidal capture of a neutron star by a red-giant or a main-sequence star whose core is now a helium dwarf companion (Verbunt 1987; Bailyn and Grindlay 1987; Miyaji 1987; Rappaport et al. 1987).

3.3 Merging of Double White Dwarfs

Merging of double white dwarfs would be an another possible site of neutron star formation by accretion-induced collapse. Two white dwarfs approaching together due to the emission of gravitational wave radiation and once the smaller mass white dwarf filles its Roche lobe, either dynamical or gravitational wave radiation-driven accretion starts at a rapid rate (Iben and Tutukov 1984; Webbink 1984; Cameron and Iben 1986). If the accretion is super-Eddington, a common envelope will be formed; futher mass transfer induces a mass loss from the outer Lagrangian point that results in the eventual merging to form a rapidly rotating white dwarf (Hachisu et al. 1986). If the accretion rate is controlled to a lower value but still larger than 2.7 x 10^{-6} $M_{\odot}$ yr^{-1}, rapid accretion ignites a mild off-center thermonuclear flashes (Kawai et al. 1987). Possible cases that relate to accretion-induced collapse are:

1) ONeMg - CO white dwarfs: Rapid accretion of CO onto the ONeMg white dwarf leads to a recurrence of weak carbon flashes and increase the mass of the ONeMg white dwarf. Eventually, the ONeMg white dwarf collapses due to electron capture on ^{24}Mg and ^{20}Ne.

2) CO - CO white dwarfs: Rapid accretion of CO ignites a mild off-center carbon flash and the subsequent propagation of the carbon burning front completely changes the CO white dwarf into an ONeMg white dwarf (Saio and Nomoto 1985; Woosley and Weaver 1986a). Further evolution would be similar to the above case 1).

3) CO + He white dwarfs: accretion of helium onto the CO white dwarf forms an extended helium envelope which might be similar to R CrB stars. The increase in the CO core mass as a result of stable helium shell burning ignites an off-center carbon burning if the CO core mass exceeds 1.1 $M_{\odot}$. As in the case 2) above, the CO core is changed into an ONeMg core.

4) He - He white dwarfs: A mild off-center helium flash is ignited by accretion of helium and a burning front propagates through the center. Since appreciable amount of helium remain unburned, the merger becomes a helium burning star (Saio and Nomoto 1987; see Nomoto 1986b). This is one of the way to form a helium main-sequence star. After exhaustion of helium, the star will become a CO white dwarf.

If the merger becomes a neutron star in cases 2 and 3, the resulting neutron star is a rapid rotator. Suppose that the magnetic field of a neutron star originates from a fossil field of the white dwarf and no enhancement of the field occurs after the neutron star formation. Then the neutron star could be weakly magnetized if the magnetic field of the white dwarf was weak. (White dwarfs in cataclysmic variables are classified into two classes, with and without strong magnetic field; King et al. 1984) The neutron star could continue to rotate rapidly for relatively long time. Furthermore, the precursor of CO or ONeMg white dwarfs are 5 -10 $M_\odot$ stars and a recoil velocity given to the binary at the explosion must be small, so that a location of the resulting neutron star may not be far from the galactic plane. These features could be consistent with the properties of the single millisecond pulsar (Baker et al. 1982). If the magnetic field is enhanced during or after collapse, a normal single pulsar would result but with no supernova remnant around it (Saio and Nomoto 1985).

In globular clusters, tidal captures of white dwarfs by normal stars and merging of the double white dwarfs would frequently occur. Therefore, we can expect more frequent accretion-induced collapse in globular clusters as discussed in detail by Grindlay (1987).

4. DISCUSSION

The evolutionary scenario leading to accretion-induced collapse could give some hints on the precursors of Type Ia supernovae becasue it is common that the accreting white dwarfs increase their mass up to the Chandrasekhar mass. One is the possibility for the white dwarf to increase its mass by the accretion slower than 10^{-9} $M_\odot$ yr^{-1} as discussed in relation to PSR 1855+09. If this is the case, a strong (even nova-like) hydrogen shell flashes should increase the white dwarf mass. Secondly, if an ultrashort period X-ray system originates from "helium star cataclysmics", $\dot{M}_{det}$ must be as low as 1 x 10^{-8} $M_\odot$ yr^{-1}. Then some helium star cataclysmics could also be precursors of Type Ia supernovae.

I would like to thank Drs. J. Grindlay, C. Bailyn, S. Rappaport, P. Joss for stimulating discussion during my visit to Cambridge.

REFERENCES

Backer, D.C., Kulkarni, S.R., Heiles, C., Davis, M.M., and Goss, W.M. 1982, Nature, **300**, 615.
Bailyn, C.D., and Grindlay, J.E. 1987, Ap. J. (Letters), submitted.
Branch, D., and Nomoto, K. 1986, Astr. Ap., **164**, L13.
Cameron, A.G.W., and Iben, I. Jr. 1986, Ap. J., **305**, 228.
Canal, R., Isern, J., and Labay, J. 1980, Ap. J. (Letters), **241**, L33.
Dewey, R.J., Maguire, C.M., Rawley, L.A., Stokes, G.H., and Taylor, J.H. 1986, Nature, **322**, 712.

Grindlay, J.E. 1987, IAU Symposium 125, The Origin and Evolution of Neutron Stars, ed. D.J. Helfand and J.H. Huang (Dordrecht: Reidel).
Hachisu, I., Eriguchi, Y., and Nomoto, K. 1986, Ap. J., **308**, 161.
Hashimoto, M., Nomoto, K., Arai, K., and kaminisi, K. 1986, Ap. J., **307**, 687.
Helfand, D.J., Ruderman, M.A., and Shaham, J. 1983, Nature, **304**, 423.
Iben, I., Jr., Nomoto, K., Tornambe, A., and Tutukov, A.V. 1987, Ap. J.
Iben, I., Jr., and Tutukov, A.V. 1984, Ap. J. Suppl., **55**, 335.
Iben, I., Jr., and Tutukov, A.V. 1987, Ap. J., submitted.
Isern, J., Labay, J., Hernanz, M., and Canal, R. 1983, Ap. J., **273**, 320.
Isern, J., Labay, J., and Canal, R. 1984, Nature, **309**, 431.
Iyetomi, H., and Ichimaru, S. 1987, private communication.
Joss, P.C., and Rappaport, S.A. 1983, Nature, **304**, 419.
Joss, P.C., Rappaport, S.A., and Lewis, W. 1987, Ap. J., in press.
Kawai, Y., Saio, H., and Nomoto, K. 1987, Ap. J., **315**, in press.
King, A.R., Frank, J., and Ritter, H. 1985, M.N.R.A.S., **213**, 181.
Kulkarni, S.R. 1986, Ap. J. (Letters), **306**, L85.
Miyaji,S. 1987, Ap. J. (Letters), submitted.
Miyaji, S., Nomoto, K., Yokoi, K., and Sugimoto, D. 1980, Pub. Astr. Soc. Japan, **32**, 303.
Nomoto, K. 1982a, Ap. J., **253**, 798.
Nomoto, K. 1982b, Ap. J., **257**, 780.
Nomoto, K. 1984, Ap. J., **277**, 791.
Nomoto, K. 1986a, Ann. NY Acad. Sci., **470**, 294.
Nomoto, K. 1986b, Prog. Part. Nucl. Phys., **17**, 249.
Nomoto, K., and Hashimoto, M. 1987, IAU Colloquium 93, Cataclysmic Variables, ed. H. Drechsel, Y. Kondo, and J. Rahe (Reidel).
Nomoto, K., and Iben, I., Jr. 1985, Ap. J., **297**, 531.
Nomoto, K., Miyaji, S., Sugimoto, D., and Yokoi, K. 1979, in IAU Colloq. 53, White Dwarfs and Variable Degenerate Stars, ed. H.M. Van Horn and V. Weidemann (Rochester: Univ. of Rochester), p.56.
Nomoto, K., Thielemann, F.-K., and Yokoi, K. 1984, Ap. J., **286**, 644.
Paczynski, B. 1983, Nature, **304**, 421.
Rappaport, S., Nelson, L.A., Ma, C.P., and Joss, P.C. 1987, Ap. J. (Letters), submitted.
Saio, H. and Nomoto, K. 1985, Astr. Ap., **150**, L21.
Saio, H. and Nomoto, K. 1987, in preparation.
Savonije, G.J. 1983, Nature, **304**, 422.
Savonije, G.J. 1987, Nature, **325**, 416.
Savonije, G.J., de Kool, M., and van den Heuvel, E.P.J. 1986, Astr. Ap., **155**, 51.
Stella, L. Priedhorsky, and White, N.E. 1987, Ap. J. (Letters), **312**, L17.
Taam, R.E. 1983, Ap. J., **270**, 694.
Taam, R.E., van den Heuvel, E.P.J. 1986, Ap. J. **305**, 235.
Tornambè, A., and Matteucci, F. 1986, M.N.R.A.S., **223**, 69.
Van den Heuvel, E.P.J. 1984, J. Ap. Astr., **5**, 209.
Verbunt, F. 1987, Ap. J. (Letters), **312**, L23.
Webbink, R.F. 1984, Ap. J., **277**, 355.
Webbink, R.F., Rappaport, S., and Savonije, G.J. 1983, Ap. J., **270**, 678.
Wood, M.A., Winget, D.E., Nather, R.E., Hessman, F.V., Liebert, J., Kurtz, D.W., Wesemael, F., and Wegner, G. 1987, Ap. J., in press.
Woosley, S.E., and Weaver, T.A. 1986a, in Nucleosynthesis and Its Implications for Nuclear and Particle Physics (Reidel), p. 145.
Woosley, S.E., and Weaver, T.A. 1986b, Ann. Rev. Astr. Ap., **24**, 205.
Wright, G.A., and Loh, E.D. 1986, Nature, **324**, 127.

STELLAR-MASS BLACK HOLES

Jeffrey E. McClintock

Harvard-Smithsonian Center for Astrophysics
Cambridge, Massachusetts

ABSTRACT

The best available evidence that black holes exist comes from dynamical studies of three X-ray binaries, Cyg X-1, LMC X-3, and A0620-00. The probable masses of the compact objects in these systems are about 10 $M_\odot$. Evidence is presented that the masses of Cyg X-1 and A0620-00 are not less than 3.4 $M_\odot$ and 3.2 $M_\odot$, respectively. It is generally believed that a neutron star with a mass above about 3 $M_\odot$ cannot exist; it would collapse through its gravitational radius. It therefore appears highly probable that these objects are black holes.

1. INTRODUCTION

The title of my talk implies that black holes exist. And, within the framework of general relativity, I will present an outline of the strongest available evidence that they do exist. But the case is not watertight (or photontight, which is a pun due to Elihu Boldt). And so, a more appropriate and humble title for this talk might be, "Do Black Holes Exist?"

Neutron stars definitely exist and, with surface potentials GM/Rc^2 ~0.1, they are the very next thing to a black hole. The first clear-cut evidence for neutron stars was the serendipitous discovery of radio pulsars in 1968 by Hewish, Bell and others, followed promptly by Tommy Gold's enlightened interpretation. Today it is plain that neutron stars are as real as solar-type stars. Consider, for example, the Crab Pulsar, a core remnant of an historical supernova that has maintained an X-ray bright synchrotron nebula for centuries. Consider the "millisecond" pulsar (PSR 1937+214), which rotates a thousand times too rapidly to be a white dwarf. And consider also compact X-ray sources with their radii of 10 km and field strengths of 10^{12} G, as deduced from observations of bursts, spin-up rates and cyclotron lines. Thus the evidence for neutron stars is indisputably direct.

It is a presumption of our time that black holes also exist. It seems that they should. After all, general relativity not only allows for the possibliity of black holes, it strongly favors their formation because the same pressure forces that forestall collapse also act as gravitating mass [1]. In a neutron star this effect operates in a significant and self-regenerative way to promote the star's collapse. This fact leads to a limit on the mass of a neutron star of 3 $M_{\odot}$ -- a limit that does not depend on the unknown properties of nuclear forces at high densities (see Section 3). Thus black holes are a logical possibility in general relativity. However, whether or not they actually exist can only be decided by observation.

Apart from theory, current models of stellar and galactic evolution provide strong circumstantial evidence that black holes exist. For example, consider first stellar-mass black holes. It is widely believed that they will be formed in the same way that neutron stars are formed, by the core collapse of massive stars[2]. After all, can a collapsing star somehow always conspire to blow off enough mass so that the core remnant is a 1-2 $M_{\odot}$ neutron star? (Moreover, even if black holes are not formed by stellar collapse, accretion flows in X-ray binaries over a Hubble time might be expected to push at least some neutron stars over into black holes.) If, as has been suggested, stars with masses > 40 $M_{\odot}$ collapse to form black holes[3-4], then there are at least several million stellar-mass black holes in the galaxy. So it seems that an abundance of black holes is a logical consequence of our present knowledge of stellar evolution. In no way, however, does this prove that black holes exist.

What about supermassive black holes? It seems plausible that they might form and grow in the cores of galaxies, a dissipative environment that contains dense condensations of stars and gas. Here Martin Rees[5], especially, has painted a compelling picture that such galactic cores often collapse to form supermassive black holes. Of course, the real driver in this argument, as Rees makes plain, is our need to account for both the compactness (~1 AU) and the prodigious luminosities ($L \sim 10^{12}\ L_{\odot}$) of active galactic nuclei (AGN). It is indeed true that the observations of AGN can be interpreted in terms of a black hole model; however, the interpretation is not at all unique[6] and plausible alternative models exist[7-9].

The strongest arguments for black holes in AGN are model dependent. For example, the appeal to a $\sim 10^{9}$-$10^{10}\ M_{\odot}$ black hole in M87[10-11] was based on a model with isotropic stellar orbits. However, no black hole is required for an equally tenable class of models that invoke an anisotropic velocity distribution[12-13]. In short, it is likely to remain exceedingly difficult to establish the certain existence of a 1-AU sized black hole in the complex environment of an AGN.

Thus, circumstantial evidence that black holes exist is provided by our current understanding of stellar and galactic evolution, the AGN phenomenon, and the nature of general relativity itself. None of this evidence, however, relieves us of the responsibility to seek out hard evidence that black holes are real, physical objects.

2. THREE BLACK-HOLE BINARIES

The best available evidence for black holes comes from dynamical studies of three X-ray binaries: Cyg X-1[14)], LMC X-3[15)], and A0620-00[16)]. This subject has been reviewed elsewhere[17)]; only an outline of the case is presented here. The argument for each binary rests on the following three pieces of evidence: (1) There exists a highly luminous ($\sim 10^{38}$ ergs s^{-1}) and variable X-ray source which is powered by gravitational infall onto a compact object. Our detailed knowledge of more than 40 X-ray binaries that pulse or burst, and therefore manifestly contain neutron stars, establishes beyond reasonable doubt that these three binaries contain compact objects that are at least as dense and massive as a neutron star. Many reviews have been published that compellingly make this point[18)]. (2) The second piece of evidence is also observational. The observed velocity of the optical star, the orbital period of the binary system, and a few particulars (such as the absence of an X-ray eclipse) imply that the mass of the compact object exceeds three solar masses. (3) The final argument rests on theory. As mentioned above, it is believed that a neutron star of three or more solar masses cannot exist; it would be crushed by its self-gravity to form a black hole.

Some of the nominal characteristics of the three leading black-hole candidates are summarized in Table 1. The X-ray luminosities are the maximum that have been observed. The visual magnitude of A0620-00 corresponds to a state of X-ray quiescence. For all three systems, the orbits are circular and X-ray eclipses are not observed.

TABLE 1
PROPERTIES OF THREE BLACK-HOLE BINARIES

	Cyg X-1	LMC X-3	A0620-00
L_x (ergs s^{-1})	2×10^{37}	3×10^{38}	1×10^{38}
MK type	09.7Iab	B3V	K5V
d(kpc)	2.5	55	1
m_v	9	17	18
$V_c \sin i$ (km s^{-1})	76±1	235±11	457±8
P(days)	5.6	1.7	0.32
$f(M/M_\odot)$	0.25±0.01	2.3±0.3	3.18±0.16

The principal observable for such single-line binaries is the optical mass function:

$$f(M) \equiv (M_x \sin i)^3/(M_x + M_c)^2 = \frac{P(V_c \sin i)^3}{2\pi G}$$

where M_x and M_c are the masses of the X-ray source and the companion star, respectively, i is the orbital inclination angle, P is the orbital period, and $v_c \sin i$ is the projected velocity semiamplitude of the optical companion. The mass function is, therefore, a cubic equation for M_x, with M_c and i as parameters.

The value of the mass function is the least possible value that M_x can take on; this limiting value of M_x corresponds to a system with a zero-mass companion ($M_c = 0$) viewed at the maximum inclination angle ($i = 90°$). Therefore, based on the value of the mass function alone, one can say that A0620-00 is a strong black-hole candidate. For Cyg X-1 and LMC X-3, one must argue in addition that the companion stars are massive and/or the systems are viewed at low inclination. These arguments and further details, which are weighed elsewhere[17], lead to the following extreme lower limits for the masses of the compact X-ray sources: M_x(Cyg X-1) > $3.4M_\odot$; M_x(LMC X-3) > 2.2 $M_\odot$ [19]; and M_x(A0620-00) > $3.2M_\odot$.

It is important to emphasize that the above values are rock-bottom limits which correspond to extreme models. For more plausible models, the masses of the compact sources are all about $10M_\odot$, as indicated in Figure 1, which is a schematic sketch to scale of the three systems.

3. UPPER BOUND TO THE MASS OF A NEUTRON STAR

The maximum mass of a neutron star is a sensitive function of the unknown equation of state for nuclear matter. The most realistic and sophisticated models of nonrotating neutron stars imply a maximum mass of 2.0 $M_\odot$, and even the stiffest equation of state predicts a mass of only 2.7 $M_\odot$ [21]. Moreover, calculations indicate that rotation cannot increase the maximum mass by more than about 20%[22].

Fortunately, there is a firm bound on the mass of a neutron star that does not depend on the details of the unknown equation of state at high densities. The key assumptions are the following: general relativity is the corrrect theory of gravity and causality is obeyed (i.e., the speed of sound is less than the speed of light). These minimal assumptions lead to a strong conclusion: a firm upper bound to the mass of a nonrotating neutron star is about 3 $M_\odot$ [23-24].

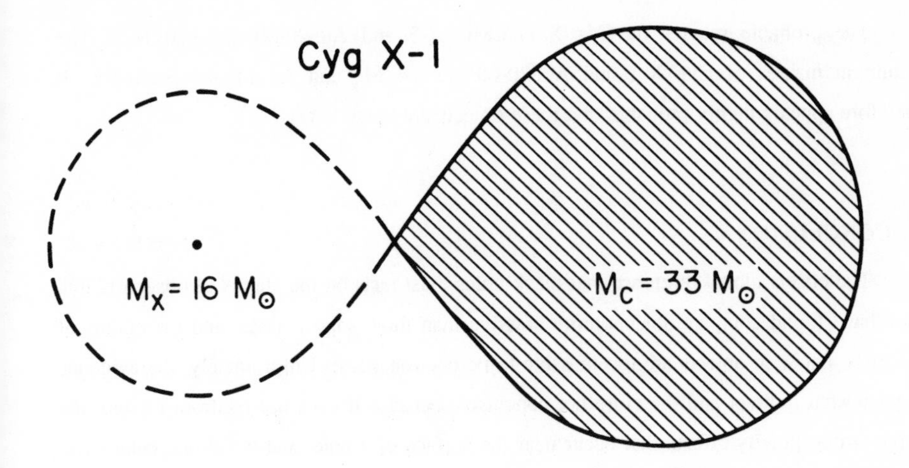

Figure 1. Cartoon comparison of three black-hole binaries[14-16]. The optical companions (shaded regions) are shown filling their Roche lobes. LMC X-1 has been suggested as a black-hole candidate; however, the dynamical evidence is less compelling and the orbital period and the optical counterpart are somewhat uncertain[20].

The probable masses of Cyg X-1, LMC X-3 and A0620-00 are ~10 $M_\odot$. The minimum masses of Cyg X-1 and A0620-00 are 3.4 $M_\odot$ and 3.2 $M_\odot$, respectively. It therefore appears highly probable that these objects are black holes.

4. CONCLUSION

The observational inference that black holes exist rests on the claims of observers that they have found compact objects more massive than three solar masses, and the claims of theorists that such objects are too massive to be neutron stars. Unfortunately, the evidence is somewhat indirect and therefore not conclusive because it does not confront us with the super-strong gravity effects that occur near the surface of a hole, and which are peculiar to black holes alone.

Nevertheless, in the past few years the observational position on the existence of black holes has been strengthened significantly by the addition of two new candidates, LMC X-3 and A0620-00. Obviously it is important to search for additional candidates, and also to strive to detect a unique black-hole signature. It is equally important to improve the theory of massive neutron stars. The limiting masses of the three candidates are perilously close to the causality limit. Further definition of this limit, which is not universally accepted[25)], and further work on the equation of state of nuclear matter would probably serve to strengthen the case that black holes exist.

ACKNOWLEDGEMENTS

I thank Roger Blandford for sending me a draft copy of his review paper entitled "Astrophysical Black Holes" and Ken Brecher for several stimulating discussions.

REFERENCES

1. Misner, C.W., Thorne, K.S., and Wheeler, J.A., Gravitation (San Francisco: Freeman), p. 605 (1973).
2. Bowers, R. and Wilson, J., Ap.J. 263 , 366 (1982), and references therein.
3. van den Heuvel, E.P.J. and Habets, G.M.H.J., Nature 309 , 598 (1984).
4. Schild, H. and Maeder, A. Astron. Astrophys. 143 , L7 (1985).
5. Rees, M., Ann. Rev. Astron. Astrophys. 22 , 471 (1984).
6. Burbidge, G., in Active Galactic Nuclei, ed. J.E. Dyson (Manchester: Manchester Univ. Press), p. 369 (1985).

7. Bailey, M.E. and Clube, S.V.M., Nature 275 , 278 (1978).
8. Flaser, F.M. and Morrison, P., Ap.J. 204 , 352 (1976).
9. Ginzburg, V.L. and Ozernoy, L.M., Astrophys. Space Sci. 50 , 23 (1977).
10. Young, P.J., Westphal, J.A., Kristian, J., Wilson, C.P., and Landauer, F.P., Ap.J. 221 721 (1978).
11. Sargent, W.L.W., Young, P.J., Boksenberg, A., Shortridge, K., Lynds, C.R., and Hartwick, F.D.A., Ap.J. 221 , 731 (1978).
12. Duncan, M.J. and Wheeler, J.C., Ap.J.(Lett.) 237 , L27 (1980).
13. Binney, J. and Mamon, G.A., MNRAS 200 , 361.
14. Gies, D.R. and Bolton C.T., Ap.J. 304 , 371 (1986).
15. Cowley, A.P., Crampton, D., Hutchings J.B., Remillard, R., and Penfold, J.E., Ap.J. 272 118 (1983).
16. McClintock, J.E. and Remillard, R.A., Ap.J. 308 , 110 (1986).
17. McClintock, J.E., in Physics of Accretion onto Compact Objects, Lecture Notes in Physics, 266 , eds. K.O. Mason, M.G Watson, and N.E. White (Heidelberg: Springer-Verlag), p. 211 (1986).
18. Joss, P.C. and Rappaport, S., Ann. Rev. Astron. Astrophys. 22 , 537 (1984).
19. Mazeh, T., van Paradijs, J., van den Heuvel, E.P.J., and Savonije, G.J., Astron. Astrophys. 157 , 113 (1986).
20. Hutchings, J.B., Crampton, D., and Cowley, A.P., Ap.J. (Lett.) 275 , L43 (1983).
21. Arnett, W.D. and Bowers, R.L., Ap.J. Suppl. 33 , 415 (1977).
22. Shapiro, S.L. and Teukolsky, S.A., Black Holes, White Dwarfs, and Neutron Stars (New York: Wiley), p. 264 (1983).
23. Rhoades, C.E. and Ruffini, R., Phys. Rev. Lett. 32 , 324 (1974).
24. Chitre, D.M. and Hartle, J.B., Ap.J. 207 , 592 (1976).
25. Caporaso, G. and Brecher, K., Phys. Rev. D 20 , 1823 (1979).

ELECTRODYNAMIC ORIGIN OF JETS FROM RAPIDLY ROTATING MATTER

R.V.E. Lovelace, J.C.L. Wang, and M.E. Sulkanen

Department of Applied Physics, Department of Physics,
and Department of Astronomy
Cornell University, Ithaca, New York 14853

ABSTRACT

Progress has been made in the understanding of the global electro-magneto-hydrodynamics of accretion disks around black holes. Rigorous solutions of the basic equations have recently been obtained which exhibit self-collimated electromagnetic jets. These jets propagate energy and angular momentum in the $\pm z$ directions away from the black hole. The methods are being extended to the treatment of jets with non-negligible inertia.

1. Introduction

The origin of the narrow radio jets seen in many extragalactic sources[1] remains a fundamental open question in the theory of accretion phenomena. Currently, a widely favored idea for the origin of the jets is based on the occurrence of narrow, essentially empty vortex funnels in the accretion flows[2-4]. An alternative proposal is that jets are electromagnetically accelerated due to the unipolar induction dynamo effect[5,6].

Here, we summarize our ongoing work on the global electro-magneto-hydrodynamics of rapidly rotating accreting matter [7-10]. We first discuss in §2 the magnetic and fluid dynamics inside a viscous resistive accretion disk. Second, in §3 we discuss the field and plasma dynamics outside the disk using ideal, relativistic magneto-hydrodynamics. The interior/exterior field solutions are matched at the disk surface. In the force-free limit, we have found[9] rigorous solutions of the basic exterior equations which correspond to self-collimated electromagnetic (or Poynting-flux) jets. These jets are oppositely directed and propagate energy and angular momentum outward along the disk rotation axis. Finally, in §4 we discuss our recent work which extends the methods of Ref. 9 to the more general case of jets with non-negligible inertia.

2. Inside the Disk

The basic equations for a stationary state of the fluid/magnetic field dynamics inside the disk are:

$$\underline{\nabla} \cdot (\rho \underline{v}) = 0 \quad , \tag{1}$$

$$\underline{\nabla} \times \underline{B} = (4\pi/c)\underline{J} \quad , \tag{2}$$

$$\underline{J} = \sigma(\underline{E} + \frac{1}{c}\,\underline{v} \times \underline{B}) \quad , \tag{3}$$

$$\frac{\partial \underline{B}}{\partial t} = 0 = \underline{\nabla} \times (\underline{v} \times \underline{B}) + \eta \nabla^2 \underline{B} \quad , \tag{4}$$

$$\rho \frac{d\underline{v}}{dt} = -\underline{\nabla} p - \rho \underline{\nabla} \Phi_{grav} + \frac{1}{c}\,\underline{J} \times \underline{B} + \underline{f}_{vis.} \quad , \tag{5}$$

$$\underline{\nabla} \cdot \underline{B} = 0 \quad , \tag{6}$$

Here, σ is the conductivity, $\eta = c^2/(4\pi\sigma)$ is the magnetic diffusivity, and $\underline{f}_{vis}$ is the viscous force. Equation (1) is mass conservation, (2) is Ampère's law, (3) is Ohm's law, (4) is Faraday's law (the induction equation), and (5) is the Navier-Stokes equation.

We consider axi-symmetric solutions of equations (1)-(6). Thus the magnetic field can be written in terms of poloidal (p) and toroidal (t) components, $\underline{B} = \underline{B}_p + B_\phi \hat{\underline{\phi}}$, where $\underline{B}_p = (B_r, B_z)$ with $B_r = -r^{-1}(\partial\Psi/\partial z)$ and $B_z = r^{-1}(\partial\Psi/\partial r)$. The function $\Psi(r,z)$ is referred to as the flux-function in that $\Psi(r,z)$ = const. describes the poloidal projection of a magnetic field line. There are two simple field symmetries: odd-symmetric where $\Psi(r,z) = -\Psi(r,-z)$; and even symmetry where $\Psi(r,z) = +\Psi(r,z)$. Figure 1 shows the nature of the corresponding electric and magnetic fields.

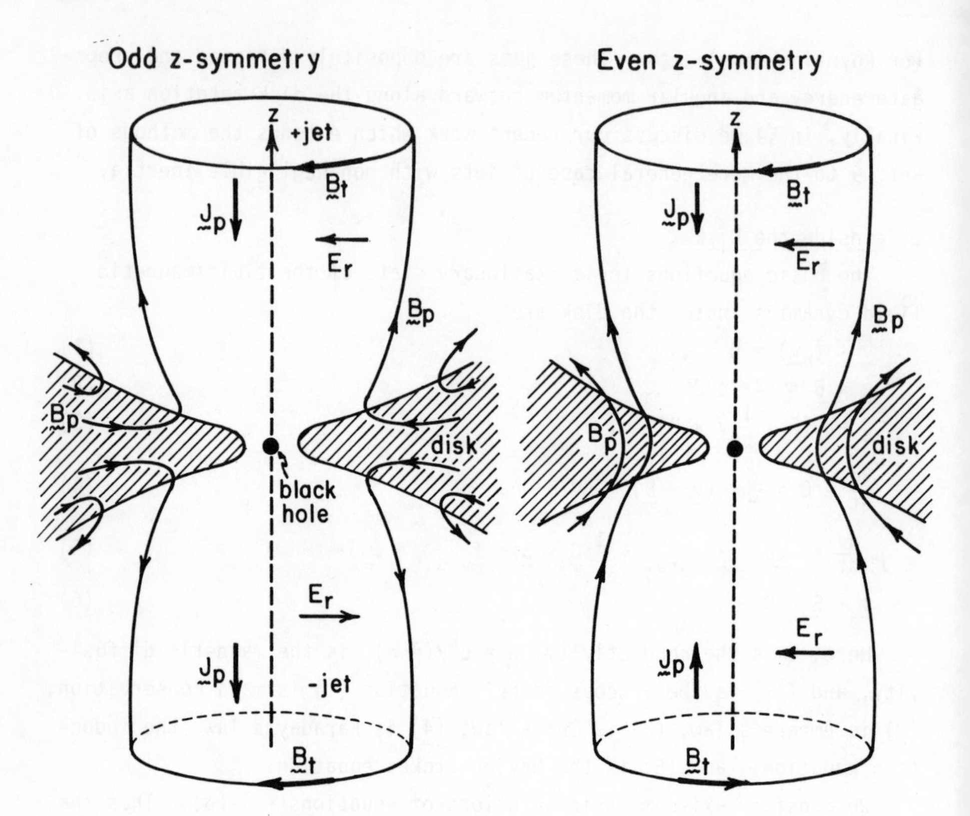

Figure 1. Field symmetries.

For thin disks, where the vertical (z) thickness 2h is small compared with the radial distance r, the induction equation (4) can be simplified to give

$$\frac{\partial\Psi(r,\zeta)}{\partial r} + D\,\frac{\partial^2\Psi}{\partial\zeta^2} = 0 \quad ,$$

$$\frac{\partial}{\partial r}(uhB_\phi) + D\,\frac{\partial^2}{\partial\zeta^2}(uhB_\phi) = \left[\frac{d}{dr}\left(\frac{v_\phi}{r}\right)\right]\frac{\partial\Psi}{\partial\zeta} \quad , \tag{7}$$

where $u \equiv -v_r > 0$ is the accretion speed, $\zeta \equiv z/h(r)$, and $D \equiv \eta/(uh^2)$. A simple example of a solution of (7) with odd z-symmetry is

$$\Psi = \frac{\Phi_p \sin(k\zeta)}{4\pi \sin(k)} \exp\left[-k^2\int_r^\infty dr' D(r')\right] \quad ,$$
$$B_\phi = \frac{\Phi_p \Omega_i \cos(k\zeta)}{4\pi uh \sin(k)} \left[\delta + \frac{k\Omega(r)}{\Omega_i}\right] \exp[\,''\,] \quad . \tag{8}$$

Here, Φ_p is the total poloidal magnetic flux entering the disk at large r, k is a dimensionless constant which determines the current flow from the disk surface, and δ is a dimensionless constant proportional to the strength of the intrinsic component of the toroidal magnetic field (the component not due to differential rotation of the disk).

3. Exterior of the Disk

In this region, the basic equations for a stationary state of the plasma and fields are taken to be

$$\underline{\nabla} \cdot (\rho \underline{v}) = 0 \quad , \tag{9}$$
$$\underline{\nabla} \times \underline{B} = (4\pi/c)\underline{J} \quad , \tag{10}$$
$$\underline{\nabla} \cdot \underline{E} = 4\pi\rho_e \quad , \tag{11}$$
$$\underline{\nabla} \times \underline{E} = 0 \quad , \tag{12}$$
$$\underline{E} + \underline{v} \times \underline{B}/c = 0 \quad , \tag{13}$$
$$\rho_e \underline{E} + \frac{1}{c} \underline{J} \times \underline{B} = \rho(\underline{v} \cdot \underline{\nabla})(\gamma \underline{v}) \quad . \tag{14}$$

Thus, outside the disk all non-ideal effects are neglected ($\sigma \to \infty$, $\underline{f}_{vis} \to 0$). However, the plasma motion may be relativistic with $(\gamma-1)$ non-negligible. Consequently the electric force $\rho_e \underline{E}$ cannot, in general, be neglected in the relativistic Navier-Stokes equation (14). For simplicity, we assume that the pressure and gravitational forces are negligible in (14).

In earlier work[9], we analyzed the force-free-limit of equations (9)-(14) where $\rho|(\underline{v}\cdot\underline{\nabla})(\gamma\underline{v})| \ll |\underline{J} \times \underline{B}|/c$. For axi-symmetric force-free situations, equations (9)-(14) boil down to a single non-linear second-order partial differential equation for $\Psi(r,z)$ – the "pulsar equation". The two unknown functions of Ψ in this equation are fixed by the condition that the Ψ and B_ϕ must match the interior field solution (8) at the disk surface ($\zeta = \pm 1$). A new Green's function method for the pulsar equation is used to obtain a general expression relating the far field $\Psi(r,z \to \infty)$ to the values of Ψ on the disk surface. The far field solutions exhibit self-collimated jets which propagate energy and angular momentum in the $\pm z$ directions. The collimation results from the magnetic pinch effect of B_ϕ which in turn results from the axial current of the jet. The energy transport is due to the Poynting flux of the jet ($\propto E_r B_\phi$). The angular momentum transport is due to the magnetic field ($\propto B_\phi B_z$). The energy and angular momentum in the jet is extracted from the accreting matter of the disk. Figure 2 shows the radial dependences of the different field components for $z \to \infty$.

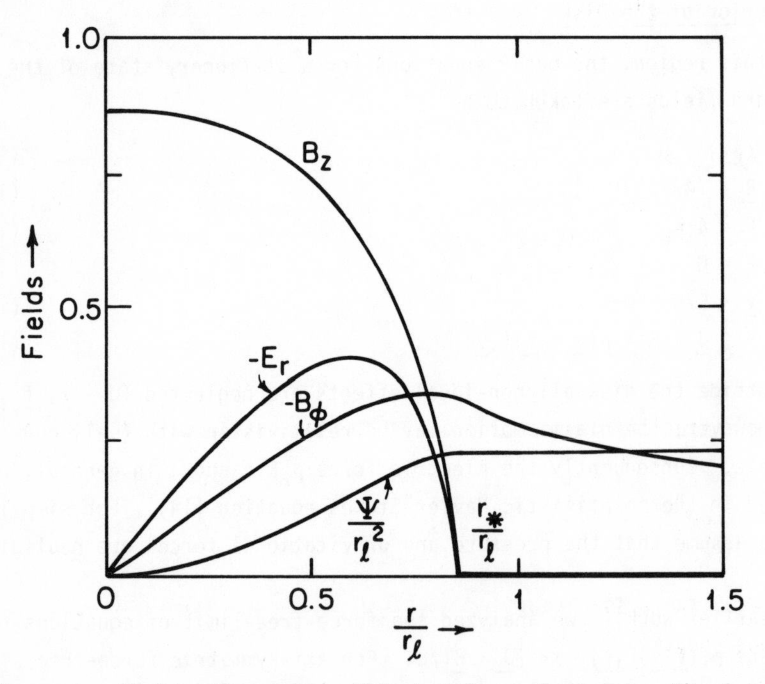

Figure 2. Fields of a self-collimated force-free jet. The jet radius is $r_* = 0.8667\ r_\ell$, where r_ℓ is the light cylinder radius.

The overall power flow in the disk-jet system has so far been fully analyzed only for the case of odd field symmetry[9]. For this symmetry the power carried away by the jets is in general small compared with the "available" accretion power ($\sim\dot{M}c^2/10$). For the even symmetry case it appears that the jets can carry away most of the available power[10]. The odd symmetry solutions may be relevant to typical quasars and Seyfert galaxies where the fraction of the total luminosity in the form of non-thermal emission from jets is small[11,12]. On the other hand, for strong radio galaxies, most of the power output is probably in the jets[13]. This may also be the case for BL Lac objects[14]. The even symmetry solutions are of evident interest for the radio galaxies and BL Lacs.

4. Self-Collimated Inertial Jets

With inertia included, the axisymmetric solutions of equations (9)-(14) involve in general five functions of Ψ which are not known a priori[7]. (In the force-free limit only two functions are important.) Because we neglect the pressure for simplicity, there are only four functions: $F(\Psi)$, which arises from the conservation of mass; $H(\Psi)$, which comes from the conservation of angular momentum; $g(\Psi)$, which results from the perfect conductivity (Eq. 13); and $I(\Psi)$, which is Bernoulli's constant expressing the conservation of the sum of the mechanical and electromagnetic energy flow on a given flux surface (Ψ = const.). A number of relations can be derived[7] from the basic equations (9)-(14):

$$rB_\phi - \gamma F r v_\phi = H \quad , \tag{15}$$

$$\gamma F - g r B_\phi / c^2 = I \quad , \tag{16}$$

$$r v_\phi - F r B_\phi/(4\pi\rho) = r^2 g \quad . \tag{17}$$

In place of the pulsar equation we now have the relativistic Grad-Shafranov equation[7]:

$$P\Delta^* \Psi - \frac{1}{2r^2}\left[\underline{\nabla}\left(\frac{r^4 g^2}{c^2}\right)\right] \cdot \underline{\nabla\Psi} - F\left[\underline{\nabla}\left(\frac{F\gamma}{4\pi\rho}\right)\right] \cdot \underline{\nabla\Psi}$$

$$= -(H + \gamma r v_\phi F)(H' + \gamma r v_\phi F') - 4\pi\rho r^2 (J' + \gamma r v_\phi g') \quad . \tag{18}$$

$$P \equiv 1 - \left(\frac{rg}{c}\right)^2 - \frac{\gamma F^2}{4\pi\rho} \quad ,$$

$$\Delta^* \equiv \frac{\partial^2}{\partial r^2} - \frac{1}{r}\frac{\partial}{\partial r} + \frac{\partial^2}{\partial z^2} \quad ,$$

$$F' \equiv \frac{dF}{d\Psi} \quad , \text{ etc.,}$$

and $J(\Psi) \equiv (c^2 I + gH)/F - c^2$.

As in the force-free case, the functions F, g, H, and I are fixed by the requirement that the exterior solution match the interior solution at the disk surface. We have generated a large number of far field jet solutions to equations (15)-(18) for cases where $F = F_0$, $g = g_0$, $H = K_H\Psi$, and $I = I_0 + \varepsilon_I K_H \Psi + D\Psi^2$, where F_0, g_0, K_H, I_0, ε_I, and D are constants. These solutions have $\Psi = \Psi(r)$ and correspond to cylindrical jets with no divergence. (The inclusion of non-ideal effects would of course lead to a slow divergence.) The solutions can be made to satisfy the relevant radial virial equation to high accuracy ($<10^{-6}$). There are at least two classes of solutions: Jets of the first class have sharp edges similar to the force-free jets (Figure 2), and are everywhere sub-Alfvénic ($P > 0$). Jets of the second class have diffuse edges as shown in Figure 3 and are sub-Alfvénic ($P > 0$) near the axis but super-Alfvénic ($P < 0$) far from the axis. As yet we have not found jet solutions which are everywhere super-Alfvénic. Nor have we found jets in which more than about 60% of the total power is carried by the particles.

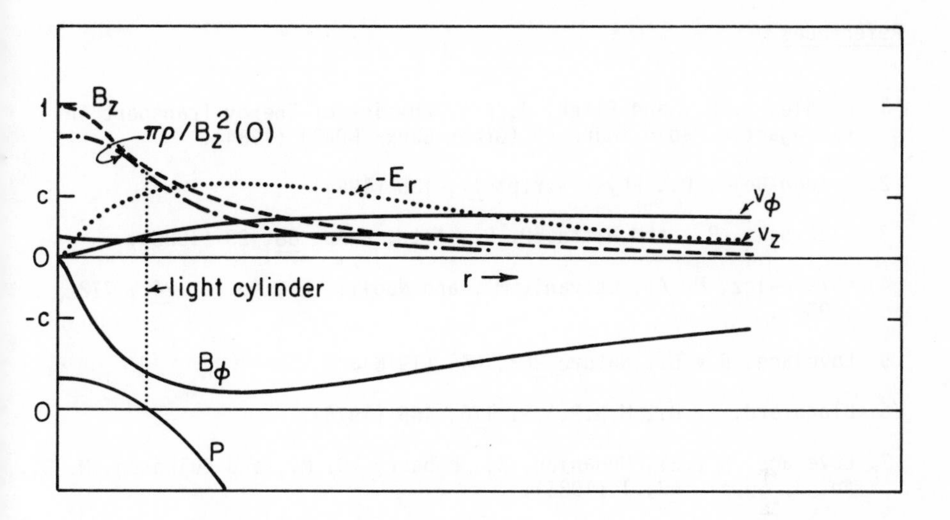

Figure 3. Fields of a self-collimated inertial jet. The vertical dotted line indicates the light-cylinder radius where $P(r) = 0$. For the case shown, 20% of the jet power is carried by the particles while the remainder is carried by the Poynting flux. The particles carry 58% of the angular momentum with the remainder carried by the B field. The average axial velocity of the matter is 0.306 c.

References

1. Bridle, A. H., and Eilek, J. A., "Physics of Energy Transport in Extragactic Radio Sources" (Green Bank: NRAO) (1984).

2. Lynden-Bell, D., Phys. Scripta 17, 185 (1984).

3. Paczynski, B., and Witta, P. S., Astr. & Ap. 88, 23 (1980).

4. Abramowicz, M. A., Calvani, M., and Nobili, L., Ap. J. 242, 772 (1980).

5. Lovelace, R.V.E., Nature 262, 649 (1976).

6. Blandford, R. D., M.N.R.A.S. 176, 465 (1976).

7. Lovelace, R.V.E., Mehanian, C., Mobarry, C. M., and Sulkanen, M. E., Ap. J. Suppl. 62, 1 (1986).

8. Mobarry, C. M., and Lovelace, R.V.E., Ap. J. 309, 455 (1986).

9. Lovelace, R.V.E., Wang, J.C.L., and Sulkanen, M. E., Ap. J. (to appear 1987).

10. Lovelace, R.V.E., Wang, J.C.L., and Sulkanen, M. E., preprint (1987).

11. Sramek, R. A., and Weedman, D. W., Ap. J. 221, 468 (1978).

12. Weedman, D. W., in "Active Galactic Nuclei" (ed. C. Hazard and S. Mitton) (Cambridge, Cambridge University Press), p. 121 (1979).

13. Strittmatter, P. A., Phys. Scripta 17, 145 (1978).

14. Wolfe, A. M. (ed.), "PittsburghConference on BL Lac Objects" (Pittsburgh: University of Pittsburgh) (1978).

HYDROGEN ABUNDANCE IN THE ULTRA-COMPACT BINARY 4U1820-30 ?

SHIGEKI MIYAJI[1,2]

Space Science Laboratory,NASA/Marshall Space Flight Center

[1]NAS/NRC Resident Research Associate

[2]On leave from Dept. of Natural History, Chiba Univ., Japan

Recent observations by EXOSAT satellite indicate a possible existence of hydrogen envelope (Miyaji 1986). Here, detailed analysis of the data (Ebisuzaki and Miyaji 1986) and other possible evidence of this hydrogen hypothesis are presented.

The result of analysis are shown in Figures 1 and 2. As shown in Figure 1, the data in cooling phase systematically differ from theoretical luminosity-color temperature relation (thick solid curve). If we accept hydrogen hypothesis and allow a transition from helium rich atmosphere (thin solid curve) to hydregen rich atmosphere (dashed curve) as like MXB1636-536, the fit becomes significantly better. Obtained mass and radius of the neutron star at 6.4 kpc distance are indicated by shadow areas in Figure 2.

REFERENCES

Ebisuzaki, T. and Miyaji, S. 1986, in this Symposium.
Miyaji, S. 1986, submitted to A. J. (Letters).

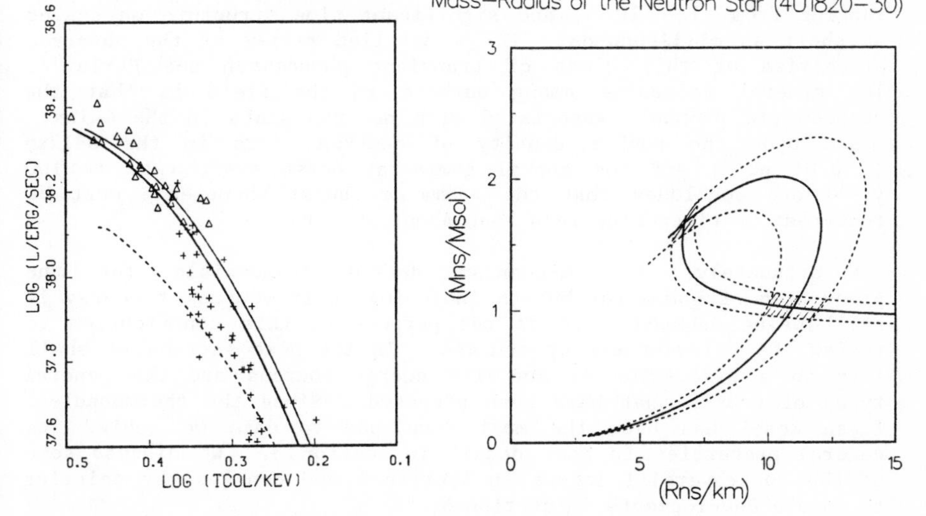

Figure 1

Figure 2

ENERGY SOURCES IN GAMMA-RAY BURST MODELS

Ronald E. Taam

Department of Physics and Astronomy
Northwestern University
Evanston, Illinois 60201

ABSTRACT

The current status of energy sources in models of gamma-ray bursts is examined. Special emphasis is placed on the thermonuclear flash model which has been the most developed model to date. Although there is no generally accepted model, if the site for the gamma-ray burst is on a strongly magnetized neutron star, the thermonuclear model can qualitatively explain the energetics of some, but probably not all burst events. The critical issues that may differentiate between the possible sources of energy for gamma-ray bursts are listed and briefly discussed.

1. Introduction

Despite the accumulation of a wealth of observational data since the discovery of cosmic gamma-ray bursts (first reported in 1973[1])) the nature and origin of the phenomenon still remains enigmatic[2,3]). Observationally, these bursts are distinguished by their hard spectrum with most of their energy emitted in the 30 - 300 keV range. For some sources the emission even extends to energies as high as 10 MeV. The burst time histories are characterized by rise times less than 0.1 to 1 s, total durations ranging from .1 - 10 s, and significant time structure on scales as short as milliseconds. For a detailed review of the observed properties of this class of transient phenomenon see Hurley[4]). The general consensus among workers in the field is that the sources are probably associated with neutron stars in the Galaxy. Based upon the number density of neutron stars in the Galaxy (~ 0.01 pc^{-3}) and the annual gamma-ray burst event rate (~ 100 yr^{-1}) one concludes that the gamma-ray burst phenomenon must be recurrent on timescales less than about 10^7 yrs.

Unfortunately, the mechanism deemed responsible for the production of gamma-ray bursts including their source of energy is still being debated. It is the purpose of this contribution to review the relevant energy sources. In the next section we shall turn to a discussion of specific energy sources and the general types of models that have been proposed. Since the thermonuclear flash model has been the most developed to date we review its general properties in some detail in section 3. We discuss some of the controversial issues in section 4 and conclude by pointing to future developments in section 5.

2. Specific Energy Sources

A key in determining the relevant energy sources is provided by an estimate of the characteristic energies of the gamma-ray burst events. From the observational data[4] the fluence, S, lies in the range between 10^{-7} and 10^{-3} ergs cm^{-2}. Since the distance to the sources is unknown it is usually assumed (based upon their nearly isotropic distribution) that they must either be very far (> 30 Mpc) or very close (< several hundred pc or in a halo population ~ 1-10 kpc).

If we adopt a characteristic fluence of 10^{-5} ergs cm^{-2} then for sources at extragalactic distances the gamma-ray burst energy exceeds 10^{47} ergs. Since a significant part of the energy is emitted in spikes of widths less than ~ 0.1 s the gamma-ray luminosities exceed 10^{48} ergs s^{-1}. A cosmological model which presumes that the sources are at redshifts of 1 or 2 (corresponding to burst energies of 10^{51} ergs, perhaps involving supernova explosions) has been recently revived by Paczynski[5]; however, constraints on the size of the emitting region obtained from arguments based on time variability and the blackbody limit may pose severe problems[2].

On the other hand, if the sources are Galactic at distances ~ 0.1 - 1 kpc then the gamma-ray burst energies are ~ 10^{36} - 10^{38} ergs. For these characteristic energies the bursts can be explained in terms of models invoking the episodic accretion of matter onto a neutron star[6-10], a readjustment of the interior structure of a neutron star[11,12], or a nuclear shell flash occurring in the accreted layers of a neutron star[13-15].

In models where the energy is derived from the accretion of matter onto the neutron star surface the energy released corresponds to Q_{grav} ~ 140 MeV per nucleon. If the conversion of gravitational energy to gamma-ray energy were 100% efficient this would correspond to an accreted mass of $10^{12}(d/pc)^2$ gm. In this context models invoking the impact of comets or asteroids[6-8] or of matter dumped (as a result of the nonlinear development of an instability in an accretion disk[9,10]) onto a neutron star have been proposed. Although these models are attractive, they are incomplete. The identification and detailed study of the relevant physical processes will be required before predictions can be made regarding for example the burst recurrent rate, the rise timescale, and burst duration.

In addition to the models involving activity on the surface of a neutron star, models have been proposed involving activity in the interior. Such models are based on the internal readjustment of the neutron star discussed either in terms of pulsar glitches[11] (where rotational energy is converted into gamma-ray energy) or phase transitions involving the formation of pion condensates[12] (where the release of gravitational energy associated with a slight contraction of the neutron star is converted into gamma-rays). At the present time, these models are out of favor[3].

Specifically, in the glitch model where the energies would be comparable to the energies observed in pulsar glitches the sources would be at distances less than 10 pc. Since the number of neutron stars within 10 pc is expected to be about 100 one would expect to observe one burst per year per neutron star. This is much too frequent since the observationally derived lower limit to the burst recurrent timescale is 10 yr[4]. In the phase transition model the formation of a pion condensate (if it exists) would only occur once in the lifetime of a neutron star and, hence, the presumed associated gamma-ray burst event would be extremely rare.

In contrast to the majority of the former models where the energy release is gravitational in origin a number of models have been constructed based upon the premise that the nuclear energy released by fusion in the accreted matter is the primary energy source[13-15]. For nuclear fusion the energy yield corresponds to $Q_{nuc} \sim 7.5$ MeV per nucleon for the conversion of H to Fe. Since $Q_{grav} > Q_{nuc}$ the accretion luminosity dominates the nuclear contribution unless the accreted fuel is stored and then burned rapidly. In particular, a prediction of such a model is that the ratio of the bolometric quiescent luminosity to the time averaged burst luminosity (where the average is taken over a recurrence time) should be equal to Q_{grav}/Q_{nuc}. This type of model has been very successful in explaining x-ray bursters as thermonuclear flashes occurring on the surfaces of non magnetized or weakly magnetized neutron stars[16,17]. Based upon its success it has been modified and extended to gamma-ray burster phenomenon[14,15]. The fundamental difference between the two models stems from the strong magnetic fields and low mass accretion rates presumed for the gamma-ray bursters. The presence of a strong magnetic field can localize the region of the accreted matter to ~ 0.1% of the surface area of the neutron star, confine the heated matter to a small region, and produce the hard non-thermal radiation which is characteristic of the gamma-ray bursts. Low accretion rates are needed for otherwise the accreted matter would be burned continuously and one would have a model for an x-ray pulsar. To produce non steady, unstable burning on the surface of strongly magnetized neutron stars mass accretion rates $< 10^{-11}$ $M_{\odot}$ yr^{-1} are required. These rates may be achieved, for example, in the case for which the neutron star accretes matter from a cloud in the interstellar medium, from the interstellar medium itself, or from a companion in a binary system. It is to the detailed discussion of the thermonuclear model that we now turn our attention.

3. Thermonuclear Model

In contrast to the model for x-ray bursts the magnetic field channels the matter to a polar cap region which leads to the result that for a given mass accretion rate, the critical accumulated masses and hence energies of outburst and recurrence timescales are reduced in comparison with accretion over the entire surface. Because the relevant mass accretion rates are low ($< 10^{-13}$ $M_{\odot}$ yr^{-1}) and the energy released by nuclear reactions (produced in the polar cap region) is radiated over the entire surface (not just in the localized region) the envelope is relatively cool (10^6 - 10^7 K) and nuclear fusion occurs at high densities (~ 10^7 g cm^{-3}). Since the mass accretion timescale is long compared with the gravitational settling timescale, the heavier ions diffuse through the hydrogen layer resulting in a nearly pure hydrogen layer overlying a layer composed of helium and heavier elements. Gamma-ray burst models also differ from x-ray burst models in their nuclear burning development since electron capture reactions are important in the former in the case where the neutron star is cool (T ~ 10^6 -10^7 K). Here the temperature in the accreted layer is much cooler than the case in which the neutron star accretes over the entire surface at the same mass accretion rate per unit area. In this case hydrogen burns via the reaction $p(e^-,\nu)n(p,\gamma)^2H(p,\gamma)^3He(^3He,2p)^4He$. In addition, at higher temperatures where hydrogen burns via the CNO cycle electron captures accelerate the rate of the beta limited CNO cycle. For mass accretion rates greater than about 10^{-15} $M_\odot$ yr^{-1} the hydrogen burning heats the envelope to temperatures of ~ 10^8 K whereupon helium ignites via the triple alpha reaction. Here, the hydrogen burning region which is temporarily stabilized by the weak interaction rates triggers a thermonuclear runaway driven by the highly temperature dependent reaction rates characteristic of helium burning in a manner similar to that found for the combined H-He flashes in x-ray burst models[18-20].

On the other hand, for hotter neutron stars where the interior temperatures exceed about 5×10^7 K, the helium flash is not triggered by the hydrogen flash[14]. Here, the accreted envelope evolves into one where the hydrogen burning reaches a steady state and a helium layer accumulates beneath the hydrogen burning shell. Eventually, when sufficient mass has been processed through the stable burning shell, the accumulated helium is burned in a shell flash. The character of the burning can be quite different than found in x-ray burst models since the density at which the matter burns is about 10^7 - 10^8 g cm^{-3} (about a factor of 10 to 100 times higher than the densities encountered in x-ray burst models). At such high densities helium burning may either produce a detonation wave (at densities of 10^8 g cm^{-3}) or a deflagration wave[14] (at densities ~ 10^7 g cm^{-3}). In the former case the matter will be ignited over the layer within a millisecond while in the latter case the matter in the polar cap will be ignited on a much longer timescale (~ several seconds). It is important to note that the energy can be transported much more rapidly in the former case than in the latter case. Because the burning occurs at such high densities the envelope temperatures rise to 4-5 x 10^9 K with most of the light nuclei converted into ^{56}Ni. As a result of the large internal nuclear luminosities (super Eddington) generated near the base of the accreted layer which are transported to the surface, the surface temperature and flux rise to exceed their local Eddington values and a wind develops from the polar cap of the neutron star. The magnetic field, important for localization during the quiescent phase, is equally important during this explosive phase which distinguishes it from x-ray burst models since (i) it provides for efficient energy transport to the surface[3] (for otherwise the rise time of the burst may be too long), (ii) it prevents the matter from spreading over the entire neutron star surface (which would otherwise degrade the hard emission) and (iii) it facilitates the acceleration of electrons to relativistic speeds as explosion energy stored in the magnetic field is released when the stressed field lines reconnect[21]. The interaction between the electrons and the soft photons by inverse Compton scattering for example, may produce the hard non thermal radiation. Since $Q_{grav} \gg Q_{nuc}$ only a small fraction of the accreted matter is ejected (≤ 0.001). For typical neutron star parameters, the recurrence timescale of the outburst is roughly 50 yrs $\Delta M(10^{20}gm)/\dot{M}\ (10^{-15}\ M_{\odot}\ yr^{-1})$.

Since the thermonuclear model has been studied in some detail a number of its predictions can be observationally tested. Among the main predictions are the following: (i) bursts should be recurrent with the energy in the burst proportional to the recurrence timescale, (ii) the recurrence timescale should be inversely proportional to the quiescent bolometric luminosity, and (iii) since hydrogen is depleted after helium (hydrogen burning involves weak interactions whereas helium burning involves charged particle interactions) the energy from hydrogen burning and that stored in the envelope is radiated away as x-rays after the explosive phase on timescales of minutes.

A detailed comparison of the first two predictions with observations cannot yet be attempted since gamma-ray bursts have not been observed to recur (except for perhaps three sources which are soft gamma-ray repeaters[22]). On the other hand, the post burst phase has been observed in x-rays in the 3 -10 keV range and there is some evidence, but not in all bursters, for a cooling x-ray tail[23]. Although this evidence provides some support for the thermonuclear model, the fact that the observed ratio of the x-ray to gamma-ray luminosity is only 0.02[23] may present serious difficulties for the model unless the gamma-rays are beamed or the total x-ray emission in the 3 - 10 keV range can be suppressed. Observations of the quiescent x-ray luminosity may also provide some constraints on the mass accretion rates, burst energies, and recurrence timescales[24] of the thermonuclear model, but they are still somewhat inconclusive (see section 5).

4. Important Issues

Two issues which would have immediate impact on the suitability of the above described models involve questions regarding the strength of the magnetic field of the neutron star and the distance to the sources. The former has important implications for the thermonuclear model which would no longer be considered viable if the magnetic field strength were not strong. Similarly, it would impact on emission models involving the synchrotron process for the production of gamma-ray[25] and cyclotron process for the optical bursts[21,26]. On the other hand, if the neutron stars were found to have strong fields it would impact on our understanding of the neutron star population and on the issue of magnetic field decay.

Source distances would also be important since they would place important constraints on the energetics (and perhaps distinguish between the various sources of energy if the accreted mass involved in the outburst could be estimated), the space density of sources, and the relevant radiation processes involved in producing the hard non thermal radiation.

5. Future Theoretical Work

At this point it is clear that the theory for gamma-ray bursts is still controversial. The majority view is that the gamma-ray burst sources are involved with neutron stars. In order to confront theory and observation in a more meaningful way more observations and much more theoretical work on the origin of matter, the mechanism for energy release, and the conversion to gamma-rays will be required before a self consistent picture can emerge. A few of the theoretical problems that must be addressed are summarized below.

1. The physics of accretion onto magnetized neutron stars should be investigated especially with regard to the spectrum emitted during the quiescent state. This study can have important implications for the thermonuclear model as it will place constraints on the mass accretion rate.

2. The physics underlying the affects of the magnetic field for energy transport and for the production of hard non-thermal radiation should be clarified for thermonuclear models and the response of the neutron star magnetosphere to episodic mass injection and the interaction of photons with the magnetic field studied for magnetospheric models.

3. Finally, the origin of the captured matter and the nature and physics of disk instabilities needs further elucidation in accretion models.

This work was supported in part by the NASA Astrophysics Theory Program under grant NAGW-768.

References

[1] Klebesadel, R. W. et al., Ap. J. Letters, 182, L85 (1973).
[2] Ruderman, M., Ann. N. Y. Acad. Sci., 262, 164 (1975).
[3] Hameury, J. M., and Lasota, J. P., Gamma Ray Bursts, AIP Conf. Proc. 141, ed. E. P. Liang and V. Petrosian, (AIP, New York), 164 (1986).
[4] Hurley, K., Accreting Neutron Stars, MPE Report 177, ed. W. Brinkmann and J. Trumper, (Garching: Max Planck Institut), 161 (1982).
[5] Paczynski, B., Ap. J. Letters, 308, L43 (1986).
[6] Harwit, M. and Salpeter, E. E., Ap. J. Letters, 186, L37 (1973).
[7] Newman, M. G., and Cox, A. N., Ap. J., 242, 319 (1980).
[8] Michel, F. C., Ap. J., 290, 721 (1985).
[9] Epstein, R. I., Ap. J., 291, 822 (1985).
[10] Tremaine, S., and Zytkow, A., Ap. J., 301, 155 (1986).
[11] Pacini, F., and Ruderman, M., Nature, 251, 399 (1974).
[12] Ramaty, R. et al., Nature, 287, 122 (1980).
[13] Woosley, S. E., and Taam, R. E., Nature, 263, 101 (1976).
[14] Woosley, S. E., and Wallace, R. K., Ap. J., 258, 716 (1982).
[15] Hameury, J. M. et al. Astr. Ap., 111, 242 (1982).
[16] Joss, P. C., and Rappaport, S. A., Ann. Rev. Astr. Ap., 22, 537 (1984).
[17] Taam, R. E., Ann. Rev. Nuc. Part. Sci., 35, 1 (1985).
[18] Taam, R. E., and Picklum, R. E., Ap. J., 224, 210 (1978).
[19] Taam, R. E., and Picklum, R. E., Ap. J., 233, 327 (1979).
[20] Ayasli, S., and Joss, P. C., Ap. J., 256, 637 (1982).
[21] Woosley, S. E., High Energy Transients in Astrophysics, AIP Conf. Proc. 115, ed. S. E. Woosley (AIP, New York), 485 (1984).
[22] Laros, J. G. et al., Bull. AAS, 18, 928 (1986).
[23] Laros, J. G. et al., High Energy Transients in Astrophysics, AIP Conf. Proc. 115, ed. S. E. Woosley (AIP, New York), 378 (1984).
[24] Pizzichini, G. et al., Ap. J., 301, 641 (1986).
[25] Liang, E. P. T., Nature, 299, 321 (1982).
[26] Hartmann, D., Woosley, S. E., and Arons, J., Bull. AAS, 18, 928 (1986).

PHYSICS OF CONTINUUM SPECTRA OF GAMMA-RAY BURSTS

ANDRZEJ A. ZDZIARSKI
Theoretical Astrophysics
California Institute of Technology
Pasadena, CA 91125, USA

ABSTRACT

The models proposed for the formation of continuum radiation of γ-ray bursts are reviewed. The current status of the X-ray and γ-ray observations is briefly presented. Compactnesses, geometries, and magnetic fields that may characterize γ-ray burst sources are discussed. The models invoking bremsstrahlung, synchrotron, Compton, and blackbody radiation, and thermal and nonthermal electron distributions are discussed and their predictions compared with observations. A particular model based on repeated Compton scatterings by power law electrons is found to fit the observed continua best. This model, however, does not predict any X-ray absorption features, possibly seen in some burst spectra.

1. INTRODUCTION

We review current theoretical models of radiative processes in γ-ray bursts (see also Ref. 20). We do not discuss the possible ways in which the radiating particles are supplied with energy (see Ref. 18). Line formation is discussed only briefly, and the models that, in our opinion, do not yield yet specific spectral predictions are not discussed.

We define the following dimensionless quantities:

$$x \equiv \frac{E}{m_e c^2}, \quad x_B \equiv \frac{B}{B_{cr}}, \quad \Theta \equiv \frac{kT}{m_e c^2}, \quad \ell \equiv \frac{L\sigma_{\rm T}}{R m_e c^3}, \tag{1}$$

where E is the dimensional photon energy, $B_{cr} = 10^{13.65}$ G, R is the source radius, L is the total luminosity, and ℓ is the "compactness parameter". We use the energy spectral index α defined by $F_E \propto E^{-\alpha}$. We denote the neutron stellar radius by R_{ns}, and the radiating area by A.

2. SUMMARY OF OBSERVATIONS

The γ-ray bursts radiate peak fluxes[34,35] $F \lesssim 10^{-4}\,{\rm erg\,s^{-1}\,cm^{-2}}$, corresponding to the luminosities, $F = 10^{-6}\,{\rm erg\,s^{-1}\,cm^{-2}}$, the corresponding luminosity is,

$$L \simeq 10^{38} \frac{F}{10^{-r4}} \left(\frac{d}{100\,{\rm pc}}\right)^2 {\rm erg\,s^{-1}} \simeq L_{\rm Edd} \frac{F}{10^{-4}} \left(\frac{M}{M_\odot}\right)^{-1} \left(\frac{d}{100\,{\rm pc}}\right)^2, \tag{2}$$

where d is the distance to the source, and $L_{\rm Edd}$ is the Eddington luminosity. The observed variability time scales are > a few ms (Ref. 50 and references therein), which implies $R \lesssim 10^8$ cm in the fastest bursts, provided no relativistic beaming occurs. The rise time of the unusual 5 March 1979 burst (GB790305) was[6] < 0.2 ms, corresponding to $R \lesssim 10^{6.7}$ cm. If bursts are connected to neutron stars or black holes, $R \gtrsim 10^6 (M/M_\odot)$ cm.

Probably the simplest description of the spectral continua of most γ-ray bursts is a low energy power law[9,21] with an energy spectral index $\alpha_X \sim 0$ and a high energy power law[31] with α_γ varying from ~ 0.5 to ~ 2, with a transition between the two regimes occuring between 50 keV and 1 MeV. The spectral form above implies that most of the γ-ray burst energy is emitted in γ-rays. The typical observed X-ray/γ-ray luminosity ratio is[25] L_X(3–10 keV)/L_γ(>100 keV) $\sim$ 0.02. The $\alpha_X \sim 0$ power law with a high-energy break is consistent with the bremsstrahlung fits (see §4.1) to the 30 keV to $\sim$ 1 MeV burst spectra observed by the Konus experiment[36,37]. The γ-ray power laws, extending to $\sim$ 5–10 MeV or more, have been observed, e.g., by the Solar Maximum Mission[31,32,46] (SMM), and Signe[23] instruments. No clearly defined high-energy cutoffs were found in any of the SMM spectra[32]. Some Konus spectra[34,35] show exponential cutoffs at $\lesssim$ 100 keV (e.g., the very strong event GB790930). Such soft events are sometimes classified as a separate class of γ-ray bursts[26].

Fig. 1 shows spectra from two strong γ-ray bursts. The scale on the vertical axis is the received power per logarithmic photon energy interval, $\propto E^2 dN/dE$, directly showing where most of the burst energy is emitted. The GB840805 spectra[46] (Fig. 1*a*) break at $\sim$ 0.5 MeV. The solid lines are least-square power law fits obtained for the energy ranges where they are plotted. The index $\alpha_X \simeq 0$, and α_γ increases from 1 to $\sim$ 1.5. The GB830801b spectrum[23] (Fig. 1*b*) was initially qualitatively similar to that of GB840805, although with a much harder low-energy part ($\alpha_X \simeq -0.7$). It then evolved in time with the break energy decreasing to $\sim$ 100 keV and α_γ increasing to $\sim$ 2.5. In general, γ-ray burst spectra seem to soften with decreasing flux[13], and with time in individual pulses[41].

Absorption features between 30 and 80 keV have been reported in about 30% of burst spectra[36,37]. Their observational status is somewhat controversial as they are sometimes attributed to errors in detector gains or found to be artifacts of the thermal bremsstrahlung spectrum assumed in spectral unfolding[10,24]. Mazets *et al.*[36,37] did not analyze the statistical significance of these features. The evidence for broad emission features around 400 keV claimed to be seen in 10% of bursts[36,37] is even weaker. Some of these features disappear if a spectrum different from thermal bremsstrahlung is used in the spectral unfolding[11]. No spectra with lines of the width < 70 keV were found in the data from 60 bursts observed by SMM[39], and no broad line detections have been reported. A broad line at $\sim$ 400 keV claimed for GB811231 by Mazetz *et al.*[36] has not been confirmed by SMM[40].

3. FEATURES OF THEORETICAL MODELS

We will discuss here some basic features characterizing theoretical models of radiative processes in γ-ray bursts.

3.1 Compactness

The time variability arguments give $R \lesssim 10^8$ cm (§2). If no relativistic beaming occurs, we find that the sources of strong bursts are compact ($\ell > 1$) unless $d \lesssim 20$ pc,

$$\ell \simeq 30 \frac{F}{10^{-4}} \left(\frac{R}{10^8\,\mathrm{cm}}\right)^{-1} \left(\frac{d}{100\,\mathrm{pc}}\right)^2 . \tag{3}$$

If $\ell \gtrsim 1$, the cooling time of a mildly relativistic electron is shorter than the light crossing time, and γ-γ pair production can affect the radiated spectra[16,49]. The optical depth to pair production $\tau_{\gamma\gamma} \propto \ell$, and the Thomson depth of the pairs in pair equilibrium is roughly[16,49] $\tau_{\rm pair} \sim \min\left[\ell/4\pi,\, (\ell/4\pi)^{1/2}\right]$. Zdziarski and Lamb[52] found that pair absorption results in a substantial steepening of γ-ray-burst-like spectra above 511 keV for $\ell \gtrsim 10^2$ (cf. Fig. 2*b*). For bursts with flat γ-ray spectra, this implies,

$$d \lesssim 2 \left(\frac{F}{10^{-6}}\right)^{-1/2} \left(\frac{R}{10^8\,\mathrm{cm}}\right)^{1/2} \mathrm{kpc}. \tag{4}$$

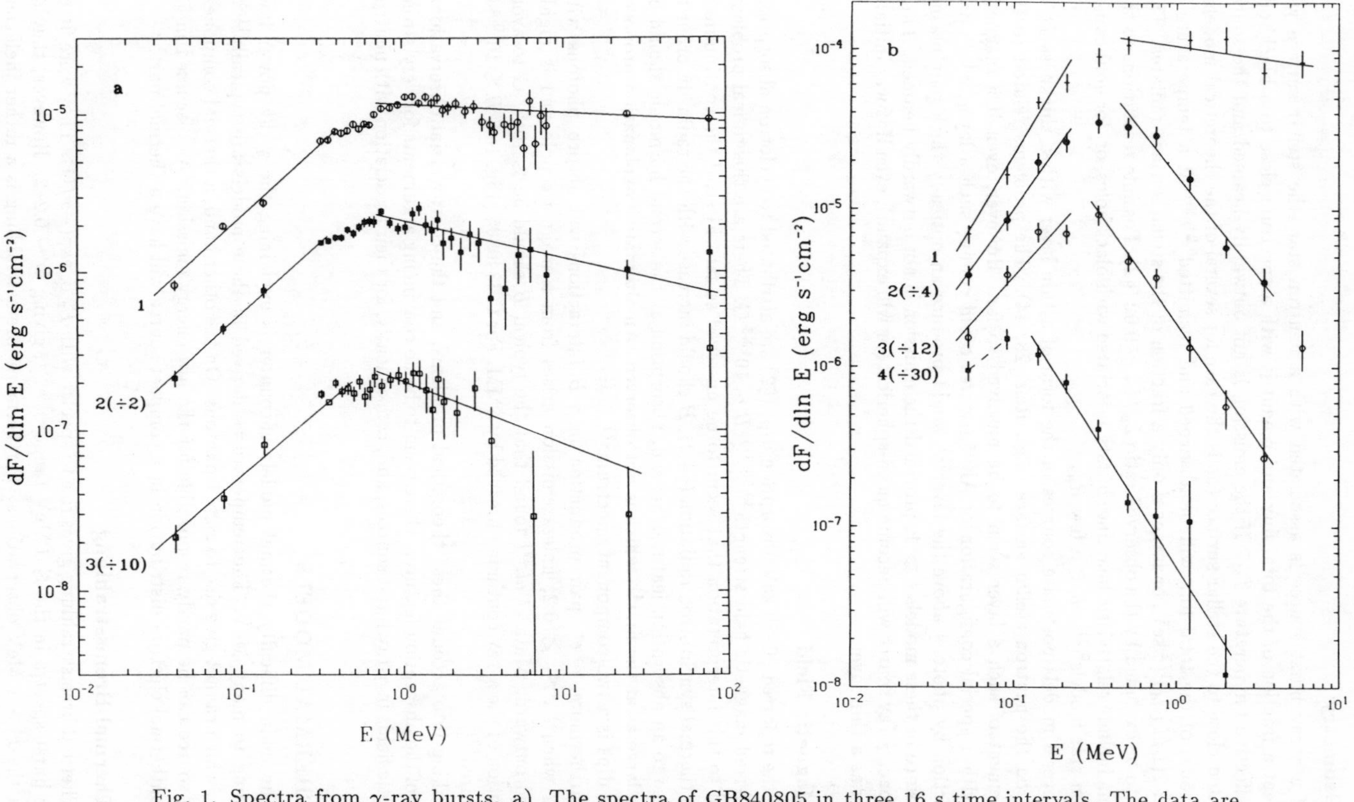

Fig. 1. Spectra from γ-ray bursts. a) The spectra of GB840805 in three 16 s time intervals. The data are from Refs. 45 and 46. b) The spectra of GB830801b in four 0.5 s time intervals. The data are from Ref. 23. See text for further details.

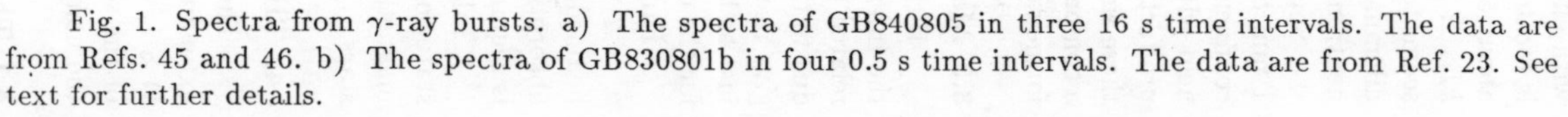

3.2 Geometry

If a γ-ray burst source is associated with a neutron star, the stellar surface will intercept a fraction of the total flux and reemit it with a spectrum close to a blackbody with effective temperature T_{bb}. If the emission is not outwardly beamed and the source is located close to the stellar surface (as is the case for several of the theoretical models), about 50% of the total flux will be absorbed and reemitted[9,21,38] at a temperature of $kT_{bb} \simeq 2(L/L_{\rm Edd})^{1/2}$ keV, or larger if only a fraction of the stellar surface radiates. This seems to be excluded by the observed ratio $L_x/L_\gamma \sim 0.02$ (§2). Possible resolutions to this dilemna include relativistic beaming, bursts associated with black holes, or, for nonbeamed neutron star models[9,21], $R \gtrsim$ a few R_{ns}.

Several models postulate sources in the form of a thin layer with a thickness $D \ll 1$ mm on the neutron stellar surface (e.g., Refs. 28, 44). The problems related to the confinement of such a layer seem to be unsolved so far. However, even if a magnetic field with a special configuration or Alfvén waves could confine such a layer, e^+e^- pair production by photons *above* the layer[51] would produce an optically thick pair plasma there since in these models $\ell \gg 1$, provided the emission is not outwardly beamed. Thus, a compact γ-ray source will become quasi-spherical (or will expand), even if it was initially formed as a thin layer.

3.3 Magnetic Field

If the reported 30–80 keV absorption dips (§2) are attributed to cyclotron absorption, the inferred magnetic field strength[30,36,37] $B \sim 10^{12.5}$ G. There are theoretical problems related to this interpretation that seem to be unsolved so far. First, if the continuum is due to thermal synchrotron radiation (§4.2), B should increase with the radius in order to give rise to an absorption feature. Second, absorption at the second harmonic should be just as large as at the first[4], which is not observed. An alternative explanation proposed for the dips is a two-component spectrum[27].

Furthermore, e^+e^- pair production in γ-B interactions will absorb photons with $x > 2$ when[8] $xx_B \gtrsim 0.2$, unless emission arises from a magnetic pole and is highly beamed outward. Matz *et al.*[32] found that the typical B should be $\lesssim 10^{11.5}$ G to avoid absorption of the γ-ray continua observed by SMM. For GB840805 (Fig. 1*a*), $B \lesssim 10^{10.5}$ G is required[19].

Taking into account these theoretical problems, and the controversial observational status of the absorption features, there seem to be no convincing arguments for very strong magnetic fields in γ-ray burst sources, and, consequently, for their association with neutron stars.

4. THERMAL MODELS

The main difficulty thermal models encounter is explaining the γ-ray power laws extending to many MeV. The temperatures derived in these models are typically[11,30] $\Theta \lesssim 1$, which cannot give rise to $x \gg 1$ photons. On the other hand, a thermal component of the source can be partly responsible for the low energy emission. As discussed in §5, the nonthermal electron distribution in a compact source will have a thermal part.

4.1 Thermal Bremsstrahlung

Thermal bremsstrahlung gives rise to spectra with $F_x \propto \exp(-x/\Theta)$. They were fitted to the burst spectra in the $\lesssim 1$ MeV range[12,36,37], giving $\Theta \sim 0.2$–2. However, they do not fit the $E > 1$ MeV observations. Furthermore, bremsstrahlung is a rather inefficient radiative process, and both synchrotron and Compton processes are more efficient for parameters considered characteristic for γ-ray bursts[20]. The same holds for nonthermal bremsstrahlung[20].

4.2 Thermal Synchrotron

For $x \ll \Theta$, the thermal synchrotron spectrum has an approximate form[43]

$$F_x \propto x \exp\left[-\frac{x}{2\Theta} - \left(\frac{9x}{2x_c}\right)^{1/3}\right]. \tag{5}$$

Here $x_c = x_B\Theta^2\langle\sin\theta\rangle$, and θ is the pitch angle. The emission peaks for $\theta = \pi/2$. For $x \ll \Theta$, the shape of this spectrum depends on x_c only, and B and Θ cannot be uniquely determined by fitting the continuum. Liang *et al.*[30] fitted Konus ($\lesssim$ 1 MeV) spectra neglecting the $x/2\Theta$ factor and obtaining $x_c \sim 10^{-2}$.

Interpreting the low-energy spectral features appearing in some Konus spectra as due to the lowest harmonic emission, Liang *et al.*[30] obtained $\Theta \sim 0.3$ and $B \sim 10^{12.7}$ G as typical for these events. This value of B is above the γ-B pair absorption limit (see §3). For bursts without low-energy emission features at $E > 30\,\mathrm{keV}$, $B \lesssim 10^{12.3}$ G is required. Inclusion of the $x/2\Theta$ factor gives[22,43] $B < 10^{12}$ G.

The thermal synchrotron fits to Konus data fall below the SMM power law spectra[43]. The thermal models with $\Theta \gg 1$ can be fitted[1,22] to some SMM spectra, but then they do not satisfy $L_X/L_\gamma \sim 0.02$.

The lack of self-absorption in most burst spectra for $E > 30\,\mathrm{keV}$ constrains the luminosity, and consequently, the distance[29]. Taking $\Theta = 0.3$, $B = 10^{12.7}$ G as typical, one obtains $L \lesssim 10^{40}(A/10^{10}\,\mathrm{cm}^{-2})\,\mathrm{erg\,s}^{-1}$, and

$$d \lesssim 10\left(\frac{F}{10^{-6}}\right)^{-1/2}\left(\frac{A}{10^{10}\,\mathrm{cm}^2}\right)^{1/2}\ \mathrm{kpc}, \tag{6}$$

(somewhat lower estimates are obtained[43] for lower values of B). In some Konus spectra, self-absorption turnovers appear to be seen[36,37] above 30 keV. In these cases, the expressions above represent the actual estimates[30] of L and d. The bursts are then highly super-Eddington and violate the pair absorption limit (eq. [4]).

In all thermal synchrotron models, an extremely efficient thermalization mechanism is required to keep electrons at higher Landau levels, as $t_{\mathrm{Coulomb}} \gg t_{\mathrm{cool}}$ (see, e.g., Ref. 20). To solve this difficulty, Hameury *et al.*[17] assumed that most electrons are at the lowest Landau levels and have a parallel temperature of $\Theta_\parallel \sim 1$. A small fraction of the electrons is at higher Landau levels due to excitation by γ-rays produced by Compton scatterings off the electrons moving along the field lines. The excited electrons are assumed to be thermal. We note here that the distribution of the excited electrons will in fact follow from balance of radiative excitations and deexcitations in the non-Planckian radiation field, and be nonthermal (see the models in §5.1).

The X-ray paucity constraint (§3.2) requires that nonbeamed radiation be produced at $R >$ a few R_{ns}. This is not satisfied, e.g., in the models in Refs. 17 and 30. Furthermore, pair equilibrium (§3.1, Ref. 48) will result in large Thomson depths for $\ell \gg 1$, whereas for thermal synchrotron models $\tau_\mathrm{T} \ll 1$ is required (e.g., Ref. 30). To conclude, we consider thermal synchrotron models unlikely for γ-ray bursts.

4.3 Thermal Comptonization

A thermal Comptonization model for γ-ray bursts was proposed by Fenimore *et al.*[11]. In the model, blackbody photons emitted by the neutron stellar surface are repeatedly upscattered in a mildly relativistic thermal corona above the surface. Spectra of GB781104 in three time intervals were fitted by this model, yielding $\tau_T \sim 1$–3, and $\Theta \sim 0.3$–0.5. For the first two intervals, Comptonization was close to saturation, yielding $\alpha \simeq 0$ and Wien features. The Wien features fill in the broad excesses at ~ 400 keV in the observed spectra, previously claimed as redshifted annihilation lines[36]. The unscattered blackbody photons may give rise to a soft component at low energies, similar to ones observed in some Konus spectra[37].

The stellar surface may be heated either by a thermonuclear explosion, similar to those in X-ray bursts, or, more likely, by the hot corona. We note here that this model implicitly assumes that most of the energy is released at $R \gg R_{ns}$. If the corona were close to the stellar surface, $\sim 50\%$ of its emission would be directed towards the surface, be reemitted as blackbody, and cool the corona. In a steady state without nuclear heating, the energy radiated by the surface must equal that reflected by the corona. For $\tau_T \sim 3$, the energy reflected by a slab approximately equals that radiated outward[15]. For the parameters of Fenimore *et al.*[11], the input luminosity is amplified 10–10^2 times, thus excluding the slab geometry. Only if $R \simeq 3$–$10R_{ns}$, the stellar surface will intercept a luminosity $\sim (R_{ns}/R)^2 L$, and the energy radiated will equal that absorbed (cf. §5.2).

4.4 Relativistic Blackbody

Paczyński[42] and Goodman[14] proposed that some γ-ray bursts are at cosmological distances. This requires a release of supernova-like energy ($10^{51}\,\mathrm{erg\,s^{-1}}$) within < 1 s, and gives an effective temperature $kT_{bb} = 3(L/10^{51}\,\mathrm{erg\,s^{-1}})^{1/4}(R/10^6\,\mathrm{cm})^{-1/2}$ MeV and $F \simeq 10^{-6}\,\mathrm{erg\,s^{-1}\,cm^{-2}}$ at $d \simeq 10^{28}$ cm. Such an event can arise, e.g., from a merger of two neutron stars. The released power drives an optically thick relativistic outflow of e^+e^- pair plasma from the source. This outflow becomes optically thin when the temperature drops to ~ 20 keV at $R \sim 10^2 R_{ns}$. Due to the Doppler blueshift, the source would look like a stationary one with $T = T_{bb}$ and $R = R_{ns}$. Pair absorption does not limit d as the escaping photons have $x < 1$ in the comoving frame of the outflow.

The model predicts no lines in the spectra. The spectra should look similar to blackbody[14], in disagreement with the observations. Perhaps a special pattern of time variability of L or a cool envelope surrounding the source could yield spectra similar to the observed ones[42].

5. NONTHERMAL MODELS

A nonthermal distribution will be formed if some process delivers a large fraction of the power to selected particles, which are then let cool. Since the cooling time of relativistic electrons in an $\ell \gtrsim 1$ source is shorter than the escape time, the nonthermal steady state distribution (typically a power law) will join onto a cooled thermal part. In strongly magnetized sources, this cooled part will consist of the electrons moving paralelly to the field lines.

Nonthermal electron distributions in γ-ray burst sources seem to be favored by the observations of hard power law spectra extending to $\lesssim 100$ MeV (§2). Extremely relativistic temperatures, which are physically unlikely, would be required in purely thermal models to give rise to such spectra.

5.1 Nonthermal Synchrotron

A nonthermal synchrotron model was proposed for GB790305 by Ramaty *et al.*[44]. In the model, monoenergetic e^+e^- pairs at $\gamma_0 = 7$ are injected into a magnetic field with $B = 10^{11}$ G. Here, γ is the Lorentz factor. The pairs cool via synchrotron emission to

$\gamma \simeq 1.2$, where they form a thermal distribution. The synchrotron spectrum from the nonthermal part of the distribution is given by[2],

$$F_x \propto x^{-1/2} \exp(-x/x_c), \tag{7}$$

where x_c is now $= (3/2)\gamma_0^2 x_B$. This fits the low-energy exponential part of the burst spectrum[33,44]. The harder emission comes from annihilation of the thermal pairs. In order for the latter emission to give rise to the flux required at the assumed 55 kpc distance[7], the emission region must form a very thin ($D \simeq 10^{-2}$ cm) layer around the neutron star.

We note here that the distance limit of $d \lesssim 1$ kpc derived for GB790305 by Liang[29] from the lack of synchrotron self-absorption at $E \geq 30$ keV does not apply for the particular model of Ramaty *et al.*[44]. Liang[29] considered the thermal[30] and nonthermal power law injection[1] cases. However, a monoenergetic injection gives much lower electron densities than that of those two models, and we obtained the self-absorption energy below 30 keV at $d = 55$ kpc, in agreement with the value given by Ramaty *et al.*[44].

The model of Ref. 44 was criticized by Zdziarski[51] as violating the pair absorption limit. The burst spectrum extends to at least 2 MeV[5] and the photon photosphere *above* the thin layer is completely optically thick to pair absorption ($\tau_{\gamma\gamma}[1\,\mathrm{MeV}] \sim 10^5$, Ref. 51). Thus, if the burst were at the distance of 55 kpc, e^+e^- pair production would give rise to a relativistic outflow such as discussed by Paczyński[42] and Goodman[14] (§4.4). The distance of 55 kpc gives $kT_{bb} \sim 70$ keV, in rough agreement with the observed spectral shape[33]. However, the distinct line[33] at ~ 400 keV is not reproduced by this model.

Bussard[2] studied models with monoenergetic electron injection at $\gamma_0 \leq 20$. For $\gamma_0 x_B \ll 1$, his Monte Carlo results conform to eq. (7). This, with $\alpha(x \ll x_c) = 0.5$, does not fit the majority of the burst spectra with $\alpha_X \sim 0$ (see §2). His models with $B = 10^{13}$ G, showing harmonic structure at $\sim 10^2$ keV and hard spectra, do not directly apply to γ-ray bursts, as hard bursts do not exhibit such structure[34,35].

Bussard[3], and Brainerd and Lamb[1] studied nonthermal power law injections, $\propto \gamma^{-\Gamma}$. The spectra they obtained fit the SMM power laws well, but do not fit the low energy $\alpha_X \sim 0$ spectra.

Sturrock[47] considered secondary synchrotron radiation from e^+e^- pairs produced by curvature γ-rays interacting with the magnetic field of a neutron star. The curvature radiation is emitted by highly relativistic electrons accelerated by magnetic field reconnection during a magnetospheric flare. The pairs radiate while crossing the dipole magnetic field lines. The resulting synchrotron spectrum has $\alpha = 0.5$ below the magnetic pair production threshold ($x x_B \sim 0.2$, cf. §3.3), and $\alpha = 2$ above it. This can fit some burst spectra if $B \sim 10^{12.5}$, but it does not explain the $\alpha_\gamma \sim 0.5$–1.5 power laws seen in many bursts (§2).

5.2 Nonthermal Compton

We are not aware of any nonthermal single Compton scattering models proposed for γ-ray bursts. Electrons injected with a power law distribution $\propto \gamma^{-\Gamma}$ would give rise to a photon power law with $\alpha = \Gamma/2$ (e.g., Ref. 49).

Repeated Compton scatterings by power law electrons were studied by Zdziarski and Lamb[52,53]. In order for repeated scatterings to dominate, the Thomson depth of the power law electrons must be $\gtrsim 1$. This occurs when the luminosity in the primary soft photons is much less then the power supplied to the injected electrons, $L_{\mathrm{soft}} \ll L_{\mathrm{el}}$. Then the relativistic electrons cool slowly and become optically thick. Solution of the kinetic equation for a power law injection, $\propto \gamma^{-\Gamma}$, gives the steady state electron distribution as

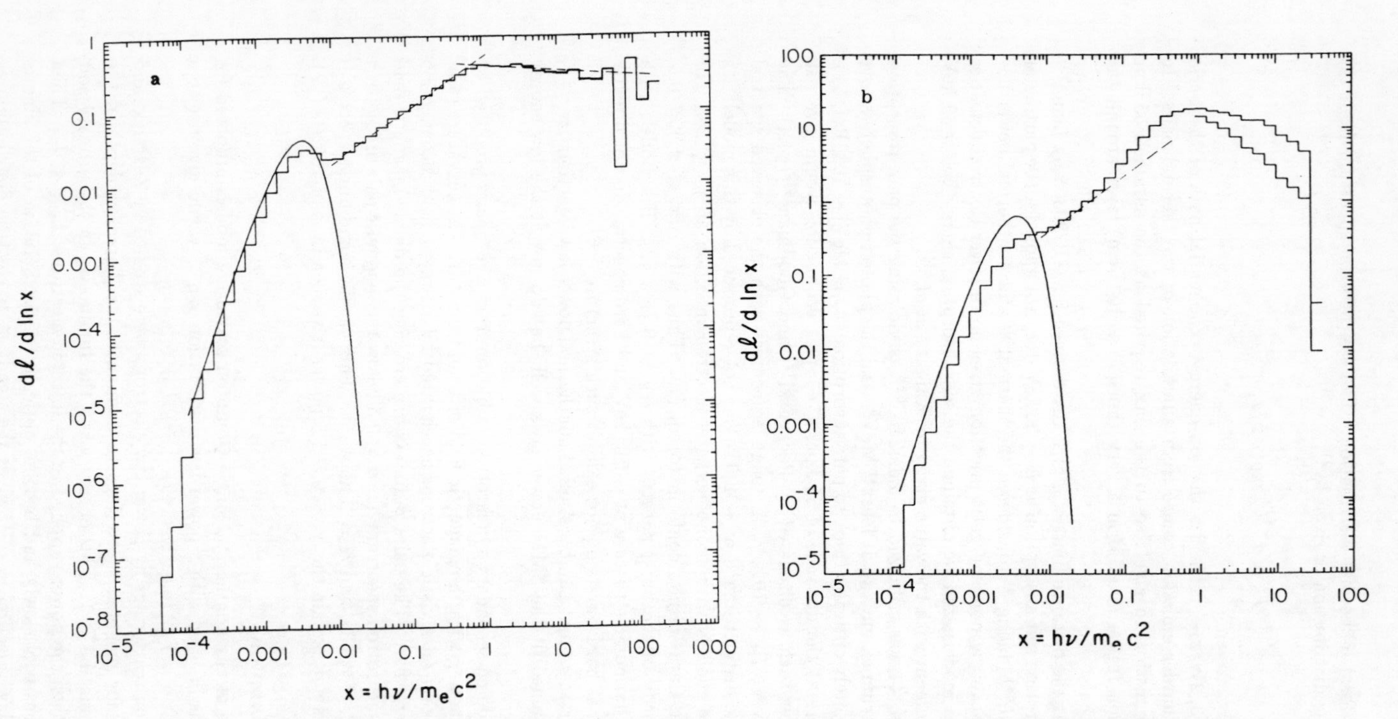

Fig. 2. Monte Carlo simulations of repeated Compton scattering (*histogram*) of blackbody photons with $kT_{bb} = 0.5\,\mathrm{keV}$ (*solid curve*) by power law electrons with the steady state distribution $\propto \gamma^{-2.5}$ and with the unit the Thomson depth, joining at low energies onto a thermal distribution. The high-energy cutoffs result from respective cutoffs in the power law electron distributions. Adopted from Ref. 53. See text for further details. a) The thermal electrons have $\tau_{\mathrm{T}} \ll 1$. This case yields $\alpha_X \simeq 0.3$ and $\alpha_\gamma \simeq 1.1$. b) The thermal electrons have $\tau_{\mathrm{T}} = 3$ and $\Theta = 0.1$. The lower histogram shows the pair-absorbed spectrum at $\ell = 70$.

a power law with another index, and a broken-power-law photon spectrum[52],

$$F_x \propto \begin{cases} x^{-\alpha_X}, & x \lesssim 1; \\ x^{-\alpha_\gamma}, & x \gtrsim 1, \end{cases} \tag{8}$$

where $\alpha_\gamma = \Gamma - 1$, α_X depends on $L_{\rm soft}/L_{\rm el}$, and $\Gamma > \alpha_X + 1$ is assumed. The break at $x \sim 1$ is due to the relativistic cutoff in the Klein-Nishina cross section. Such a spectrum can fit some γ-ray burst spectra, e.g., those of GB840805 (Fig. 1*a*). Fig. 2*a* shows an example[53] of the theoretical spectrum obtained using a Monte Carlo method.

If $\ell \gg 1$, there is an additional effect due to the cooled e^+e^- pairs forming an optically thick *thermal* component of the electron distribution. The spectrum retains its approximate broken power law form, but the break occurs now at an energy $x < 1$, decreasing with increasing thermal optical depth (cf. the spectra in Fig. 1*b*). Fig. 2*b* shows a spectrum[53] with the break at $x \sim$ 0.3–0.4.

The ratio $L_{\rm soft}/L_{\rm el} \ll 1$ and $\alpha_X \sim 0$ can be obtained, for example, if the primary power $L_{\rm el}$ is released at $R \gg R_{ns}$ (and if $B \lesssim 10^6$ G there). Then the stellar surface absorbs and reemits the luminosity $L_{\rm soft} \sim (R_{ns}/R)^2 L_{\rm el}$ as a blackbody flux. The expected initial spectrum is very hard, softening after the thermal flux cools the injected power law electrons. A hard-to-soft spectral evolution seems to be typical for γ-ray bursts[41] (cf. §2).

6. CONCLUSIONS

None of the discussed models can explain the two parts, low and high energy, of the γ-ray burst continua and the spectral features. The model by Zdziarski and Lamb[52,53] explains the entire continuum, but predicts no absorption dips. Two or multi-component models could most likely explain the continuum as well as the spectral features, but succesful models have not been proposed yet. In nonbeamed neutron star models, the $E \gtrsim 1\,\text{MeV}$ radiation has to originate at $R >$ a few R_{ns} and in a relatively weak magnetic field. The most important observational question concerns the reality of the absorption dips, whose existence would imply very large magnetic fields in γ-ray bursters.

I thank Gerald Share for providing me with the numerical data on GB840805, and Don Lamb for comments. I benefited from discussions with Włodek Kluźniak and Edison Liang. This research was supported in part by the NSF grants AST 84–15355 and AST 84–57125.

REFERENCES

1. Brainerd, J. J., and Lamb, D. Q., *Ap. J.*, **313**, in press (1987).
2. Bussard, R. W., *Ap. J.*, **284**, 357 (1984).
3. Bussard, R. W., in *Gamma-Ray Bursts*, ed. E. P. Liang and V. Petrosian, (New York: AIP), p. 147 (1986).
4. Bussard, R. W., and Lamb, F. K., in *Gamma Ray Transients and Related Astrophysical Phenomena*, ed. R. E. Lingefelter *et al.*, (New York: AIP), p. 196 (1982).
5. Cline, T. L., in *High Energy Transients in Astrophysics*, ed. S. E. Woosley (New York: AIP), p. 333 (1984).
6. Cline, T. L., *et al.*, *Ap. J. (Letters)*, **237**, L1 (1980).
7. Cline, T. L., *et al.*, *Ap. J. (Letters)*, **255**, L45 (1982).
8. Daugherty, J. K., and Harding, A. K., *Ap. J.*, **273**, 761 (1983).
9. Epstein, R. I., in *Radiation Hydrodynamics in Stars and Compact Objects*, ed. K.-H. Winkler and D. Mihalas (Berlin: Springer), p. 305 (1986).
10. Fenimore, E. E. *et al.*, in *Gamma Ray Transients and Related Astrophysical Phenomena*, ed. R. E. Lingefelter *et al.*, (New York: AIP), p. 20 (1982).
11. Fenimore, E. E., *et al.*, *Nature*, **277**, 665 (1982).
12. Gilman, D., *et al.*, *Ap. J.*, **236**, 951 (1980).

13. Golenetskii, S. V., Ilyinskii, V. N., and Mazets, E. P., *Nature*, **307**, 41 (1984).
14. Goodman, J., *Ap. J. (Letters)*, **308**, L47 (1986).
15. Górecki, A., and Wilczewski, W., *Acta Astr.*, **34**, 141.
16. Guilbert, P. W., Fabian, A. C., and Rees, M. J., *M. N. R. A. S.*, **205**, 593 (1983).
17. Hameury, J. M., *et al., Ap. J.*, **293**, 56 (1985).
18. Hameury, J. M., and Lasota, J. P., "Energy Sources", in *Gamma-Ray Bursts*, ed. E. P. Liang and V. Petrosian, (New York: AIP), p. 164 (1986).
19. Harding, A. K., Petrosian, V., and Bussard, R. W., in *Gamma-Ray Bursts*, ed. E. P. Liang and V. Petrosian, (New York: AIP), p. 127 (1986).
20. Harding, A. K., Petrosian, V., and Teegarden, B. J., "Spectra and Emission Mechanisms", in *Gamma-Ray Bursts*, ed. E. P. Liang and V. Petrosian, (New York: AIP), p. 75 (1986).
21. Imamura, J. M., and Epstein, R. I., *Ap. J.*, **313**, in press (1987).
22. Imamura, J. M., Epstein, R. I., and Petrosian, V., *Ap. J.*, **296**, 65 (1985).
23. Kuznetsov, A. V., *et al., Pisma Astr. Zh.*, **12**, 755 (Sp. Res. Inst. preprint No. 1076) (1986).
24. Laros, J. G., *et al., Astr. Sp. Sci*, **88**, 243 (1982).
25. Laros, J. G., *et al., Ap. J.*, **286**, 681 (1984).
26. Laros, J. G., *et al., Nature*, **322**, 152 (1986).
27. Lasota, J. P., and Belli, B. M., *Nature*, **304**, 139 (1983).
28. Liang, E. P., *Nature*, **310**, 121 (1984).
29. Liang, E. P., *Ap. J. (Letters)*, **308**, L17 (1986).
30. Liang, E. P., Jernigan, T. E., and Rodrigues, R., *Ap. J.*, **271**, 766 (1983).
31. Matz, S. M., Ph.D. thesis, University of New Hampshire (1986).
32. Matz, S. M., *et al., Ap. J. (Letters)*, **288**, L37 (1985).
33. Mazets, E. P., *et al., Nature*, **282**, 587 (1979).
34. Mazets, E. P., *et al., Astr. Sp. Sci.*, **80**, 3 (1981).
35. Mazets, E. P., *et al., Astr. Sp. Sci.*, **80**, 85 (1981).
36. Mazets, E. P., *et al., Nature*, **290**, 378 (1981).
37. Mazets, E. P., *et al., Astr. Sp. Sci.*, **82**, 261 (1982).
38. Nishimura, J., Fujii, M., and Yamagani, T., *Astr. Sp. Sci.*, **93**, 87 (1983).
39. Nolan, P. L., *et al.*, in *Positron-Electron Pairs in Astrophysics*, ed. M. L. Burns *et al.*, (New York: AIP), p. 59 (1983).
40. Nolan, P. L., *et al., Nature*, **311**, 360 (1984)
41. Norris, J. P., *et al., Ap. J.*, **301**, 213 (1986).
42. Paczyński, B., *Ap. J. (Letters)*, **308**, L43 (1986).
43. Pavlov, G. G., and Golenetskii, S. V., *Astr. Sp. Sci.*, **128**, 341 (1986).
44. Ramaty, R., Lingenfelter, R. E., and Bussard, R. W., *Astr. Sp. Sci.*, **75**, 193 (1981).
45. Share, G. H., private communication (1987).
46. Share, G. H., *et al., Adv. Sp. Res.*, in press (1987).
47. Sturrock, P. A., *Nature*, **321**, 47 (1986).
48. Svensson, R., *M. N. R. A. S.*, **209**, 175 (1984).
49. Svensson, R., *M. N. R. A. S.*, in press (1987).
50. Wood, K., *et al.*, in *Gamma-Ray Bursts*, ed. E. P. Liang and V. Petrosian, (New York: AIP), p. 4 (1986).
51. Zdziarski, A. A., *Astr. Ap.*, **134**, 301 (1984).
52. Zdziarski, A. A., and Lamb, D. Q. *Ap. J. (Letters)*, **309**, L79 (1986).
53. Zdziarski, A. A., and Lamb, D. Q., *Ap. J.*, submitted (1987).

A THIRD CLASS OF HIGH ENERGY BURSTER

J. G. Laros, E. E. Fenimore, and R. W. Klebesadel

Los Alamos National Laboratory

Los Alamos, NM 87544

Until recently, only two cosmic high-energy burster classes--x-ray bursters and gamma-ray bursters (GRBs)--were recognized. However, as early as 1979 it was realized that events similar to GRBs, but which did not fit comfortably into that class, occasionally occurred. This realization arrived in the form of the famous "March 5 Event" of 1979, which had, among its many unique attributes, typical photon energies of tens of keV--that is, midway between x-ray burster and GRB energies. Subsequently, the source was seen to repeat sporadically about a dozen times, with about one day separating the closest recurrences.[1] Eventually, a few other bursters unlike normal GRBs and possibly related to the March 5 source were found. All had similar average photon energies of tens of keV, and most had durations of tenths of a second. One of them, first seen on 1979 March 24, repeated twice within three days.[2] Unfortunately, this suspected third class of high-energy burster may be one of the least well studied phenomena in astrophysics. This is largely because most of the photons are emitted at "awkward" energies, in the gap between the regions of coverage provided respectively by most of the x-ray and gamma-ray instrumentation in space.

Fortuitously, the International Cometary Explorer (ICE) spacecraft contains an experiment, built by a UC Berkeley/Los Alamos collaboration, that is reasonably well suited for the study of this type of event. However, the detector field-of-view (a 10° FWHM band centered on the ecliptic plane) always includes the same 15% of the sky, with the remaining 85% being forever inaccessible. Neither of the

1979 repeating sources was favorably located; but, an event with similar characteristics occurred on 1979 January 7 in the proper part of the sky to be observed by ICE. The results of this observation were recently published.[3)]

Then, in the Spring of 1986, K. Hurley and colleagues in Toulouse, France, using data from their experiment on the Prognoz 9 (P9) satellite, informed us of possible source activity similar to that seen from the 1979 March repeaters. Specifically, in a 6-month period they observed four discrete groups of two or three short events (for a total of 9 events). Within a group, the bursts were separated by one to three days. No other information on the events was available at that time. Using our ICE data, we quickly found that four of the P9 events were located within the ICE field of view and were spectrally and temporally very similar to the 1979 January 7 burst. Thus, it appeared that the January 7 source might indeed be in the same class as the 1979 March repeaters. Further, it seemed that these objects could be dormant for long periods of time, but that they would eventually repeat as often as once per day for a few days.

At this point it occurred to us that, given the superior ICE temporal coverage and sensitivity to sources near the ecliptic plane, a systematic search through the ICE data base might turn up one or two additional source repetitions. What we actually found was completely unexpected. During a single day, 1983 November 16, a total of 18 events were detected; 10 occurred within an hour, with some separated by minutes or even seconds. During the month of November ICE recorded 63 events. A laborious search covering nearly eight years of ICE data revealed 48 additional repetitions, for a total of 111. Approximately 20 of these also were observed by other spacecraft. Of particular importance were three detected by the Pioneer Venus Orbiter Gamma Burst Detector. These three, which could then be located accurately by interplanetary arrival-time analysis, provided the firm link between GB790107 and the 1983 November activity. The Figure, which shows number of events vs. time for eight years, at 0.5-month time resolution, summarizes our results. We had clearly discovered a new and unique kind of repetitive behavior in a high-energy burster.

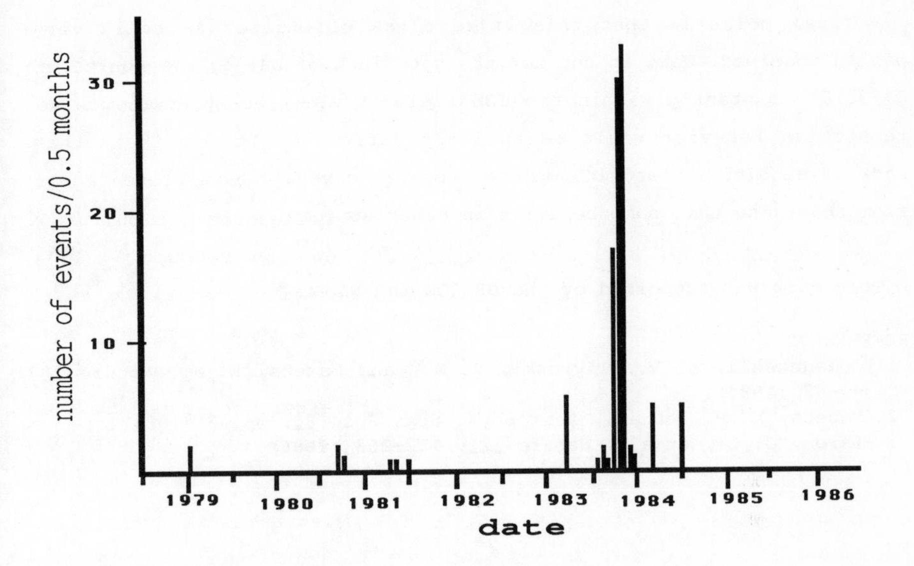

Figure. Events vs. time from August 1978 through June 1986 at 0.5-month time resolution.

The nature of the source is still being investigated. At this point we know that the repetitions are similar to one another in every respect except intensity, where a range of at least a factor of 30 is seen. The durations of the events are typically 0.1 s, plus or minus about a factor of 2. The characteristic photon energy is about 30 keV, and there is evidence for a deficiency of photons below 10 keV, relative to a simple thermal bremsstrahlung model. Perhaps the most amazing feature of the repetitions is that they can be separated by anywhere from seconds to years, and there is almost no correlation between intensity and separation. (This is in contrast to the Rapid X-ray Burster, where the intensity of a burst determines the time interval to the next burst.) No periodicities have been found. The only convincing pattern to the repetitions is their tendency to cluster, evident in the Figure. This clustering seems to persist on time scales as short as hours.

A final point is that this third class of burster is still very poorly observed. Were it not for the good luck of having the source of GB790107 constantly within the ICE field-of-view, our picture of its repetitive behavior would be entirely different. It is likely that lack of sensitivity and incomplete temporal coverage have prevented us from observing this same behavior in other sources.

This work was supported by the US DOE and NASA.

REFERENCES

1. Golenetskii, S. V., Ilyinski, V. N., and Mazets, E. P., Nature 307, 41-42 (1984).
2. Mazets, E. P., et al., Astrophys. Spa. Sci. 80, 3-143 (1981).
3. Laros, J. G., et al., Nature 322, 152-153 (1986).

OBSERVATIONS OF PEV γ-RAYS

Jerome W. Elbert
Physics Department, University of Utah, Salt Lake City, UT 84112

ABSTRACT

Nineteen observations of ultra-high energy (~10^{15} eV) γ-rays are briefly described. Observations have been made of Cygnus X-3, the Crab Nebula, Vela X-1, Hercules X-1, LMC X-4, and Centaurus X-3. Most observations are of low statistical significance and some of the detections may be spurious. A strong case can now be made for emission from Cygnus X-3, with 11 observations by 8 different research groups. The possible abundant production of low energy muons in the showers of "γ-ray" origin is under investigation. Improvements of detector sensitivities by ~2 orders of magnitude appear to be possible. Studies of unexpected muon production, particle acceleration mechanisms, galactic cosmic ray production, new galactic and extragalactic sources, and γ-ray absorption by 3^0K photons may be possible with the improved detector systems.

1. INTRODUCTION

In this paper I will briefly summarize the current results for those who are developing an interest in this rapidly developing field. Some definitions and background are needed. One PeV is equal to 10^{15} eV, 1000 TeV, or 1,602 ergs. I will describe results in the energy range 0.03 - 30 PeV.

The fluxes of PeV γ-rays are so small that direct measurements of the primary γ-rays within detectors are not possible. Instead, the "small" extensive air showers produced by the interactions of the γ-rays in the atmosphere are studied. These showers are cascades, produced by electromagnetic interactions in which γ-rays produce electron-positron pairs by interacting with the air, followed by the production of new γ-rays by bremsstrahlung by the electrons and positrons, etc. Relatively few energetic strongly interacting particles are expected in these cascades. Thus the γ-ray cascades should be muon-poor, i.e., have relatively few associated muons produced by decays of the strongly interacting mesons.

The γ-ray showers are observed in the presence of a large, nearly isotropic background of showers from cosmic ray nuclei. The showers from nuclei include large numbers of muons, and some observations attempt to reject showers which are not muon-poor. The PeV γ-ray sources are detected by observing excesses in the numbers of small air showers in certain directions at certain times. The belief that the excesses are due to γ-rays is based on the fact that no other known particles have the necessary properties to produce the excesses.

Two types of detector systems are employed in the search for PeV γ-ray sources. One type uses a large area array of detectors to sample the number of shower particles reaching the Earth's surface. Typically the detectors are 1 m^2 scintillators with 10-50 m separation between units. The second kind of detector systems are optical detectors. These detect brief flashes of Cherenkov light emitted by the showers passing through the atmosphere. The light is emitted in nearly the same direction as that of the primary particle which initiated the shower and is abundant at separations up to about 100 m from the shower core.

2. OBSERVATIONS OF CYGNUS X-3

Cygnus X-3 has 11 reported detections, more than the total number of PeV detections of other objects. It has been observed by 8 different research groups. The Haverah Park, Akeno, and Utah groups have each detected it twice. Its first PeV detection by Samorski and Stamm at Kiel[1]] in 1983, followed by the confirming detection by Lloyd-Evans et al. at Haverah Park in England,[2]] produced a strong motivation for other groups to search for PeV γ-ray signals.

The Kiel data were taken by an array of effective area about 3000 m^2, angular resolution 1^0, and during 3,838 hours in which Cygnus X-3 was within 30^0 of the zenith. Showers with ages >1.1 were selected (large age showers were identified by their broad lateral distributions). The data were taken between March 1976 and January 1980. In a 3^0 square box centered on Cygnus X-3, 31 showers were observed, with only 14.4 ± 0.4 expected from cosmic ray nuclei. The probability of this happening by chance was near 10^{-4}.

The Kiel data were also binned according to their phase (range 0-1) in the 4.8 hour X-ray period of Cygnus X-3. One of 10 phase bins had 13 events, with 1.44 expected. Thus the data displayed the characteristic periodic signature of Cygnus X-3. The γ-ray flux above 2 PeV was $7.4 \pm 3.2 \times 10^{-14} cm^{-2} s^{-1}$.

A muon detector of effective area 21.5 m^2 was located near the center of the Kiel array. When the muon content of each shower from the Cygnus X-3 bin was examined it was expected that some muon-poor showers (from γ-rays) would be present as well as some showers with the typical muon content of showers from cosmic ray nuclei. However, the measurements did not show a separate class of muon-poor showers. This puzzling result is still not understood at present.

The Haverah Park observation of Cygnus X-3 yielded a chance probability of $<1.8 \times 10^{-3}$ in the 4.8 hour periodicity test, corresponding to a flux of $1.5 \times 10^{-14} cm^{-2} s^{-1}$ above 3 PeV. When the Kiel and Haverah Park data were binned using the same ephemeris (van der Klis and Bonnet-Bidaud, 1981), the phases of the enhancements both occurred near 0.25.

With a high altitude counter array, Morello et al.[3]] detected a 2.8σ signal above 30 TeV from Cygnus X-3. Using the University of Utah Fly's Eye, an optical detector system, Baltrusaitis et al.[4]] detected a 3.5σ signal during July, 1983. A very high flux was obtained by Bhat et al.[5]] in an analysis of data taken much earlier with two large photomultiplier tubes used without mirrors.

A second detection of Cygnus X-3 by the Haverah Park group[6]] (Lambert et al, 1985) was achieved in 1984. They concluded that the presence of a signal at phase 0.63 in 1984 which was not present in their earlier data is evidence for time variability of the signal.

By selecting muon-poor showers at Akeno, Japan, Kifune et al.[7]] detected a small flux from Cygnus X-3. More recently, the same group detected a larger flux (without the low muon cut) following the 1985 radio outburst of Cygnus X-3. They also observed Cygnus X-3 in October and November, 1985, just after the radio outburst. They detected 4 showers during this time which appear to be about as muon-poor as expected for γ-ray showers. Of these 4 showers, 3 have 4.8 hour phases near 0.6.

Data taken in the summer of 1985 by the Utah Fly's-Eye[8]] show a 3.9σ signal in the phase interval 0.65-0.7. During about 4 hours of observations on June 17, 1985 (U.T.), an apparent sporadic outburst was observed. This signal was not observed on the preceding or following nights. For a part of the signal with energies above 250 TeV, an excess is present near phase 0, when some models have the neutron star in eclipse behind the companion star.

The "Carpet" scintillator array at Baksan showed a low significance enhancement near phase 0.6 in data taken between July 1984 and June 1985 (Alexeenko et al., 1986).[9]] The flux level of this observation is quite low, giving further evidence of variability. They also reported a strong sporadic outburst on 14-16 October 1985 shortly after a major radio outburst. The γ-ray signal did not show a narrow peak in the 4.8 hour phase.

The "Cygnus" collaboration,[10]] working with a scintillator array at Los Alamos and using neutrino detectors to observe muons, has presented preliminary results in recent talks that show a detection of Cygnus X-3 in >0.1 PeV data taken in April and May, 1986. The signal was observed in muon poor showers as a 3.5σ excess in the Cygnus X-3 direction and an excess near 4.8 hour phase 0.8 of comparable significance. The phase was computed using the" Mason ephemeris", but was reported to be equivalent to 0.7 according to the Van der Klis and Bonnet-Bidaud[11]] ephemeris.

3. OBSERVATION OF SOURCES OTHER THAN CYGNUS X-3

Besides the PeV observations of Cygnus X-3, there have been 3 observations of the Crab nebula, 2 observations of Vela X-1, and 1 observation each of Hercules X-1, LMC X-1, and Centaurus X-3. The first of the Crab PeV observations was by Dzikowski et al.[12]] In 1983 they presented results from data taken from 1975-1982 at Lodz, Poland.[13]] An electron array with 10^0 resolution showed a 5σ excess in the broad angular bins centered on the Crab nebula. The showers were not nearly as muon-poor as expected.

The Utah group made an observation of the Crab on a night in December, 1980.[14]] A weak (3.1σ) excess of 1 PeV events was observed. A second attempt to observe the Crab on three nights in February, 1981 showed no signal. This implies that the Crab signal, if real, is variable in flux.

During more than 17,000 hours of operation from February, 1974 to October, 1982, the Tien-Shan array in the U.S.S.R. collected muon-poor showers from the Crab direction. Analysis of showers above 0.55 PeV showed a 4.2σ excess flux of $1.9\pm0.7 \times 10^{-13}cm^{-2}s^{-1}$.[15]] Since all 3 of the observations had poor angular resolution and did not use accurate fast timing, the association of the excesses with the Crab Pulsar is uncertain.

The observations of Vela X-1 were by Protheroe, Clay, and Gerhardy[16]] using the Buckland Park air shower array in Australia and by the BASJE group from Mt. Chacaltaya in Bolivia.[17] Data from the Buckland Park array were taken in 1979-1981 with a 3 PeV threshold. A time averaged flux of $9.3 \pm 3.4 \times 10^{-15} cm^{-2}s^{-1}$ was detected in one of 50 phase bins (phase 0.63) of the 8.96 day orbital period. The chance probability of the observation was estimated to be $\sim 10^{-4}$.

A reanalysis of muon-poor showers observed from 1964-1966 on Mt. Chacaltaya shows an excess of events within about 5^0 of Vela X-1 and a non-uniform distribution of events during the orbital period.[17] The combined chance probability of the excess and the periodic effect was estimated to be $\sim 10^{-3}$. It was reported in the same paper that a similar analysis of 4 events from the Cen X-3 vicinity showed a chance probability of 0.6%.

A 40 minute long signal from 0.5 PeV γ-rays from Hercules X-1 was reported by the Utah group.[18] A periodic effect (chance probability 2×10^{-4}) was observed in the 1.24 second rotational period of Hercules X-1. Optical and X-ray data obtained from nearly the same time showed that the accretion disk may have thickened and blocked the X-ray line of sight from the earth to the neutron star. The same material may have served as a target on which multi-PeV protons produced mesons which decayed to yield the observed γ-rays.

From Buckland Park data, Protheroe and Clay (1985)[19] reported a weak detection of 1.4 day periodicity in the direction of LMC X-4, a binary X-ray source in the Large Magellanic Cloud. The observation is interesting because it is the first extragalactic object detected in the PeV energy region.

4. DISCUSSION

The observations described in Section 2 support the conclusion that Cygnus X-3 is a PeV particle source. It is probable that the periodic flux is variable on a time scale of months. Shorter term sporadic effects may also be present. The PeV signal's phase appears to have switched from near 0.25 for earlier times to near 0.6 during 1984-1986. There is some evidence for PeV emission from the Crab, but it is not well established. Other objects are not adequately confirmed at present. The experiments have not achieved sufficient sensitivity to allow typical Cygnus X-3 flux levels to be detected with unquestionable statistical significance.

Except for the Crab, the observed PeV sources are all probable or certain binary X-ray sources. The previously known periodicity of the sources has helped in the detection of the binary objects. Except for the Crab and Hercules X-1, the orbital periods have been detected. Because they are easier to detect, there may be some bias toward detection of periodic sources. However, a broad northern hemisphere survey has been performed using the Fly's Eye,[4] and it set flux limits of 3×10^{-13} to 3×10^{-12} $cm^{-2}s^{-1}$ for point sources of showers above 1 PeV. This indicates that undiscovered sources are not likely to be observed at flux levels 1-2 orders of magnitude higher than Cygnus X-3 fluxes.

Because of synchrotron losses, it is not considered to be feasible to produce electrons and positrons with high enough energies to be the sources of PeV gamma rays. The probable process is acceleration of protons and nuclei, followed by collisions which generate neutral pions. The neutral pions decay almost immediately, producing the γ-rays. It is likely that the accelerated particles do not always strike a target. The escaping energetic particles would contribute heavily to the galactic PeV cosmic ray fluxes. Hillas has given a model in which Cygnus X-3 would produce enough 100 PeV cosmic rays to exceed the estimated galactic escape rate of cosmic rays at that energy.[20]]

The detector systems which have been used in most of the observations done so far were designed for cosmic ray studies and are not optimized for γ-ray detection. Advances by about 2 orders of magnitude in sensitivity are expected in a collaborative effort by the Universities of Chicago, Michigan, and Utah. The plans include a 300 m x 300 m array of scintillators at 10 m separation. Angular resolution is expected to be ~ $1/2^0$ for this array. In addition, 1100 m^2 of detectors buried under 10 feet of earth will allow muon poor showers to be selected. Such improved experiments should uncover a number of additional sources in our galaxy. Perhaps more importantly, such detectors may give more detailed information on such sources as Cygnus X-3, Hercules X-1, and the Crab Nebula so that the acceleration mechanism (or mechanisms) may be understood.

Such detector systems as the proposed system mentioned above and the Cygnus project at Los Alamos will have a significant capability of measuring the degree to which the signal from point sources involves muon-poor air showers. Since the Chicago-Michigan-Utah project will be located at Dugway, Utah, it will also have the capability of making simultaneous observations with the Los Alamos experiments. This will allow tests for the consistency of results obtained at the two sites. The simultaneous observations make possible tests of "reproducibility" of data from variable sources such as Cygnus X-3. The experiments will also be able to correlate the results with TeV observations made from the Whipple Observatory in Arizona.

For Cygnus X-3, the TeV-PeV γ-ray energy flux is ~25% of the X-ray energy flux.[21]]. With improved detector systems, some active galaxies such as NGC 4151 would be detectable in PeV γ-rays if a similar ratio of TeV-PeV to X-ray energy fluxes exists. Although 1 PeV γ-rays are absorbed by the 3^0K black body radiation on distance scales of ~10 kpc, 0.03 PeV γ-rays have long enough mean free paths that such extragalactic observations may be possible.

5. ACKNOWLEDGEMENTS

This research was sponsored by the National Science Foundation under grants PHY8515265 and PHY8415294.

6. REFERENCES

1. Samorski, M. and Stamm, W., Ap. J. (Letters), 268, L17 (1983).
2. Lloyd Evans, J. et al., Nature 305, 784 (1983).
3. Morello, C., Navarra, G., and Vernetto, S., Proc. 18th Int. Cosmic Ray Conf. (Bangalore) 1, 127 (1983).
4. Baltrusaitis, R.M. et al., Ap. J. 297, 145 (1985).
5. Bhat, C.L., Sapru, M.L., and Razdan, H., Bhabha Atomic Research Center (Srinagar, India) preprint (1985).
6. Lambert, A. et al., Proc. 19th Int. Cosmic Ray Conf. (La Jolla) 1, 71 (1985).

7. Kifune, T. et al., Ap. J. 301, 230 (1986).
8. Baltrusaitis, R.M. et al., University of Utah preprint (1987).
9. Alexeenko, V.V. et al.,University of Moscow preprint (1986).
10. Nagle, D.E. et al., "Techniques in Ultra High Energy Gamma Ray Astronomy", Edited by R.J. Protheroe and S.A. Stephens (University of Adelaide, 1985) 66.
11. Van der Klis, M., and Bonnet-Bidaud, J.M., Astr. Ap.95, L5 (1981).
12. Dzikowsky, T. et al., Proc. 17th Int. Cosmic Ray Conf. (Paris) 1, 8 (1981).
13. Dzikowsky, T. et al., Proc. 18th Int. Cosmic Ray Conf. (Bangalore) 2, 132 (1983).
14. Boone, J. et al., Ap. J. 285, 264 (1984).
15. Kirov, I.N. et al., Proc. 19th Int. Cosmic Ray Conf. (La Jolla) 1, 135 (1985).
16. Protheroe, R.J., Clay, R.W., and Gerhardy, P.R., Ap. J. (Letters) 280, L47 (1984).
17. Suga, K. et al., "Techniques in Ultra High Energy Gamma Ray Astronomy", Edited by R.J. Protheroe and S.A. Stephens (University of Adelaide, 1985) 48.
18. Baltrusaitis, R.M. et al, Ap. J. (Letters), 293, L69 (1985).
19. Protheroe, R.J. and Clay, R.W., Nature 315, 205 (1985).
20. Hillas, A.M., Nature 312, 50 (1984).
21. Vestrand, W. Thomas and Eichler, David, Ap. J. 261, 251 (1982).

Galactic Cosmic Ray Isotope Spectroscopy: Status and Future Prospects

R. A. Mewaldt
California Institute of Technology
Pasadena, CA 91125

Abstract: A brief review of cosmic ray isotope spectroscopy is presented, focusing on cosmic ray clocks and the composition of cosmic ray source material. Some of the goals and prospects for future cosmic ray isotope spectrometers are discussed.

1. Introduction

The relative abundances of the isotopes of galactic cosmic rays represent a record of the nuclear history of a sample of matter from other regions of the galaxy, including its synthesis in stars, and its subsequent nuclear interactions with the interstellar gas. Recent progress in cosmic ray isotope spectroscopy has revolutionized our views of both cosmic ray origin and propagation. In this paper the present status of cosmic ray isotope measurements and prospects for future progress are briefly reviewed, including possibilities for improving on the yield and energy coverage of present experiments by more than two orders of magnitude. For more complete discussion and references on many of these topics, there are a number of other recent reviews[1,2,3,4,5].

2. Cosmic Ray Clocks and Secondary Nuclei

There are several long-lived radioactive isotopes produced as "secondaries" in cosmic ray interactions with the interstellar medium (ISM) that can be used to measure the average lifetime of cosmic ray propagation in the galaxy. Examples are ^{10}Be, ^{14}C, ^{26}Al, ^{36}Cl, and ^{54}Mn. Figure 1 shows measurements of two of these cosmic ray clocks: ^{10}Be (half-life = 1.6 x 10^6 yr) and ^{26}Al (half-life = 9 x 10^5 yr), along

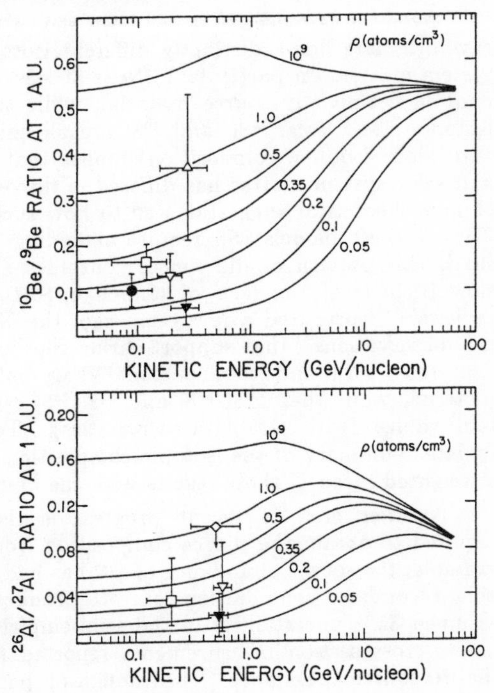

Figures 1 and 2: The ^{10}Be/^{9}Be and ^{26}Al/^{27}Al ratios vs. energy per nucleon, with the Calculations by Guzik and Wefel[6], parameterized by the density of the propagation region. References to measurements (by Berkeley, Chicago, Goddard, Minnesota, and New Hampshire) can be found in 1); recent data is from references 7) and 8).

with calculations parameterized by the density of the propagation region. Note that both ^{10}Be and ^{26}Al suggest a density of ~0.2 atoms/cm^3, substantially less than the average density of matter in the galactic plane. For a mean free path of 6 g/cm^2 for escape from the galaxy the "lifetime" is ~2 x 10^7 years, much younger than solar system matter. The similar lifetimes for ^{10}Be and ^{26}Al, suggest that their parents share a common history.

With improved precision, measurements of clocks with different half-lives can determine the *uniformity* of the matter density traversed, and thereby test whether some of the matter is in the source region. In addition, measurements of clocks over a broad energy interval can test whether continuous acceleration is occurring in the galaxy as a result of encounters with supernova shock waves. Finally, studies of Fe, Ni, and Co electron-capture isotopes can determine the time-delay between nucleosynthesis and cosmic ray acceleration.

There has been a great deal of recent interest in the secondary isotopes ^{2}H and ^{3}He, as a result of indications that cosmic ray H and possibly He may have had a different origin and history from that of heavier nuclei. In particular, a number of models [see, e.g., ref 9)] designed to explain the excess of antiprotons observed at GeV energies have suggested that at least some ^{1}H and ^{4}He nuclei have traversed considerably more matter than heavier nuclei, which would also produce an excess of ^{2}H and ^{3}He. Although recent studies[10,11,12)] find no evidence for an excess of ^{3}He beyond that expected from standard models, the ^{2}H situation is less clear, and most observations are at low energies (<300 MeV/nucleon), far below the threshold for antiproton production.

3. Source Abundances of Cosmic Ray Isotopes

Over the last decade it has been shown that the matter from which cosmic rays originate has a distinctly different isotopic composition from typical solar system matter. In particular, ^{22}Ne is at least a factor of three times more abundant in cosmic ray source material, while the abundances of the neutron-rich isotopes ^{25}Mg, ^{26}Mg, ^{29}Si, and ^{30}Si are all enhanced by a factor of ~1.5. This anomalous isotopic composition implies that the nucleosynthesis of cosmic ray and solar system matter has differed, a discovery that has stimulated a number of new theoretical suggestions as to how such differences might have occurred. To test these models will require extending observations to other elements to see if this pattern of differences continues. Unfortunately, progress has been slow because of the lack of launch opportunities for new instruments. Since reviews[1,2,3)] prepared a few years ago, the New Hampshire group has reported new observations[8)] that support earlier reports of a very low ^{14}N abundance [see, e.g., refs. 2,5)] and confirm the ^{25,26}Mg enhancements in cosmic ray source material, while their ^{29}Si/^{28}Si and ^{30}Si/^{28}Si ratios are consistent with solar system values (with sizable uncertainties). Table 1 and Figure 2 present an updated summary of the isotopic composition of the cosmic ray source, based on a weighted mean of those results with the best mass resolution [see also 5)].

Another area of recent progress is the measurement of cross sections required to obtain the source composition from the measured composition. For example, the source abundance of ^{13}C has had a large uncertainty due to the sizable secondary contribution of ^{13}C produced during propagation, and the assumed 35% uncertainity in the semi-empirical cross sections used[14)]. Preliminary cross section measurements reported recently allow an improvement in the ^{13}C source abundance[15)]. Although propagation uncertainities still dominate

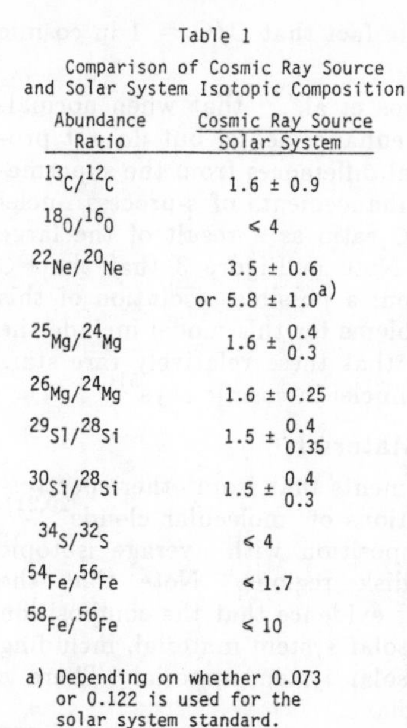

Table 1

Comparison of Cosmic Ray Source and Solar System Isotopic Composition

Abundance Ratio	Cosmic Ray Source / Solar System
$^{13}C/^{12}C$	1.6 ± 0.9
$^{18}O/^{16}O$	≤ 4
$^{22}Ne/^{20}Ne$	3.5 ± 0.6 or 5.8 ± 1.0 [a]
$^{25}Mg/^{24}Mg$	$1.6 {}^{+0.4}_{-0.3}$
$^{26}Mg/^{24}Mg$	1.6 ± 0.25
$^{29}Si/^{28}Si$	$1.5 {}^{+0.4}_{-0.35}$
$^{30}Si/^{28}Si$	$1.5 {}^{+0.4}_{-0.3}$
$^{34}S/^{32}S$	≤ 4
$^{54}Fe/^{56}Fe$	≤ 1.7
$^{58}Fe/^{56}Fe$	≤ 10

a) Depending on whether 0.073 or 0.122 is used for the solar system standard.

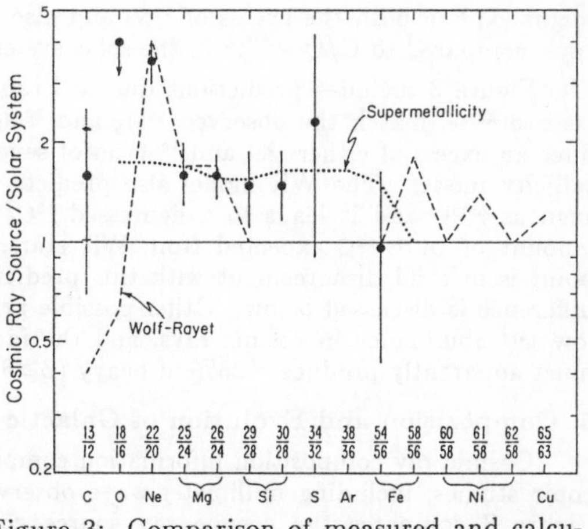

Figure 3: Comparison of measured and calculated cosmic ray source and solar system[13] compositions. The data points are based on a weighted mean of selected measurements, updated from 2). Dotted extensions to the ^{13}C error bar indicate the propagation uncertainty.

the revised $^{13}C/^{12}C$ ratio in Table 1 and Figure 3, and it is not yet possible to tell if there is an excess of ^{13}C, the uncertainity may be reduced further in the next year or two. Recent cross section measurements[16] have also supported the conclusion that cosmic ray source material is deficient in ^{14}N, and they have confirmed that the enhancements of ^{22}Ne, $^{25,26}Mg$, and $^{29,30}Si$ reported earlier could not be the result of secondary production during propagation.

4. Interpretation of the Cosmic Ray Source Composition

Of the models proposed to explain the observed excess of neutron-rich isotopes in cosmic rays, the most quantitative are the so-called "supermetallicity" model of Woosley and Weaver[17], and the Wolf-Rayet model proposed by Casse and Paul[18]. Woosley and Weaver pointed out that the production of neutron-rich isotopes in massive stars is proportional to the initial "metallicity" (the fraction of $Z>2$ elements) of the material from which the star formed, and proposed that the excess of neutron-rich isotopes might result if cosmic rays originate in regions of the galaxy that are metal-rich compared to the solar system. They predicted similar enhancements for a number of other neutron-rich nuclei including ^{18}O, ^{34}S, ^{38}A, and ^{54}Fe, as shown in Figure 3. Note that this model is consistent with the approximately equal enhancements observed for the Mg and Si isotopes, but it would apparently require an additional source of ^{22}Ne to explain the large $^{22}Ne/^{20}Ne$ ratio in cosmic rays.

In the Wolf-Rayet (WR) model it is assumed that a fraction of heavy cosmic rays originate from the material that has been expelled by Wolf-Rayet stars. These massive stars are undergoing significant mass loss by means of high-velocity stellar winds. As a result they have been stripped of their hydrogen envelopes, and helium-burning products including ^{12}C, ^{16}O, and ^{22}Ne have been exposed and are being expelled from their surface. Casse and Paul suggested that if a fraction of cosmic rays were to originate from WR material, this might explain both the excess of ^{22}Ne and also the fact that C/O ≃ 1 in cosmic rays, compared to C/O ≃ 0.5 in the solar system.

Figure 3 includes predictions due to Prantzos et al.[19], that when normalized to ^{22}Ne, match the observed ^{25}Mg and ^{26}Mg enhancements, but do not produce an excess of either ^{29}Si and ^{30}Si; one of several differences from the supermetallicity model. The WR model also predicts enhancements of s-process nuclei such as ^{58}Fe, and it leads to a depressed ^{13}C/^{12}C ratio as a result of the large amount of pure ^{12}C expected from WR stars. Note in Figure 3 that the ^{13}C point is in mild disagreement with this prediction; a possible resolution of this difference is discussed below. Other possible problems for this model include the low ^{14}N abundance in cosmic rays, and the fact that these relatively rare stars must apparently produce ~25% of heavy (Z≥6) nuclei in cosmic rays[5].

5. Composition and Evolution of Galactic Material

Cosmic ray composition information complements that from other spectroscopic studies, including millimeter-wave observations of molecular clouds[20,21]. Figure 3 compares the cosmic ray source composition with average isotopic ratios in the galactic center and galactic disk regions. Note that the millimeter-wave observations have produced clear evidence that the composition in the galactic center region differs from that of solar system material, including variations much greater than typically found in solar system material. There is also evidence for such differences in the galactic plane.

Recently Hawkins and Jura[23] reported optical measurements of the ^{13}C/^{12}C ratio in four different directions that yield a ^{13}C/^{12}C ratio in the local neighborhood a factor of 2.1 ± 0.2 greater than the solar system value. These new observations, which are consistent with several galactic chemical evolution models, provide proof that the composition of the local neighborhood has evolved since the formation of the Sun. It is interesting that if WR material is mixed with material characteristic of the present-day local ISM, as would seem appropriate in this model, the difference between the measured and predicted ^{13}C/^{12}C ratios (Figure 3) would disappear. Although the rate of evolution of the local ^{13}C/^{12}C ratio does not necessarily imply similar changes in heavier neutron-rich isotopes because of differences in their origin[24], it does show that evolutionary effects are important

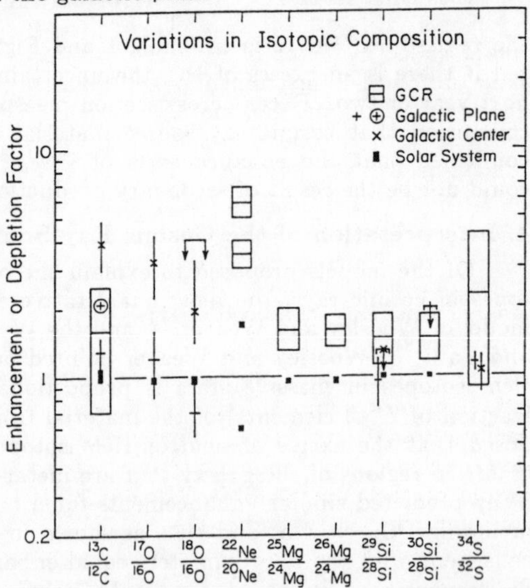

Figure 4: Comparison of variations in isotopic composition detected in galactic cosmic rays with those detected in the galactic plane and center regions using other techniques,[20,21] and with those found in solar system material.[22] The circled cross is the recent observation by Hawkins and Jura.[23]

locally on relatively recent time scales, and it underscores the need for both theoretical modeling and further measurement of galactic isotope ratios via cosmic ray and other observations.

6. Future Prospects

Our knowledge of the isotopic composition of cosmic ray source material is still very limited. Only the Ne, Mg, and Si isotopes have had their source abundance determined to an accuracy of ~30% or better, and in each case differences from the solar system composition have been found. If such observations are to be extended to other elements such as Fe and Ni, it will be necessary to expose larger instruments in space. The techniques for resolving isotopes have now been proven, they need only be applied on a larger scale. In the next few years two instruments built by the University of Chicago for the Ulysses and CRRES missions (a similar Goddard experiment is planned for ISTP/WIND) should collect a factor of ~10 more data than the ISEE-3 instruments, determining composition of the more abundant isotopes from ~100 to ~300 MeV/nucleon. At somewhat higher energies, further results can be expected from balloon-borne experiments. Beyond this, the Cosmic Ray Program Working Group[25] in its "Particle Astrophysics Program for 1985-1995" identified the need for two major new projects that would improve on existing observations by more than two orders of magnitude: a superconducting magnet facility for particle astrophysics on the Space Station, and a cosmic ray composition Explorer. To meet this need two possible missions have been proposed.

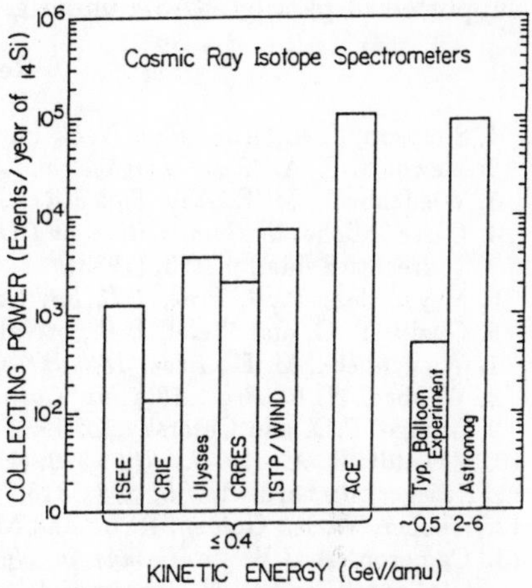

Figure 5: A comparison of the collecting power of previous cosmic ray isotope spectrometers (ISEE, CRIE) with that of planned or proposed instruments.

The Advanced Composition Explorer (ACE) would measure the isotopic and elemental composition of several samples of matter with unprecedented resolution and collecting power, including galactic cosmic rays (30-400 MeV/nucleon), the "anomalous" cosmic ray component (thought to represent a sample of the neutral ISM), energetic particles accelerated in solar flares, and the solar wind. *Astromag,* the superconducting magnet facility being planned for the Space Station[26], would extend cosmic ray isotope measurements to energies from ~2 to as high as ~50 GeV/nucleon, allowing clocks such as ^{10}Be to be read over a wide range of time-dilation factors, and extending measurements of the source composition up to high energies. *Astromag* would also measure the spectra of antiprotons, electrons, positrons, and heavy nuclei, and make a sensitive search for antinuclei. As indicated in Figure 4, the collecting power of both ACE and *Astromag* would be more than an order of magnitude greater than that of any previous or planned experiment, sufficient to obtain definitive measurements of even rare species.

Acknowledgements: I am grateful to E. C. Stone, W. R. Webber, J. P. Wefel, M. E. Wiedenbeck, and S. E. Woosley for helpful discussions. This work was supported in part by NASA under grants NAG5-722 and NGR 05-002-160.

References

1. Simpson, J. A., *Ann. Rev. Nucl. Part. Sci.* 33, 323, (1983).
2. Mewaldt, R. A., *Rev. Geophys. and Sp. Phys.* 21, 295, (1983).
3. Wiedenbeck, M. E. *Adv. Space Res.* 4, 15, (1984).
4. Casse, Michel in *Composition and Origin of Cosmic Rays,* ed. Shapiro, M. M. (Dordrecht: Reidel) p. 193, (1983).
5. Meyer, Jean-Paul, *Proc. 19th Int. Cosmic Ray Conf. (La Jolla)* 9, 141, (1985).
6. Guzik, T. G. and Wefel, J. P., private communication.
7. Wiedenbeck, M. E., *Proc. 18th Int. Cosmic Ray Conf. (Bangalore)* 9, 147, (1983).
8. Webber, W. R. *Proc. 19th Int. Cosmic Ray Conf. (Bangalore)* 2, 88, (1985).
9. Lagage, P.O. and Cesarsky, C. J. *Astron. Astrophys.* 147, 127, (1985).
10. Mewaldt, R. A., *Ap. J.* 311, 979, (1986)
11. Kroeger, Richard, *Ap. J.,* 303, 816, (1986).
12. Webber, W. R., Golden, R. L., and Mewaldt, R. A., *to be published in Ap.J. (1987).*
13. Cameron, A.G.W. in *Essays in Nuclear Astrophysics*, eds. C. A. Barnes, D. D. Clayton, and D. N. Schramm, (Cambridge Univ. Press, Cambridge) p. 23 (1982).
14. Wiedenbeck, M. E. and Greiner, D. E., *Ap. J.* 247, L119, (1981).
15. Based on data from references 14) and 27) with calculations and cross sections from Guzik, T. G., Wefel, J. P., Crawford, H. J., Greiner, D. E., and Lindstrom, P. J. *Proc. 19th Int. Cosmic Ray Conf. (La Jolla),* 2, 80, (1985).
16. Webber, W. R. and Gupta, M., *submitted to the 20th Int. Cosmic Ray Conf., (Moscow),* paper OG 7.2-4 (1987), and private communication.
17. Woosley, S. E. and Weaver, Thomas A., *Ap. J.* 243, 561, (1981).
18. Casse, M. and Paul, J. A., *Ap. J.* 258, 860, (1982).
19. Prantzos, N., Arnould, M., Arcoragi, J.P., and Casse, M., *Proc. 19th Int. Cosmic Ray conf. (La Jolla).* 3, 167, (1985).
20. Penzias, Arno A. *Science* 208, 663, (1983).
21. Wannier, P. G., *Ann. Rev. Astron. Astrophys.* 18, 399, (1980).
22. Podosek, F. A., *Ann. Rev. Astron. Astrophys.* 16, 293, (1978).
23. Hawkins, Isabel and Jura, Michael *Submitted to Ap. J. (1986).*
24. Woosley, S. E., private communication.
25. *The Particle Astrophysics Program for 1985 - 1995,* Report of the NASA Cosmic Ray Program Working Group (Israel, M. H. and Ormes, J. F., co-chairmen), December, (1985).
26. *The Particle Astrophysics Magnet Facility ASTROMAG,* (ed. Ormes, J. F., Israel, M., Wiedenbeck, M., and Mewaldt, R. A.), August, (1986).
27. Mewaldt, R. A., Spalding, J. D., Stone, E. C., and Vogt, R. E. *Ap. J.* 251, L27, (1981).

HIGH-ENERGY GAMMA RAYS PROBING COSMIC-RAY PROPAGATION

Hans Bloemen

Astronomy Department, University of California, Berkeley, CA 94720

Recent analyses of the *COS-B* γ-ray observations (about 50 MeV – 5 GeV) have given a new view on the galactic cosmic-ray (CR) distribution[1]. It turned out, surprisingly, that there is barely a galacto-centric CR density gradient (at ~1-10 GeV/nucl), at least much weaker than the roughly 5-kpc radial scale length for candidate CR sources. In addition, an energy dependence in the γ-ray data was firmly established, but the physical explanation for this could not be determined unambiguously. In fact, the interpretation that appeared to describe best the data (I will not repeat it here because the results described below change the situation), and the lack of a strong CR gradient, seemed theoretically so uncomfortable that some have questioned the whole approach. There is, however, far more spectral information in the *COS-B* data which can provide further insight. The results of such a detailed spectral analysis[2] are summarized here and seem to clarify the puzzling situation.

Fig. 1 shows γ-ray spectra of the inner ($310° < \ell < 50°$; $|b| < 30°$) and outer Galaxy ($90° < \ell < 270°$), derived from a combination of all the data obtained during the almost 7 years of the *COS-B* mission. The isotropic background contribution (instrumental + extragalactic) for each energy interval was determined at $|b| > 30°$ (combining all high-latitude observations) and subtracted from the galactic-plane intensities. The spectra in Fig. 1 show a remarkable difference (further clarified in Fig. 2): the outer-Galaxy γ-ray spectrum ***above a few hundred MeV*** is flatter than the inner-Galaxy spectrum. It is very suggestive to find this effect at such high energies because the emission at these energies is expected to be largely due to π°-decay and the slope of the π°-decay spectrum converges at ~1 GeV to that of the CR spectrum above a few GeV/nucl. The data points above 300 MeV in Fig. 2 can be fitted by a power law $E^{-\beta}$ with best-fit spectral index $\beta = 0.4 \pm 0.1$. Although, in principle, the high formal statistical significance of this decrease (chance probability 3×10^{-5}) should be judged with care because of possible systematic errors, all these potential uncertainties turn out to be essentially eliminated in this approach[2]. Fig. 2 shows the impact of arbitrarily varying the isotropic background, which is probably the major potential source of uncertainty; the decrease of the ratio is barely affected. Fig. 1 shows that the inner-Galaxy γ-ray spectrum is consistent with the expectation from the local CR electron and proton spectra (with some scaling factor, the same for both).

It seems inevitable to conclude that the CR spectrum (more precisely, the CR proton spectrum, between a few GeV and a few tens of GeV) is flatter in the outer Galaxy than in the solar vicinity and inner Galaxy (approximately $\propto E^{-2.3}$, and maybe flatter, vs. $\propto E^{-2.7}$), suggesting that the energy dependence of the CR path length is flatter in the outer Galaxy, unless the CR source spectrum is flatter. In the framework of the simple CR propagation model considered by Jones[3] (more detailed modelling gives essentially the same result[4,5]), this can be understood if the transition between diffusion-dominated and convection-dominated propagation occurs in the outer Galaxy at a higher energy ($\gtrsim 10$ GeV/nucl) than estimated from CR secondary-to-primary ratios for the local region of the Galaxy (~1 GeV/nucl). An increase of the transition energy from some value E_1 to E_2 requires in this model, where the particles are produced in a thin disk and propagate outwards in a halo with total thickness $2D$, with convection velocity V and diffusion coefficient $\kappa = \kappa_\circ E^{-\delta}$ (with $\delta \simeq 0.6$ in the solar vicinity), an increase of the quantity $VD/\kappa_\circ$ by about $(E_2/E_1)^\delta$ (so $\gtrsim 3$ between the outer Galaxy and locally). In view of the weak CR gradients (even few-GeV protons appear to have a radial exponential scale length of ~10-15 kpc), it seems that this should be attributed to a smaller diffusion velocity ($\propto \kappa_\circ/D$) for the outer Galaxy, rather than a larger V, because the CR density is inversely proportional to the outward velocity (i.e. $\propto 1/V$ in the convective regime).

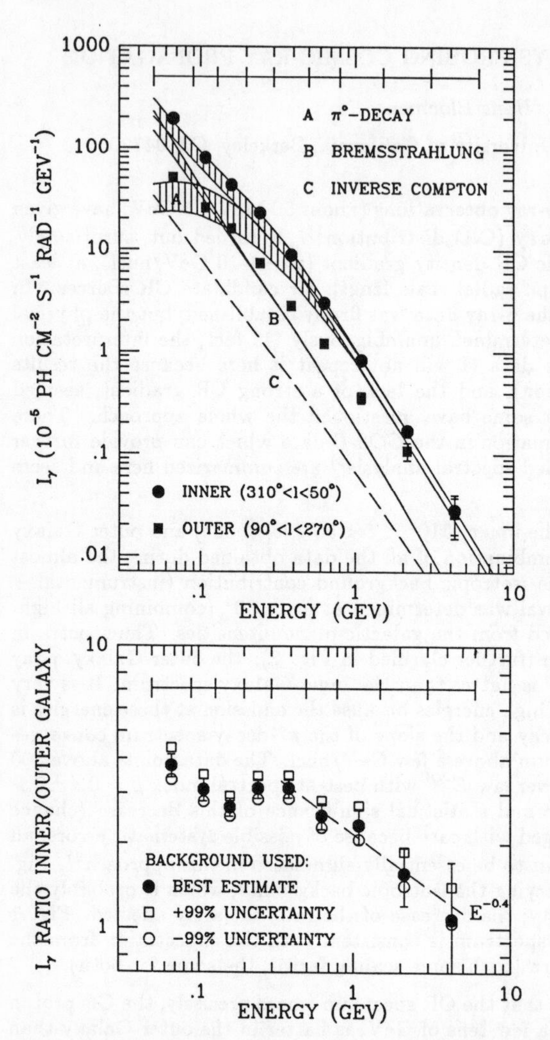

Fig. 1: COS-B γ-ray spectra towards the inner and outer Galaxy ($|b| < 30°$) and γ-ray intensities derived from the local interstellar CR spectra (the sum of the components is normalized to the inner-Galaxy spectrum at ~1 GeV). The error-bars are smaller than the dots for most data points and include the statistical uncertainties of the subtracted background. The upper bound of the π^0-decay contribution corresponds to the 'M_U' curve of Stephens and Badhwar[7] and the lower bound to the spectrum given by Dermer[8]. The bremsstrahlung spectrum is determined from the electron spectrum of Webber[9]; the upper and lower bound correspond to an $E_e^{-2.4}$ and $E_e^{-2.1}$ electron spectrum for $E_e \lesssim 200$ MeV.

Fig. 2: Ratio between the γ-ray spectra towards the inner and outer Galaxy. The black dots and error bars correspond to the spectra shown in Fig. 1; the best power-law fit above 300 MeV, $E^{-0.4}$, is indicated. The squares and circles illustrate the impact of arbitrarily varying the high-latitude measurements of the isotropic background by their 99% uncertainties.

It is interesting that a recent spectral-index study[6] of radio-continuum surveys of the Galaxy at 408 MHz and 1420 MHz has led to essentially the same result for CR electrons.

I gratefully acknowledge receipt of a Miller Fellowship and thank Drs. F.C. Jones and R. Schlickeiser for valuable discussions and comments during the meeting.

REFERENCES

1. Bloemen, J.B.G.M. *et al.* 1986, *Astr. Ap.*, **154**, 25.
2. Bloemen, J.B.G.M. 1987, *Ap. J. Letters*, submitted.
3. Jones, F.C. 1979, *Ap. J.*, **229**, 747.
4. Kóta, J., and Owens, A.J. 1980, *Ap. J.*, **237**, 814.
5. Lerche, I., and Schlickeiser, R. 1982, *M.N.R.A.S.*, **201**, 1041.
6. Reich, P. 1986, Ph. D. Thesis, University of Bonn.
7. Stephens, S.A., and Badhwar, G.D. 1981, *Ap. Space Sci.*, **76**, 213.
8. Dermer, C.D. 1986, *Astr. Ap.*, **157**, 223.
9. Webber, W.R. 1983, in *Composition and Origin of Cosmic Rays*, ed. M.M. Shapiro, Reidel, Dordrecht p.83.

PULSARS AS COSMIC RAY PARTICLE ACCELERATORS - ACCELERATION BOUNDARY FOR PROTONS

K.O. Thielheim
Institut für Reine und Angewandte Kernphysik
Abteilung Mathematische Physik
University of Kiel, FRG

Summary

Systematic investigations on the dynamics of protons (and other types of charged particles) in the vacuum field of a magnetic dipole rotating with its vector $\underline{\omega}$ of angular velocity perpendicular to its vector $\underline{\mu}$ of dipole moment, using a standard set of parameters $\omega = 20\pi\ \mathrm{sec}^{-1}$ and $\mu = 10^3\ \mathrm{Gcm}^3$ (as well as other parameter values) have revealed some features which may be relevant to the functioning of pulsars as cosmic ray particle accelerators[1,2,3], as for example the 'critical surface' and the 'acceleration boundary'. The limited space available here, does not allow more than a few remarks on the latter.

Obviously the ability of the orthogonal rotator to accelerate protons to very high energies and to propagate them to very large distances must break down beyond a certain limit of radial distance[4]. This is what I have called the 'acceleration boundary' r_B. A 'first estimation' on its value in the equatorial plane is obtained as follows: If r_0 is the initial radial coordinate value, $x_P = \pi r_T^4 / r_L r_0$, $r_L = c/\omega$ and $r_T = (e\mu/mc^2)^{1/2}$, the suggestion $r_B = x_P = r_0$ is leading to $r_B / r_L = (r_T / r_L)^{4/3} = (e\mu/mc^2)^{2/3} \cdot (\omega/c)^{4/3}$, which for the standard set of parameters gives $r_B / r_L \cong 10^4$.

Numerical integrations of the equations of motion for protons performed by one of my students and myself[5] shed some light on features of particle dynamics associated with the phenomenon of the acceleration boundary. The latter is defined here through a deviation observed from the monotonous increase of particle energy over a certain range of radial distance (from r_0 to $10 \cdot r_0$ in this case) when r_0 passes the acceleration boundary from inside as is illustrated in figure 1. The results of these numerical calculations confirm that for the standard set of parameters the acceleration boundary is about $r_B = 10^4 \cdot r_L$.

The appearance of the acceleration boundary is also reflected in the phase development of protons. Those originating from positions well inside the acceleration boundary (though still inside the wave zone) appear to be well stabilized inside one period of the propagating wave over the range of radial distance considered here (r_o to $10 \cdot r_o$), while others originating from positions well outside the acceleration boundary increasingly exhibit a tendency for slipping through from one period to the other on this range of radial distance.

For illustration a graphical representation of the shape of the acceleration boundary, using radial and latitudinal coordinates together with the phase angle (instead of the longitudinal angle) is given in figure 2.

ENERGY DEVELOPMENT OF PROTONS ORIGINATING FROM INITIAL POSITIONS NEAR TO THE ACCELERATION BOUNDARY

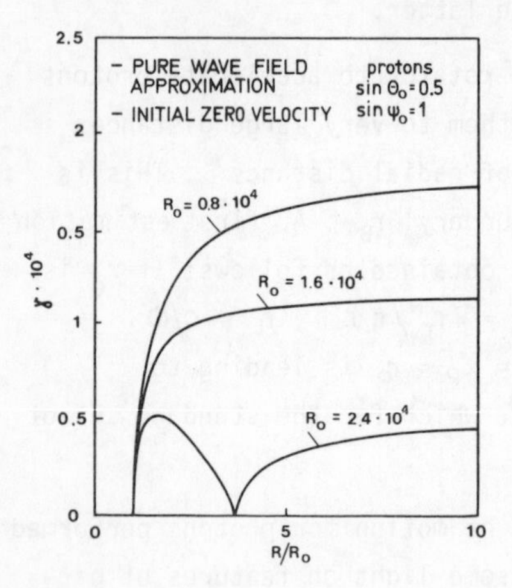

Figure 1

PROTONS ACCELERATION BOUNDARY

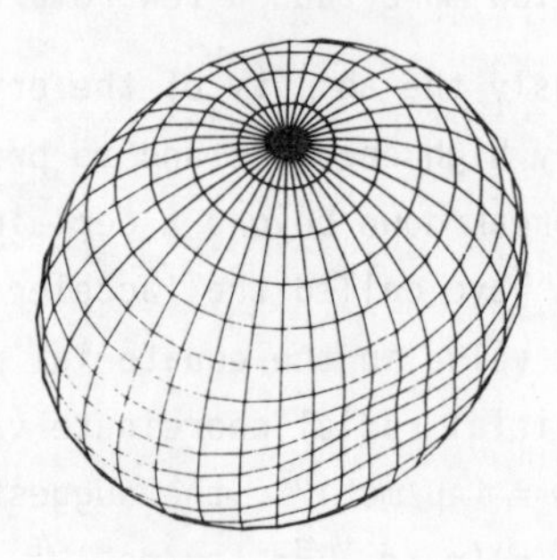

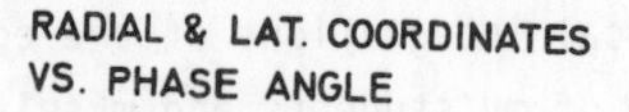

Figure 2

References

1. K.O. Thielheim, Proc. 2nd ESO/CERN Symposium, Munich 17-21 March 1986, 317-320
2. K.O. Thielheim, Proc. IAU-Symposium No. 125, Nanjing 26-26 May 1986
3. H. Laue and K.O. Thielheim, ApJ, Suppl. Ser. 61, 465-478 (1986)
4. K.O. Thielheim, Journ. Phys. A: Math. Gen. 20 (1987)
5. R. Leinemann and K.O. Thielheim, to be published

DISTRIBUTED REACCELERATION OF COSMIC-RAYS

Amri Wandel

Center for Space Science and Astrophysics
Stanford University, ERL, Stanford, California 94305

ABSTRACT

Continuous acceleration of cosmic rays during their propagation through the galaxy would alter the cosmic ray spectrum. A model is proposed, in which finite, power-law redistribution accelerating process (e.g. shock acceleration) is distributed over the cosmic ray propagation, and combined with escape from the galaxy. Comparing the calculated secondary/primary ratio to the observed boron/carbon (B/C) ratio, we find constraints on the amount and form of reacceleration consistent with the data. It is found that including reacceleration , the B/C data can be fit with escape rates quite higher than the standart "leaky box" model. It is also shown that reacceleration by strong shocks causes the secondary/primary ratio to become constant at sufficiently high energies, an effect which may be seen in current B/C data.

The observed spectra of cosmic ray above an energy of a few GeV/nucleon have a power law form in momentum (which is equivalent to kinetic energy for relativistic particles), $J(p) \propto p^{-q}$. The power is $q \approx 2.7 \pm 0.1$ for protons, He, and medium nuclei such as C, N, O, (Engelmann 1985; Ormes and Protheroe 1983), and somewhat steeper ($p^{q+\alpha}$; $\alpha \sim 0.5 \pm 0.1$) for light nuclei (Li, Be, B), which are mostly secondary cosmic rays , produced by spallation of medium nuclei on interstellar hydrogen. These observations are interpreted in the energy dependent "leaky box" model: cosmic rays are produced in the galaxy with a source spectrum $J_0(p) \propto p^{-q_0}$ with $q_0 \approx 2.1$, and escape after having traversed a column of interstellar matter (path length), which at a few GeV/nucleon has the mean value of ≈ 9 g cm^{-2}. From the decrease in the secondary to primary ratio with energy, it is inferred that the escape rate, which is usually expressed in terms of inverse path length, increases with energy (or momentum) per nucleon, $R \propto p^{\alpha}$, $\alpha \approx 0.6$.

It has been suggested that in addition to their inital acceleration, cosmic rays are continuosly accelerated during their propagation in the galaxy (Blandford and Ostriker 1980; Fransson and Epstein 1980). Silberberg *et. al.* (1983) have suggested reaaceleration at sub-GeV/n energies in order to explain the observed anomalies in the relative abundances of sub-iron secondary cosmic rays. However, reacceleration alters the cosmic ray specrum, and in particular the enrgy dependence of the secondary-to-primary cosmic ray ratio. It has been shown (e.g. Eichler 1980; Cowsik 1986) that observations of this ratio strongly constrain the acceptable amout of reacceleration .

In the presence of a reacceleration process which transforms a given initial distribution into a power law distribution, p^{-q}, the steady state differential distribution of cosmic rays can be described by the equation (Wandel *et. al.* , hereafter WELST)

$$J_0(p) - (R+S)J(p) = BJ(p) - (q-1)B\int_{p_0}^{p}(\frac{p'}{p})^q J(p')dp' \equiv B\ L_q(J) \tag{1}$$

where R, B, and S are the rates of escape, reacceleration and spallation (which may be a function of momentum), and p_0 is a cutoff due to ionization losses at low energies. The operator L_q defined on the right hand side describes the change in the distribution due to reacceleration .

The distribution equation for secondary cosmic rays is analogous to eq. (1), with the source term J_0 replaced by the formation rate of the secondary species. For simplicity, let us assume the secondary species 2 is produced entirely due to spalation of the primary species 1. If the escape and reacceleration reates for the two species are the same, their distribution equations are given by

$$J_0(p) - (R+S_1)J_1(p) = BL_q(J_1), \tag{2.a}$$

and

$$f_{12}S_1J_1 - (R+S_2)J_2(p) = BL_q(J_2), \tag{2.b}$$

where S_i are the respective spallation rates, f_{12} is the fraction of spallated nuclei of species 1 which go into species 2, and L_q is the reacceleration operator defined in eq. (1). If more species are involved, the secondary source term must take into account contributions from several primary species. WELST have soved this set of integral equations, for spallation rates $S_1 = S_2$ independent of momentum, and an escape law $R = R_1 p_{GeV}^{\alpha}$, where p_{GeV} is the momentum in unuits of GeV/c/nucleon. The ratio J_2/J_1 was then normalized to the boron/carbon (hereafter B/C) ratio by replacing f_{12} by an effective value, which takes into account the production rates of boron from other medium primary nuclei (mainly ^{16}O), as well as secondary carbon. These approximations are reasonable, since the spallation rate is proportional to $A^{-0.7}$ (A being the atomic mass number), so that elements of similar atomic masses (^{11}B, ^{12}C, ^{16}O) have similar spallation rates, and hence similar primary spectra.

The B/C ratio has been calculated for various combinations of the parameters α, R_1, B, and q, and compared to the observations. It was found that if the "traditional" leaky box ($\alpha = 0.6$, $R_1 = 0.1\mathrm{gm}^{-1}\mathrm{cm}^2$) is assumed, only a neglible amount of reacceleration is consistent with the data, $B/R_1 \leq 0.1$ (see Fig. 1). However, it turns out that in the context of distributed reacceleration , the data can be fitted with escape laws quite different from the traditional one, in partibular the escape rate can be significantly larger then previously assumed (see Fig. 2). For $\alpha = 0.6$ WELST find that the parameter sets that give the best fit to the B/C data obey the empirical relation

$$\frac{B}{R_1} = (1.67 \pm 0.1)(q-2)R_1, \quad 0.15 \leq R_1 \leq 0.4, \tag{3}$$

where R_1 is in units of $\mathrm{g}^{-1}\mathrm{cm}^2$. For example, the parameters which best fit the high quality data of the HEAO-3 experiament (Engelmann *et. al.* 1985), $R_1 = 0.2, q = 4$, yield $B/R_1 = 0.7$

It is possible to show that without reacceleration , or with reacceleration by weak shocks, the secondary/primary ratio at high energies decreases as $p^{-\alpha}$, as observed. WELST have shown that if the reaccelerating shocks are strong enough, $q < q_0 + \alpha$, the secondary/primary ratio at high energies will be flatter, having a slope of $p^{-(q-q_0)}$. In particular, if $q \approx q_0$, the secondary/primary ratio flattens assymptotically, the transition energy being determined by the ratio B/R_1 (Fig. 3). There may be a hint to this effect in the present B/C data from the HEAO-3 experiament, and higher energy measurments could definetely confirm or reject it.

The strong shock reacceleration provided by young supernova remnants in the interstellar medium is more than enoung to explain the amount of strong shock reacceleration required to fit the B/C data (Wandel 1987), and the latter may actually constrain the expansion and/or frequency of supernova remnants in the galaxy.

REFERENCES

Blanford, R.D. and Ostriker, J.P. 1980, *Ap. J.*, **237**, 793.
Cowsik, R. 1986, *Astr. Ap.*, **155**, 344.
Eichler, D.S. 1980, *Ap. J.*, **237**, 809.
Fransson, C. and Epstein, R.I. 1980, *Ap. J.*, **242**, 411.
Engemmann, J.J., *et. al.* 1983, *Proc. 18th Internat. Cosmic Ray Conf.* (Bangalore), **2**, 17.
Ormes, J.F., and Protheroe, R.J. 1983,, *Proc. 18th Internat. Cosmic Ray Conf.* (Bangalore), **2**, 17.
Silberberg, R., Tsao, C.H., Letaw, J.R. and Shapiro, M.M. 1983, *Phys. Rev. Lett.*,**51**, 1217.
Wandel, A., Eichler, D.S., Letaw, J.R., Silberberg, R., and Tsao, C.H. 1987, *Ap. J.*, **317**, (WELST).
Wandel, A. 1987, in preparation.

Fig. 1 secondary/primary ratio (normalized to the Boron/carbon ratio) in the standart "leaky box" model ($R = 0.11$, $\alpha = 0.6$ for various amounts of reacceleration (B). The data points are from Engelmann (1983).

Fig. 2 Boron/carbon ratio for an increased escape rate. The curves correspond to no reacceleration ($B = 0$), and B=0.03, 0.085, 0.2 and 0.4 in increasing order.

Fig. 3 secondary/primary ratio for strong shock reacceleration ($q = q_0 = 2.1$), for an enhanced escape law ($R = 0.2$). The curvese are labelled by the reacceleration rate.

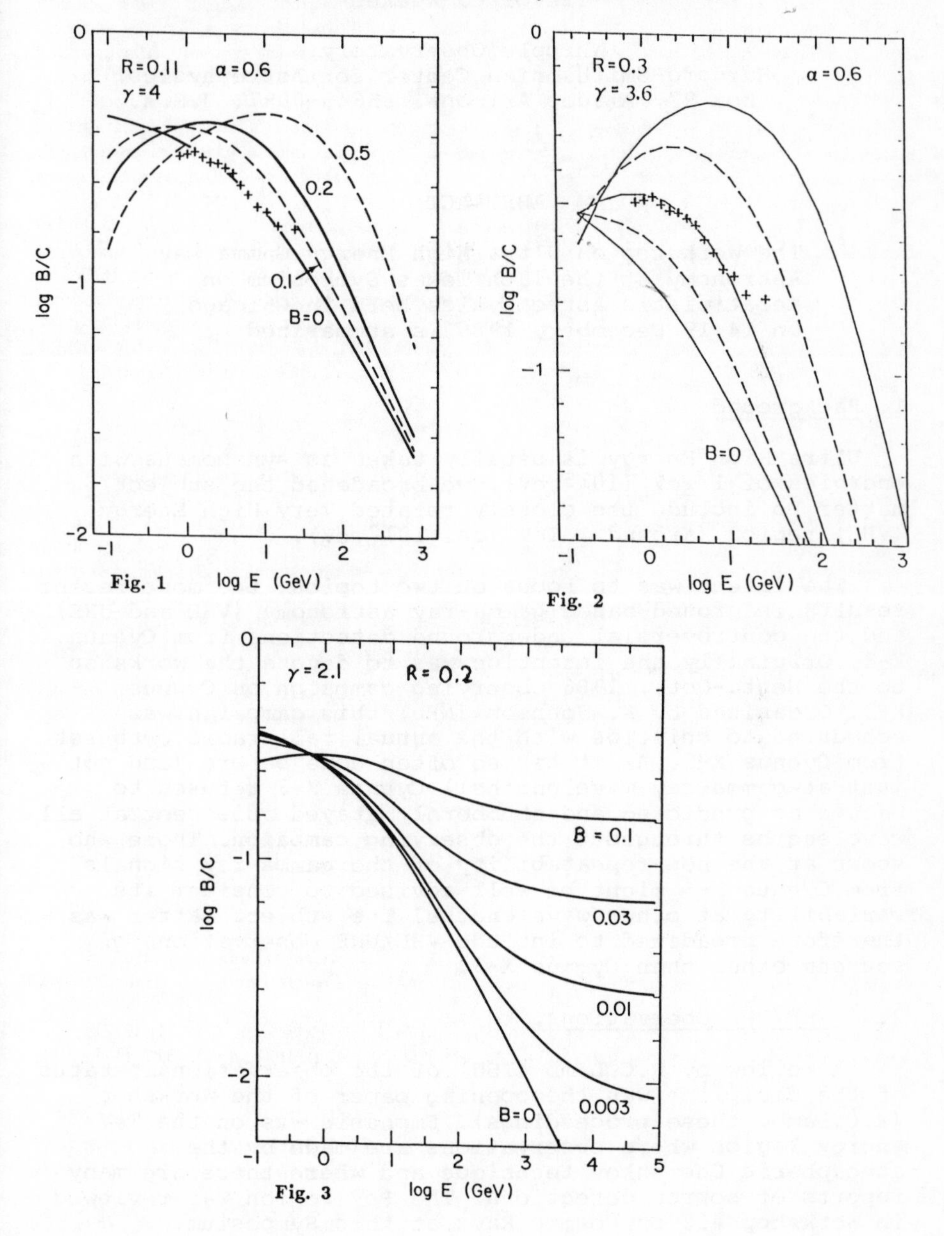

ULTRA HIGH ENERGY GAMMA RAY WORKSHOP SUMMARY

Trevor C. Weekes

Whipple Observatory,
Harvard-Smithsonian Center for Astrophysics,
Box 97, Amado, Arizona. 85645-0097. U.S.A.

ABSTRACT

The Workshop on Ultra High Energy Gamma Ray Astronomy at the 13th Texas Symposium on Relativistic Astrophysics held in Chicago on 14-19 December, 1986 is summarized.

1. Background

Ultra High Energy is usually taken as synonomous with energies of 1 PeV (10^{15} eV); we broadened the subject matter to include the closely-related Very High Energy (VHE) region (around 1 TeV i.e. 10^{12} eV).

The intent was to focus on two topics: the most recent results in ground-based gamma-ray astronomy (VHE and UHE) and the controversial underground detections from Cygnus X-3. Originally the intention was to devote the workshop to the Sept.-Oct., 1986 observing campaign on Cygnus X-3. Organised by K. Johnson (NRL) this campaign was scheduled to coincide with the annual fall radio outburst from Cygnus X-3. As it has so often done before (and not just at gamma-ray wavelengths!) Cygnus X-3 refused to behave as predicted and stubbornly stayed quiescent at all wavelengths throughout the observing campaign. Those who scorn at the non-repeatability of the gamma-ray signals from Cygnus X-3 might be well-advised to consider its variability at other wavelengths! The subject matter was therefore broadened to include VHE/UHE observations of sources other than Cygnus X-3.

2. VHE/UHE Obervations.

A review by R.C.Lamb (ISU) of the observational status of the dicipline was the opening paper of the Workshop (R.C.Lamb, these proceedings). Emphasis was on the TeV energy region where observations are made by the atmospheric Cherenkov technique and where there are many reports of source detections. The PeV region was reviewed in Workshop #12 on Cosmic Rays at this Symposium.

There were then a series of six invited short contributions from gamma-ray observatories that are active in the Northern Hemisphere. The first of these was from the Halaekala Collaboration (University of Athens, University of Hawaii, Purdue University and University of Wisconsin) and presented by A. Kenter; it concerned the detection of episodes of periodic emission of TeV gamma rays from the Crab pulsar. The Halaekala Gamma Ray Telescope consists of six spherical mirrors, each of 1.5 m aperture, on a single mount. Three episodes of emission lasting 2000 seconds were detected. The light-curves were broad and the absolute phase was not determined.

The newest gamma-ray observatory, with potentially the lowest energy threshold, was represented by C.Akerlof (U.Mich.). This new project is a collaboration between the University of California at Riverside, the Jet Propulsion Laboratory, the University of Michigan and the Sandia Laboratory; it is centered arounded the use of two 11 m Solar Energy Collectors at the Kirkland Air Force Base in Albuquerque, New Mexico. The time of arrival difference between the shower front striking the two detectors, which are 41 m apart, is used to determine the arrival direction. No source observations have yet been made but these are expected in the spring of 1987.

A new observation of emission from Hercules X-1 was presented by M.Cawley (ISU) representing the Whipple Observatory Collaboration (University College,Dublin, the University of Leeds, Iowa State University and the Smithsonian Astrophysical Observatory). This group uses a 10 m Optical Reflector at the Whipple Observatory on Mount Hopkins in Southern Arizona. This report concerned the detection of a periodic signal at orbital phase 0.7 at the putative turn-on point of the 35 day cycle.

Switching to higher energies (>0.1 PeV) D.Nagle (LANL) descibed a new detection of Cygnus X-3 by the CYGNUS Collaboration at Los Alamos (D.Nagle, these proceedings). There was much interest at the workshop on the muon content of the detected signal since many new experiments plan to exploit this difference between the gamma rays and the hadrons that constitute the background.

There are two UHE experiments in India, both operated by the Tata Institute for Fundemental Research in Bombay. The status of these experiments was summarised by B.V.Sreekantan (TIFR). The newest and largest of these experiments is the array at the Kolar Goldfield; data from this experiment has not yet been fully analysed but the angular resolution has been measured to be 0.5^{o}. Results from the Ooty Array indicates a new detection of Cygnus X-3 in data taken since 1984.

The Fly's Eye Experiment was represented by B.Fisk (U.Utah) who described observations of Cygnus X-3; a signal was seen in 1985 but not in 1986.

3. Underground Detection of Cygnus X-3?

The second part of the Workshop was devoted to the controversial detection of Cygnus X-3 in underground muon proton-decay experiments. The detections reported by the Soudan Mine and NUSEX Experiments were summarised by K.Ruddick (U.Minn.). K.Lande (U.Penn.) gave the opposing view, describing the null results from the Homestake Mine, from FREJUS, from Kamioka and from IMB. Both papers are in these proceeding. A new result from the IMB experiment, in which a signal is seen from Cygnus X-3 during a 1983 radio outburst, was presented by J.Learned (U.Hawaii).

In his summary T.Gaisser (Bartol), who acted as moderator, outlined the difficulties of explaining this result in terms of conventional physics. He emphasised the distinction between the surface experiments, whose results can be explained in terms of gamma rays and which involve no new physics, and the underground experiments, which have no such easy interpretation. He noted that the statistical basis of both sets of experiments is not high.

4. Prospect

In summary VHE/UHE gamma-ray astronomy may be seen to be at a cross-roads, with considerable uncertainty as to its future. In the past year the number and quality of gamma-ray telescopes has increased giving some hope that the quality of observations may also improve. The current uncertainty may therefore be short-lived.

For the moment it is difficult not to believe in the existence of some sources at energies above 1 TeV. Cyg X-3, Her X-1, Vela X-1 and the Crab have been seen by a number of groups; the emission of gamma rays at varying intensities from these objects, at least, seems quite credible. However there is still a large number of unconfirmed sources and it would be unwise to put too much faith in a source whose existence rests on a single detection.

This Workshop did not answer any of the outstanding questions in gamma-ray astronomy but it did provide a standing-room only attendance an opportunity to see what is happening at the frontiers of this new dicipline and to get some flavor of what is to come with the advent of a new generation of detectors.

GROUND-BASED GAMMA-RAY ASTRONOMY: AN OVERVIEW

R. C. Lamb

Iowa State University
Ames, Iowa 50011

ABSTRACT

A brief overview of ground-based gamma ray astronomy is given. Seven sources have been confirmed at energies greater than 1 TeV. Emphasis is given to the Crab Pulsar, Cygnus X-3 and Hercules X-1.

I. INTRODUCTION

At energies above about 100 GeV cosmic photons may be detected on the ground by virtue of the air showers they create in the atmosphere. In the region 10^{11} to 10^{14} eV (the "very high energy" or "TeV" region for short) the detection is generally, by means of the Cherenkov light, emitted by the relativistic charged-particle secondaries. At energies $\geq 10^{14}$ eV ("ultra high energy" or "PeV") the detection is generally by means of the charged particle secondaries themselves.

This paper is a brief overview of the field. For a comprehensive review the reader is referred to recent articles by Watson[1)], Hillas[2)], Turver[3)], Weekes[4)] and Ramana-Murthy[5)].

Why is this field important? In the first place, it represents an underdeveloped and still largely unexplored frontier of the cosmic electromagnetic spectrum. There may be surprises. The observation of TeV and PeV emission from some of the X-ray binaries is one such surprise and there may be others. Certainly one cannot claim to understand any particular celestial object unless it has been observed at all wavelengths at which significant energy is emitted. Furthermore, observations of cosmic photons at these high energies give direct information regarding the sites of cosmic rays and thus play an important, perhaps decisive, role in answering long-standing questions about the origins of cosmic rays. Finally, the localized sources that have been discovered are natural "laboratories" for the exploration of physical phenomena which are far beyond the capabilities of terrestrial laboratories.

The first clear source observations at TeV[6] and PeV[7] energies are shown in figure 1. The source, Cygnus X-3 in both cases, appears as a relatively narrow angular enhancement above an otherwise smooth isotropic background which arises from charged-particle cosmic ray air showers.

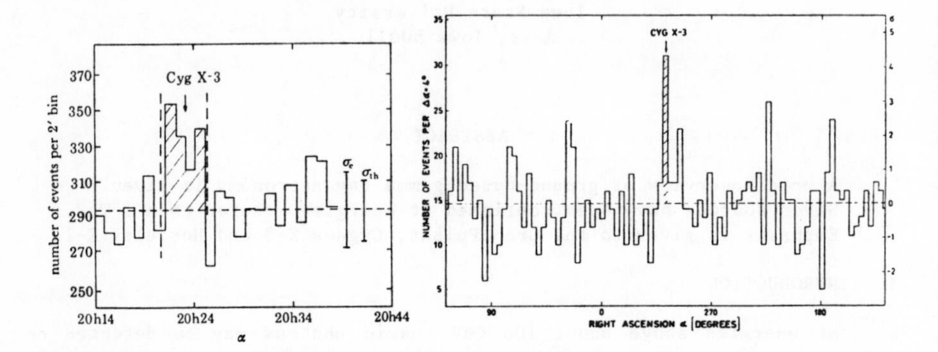

Figure 1. First TeV and PeV source detections. 1a: Crimean Astrophysical observations in 1972 of Cygnus X-3 at energies above 1 TeV[6]. 1b: Kiel observations from 1976-80 of Cygnus X-3 at energies above 2 PeV.[7]

The presence of this cosmic ray background is the major obstacle to rapid progress in this area of astronomy, since photon signal rates, particularly in the TeV region, are adequate, giving for a "bright" source counting rates in excess of 1 per minute. The cosmic ray background may be reduced by improving angular resolution and by exploiting the differences which exist between hadron-induced air showers and photon-induced showers. However, tests and development of various schemes proposed to reduce the cosmic ray background have been handicapped by the absence of a bright "standard candle". Thus, the results presented here are the result of "first generation" detectors which have not begun to exploit, in any significant way, methods to reduce the cosmic ray background.

In the main section of this paper we will present the source list, giving attention particularly to the Crab Pulsar, Cygnus X-3, and Hercules X-1.

II. THE SOURCE LIST

The table shows a list of all sources as of December 1986 which have been reported in the literature and for which there are no contradictory observations. The table is divided into four categories: radio pulsars, binary X-ray pulsars, supernova remnants and radio galaxies. For the first two categories either a neutron star spin period (in the range of ms to s) or a binary orbital period (hours to days) is associated with the source and has aided its detection. For those cases where there are two or more detections the stated flux values are rough averages scaled generally to 1 TeV or 1 PeV. With one exception, which is indicated, the fluxes are determined from observations which span at least several days. The luminosities quoted assume 4π geometry; for some beam

TeV & PeV Gamma-Ray Source Catalog

Source	Period	Energy Range (# of detections)	Integral Flux ($cm^{-2}s^{-1}$)	Luminosity (erg/s)
Radio Pulsars				
Crab Pulsar	33 ms	TeV (5)	8×10^{-12} (>1 TeV)	2×10^{34}
Vela Pulsar	89 ms	TeV (2)	7×10^{-12} (>1 TeV)	5×10^{32}
PSR 1953+29	6.1 ms	TeV (1)	1×10^{-12} (>1 TeV)	6×10^{35}
PSR 1802-23	112 ms	TeV (1)	2×10^{-10} (>1 TeV)	3×10^{35}
PSR 1937+21	1.6 ms	TeV (1)	2×10^{-11} (>1 TeV)	2×10^{35}
Cygnus X-3	4.8 h	TeV (≥ 10)	5×10^{-11} (>1 TeV)	3×10^{36}
	(12.6 ms)	PeV (2)	2×10^{-14} (>1 PeV)	10^{36}
Binary X-Ray Pulsars				
Hercules X-1	1.24 s	TeV (8)	3×10^{-11} (>1 TeV)	3×10^{35}
		PeV (1)	3×10^{-12} (>1 PeV)*	2×10^{37}*
4U0115+63 (Cas γ-1)	3.6 s	TeV (5)	7×10^{-11} (>1 TeV)	6×10^{35}
Vela X-1 (4U1700-37)	283 s	TeV (1)	2×10^{-11} (>1 TeV)	3×10^{34}
	9.0 d	PeV (1)	9×10^{-15} (>3 PeV)	2×10^{34}
LMC X-4	1.4 d	PeV (1)	5×10^{-15} (>10 PeV)	10^{38}
Supernova Remnants				
Crab Nebula		TeV (2)	10^{-11} (>1 TeV)	2×10^{34}
Radio Galaxies				
Centaurus A		TeV (1)	10^{-11} (>1 TeV)	3×10^{40}

*40 minute outburst[8)]

geometries this may be a significant overestimate. The luminosity associated with the charged particles responsible for the radiation may be estimated as follows: If the charged particles are electrons, their luminosity should be comparable to the photon luminosity, whereas a hadron beam luminosity will be perhaps two orders of magnitude greater. For a complete bibliography, see references 1-5.

Radio Pulsars. The Crab and Vela pulsars have been reported by several groups whereas the other radio pulsars listed are yet to be confirmed. In figure 2a, the most convincing observation of the Crab pulsar at TeV energies is shown. This detection, by the Durham group,[9] is based on 103 hours of observation. For comparison the COS-B satellite detection for energies above 100 MeV[10] is shown in figure 2b. The gamma-ray luminosity above 1 TeV for both the Crab and Vela pulsars is of order 10^{-4} of their estimated total spin-down luminosities. Thus the high energy gamma ray emission of the radio pulsars (if the Crab and Vela pulsars are typical) is not as important as it is for the binary X-ray pulsars where the luminosity above 1 TeV is $\sim 10^{-2}$ or more of the total luminosity.

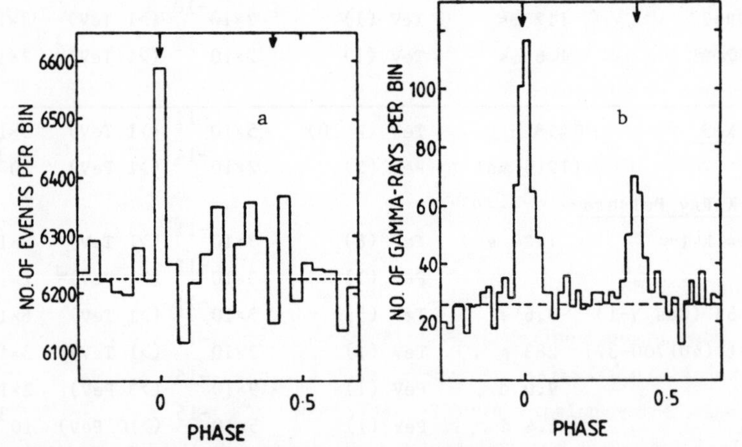

Figure 2. Observations of the Crab Pulsar. a) at energies above 1 TeV[9]; b) at energies above 100 MeV[10].

Cygnus X-3. Although Cygnus X-3 is generally regarded as some sort of a binary source, there is, as yet, no convincing demonstration that the collapsed member of the presumed binary system is, in fact, a spinning neutron star. I have therefore separated it from the list of X-ray binary pulsars. It should be noted that the reported[11] 12.6 ms pulsation of Cygnus X-3 is yet to be confirmed. Two observatories have searched for this periodicity with negative results[12,13]. Recent observations of Cygnus X-3 at both TeV and PeV energies have relied on the observation of rather narrow phase peaks ($\leqslant$ 10% of the assumed 4.8 hour orbital cycle) above the cosmic ray background level (cf. figure 1, reference 1). There is some evidence for clustering of emission for orbital phases near 0.6 and also 0.2 (cf. figure 3, reference 1). Time variability for Cygnus X-3 has been established in the TeV region on a monthly basis[14] as well as on time scales of minutes[15,16].

Binary X-Ray Pulsars. Of the three confirmed X-ray binary pulsars: Hercules X-1, 4U0115+63, and Vela X-1, the observations of Hercules X-1 are the most extensive. The source was discovered by the Durham group[17]. A Fourier analysis of a 3-minute excess of showers coming from the direction of Hercules X-1 showed the characteristic 1.24 s spin period of the neutron star member of the binary system. The Fly's Eye group has reported[8] a 40-minute excess at energies above 500 GeV. The Whipple Observatory has made observations in 1984[18] and 1985[19,20] which show that 5 to 10% of the time Hercules X-1 emits TeV γ-rays during a small fraction of its 1.7 day orbital cycle. In figure 3, the observations of Hercules X-1 are summarized. Note that there are no episodes of emission during the OFF states of the 35-day cycle; and, although there is no clearly preferred orbital phase, five of the nine episodes occur between orbital phase 0.6 and 0.8 (phase 1.0 corresponds to mid-eclipse).

III. CONCLUSIONS

Of the seven confirmed sources listed in the catalog, three are binary X-rays pulsars (and perhaps Cygnus X-3 should be a fourth). Thus, some binary X-ray pulsars have a significant portion (a few percent) of their luminosity at energies above 1 TeV. If the primary "beam" responsible for the radiation is hadronic, then its energy dominates the energy output of these sources. Second generation detectors in both the TeV and PeV region promise sensitivity improvements of more than an order-of-magnitude. Thus, 50 to 100 sources should be detected with such instruments and brighter sources studied on time scales which are inaccessible with present detectors.

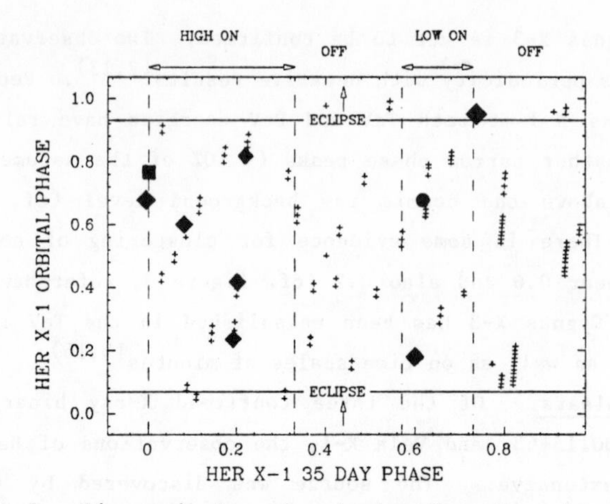

Figure 3. Nine episodes of emission from Hercules X-1 are shown as a function of orbital phase (vertically) and 35 day phase (horizontally). The seven diamonds mark the positive detections by the Whipple Observatory during 1984-85[20] the crosses indicate times of no detectable emission. The square marks the discovery detection by the Durham group[17], the circle the Fly's Eye detection[8].

REFERENCES

1. Watson, A. A., 19th ICRC (La Jolla), 9, 111 (1985).
2. Hillas, A. M., 19th ICRC (La Jolla), 9, 407 (1985).
3. Turver, K. E., 19th ICRC (La Jolla), 9, 399 (1985).
4. Weekes, T. C., Proceedings of the VIth Astrophysics Meeting of the XXIth Rencontre de Moriond (1986) in press.
5. Ramana-Murthy, P. E., NATO Advanced Workshop on Very High Energy Gamma-Ray Astronomy (Durham), 1986) in press.
6. Vladimirsky, B. M., et al., 13th ICRC (Denver) 1, 456 (1973).
7. Samorski, M. and Stamm, W., Ap. J. Letters, 268, L17 (1983).
8. Baltrusaitis, R. M., et al., Ap. J. Letters, 293, L17 (1983).
9. Dowthwaite, J. C., et al., Nature, 309, 691 (1984).
10. Wills, R. D., et al., Nature, 296, 723 (1982).
11. Chadwick, P. M., et al., Nature, 318 642 (1985).
12. Resvanis, L., et al., NATO Advanced Workshop on Very High Energy Gamma-Ray Astronomy (Durham), (1986) in press.
13. Fegan, D. J., et al., NATO Advanced Workshop on Very High Energy Gamma-Ray Astronomy (Durham), (1986) in press.
14. Cawley, M. F., et al., Ap. J., 296, 185 (1985).
15. Weekes, T. C., Astr. Ap., 121, 232 (1983).
16. Dowthwaite, J. C., et al., Ap. J. 286, L35 (1984).
17. Dowthwaite, J. C., et al., Nature, 309, 691 (1984).
18. Gorham, P. W., et al., Ap. J., 309, 114 (1986).
19. Gorham, P. W., et al., Ap. J. Letters, 308, L11 (1986).
20. Gorham, P. W., Ph.D. Thesis, University of Hawaii, unpublished (1986).

DETECTION OF 10^{14}-eV GAMMA RAYS FROM CYGNUS X-3 DURING 1986

D.E. Nagle, R.D. Bolton, R.L. Burman, K.B. Butterfield, R. Cady, R.D. Carlini, V.D. Sandberg, R.A. Williams, C. Wilkinson; Los Alamos National Laboratory; Los Alamos, NM 87545

C.-Y. Chang, B. Dingus, J.A. Goodman, R.L. Talaga, G.B. Yodh, D.A. Krakauer; University of Maryland; College Park, MD 20742

S. Gupta; TATA Institute; Bombay, INDIA

R.W. Ellsworth; George Mason University; Fairfax, VA 22030

J. Linsley; University of New Mexico; Albuquerque, NM 875131

R.C. Allen; University of California; Irvine, CA 92717

In this workshop we are asked to emphasize recent observations. However, a brief description of the system is probably in order. (A more complete one is given in the La Jolla Workshop (1985), R. Protheroe, ed.). An array of scintillation counters at Los Alamos, NM (Latitude 35°52′N) has been in operation since March 1986. During the period reported on the array consisted of 50 counters of area 1 m^2, which sampled cosmic ray air showers over an area approximately 10^4 m^2. The direction of the shower is determined from the scintillators with a resolution of better than 1°. The system is augmented with a high-resolution track detector for observing muons associated with the shower, which is used in the so-called LAMPF E225 experiment on neutrino-electron scattering. The anticoincidence shield array of E225 consists of three layers of MWPC anticoincidence counters below a shield of 1.4 kg/cm^2 of Fe, corresponding to muon energies greater than 3 GeV.

For our purposes we use the signals from the anticounters and the central detector to register cosmic-ray muons associated with an extensive air shower detected by our scintillator array. Data are collected within a 30° cone centered about the zenith. About 20,000 showers per day are recorded.

The following are our preliminary results: Figure 1 represents a scan in right ascension thru the region containing Cyg X-3. The bin width is 2.472° in right ascension, and 2.806° in declination. One bin is centered on Cyg X-3. The period of observation was 38 days during April and May 1986. A peak is observed in the Cygnus bin. The null hypothesis would require a chance fluctuation of 4σ. A cut requiring (0 or 1) muons was imposed.

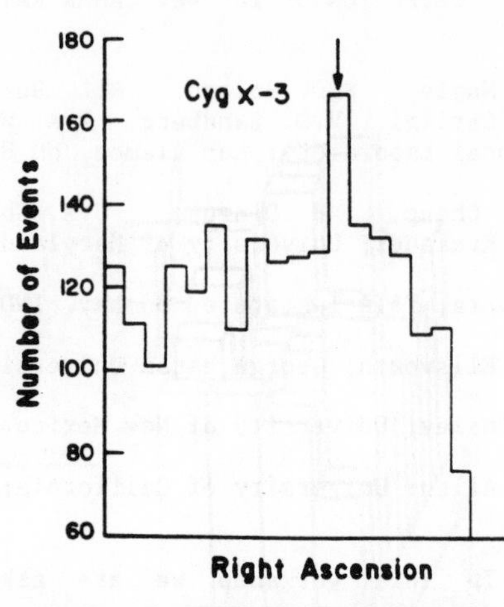

Figure 1. Right ascension scan of Cyn X-3 region.

In Figure 2 the events in the Cyg bin are folded into the nominal 4.8-hour period using the ephemeris of Mason (1985).[1] A peak is seen in the phase (0.7-0.8) bin. A phase diagram for the events of Fig. 1 other than in the Cyg bin shows no obvious peak. Note that the Vander Klis and Bonnet-Bidand ephemeris would give a phase during this time of 0.08 less than the Mason one.

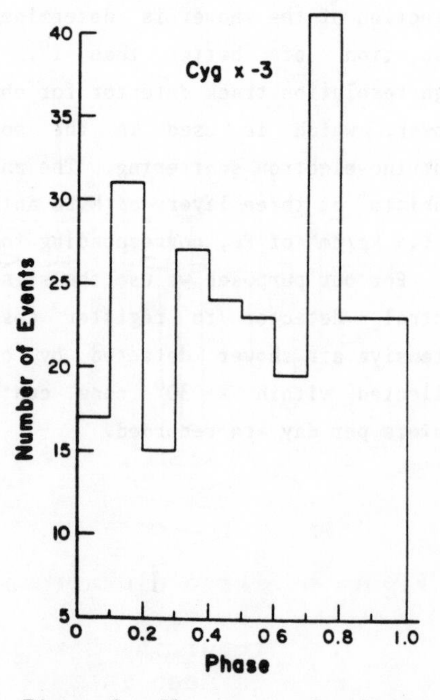

Figure 2. Phasogram: Cyg Events.

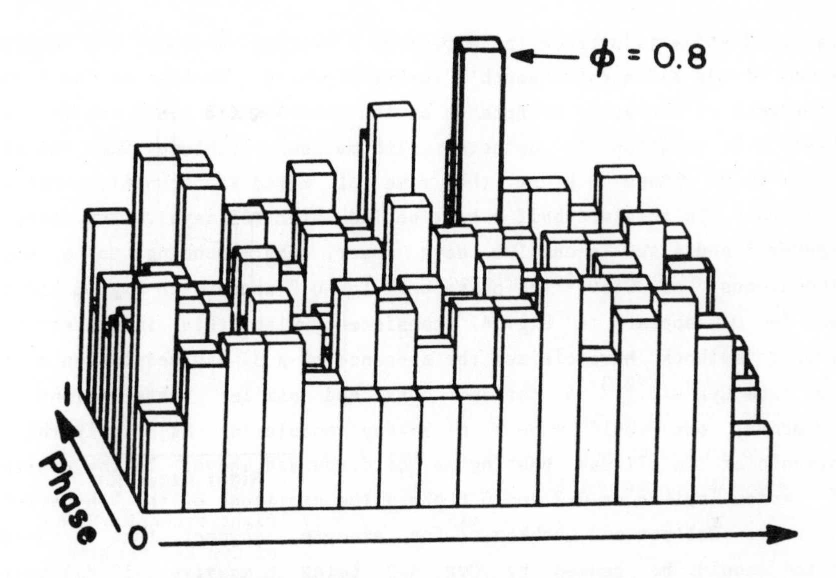

Figure 3. Three-dimensional plot of counts during the April-May '86 interval. A peak is seen in the (0.7-0.8) phase bin in the Cygnus direction.

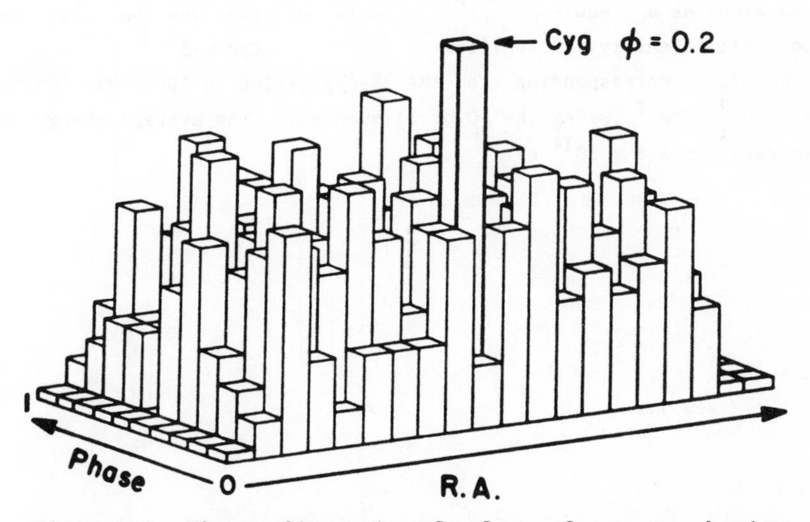

Figure 4. Three-dimensional plot of events during the period 19 June-27 September. A cut requiring zero muons was used. A peak is seen in the (0.1-0.2) phase bin.

Figures 3 and 4 illustrate the effect of a beating between the orbital period of Cyg X-3 and the earth's rotation period. Because of the strong dependence of the array acceptance on the zenith angle end($\propto \cos^7\Theta$), the phase 0.8 emission is detected with maximum efficiency when the star system is at phase 0.8 at the time of meridian transit. Minimum efficiency is obtained half a beat period later (85 days). The dates of Figures 3 and 4 are taken 100 days apart, corresponding to a phase displacement of 0.6. The phase peak for 0.2 appears in Fig. 3 and the peak for 0.8 appears in Fig. 4, consistent with this interpretation. Several authors have claimed the presence of a 34-day modulation of the flux from Cyg X-3.[2,3] As far as ours and similar installations are concerned, one would expect a 34-day modulation as simply the 5th harmonic of the 171-day beating period discussed above. In the presence of narrow spikes at 0.2 and 0.6 phase the amplitude of the 5th harmonic is large. Gal'per and Luchkov,[3] for example, suggest that a 34-day period could be caused by Cyg X-3 being a massive ($\sim 10^3 M_\odot$) system undergoing relativistic precession. A less drastic explanation is the simple beating phenomenon discussed here. In general, many groups have reported emission at a low phase (0.2) and/or at a high phase (0.65), but in many publications both phases have not been reported.[4] This may also be understood as a result of the effects of the beating plus the non-periodic intensity fluctuations.[5]

The flux corresponding to the 38-day period in April-May '86 was $(7\pm2) \times 10^{-9}\ m^{-2}s^{-1}$, using the (0 or 1) muon cut. The average energy of the showers was 1.6×10^{14} eV.

References:

1. Mason, K., private communication.
2. Morello, C., Navarra, G., and Vernetto, S., Proc. 18th Intl. Cosmic Ray Conf. 1, 127 (1983) XG-4-13 p. 127.
3. Gal'per, A.M., and Luchkov, B., Sov. Astron. Lett. 10, (3), May-June 1984, p. 150.
4. Watson, A.A., Proc. 19th Int. Conf. on Cosmic Rays, 9, p. 111 (1985).
5. Priedhorsky, W., and Terrell, J., Ap. J. 301, 886 (1986).

UNDERGROUND MUONS FROM CYGNUS X-3: THE CASE FOR.

K. Ruddick

School of Physics and Astronomy,
University of Minnesota, Minneapolis, MN 55455.

ABSTRACT

The principal case for the underground muons associated with Cygnus X-3 still rests on the original simultaneous observations in the Soudan 1 and NUSEX detectors. Recent reports of the non-observation of the effect do not necessarily negate the effect. Rather, they may emphasize the high variability of the system. The data may be used to deduce the properties of the particles and interactions responsible for the phenomenon.

The observation of muons deep underground apparently originating from the direction of the Cygnus X-3 system was first reported in early 1985 in two different experiments: the Soudan 1 experiment situated in a mine in northern Minnesota[1], and the NUSEX experiment situated below Mont Blanc[2]. The principal case for the existence of the effect is still primarily based on these original simultaneous observations, which were made during 1982-83. Since that time, confirmatory evidence has been obtained at Soudan[3], coincident with a large radio burst from Cygnus X-3 during September-October, 1985, but no other experiment has observed the effect. On the contrary, some other underground experiments have published apparently contradictory evidence[4,5]. These latter data, plus the apparent inability of conventional physics, "the standard model", to explain the phenomenon, have tended to discount the original data. However, in this paper, I shall show that a self-consistent picture of the phenomenon is indeed possible.

The muon data were obtained using the very accurately known time modulation of the Cygnus X-3 source which has been presumed to be due to the 4.79 hour orbital period of a neutron star about a large companion star. The muons occur only within a characteristic range of phases during this period, between 0.65 and 0.9 at Soudan, and from 0.8 to 0.9 at NUSEX. They appear to originate from directions not directly from Cygnus X-3, but, rather, within a range of angles up to 3 to 5 degrees from the source. The average flux at Soudan was $7\text{x}10^{-11}/\text{cm}^2/\text{sec}$ and was approximately 10 times lower at NUSEX. The depths of the two experiments are $1.8\text{x}10^5$ and $5\text{x}10^5$ gm/cm^2, respectively.

Superficially, these muon data appear very similar to data obtained at the surface of the earth for air showers initiated in the atmosphere by TeV and PeV cosmic rays, assumed to be photons originating from Cygnus X-3. A natural explanation of the muon data, then, would have the muons produced by the same photons which are responsible for the air showers. However, if the muons are produced in the atmosphere, they require energies at least 0.7 and 5.0 TeV, respectively, to penetrate to the depths of the two experiments, while the muon fluxes are very close to the photon fluxes reported at these same energies. But photons do not produce muons this efficiently. The muon flux is too high by a factor 200 to 300 to have been produced by the photons responsible for the atmospheric shower data[6)].

Neutrons originating at Cygnus X-3 have too short a lifetime to explain the signal, while neutrinos would produce muons at mostly horizontal angles, where none are seen. Neutral atoms would be stripped in the galactic medium. Charged primaries, would, of course, be bent in the galactic magnetic field and not point directly at their source.

During October, 1985, radio-astronomers detected a very large outburst from Cygnus X-3. During September 18 to October 18, a significant signal of 16 muons from the direction of Cygnus X-3 were observed in the Soudan 1 detector, over an estimated background of 6 events[3]. The muons appear to precede the radio signal by approximately one week. They arrive within a very narrow phase range, 0.725 to 0.745. This is the only observation of Cygnus X-3 which has been made at Soudan or in any other detector since the original 1982-83 observation. If the data are correct, Cygnus X-3 has been quiescent since 1983, except for the brief period during September-October 1985.

Data have recently been published from two other underground experiments: Kamioka, a large water detector in Japan[4], and Frejus, another large tracking detector beneath the Alps[5]. They have not observed any signal from the direction of Cygnus X-3. Unfortunately, the former experiment was not operating at the time when the original signal was observed at Soudan and NUSEX. Of some more significance is the Frejus experiment, which overlapped the Soudan experiment during the 1985 radio burst. They observed no obvious signal at the Soudan phase. On a background of about two events per phase interval, they should have observed 6+-3 events based on the Soudan rate extrapolated to the Frejus depth (using the relative rates between Soudan and NUSEX). The statistical and systematic uncertainty of this result is inadequate to nullify the Soudan data.

At this symposium, new data have been presented by the IMB group, who occasionally observe muons in their large water cerenkov nucleon decay detector[7]. This detector is at a depth slightly shallower than Soudan. None of their data, which overlap the original 1982-83 observations, set limits below the Soudan flux. However, in a search for active Cygnus periods, coinciding with previous radio outbursts, they have observed an enhanced signal during a burst in October, 1983. This signal occurs at phase 0.5 to 0.6, consists of approximately 20 events over a background of 36 in that phase interval, and is quoted as having a statistical significance of 4.3 standard deviations. A similar number of events should have been seen in the Soudan 1 detector. None were observed. The significance of this result is presently unclear: account must be taken of the number of data sets which have been examined in the analysis.

In order to accommodate all the observations, (and non-observations), of Cygnus X-3 deep underground, it is necessary to accept the fact that the signal is highly variable in amplitude, and also somewhat variable in the phase of the emission. It is not surprising that what is possibly the most energetic system in the galaxy is highly unstable. If Cygnus X-3 presently is in a long period of quiescence, there may be only a small chance to test the data, and it is necessary to investigate other astronomical objects for possible evidence, and to operate other large precision tracking devices at Soudan depths, or less. There are indications that the effect occurs in other X-ray emitting binary systems, but the data are not absolutely conclusive so far[3,8]. The Soudan 2 detector will soon be operational at a depth only slightly greater than Soudan 1 and should be able to receive data at 10 times the rate of Soudan 1.

Numerous efforts have been made to explain the data[9]. No explanation of the phenomenon has been possible in terms of known particles and interactions, or even those particles postulated in possible extensions of the "standard model". However, the data themselves can be used to infer the necessary properties of the particles and their interactions.

From the known distance to Cygnus X-3 (>11kpsec) it is easily shown that the mass of any particle carrying the signal from Cygnus X-3 must be less than about 20 MeV/c^2 in order to maintain the narrow peak observed in the phase distribution[6,9]. This condition on the maintenance of phase coherence actually limits only the velocities of the particles. In order to infer the mass, it is necessary to use the additional fact that the muons have energies of at least tens of GeV. These neutral particles, which have come to be called "cygnets"[10], must be either some previously unknown particles, or they could be neutrinos with some new interaction not observed at the energies of present particle accelerators.

It is very significant that the source apparently has an angular size of 3 to 5 degrees at both Soudan and NUSEX. These angles must be produced locally, originating in the interaction of the cygnet particles. From the angles we are able to deduce that the cygnets must interact in the rock, not in the atmosphere, that they interact with nucleons, not electrons, and that the interaction must involve a new mass scale of tens of GeV[9,11].

That the interaction occurs in the rock follows immediately from the fact that the angular spread of the source is similar at both Soudan and NUSEX. It has already been pointed out that muons produced in the atmosphere require minimum energies of 0.7 and 5.0 TeV to reach the two detectors. For the same characteristic momentum transfer produced in the cygnet interaction, one would then expect the NUSEX angular spread to be lower than that at Soudan by approximately the ratio of these energies, for any reasonable picture of the process. The difference in rates between Soudan and NUSEX must be due to attenuation of the primary cygnet beam, and corresponds to an interaction length on the order of 10^5 gm/cm^2, a cross-section of 10 to 20 microbarns/nucleon.

The mass scale of the particles involved in the process is set by the magnitude of the muon angular spread. We assume that the muons are decay products of a massive particle produced in the primary cygnet interaction in the rock. The median emission angle of the muons will be approximately M/E where M is the mass of this new particle and E its energy. But for a process occurring near threshold, E and M are related through the requirement that the square of the center of mass energy M^2, approximately, should equal 2Em where m is the mass of the target particle with which the cygnet interacts. E and M are thus determined directly from the measured angle. The resulting muon energies are much too low for the process to have occurred off electrons. The cygnets must interact with nuclear matter and produce particles of mass 20 to 40 GeV/c^2 with energies of hundreds of GeV, consistent with the observations. A slightly different scenario has the cygnets interacting to produce the muons directly in a new interaction with a mass scale in this same region[11]. This mass scale may be almost within reach of the new energy regime at particle accelerators.

From a practical point of view, the existence of a new particle with an interaction length about 10^5gm/cm^2 would provide a very useful probe of the more dense regions of gas around binary X-ray systems. It is likely that photons and cygnets are produced by a beam of protons accelerated in the neighbourhood of the neutron star to energies of order 10^{17}eV, and which interact in the gaseous matter which surely exists between the neutron star and companion, perhaps in the form of an accretion disc[12]. Photons are produced and fully absorbed within thicknesses approximately 100gm/cm^2. Thus, the observation of both photons and cygnets from a source would significantly enhance our ability to investigate the structure of such systems.

In conclusion, no strong new evidence has been presented for the existence of underground muons from the direction of Cygnus X-3. On the other hand, no strong new evidence against the effect has been presented. The high variability of Cygnus X-3, and presumably similar X-ray emitting binary systems, makes the situation problematic. While no theory with conventional particles is able to explain the effect, it is possible to construct a simple model which is able to explain all features of the data. The particles and interactions required are novel, and the ultimate test of the ideas may come at the next generation of particle accelerators. Let us hope that Cygnus X-3 shines brightly again before that time.

REFERENCES.

1. M.L.Marshak et al., Phys.Rev.Lett. 54,2079(1985),55,1965(1985)
2. G.Battistoni et al., Phys.Lett. 155B,465(1985)
3. D.Ayres, Proc. 2nd Conf. on Int. bet. Part.and Nuc.Phys. (AIP Conf.Proc. 150,1986), M.L.Marshak, 23rd ICHEP, Berkeley(1986)
4. K.Oyama et al., Phys.Rev.Lett. 56,991(1986)
5. Ch.Berger et al., Phys.Lett. 174B,118(1986)
6. M.V.Barnhill et al., Nature 317,409(1985)
7. J.Learned, this meeting.
8. K.Ruddick, NATO Adv.Wkshp. on H.E. Gam Astr., Durham (1986)
9. K.Ruddick, Phys.Rev.Lett. 57,531(1986)
10. G.Baym et al., Phys.Lett. 160B,181(1985)
11. J.Collins et al., 23rd ICHEP, Berkeley(1986)
12. A.M.Hillas, Nature 312,50(1984)

Superconducting Cosmic Strings

EDWARD WITTEN*

Joseph Henry Laboratories

Princeton University

Princeton, New Jersey 08544

The early universe, as it froze, may — depending on assumptions about particle physics — have produced a variety of topological defects. Such defects may have dimension zero, one, or two. They may be, in other words, points (monopoles), lines (cosmic strings), or sheets (domain walls). As we go up in "dimension", the effects of such defects become more drastic. Monopoles — defects of dimension zero — probably can only be detected if they pass through the laboratory. (There is a slight chance of of observing the collective gravity of monopoles if they are the "missing mass". Likewise macroscopic plasma effects of monopoles might be observable.) Domain walls — defects of dimension two — have such tremendous gravitational effects, because of their great mass, that we can be pretty sure they do not exist in our present-day universe, at least inside our field of view. The fascinating intermediate case is the defect of dimension one, the cosmic string.[1] Stretching across a galaxy, perhaps across the whole sky, the effects of a cosmic string are subtle enough that it might elude discovery, yet conspicuous enough that such a string might be visible at an astronomical distance.

---

* Supported in part by NSF Grants PHY80-19754 and 86-16129

How indeed can we hope to observe cosmic strings? One property that strings certainly have is mass. A string that forms due to symmetry breaking (or confinement) at an energy scale μ has a mass per unit length

$$\frac{dM}{dL} \sim \mu^2$$

For instance, here are some typical values:

μ	dM/dL
10^3 Gev (electroweak scale)	10^{-6} gm/cm
10^{10} Gev ("Intermediate mass scale")	10^{10} gm/cm
10^{16} Gev (traditional GUT scale)	10^{22} gm/cm
10^{19} Gev (Planck mass)	10^{28} gm/cm

Now, near the Planck mass, strings are so heavy that their gravitational effects are just too big. Light deflection from such strings would involve a lensing angle of order π. Lensing by such strings would produce a kaleidoscopic universe, with many images of the same objects, giant jumps in the apparent microwave temperature, etc. On the other hand, much below the traditional grand unified mass $M_{GUT} \sim 10^{16} Gev$, the gravitational effects of strings are negligible in astronomical terms. For instance, for $\mu = 10^{10} Gev$, a string crossing the Milky Way weighs about one tenth of the mass of the sun, while a string crossing the solar system weighs about 10^{-4} of the mass of the earth. We would certainly not notice the intrusion of .1 solar masses in our galaxy, or 10^{-4} terrestrial masses in our solar system.

On the other hand, just near $\mu \sim 10^{16} Gev$, the gravitational effects of strings are interesting. Such strings may well be observed as gravitational lenses, gravitational wave sources, or seeds for galaxy formation. Most of these possibilities have been reviewed in [1].

We cannot do much about the fact that strings presumably cannot exist (in today's universe) above $\mu \sim 10^{16} Gev$. However, it is a shame to dismiss strings below $\mu \sim 10^{16} Gev$ as undetectable. And even if strings do exist with $\mu \sim 10^{16} Gev$, their gravitational effects effects may not be the whole story.

This then leads us to think about non-gravitational effects of cosmic strings. The possibility we will consider here[2] is that strings might behave as current-carrying wires, in fact superconducting wires, and produce large scale elecromagnetic effects. The superconductivity arises because of charges trapped on the string, which may be either bosons or fermions. There have been several recent suggestions about astrophysical applications of superconducting cosmic strings, in connection with certain galactic objects,[3] galaxy formation,[4] high energy cosmic rays,[5] and quasars.[6] Here, I will not enter into such matters (except briefly at the end), and will instead sketch the microphysics of how a string may come to be a superconductor. A more general scenario was recently considered in [7].

First, I will consider the reasoning that leads to superconductivity with fermion charge carriers. Dirac's equation for fermions was originally

$$(i \not{D} - m)\psi = 0 \tag{1}$$

In contemporary gauge theories, the constant m is replaced by a coupling to a Higgs field

$$(i \not{D} - g\phi)\psi = 0 \tag{2}$$

Here g is a coupling constant, and ϕ is the Higgs field. I have hidden various indices in (2), but an essential point is that there is parity violation in the appearance of ϕ rather than its complex conjugate ϕ^* in (2). In fact, parity violation is implicit in the mechanism by which the quarks and leptons are believed to get mass.

Now, we must recall that the whole idea of cosmic strings — at least, those cosmic strings that arise as vortex lines analogous to the vortex lines in superconductors — is that in circling the string, the Higgs field undergoes a "twist", and it vanishes in the core of the string. Thus, in the core of the string, the fermion is effectively massless.

To understand the implications of this, we will solve the Dirac equation (1) by separation of variables. We consider a string running in the z direction, so that x and y are the directions transverse to the string, while z and t (= time) are tangent to the world-sheet cut out by the string as it moves in space-time. We have then

$$i\,\not{D} \;=\; i\sum_{\mu=1}^{4}\Gamma^{\mu}D_{\mu} \;=\; i\,\not{D}_T + i\,\not{D}_L \tag{3}$$

where $\not{D}_T$ and $\not{D}_L$, the transverse and longitudinal Dirac operators, are defined by

$$\begin{aligned}\not{D}_T \;&=\; \sum_{i=x,y}\Gamma^{i}D_{i}\\ \not{D}_L \;&=\; \sum_{j=z,t}\Gamma^{j}D_{j}\end{aligned} \tag{4}$$

So our Dirac equation is

$$((i\,\not{D}_T - g\phi) + i\,\not{D}_L)\psi \;=\; 0.$$

Here I have included the Higgs coupling $g\phi$ with the transverse Dirac operator, since ϕ has a non-trivial dependence on the transverse coordinates.

Solving (5) by separation of variables, we must consider the transverse eigenvalue problem

$$(i\,\not{D}_T - g\phi)\psi \;=\; \lambda\psi \tag{6}$$

Now, here is where something interesting happens. In general, eigenfunctions and eigenvalues of an operator like that on the left of (6) depend on various details — in this case the precise form of the Higgs field ϕ, and of the gauge field A buried in the symbol "$\not{D}$". In the case under study, though, something nice happens (if the particle physics is chosen properly). (6) has a zero mode solution, a normalizable solution with $\lambda = 0$, purely for topological reasons related to parity violation in the underlying gauge theory, and independent of all the details.

Looking back at our four dimensional Dirac equation (5), we see that a $\lambda = 0$ solution of (6) must also obey

$$i \not{D}_L \psi = 0 \tag{7}$$

(7) is the equation that will govern the motion of our fermion *along* the string, and as (7) is the massless Dirac equation in two dimensions, we see that a fermion in the transverse zero mode will travel along the string at the speed of light.

To be more specific about this, we start in four dimensions with fermions, e.g. e_L^+ or $\binom{\nu}{e^-}_L$, of definite chirality, i.e. definite eigenvalue of

$$\overline{\Gamma} = \Gamma^0 \Gamma^1 \Gamma^2 \Gamma^3 \tag{8}$$

This a fact of life in electroweak theory, and it is the reason we obtained topological zero modes. For, e.g., e_L^+ one has

$$-1 = \overline{\Gamma} = \Gamma^0\Gamma^1\Gamma^2\Gamma^3 \tag{9}$$

so

$$\Gamma^1\Gamma^2 = \Gamma^0\Gamma^3 \tag{10}$$

The topological argument that predicts the existence of the transverse zero mode also predicts that it has a definite chirality, e.g.

$$\Gamma^1\Gamma^2 = +1 \tag{11}$$

So it must obey

$$\Gamma^0\Gamma^3 = +1. \tag{12}$$

Now, what does it mean to have a solution of the longitudinal Dirac equation

$$0 = i \not{D}_L \psi = i \left(\Gamma^0 \frac{\partial}{\partial t} + \Gamma^3 \frac{\partial}{\partial z} \right) \psi \tag{13}$$

which also obeys

$$\psi = \Gamma^0\Gamma^3\psi ? \tag{14}$$

With $(\Gamma^0)^2 = 1$, (13) becomes

$$0 = \left(\frac{\partial}{\partial t} + \Gamma^0\Gamma^3\frac{\partial}{\partial z}\right)\psi = \left(\frac{\partial}{\partial t} + \frac{\partial}{\partial z}\right)\psi \tag{15}$$

and the solution is, of course, $\psi(t,z) = \psi(t-z)$.

This describes a wave that travels in the $+z$ direction at the speed of light. The energy-momentum relation of such a particle is the usual relation

$$E = p \tag{16}$$

of massless particles. This is remarkable. In vacuum the electron, say, has a minimum rest energy $mc^2 = 511$ kev, but in the string it behaves like a massless particle, whose energy can be arbitrarily small, depending on p.

The result that the electron can only travel in one direction along the string may also seem rather odd. This odd result is possible because parity violation was built into our reasoning at several steps along the way. In a realistic model, while electrons will travel one way, other fermions will travel backwards. It all depends on the "handedness" of the coupling to the Higgs field. In some of the simplest models, for instance, things work out as indicated in figure (2). Electrons and down quarks travel one way along the string and up quarks travel backwards. (It should be pointed out that particles and the corresponding antiparticles always travel in the same direction along the string.)

Now, we can see without further ado why such a string is superconducting. Consider the following. A hydrogen atom basically consists of $uude^-$. In vacuum, it weighs a little over 935 Mev. If it falls into the string, the individual quarks and leptons behave like massless particles. Thus, we have a spontaneous pròcess

$$H\,atom\ +\ string\ \Rightarrow\ (\text{string with fermions on it}) + 935\text{Mev}. \tag{17}$$

Notice, though, from the figure, that once the atom falls into the string, the negative changes all travel one way, and the positive charges travel backwards. Thus there is a net current. (It can be shown via bosonization of fermions or in various other ways that the current carrying mode is really a color singlet collective mode; thus we need not worry about color confinement or other effects of QCD.)

There is no way for the current we just encountered to stop, as long as our quarks and leptons are trapped in the transerse zero modes. The only way to stop the current is to pump in 935 Mev of energy and eject the fermions from the string. Thus, the string is a superconductor. Left to its own, without any energy supplied to stop it, the current in the string will persist indefinitely.

Of course, we are not really interested in a situation in which just a single atom is trapped on the string. More interesting is the situation in which one feeds the string with a whole macroscopic supply of gas. In this case, one will build up a whole fermi gas of trapped charge carriers. Fermions trapped in transverse zero modes can be characterized by the longitudinal momentum p of the fermion parallel to the string. At zero temperature, after supplying the string with a macroscopic supply of gas, every species of fermion will behave as a fermi gas with all the states filled up to some fermi momentum P_F.

To gain some insight into how such a system will behave, let us consider some processes characteristic of superconductivity. We consider a loop of string which is initially in a magnetic field B of, perhaps, some external origin (figure (3)). We turn off the sources of the field. What will happen when some of the magnetic flux tries to escape?

We recall that the electric and magnetic field $\vec{E}$ and $\vec{B}$ obey

$$\vec{\nabla} \times \vec{E} = \frac{d\vec{B}}{dt} \tag{18}$$

Thus, if $\Phi = \int_S B \cdot n$ is the flux threading the loop, we have

$$\frac{d\Phi}{dt} = \int_S \frac{dB}{dt} \cdot n = \int_S \nabla \times E \cdot n$$
$$= \int_L E \cdot d\ell . \tag{19}$$

where S is a surface spanning the loop L.

Thus, when magnetic flux escapes, there is an $\vec{E}$ field along the string. So we should think about a string in an $\vec{E}$ field.

An electron trapped in the string obeys the Lorentz force law

$$\frac{dp}{dt} = e\,E \tag{20}$$

with E the component of $\vec{E}$ along the string and p the component of momentum along the string. Usually, the Lorentz force law contains the magnetic field as well, but that is irrelevant here, since fermions trapped in transverse zero modes cannot be deflected off the string.

Now we want to apply (20) to the fermion on top of the fermi sea. Its momentum is just the fermi momentum P_F, so we can interpret (20) as a formula for the time dependence of the fermi momentum,

$$\frac{dP_F}{dt} = e\,E \tag{21}$$

(Note that when the particle on top of the fermi sea changes its momentum, so do all the others, and by the same amount. Thus we get, as in figure (4), a filled fermi gas of different momentum.)

This is actually a change in the net number of particles (minus antiparticles).

For a 1 + 1 dimensional fermi gas, the particle density is $n = \frac{p_F}{2\pi}$. Thus

$$\frac{dn}{dt} = \frac{eE}{2\pi} \tag{22}$$

The created particles are in essence flowing in from the Dirac sea at $p = -\infty$. But they are true, physical particles. You could literally kick them off the string.

In the situation considered earlier, while creating electrons at the rate (22), we will create d quarks at the same rate, and twice as many u quarks. So the produced particles have quantum numbers $uude^-$, i.e. H atoms. If B is the total number of atoms (or baryons) integrated along the string, then the time rate of charge of B is

$$\begin{aligned} \frac{dB}{dt} &= \oint_L d\ell \frac{dn}{dt} = \oint_L d\ell \frac{eE}{2\pi} \\ &= e \int_S dS \frac{\nabla \times E}{2\pi} = \frac{e}{2\pi} \frac{d\Phi}{dt} \end{aligned} \tag{23}$$

where Φ is the magnetic flux through the loop; dn/dt is the time rate of change of the baryon density per unit length.

From this we see that in the interaction of matter with strings and magnetic fields, the conserved quantity is not the baryon number B, but rather is

$$\hat{B} = B - \frac{e}{2\pi}\Phi \tag{24}$$

What limits the above described process is that the fermi momentum of the fermions reaches

$$p_F^c = m_F c^2 \tag{25}$$

then it is no longer energetically favored for them to be trapped in the string. If the charge carriers are ordinary quarks and leptons, this arises at a current of

$\sim 10^4$ amps, Beyond that point, passing the string over magnetic field lines causes particle creation, at a rate of order

$$10^{11}\ \frac{atoms}{cm \cdot sec} \cdot \left(\frac{B}{10^{-6}\text{gauss}}\right) \cdot N_{gen} \tag{26}$$

Here N_{gen} is the number of fermion generations. Actually, the produced particles consist of baryons and leptons or antibaryons and antileptons depending on which way the string is moving relative to the magnetic field.

So far I have described superconducting strings in which the charge carriers are fermions. It is also possible to make a superconducting string with charge carriers that are bosons. Consider a $U(1) \times U(1)'$ gauge theory, with $U(1)$ being electromagnetism and $U(1)'$ being some other interaction which will be responsible for the existence of cosmic strings. We introduce Higgs fields ϕ and σ of charges $(e,0)$ and $(0,g)$ respectively; thus ϕ carries ordinary electric charge and σ carries the $U(1)'$ interaction. In vacuum, we suppose $\phi = 0$ but $\sigma \neq 0$. Thus electromagnetism is unbroken, but $U(1)'$ is spontaneously broken, producing strings in whose core $U(1)'$ is restored. Whether electromagnetism is broken or unbroken in the core of the string depends on details of the dynamics. Thus, consider a Higgs potential of the form

$$V(\phi,\sigma) \;=\; \lambda(|\sigma|^2 - a^2)^2 \;+\; |\phi|^2(-M^2 + g^2|\sigma|^2) \;+\; \lambda'|\phi|^4 \tag{27}$$

Denoting the vacuum expectation value of σ as $<\sigma>$, we suppose $-M^2+g^2|\sigma|^2 > 0$, so $<\phi>= 0$ in vacuum, but $-M^2 < 0$, so $<\phi>\neq 0$ in the core of the string. Thus, electromagnetism is spontaneously broken in the core of the string, and the string is a superconductor. The current carrying mode is essentially the phase of the ϕ field in the core of the string. I will not enter here into the mathematics of superconducting strings with bosonic charge carriers, except to remark that it is very similar to the mathematics of an ordinary laboratory superconducting wire.

Superconductivity with boson charge carriers isn't topological, so it can't be "predicted" (as of 1986). On the other hand, it has the virtue that it can arise in practically any cosmic string model, and the critical current can be enormous, of order

$$J_{crit} \sim \frac{e\mu c^2}{\hbar} \sim 10^{20} amps \cdot \frac{\mu}{10^{16} Gev} \tag{28}$$

where μ is the mass scale at which the string forms. By "the critical current," I mean the current at which the superconducting state is no longer energetically favored, and superconductivity is lost. At that point a bosonic superconductor undergoes a catastrophic phase transition, analogous to occasional "quenching" of Fermilab magnets.

I will conclude by briefly commenting on some astronomical effects. Consider a string moving in a maganetic field. As the string moves, there is an $\vec{E}$ field in its rest frame, which will excite currents. As it moves upstream (figure (5)), the string will tend to move the field lines upstream via the Meissner effect. After crossing a galaxy ($B = 10^{-6}$ gauss, $R \simeq 10^4$ parsecs), the string reaches a current of 10^{16} amps (!!). Its magnetic field is enormous:

$$B \;=\; 3 \times 10^{14}\text{gauss} \cdot \left(\frac{1\text{cm}}{\text{r}}\right) \cdot \left(\frac{\text{R}}{10^4\text{parsecs}}\right) \cdot \left(\frac{B_0}{10^{-6}\text{gauss}}\right) \tag{29}$$

Here r is the distance from the string. This field will dominate the galactic field at a distance of 200 parsecs. These numbers should illustrate the fact that astronomical effects of superconducting strings can well be quite interesting. Several proposals concerning such effects have been made recently, including a model of galaxy formation by J. Ostriker, C. Thompson and me which has been reported at this conference by Thompson. In this proposal, we assume a primordial magnetic field with about 10^{-9} of the cosmic energy density; extrapolated to today's universe, this corresponds to an intergalactic magnetic field of roughly 10^{-9} gauss. If strings are born in such a field, they will develop huge currents when the expanding universe thins out the primdordial magnetic flux.

The oscillating loop then emits vast amounts of extremely long wavelength magnetic dipole radiation. Such waves cannot propagate through the primadial plasma (the plasma frequency is far too high), so they tend to exert a pressure on the plasma and to expel it from the region around the string. This behaves rather like an explosion. A void forms around the string, the expanding wall of the void is a sort of shock wave, and galaxies will form in sheets where these walls collide. All of this has at least a familial resemblance to recent observations of large scale structure in the universe, as reported in various talks at this conference.

REFERENCES

1. For a review, see A. Vilenkin, *Phys. Rept.* **121** (1985), 263.
2. E. Witten, *Nucl. Phys.* **B246** (1985), 557.
3. E. M. Chudnovsky, G. B. Field, D. N. Spiegel, and A. Vilenkin, *Phys. Rev.* **D34** (1986), 944.
4. C. T. Hill, D. N. Schramm, and T. P. Walker, Fermilab Pub-86/146-T.
5. J. Ostriker, C. Thompson, and E. Witten, *Phys. Lett.* **B180** (1986), 231.
6. A. Vilenkin and G. B. Field, "Quasars and Superconducting Cosmic Strings" (Preprint, 1986).
7. C. T. Hill and L. M. Widrow, FERMILAB-Pub-86/162-T.

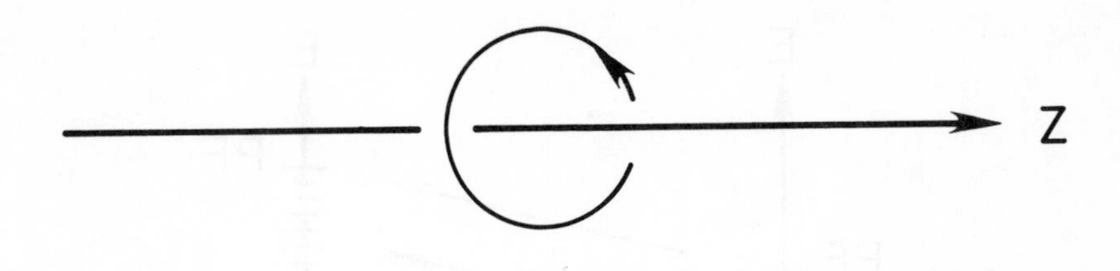

Fig. 1

A string running along the z axis. The phase of the Higgs field changes by 2π in circling the string.

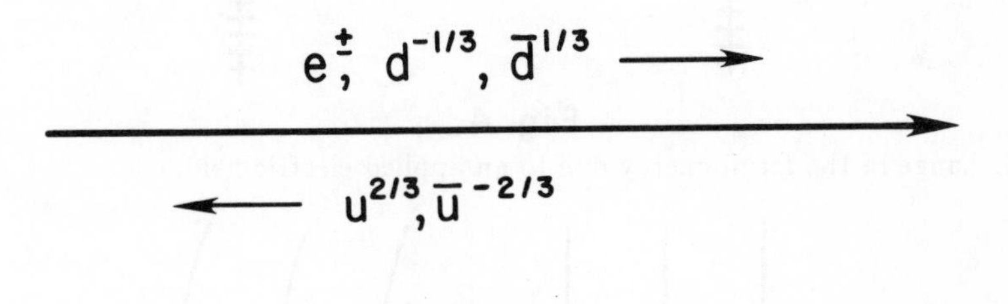

Fig. 2

A superconducting string with electrons and d quarks moving to the right, and u quarks to the left.

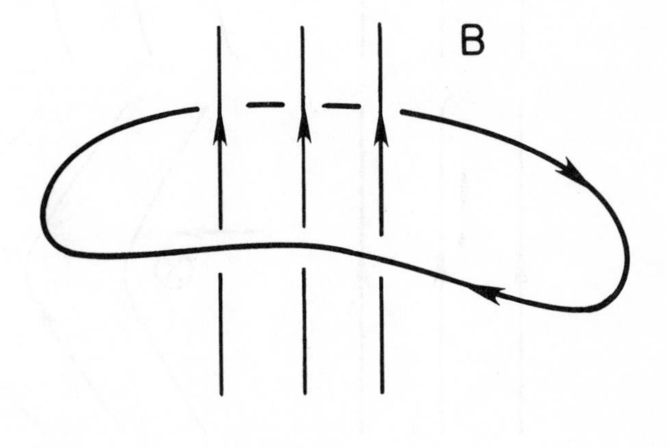

Fig. 3

A $\vec{B}$ field threading a loop of string.

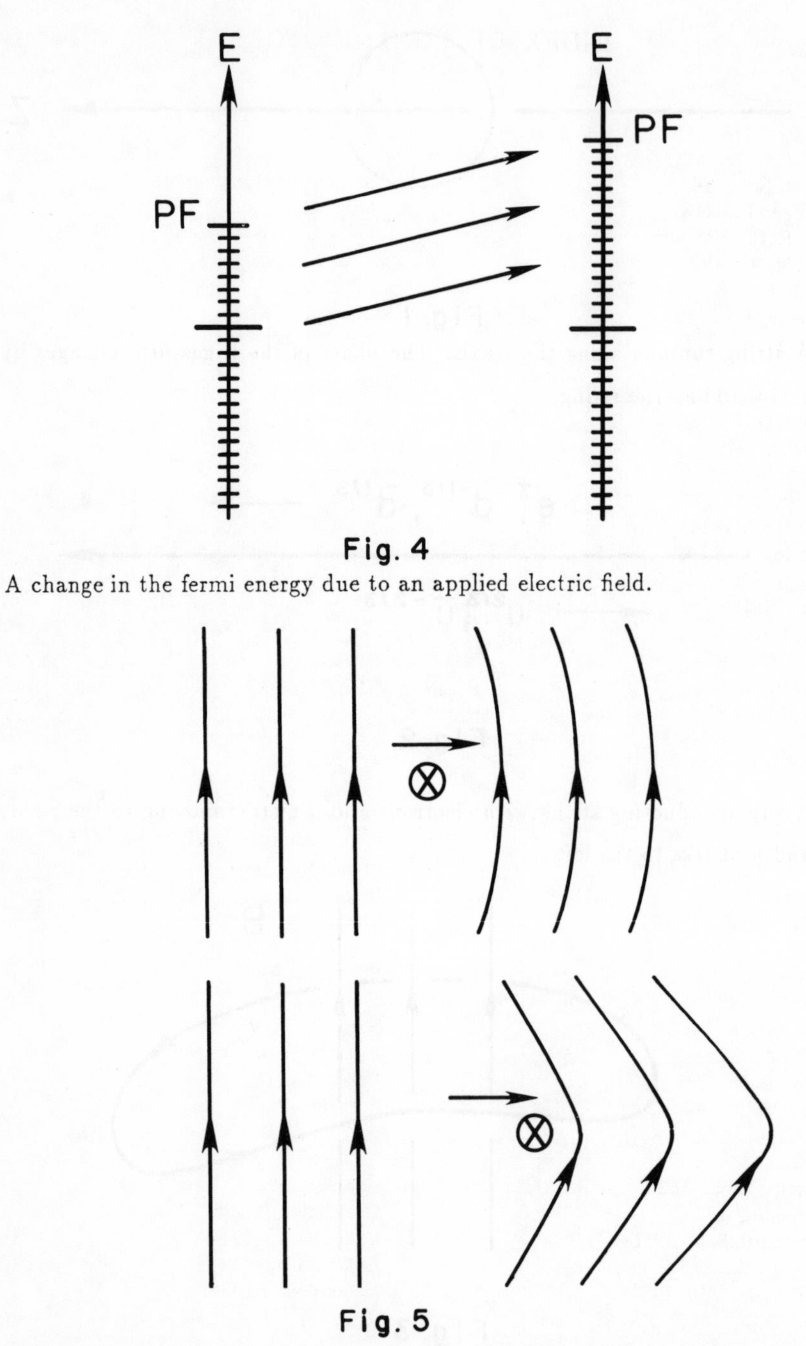

Fig. 4

A change in the fermi energy due to an applied electric field.

Fig. 5

A string (pointing into the paper and denoted by ⊗) moving across field lines.

INDEX OF CONTRIBUTORS

LIST OF PARTICIPANTS

NAME	INSTITUTION
ABRAHAMS, ANDREW	UNIVERSITY OF ILLINOIS
ACCETTA, FRANK	UNIVERSITY OF CHICAGO
AKERLOF, CARL W.	UNIVERSITY OF MICHIGAN
ALBRECHT, ANDREAS	FERMILAB
ALPAR, M. ALI	MPI FOR ASTROPHYSICS
ALSCHULER, MATTHEW B.	
AMALDI, EDOARDO	UNIVERSITY OF ROME
AMES, SUSAN	UNIVERSITY OF BONN
AMSTERDAMSKI, PIOTR	UNIVERSITY OF TEXAS
ANDERSON, CHRIS	IOWA STATE UNIVERSITY
ANDERSON, MARTHA	CARLETON COLLEGE
ANDERSON, PAUL R.	MONTANA STATE UNIVERSITY
ARBEITER, LARRY	UNIVERSITY OF CHICAGO
ARNETT, W. DAVID	UNIVERSITY OF CHICAGO
ASSEO, ESTELLE	ECOLE POLYTECHNIQUE
AURIEMMA, GIULIO	UNIVERSITY OF ROME
AXENIDES, MINOS	UNIVERSITY OF WASHINGTON
BABUL, ARIF	PRINCETON UNIVERSITY
BAHCALL, JOHN N.	INST FOR ADVANCED STUDY
BAND, DAVID L.	LAWRENCE LIVERMORE NAT LAB
BARBON, R.	ASTROPHYS OBSERV, PADOVA
BARDEEN, JAMES M.	UNIVERSITY OF WASHINGTON
BARDEEN, WILLIAM A.	FERMILAB
BARING, MATTHEW	UNIVERSITY OF CAMBRIDGE
BARNOTHY, JENO M.	
BARNOTHY, MADELEINE F.	
BATUSKI, DAVID J.	SPACE TELESCOPE SCI INST
BEATTY, JAMES J.	BOSTON UNIVERSITY
BECKLIN, ERIC E.	UNIVERSITY OF HAWAII
BEERS, TIMOTHY C.	MICHIGAN STATE UNIVERSITY
BEGLEY, SHARON	NEWSWEEK
BEKENSTEIN, JACOB	CITA, MCLENNAN LABS
BELL, THEODORE	UNIVERSITY OF CHICAGO
BELY, PIERRE	SPACE TELESCOPE SCI INST
BEMPORAD, CARLO	UNIVERSITY OF PISA
BENAROYA, MARTHA	ADLER PLANETARIUM
BENNETT, DAVID	FERMILAB
BENTLEY, ALAN F.	EASTERN MONTANA COLLEGE
BERGER, BEVERLY K.	LAWRENCE LIVERMORE NAT LAB
BERLEY, DAVID	NATL SCIENCE FOUNDATION
BERNSTEIN, DAVID	UNIVERSITY OF ILLINOIS
BERNSTEIN, GARY	UNIV OF CALIF, BERKELEY
BERSHADY, MATTHEW	UNIVERSITY OF CHICAGO
BERTOLA, FRANCESCO	PADOVA OBSERVATORY
BERTSCHINGER, EDMUND W.	MASS. INST OF TECHNOLOGY
BHAVSAR, SUKETU P.	UNIVERSITY OF KENTUCKY
BLAIR, DAVID	U. OF WESTERN AUSTRALIA
BLOEMEN, HANS	U OF CALIF, BERKELEY
BLONDIN, JOHN M.	UNIVERSITY OF CHICAGO
BLUDMAN, SIDNEY A.	UNIVERSITY OF PENNSYLVANIA
BODO, GIANLUIGI	TORINO OBSERVATORY
BOND, JOHN R.	CITA, MCLENNAN LABS
BONIFAZI, PAOLO	INFN, FRASCATI
BORRA, ERMANNO	LAVAL UNIVERSITY
BOUCHET, FRANCOIS	LAWRENCE LIVERMORE NAT LAB
BOUGHN, STEPHEN	HAVERFORD COLLEGE

NAME	INSTITUTION
BOWER, CHARLES R.	INDIANA UNIVERSITY
BOYD, RICHARD N.	OHIO STATE UNIVERSITY
BRAATEN, ERIC	NORTHWESTERN UNIVERSITY
BRANCH, DAVID R.	UNIVERSITY OF OKLAHOMA
BRECHER, KENNETH	BOSTON UNIVERSITY
BROWN, JOHN C.	UNIVERSITY OF GLASGOW
BROWN, LAWRENCE	UNIVERSITY OF CHICAGO
BROWN, STANLEY G.	PHYSICAL REVIEW LETTERS
BRUENN, STEPHEN	FLORIDA ATLANTIC UNIV
BUCKBY, MARGARET	RYERSON POLYTECHNICAL INST
BURNS, JACK O.	UNIVERSITY OF NEW MEXICO
BURROWS, ADAM S.	UNIVERSITY OF ARIZONA
BUSSARD, ROGER W.	MARSHALL SPACE FLIGHT CTR
CADEZ, ANDREJ	CALTECH
CAI, SHEN-OU	NORTHWESTERN UNIVERSITY
CANDY, BERNARD	U. OF WESTERN AUSTRALIA
CAPACCIOLI, MASSIMO	UNIVERSITY OF PADOVA
CARLSON, ERIC D.	ADLER PLANETARIUM
CARNEY, BRUCE W.	UNIVERSITY OF N. CAROLINA
CARRIGAN, RICHARD A.	FERMILAB
CASEY, SEAN	YERKES OBSERVATORY
CAULET, ADELINE M.	UNIVERSITY OF CHICAGO
CAVALIERE, A. G.	UNIVERSITY OF ROME
CAWLEY, MICHAEL F.	IOWA STATE UNIVERSITY
CENTRELLA, JOAN	DREXEL UNIVERSITY
CHAKRABARTI, SANDIP K.	CALTECH
CHALLENER, SHARON L.	CASE WESTERN RESERVE UNIV
CHARLTON, JANE C.	UNIVERSITY OF CHICAGO
CHASE, SIMON T.	IMPERIAL COLLEGE
CHENG, KWONG-SANG	UNIVERSITY OF ILLINOIS
CHERRY, MICHAEL L.	UNIVERSITY OF PENNSYLVANIA
CHRISTENSEN, STEVEN M.	UNIVERSITY OF ILLINOIS
CHUN, ELAINE L.	UNIVERSITY OF CHICAGO
CLARKE, DAVID	UNIVERSITY OF NEW MEXICO
CLINE, DAVID B.	UNIVERSITY OF WISCONSIN
COLEY, ALAN	DALHOUSIE UNIVERSITY
COLGATE, STIRLING A.	LOS ALAMOS NATIONAL LAB
CONWELL, JAMES C.	EASTERN ILLINOIS UNIV
COOPERSTOCK, FRED I.	UNIVERSITY OF VICTORIA
COPELAND, EDMUND	IMPERIAL COLLEGE
CORN, PHILIP B.	OHIO STATE UNIVERSITY
CORSO, GEORGE J.	LOYOLA UNIVERSITY
COTTINGHAM, DAVID A.	PRINCETON UNIVERSITY
COUCHMAN, H. M. P.	CITA, TORONTO
COURVOISIER, THIERRY J.-L.	SPACE TELESCOPE, MUNICH
COWAN, JOHN J.	UNIVERSITY OF OKLAHOMA
CRANE, PHILLIPE	EUROPEAN SOUTHERN OBS
CRANNELL, CAROL JO	GODDARD SPACE FLIGHT CTR
CRANNELL, HALL L.	CATHOLIC UNIV OF AMERICA
CROMIE, WILLIAM	
CRONIN, JAMES W.	ENRICO FERMI INSTITUTE
CUTLER, CURT	UNIVERSITY OF CHICAGO
D'AMICO, NICOLO'	UNIVERSITY OF PALERMO
DALY, RUTH A.	BOSTON UNIVERSITY
DANNEHOLD, TERRY L.	AT&T BELL LABORATORIES
DAS, ANADI J.	SIMON FRASER UNIVERSITY

NAME	INSTITUTION
DAVIDS, CARY N.	ARGONNE NATIONAL LAB
DAVIES, R. D.	NUFFIELD ASTRONOMY LABS
DAVIS, MARC	U OF CALIF, BERKELEY
DAVIS, RAYMOND	UNIVERSITY OF PENNSYLVANIA
DE KOOL, MARTHIJN	UNIVERSITY OF AMSTERDAM
DE LAPPARENT, VALERIE	CENTER FOR ASTROPHYSICS
DE YOUNG, DAVID S.	KITT PEAK NATL OBSERVATORY
DETWEILER, STEVEN	UNIVERSITY OF FLORIDA
DIXIT, VIJAI V.	PARKS COLLEGE/ST. LOUIS U.
DOYLE, LARRY	UNITED PRESS INTERNATIONAL
DRESSLER, ALAN	MT. WILSON OBSERVATORY
DREVER, RONALD W. P.	CALTECH
DROEGE, WOLFGANG	MPI F. RADIOASTRONOMY
DRUKIER, ANDRZEJ K.	CENTER FOR ASTROPHYSICS
DULL, JAMES D.	NORTHWESTERN UNIVERSITY
EBISUZAKI, TOSHIKAZU	MARSHALL SPACE FLIGHT CTR
EDELSON, RICHARD	CALTECH
ELBERT, JEROME W.	UNIVERSITY OF UTAH
ENGVIST, KARI	UNIVERSITY OF WISCONSIN
ENSMAN, LISA	LICK OBSERVATORY
EVANS, CHARLES R.	CALTECH
EVANS, MARK	UNIVERSITY OF PENNSYLVANIA
EWAN, GEORGE T.	QUEEN'S UNIVERSITY
FALL, S. MICHAEL	SPACE TELESCOPE SCI INST
FAUCI, FRANCESCO	UNIVERSITY OF PALERMO
FENYVES, ERVIN J.	UNIVERSITY OF TEXAS
FERLAND, GARY J.	OHIO STATE UNIVERSITY
FERRARI, ATTILIO	UNIVERSITY OF TURIN
FICK, BRIAN E.	UNIVERSITY OF UTAH
FIEDLER, RALPH	NATIONAL RESEARCH LAB
FIELD, GEORGE B.	HARVARD ASTROPHYSICS CTR
FIGER, DONALD F.	NORTHWESTERN UNIVERSITY
FINZI, ARRIGO	TECHNION
FIREMAN, EDWARD L.	CENTER FOR ASTROPHYSICS
FISCHER, MARC	UNIV OF CALIF, BERKELEY
FISHBONE, LESLIE G.	BROOKHAVEN NATIONAL LAB
FITCHETT, MICHAEL	CITA, TORONTO
FITZPATRICK, RICHARD	UNIVERSITY OF SUSSEX
FOMALONT, EDWARD B.	NRAO, CHARLOTTESVILLE
FORD, LAWRENCE H.	TUFTS UNIVERSITY
FORTNER, BRAND	UNIVERSITY OF ILLINOIS
FORWARD, ROBERT L.	HUGHES RESEARCH LABS
FOWLER, JAMES R.	YERKES OBSERVATORY
FRANKLIN, MICHIKO	UNIVERSITY OF ILLINOIS
FRANSSON, CLAES	STOCKHOLM OBSERVATORY
FREESE, KATHERINE	U OF CALIF, SANTA BARBARA
FRIEDMAN, JOHN L.	UNIVERSITY OF WISCONSIN
FRIEMAN, JOSHUA A.	SLAC
FRUCHTER, ANDREW S.	PRINCETON UNIVERSITY
FRY, JAMES N.	UNIVERSITY OF FLORIDA
FRYXELL, BRUCE A.	UNIVERSITY OF CHICAGO
FU, ALBERT	UNIVERSITY OF CHICAGO
FULLER, GEORGE M.	LAWRENCE LIVERMORE NAT LAB
FUSHIKI, IKKO	UNIVERSITY OF ILLINOIS
GABRIEL, MARCUS D.	PURDUE UNIVERSITY
GAIDOS, JAMES A.	PURDUE UNIVERSITY

NAME	INSTITUTION
GAISSER, THOMAS K.	BARTOL RESEARCH FOUNDATION
GALLAGHER, JOHN S.	LOWELL OBSERVATORY
GARFINKLE, DAVID	WASHINGTON UNIVERSITY
GARWIN, CHARLES A.	
GATES, EVALYN	CASE WESTERN RESERVE UNIV
GAUTREAU, RONALD	NEW JERSEY INST OF TECH
GEGENBERG, JACK D.	UNIV OF NEW BRUNSWICK
GELDZAHLER, BARRY	APPLIED RESEARCH CORP
GELLER, MARGARET J.	CENTER FOR ASTROPHYSICS
GELMINI, GRACIELA B.	ENRICO FERMI INSTITUTE
GENZEL, REINHARD	MPI PHYSICS & ASTROPHYSICS
GERBAL, DANIEL	PARIS OBSERVATORY, MEUDON
GERBIER, GILLES	U OF CALIF, BERKELEY
GERLACH, ULRICH H	OHIO STATE UNIVERSITY
GHABOUSSI, FARHAD	UNIVERSITY OF KONSTANZ
GIACCONI, RICCARDO	SPACE TELESCOPE SCI INST
GLASS, EDWARD	UNIVERSITY OF WINDSOR
GLEISER, MARCELO	FERMILAB
GOERNITZ, THOMAS	STARNBERG
GONATAS, DINOS P.	UNIVERSITY OF CHICAGO
GRABELSKY, DAVID A.	NORTHWESTERN UNIVERSITY
GRAHAM, JAMES R.	U OF CALIF, BERKELEY
GRAZIANI, FRANK R.	UNIVERSITY OF MINNESOTA
GRIFFITHS, RICHARD E.	SPACE TELESCOPE SCI INST
GRILLO, AURELIO F.	INFN, FRASCATI
GRINDLAY, JONATHAN E.	HARVARD OBSERVATORY
GROOM, DONALD E.	LAWRENCE BERKELEY LAB
GRUNSFELD, JOHN	UNIVERSITY OF CHICAGO
GURSKY, HERBERT	NAVAL RESEARCH LABORATORY
GUSH, HERBERT P.	UNIV OF BRITISH COLUMBIA
GUTH, ALAN H.	MASS. INST OF TECHNOLOGY
GUZIK, T. GREGORY	LOUISIANA STATE UNIVERSITY
HABISOHN, CHRIS	UNIVERSITY OF CHICAGO
HAGGIN, JOSEPH	CHEM & ENGINEERING NEWS
HAIM, DAN	UNIVERSITY OF ILLINOIS
HAJICEK, PETR	UNIVERSITY OF BERNE
HALL RENO, MARY	FERMILAB
HAMILTON, JEFFREY J.	UNIVERSITY OF MARYLAND
HAMILTON, WILLIAM O.	LOUISIANA STATE UNIVERSITY
HANSELL, TOM	UNIVERSITY OF TEXAS
HANSEN, PETER M.	THE AEROSPACE CORPORATION
HARARI, DIEGO	UNIVERSITY OF FLORIDA
HARKNESS, ROBERT	UNIVERSITY OF TEXAS
HARTLE, JAMES B.	U OF CALIF, SANTA BARBARA
HARTMANN, DIETER	LICK OBSERVATORY
HARVEY, JEFFREY A.	PRINCETON UNIVERSITY
HAUGAN, MARK	PURDUE UNIVERSITY
HAWKING, STEPHEN W.	UNIVERSITY OF CAMBRIDGE
HAWLEY, JOHN F.	CALTECH
HAYNES, MARTHA P.	CORNELL UNIVERSITY
HEGYI, DENNIS J.	UNIVERSITY OF MICHIGAN
HEHNER, NED	
HEINZ, RICHARD M.	INDIANA UNIVERSITY
HEISE, JOHN	SPACE RESEARCH UTRECHT
HELFAND, DAVID J.	COLUMBIA UNIVERSITY
HEWITT, JACQUELINE N.	HAYSTACK OBSERVATORY

NAME	INSTITUTION
HIGDON, HAL	
HIGUCHI, ATSUSHI	UNIVERSITY OF WISCONSIN
HILDEBRAND, ROGER H.	ENRICO FERMI INSTITUTE
HILL, CHRISTOPHER T.	FERMILAB
HILL, HENRY A.	UNIVERSITY OF ARIZONA
HINDMARSH, MARK	LOS ALAMOS NATIONAL LAB
HISCOCK, WILLIAM A.	MONTANA STATE UNIVERSITY
HOBBS, LEWIS M.	UNIVERSITY OF CHICAGO
HOBILL, DAVID W.	STEVENS INST OF TECHNOLOGY
HODGES, HARDY M.	UNIVERSITY OF CHICAGO
HOGAN, CRAIG J.	STEWARD OBSERVATORY
HOLCOMB, KATHERINE	DREXEL UNIVERSITY
HOLMAN, RICHARD F.	FERMILAB
HORWITZ, RICHARD	ASSOCIATED PRESS
HOUGH, JAMES	UNIVERSITY OF GLASGOW
HOWARD, MICHAEL	LAWRENCE LIVERMORE NAT LAB
HUCHRA, JOHN P.	CENTER FOR ASTROPHYSICS
HUDSPETH, DAVID S.	UNIV OF ILLINOIS, CHICAGO
HUGHSTON, L. P.	LINCOLN COLLEGE
HUNTER, DEIDRE A.	LOWELL OBSERVATORY
HURLEY, KEVIN	CESR, TOULOUSE
ICKE, VINCENT	LEIDEN OBSERVATORY
INCANDELA, JOSEPH R.	BOSTON UNIVERSITY
IPSER, JAMES R.	UNIVERSITY OF FLORIDA
IWAMOTO, NAOKI	UNIVERSITY OF TOLEDO
JAFFE, DANIEL	UNIVERSITY OF TEXAS
JAMES, PHILIP A.	IMPERIAL COLLEGE
JIN, LIPING	UNIVERSITY OF CHICAGO
JOHNSON, DAVID G.	PRINCETON UNIVERSITY
JOHNSON, WARREN W.	LOUISIANA STATE UNIVERSITY
JONES, FRANK C.	GODDARD SPACE FLIGHT CTR
JORDAN, THOMAS F.	UNIVERSITY OF MINNESOTA
JUSZKIEWICZ, ROMAN	UNIV OF CALIF, BERKELEY
KANTOWSKI, RONALD	UNIVERSITY OF OKLAHOMA
KATES, RONALD	MPI FOR ASTROPHYSICS
KATGERT, PETER	LEIDEN OBSERVATORY
KAUFMANN, WILLIAM J.	SAN DIEGO STATE UNIV
KAWANO, LAWRENCE	UNIVERSITY OF CHICAGO
KEMBHAVI, AJIT K.	TATA INST OF FUND RESEARCH
KESSLER, DAN	CARLETON UNIVERSITY
KIBBLE, THOMAS W. B.	IMPERIAL COLLEGE
KIELKOPF, JOHN F.	UNIVERSITY OF LOUISVILLE
KIM, SUNG K.	MACALESTER COLLEGE
KLUZNIAK, WLODZIMIERZ	STANFORD UNIVERSITY
KNUTSEN, HENNING	NORDITA
KOCHANEK, CHRISTOPHER S.	CALTECH
KOLB, EDWARD W.	FERMILAB
KOLTICK, DAVID S.	PURDUE UNIVERSITY
KONDOYANNIDIS, NIKOS	UNIVERSITY OF WISCONSIN
KONIGL, ARIEH	UNIVERSITY OF CHICAGO
KOO, DAVID C.	SPACE TELESCOPE SCI INST
KORFF, SERGE A.	NEW YORK UNIVERSITY
KOTULAK, RONALD	CHICAGO TRIBUNE
KRAUSE, DENNIS	UNIVERSITY OF WISCONSIN
KUEBEL, DAVID J.	UNIVERSITY OF CHICAGO
KUHN, JEFF	MICHIGAN STATE UNIVERSITY

NAME	INSTITUTION
KUNDT, WOLFGANG	UNIVERSITY OF BONN
KUNIEDA, HIDEYO	GODDARD SPACE FLIGHT CTR
KUNSTATTER, GABOR	UNIVERSITY OF WINNIPEG
KURKI-SUONIO, HANNU	DREXEL UNIVERSITY
L'HEUREUX, JACQUES	UNIVERSITY OF CHICAGO
LAMB, DON Q.	UNIVERSITY OF CHICAGO
LAMB, FREDERICK K.	UNIVERSITY OF ILLINOIS
LAMB, RICHARD C.	IOWA STATE UNIVERSITY
LAMBAS, DIEGO GARCIO	PRINCETON UNIVERSITY
LANDE, KENNETH	UNIVERSITY OF PENNSYLVANIA
LANDSMAN, N. P.	UNIVERSITY OF AMSTERDAM
LANGE, ANDREW E.	U OF CALIF, BERKELEY
LAROS, JOHN	LOS ALAMOS NATIONAL LAB
LARSEN, KRISTINE	UNIVERSITY OF CONNECTICUT
LARSON, RICHARD B.	YALE UNIVERSITY
LATTIMER, JAMES M.	SUNY, STONY BROOK
LEAHY, DENIS A.	UNIVERSITY OF CALGARY
LEARNED, JOHN G.	UNIVERSITY OF HAWAII
LEAVER, EDWARD W.	THE BDM CORPORATION
LEE, KIMYEONG	FERMILAB
LEISING, MARK D.	NAVAL RESEARCH LABORATORY
LEON, JUAN	FERMILAB
LESKE, RICHARD A.	UNIVERSITY OF CHICAGO
LEUCHS, GERD	MPI FOR QUANTUM OPTICS
LEVENTHAL, MARVIN	AT&T BELL LABORATORIES
LEWIS, DAVID A.	IOWA STATE UNIVERSITY
LIANG, EDISON P.	LAWRENCE LIVERMORE NAT LAB
LIGHTMAN, ALAN P.	CENTER FOR ASTROPHYSICS
LINDBLOM, LEE A.	MONTANA STATE UNIVERSITY
LINK, BENNETT	UNIVERSITY OF ILLINOIS
LIVIO, MARIO	UNIVERSITY OF ILLINOIS
LO, KWOK-YUNG	UNIV OF ILLINOIS/CALTECH
LOH, EDWIN D.	PRINCETON UNIVERSITY
LOKEN, CHRIS	QUEEN'S UNIVERSITY
LOREDO, TOM	UNIVERSITY OF ILLINOIS
LOVELACE, RICHARD V. E.	CORNELL UNIVERSITY
LOWDER, DOUGLAS M.	U OF CALIF, BERKELEY
LUBIN, PHILIP	LAWRENCE BERKELEY LAB
LUCCHIN, FRANCESCO	UNIVERSITY OF PADOVA
LUKASZEK, DONALD	UNIVERSITY OF CONNECTICUT
LUMINET, JEAN-PIERRE	PARIS OBSERVATORY, MEUDON
MACOMB, DARYL	IOWA STATE UNIVERSITY
MAISCHBERGER, KARL	MPI FOR QUANTUM OPTICS
MAJEWSKI, STEVEN R.	YERKES OBSERVATORY
MALLETT, RONALD L.	UNIVERSITY OF CONNECTICUT
MANN, ALFRED K.	UNIVERSITY OF PENNSYLVANIA
MARCK, JEAN-ALAIN	MEUDON OBSERVATORY
MARKS, DENNIS W.	VALDOSTA STATE COLLEGE
MARSHAK, ROBERT E.	VIRGINIA POLYTECH
MARTINEZ, JUAN	UNIVERSITY OF ILLINOIS
MASI, SILVIA	UNIVERSITY OF ROME
MATHER, JOHN C.	GODDARD SPACE FLIGHT CTR
MATHEWS, GRANT J.	LAWRENCE LIVERMORE NAT LAB
MATSUI, YUTAKA	NORTHWESTERN UNIVERSITY
MATZNER, RICHARD A.	UNIVERSITY OF TEXAS
MC CLINTOCK, JEFFREY E.	CENTER FOR ASTROPHYSICS

NAME	INSTITUTION
MC DERMOTT, PATRICK N.	NORTHWESTERN UNIVERSITY
MC DONALD, KIM A.	CHRON OF HIGHER EDUCATION
MCLERRAN, LARRY D.	FERMILAB
MELIA, FULVIO	UNIVERSITY OF CHICAGO
MELOTT, ADRIAN L.	UNIVERSITY OF KANSAS
MESSINA, ANTONIO	UNIVERSITY OF BOLOGNA
MESTEL, LEON	UNIVERSITY OF SUSSEX
MEWALDT, RICHARD	CALTECH
MEYER, BRADLEY S.	UNIVERSITY OF CHICAGO
MEYER, DAVID M.	UNIVERSITY OF CHICAGO
MEYER, DONALD I.	UNIVERSITY OF MICHIGAN
MEYER, PETER	UNIVERSITY OF CHICAGO
MICHELSON, PETER	STANFORD UNIVERSITY
MILLER, RICHARD H.	UNIVERSITY OF CHICAGO
MINN, HOKEE	SEOUL NATIONAL UNIVERSITY
MIYAJI, SHIGEKI	MARSHALL SPACE FLIGHT CTR
MORRIS, DANIEL J.	UNIV OF NEW HAMPSHIRE
MORRIS, MICHAEL S.	CALTECH
MORRISON, MIKE	
MOSKOWITZ, BRUCE	UNIVERSITY OF ROCHESTER
MOSS, IAN G.	UNIVERSITY OF NEWCASTLE
MOTTOLA, EMIL	LOS ALAMOS NATIONAL LAB
MUELLER, DIETRICH	UNIVERSITY OF CHICAGO
MUELLER, EWALD	MPI FOR ASTROPHYSICS
MUFSON, STUART L.	INDIANA UNIVERSITY
MUKERJEE, MADHUSREE	ENRICO FERMI INSTITUTE
MULLER, RAFAEL J.	HUMACAO UNIVERSITY COLLEGE
MUNN, JEFF	YERKES OBSERVATORY
MUSSER, JAMES	UNIVERSITY OF MICHIGAN
NAGLE, DARRAGH E.	LOS ALAMOS NATIONAL LAB
NAKAMURA, TAKASHI	KYOTO UNIVERSITY
NAMBU, YOICHIRO	ENRICO FERMI INSTITUTE
NANDKUMAR, RADHA	UNIVERSITY OF ILLINOIS
NASH, MADELEINE	TIME MAGAZINE
NEGROPONTE, JOHN	T.E.S.R.E - CNR, BOLOGNA
NEMIROFF, ROBERT J.	UNIVERSITY OF PENNSYLVANIA
NEUHAUSER, BARBARA	STANFORD UNIVERSITY
NEWPORT, BRIAN	UNIVERSITY OF CHICAGO
NEZRICK, FRANK A.	FERMILAB
NOMOTO, KENICHI	UNIVERSITY OF TOKYO
NORDSTROM, DENNIS	PHYSICAL REVIEW D
NORMAN, COLIN A.	JOHNS HOPKINS UNIVERSITY
NORMAN, ERIC B.	LAWRENCE BERKELEY LAB
NORMAN, M. L.	UNIVERSITY OF ILLINOIS
O'BRIEN, P. T.	UNIVERSITY COLLEGE LONDON
OCCHIONERO, FRANCO	UNIVERSITY OF ROME
OLCZYK, THADDEUS L.	UNIVERSITY OF ILLINOIS
OLINTO, ANGELA V.	MASS. INST OF TECHNOLOGY
OLSON, TIMOTHY S.	UNIVERSITY OF MINNESOTA
OVERBYE, DENNIS	FIRST LIGHT PRODUCTIONS
OZERNOY, LEONID M.	CENTER FOR ASTROPHYSICS
OZSVATH, ISTVAN	UNIVERSITY OF TEXAS
PACINI, FRANCO	ARCETRI ASTROPHYSICS OBS
PAKEY, DONALD	UNIVERSITY OF ILLINOIS
PARK, HYE-SOOK	U OF CALIF, BERKELEY
PARKER, EUGENE N.	UNIVERSITY OF CHICAGO

NAME	INSTITUTION
PARKER, LEONARD E.	UNIVERSITY OF WISCONSIN
PAYNE, DAVID G.	JET PROPULSION LABORATORY
PEARSON, MARGARET M.	FERMILAB
PENNYPACKER, CARL	LAWRENCE BERKELEY LAB
PENPRASE, BRYAN	UNIVERSITY OF CHICAGO
PERLMUTTER, SAUL	LAWRENCE BERKELEY LAB
PETERS, PHILIP C.	UNIVERSITY OF WASHINGTON
PETERSON, BRADLEY M.	OHIO STATE UNIVERSITY
PETERSON, J.	PRINCETON UNIVERSITY
PETRAKIS, JOHN P.	INDIANA UNIVERSITY
PETROSIAN, VAHE	STANFORD UNIVERSITY
PIM, RICK	QUEEN'S UNIVERSITY
PINES, DAVID	UNIVERSITY OF ILLINOIS
PINTO, PHILIP A.	LICK OBSERVATORY
PIRAN, TSVI	INST FOR ADVANCED STUDY
PIZZELLA, GUIDO	UNIVERSITY OF ROME
POIRIER, JOHN A.	UNIVERSITY OF NOTRE DAME
POKORNY, BRAD	BOSTON GLOBE
PRESS, WILLIAM	HARVARD UNIVERSITY
PRICE, P. BUFORD	U OF CALIF, BERKELEY
PRIEDHORSKY, WILLIAM C.	LOS ALAMOS NATIONAL LAB
PRINSTER, SCOTT	PURDUE UNIVERSITY
PUCACCO, GIUSEPPE	UNIVERSITY OF ROME
QUIGG, CHRIS	FERMILAB
RABINOWITZ, DAVID	UNIVERSITY OF CHICAGO
RAFFELT, GEORG	LAWRENCE LIVERMORE NAT LAB
RAGHAVAN, R. S.	A T & T BELL LABS
REES, MARTIN J.	CAMBRIDGE UNIVERSITY
REFSDAL, SJUR	HAMBURG OBSERVATORY
RENSBERGER, BOYCE	WASHINGTON POST
RESSELL, M. TED	UNIVERSITY OF CHICAGO
RICHARDS, PAUL L.	U OF CALIF, BERKELEY
RINGWALD, ANDREAS	UNIVERSITY OF HEIDELBERG
RITTER, JAMES	CHICAGO SUN-TIMES
ROBERTSON, NORNA A.	UNIVERSITY OF GLASGOW
ROBINSON, IVOR	UNIVERSITY OF TEXAS
RODDIER, F.	KITT PEAK NATL OBSERVATORY
ROEDER, ROBERT C.	SOUTHWESTERN UNIVERSITY
ROGERS, JOSEPH T.	UNIVERSITY OF ROCHESTER
ROMAN, THOMAS A.	CENT. CONN. STATE UNIV.
ROMANI, ROGER W.	CALTECH
ROSEN, S. PETER	LOS ALAMOS NATIONAL LAB
ROSENZWEIG, CARL	SYRACUSE UNIVERSITY
ROSNER, JONATHAN L.	ENRICO FERMI INSTITUTE
ROTH, KATHERINE	NORTHWESTERN UNIVERSITY
ROTHMAN, TONY	DISCOVER MAGAZINE
ROYCROFT, KELLY J.	UNIVERSITY OF CHICAGO
RUDERMAN, MALVIN A.	COLUMBIA UNIVERSITY
RUFFOLO, DAVID J.	UNIVERSITY OF CHICAGO
SADOULET, BERNARD	UNIV OF CALIF, BERKELEY
SALAMON, MICHAEL	U OF CALIF, BERKELEY
SALATI, PIERRE	LAPP, ANNECY-LE-VIEUX
SALE, KEN	LAWRENCE LIVERMORE NAT LAB
SARGENT, WALLACE L. W.	CALTECH
SAULSON, PETER R.	MASS. INST OF TECHNOLOGY
SCHAEFFER, RICHARD	CEN/SACLAY

NAME	INSTITUTION
SCHECHTER, BRUCE M.	OMNI MAGAZINE
SCHERRER, ROBERT J.	HARVARD UNIVERSITY
SCHINDER, PAUL	UNIVERSITY OF PENNSYLVANIA
SCHLEICH, KRISTIN	U OF CALIF, SANTA BARBARA
SCHLICKEISER, R.	MPI FOR RADIOASTRONY
SCHLUTER, ROBERT A.	NORTHWESTERN UNIVERSITY
SCHMIDT, MARTEEN	CALTECH
SCHRAMM, DAVID N.	UNIVERSITY OF CHICAGO
SCHWARZSCHILD, BERTRAM	PHYSICS TODAY
SEIDEL, H. EDWARD	YALE UNIVERSITY
SEMBROSKI, GLENN	PURDUE UNIVERSITY
SHANKAR, ANURAG	UNIVERSITY OF ILLINOIS
SHANKS, THOMAS	UNIVERSITY OF DURHAM
SHAPIRO, MAURICE M.	
SHAPIRO, STUART L.	CORNELL UNIVERSITY
SHAVER, ERIC	QUEEN'S UNIVERSITY
SHELDON, WILLIAM R.	UNIVERSITY OF HOUSTON
SHELLARD, E. PAUL	CAMBRIDGE UNIVERSITY
SHEN, TSUNG-CHENG	S.I.S.S.A., TRIESTE
SHIBAZAKI, NORIAKI	MARSHALL SPACE FLIGHT CTR
SHIELDS, GREGORY A.	UNIVERSITY OF TEXAS
SILK, JOSEPH I.	U OF CALIF, BERKELEY
SIMPSON, JOHN A.	UNIVERSITY OF CHICAGO
SIRONI, GIORGIO	UNIVERSITY OF MILAN
SKAGERSTAM, BO-STURE	UNIVERSITY OF GOTEBORG
SMARR, LARRY L.	NAT CTR FOR SUPERCOMPUTERS
SMITHER, ROBERT K.	ARGONNE NATIONAL LAB
SMOLIN, LEE	YALE UNIVERSITY
SOIFER, B.	CALTECH
SOLOMON, PHILIP	SUNY, STONY BROOK
SONG, YI-QIAO	NORTHWESTERN UNIVERSITY
SPERGEL, DAVID N.	INST FOR ADVANCED STUDY
SPINELLI, GIANCARLO	MILAN POLYTECH
SREEKANTAN, B. V.	TATA INST OF FUND RESEARCH
STEBBINS, ALBERT	FERMILAB
STEIGMAN, GARY	OHIO STATE UNIVERSITY
STEIN-SCHABES, JAIME	FERMILAB
STELLE, KELLOGG S.	INST FOR ADVANCED STUDY
STRAUMANN, NORBERT	UNIVERSITY OF ZURICH
SUEN, WAI-MO	UNIVERSITY OF FLORIDA
SULAK, LAWRENCE R.	BOSTON UNIVERSITY
SULLIVAN, WALTER	NEW YORK TIMES
SUSON, DANIEL J.	UNIVERSITY OF TEXAS
SUTHERLAND, PETER G.	UNIVERSITY OF COLORADO
SWARTZ, DOUGLAS	UNIVERSITY OF TEXAS
SWEETLAND, EMMA	PURDUE UNIVERSITY
SWORDY, SIMON	UNIVERSITY OF CHICAGO
SZAMOSI, GEZA	UNIVERSITY OF WINDSOR
TAAM, RONALD	NORTHWESTERN UNIVERSITY
TAMMANN, GUSTAV A.	ASTRONOMY INST, BINNINGEN
TANGHERLINI, FRANK R.	COLLEGE OF THE HOLY CROSS
TARLE, GREGORY	UNIVERSITY OF MICHIGAN
TAVANI, MARCO	COLUMBIA UNIVERSITY
TAYLOR, JOSEPH	PRINCETON UNIVERSITY
TERRELL, JAMES	LOS ALAMOS NATIONAL LAB
TERZIAN, YERVANT	CORNELL UNIVERSITY

NAME	INSTITUTION
TESCHE, CLAUDIA D.	IBM
TEUKOLSKY, SAUL A.	CORNELL UNIVERSITY
THIELEMANN, FREDERICH K.	CENTER FOR ASTROPHYSICS
THIELHEIM, KLAUS O.	KIEL UNIVERSITY
THOMPSON, CHRIS	PRINCETON UNIVERSITY
THOMSEN, DIETRICK E.	SCIENCE NEWS
THORNE, KIP S.	CALTECH
TILANUS, REMO P. J.	UNIVERSITY OF ILLINOIS
TIPLER, FRANK J.	TULANE UNIVERSITY
TITUS, TIMOTHY N.	IOWA STATE UNIVERSITY
TOLMAN, BRIAN W.	UNIVERSITY OF TEXAS
TRASCHEN, JENNY	UNIVERSITY OF CHICAGO
TREFIL, JAMES S.	UNIVERSITY OF VIRGINIA
TRIMBLE, VIRGINIA	UNIVERSITY OF MARYLAND
TRINKLEIN, GRETCHEN	WANG LABORATORIES
TRURAN, JAMES W.	UNIVERSITY OF ILLINOIS
TSCHIRHART, ROBERT S.	UNIVERSITY OF MICHIGAN
TSE, SZEMAN	NORTHWESTERN UNIVERSITY
TSURUTA, SACHIKO	MONTANA STATE UNIVERSITY
TUCKER, GREGORY S.	PRINCETON UNIVERSITY
TURNER, EDWIN L.	PRINCETON UNIVERSITY
TURNER, MICHAEL S.	FERMILAB/U. OF CHICAGO
TUROK, NEIL	IMPERIAL COLLEGE
TYTLER, DAVID	COLUMBIA UNIVERSITY
ULMER, MELVILLE P.	NORTHWESTERN UNIVERSITY
VAN DALEN, ANTHONY	UNIVERSITY OF CHICAGO
VAN DER BIJ, J. J.	FERMILAB
VAN LEEUWEN, WILLEM A.	UNIVERSITY OF AMSTERDAM
VANDER VELDE, JACK	UNIVERSITY OF MICHIGAN
VANDERVOORT, PETER O.	UNIVERSITY OF CHICAGO
VEERARAGHAVAN, SHOBA	U OF CALIF, BERKELEY
VERMEULEN, RENE C.	STERREWACHT LEIDEN
VILENKIN, ALEX	TUFTS UNIVERSITY
VINCENT, DWIGHT	UNIVERSITY OF WINNIPEG
VISNOVSKY, KAREN	PURDUE UNIVERSITY
VITTORIO, NICOLA	UNIVERSITY OF ROME
VOGEL, SHAWNA	DISCOVER MAGAZINE
VOSS, DAVID	SCIENCE MAGAZINE-AAAS
WADDINGTON, CECIL J.	UNIVERSITY OF MINNESOTA
WAGONER, ROBERT V.	STANFORD UNIVERSITY
WALD, ROBERT M.	UNIVERSITY OF CHICAGO
WALDROP, M. MITCHELL	SCIENCE MAGAZINE
WALLIN, JOHN	IOWA STATE UNIVERSITY
WANDEL, AMRI	STANFORD UNIVERSITY
WARD, RICHARD A.	LAWRENCE LIVERMORE NAT LAB
WASSERMAN, IRA	CORNELL UNIVERSITY
WAUGH, BRAD	QUEEN'S UNIVERSITY
WEAVER, THOMAS A.	LAWRENCE LIVERMORE NAT LAB
WEBER, JOSEPH	UNIVERSITY OF MARYLAND
WEBSTER, RACHEL L.	UNIVERSITY OF TORONTO
WEEKES, TREVOR C.	MT. HOPKINS OBSERVATORY
WEFEL, JOHN P.	LOUISIANA STATE UNIVERSITY
WEILER, THOMAS J.	VANDERBILT UNIVERSITY
WEINBERG, ERICK J.	COLUMBIA UNIVERSITY
WEYMANN, RAY J.	MT. WILSON OBSERVATORY
WHEELER, J. CRAIG	UNIVERSITY OF TEXAS

NAME	INSTITUTION
WHEELER, JOHN A.	UNIVERSITY OF TEXAS
WHITE, SIMON D. M.	STEWARD OBSERVATORY
WIDROW, LARRY	UNIVERSITY OF CHICAGO
WIEDENBECK, MARK E.	UNIVERSITY OF CHICAGO
WIITA, PAUL J.	GEORGIA STATE UNIVERSITY
WILKINSON, DAVID T.	PRINCETON UNIVERSITY
WILL, CLIFFORD M.	WASHINGTON UNIVERSITY
WILSON, JAMES R.	LAWRENCE LIVERMORE NAT LAB
WIRINGA, ROBERT B.	ARGONNE NATIONAL LAB
WITTEN, EDWARD	PRINCETON UNIVERSITY
WOLFE, ARTHUR	UNIVERSITY OF PITTSBURGH
WOLFSBERG, KURT	LOS ALAMOS NATIONAL LAB
WOOSLEY, STANLEY E.	LICK OBSERVATORY
WORTHY, WARD	CHEM & ENGINEERING NEWS
WRIGHT, GREGORY A.	PRINCETON UNIVERSITY
WRIGHT, JAMES P.	NATL SCIENCE FOUNDATION
WUNNER, GUENTER	UNIVERSITY OF TUBINGEN
WYSE, ROSEMARY F. G.	U OF CALIF, BERKELEY
YANG, JONGMANN	EWHA WOMANS UNIVERSITY
YANNY, BRIAN	UNIVERSITY OF CHICAGO
YODH, GAURANG B.	UNIVERSITY OF MARYLAND
YORK, DONALD G.	UNIVERSITY OF CHICAGO
ZDZIARSKI, ANDRZEJ	CALTECH
ZHANG, HUAI	UNIVERSITY OF WISCONSIN
ZHANG, XIAO-HE	CALTECH
ZHANG, YANG	UNIVERSITY OF WISCONSIN
ZHOU, ZIYE	PURDUE UNIVERSITY
ZUREK, WOJCIECH H.	LOS ALAMOS NATIONAL LAB

UNREGISTERED PARTICIPANTS

NAME	INSTITUTION
FREEDMAN, STUART	ARGONNE NATIONAL LAB
GEROCH, ROBERT P.	UNIVERSITY OF CHICAGO
GUNNARSON, LAWRENCE	UNIVERSITY OF CHICAGO
HARPER, D. A.	UNIVERSITY OF CHICAGO
KATZ, JONATHON	WASHINGTON UNIVERSITY
MARQUES, SIRLEY	FERMILAB
OSTRIKER, JEREMIAH P.	PRINCETON UNIVERSITY